Ernest F. Brickell (Ed.)

Advances in Cryptology – CRYPTO '92

12th Annual International Cryptology Conference
Santa Barbara, California, USA
August 16-20, 1992
Proceedings

Springer-Verlag

Berlin Heidelberg New York
London Paris Tokyo
Hong Kong Barcelona
Budapest

Series Editors

Gerhard Goos
Universität Karlsruhe
Postfach 69 80
Vincenz-Priessnitz-Straße 1
D-76131 Karlsruhe, Germany

Juris Hartmanis
Cornell University
Department of Computer Science
4130 Upson Hall
Ithaca, NY 14853, USA

Volume Editor

Ernest F. Brickell
Department 1423, Sandia National Laboratories
PO Box 5800, Albuquerque, NM 87185, USA

CR Subject Classification (1991): E.3-4, D.4.6, G.2.1

ISBN 3-540-57340-2 Springer-Verlag Berlin Heidelberg New York
ISBN 0-387-57340-2 Springer-Verlag New York Berlin Heidelberg

Typesetting: Camera-ready by author
Printing and binding: Druckhaus Beltz, Hemsbach/Bergstr.
45/3140-543210 - Printed on acid-free paper

Preface

Crypto'92 took place on August 16-20, 1992. It was the twelfth in the series of annual cryptology conferences held on the beautiful campus of the University of California, Santa Barbara. Once again, it was sponsored by the International Association for Cryptologic Research, in cooperation with the IEEE Computer Society Technical Committee on Security and Privacy. The conference ran smoothly, due to the diligent efforts of the general chair, Spyros Magliveras of the University of Nebraska.

One of the measures of the success of this series of conferences is represented by the ever increasing number of papers submitted. This year, there were 135 submissions to the conference, which represents a new record. Following the practice of recent program committees, the papers received anonymous review. The program committee accepted 38 papers for presentation. In addition, there were two invited presentations, one by Miles Smid on the Digital Signature Standard, and one by Mike Fellows on presenting the concepts of cryptology to elementary-age students. These proceedings contains these 40 papers plus 3 papers that were presented at the Rump Session. I would like to thank all of the authors of the submitted papers and all of the speakers who presented papers.

I would like to express my sincere appreciation to the work of the program committee: Ivan Damgard (Aarhus University, Denmark), Oded Goldreich (Technion, Israel), Burt Kaliski (RSA Data Security, USA), Joe Kilian (NEC, USA), Neal Koblitz (University of Washington, USA), Ueli Maurer (ETH, Switzerland), Chris Mitchell (Royal Holloway, UK), Kazuo Ohta (NTT, Japan), Steven Rudich (Carnegie Mellon, USA), and Yacov Yacobi (Bellcore, USA). I would also like to thank Joan Boyar for agreeing to chair one of the sessions.

Ernest Brickell
Albuquerque, NM
August, 1993

PROCEEDINGS OF CRYPTO '92

TABLE OF CONTENTS

Session IX: Cryptography Education

Session X: Theory II

Session XI: Key Distribution

Session XII: DES

Provably Unforgeable Signatures

Jurjen N.E. Bos*
David Chaum†

Abstract. Very strong definitions of security for signature schemes have been proposed in the literature. Constructions for such schemes have been proposed, but so far they have only been of theoretical interest and have been considered far too inefficient for practical use.

Here we present a new scheme that satisfies these strongest definitions and uses essentially the same amount of computation and memory as the widely applied RSA scheme. The scheme is based on the well known RSA assumption.

Our signatures can be thought of as products resulting from a two-dimensional Lamport scheme, where one dimension consists of a list of public constants, and the other is the sequence of odd primes.

Introduction

One of the greatest achievements of modern cryptography is the digital signature. A digital signature on a message is a special encryption of the message that can easily be verified by third parties. Signatures cannot be denied by the signer nor falsified by other parties.

This article introduces a new signature scheme that combines the strength of the strongest schemes with the efficiency of RSA.

Signing a message of 245 bits in our scheme is possible in roughly 910 multiplications, and verifying it costs about 152 multiplications. In comparison, RSA, using the ISO/IEC standard 9796 redundancy scheme, takes roughly 768 multiplications (or 610 using addition chains) for signing, and 3 (or optionally 17) for verification. RSA signatures are 512 bits long, while ours requires an additional message counter. Thus, 16 extra bits give a scheme that allows 65,536 signatures per public key.

A variation involving pre-computation, signs short messages (64 bits) in 33 multiplications (not counting precomputation) and verifies in 35 multiplications.

After the introduction, we discuss other signature schemes relevant to this work. We discuss the Lamport signature scheme, on which this signature scheme is based, in detail. Then, the new scheme is explained, and the possible choices for parameter values are shown.

---

* This article is adapted from the dissertation "Practical Privacy" of Jurjen N.E. Bos, written while he was at CWI (the Dutch nationally funded centre for Mathematics and Computer Science). He is currently affiliated with Irdeto (a pay TV company) in Hoofddorp, Netherlands.

† David Chaum is affiliated both with CWI and DigiCash (innovators in electronic money systems).

Signature scheme

An overview of signature schemes, comparing securities, can be found in the paper mentioned earlier [GMR88]. We use their notation. They define a signature scheme as consisting of the following components:

- A *security parameter k*, that defines the security of the system, and that may also influence performance figures such as the length of signatures, running times and so on.
- A *message space* **M**, that defines on which messages the signature algorithm may be applied.
- A *signature bound b*, that defines the maximal number of signatures that can be generated without reinitialization. Typically, this value depends on k, but it can be infinite.
- A *key generation algorithm* **G**, that allows a user to generate a pair of corresponding public and secret keys for signing. The secret key S is used for generating a signature, while the public key P is used to verify the signature.
- A *signature algorithm* σ, that produces a signature, given the secret key and the message to be signed.
- finally, a *verification algorithm*, that produces **true** or **false** on input of a signature and a public key. It ouputs **true** if and only if the signature is valid for the particular public key.

Some of these algorithms may be *randomized*, which means that they may use random numbers. Of course, **G** must be randomized, because different users must produce different signatures. The signing algorithm σ is sometimes randomized, but this tends to produce larger signatures. The verification algorithm is usually not randomized.

A simple example of a signature scheme is a trapdoor one-way function f. The function f is used for verification by comparing the function value of the signature with the message to be signed, and σ is the trapdoor of f. The main problem with such a scheme is that random messages $f(x)$ can be signed by taking a random signature value x. A simple solution is to let **M** be a sparse subset of a larger space, so that the probability that $f(x)$ is a valid message for random x is low. An example of a sparse subset is the set of "meaningful" messages.

Related work

The notion "digital signature" was introduced in [DH76]. This paper, which can be considered the foundation of modern cryptography, discusses the possibility of digital signatures and the use of a trapdoor one-way function to make them.

[RSA78] is the original article on the RSA scheme. It introduces the famous RSA trapdoor one-way function. This function is still widely in use and is applied frequently. A well-known weakness of RSA is that it is multiplicative: the product of two signatures is the signature of the product. This potential problem can be prevented as above by choosing an appropriate sparse message space.

Since then, an enormous number of signature schemes have been proposed [Rab77, MH78, Sha78, Rab79, Lie81, DLM82, GMY83, Den84, GMR84, OSS84, ElG85, OS85, FS86, BM88, GMR88, CA89, EGL89, EGM89, Mer89, Sch89, SQV89, BCDP90, Cha90, CR90, Hay90, CHP91], applied [Wil80, Cha82, Gol86, Bet88], and broken [Yuv79, Sha82, Tu84, BD85, EAKMM85, Roo91]. We will not discuss all these schemes here; we only discuss the ones that are interesting to compare with the new scheme.

The schemes [Rab79, GMY83, GMR84, GMR88] are steps towards a provably secure signature scheme. The scheme described in the last article is secure in a very strong way: it is "existentially unforgeable under an adaptive chosen-message attack" with probability smaller than $1/Q(k)$ for every polynomial Q. This means that generating a new signature is polynomially hard if signatures on old messages are known, even if the old signatures are on messages chosen by the attacker.

The scheme in [GMR88] is based on factoring. While our scheme is based on the slightly stronger RSA assumption, it is much more efficient. The signature scheme of [GMR88] uses a large amount of memory for the signer, and quite a lot of computation. Our scheme uses no memory at all, except for a counter and the public values, and signing and verifying takes about as much computation as RSA does, depending on the parameters.

The Lamport Scheme

Our scheme can be thought of an optimization for both security and efficiency of [GMY83]. To explain the new system, we compare it to the earlier *Lamport scheme* (explained already in [DH76, page 650]). To make a signature in this scheme, the signer makes a secret list of $2k$ random numbers

$$\mathbf{A} = a_{1,0}, a_{1,1}, a_{2,0}, a_{2,1}, ..., a_{k,0}, a_{k,1},$$

applies a one-way function f to all elements, and publishes the result $\mathbf{B}$:

$$\mathbf{B} = \begin{cases} f(a_{1,0}), f(a_{2,0}), ..., f(a_{k,0}) \\ f(a_{1,1}), f(a_{2,1}), ..., f(a_{k,1}) \end{cases}$$

The signature consists of the numbers $a_{1,m_1}, a_{2,m_2}, ..., a_{k,m_k}$ from the list $\mathbf{A}$ (one from each "column"), where $m_1, m_2, ..., m_k$ are the bits of the message to be signed. The lists $\mathbf{A}$ and $\mathbf{B}$ cannot be used again.

The properties of Lamport's scheme are easy to verify:
- Signing a message is only the publication of the proper elements of **A**.
- To forge a signature, one needs to find certain values from the list **A**. How hard this is, depends on the security of the one-way function f.
- If the values **A** are only used for one signature, new signatures cannot be made from old ones.
- Verification of a signature consists of applying the one-way function to the signature values, and comparing them to the public values determined by the signed message.

The new system uses the same idea, with three important differences. first, the list **B** is replaced by another list that can be used for all signatures. Second, the list **A** is constructed from two lists so that less memory is needed to define it. Third, the elements of **A** in the signature can be combined into a single number.

A small optimization

There is a trivial optimization of Lamport's scheme that reduces the number of public function values to almost half, that we could not find in the literature. This optimization is independent of the signature scheme as such. Basically, the signer signs by publishing a k-element subset of the $2k$ secret numbers. Lamport's scheme chooses a particular set of subsets of the set of $2k$ elements, as shown above. The necessary property of this set of subsets is that no subset includes another.

There are other sets of subsets with the property that no subsets includes another. A largest set of subsets with this property is the set of all k-element subsets (a well-known result from lattice theory). For these sets, it is easy to see that no subset includes another.

For example, in Lamport's scheme, the list of 6 elements
$$A = a_{1,0}, a_{1,1}, a_{2,0}, a_{2,1}, a_{3,0}, a_{3,1}$$
allows us to sign messages of 3 bits. If we renumber **A** as $a_1, a_2, a_3, a_4, a_5, a_6$, we get the set of 20 three-element subsets of **A**:
$$\{a_1, a_2, a_3\}, \{a_1, a_2, a_4\}, \{a_1, a_2, a_5\}, \{a_1, a_2, a_6\}, \{a_1, a_3, a_4\},$$
$$\{a_1, a_3, a_5\}, \{a_1, a_3, a_6\}, \{a_1, a_4, a_5\}, \{a_1, a_4, a_6\}, \{a_1, a_5, a_6\},$$
$$\{a_2, a_3, a_4\}, \{a_2, a_3, a_5\}, \{a_2, a_3, a_6\}, \{a_2, a_4, a_5\}, \{a_2, a_4, a_6\},$$
$$\{a_2, a_5, a_6\}, \{a_3, a_4, a_5\}, \{a_3, a_4, a_6\}, \{a_3, a_5, a_6\}, \{a_4, a_5, a_6\};$$
this allows us to sign one of 20 messages, which is equivalent to more than 4 bits.

In general, there are
$$\binom{2k}{k}, \text{ or about } \frac{2^{2k}}{\sqrt{k\pi}},$$
k-element subsets, so that we can sign messages of about $2k - \frac{1}{2}\log_2(k\pi)$ bits. The original Lamport scheme allowed messages of only k bits, so that we get almost

a doubling of the message size for the same size of the list **B**. This simple improvement can also be used in our new signature scheme.

To encode a signature, a mapping needs to be defined between messages and these subsets:

$$s(message) = subset.$$

The simplest mapping just enumerates messages (interpreted as numbers from 0 onwards) to sets (seen as binary strings that denote 1 for presence and 0 for absence) in order. Such a mapping is easily and efficiently computed by the algorithm shown in figure 1. The binomial coefficients do not need to be computed by repeated multiplication and division. The first binomial coefficient is always the same, so it can be precomputed, and the others can be computed by one multiplication and one division by small numbers using the properties:

$$\binom{t-1}{e} = \binom{t}{e} \cdot \frac{t-e}{t} \text{ and } \binom{t-1}{e-1} = \binom{t}{e} \cdot \frac{e}{t}.$$

The algorithm outputs ones and zeros corresponding to the elements in the resulting set.

Note that the Lamport scheme uses another mapping that maps numbers onto k-element subsets, but that only a small number of these sets are used.

Let n, the message, be a number in the range $0 \ldots \binom{2k}{k} - 1$.

Put $2k$ in t and k in e.

While $t > 0$:

    Put $t-1$ in t.

    If $n \geq \binom{t}{e}$, put $n - \binom{t}{e}$ in n, $e-1$ in e, and output a 1 (this t is in the set).

    Else, output a 0 (this t is not in the set).

Fig 1. Algorithm for the mapping s.

The New Signature Scheme

The new signature scheme replaces the list **A** of the Lamport scheme by a list of numbers that can be organized in a matrix. Instead of using a new list **B** for every signature, a fixed list called **R** is used for all signatures and all participants. The one-way function f is replaced by a set of trapdoor one-way functions, that changes per signature. For the trapdoor one-way functions, we use the modular root function of [RSA78].

The construction allows us to sign long messages using only a few numbers to define the set **A**. In the example of figure 2, the set **A** of 12 elements is constructed from three primes p_1, p_2, p_3 (used only for this signature) and four public values r_1, r_2, r_3, r_4 (that can be used again). This set allows us to sign messages of 9 bits, since there are $924 > 2^9$ possible 6-element subsets of **A**. Signing messages of 9 bits in the original Lamport scheme takes 18 public values that can be used only once.

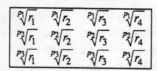

Fig. 2. Example list **A** of the new scheme.

The numbers a_i of **A** are secret encryptions of the numbers r_i of **R**, and the corresponding decryption exponents are public. The multiplicative property of RSA allows us to multiply the values of the signature to form one number. Verification of a signature can be done using a simple computation, without having to compute the separate factors.

The public values of the new system are:
- One modulus per signer;
- The system-wide list **R**. This list is used by all users, and that it does not change often, so that distribution does not require much traffic. The numbers in **R** are smaller than the smallest modulus used by the signers.
- A list of sets of primes that may be used for signing. For security reasons, the sets may not overlap each other, and the signers may only use these sets of primes.

A signature consists of the original message signed, the signature proper (an integer smaller than the modulus of the signer), and a description of the prime set.

In the language of [GMR88]:
- The security parameter determines the size of the RSA modulus. This modulus can vary per user.
- The message space **M** is (equivalent to) the set of subsets of **A** that include half the elements.
- The size of the public list of sets of primes determines the signature bound b.
- Key generation is a matter of generating an RSA modulus, and computing exponents for the modular root extractions.
- Signing and verification are defined below.

Signing

For the list **A** of a signature, the set of RSA encryptions

$$A = \left\{ \sqrt[p]{r} \bmod n \,\middle|\, p \in \mathbf{P}; r \in \mathbf{R} \right\}$$

is used, where:

- **P** a set of primes from the public list;
- **R** is the public list of verification values;
- n is the RSA modulus of the signer.

As explained above, a signature is constructed from a subset determined by $s(m)$ of half these numbers. The constant k used in the algorithm that maps s is equal to $\left\lfloor \frac{\#P \cdot \#R}{2} \right\rfloor$. This allows us to sign a message of almost $\#A = \#P \cdot \#R$ bits. The product of the elements of **A** in this subset is the signature. Since this is a single number, the signature is much more compact than in Lamport's scheme.

Thus, signing a message consist of the following steps:

- Choose the set **P** of primes that is to be used for this signing from the public list. This determines **A**:

$$A = \left\{ \sqrt[p_i]{r_j} \bmod n \,\middle|\, i, j \in \{1, \ldots, \#P\} \times \{1, \ldots \#R\} \right\}.$$

Like the sets **A** and **B** in Lamport's scheme, the set **P** can be used only once. The list **A** need not be computed.

- Determine the message m to sign. This could be a message, or a public hash function value of that message, for example.
- Compute the subset **M** of index pairs from $\{1, \ldots, \#P\} \times \{1, \ldots, \#R\}$ from the message m with the algorithm described above:

$$M = s(m)$$

- Compute the signature proper:

$$S = \prod_{i, j \in M} \sqrt[p_i]{r_j} \pmod{n},$$

and send m, **P**, and S to the recipient.

There are two ways to increase the efficiency of signing. If there is time to do a precomputation, the entire set **A** can be computed before the value of m is known. Although this takes quite a while, signing becomes much faster, since signing consists only of multiplying the proper values of **A** together. If precomputation is not possible, the computation of S can be speeded up with a *vector addition chain* [Bos92].

Verification

Instead of trying to compute individual factors of the signature, the number S can be verified in a single computation. To see this, we note that the power of the signature

$$\frac{\prod_{k \in P} p_k}{S}$$

should be equal to the following product that can be computed from public values:

$$\prod_{i,j \in M} r_i^{\prod_{k \in P} p_k / p_j} .$$

The lower product can be computed with a vector addition chain. Verification of a signature consists of checking that these two values are the same. The verification can be performed with a single vector addition chain, if the inverse of the signature is computed first:

$$(S^{-1})^{\prod_{k \in P} p_k} \cdot \prod_{i,j \in M} r_i^{\prod_{k \in P} p_k / p_j} ,$$

which must evaluate to 1 (mod n). To increase the efficiency of the verification, the signer could send $1/S$ instead of S, so that the inversion is performed only once by the signer, and not by every verifier.

If not all prime numbers from **P** occur as exponents in the set **M**, it is possible to verify a signature using slightly fewer multiplications by raising S to only the occurring primes. Unfortunately, this optimization is only applicable in the less interesting cases where verification requires a lot of multiplications.

The verifier must also check whether **P** occurs in the public list. If **P** is described as an index number in this list, this is of course unnecessary.

Parameters

In practice, the following parameter values could be used:

- A modulus size big enough to make factorization hard (200 digits, or 668 bits).
- **R** a list of 50 numbers.
- The sets **P** consisting of the $(5n+1)^{th}$ to the $(5n+5)^{th}$ odd prime number, where $n \in \{0,\ldots,16404\}$ is the sequence number of the signature. This uses the primes of up to 20 bits.

With these parameters, we have sets **A** of 250 elements, so that a message of 245 bits (30 bytes) can be signed. A signature consists of the message, the signature product (668 bits, or 84 bytes), and the index number of the prime set (15 bits, or 2 bytes). Computing a signature takes about 1512 modular multiplications, and verification about 272; both these numbers are obtained using vector addition chains.

The list of the odd primes up to 20 bits (the highest being 1048557) can easily be stored; it would need only 64 K bytes of storage (using a bit table of the odd numbers) and contain 82025 primes. Such a list can easily be stored in a ROM chip. When all primes are used up, the user can choose a new modulus and start again. Another solution is to change the list **R** often enough so that users do not run out of primes. To make it possible to verify old signatures, old values of **R** and the user moduli must be saved.

The list **R** can be computed from a seed number using a public hash function. This way, only one seed number is needed to define **R**. This allows us one to use a long list **R** while using small amounts of data to distribute it. Also, less data is needed to save old lists.

Figure 3 shows the performance of the algorithm for several sizes of **R** and **P**. For each of the entries in the table, the modulus is 668 bits (200 decimal digits), and the size of the primes in **P** is 20 bits. The entries are computed by averaging random number approximations. The entries marked by * have an estimated standard deviation higher than 10, so that the last digits are likely to be inaccurate.

Powers and products were computed using addition chains and sequences; see [Bos92, chapter 4]. The products were computed collecting the base numbers; for example, the product

$$b_2^{e_1} \cdot b_3^{e_1} \cdot b_1^{e_2} \cdot b_3^{e_2} \cdot b_4^{e_2} \cdot b_2^{e_3}$$

would be computed as

$$b_1^{e_2} \cdot b_2^{e_1+e_3} \cdot b_3^{e_1+e_2} \cdot b_4^{e_2}$$

using a vector addition chain algorithm. In the cases were a single power was to be computed, the "window method" of [Bos92] was applied.

The table shows that in the general case, where verification is done more often than signing, it is advantageous to use a small **P**, possibly of only one element. The length of the list **R** is not a problem if it is generated from a seed, as suggested above. Another advantage of using a small set **P** is that the list **R** has to change less often.

#R	#P	message	sign	verify
250	1	245	910	152
50	5	245	1512	272
5	50	245	1451	2048*
1	250	245	796	7123*
500	1	495	1035	278
50	10	495	2964*	1372*
68	1	64	819	61
17	4	64	1317	162
4	17	64	1301	659*

Fig. 3. Performance for different size of R and P.

The influence of the modulus size and prime size on the performance is shown is

Figure 4. In this table, the size of **R** is set to 50 elements, while the sets **P** contain 5 elements each. The number of multiplications for signing depends on the size of the modulus only, while the number of multiplications for verifying depends on the size of the prime numbers only. Although it saves a little time during the signing to use a shorter modulus, we suggest using a modulus of 668 bits, since the current technology already allows factoring numbers of up to 351 bits.

The size of the primes in the sets **P** determines the verification time. Choosing smaller primes increases the speed of verification, but allows fewer signatures before a new list **R** is needed.

modulus size	signing
512	1172
668	1512

prime size	verifying
10	171
20	272
30	381

Fig. 4. Performance for different sizes of modulus and primes.

If the elements of **A** are precomputed, signing takes #A/2−1 multiplications. The precomputation takes about 796·#A multiplications, so precomputation is only effective if there is plenty of time for doing it.

For extremely fast verification of signatures, we choose a list **R** of 68 elements, generated from a seed number that is part of the signature, and **P** ={3}. For these parameters, the message to be signed is 64 bits (8 bytes). This allows verification of a signature in only 35 modular multiplications, plus the time to generate the elements of **R**. Signing takes about 819 multiplications. Using precomputation, signing takes 33 multiplications, but about 55000 multiplications for the precomputation.

Proof of unforgeability

We prove that the signature scheme is "existentially unforgeable under an adaptive chosen-message attack". This means that, under the RSA assumption, if an attacker can influence the signer to sign any number of messages of his liking, he cannot forge new signatures in polynomial time, even if the messages depend on the signatures on earlier messages.

The main theorem used to prove unforgeability of the signature system is proved by Jan-Hendrik Evertse and Eugène van Heijst in [EH90], and is a generalization of a theorem by Adi Shamir [Sha83]. The theorem is about computing a product of RSA roots with a given modulus if a set of products of signatures is known. Under the RSA assumption, the theorem states that if a set of products of roots is known, the only new products of roots that can be constructed in polynomial time are those that can be computed using multiplication and division.

One assumption we make is that the attacker cannot combine the signatures of different participants, because they have different moduli. This is still an open problem. This assumption allows us to use the results of [EH90].

In our situation, we assume an attacker who knows many signature products S from a participant. These products can be written as products of roots of elements of **R**:

$$r_1^{x_1} r_2^{x_2} r_3^{x_3} \cdots r_{\#\mathbf{R}}^{x_{\#\mathbf{R}}},$$

where the numbers x_i are rational numbers. The theorem of [EH90] states that if we interpret the x as vectors, the only new products that can be computed by the attacker correspond to linear combinations of these vectors. What remains to be proved is that linear combinations of these vectors do not give products that the attacker can use for new signatures.

The denominators of the rational numbers x_i are products of primes from the set **P** of the corresponding signature, since the x_i are sums of the form $\frac{1}{p_1} + \frac{1}{p_2} + \cdots$, where $p_i \in \mathbf{P}$. This means that we can speak of "the set of primes in a vector", meaning both the set of primes that occur in the denominators of the elements, and the set **P** used for generating the signature. Every signature uses another **P**, and the sets **P** do not overlap, so the sets of primes in the vectors also do not overlap. A linear combination of vectors will contain only primes that occurred in the original vectors. From this we see that combining signatures with multiplication and division will not produce a signature with a set **P** that is not used before.

For a set **P** that has already been used, the only linear combination of vectors that contains the primes of **P** is a multiple of the corresponding vector, because any other linear combination of vectors contains primes not in **P** . This means that other signature products do not help compute a new signature product with a given set **P**. From the definition of the signature product, we see that a power of a product cannot be a signature on another message, so this method also yields no new signatures for the attacker.

Note that if m is a one-way hash function of a message, signatures on other messages can be forged if the hash function is broken. This is of course a separate problem from the security of the signature scheme.

From the above we conclude that an attacker cannot, under the RSA assumption, produce a signature product that is not already computed by the signer. This finishes the proof that the signature scheme is secure.

Conclusion

It was already known that a signature with provable unforgeability existed under the factoring assumption. Our scheme, based on the modular root assumption, improves on the scheme in the literature on several points: signatures are smaller, while signing and verification use much less memory and computation. The new scheme has a large degree of flexibility, allowing the signing of both long and short messages by varying the parameters.

References

[BCDP90] J. F. Boyar, D. Chaum, I. B. Damgård and T. Pedersen: *Convertible Undeniable Signatures*, Advances in Cryptology: Proc. Crypto '90 (Santa Barbara, CA, August 1990), to be published.

[BD85] E. F. Brickell and J. M. DeLaurentis: *An Attack on a Signature Scheme proposed by Okamoto and Shiraishi*, Advances in Cryptology: Proc. Crypto '85 (Santa Barbara, CA, August 1985), pp. 28-32.

[Bet88] T. Beth: *A fiat-Shamir-like Authentication Protocol for the ElGamal Scheme*, Advances in Cryptology: Proc. Eurocrypt '88 (Davos, Switzerland, May 1988), pp. 77-86.

[BM88] M. Bellare and S. Micali: *How to Sign Given any Trapdoor Function*, Advances in Cryptology: Proc. Crypto '88 (Santa Barbara, CA, August 1988), pp. 200-215.

[Bos92] J. N. E. Bos: *Practical Privacy*, dissertation of the Eindhoven University of Technology, march 1992.

[CA89] D. Chaum and H. van Antwerpen: *Undeniable Signatures*, Advances in Cryptology: Proc. Crypto '89 (Santa Barbara, CA, August 1989), pp. 212-216.

[Cha82] D. Chaum: *Blind Signatures for Untraceable Payments*, Advances in Cryptology: Proc. Crypto '82 (Santa Barbara, CA, August 1982), pp. 199-203.

[Cha90] D. Chaum: *Zero-knowledge Undeniable Signatures*, Advances in Cryptology: Proc. Eurocrypt '90 (Århus, Denmark, May 1990), pp. 458-464.

[CHP91] D. Chaum, E. van Heijst, and B. Pfitzmann: *Cryptographically Strong Undeniable Signatures, Unconditionally Secure for the Signer*, Advances of Cryptology: Proc. Crypto '91 (Santa Barbara, August 1991), to be published.

[CR90] D. Chaum and S. Roijakkers: *Unconditionally Secure Digital Signatures*, Advances in Cryptology: Proc. Crypto '90 (Santa Barbara, CA, August 1990), pp. 209-217.

[Den84] D. E. R. Denning: *Digital Signatures with RSA and Other Public-Key Cryptosystems*, Comm. ACM **27** (No. 4, April 1984), pp. 388-392.

[DH76] W. Diffie and M. E. Hellman: *New Directions in Cryptography*, IEEE Trans. Information Theory **IT-22** (No. 6, November 1976), pp. 644-654.

[DLM82] R. DeMillo, N. Lynch, and M. Merritt: *Cryptographic Protocols*, Proc. 14th ACM Symp. Theory of Computing (San Fransisco, CA, May 1982), pp. 383-400.

[EAKMM85] D. Estes, L. M. Adleman, K. Kompella, K. McCurley, and G. L. Miller: *Breaking the Ong-Schnorr-Shamir Signature Scheme for Quadratic Number fields*, Advances in Cryptology: Proc. Crypto '85 (Santa Barbara, CA, August 1985), pp. 3-13.

[EGL89]   S. Even, O. Goldreich, and A. Lempel: *A Randomized Protocol for Signing Contracts*, Advances in Cryptology: Proc. Crypto '89 (Santa Barbara, CA, August 1989), pp. 205-210.

[EGM89]   S. Even, O. Goldreich, and S. Micali: *On-line/Off-line Digital Signatures*, Advances in Cryptology: Proc. Crypto '89 (Santa Barbara, CA, August 1989), pp. 263-275

[EH90]   J-H. Evertse and E. van Heyst: *Which RSA Signatures can be Computed from Some Given Signatures?*, Advances in Cryptology: Proc. Eurocrypt '90 (Århus, Denmark, May 1990), pp. 83-97.

[EH91]   J-H. Evertse and E. van Heyst: *Which RSA Signatures can be Computed from Certain Given Signatures?*, Report W 91-06, February 1991, Mathematical Institute, University of Leiden.

[ElG85]   T. ElGamal: *A Public Key Cryptosystem and a Signature Scheme Based on Discrete Logarithm*, IEEE Trans. Information Theory IT-31 (No. 4, July 1985), pp. 469-472.

[FS86]   A. fiat and A. Shamir: *How to Prove Yourself: Practical Solutions of Identification and Signature Problems*, Advances in Cryptology: Proc. Crypto '86, (Santa Barbara, CA, August 1986), pp. 186-194.

[GMR84]   S. Goldwasser, S. Micali, and R. L. Rivest: *A "Paradoxical" Solution to the Signature Problem*, Proc. 25th IEEE Symp. Foundations of Computer Science (Singer Island, 1984), pp. 441-448.

[GMR88]   S. Goldwasser, S. Micali, and R. L. Rivest: *A Digital Signature Scheme Secure Against Adaptive Chosen-Message Attacks*, SIAM Journal on Computing 17 (No 2, April 1988), pp. 281-308.

[GMY83]   S. Goldwasser, S. Micali, and A. Yao: *Strong Signature Schemes*, Proc. 15th ACM Symp. Theory of Computing (Boston, MA, April 1983), pp. 431-439.

[Gol86]   O. Goldreich: *Two Remarks Concerning the Goldwasser-Micali-Rivest Signature Scheme*, Advances in Cryptology: Proc. Crypto '86 (Santa Barbara, CA, August 1986), pp. 104-110.

[Gol86a] O. Goldreich: *Two Remarks Concerning the Goldwasser-Micali-Rivest Signature Scheme*, Report MIT/LCS/TM-315, Massachusetts Institute of Technology.

[Hay90]   B. Hayes: *Anonymous One-Time Signatures and flexible Untraceable Electronic Cash*, Advances in Cryptology: Proc. Auscrypt '90 (Sydney, Australia, January 1990), pp. 294-305.

[Lie81]   K. Lieberherr: *Uniform Complexity and Digital Signatures*, Theoretical Computer Science 16 (1981), pp. 99-110.

[Mau91]   U. Maurer: *Non-interactive Public Key Cryptography*, Advances in Cryptology: Proc. Eurocrypt '91 (Brighton, United Kingdom, April 1991), to be published.

[Mer89]   R. C. Merkle: *A Certified Digital Signature*, Advances in Cryptology: Proc. Crypto '89 (Santa Barbara, CA, August 1989), pp. 218-238.

[MH78]   R. C. Merkle and M. E. Hellman: *Hiding Information and Signatures in Trapdoor Knapsacks*, IEEE Trans. Information Theory IT-24 (No. 5, September 1987), pp. 525-530.

[Oka88]   T. Okamoto: *A Digital Multisignature Scheme Using Bijective Public-Key Cryptosystems*, ACM Trans. Computer Systems 6 (No. 8, November 1988), pp. 342-441.

[OS85]   T. Okamoto and A. Shiraishi: *A Fast Signature Scheme Based on Quadratic Inequalities*, Proc. 1985 Symp. Security and Privacy (Oakland, CA, April 1985), pp. 123-132.

[OSS84]   H. Ong, C. P. Schnorr, and A. Shamir: *Efficient Signature Schemes based on Polynomial Equations*, Advances in Cryptology: Proc. Crypto '84 (Santa Barbara, August 1984), pp. 37-46.

[Rab77]   M. O. Rabin: *Digitalized Signatures*, Foundations of Secure Computations 1977 (Atlanta, GA, October 1977), pp. 155-168.

[Rab79]   M. O. Rabin: *Digitalized Signatures and Public-key Function as Intractable as Factorization*, Report MIT/LCS/TR-212, Massachusetts Institute of Technology.

[Roo91]   P. J. N. de Rooij: *On the security of the Schnorr Scheme using Preprocessing*, Proc. Eurocrypt '91 (Brighton, United Kingdom), to be published.

[RSA78]   R. L. Rivest, A. Shamir, and M. Adleman: *A Method for Obtaining Digital Signatures and Public Key Cryptosystems*, Comm. ACM 21 (No 2, February 1978), pp. 120-126.

[Sch89]   C. P. Schnorr: *Efficient Identification and Signatures for Smart Cards*, Advances in Cryptology: Proc. Crypto '89 (Santa Barbara, CA, August 1989), pp. 239-251.

[Sha78]   A. Shamir: *A Fast Signature Scheme*, Report MIT/LCS/TR-107, Massachusetts Institute of Technology.

[Sha82]   A. Shamir: *A polynomial Time Algorithm for Breaking the Basic Merkle-Hellman Cryptosystem*, Proc. 23rd IEEE Symp. Foundations of Computer Science (Chicago, IL, 1982), pp. 145-152.

[Sha83]   A. Shamir: *On the Generation of Cryptographically Strong Pseudorandom Sequences*, ACM Trans. Computer Systems 1 (No. 1, February 1983), pp. 38-44.

[Sha84]   A. Shamir: *Identity-based Cryptosystems and Signature Schemes*, Advances in Cryptology: Proc. Crypto '84 (Santa Barbara, CA, August 1984), pp. 47-53.

[SQV89]   M. de Soete, J.-J. Quisquater, and K. Vledder: *A Signature with Shared Verification Scheme*, Advances in Cryptology: Proc. Crypto '89 (Santa Barbara, CA, August 1989), pp. 253-262.

[Tu84]   Y. Tulpan: *Fast Cryptanalysis of a Fast Signature System*, Master's thesis in Applied Mathematics, Weizmann Institute, Israel, 1984.

[Wil80]   H. C. Williams, *A Modification of the RSA Public-Key Encryption Procedure*, IEEE Trans. Information Theory IT-26, (No. 6, November 1980), pp. 726-729.

[Yuv79]   G. Yuval: *How to Swindle Rabin*, Cryptologia 3 (No. 3, July 1979), pp. 187-189.

New Constructions of Fail-Stop Signatures and Lower Bounds

(Extended Abstract)

Eugène van Heijst[1], Torben Pryds Pedersen[2], Birgit Pfitzmann[3]

Abstract. With a fail-stop signature scheme, the supposed signer of a forged signature can prove to everybody else that it was a forgery. Thus the signer is secure even against computationally unrestricted forgers. Until recently, efficient constructions were only known for restricted cases, but at Eurocrypt '92, van Heijst and Pedersen presented an efficient general scheme, where the unforgeability is based on the discrete logarithm.

We present a similar scheme based on factoring: Signing a message block requires approximately one modular exponentiation, and testing it requires a little more than two exponentiations. It is useful to have such alternative constructions in case one of the unproven assumptions is broken.

With all fail-stop signatures so far, the size of the secret key is linear in the number of messages to be signed. In one sense, we prove that this cannot be avoided: The signer needs so many secretly chosen random bits. However, this does not imply that these bits ever have to be secretly stored at the same time: We present a practical construction with only logarithmic secret storage and a less practical one where the amount of secret storage is constant.

We also prove rather small lower bounds for the length of public keys and signatures. All three lower bounds are within a small factor of what can be achieved with one of the known schemes.

Finally, we prove that with unconditionally secure signatures, like those presented by Chaum and Roijakkers at Crypto '90, the length of a signature is at least linear in the number of participants who can test it. This shows that such schemes cannot be as efficient as fail-stop signatures.

1   Introduction and Overview over the Results

Ordinary and Fail-Stop Signatures

Ordinary digital signatures, as introduced in [DH76] and formally defined in [GMR88], allow a person who knows a secret key to make signatures that everybody else can verify with a corresponding public key. Such signatures can only be computationally secure: A forger with unrestricted computing power can always forge signatures of other persons. The security of the schemes relies on the fact that a realistic forger has not enough time to carry out brute-force search and the assumption that there is no really efficient algorithm to compute forgeries.

---

[1] CWI, Kruislaan 413, NL-1098 SJ Amsterdam
[2] Aarhus Universitet, Computer Science Department, Ny Munkegade, DK-8000 Aarhus C, tppedersen@daimi.aau.dk
[3] Institut für Informatik, Universität Hildesheim, Samelsonplatz 1, W-3200 Hildesheim, pfitzb@informatik.uni-hildesheim.de

With fail-stop signatures, introduced in [WP90] and formally defined in [PW90], unforgeability also relies on a computational assumption. If nevertheless a signature is forged, the alleged signer can prove that the signature is a forgery. More precisely, she can prove that the underlying computational assumption has been broken. This proof may fail with an exponentially small probability, but the ability to prove forgeries does not rely on any cryptographic assumption and is independent of the computing power of the forger. Thus a polynomially bounded signer can be protected from an all-powerful forger. Moreover, after the first forgery, all participants, or the system operator, know that the signature scheme has been broken, so that it can be stopped. This is where the name "fail-stop" comes from.

For more details about possible benefits of fail-stop signatures in applications, e.g., in electronic payment systems, and possible advantages for the acceptability of digital signatures in law, see [PW91, P91].

Previous Constructions

So far, there have been three significantly different results about fail-stop signatures.

Theoretically, fail-stop signature schemes are known to exist if claw-free pairs of permutations (not necessarily with trap-door) exist; see [BPW91, PW91] for descriptions and [PW90] for a proof. In particular, this shows that fail-stop signatures exist if factoring large integers or computing discrete logarithms is hard. The construction uses one-time signatures, similar to [L79], i.e., messages are basically signed bit by bit. Therefore, although messages can be hashed before signing and tree-authentication is used (similar to [M80]), this general construction is not very efficient.

There is an efficient variant especially suited for making clients unconditionally secure in on-line payment systems, see [P91]. However, in this scheme, all signatures by one client (with one key) must have the same recipient, like the bank in a payment system. Furthermore, signing is a 3-round protocol between the signer and the recipient.

The first efficient general fail-stop signature scheme was presented in [HP92]. The unforgeability relies on a discrete logarithm assumption. Signatures for one message block are about as efficient as with RSA. Messages can be hashed before signing. In contrast to RSA, the signer needs some new random bits for each new signature, and tree authentication is needed to keep the public keys short. However, fast hash functions can be used without reducing the security of the signer.

Related Types of Systems

In [CHP92], unconditional security for the signer was achieved in undeniable signatures (cf. [CA90]). The construction was the first not to use bit-by-bit signing. Apart from the usual differences between ordinary and undeniable signatures, this scheme differs from efficient fail-stop signatures in two ways: First, although the signatures themselves are efficient, the verification protocol requires quite a lot of computation, because it needs σ challenges (similar to signatures) to achieve an error probability of $2^{-\sigma}$. Secondly, if the computational assumption is broken, signers can disavow signatures, but there is no way for the recipient to prove to a third party that this is due to cheating (whereas with fail-stop signatures, third parties can distinguish whether the signatures just don't pass the test, or whether they are disavowed due to a proof of forgery). In particular, one cannot stop the scheme as soon as this happens.

In [CR91], unconditionally secure signatures were introduced, i.e., signature-like schemes where both the signer and the recipient are unconditionally secure. In [PW92], a transferable version was presented, i.e., signatures can be passed on from one recipient to another, and security against active attacks on recipients was achieved; such attacks must be considered because the recipients, too, have secret information in such schemes. With these extensions, unconditionally secure signatures could in principle replace other signatures in many applications. So far, however, they are too inefficient to be used in practice: They require a complicated interactive key generation protocol in many rounds, and signatures are very long. Hence they cannot replace ordinary or fail-stop signatures at present.

Overview over the New Results

We present two new constructions of efficient fail-stop signatures (Ch. 3 and 5) and some general lower bounds (Ch. 4).

The first construction has similar properties to that from [HP92], but the unforgeability is based on factoring instead of the discrete logarithm. Signing a message block requires about one modular exponentiation, testing a little more than two. Key exchange is in general more complicated than for the discrete logarithm scheme. Nevertheless, with all types of cryptographic systems it is useful to have alternative constructions, in case one of the unproven assumptions is broken.

The second construction and the first lower bound deal with the fact that in all fail-stop signature schemes so far, the size of the secret key is linear in the number of messages to be signed. We show that in the sense of secret storage needed, this can be avoided, whereas in the sense of choosing secret random bits, it cannot.

Constructions with small secret storage may be important since secret storage is quite hard to realize: One needs a more or less tamper-proof device. In contrast, information can quite easily be stored just securely, since one can distribute several copies. (Note that even ordinary digital signatures assume that a lot of information can be stored securely, since all signatures must usually be stored by their recipients.) In Ch. 5, we present an efficient construction where the size of the secret storage space is logarithmic in the number of messages to be signed, and an otherwise less efficient variant where this size is constant.

For the lower bounds, we assume that the probability that a forgery cannot be proved is smaller than $2^{-\sigma}$ for some security parameter σ, and that the recipient wants a similar level of security at least against simple brute-force forging algorithms. Then the most important result we obtain about fail-stop signatures is:

- If N messages are to be signed, the signer needs at least $(N + 1)(\sigma - 1)$ secretly chosen random bits. More precisely, this is a lower bound for the entropy of her secrets, given the public key.

Additionally, we show two more lower bounds for fail-stop signatures. They are not much larger than similar bounds for ordinary digital signatures would be, since they concern parameters where the difference between current fail-stop signatures and ordinary digital signatures is already quite small.

- The entropy, and hence the length, of a signature is at least $2\sigma - 1$, and the entropy of the public key is at least σ, even if a prekey is already given, i.e., some information trusted by recipients and chosen before the signer chooses her actual keys.

Finally, we show that unconditionally secure signatures cannot be as efficient as fail-stop signatures:

- The entropy (and thus the length) of each unconditionally secure signature that can be tested by M participants, including those that only have to settle disputes, is at least $M \cdot \sigma$.

2   Brief Sketch of Definitions

Like an ordinary digital signature scheme, a fail-stop signature scheme contains a method to generate secret and public keys and algorithms *sign* for signing messages and *test* for testing signatures. Additionally, there is an algorithm *prove*, which the signer uses to produce a proof of forgery from a forged signature, and an algorithm *proof_test*, which everybody else uses to test if something really is a proof of forgery.

A secure fail-stop signature scheme has the following properties, where 2. is a consequence of the others:

1. If the signer signs a message correctly, then the recipient accepts the signature.

2. A polynomially bounded forger cannot make signatures that pass the signature test.

3. If an unrestricted forger succeeds in constructing a signature that passes the signature test, then with "overwhelming" probability, the signer can produce a proof of forgery that convinces any third party that a forgery has in fact occurred (i.e., the output of *prove* passes *proof_test*).

4. A polynomially bounded signer cannot make a (false) signature that she can later prove to be a forgery.

The basic idea to achieve these properties is that (exponentially) many secret keys correspond to each public key, and different secret keys give different signatures on the same message. The signer knows exactly one of these secret keys and can only construct one of the possible signatures on a given message. However, even an arbitrarily powerful forger does not have sufficient (Shannon) information to determine which of the many possible signatures the signer can construct on a new message. Consequently, with very high probability a forged signature will be different from the signature that the signer would have constructed. The knowledge of two different signatures on the same message then yields a proof of forgery.

Since there must be security for both signers (see 3.) and recipients (see 4.), both take part in key generation. Usually, the recipient (or all possible recipients together, or a device trusted by all recipients) chooses a value called prekey, such as a number that the signer cannot factor, and then the signer chooses the real secret and public key based on this prekey. However, we prove the lower bounds for an arbitrary key generation protocol.

There are also two security parameters: σ determines that the probability of unprovable forgeries is smaller than $2^{-\sigma}$, and k is the parameter for the cryptographic security of the recipient. Usual choices of σ may be between 40 and 100, whereas k, if it is the binary length of numbers that should be hard to factor in Ch. 3, must be larger than 500.

Remark: Note that it is not a matter of the definition how one acts if a proof of forgery occurs. In particular, instead of making signers unconditionally secure by invalidating signatures after proofs of forgery, one could leave the responsibility with the signer. Then one has all the properties of an ordinary digital signature scheme, plus the possibility to stop after forgeries. (This shows that fail-stop signatures are a strictly stronger notion.)

Furthermore, the current definition does not specify for how much of a system a particular proof of forgery is valid. As long as forging even one signature is provably as hard as, say, factoring, one should stop the whole scheme after any forgery, because if one signature has been forged, the same forger can probably forge them all. Therefore, the constructions usually assume that there is just one type of proof of forgery. However, it is no problem to make proofs of forgery specific to the keys of individual signers or even (although currently with some loss in efficiency) to each particular signature. ◆

For a complete formal definition, see [PW90]. In this abstract, we will only make those parts more precise that are actually needed in the proofs of the lower bounds.

3 Efficient Fail-Stop Signatures based on Factoring

This section presents a fail-stop signature scheme based on the assumption that it is infeasible to factor large integers. To emphasize the generality of the construction, the scheme is first described in general terms. Like in [HP92], we first present a version for signing just one message block.

3.1 General Structure of the Construction

The following construction generalizes that from [HP92]. We base it on so-called **bundling homomorphisms**, i.e., functions h with the following properties:
1. h is a homomorphism between two Abelian groups.
2. Given an image $h(a)$, there exist at least 2^τ possible preimages.
3. It is infeasible to find collisions, i.e., two different values that are mapped to the same value by h.

More precisely, there must be a family of such functions and groups, and a key generation algorithm that selects a particular function h, given τ and a security parameter k. One also needs efficient algorithms for the group operations and to choose random elements.

Now we define all the components of a fail-stop signature scheme (cf. Ch. 2):

- Prekey: The recipient selects a function h from the family. Let the domain be G and the range H.

- Prekey test: The recipient must prove that his choice of h was correct, or at least that his h is in a set of functions with Properties 1 and 2 (which are needed for the security of the signer).

- Secret key: $sk := (sk_1, sk_2)$, where sk_1 and sk_2 are chosen at random from G.

- Public key: $pk := (pk_1, pk_2)$, where $pk_i = h(sk_i)$ for $i = 1, 2$.

- Signing: $sign(sk, m) = sk_1 \cdot sk_2{}^m$ for messages m from a subset (to be defined) of $\mathbb{Z}$.

- Test: $test(pk, m, s) = ok \; :\Leftrightarrow \; pk_1 \cdot pk_2{}^m = h(s)$.

- Proving forgeries: Given a forged signature sf on a message m^*, the signer computes $s = sign(sk, m^*)$, and if $s \neq sf$, she uses the pair (s, sf) as a proof of forgery.

- Testing proofs of forgery: Given two elements of $\acute{G}$, verify that they collide under h.

Theorem 1: Independently of the choice of h, the following holds:

1. Correct signatures pass the test: $h(s) = h(sk_1 \cdot sk_2{}^m) = pk_1 \cdot pk_2{}^m$.

2. A polynomially bounded signer cannot construct a signature and a proof that it is a forgery.

3. If sf is a forged signature on m^* and $sf \neq sign(sk, m^*)$, then the signer obtains a valid proof of forgery. ◆

Proof: Follows easily from the definitions. ☐

This theorem shows that the general scheme is secure for the recipients, and that it is also secure for the signer if even an all-powerful forger cannot guess a correct signature $s = sign(sk, m^*)$, except with a very small probability. In order to estimate the probability with which a forger can guess s, first note that the public key contains no information about which of at least $2^{2\tau}$ possible secret keys the signer actually has. However, after having received a signature on a message m, the forger has more information about sk. Theorem 2 gives a condition for when this information is not sufficient to construct new signatures that the signer cannot repudiate:

Theorem 2: Let pk, a signature $s = sign(sk, m)$, and a message $m^* \neq m$ be given, and let $m' := m^* - m$. Whatever value sf an all-powerful forger selects as a forged signature on m^*, the probability that it is correct is at most $|T| / 2^\tau$, where

$$T := \{d \in G \mid h(d) = 1 \wedge d^{m'} = 1\} = \{d \mid h(d) = 1 \wedge \operatorname{ord}(d) \mid m'\}.$$

(The probability is given by the secret keys that are still possible when pk and s are known.)
◆

Proof: The set of possible secret keys is

$$SK^* := \{(sk_1, sk_2) \in G \times G \mid h(sk_1) = pk_1 \wedge h(sk_2) = pk_2 \wedge sk_1 \cdot sk_2{}^m = s\}$$
$$= \{(s / sk_2{}^m, sk_2) \mid h(sk_2) = pk_2\},$$

because of the homomorphism property; and the size of this set is 2^τ. The attacker is successful if

$$sk_1 \cdot sk_2{}^{m^*} = sf.$$

For keys from SK^*, this equation is equivalent to $sk_2{}^{m'} = sf / s$. This equation may be unsolvable, but if there is any solution sk_2^*, then the set of all solutions in SK^* is

$$\{(s / sk_2{}^m, sk_2) \mid h(sk_2 / sk_2^*) = 1 \wedge (sk_2 / sk_2^*)^{m'} = 1\}.$$

Hence the number of solutions is $|T|$, and the attacker is successful with the claimed probability. ☐

Consequently, in order to estimate the probability of successful forgeries we must find the size of T. This size depends on the chosen family of homomorphisms.

3.2   The Special Case with Factoring

Our family of bundling homomorphisms was defined in [BPW91], using ideas from [GMR88, G87]: A member of the family is characterized by τ and a k-bit integer $n = pq$, with p, q prime and $p \equiv 3$ and $q \equiv 7 \bmod 8$. We omit τ and n in the following. The groups are

$$H = \pm QR/\{\pm 1\}, \text{ and } G = \mathbb{Z}_{2^\tau} \times H,$$

where QR denotes the group of quadratic residues modulo n, and the operation on G is given by

$$(a, x) \circ (b, y) := ((a + b) \bmod 2^\tau, x \cdot y \cdot 4^{(a + b) \operatorname{div} 2^\tau}).$$

Elements of H are represented by numbers between 0 and $n/2$; H is used instead of QR because membership can be tested efficiently. The unit element of G is $(0,1)$. The homomorphism is given by

$$h((a, x)) = \pm (4^a \cdot x^{2^\tau}).$$

Theorem 3: The construction described above is a family of bundling homomorphisms. Properties 1 and 2 even hold for any odd n. Furthermore, if n is chosen correctly or at least as $n = p^r q^s$ where p and q are correct and r, s odd, then for any a, z, there exists exactly one x so that $h((a, x)) = z$.   ♦

Proof: See [BPW91]. The last sentence is only proved for correctly chosen n there, but the same proof is valid for the more general form.   □

To use these homomorphism in a fail-stop signature scheme according to Sect. 3.1, let the message space be $\{0, ..., 2^\rho - 1\}$ for some ρ and $\tau := \rho + \sigma$. As an efficient prekey test, we use the protocol from [GP88] and a test that $n \equiv 5 \bmod 8$. Actually, this does not completely prove that n is of the correct form, but it ensures that $n = p^r q^s$ where p and q are correct and r, s odd.

Theorem 4: With the definitions made above, the probability of undetected forgery is at most $2^{-\sigma}$.   ♦

Proof: According to Th. 1 and 2, it only remains to prove $|T| \leq 2^\rho$. Note that in G

$$(a, x)^{m'} = (0, 1) \implies m' \cdot a \bmod 2^\tau = 0 \implies \operatorname{ord}(a) \mid m'.$$

Hence $T \subseteq \{(a, x) \in G \mid h((a, x)) = 1 \wedge \operatorname{ord}(a) \mid m'\}.$

According to Th. 3, for each a, there is exactly one x such that $h((a, x)) = 1$. Thus

$$|T| \leq |\{a \in \mathbb{Z}_{2^\tau} \mid \operatorname{ord}(a) \mid m'\}| = \gcd(2^\tau, m'). \tag{3}$$

By the choice of message space, every pair of messages $m \neq m^*$ satisfies $|m - m^*| < 2^\rho$ and therefore $\gcd(2^\tau, m - m^*) < 2^\rho$.   □

As to efficiency, first note that a multiplication in G is mainly one modular multiplication, since the exponent of 4 is 0 or 1, and a multiplication by 4 can be replaced by shifts and subtractions. We can choose any fixed message length ρ; long messages are hashed before signing. Since even the hash functions as secure as factoring from [D88] take only one multiplication per message bit, i.e., not more than signing or testing, one should always hash messages as short as possible. Thus ρ is determined by the size of the output of the hash function. In the following table of the efficiency of signing one message block, we assume $\rho = k$. If one trusts a faster hash function, or in applications where only short messages are signed, one can still gain efficiency by making ρ smaller.

sign:	k multiplications
test:	$2k + \sigma$ multiplications
Length of *pk*:	$2k$
Length of *sk*:	$4k + 2\sigma$
Signature length:	$2k + \sigma$

To sign several messages, one can use tree authentication as in [PW91, HP92], after [M80]. Note that key exchange is more efficient in [HP92] because the choice of the prekey is just a choice of random numbers, and no prekey test is necessary even if there is no trusted device to choose the prekey.

4   Lower Bounds

The idea of each of our proofs will first be described informally. For the formal sketches, we assume the reader knows the notions of conditional entropy, $H(X \mid Y)$, and mutual information, $I(X; Y)$; see [S49, G68 Sect. 2.2, 2.3]. Like in [G68], we use capital letters for random variables and small letters for corresponding values, and abbreviate $P(X = x)$ by $P(x)$ etc. The formula we need most is the **chain rule** to add entropies:

$$H(Y, Z \mid X) = H(Y \mid X) + H(Z \mid Y, X).$$

Additionally, when we know that the probability that something can be guessed correctly is small, and want to derive that a conditional entropy is large, we often need **Jensen's inequality** for the special case of the logarithm [F71]: If $p_i \geq 0$ and $x_i > 0$ for all i, and the sum of the p_i's is 1, then

$$\log\left(\sum_i p_i x_i \right) \geq \sum_i p_i \log(x_i).$$

4.1   Secret Keys, or Rather, Secret Random Choices

The basic reason why the signer needs a lot of secretly chosen random bits is:

1. Even an arbitrarily powerful forger must not be able to guess the signer's correct signatures.

2. Since this holds for each additional signature, even when some signatures are already known, the entropy of each new signature must be large, and therefore the overall entropy of the signer's secrets is large.

However, sometimes the forger *does* know correct signatures on new messages. For instance, in schemes with message hashing, the forger knows the signatures on all messages with the same hash value as the original message. (Then the collision counts as a proof of forgery.) Hence Statement 1 does not hold absolutely. Instead, we will derive an average version as follows:

1.1 With high probability, the signer should not obtain proofs of forgery if she applies *prove* to her correct signatures; otherwise she could cheat the recipient. (The probability is over the choice of the keys; we will see that we can leave the messages fixed.)

1.2 Thus, on average, even an all-powerful forger must not be able to guess those correct signatures.

In 1.1, the recipient's security is needed. (Note that the desired theorem cannot possibly be proved from the signer's security alone. As a counterexample consider that the signer were allowed to disavow all signatures in an ordinary digital signature scheme; then she would be unconditionally secure without many random bits, but the recipient would not be secure at all.) This is a problem, since the recipient's security, like all computational cryptographic definitions, is only defined asymptotically. It says: For any polynomial-time algorithm $\tilde{A}$ and any c, there exists k_0 so that the probability that $\tilde{A}$ successfully cheats the recipient is smaller than k^{-c} for all $k \geq k_0$. Thus, in a certain sense, we can only derive lower bounds for $k \geq k_0$, for an unknown k_0. This may seem unsatisfactory: Nobody would have doubted that we need arbitrarily long keys if we make k sufficiently large.

However, note that the real purpose of our lower bounds is to say "whenever we have certain requirements on the security, then we have to pay the following price in terms of efficiency". In this section, this is more precisely: "If the signers want the probability of unprovable forgery to be at most $2^{-\sigma}$, and the recipients want some security, too, then at least the following number of random bits is needed (as a function of σ and the security of the recipients)".

To quantify the security of the recipient, it suffices for our purpose to consider the case of Statement 1.1 above, i.e., we consider the probability with which the signer can prove that her own correct signatures are forgeries just by applying the algorithm *prove* to them. In practice, one has to require this probability to be at most, say, 2^{-20}, or, more generally, $2^{-\sigma^*}$ for some σ^*. We will prove the lower bounds as a function of this parameter σ^* (in addition to the σ from the signer's security). To formulate the theorem precisely, we need some more notation and partial definitions:

- **Key exchange and probability space:** Key exchange is a protocol G with inputs σ, k, and the number of messages to be signed, N, all in unary. The output is a pair (sk, pk) of a secret and a public key. For the lower bounds, we only need the case where all parties execute G honestly, and we always consider a fixed triple of parameters. Then the probabilities of sk and pk are uniquely determined, and we can define corresponding random variables SK and PK.

  Without loss of generality, we assume that all random bits that the signer needs are already in sk, so that *sign* is deterministic, and so are *test*, *prove*, and *proof_test*. Thus, the underlying probability space for all probabilities is that of the secret random bits used in the key exchange.

- **Signing:** We make the lower bounds quite general by permitting the signer to use memory in a general way, i.e., signatures may depend on all previously signed messages. We even allow testing to be equally general, although this is only useful when there is a single recipient.

- **Probability that the signer can disavow her correct signatures:** For every message sequence $\underline{m} = (m_1, \ldots, m_{N+1})$, we define a polynomial-time algorithm $\tilde{A}_{\underline{m}}$ to describe what a dishonest signer would do to disavow her own signatures: After executing G correctly, i.e., on input sk, she first signs m_1, ..., m_N correctly. Then, since m_{N+1} is one message too much, she signs it as if she had not signed m_N. From each of these signatures, together with sk and the history of preceding signatures, she tries to compute a proof of forgery using *prove*. (This algorithm should be rather useless!)

If the fail-stop signature scheme, N, and σ are fixed, we say that k is large enough to provide the security level σ^* for the recipient against $\tilde{A}_m$ if the success probability of $\tilde{A}_m$ is at most $2^{-\sigma^*}$.[4]

The formal version of the theorem is therefore:

Theorem 5: Let a fail-stop signature scheme with actual parameters σ and N and a security level σ^* be given. Let $\sigma' := \min(\sigma, \sigma^*)$. Then for all k sufficiently large to provide the security level σ^* for the recipient against an algorithm $\tilde{A}_m$ for any sequence m of $N+1$ pairwise distinct messages[5],

$$H(SK \mid PK) \geq (N+1)(\sigma'-1). \qquad \blacklozenge$$

Since m is fixed within the theorem, we can omit it in the proof. Let S_i denote the random variable of the signature on the i-th message of m, and $Hist_i$ that of the history of the first i signatures. The following lemma formalizes that on average, correct signatures cannot be guessed:

Lemma 1: With the same notation as in Th. 5, for each $i \leq N+1$:

$$H(S_i \mid PK, Hist_{i-1}) \geq \sigma' - 1. \qquad \blacklozenge$$

We must omit the proof of Lemma 1 in this abstract. However, it proceeds along the informal description, exploiting the difference that correct signatures can usually not be disavowed, whereas guessed ones can, with an application of Jensen's inequality at the end.

Proof of Th. 5: First we use Lemma 1 to show by induction over i that the entropy of all signatures together is large. Remember $Hist_i = (S_1, ..., S_i)$. Hence, we show for all $i \leq N+1$:

$$H(Hist_i \mid PK) \geq i \cdot (\sigma' - 1). \qquad (1)$$

For $i = 1$, (1) is just Lemma 1. And if (1) has already been proved for $i-1$, then it holds for i because

$$H(Hist_i \mid PK) = H(S_i \mid PK, Hist_{i-1}) + H(Hist_{i-1} \mid PK)$$
$$\geq (\sigma'-1) + (i-1)(\sigma'-1) = i \cdot (\sigma'-1).$$

We now use that signing is deterministic, i.e., SK uniquely determines $Hist_{N+1}$. This implies $H(Hist_{N+1} \mid PK, SK) = 0$, and therefore with the chain rule

$$H(SK \mid PK) = H(SK, Hist_{N+1} \mid PK) - H(Hist_{N+1} \mid PK, SK) = H(SK, Hist_{N+1} \mid PK)$$
$$\geq H(Hist_{N+1} \mid PK) \geq (N+1)(\sigma'-1). \qquad \square$$

4.2 Signatures and Public Keys

Signatures and public keys are not much longer in current fail-stop signature schemes than in ordinary signature schemes. Hence the lower bounds are very small, too.

The basic idea about the length of a signature is:

---

[4] The formal definition of the recipient's security immediately implies the existence of k_0 such that all $k \geq k_0$ have this property. We have now bypassed the problem that we do not know how large k_0 is because we just know that it must be large enough in a practical application.

[5] Note that we only require security against $\tilde{A}_m$ for one message sequence m. The contrary is that *all* these algorithms work.

a) First, there must be at least 2^σ acceptable signatures; otherwise the correct signature could be guessed too easily.

b) Secondly, it must be hard for a forger to guess signatures at all. Thus the density of the set of acceptable signatures within the signature space should be small, e.g., at most $2^{-\sigma^*}$.

Hence we expect the size of the signature space to be at least $2^{\sigma+\sigma^*}$. Indeed, we prove more generally that the entropy of each signature is at least $\sigma+\sigma^*$. What has to be done is:

• Since the forger in (b) is computationally restricted, we must show that he could guess acceptable signatures *efficiently* if their density was too high.

• As in 4.1, we must require that k is sufficiently large so that a concrete version of the asymptotic security against forgery holds.

• We must express the idea with the density in information-theoretic terms.

For this, we first define a simple algorithm $\tilde{F}_m$ that tries to guess signatures on a message m (in a rather stupid way): $\tilde{F}_m$ just chooses its own key pair (sk^*, pk^*) and signs m with sk^*.

Theorem 6: Assume a fail-stop signature scheme with actual parameters k, σ, N provides the security level σ^* against forgery by an algorithm $\tilde{F}_m$.

1. Let S be the random variable of the signature. Then
$$H(S) \geq \sigma' + \sigma^* - 1.$$

(If the scheme is not memory-less, we obtain the same result for later messages by using the last message of a message sequence $\underline{m}$.)

2. $\qquad\qquad\qquad\qquad\qquad H(PK) \geq \sigma^*.$ ♦

The following lemma formalizes the density argument. The fact that the number of possible signatures, given the public key, is much smaller than the complete signature space is generalized as follows: The public key contains a lot of information about the correct signature.

Lemma 2: With the same notation as in Th. 6,
$$I(S; PK) \geq \sigma^*.$$ ♦

The proof must be omitted in this abstract.

Proof of Th. 6: Lemma 2 means $H(S) - H(S \mid PK) \geq \sigma^*$; and a special case of Lemma 1 is $H(S \mid PK) \geq \sigma' - 1$. Consequently, $H(S) \geq H(S \mid PK) + \sigma \geq \sigma' + \sigma^* - 1$. Furthermore, $I(S; PK) \leq H(PK)$. □

For the case with a prekey (cf. Ch. 2), we obtain the same results with an additional condition over K, i.e., $H(S \mid K) \geq \sigma' + \sigma^* - 1$ and $H(PK \mid K) \geq \sigma^*$. If, as usual in such schemes, PK is a function of SK, we obtain one more result by applying the chain rule to the last formula and Th. 5:

Theorem 5*: In a fail-stop signature scheme with prekey, and where the public key is a function of the secret key, and with the same notation as in Th. 5 and 6,
$$H(SK \mid K) \geq (N+1)(\sigma'-1) + \sigma^* \geq (N+2)(\sigma'-1).$$ ♦

4.3   Unconditionally Secure Signatures

Unconditionally secure signature schemes could be achieved by replacing the globally known public key pk (which implied that an all-powerful forger could find acceptable signatures by brute-force search) by different test keys t_x for each recipient x. So far, this has made key exchange complicated and signatures long.

Essentially, we prove that such signatures must indeed be at least as long as if they consisted of an independent part for each test key, i.e., they cannot be shortened by a suitable combination. Assume M people may want to test a signature (as a recipient, or to settle a dispute), and that the probability for successful forgeries is to be $\leq 2^{-\sigma}$. The basic idea is: If some participants want to forge a signature on m, they can determine the set of signatures acceptable under all their test keys. Still, within this set, the density of signatures that another participant accepts must not exceed $2^{-\sigma}$. Inductively, this implies that the size of the original signature space must be at least $2^{M\sigma}$.

In Theorem 7, we generalize this to entropies, and we show that it holds for every signature, even if signing is not memory-less.

Theorem 7: Consider an unconditionally secure signature scheme with M recipients where N messages can be signed and the probability of successful forgery is $\leq 2^{-\sigma}$. For any given message sequence $\underline{m}$, let S_i denote the random variable of the signature on the i-th message of $\underline{m}$, and $Hist_i$ that of the history of the first i signatures. Then
$$H(S_i \mid Hist_{i-1}) \geq M\sigma. \qquad \blacklozenge$$

The basic idea for the proof is formalized similar to Lemma 2: Even when some test keys are known, any other test key still gives a lot of information about the correct signature.

Lemma 3: With the same notation as in Th. 7: For any set X of participants and $y \notin X$, if T_X denotes the joint random variable of the test keys of X:
$$I(S_i; T_y \mid T_X, Hist_{i-1}) \geq \sigma. \qquad \blacklozenge$$

Again, we must omit the proof in this abstract.

Proof of Th. 7: Lemma 3 means $H(S_i \mid T_X, Hist_{i-1}) \geq \sigma + H(S_i \mid T_{X \cup \{y\}}, Hist_{i-1})$. With induction over the size of X, one easily obtains the desired result. $\qquad \square$

From Th. 7 and Lemma 3, with induction over i similar to that in Th. 5, we can also obtain
$$H(SK) \geq (N+1)M\sigma \ \wedge \ H(T_y \mid T_X) \geq (N+1)\sigma.$$

5   Fail-Stop Signatures with Small Secret Storage

To show that the signer needs far less secret storage than the number of secret bits she must choose according to Th. 5, we proceed in two steps: First we show a simple construction where only a small amount of secret storage is needed at the start, i.e., directly after key exchange. Then we add additional measures so that the amount of secret storage is small all the time.

The basis of this section is a fail-stop signature scheme for signing just one message of arbitrary length. We use the scheme described in Section 3.1 combined with message hashing. Hence the construction works for the schemes from [HP92] and Section 3.2.

(a) Small amount of secret information at the start: Use "top-down" tree-authentication similar to [M88, GMR88]. (Note that a different "bottom-up" version, which is a little more efficient if one does not consider secret storage space, was normally used with fail-stop signatures so far.) Let a prekey, i.e., a bundling homomorphism h, be given. The signer starts with one pair of a secret and a public key at the root of the tree. Then she creates two children, each with a new key pair, and uses the old secret key to sign a message containing the two new public keys. For each of the two new nodes, she again constructs two children in the same way, and so on. Messages are signed using the secret keys at the leaves of the tree, and a complete signature is one branch of these original signatures.

During key exchange, only the root of the tree has to exist, and to sign the first message, only the keys on the left-most branch and their immediate other children have to be generated. Figure 1 shows the situation after the first message, $m_{0...0}$, has been signed.

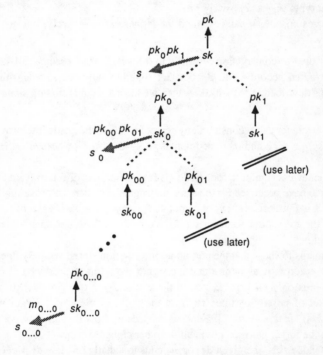

Figure 1   Fail-stop signature scheme with "top-down" tree authentication.
Thin black arrows denote the computation of a public key from a secret key in a basic scheme to sign just one message (like in Ch. 3, together with message hashing), broad grey arrows denote signatures in the basic scheme, and dotted lines just indicate a tree, but are not related to a computation.

At any time, just one branch of the tree has to be stored for signing. However, so far, the individual secret keys sk_j that are used up, i.e., that are no longer needed for signing, must be stored until the end so that forgeries at any node can be proved.

(b) Small amount of secret storage altogether: The basic idea to reduce secret storage further is to store values sk_j that are used up in encrypted form and to store just the key secretly. However, information-theoretically secure encryption is needed, and a one-time pad is of no use because the key would be just as long as the encrypted message. Hence special care must be taken that each individual sk_j is still secret enough, although information about the ensemble of sk_j's may become known.

If the individual sk_j's are formed according to Section 3.1, this is achieved by the following additional steps:

1. Initially, the signer chooses a value $e \in G$ randomly as an encryption key. She keeps e secret all the time.

2. Whenever the signer has used up a value $sk_j = (sk_{j,1}, sk_{j,2})$ by signing a message m_j, she proceeds as follows:
   - She encrypts $sk_{j,2}$ as $c_j := sk_{j,2} \cdot e$.
   - She stores m_j, the signature s_j, and the ciphertext c_j securely, but not necessarily secretly.

Theorem 8: If the tree construction described in (a) is applied to a secure fail-stop signature scheme constructed according to Section 3.1 together with message hashing, and the additional steps described in (b) are taken, then we have a secure fail-stop signature scheme again. ◆

Proof: First, the signer can reconstruct any secret key sk_j if she needs it to prove a forgery: She decrypts $sk_{j,2} = c_j / e$ and then recomputes $sk_{j,1} = s_j / sk_{j,2}^m$, where m is the hash value of m_j.

Hence, whenever a signature for a node j is forged and it is different from the signature the signer would have produced for the same message, the signer can prove this forgery just as in Section 3.1. Furthermore, every complete forgery sf (i.e., a branch of the tree) must be linked into the correct tree somewhere, i.e., it contains at least one such forgery at a node j for the correct pk_j.

Thus it remains to show that the additional information stored securely does not help a forger to find exactly the signature that the signer would have produced at node j. This signature depends only on sk_j (i.e., not on the values sk_l at other nodes). In the original scheme, the set of possible values sk_j from the point of view of a forger was $SK_j^* = \{(s_j / sk_{j,2}^m, sk_{j,2}) \mid h(sk_{j,2}) = pk_{j,2}\}$. Hence it suffices to show that all these values are still possible when the forger has seen c_j and all the other ciphertexts c_l.

Let such a value $sk^*_{j,2}$ be given. It corresponds to exactly one key $e^* = c_j / sk^*_{j,2}$. This implies that the other plaintexts must be $sk^*_{l,2} = c_l / e^* = sk^*_{j,2} \cdot c_l / c_j$. The only question is if these are possible plaintexts, i.e., if $h(sk^*_{l,2}) = pk_{l,2}$. On the one hand, $h(sk^*_{l,2}) = h(sk^*_{j,2}) \cdot h(c_l) / h(c_j) = pk_{j,2} \cdot h(c_l) / h(c_j)$. On the other hand, $h(c_l) = h(sk_{l,2}) \cdot h(e) = pk_{l,2} \cdot h(e)$ and $h(c_j) = pk_{j,2} \cdot h(e)$, hence $h(c_l) / h(c_j) = pk_{l,2} / pk_{j,2}$. This yields $h(sk^*_{l,2}) = pk_{l,2}$. □

Consequences: If this construction is applied to a usual complete tree, then it is very practical, and at any time, only e and the secret keys that have been marked "use later", i.e., at most one per level of the tree, must be stored secretly. This is a logarithmic amount.

If we use a list-like tree, i.e., the left child of each node is a real message, we only need two sk_j's at any time. However, later signatures are very long. Thus the list-like version should only be used with a fixed recipient, who can store the part of the list he already received, like in [P91].

One can also use trees of other forms or combine it with other methods to sign several messages from [HP92].

6   Conclusion

We have constructed efficient fail-stop signatures based on the assumption that factoring large integers is hard, giving an alternative to the previous scheme based on a discrete logarithm assumption. We also presented a construction which only needs a small amount of secret storage space, whereas in all previous constructions, a secret key whose length was linear in the number of signatures to be issued was stored all the time.

On the other hand, we proved that there is a definite difference to ordinary digital signatures in that the signer must choose an amount of random bits linear in the number of signatures to be issued. Finally, we showed that there is no hope that unconditionally secure signatures can become as efficient as fail-stop signatures, because the length of each unconditionally secure signature is linear in the number of participants who can test it, whereas the length of a fail-stop signature (or an ordinary digital signature) does not depend on this number.

Acknowledgements

It is a pleasure to thank Joachim Biskup, Gerrit Bleumer, David Chaum, Andreas Pfitzmann, and Michael Waidner for interesting discussions.

References

[BPW91] Gerrit Bleumer, Birgit Pfitzmann, Michael Waidner: A remark on a signature scheme where forgery can be proved; Eurocrypt '90, LNCS 473, Springer-Verlag, Berlin 1991, 441-445.

[CA90] David Chaum, Hans van Antwerpen: Undeniable signatures; Crypto '89, LNCS 435, Springer-Verlag, Heidelberg 1990, 212-216.

[CHP92] David Chaum, Eugène van Heijst, Birgit Pfitzmann: Cryptographically Strong Undeniable Signatures, Unconditionally Secure for the Signer; Crypto '91, LNCS 576, Springer-Verlag, Berlin 1992, 470-484.

[CR91] David Chaum, Sandra Roijakkers: Unconditionally Secure Digital Signatures; Crypto '90, LNCS 537, Springer-Verlag, Berlin 1991, 206-214.

[D88] Ivan Bjerre Damgård: Collision free hash functions and public key signature schemes; Eurocrypt '87, LNCS 304, Springer-Verlag, Berlin 1988, 203-216.

[DH76] Whitfield Diffie, Martin E. Hellman: New Directions in Cryptography; IEEE Transactions on Information Theory 22/6 (1976) 644-654.

[F71] William Feller: An Introduction to Probability Theory and Its Applications, Vol. II (2nd. ed.); John Wiley & Sons, New York 1971.

[G68] Robert G. Gallager: Information Theory and Reliable Communication; John Wiley & Sons, New York 1968.

[G87]   Oded Goldreich: Two Remarks Concerning the Goldwasser-Micali-Rivest
          Signature Scheme; Crypto '86, LNCS 263, Springer-Verlag, Berlin 1987, 104-
          110.

[GMR88]   Shafi Goldwasser, Silvio Micali, Ronald L. Rivest: A Digital Signature Scheme
          Secure Against Adaptive Chosen-Message Attacks; SIAM J. Comput. 17/2
          (1988) 281-308.

[GP88]   Jeroen van de Graaf, René Peralta: A simple and secure way to show the validity
          of your public key; Crypto '87, LNCS 293, Springer-Verlag, Berlin 1988, 128-
          134.

[HP92]   Eugène van Heijst, Torben Pryds Pedersen: How to Make Efficient Fail-stop
          Signatures; Eurocrypt '92, Extended Abstracts, 24.-28. 5. 1992, Balatonfüred,
          Hungary, 337-346.

[L79]   Leslie Lamport: Constructing Digital Signatures from a One-Way Function; SRI
          Intl. CSL-98, Oct. 1979.

[M80]   Ralph C. Merkle: Protocols for Public Key Cryptosystems; Proc. 1980
          Symposium on Security and Privacy, Oakland 1980, 122-134.

[M88]   Ralph C. Merkle: A digital signature based on a conventional encryption
          function; Crypto '87, LNCS 293, Springer-Verlag, Berlin 1988, 369-378.

[P91]   Birgit Pfitzmann: Fail-stop Signatures; Principles and Applications; Proc.
          Compsec '91, 8th world conference on computer security, audit and control,
          Elsevier, Oxford 1991, 125-134.

[PW90]   Birgit Pfitzmann, Michael Waidner: Formal Aspects of Fail-stop Signatures;
          Fakultät für Informatik, University Karlsruhe, Report 22/90, Dec. 1990.

[PW91]   Birgit Pfitzmann, Michael Waidner: Fail-stop Signatures and their Application;
          Securicom 91, Paris, 19.-22. March 1991, 145-160.

[PW92]   Birgit Pfitzmann, Michael Waidner: Unconditional Byzantine Agreement for any
          Number of Faulty Processors; STACS 92, LNCS 577, Springer-Verlag, Berlin
          1992, 339-350.

[S49]   Claude E. Shannon: Communication in the Presence of Noise; Proceedings of
          the Institute of Radio Engineers 37/1 (1949) 10-21.

[WP90]   Michael Waidner, Birgit Pfitzmann: The Dining Cryptographers in the Disco:
          Unconditional Sender and Recipient Untraceability with Computationally Secure
          Serviceability; Eurocrypt '89, LNCS 434, Springer-Verlag, Berlin 1990, 690.

          (Full version: Unconditional Sender and Recipient Untraceability in spite of
          Active Attacks – Some Remarks; Fakultät für Informatik, University Karlsruhe,
          Report 5/89, March 1989.)

Provably Secure and Practical Identification Schemes and Corresponding Signature Schemes

Tatsuaki Okamoto

NTT Laboratories
Nippon Telegraph and Telephone Corporation
1-2356, Take, Yokosuka-shi, Kanagawa-ken, 238-03 Japan
Email: okamoto@sucaba.ntt.jp

Abstract. This paper presents a three-move interactive identification scheme and proves it to be as secure as the discrete logarithm problem. This provably secure scheme is almost as efficient as the Schnorr identification scheme, while the Schnorr scheme is not provably secure. This paper also presents another practical identification scheme which is proven to be as secure as the factoring problem and is almost as efficient as the Guillou-Quisquater identification scheme: the Guillou-Quisquater scheme is not provably secure. We also propose practical digital signature schemes based on these identification schemes. The signature schemes are almost as efficient as the Schnorr and Guillou-Quisquater signature schemes, while the security assumptions of our signature schemes are weaker than those of the Schnorr and Guillou-Quisquater signature schemes. This paper also gives a theoretically generalized result: a three-move identification scheme can be constructed which is as secure as the random-self-reducible problem. Moreover, this paper proposes a variant which is proven to be as secure as the difficulty of solving both the discrete logarithm problem and the specific factoring problem simultaneously. Some other variants such as an identity-based variant and an elliptic curve variant are also proposed.

1 Introduction

Public-key based identification schemes and digital signature schemes are very useful and fundamental tools in many applications such as electronic fund transfer and online systems for preventing data access by invalid users and proving the authenticity of messages.

Identification schemes are typical applications of zero-knowledge interactive proofs [GMRa], and several practical zero-knowledge identification schemes have been proposed [Bet, FiS, FFS, OhO1]. However, the zero-knowledge identification schemes have the following shortcomings in practice, where we simply call "black-box simulation zero-knowledge" "zero-knowledge", since we do not know of any effective measure to prove zero-knowledgeness except the black-box simulation technique, although "auxiliary-input zero-knowledge" is more general than "black-box simulation zero-knowledge":

- A zero-knowledge identification scheme requires more than three interactions (three-moves [1]) from Goldreich et.al.'s result [GK] unless the language for the proof is trivial. A zero-knowledge protocol is less practical than the corresponding (three-move) parallel version since interaction over a network often requires more time than taken by the calculation in these identification schemes. Although four-move and five-move zero-knowledge proofs have been proposed [BMO1, FeS2], these protocols impose fairly big additional communication and computation overheads compared to the three-move parallel versions (especially Type 2 below).
  Note: Here, the "(three-move) parallel version" denotes two types of protocols. One (Type 1) is just the parallel execution of a zero-knowledge protocol (e.g., the three-move version of the Fiat-Shamir scheme with $k = 1$ and $t = Poly(|n|)$ [FiS]). The other (Type 2) is a protocol which can be converted to zero-knowledge by executing the protocol repeatedly many times and setting the security parameter of one repetition to be constant (e.g., the three-move and higher-degree version of the Fiat-Shamir scheme [GQ, OhO1]). The communication complexity of the Type 1 protocol is the same as that of the original zero-knowledge protocol. Usually, the communication complexity of the Type 2 protocol is much less than that of the corresponding zero-knowledge protocol (or Type 1).
- No zero-knowledge identification can be converted into a signature scheme using Fiat-Shamir's technique [FiS], which is a truly practical way of converting an identification scheme into a signature scheme with a one-way hash function. This is because: if the identification protocol is zero-knowledge, the signature converted from this identification protocol through Fiat-Shamir's technique can be forged by using the same algorithm as the simulation for proving the zero-knowledgeness of the identification protocol. Therefore, for example, the above-mentioned four-move and five-move zero-knowledge proofs [BMO1, FeS2] cannot be used to construct a signature scheme.

In contrast, the three-move identification schemes [Bet, BM1, FiS, FFS, GQ, OhO1, Sch], which are the parallel version (Type 2) of zero-knowledge proofs, have the following merits in practice.

- The communication and computation overheads are smaller than those of the zero-knowledge identification schemes.
- The three-move identification schemes can be converted into practical signature schemes by using Fiat-Shamir's technique.

How then can we prove the security of the three-move identification schemes? As mentioned above, the zero-knowledge notion seems to be ineffective for this purpose. Feige, Fiat and Shamir [FFS] have developed an effective measure called "no-useful information transfer" to prove the security of their three-move identification scheme. Ohta and Okamoto [OhO1] have proposed a variant called

---

[1] A scheme is called "one-move" if prover A only sends one message to verifier B, and is called "two-move" if B sends to A and then A sends to B. "j-move" is defined in the obvious way.

"no transferable information with (sharp threshold) security level," which characterizes the security level theoretically. Therefore, only "no-useful information transfer" [FFS] and its variant [OhO1] have been known to be effective to prove the security of three-move identification schemes.

Only three three-move identification schemes [FFS, OhO1, BM1] have been proven to be secure assuming reasonable primitive problems, in the sense of [FFS, OhO1]. The Feige-Fiat-Shamir identification scheme [FFS], based on square root mod n, has been proven to be as secure as the factoring problem. The Ohta-Okamoto scheme [OhO1], which is the higher (the L-th) degree modification of the Feige-Fiat-Shamir scheme, has been proven to be as secure (with sharp threshold security level $1/K$) as factoring, where $v^{1/L}$ mod n has at least K solutions (e.g., $\gcd(L, p-1) = K$; see [OhO1] for more detail conditions). The Brickell-McCurley scheme [BM1], which is a modification of the Schnorr scheme [Sch], has been proven to be secure assuming that it is intractable to find a factor, q, of $p-1$, given additional information g whose order is q in $\mathbf{Z}_p^*$, although the security of their scheme also depends on the discrete logarithm.

Therefore, there is no existing alternative that is "provably secure" and "three-move" practical identification if factoring intractability fails in the future, since the security of all these provably secure schemes depends on the factoring assumption. In addition, although their schemes are efficient, they have some shortcomings in practice: the transmitted information size and memory size cannot be small simultaneously [FFS], and a priori fixed value v (e.g., v is the identity of a user) cannot be used as a public key [OhO1], (or the identity based scheme [Sha] cannot be constructed on this scheme). In addition, the security assumption of [BM1] is fairly stronger than the ordinary factoring problem (or the level of the provable security is lower than those of [FFS, OhO1]).

In contrast, other previously proposed practical three-move identification schemes, the Schnorr [Sch] and Guillou-Quisquater [GQ] schemes, have some merits compared to [FFS, OhO1, BM1]: The security of the Schnorr scheme depends on the discrete logarithm, which is a promising alternative if factoring becomes tractable, since we have several different types of discrete logarithms such as elliptic curve logarithms which seem to be more intractable than factoring. Moreover, the transmitted information size and memory size with these schemes can be small simultaneously, while it is impossible in [FFS]. The Schnorr scheme is more efficient than [BM1]. In addition, in the Guillou-Quisquater scheme, a priori fixed value v can be used as the public key. Unfortunately, the Schnorr and Guillou-Quisquater schemes are not provably secure. The difficulty of proving the security of these schemes resides in the fact that the discrete logarithm and RSA inversion have single solutions in restricted domains, that is, $\log_g x$ mod p has a single solution (x is in the restricted domain, $\{0, 1, \ldots, \mathrm{ord}(g) - 1 \}$), and $x^{1/e}$ mod n has also a single solution ($\gcd(e, \phi(n)) = 1$, ϕ is the Euler function).

In this paper, we propose three-move identification schemes that are proven to be as secure as the discrete logarithm or RSA inversion. We also propose a variant which is proven to be as secure as the factoring problem. Our new schemes inherit almost all the merits of the Schnorr and Guillou-Quisquater schemes even though they are provably secure. That is, these schemes are almost as efficient as the Schnorr and Guillou-Quisquater identification schemes from all practical viewpoints such as communication overhead, interaction number, required memory size, and processing speed. In addition, the new schemes duplicate the other advantage of the Guillou-Quisquater scheme: the identity based schemes can be constructed on these schemes.

This paper also develops new practical digital signature schemes from the proposed provably secure three-move identification schemes. The signature schemes are almost as efficient as the Schnorr and Guillou-Quisquater signature schemes, while the security assumptions of our schemes are weaker than those of the Schnorr and Guillou-Quisquater signature schemes. That is, the security (existentially unforgeable against adaptive chosen message attacks [GMRi]) of our new signature schemes only depends on just one reasonable assumption about the one-way hash function (or the existence of a "correlation-free one-way hash function") as well as the primitive assumption (e.g., the intractability assumption of the discrete logarithm).

We also extend these specific and practical results to a more general and theoretical result. We show that any random-self-reducible problem [TW] can lead to a provably secure and three-move identification scheme.

We also construct some variants of our new identification and signature schemes. One is a variant of our identification scheme based on the discrete logarithm using the idea of the Brickell-McCurley scheme [BM1]. This variant is proven to be as secure as the difficulty of solving both the discrete logarithm and the specific factoring problem (or the finding order problem) simultaneously, while, as mentioned above, the Brickell-McCurley scheme is proven to be secure assuming the intractability of the finding order problem, although the security of their scheme also depends on the discrete logarithm. Some other variants of our scheme, identity-based and certification-based versions, and an elliptic curve version, are also proposed. The elliptic curve variant has the significant property that it is proven to be secure assuming the intractability of the (non-supersingular) elliptic curve logarithms against which only exponential-time attacks are known so far.

2 Definition of Secure Identification

2.1 Identification

Definition 1. An *identification* scheme consists of two stages:

1. Initialization: In this stage, each user (e.g., A) generates a secret key (e.g., SK_A) and a public key (e.g., PK_A) by using probabilistic polynomial-time generation algorithm G on input of the key size. A link between each user and its public key is established. Note that in some schemes a part of the public key can be commonly shared among all users as a system parameter.
2. Operation: In this stage any user (e.g., A) can demonstrate its identity to a verifier by performing some identification protocol related to its public key

(e.g., PK_A), where the input for the verifier is the public key (e.g., PK_A). At the conclusion of this stage, the verifier either outputs "accept" or "reject".

2.2 Security of Identification schemes

We define a *secure* identification scheme based on the definition (the "no useful information transfer") given by Feige et. al. [FFS].

Definition 2. A prover A (resp. verifier B) is a "good" prover denoted by $\overline{A}$ (resp. "good" verifier denoted by $\overline{B}$), if it does not deviate from the protocols dictated by the scheme. Let $\tilde{A}$ be a fraudulent prover who does not complete the Initialization stage of Definition 1 as A and may deviate from the protocols (so another person/machine can simulate $\tilde{A}$). $\tilde{B}$ is not a good B. $\tilde{A}$ and $\tilde{B}$ are assumed to be polynomial time bounded machines, which may be nonuniform.

An identification scheme (A, B) is *secure* if

1. $(\overline{A}, \overline{B})$ succeeds with overwhelming probability.
2. There is no coalition of $\tilde{A}, \tilde{B}$ with the property that, after a polynomial number of executions of $(\overline{A}, \overline{B})$ and relaying a transcript of the communication to $\tilde{A}$, it is possible to execute $(\tilde{A}, \overline{B})$ with nonnegligible probability of success. The probability is taken over the distribution of the public key and the secret key as well as the coin tosses of $\overline{A}$, $\tilde{B}$, $\tilde{A}$, and $\overline{B}$, up to the time of the attempted impersonation.

Remark: When an identification scheme is "witness hiding" [FeS1] and an interactive proof of "knowledge" [FFS], this scheme is secure in the sense of Definition 2. This is roughly because if there exists $(\tilde{A}, \tilde{B})$ with nonnegligible probability of success, we can construct a knowledge extractor (from the "knowldge soundness"), which leads to contradiction with "witness hiding". Thus there are two ways to prove the security of Definition 2: One is to prove it directly as in [FFS, OhO1], and the other way is to prove that a scheme is "witness hiding" and an interactive proof of "knowledge". Some schemes such as [OhO1] seem to be proven only in the former way, since the knowledge soundness is sometimes hard to prove (e.g., [OhO1]). In this paper, we will prove our schemes in the former way, since it is compatible with the way to prove it by a variant of Definition 2, [OhO1], to be described below, although we can prove them in the latter way.

In the Appendix A, we introduce a variant of the "no useful information transfer" given by Ohta and Okamoto [OhO1], called "no transferable information with (sharp threshold) security level". This notion does not guarantee the security guaranteed by [FFS] i.e., the success probability of cheating by any adversary $(\tilde{A}, \tilde{B})$ is negligible in an asymptotic sense. However, the notion sheds light on another aspect of the security of identification schemes, the *security level* in a non-asymptotic sense. In practice, the security parameter is fixed in a system (e.g., the values of k and t of the Fiat-Shamir scheme [FiS]). Then we can assume a fixed security level for the system. The definition [OhO1] guarantees that such a fixed security level has theoretical significance [2]. Note that

---

[2] An asymptotic extension of the security level is recently studied in [CD]

this notion is defined essentially in an asymptotic manner although the security level is characterized in a non-asymptotic manner. The provable security of an identification scheme can be guaranteed by both these notions.

3 Proposed Three-Move Identification Schemes

3.1 Identification Scheme as Secure as the Discrete Logarithm

In this subsection, we propose a new scheme which is almost as efficient as the Schnorr identification scheme [Sch], and prove that it is as secure as the discrete logarithm problem.

A user generates a public key (p, q, g_1, g_2, t, v) and a secret key (s_1, s_2) and publishes the public key. Here, if g_2 is calculated by $g_2 = g_1^\alpha \bmod p$, α can be discarded after publishing g_2.

- primes p and q such that $q|p - 1$. (e.g., $q \geq 2^{140}$, and $p \geq 2^{512}$.)
- g_1, g_2 of order q in the group Z_p^*, and an integer $t = O(|p|)$. (e.g., $t \geq 20$.)
- random numbers s_1, s_2 in Z_q, and $v = g_1^{-s_1} g_2^{-s_2} \bmod p$.

Remark: (p, q, g_1, g_2, t) can be published by a system manager and used commonly by all system users as a system parameter. The system manager should then also publish some information to confirm to users that these parameters were selected honestly. For example, (s)he publishes some witness that no trapdoor exists in p, g_1, g_2, or that these values are generated honestly. Since the primality test for p and q is fairly easy for users, they can confirm for themselves that g_1 and g_2 are both of order q. When, as described above, the system parameter is generated and published by each user individually, (s)he does not need to publish such information.

We now describe our new identification scheme (Identification scheme 1) by which party A (the prover) can prove its identity to B (the verifier).

Protocol: Identification scheme 1

Step 1 A picks random numbers $r_1, r_2 \in Z_q$, computes

$$x = g_1^{r_1} g_2^{r_2} \bmod p,$$

and sends x to B.

Step 2 B sends a random number $e \in Z_{2^t}$ to A.

Step 3 A sends to B (y_1, y_2) such that

$$y_1 = r_1 + es_1 \bmod q, \quad \text{and} \quad y_2 = r_2 + es_2 \bmod q.$$

Step 4 B checks that

$$x = g_1^{y_1} g_2^{y_2} v^e \bmod p.$$

If it holds, B accepts, otherwise rejects.

Next, we prove the security of the above identification scheme. First, we show a definition and lemma in preparation.

Definition 3. Let RA denote $\widetilde{A}$'s random tape, and RB denote $\overline{B}$'s random tape. The possible outcomes of executing $(\widetilde{A}, \overline{B})$ can be summarized in a large Boolean matrix H whose rows correspond to all possible choices of RA. Its columns correspond to all possible choices e of RB, and its entries are 1 if $\overline{B}$ accepts $\widetilde{A}$'s proof, and 0 if otherwise.

When the success probability of $\widetilde{A}$ is ε (or the rate of 1-entries in H is ε), we call a row *heavy* if its ratio of 1's is at least $\varepsilon/2$.

Lemma 4. *If, given A's public key (p, q, g_1, g_2, t, v), the success probability, ε, of $\widetilde{A}$ is greater than 2^{-t+1}, then there exists a probabilistic algorithm which runs in expected time $O(\|\widetilde{A}\|/\varepsilon)$ and outputs the history of two accepted executions of $(\widetilde{A}, \overline{B})$, (x, e, y_1, y_2) and (x, e', y_1', y_2'), where $e \neq e'$. Here, $\|\widetilde{A}\|$ denotes the time complexity of $\widetilde{A}$. The success probability ε is taken over the coin tosses of $\widetilde{A}$ and $\overline{B}$.*

Sketch of Proof:

Assume that at least $1/2$ of the 1's in H are not located in heavy rows. Then the fraction of non-heavy rows in H, which we denote τ, is estimated as follows: $\tau \geq \frac{2^t \varepsilon/2}{2^t \varepsilon/2 - 1} > 1$. This is a contradiction. Therefore, at least $1/2$ of the 1's in H are located in heavy rows. Since ε is greater than 2^{-t+1} and the width of H is 2^t, a heavy row contains at least two 1's. To find two 1's in the same row, we thus adopt the following strategy:

1. Probe $O(1/\varepsilon)$ random entries in H (or pick (RA, e) randomly and check it, and repeat this until successful).
2. After the first 1 is found (or accepted (x, e, y_1, y_2) with RA is found), probe $O(1/\varepsilon)$ random entries along the same row (or probe (x, e', y_1', y_2') with the same RA).

Since at least $1/2$ of the 1's in H are located in heavy rows, this strategy succeeds with constant probability in $O(1/\varepsilon)$ probes. □

Definition 5. The discrete logarithm is (nonuniformly) intractable, if any family of boolean circuits, which, given properly chosen (g_1, g_2, p, q) in the same distribution as the output of key generator G, can compute the discrete logarithm $\alpha \in Z_q$ ($g_2 = g_1^\alpha \mod p$) with nonnegligible probability, must grow at a rate faster than any polynomial in the size of the input, $|p|$.

Remark The discrete logarithm above might be less intractable than that when the order of g_1 is greater than q (e.g., $p - 1$), although no attack has yet been reported when q is appropriately large (considering an attack, [PH]).

Theorem 6. *Identification scheme 1 is secure if and only if the discrete logarithm is intractable.*

Sketch of Proof:

(Only if:)

Suppose that the discrete logarithm is not intractable. Clearly a (nonuniform) polynomial time machine can calculate (s_1', s_2') satisfying $v = g_1^{-s_1'} g_2^{-s_2'} \bmod p$ with nonnegligible probability. Thus Identification scheme 1 is not secure.

(If:)

To prove the "If" part, we show that if Identification scheme 1 is not secure, then, given (g_1, g_2, p, q) with the same distribution as the output of key generator G, the discrete logarithm $\alpha \in Z_q$ ($g_2 = g_1^\alpha \bmod p$) can be computed by a polynomial time machine P with non-negligible probability.

Assume that Identification scheme 1 is not secure. Then $(\tilde{A}, \overline{B})$ can be accepted with nonnegligible probability ε after $O(|p|^c)$ executions of $(\overline{A}, \tilde{B})$. The complete history of the executions of $(\overline{A}, \tilde{B})$ and $(\tilde{A}, \overline{B})$ can be simulated by one polynomial time procedure P, which may be nonuniform, if P knows $\overline{A}$'s secret key.

To calculate the discrete logarithm $\alpha \in Z_q$ ($g_2 = g_1^\alpha \bmod p$), given (g_1, g_2, p, q), P firstly chooses $s_1^*, s_2^* \in Z_q$ randomly, and calculates $v = g_1^{-s_1^*} g_2^{-s_2^*} \bmod p$.

Then, using (s_1^*, s_2^*) as $\overline{A}$'s secret key, P simulates $(\overline{A}, \tilde{B})$ as well as $(\tilde{A}, \overline{B})$. So, for (v, g_1, g_2, p, q), after simulating $O(|p|^c)$ executions of $(\overline{A}, \tilde{B})$, P tries to find two accepted interactions of $(\tilde{A}, \overline{B})$, (x, e, y_1, y_2) and (x, e', y_1', y_2') ($e \neq e'$). From Lemma 4, this is possible with overwhelming probability, since ε is nonnegligible i.e. greater than 2^{-t+1}.

P can then calculate $(s_1, s_2) = ((y_1 - y_1')/(e - e') \bmod q, (y_2 - y_2')/(e - e') \bmod q)$ by

$$y_1 = r_1 + es_1 \bmod q, \quad y_2 = r_2 + es_2 \bmod q,$$

$$y_1' = r_1 + e's_1 \bmod q, \quad y_2' = r_2 + e's_2 \bmod q.$$

There are q solutions of (s_1, s_2) which satisfy $v = g_1^{-s_1} g_2^{-s_2} \bmod p$, given (v, g_1, g_2, p, q). Even an infinitely powerful $\tilde{B}$ cannot determine from x's, y_1's, and y_2's sent by $\overline{A}$ during the execution of $(\overline{A}, \tilde{B})$ which (s_1, s_2) satisfying $v = g_1^{-s_1} g_2^{-s_2} \bmod p$ actually uses. To prove this, for two different solutions, (s_1, s_2) and (s_1^*, s_2^*) satisfying $v = g_1^{-s_1} g_2^{-s_2} \equiv g_1^{-s_1^*} g_2^{-s_2^*} \pmod{p}$, we show that even an infinitely powerful $\tilde{B}$ cannot determine which solution was used from x's, y_1's, and y_2's. When $r_1^* = r_1 + e(s_1 - s_1^*) \bmod q$ and $r_2^* = r_2 + e(s_2 - s_2^*) \bmod q$, the following three equations hold.

$$x = g_1^{r_1} g_2^{r_2} \equiv g_1^{r_1^*} g_2^{r_2^*} \pmod{p},$$

$$y_1 = r_1 + es_1 \equiv r_1^* + es_1^* \pmod{q},$$

$$y_2 = r_2 + es_2 \equiv r_2^* + es_2^* \pmod{q}.$$

In addition, the distributions of (r_1, r_2) and (r_1^*, r_2^*) are exactly equivalent even if they satisfy the above relation. Hence, although P knows (s_1^*, s_2^*), (s_1, s_2), which is calculated by P by simulating the operations of $(\overline{A}, \widetilde{B})$ and $(\widetilde{A}, \overline{B})$, is independent from (s_1^*, s_2^*).

Therefore, (s_1^*, s_2^*) which was randomly chosen by P at first is different with probability $(q-1)/q$ from (s_1, s_2). Thus, α can be calculated with probability $(q-1)/q$ from (s_1, s_2) and (s_1^*, s_2^*) such that $\alpha = (s_1 - s_1^*)/(s_2^* - s_2) \bmod q$. The total success probability of P is nonnegligible.

This contradicts the intractability assumption of the discrete logarithm. $\square$

Theorem 7. *Let $t = O(1)$. Identification scheme 1 is secure with sharp threshold security level $1/2^t$ if and only if the discrete logarithm is intractable.*

The proof of Theorem 7 is similar to that of Theorem 6. It is shown in the final version.

3.2 Identification Scheme as Secure as RSA Inversion

This subsection proposes another practical identification scheme which is almost as efficient as the Guillou-Quisquater identification scheme [GQ], and proves that it is as secure as RSA inversion.

A user generates a public key (a, k, n, v) and a secret key (s_1, s_2) and publishes the public key. Here, p, q can be discarded after publishing n. Note that (a, k) can be common among users as the system parameter.

- primes p, q, $n = pq$, and prime k such that $\gcd(k, \phi(n)) = 1$ and $|k| = O(|n|)$, where $\phi(n) = \mathrm{lcm}(p-1, q-1)$. (e.g., $k \geq 2^{20}$, $n \geq 2^{512}$)
- random number $s_1 \in \mathbf{Z}_k$, and random numbers $a, s_2 \in \mathbf{Z}_n^*$, and $v = a^{-s_1} s_2^{-k} \bmod n$.

We now describe our new identification scheme (Identification scheme 2) by which party A (the prover) can prove its identity to B (the verifier).

Protocol: Identification scheme 2

Step 1 A picks random numbers $r_1 \in \mathbf{Z}_k$ and $r_2 \in \mathbf{Z}_n^*$, computes

$$x = a^{r_1} r_2^k \bmod n,$$

and sends x to B.

Step 2 B sends a random number $e \in \mathbf{Z}_k$ to A.

Step 3 A sends to B (y_1, y_2) such that

$$y_1 = r_1 + e s_1 \bmod k, \quad y_2 = a^{\lfloor (r_1 + e s_1)/k \rfloor} r_2 s_2^e \bmod n.$$

Step 4 B checks that $x = a^{y_1} y_2^k v^e \bmod n$.

Definition 8. RSA inversion is (nonuniformly) intractable, if any family of boolean circuits, which, given properly chosen (a, k, n) in the same distribution as the output of key generator G, can compute $a^{1/k} \bmod n$ with nonnegligible probability, must grow at a rate faster than any polynomial in the size of the input, $|n|$.

Theorem 9. *Identification scheme 2 is secure if and only if RSA inversion is intractable.*

Sketch of Proof:

(Only if:)

Suppose that the RSA inversion is not intractable. Clearly a (nonuniform) polynomial time machine can calculate (s'_1, s'_2) satisfying $v = a^{-s'_1} s'^{-k}_2 \bmod n$ with nonnegligible probability. Thus Identification scheme 2 is not secure.

(If:)

To prove the "If" part, we can prove this in a manner similar to the "if" part proof of Theorem 6. So we only sketch the different points here.

First, P chooses $s^*_1 \in Z_k$, and $s^*_2 \in Z^*_n$ randomly, and calculates $v = a^{-s^*_1} s^{*-k}_2 \bmod n$.

Then, for (a, k, n, v), P finds (x, e, y_1, y_2) and (x, e', y'_1, y'_2) $(e \neq e')$ by the technique of Lemma 4.

Next P calculates $s_1 = (y_1 - y'_1)/(e - e') \bmod k$, and $r_1 = y_1 - e s_1 \bmod k$. P then calculates X, Y as follows:

$$X = \frac{y_2/a^{\lfloor (r_1 + e s_1)/k \rfloor}}{y'_2/a^{\lfloor (r_1 + e' s_1)/k \rfloor}} \bmod n \ (= s_2^{e-e'} \bmod n),$$

$$Y = 1/(v a^{s_1}) \bmod n \ (= s_2^k \bmod n).$$

Since $\gcd(k, e - e') = 1$ (as k is prime), P can compute α, β satisfying $\alpha(e - e') + \beta k = 1$ by the extended Euclidean algorithm. Hence P calculates $s_2 = X^\alpha Y^\beta \bmod n$.

There are k solutions of (s_1, s_2) which satisfy $v = a^{-s_1} s_2^{-k} \bmod n$, given (v, n, a, k). Even an infinitely powerful $\tilde{B}$ cannot determine from x's, y_1's, and y_2's which (s_1, s_2) was actually used.

P then obtains (s_1, s_2), (s^*_1, s^*_2) $(s_i \neq s^*_i)$ such that $v = a^{s_1} s_2^k \equiv a^{s^*_1} s_2^{*k}$ (mod n), so $a^{(1/k)(s_1 - s^*_1)} \equiv s^*_2/s_2$ (mod n). After repeating the above procedure, P obtains another (s'_1, s'_2), (s'^*_1, s'^*_2) $(s'_i \neq s'^*_i)$ such that $a^{(1/k)(s'_1 - s'^*_1)} \equiv s'^*_2/s'_2$ (mod n) with nonnegligible probability. If $\gcd(s_1 - s^*_1, s'_1 - s'^*_1) = 1$, then P can calculate $a^{1/k} \bmod n$. The probability that $\gcd(s_1 - s^*_1, s'_1 - s'^*_1) = 1$ is more than a constant, since s^*_1, s'^*_1 is selected randomly and s_1, s'_1 is independent from s^*_1, s'^*_1. Thus, the total success probability of P is nonnegligible.

This contradicts the intractability assumption of RSA inversion. □

Theorem 10. *Let $|k| = O(1)$. Identification scheme 2 is secure with sharp threshold security level $1/k$ if and only if RSA inversion is intractable.*

3.3 Identification Scheme as Secure as Factoring

In this subsection, we show a slight variant of the previous identification scheme (Identification scheme 2), which is as secure as factoring, while Identification scheme 2 is as secure as the inversion of the RSA function. The protocol of this variant (Identification scheme 3) is exactly same as Identification scheme 2. The only difference is that the value of k is selected so that $\gcd(k, \phi(n)) = 2$ and $k/2$ is prime, while $\gcd(k, \phi(n)) = 1$ and k is prime in Identification scheme 2.

Definition 11. Factoring is (nonuniformly) intractable, if any family of boolean circuits, which, given properly chosen (n) in the same distribution as the output of key generator G, can factor n with nonnegligible probability, must grow in a rate faster than any polynomial in the size of the input, $|n|$.

Theorem 12. *Identification scheme 3 is secure if and only if factoring is intractable.*

Theorem 13. *Let $|k| = O(1)$. Identification scheme 3 is secure with sharp threshold security level $1/k$ if and only if factoring is intractable.*

4 Generalization to Random-Self-Reducible Problems

This section shows that any random self-reducible problem [TW] leads to provably secure and three-move identification.

Definition 14. Let $\mathcal{N}$ be a countable infinite set. For any $N \in \mathcal{N}$, let $|N|$ denote the length of a suitable representation of N, and denote the problem size. For any $N \in \mathcal{N}$, let X_N, Y_N be finite sets, and $R_N \subseteq X_N \times Y_N$ be a relation. Let

$$dom R_N = \{x \in X_N \mid (x, y) \in R_N \text{ for some } y \in Y_N\}$$

denote the *domain* of R_N,

$$R_N(x) = \{y \mid (x, y) \in R_N\}$$

the *image* of $x \in X_N$.

R is *random self-reducible (RSR)* if and only if there is a polynomial time algorithm A that, given any inputs $N \in \mathcal{N}$, $x \in dom R_N$, and a source $r \in \{0, 1\}^{\omega}$, outputs $x' = A(N, x, r) \in dom R_N$ satisfying the following seven properties.

1. If r is randomly and uniformly chosen on $\{0, 1\}^{\omega}$, then x' is uniformly distributed over $dom R_N$.
2. There is a polynomial time algorithm that, given N, x, r, and any $y' \in R_N(x')$, outputs $y \in R_N(x)$.
3. There is a polynomial time algorithm that, given N, x, r, and any $y \in R_N(x)$, outputs some $y' \in R_N(x')$. If, in addition, the bits of r is random, uniform, and independent, then y' is uniformly distributed over $R_N(x')$.
4. There is an expected polynomial time algorithm that, given N, x', and y', determines whether $(x', y') \in R_N$.
5. There is an expected polynomial time algorithm that, given N, outputs random pairs $(x', y') \in R_N$ with x' uniformly distributed over $dom R_N$ and y' uniformly distributed over $R_N(x')$.

6. There is an expected polynomial time algorithm that, given N, x_0, x_1, x_2, r_1, r_2 satisfying $x_i = A(N, x_0, r_i)$ $(i = 1, 2)$, outputs r^* satisfying $x_2 = A(N, x_1, r^*)$.

7. There is an expected polynomial time algorithm that, given N, x_1, x_2, y_1, y_2 satisfying $(x_i, y_i) \in R_N$ $(i = 1, 2)$, outputs r^* satisfying $x_2 = A(N, x_1, r^*)$.

Next we construct a three-move identification scheme based on random self-reducible problem R (Identification scheme 4).

A user generates a public key (N, a, t, v) and a secret key (s_i) $(i = 0$ or $1)$ and publishes the public key.

- A random bit $i \in \{0, 1\}$, $N \in \mathcal{N}$, $a \in dom R_N$, and an integer $t = O(|N|)$.
- When $i = 0$, random bits $s_0 \in \{0, 1\}^\omega$, and $v = A(N, a, s_0)$.
- When $i = 1$, a random pair $(v, s_1) \in R_N$.

Protocol: Identification scheme 4

Step 1 A generates random bits $y_{j0} \in \{0, 1\}^\omega$, and $x_{j0} = A(N, a, y_{j0})$, $(j = 1, \ldots, t)$. A also generates random pairs $(x_{j1}, y_{j1}) \in R_N$, $(j = 1, \ldots, t)$. A sets $x_j = (x_{jb_j}, x_{j(1-b_j)})$ with a random bit $b_j \in \{0, 1\}$, and sends $(x_1, x_2, \ldots, x_t)$ to B.

Step 2 B sends random bits $(e_1, \ldots, e_t)$ to A.

Step 3 A sends $(z_1, z_2, \ldots, z_t)$ to B. Here, if $e_j = 0$, $z_j = (y_{j0}, y_{j1})$. If $e_j = 1$ and $i = 0$, then $z_j = r_0$ such that $x_{j0} = A(N, v, r_0)$ (r_0 can be computed from property 6). If $e_j = 1$ and $i = 1$, then $z_j = r_1$ such that $x_{j1} = A(N, v, r_1)$ (r_1 can be computed from property 7).

Step 4 B checks the validity of the messages received from A.

Definition 15. The random self-reducible problem R is (nonuniformly) intractable, if any family of boolean circuits, which, given properly chosen (N, a) in the same distribution as the output of key generator G, can compute α satisfying $(a, \alpha) \in R_N$ with nonnegligible probability, must grow at a rate faster than any polynomial in the size of the input, $|p|$.

Theorem 16. *Identification scheme 4 is secure if and only if the random self-reducible problem R is intractable.*

The basic techniques to prove this theorem are similar to those shown in Section 3. Scheme 4 is much less efficient than the schemes in Section 3, since the schemes in Section 3 are Type 2 of the parallel versions (see Section 1), while this scheme is Type 1.

Because of space limitations, we omit the proof of this theorem in this extended abstract.

5 Variants of the Proposed Identification Schemes

5.1 Identification Scheme as Secure as the Discrete Logarithm and Factoring Simultaneously

This subsection introduces a variant of Identification scheme 1 (Identification scheme 5) using the idea of the Brickell-McCurley scheme [BM1]. This variant is proven to be as secure as the difficulty of solving both the discrete logarithm and the specific factoring problem (or the finding order problem) simultaneously.

In this identification scheme, a user generates a public key (p, g_1, g_2, v) and secret key (s_1, s_2) and publishes the public key. (q, w) can be discarded after publishing the public key. (p, g_1, g_2) can be a system parameter, which is commonly used by all users.

- primes p, q and w such that $qw|p-1$ (e.g., $q \geq 2^{140}$, $p \geq 2^{512}$, and $qw \geq 2^{512}$).
- g_1 and g_2 of order q in the group Z_p^*.
- random numbers s_1, s_2 in Z_{p-1}.
- $v = g_1^{-s_1} g_2^{-s_2} \bmod p$.

We now describe our new identification scheme (Identification scheme 5).

Protocol: Identification scheme 5

Step 1 A picks random numbers $r_1, r_2 \in Z_{p-1}$, computes

$$x = g_1^{r_1} g_2^{r_2} \bmod p,$$

and sends x to B.

Step 2 B sends random numbers $e \in Z_{2^t}$ to A.

Step 3 A sends to B (y_1, y_2) such that

$$y_1 = r_1 + ex_1 \bmod p - 1, \quad \text{and} \quad y_2 = r_2 + ex_2 \bmod p - 1.$$

Step 4 B checks that

$$x = g_1^{y_1} g_2^{y_2} v^e \bmod p.$$

Definition 17. The finding order problem is (nonuniformly) intractable, if any family of boolean circuits, which, given properly chosen (p, g_1) in the same distribution as the output of key generator G, can compute the order of g_1 in the group Z_p^* with nonnegligible probability, must grow at a rate faster than any polynomial in the size of the input, $|p|$.

Remark This problem is more tractable than the factoring problem (Definition 11), since if there exists an polynomial time algorithm to solve the factoring problem, then the finding order problem can be solved by factoring $p-1$. So, the finding order problem can be considered a subproblem of the factoring problem.

Theorem 18. *Identification scheme 5 is secure if and only if the problem to solve both the discrete logarithm and the finding order problem simultaneously is intractable.*

5.2 Identity-Based and Certification-Based Variants

There are two methods of eliminating the public key directory from the conventional public key schemes: one is the identity-based method and the other is the certification-based method.

In the certification-based method, a trusted center (key authentication center, or certification authority) publishes its public key and gives a user A its signature S for the pair of identity Id_A and public key PK_A of A. The user A sends (Id_A, PK_A, S) to the verifier, who checks the validity of PK_A by verifying the trusted center's signature S for (Id_A, PK_A) in place of retrieving PK_A through Id_A from the public key directory.

In the identity-based method, proposed by Shamir [Sha] and independently by Okamoto [Oka], the public key is replaced by the identity related value of a user.

The difference between the certification-based method and identity-based method is as follows:

- Any public-key system can be converted into the certification-based variant by the same technique, while each public-key system needs a peculiar technique to convert to the identity-based variant.
- The trusted center of the certification-based method does not know each user's secret key, while the trusted center of the identity-based method generates and knows each user's secret key.
- The size of the public key that a user keeps and sends to the verifier in the certification-based method is longer than that in the identity-based method.

In this extended abstract, only two examples, identity-based variants of Identification schemes 1 and 2, are introduced briefly. In particular, we show a new construction technique to realize the identity-based variant of a scheme which is based on the discrete logarithm (e.g., Identification scheme 1), although the identity-based scheme based on the discrete logarithm is usually difficult to construct. Our technique is similar to Beth's idea [Bet], but, ours seems to be more natural, since we use the digital signature corresponding to the identification (Section 6), while the ElGamal scheme is used in [Bet]. (Our technique can be also applied to the Schnorr scheme: See Appendix B.)

Identity-Based Variant of Identification scheme 1 A trusted center T (or key authentication center) generates a public key (p, q, g_1, g_2, t, v_T) and its secret key (s_{T1}, s_{T2}), and publishes the public key as a system parameter. T generates T's digital signature, (e_A, y_{A1}, y_{A2}), of A's identity, Id_A, by using its secret key. So, $e_A = h((g_1^{y_{A1}} g_2^{y_{A2}} v_T^{e_A} \bmod p), Id_A)$ (see Section 6). T gives A A's secret key (s_{A1}, s_{A2}) and e_A, where $(s_{A1}, s_{A2}) = (q - y_{A1}, q - y_{A2})$. Then A generates A's public key $v_A = g_1^{-s_{A1}} g_2^{-s_{A2}} \bmod p$ from the secrete key given by T.

In this identity-based identification protocol, A first sends (Id_A, v_A, e_A) to verifier B along with x (same as x in the first step of Identification scheme 1). B checks the validity of Id_A and v_A by checking whether $e_A = h((v_A v_T^{e_A} \bmod$

$p), Id_A$) holds or not. If the check passes, the remainig protocol is the same as Identification scheme 1 (or B sends A e, A sends B (y_1, y_2), and B checks it). So, B does not need to retrieve v_A from the public-key directory. Here, the communication overhead except (Id_A, v_A) is just e_A, whose size is much smaller than those of v_A and x.

Identity-Based Variant of Identification scheme 2 A trusted center (or key authentication center) generates a public key (a, k, n) and gives user A its secret key (s_{A1}, s_{A2}), where $Id_A = a^{-s_{A1}} s_{A2}^{-k} \bmod n$. (First $s_{A1} \in Z_k$ is randomly determined, then $s_{A2} = (Id_A a^{s_{A1}})^{-1/k} \bmod n$ is calculated. Id_A can be replaced by $h(Id_A)$ with a one-way function.)

In this identity-based identification protocol, Id_A is used in place of v in Identification scheme 2. In a manner similar to the above-mentioned identity-based protocol, Id_A is sent to B along with x in the first step and the remaining part is the same as Identification scheme 2. So, B does not need to retrieve v from the public-key directory.

5.3 Elliptic Curve Version

Some techniques to construct cryptosystems based on the elliptic curve logarithm over a finite field [HMV, Kob1, Kob2, Mil, Miy] can be straightfowardly applied to our Identification scheme 1.

The elliptic curve variant of Identification scheme 1 has the significant property that three-move practical identification is proven to be secure assuming the intractability of the (non-supersingular) elliptic curve logarithms against which only exponential-time attacks have been reported so far [MOV, Kob2].

6 Signature Schemes

This section describes digital signature schemes converted from the identification schemes given in the previous sections. We also prove the security (existentially unforgeable against adaptive chosen message attacks [GMRi]) of our new signature schemes assuming one reasonable assumption about the one-way hash function (correlation-free one-way hash function) as well as a primitive assumption.

Since this conversion [FiS] is very simple, in this extended abstract, we only show one example (Signature scheme 1) based on Identification scheme 1. Other signature schemes (Signature schemes 2 to 5, and others) can be realized in the same way based on Identification schemes 2 to 5, and the variants described in subsections 5.2 and 5.3.

6.1 Signature Scheme Based on Identification Scheme 1

Signature scheme 1 is almost as efficient as the Schnorr signature scheme and DSA (see Section 7), while the security [GMRi] assumption of our scheme is weaker and more reasonable than those of the Schnorr signature scheme and DSA.

A public key (p, q, g_1, g_2, t, v) and secret key (s_1, s_2) of each user are determined in the same manner as Identification scheme 1. h is a one-way hash function.

We now describe our new signature scheme (Signature scheme 1) by which party A (the signer) generates a signature (e, y_1, y_2) of a message m, and sends (m, e, y_1, y_2) to B (the verifier).

Protocol: Signature scheme 1

Step 1 A (signer) picks random numbers $r_1, r_2 \in Z_q$, computes $x = g_1^{r_1} g_2^{r_2}$ mod p. A computes $e = h(x, m) \in Z_{2^t}$ and (y_1, y_2) such that $y_1 = r_1 + es_1 \bmod q$, and $y_2 = r_2 + es_2 \bmod q$.

Step 2 A sends to B (e, y_1, y_2) along with message m.

Step 3 B computes $x = g_1^{y_1} g_2^{y_2} v^e \bmod p$, and checks that $e = h(x, m)$.

6.2 Security of Signature Schemes

In this subsection, we discuss the security of our signature schemes in the sense of "existentially unforgeable against adaptive chosen message attacks" defined by [GMRi]. Fiat and Shamir [FiS] have shown that the existence of an "ideal random function" as well as factoring assumption is sufficient to prove the security of the Fiat-Shamir signature scheme. However, their assumption, the existence of an ideal random function, can never be realized in the real world, and to realize the "pseudo-random function" [GGM] as a common function requires a tamper-free device.

In this paper, we clarify a reasonable assumption to prove the security of the Fiat-Shamir type signature schemes. We introduce a new class of one-way hash functions, *correlation-free one-way hash functions*, and show that the existence of a "correlation-free one-way hash function", as well as a primitive assumption, is sufficient to prove the security of our schemes. Although the existence of a correlation-free one-way hash function seems to be a stronger assumption than those of universal one-way hash function, claw-free pair of functions and collision-free hash function, we highly believe that carefully designed practical one-way hash functions such as MD5 and SHA are correlation-free one-way hash functions with any number theoretic predicate.

Definition 19. A family of *correlation-free one-way hash functions* with F is a set of hash functions, $H = \{H_n\}$ (H_n is a subset of H with security parameter n), with the following properties:

- **Poly-time indexing:** Each function in H_n has a unique n bit index, σ_n, associated with it: $H_n = \{h_{\sigma_n} \mid \sigma_n \in \{0,1\}^n, h_{\sigma_n} : \{0,1\}^{p(n)} \times \{0,1\}^{s(n)} \to \{0,1\}^{q(n)}\}$, where $p(n)$, $s(n)$, and $q(n)$ are polynomial in n. There is a probabilistic polynomial time algorithm, which, on input n, selects uniformly and randomly σ_n in $\{\sigma_n\}$.

- **Poly-time evaluation:** There exists a polynomial time algorithm that (for all $n \geq 1$), upon input of an index σ_n and an argument $(x, m) \in \{0,1\}^{p(n)} \times \{0,1\}^{s(n)}$, computes $h_{\sigma_n}(x, m)$.
- **Correlation-freeness:** Let $F = \{F_n \mid F_n = \{f_{\delta_n}\}\}$ be a poly-time indexing (δ_n) and poly-time evaluation predicate family such that $f_{\delta_n} : \{0,1\}^{p(n)} \times \{0,1\}^{q(n)} \times \{0,1\}^{r(n)} \to \{0,1\}$, where $r(n)$ is polynomial in n. Suppose that any family of boolean circuits, which, given δ_n, can compute x and (e_i, y_i) $(i = 1, \ldots, t(n))$ $(t(n)$ is polynomial in $n)$ with nonnegligible probability such that $f_{\delta_n}(x, e_i, y_i) = 1$, must grow at a rate faster than any polynomial in n. Then, any family of boolean circuits, which, given σ_n, and δ_n, can compute (x, e, y, m) with nonnegligible probability such that $h_{\sigma_n}(x, m) = e$ and $f_{\delta_n}(x, e, y) = 1$, must grow at a rate faster than any polynomial in n.
- **One-wayness:** Any family of boolean circuits, which, given (x, m), can compute m' $(m' \neq m)$ with nonnegligible probability such that $h_{\sigma_n}(x, m') = h_{\sigma_n}(x, m)$, must grow at a rate faster than any polynomial in n.

Theorem 20. *Signature scheme 1 is existentially unforgeable against any adaptive chosen message attacks if the discrete logarithm problem is intractable and h is a correlation-free one-way hash function with $F = \{f_{(g_1, g_2, p, v)}\}$, where $f_{(g_1, g_2, p, v)}(x, e, (y_1, y_2)) = 1$ if and only if $x = g_1^{y_1} g_2^{y_2} v^e \bmod p$ holds.*

Sketch of Proof:

Assume that there exists an adaptive chosen message attacker, P, to Signature scheme 1. We also assume that the discrete logarithm problem is intractable. Then we will show a contradiction with the assumption that h is a correlation-free one-way hash function with $F = \{f_{(g_1, g_2, p, v)}\}$.

First, assume that P can find $(x, e, y_1, y_2, e', y_1', y_2')$ $(e \neq e')$ with nonnegligible probability such that $x = g_1^{y_1} g_2^{y_2} v^e \bmod p$ and $x = g_1^{y_1'} g_2^{y_2'} v^{e'} \bmod p$, after adaptive chosen message attacks. Since, given (g_1, g_2, p), P can exactly simulate the valid signer by generating his/her secret key (s_1, s_2) and following signer's valid procedure, P can calculate the discrete logarithm α $(g_2 = g_1^{\alpha} \bmod p)$ by the technique described in the proof of Theorem 6. This contradicts the intractability assumption of the discrete logarithm problem. Therefore, P can find $(x, e, y_1, y_2, e', y_1', y_2')$ $(e \neq e')$ with negligible probability.

On the other hand, from the assumption that P is an adaptive chosen message attacker, P can find (x, e, y_1, y_2, m) with nonnegligible probability such that $h(x, m) = e$ and $x = g_1^{y_1} g_2^{y_2} v^e \bmod p$. This contradicts the assumption that h is a correlation-free hash function with $F = \{f_{(g_1, g_2, p, v)}\}$.

Thus, any attacker P cannot find a valid signature message (x, e, y_1, y_2, m) with nonnegligible probability after adaptive chosen message attacks. $\quad\square$

6.3 Two-Move and One-Move Identification Schemes

In this subsection, we briefly introduce two-move and one-move identification schemes by using secure signature schemes above, which are almost as efficient as the proposed three-move identification schemes.

Two-move secure identification scheme can be trivially constructed using a secure (existentially unforgeable against any adaptive chosen message attacks) signature scheme as follows: First, verifier B sends a random message x to prover A, then A generates and sends A's signature of message x to B, finally B checks the validity of A's signature.

We can easily convert a two-move identification scheme into a one-move identification by changing challenge message x into time-stamp t, which both A and B share. That is, first A sends A's signature of message t to B, then B checks it.

6.4 Multi-Signature and Blind Signature

The multi-signature and blind signature schemes of our proposed signature schemes (Signature schemes 1 to 5 and the variants) can be constructed. The multi-signature schemes are constructed in a manner similar to [OhO2], and the blind signature schemes are constructed based on the idea shown in [OkO].

Blind Signature for Signature Scheme 1 Here, we present only one example of the blind signature schemes, based on Signature scheme 1. The other blind signature schemes are constructed in the same way using the idea shown in [OkO]. (The blind signature scheme based on the Schnorr scheme is shown in Appendix B.)

In the blind signature scheme, which was originally proposed by Chaum [Cha] based on the RSA scheme, a client, Bob, generates a blinded message $b(m)$ from a message m, and sends $b(m)$ to a blind signer, Alice. She generates her signature $s_A(b(m))$ of $b(m)$, and sends it to Bob. He calculates Alice's signature $s_A(m)$ of message m from $s(b(m))$. Here, Alice has no information of m, and Bob has no information of Alice's secret key.

We now describe our blind signature scheme based on Signature scheme 1. Alice's public key is (p, q, g_1, g_2, t, v) and her secret key is (s_1, s_2), which are those of Signature scheme 1.

Protocol: Blind signature based on Signature scheme 1

Step 1 Alice (blind signer) picks random numbers $r_1, r_2 \in Z_q$, computes $x = g_1^{r_1} g_2^{r_2} \bmod p$, and sends x to Bob (client).

Step 2 Bob picks random numbers $d, u_1, u_2 \in Z_q$, and computes

$$x^* = g_1^{u_1} g_2^{u_2} v^{-d} x \bmod p, \quad e^* = h(x^*, m), \quad e = e^* + d \bmod q.$$

Bob sends e to Alice. Here, m is a message to be signed.

Step 3 Alice computes (y_1, y_2) such that $y_1 = r_1 + es_1 \bmod q$, and $y_2 = r_2 + es_2 \bmod q$, and sends (y_1, y_2) to Bob.

Step 4 Bob computes $y_1^* = y_1 + u_1 \bmod q$, $\quad y_2^* = y_2 + u_2 \bmod q$.

(e^*, y_1^*, y_2^*) is Alice's signature of message m.

Note: e is distributed on Z_q, while e^* is distributed on Z_{2^t}. The difference is no problem in the blind signature scheme, since even an infinite power attacker cannot find any linkage between e and e^*.

7 Performance

This section compares the computation amount of our schemes against those of the previous practical schemes in the light of the required number of modular multiplications, and also compare the key and signature lengths.

We assume that moduli p and q for our scheme 1, Schnorr are 512 bits and 140 bits respectively, p and q for DSA are 512 bits and 160 bits, and the modulus n for our scheme 3, Guillou-Quisquater (GQ), Ohta-Okamoto (OO) and Feige-Fiat-Shamir (FFS) is 512 bits. The security parameter for the identification schemes is assumed to be 20, or e (the challenge from the verifier) is 20 bits. The security parameter for the signature schemes is assumed to be 128, or e (the output of the hash function of x and a message) is 128 bits, since the output size of many typical hash functions such as MD5 is 128 bits. We also assume that the parameters for Feige-Fiat-Shamir are $k = |e|$ and $t = 1$.

Here, we estimate the performance of unsophisticated implementations, since the purpose of this comparison is to relatively compare some schemes with the same primitive problem (e.g., our scheme 1 and Schnorr), and many sophisticated techniques (e.g., [Mon, BGMW]) can be fairly evenly applied to the schemes with the same primitive problem. We assume the standard binary method and the extended binary method (4.6.3 ex.27 in [Kun]) for the modular exponentiation.

Table 1. Comparison of Identification Schemes

	Proposed Scheme 1	Schnorr	Proposed Scheme 3	GQ	OO	FFS
Provably secure?	Yes	No	Yes	No	Yes	Yes
Primitive problem	Disc.log.	Disc.log.	Fact.	RSA	Fact.	Fact.
ID-based variant	Possible	Possible	Possible	Possible	Hard	Possible
System parameter size (bits)	1676	1164	532	20	20	0
Public key size (bits)	512	512	1024	1024	1024	10240
Secret key size (bits)	280	140	532	512	512	10240
Communication amount (bits)	812	672	1064	1044	1044	1044
Preprocessing (Prover) (# of 512-bit modular multiplications)	245	210	35	30	30	1
On-line processing (Prover) (# of 512-bit modular multiplications)	almost 0	almost 0	32	31	31	10
On-line Processing (Verifier) (# of 512-bit modular multiplications)	248	210	38	35	35	11

Table 2. Comparison of Signature Schemes

	Proposed Scheme 1	Schnorr	DSA	Proposed Scheme 3	GQ	OO	FFS
Assumption	Weak	Strong	Strong	Weak	Strong	Weak	Weak
Primitive problem	Disc.log.	Disc.log.	Disc.log.	Fact.	RSA	Fact.	Fact.
ID-based variant	Possible	Possible	Possible	Possible	Possible	Hard	Possible
Multi-signature	Possible	Possible	Hard	Possible	Possible	Possible	Possible
Blind signature	Possible	Possible	Hard	Possible	Possible	Possible	Possible
System parameter size (bits)	1676	1164	1164	640	128	128	0
Public key size (bits)	512	512	512	1024	1024	1024	66048
Secret key size (bits)	280	140	160	640	512	512	65536
Signature size (bits)	408	268	320	768	640	640	640
Preprocessing for signing (# of 512-bit modular multiplications)	245	210	237	224	192	192	1
Signing (# of 512-bit modular multiplications)	almost 0	almost 0	almost 0	194	193	193	65
Verifying (# of 512-bit modular multiplications)	261	242	277	240	224	224	66

Acknowledgments

The author would like to thank Kouichi Sakurai for his valuable comments and suggestions especially on the relationship between "no useful information transfer" and "witness hiding". He would also like to thank an anonymous referee for his/her useful comments on the preliminary manuscript.

References

[Bet]   T.Beth, "Efficient Zero-Knowledge Identification Scheme for Smart Cards," Proceedings of Eurocrypt '88, LNCS 330, Springer-Verlag, pp.77-86 (1988).

[BGMW] E.F.Brickell, D.M.Gordon, K.S.McCurley, and D.Wilson, "Fast Exponentiation with Precomputation", to appear in the Proceedings of Eurocrypt'92.

[BM1]   E.F.Brickell, and K.S.McCurley, "An Interactive Identification Scheme Based on Discrete Logarithms and Factoring," Journal of Cryptology, Vol.5, No.1, pp.29-39 (1992).

[BM2]   E.F.Brickell, and K.S.McCurley, "Interactive Identification and Digital Signatures," AT&T Technical Journal, pp.73-86, November/December (1991).

[BMO1] M.Bellare, S.Micali and R.Ostrovsky, "Perfect Zero-Knowledge in Constant Rounds," Proceedings of STOC, pp.482-493 (1990).

[BMO2] M.Bellare, S.Micali and R.Ostrovsky, "The (True) Complexity of Statistical Zero-Knowledge," Proceedings of STOC, pp.494-502 (1990).

[Cha]   D.Chaum, "Security without Identification: Transaction Systems to Make Big Brother Obsolete," Comm. of the ACM, 28, 10, pp.1030-1044 (1985).

[CD]   L.Chen, I.Damgård, "Security Bounds for Parallel Versions of Identification Protocols," Manuscript (1992).

[FeS1]   U.Feige and A.Shamir, "Witness Indistinguishable and Witness Hiding Protocols," Proceedings of STOC, pp.416-426 (1990).

[FeS2]   U.Feige and A.Shamir, "Zero Knowledge Proofs of Knowledge in Two Rounds," Proceedings of Crypto'89, LNCS 435, Springer-Verlag, pp.526-544 (1990).

[FFS]   U.Feige, A.Fiat and A.Shamir, "Zero Knowledge Proofs of Identity," Proceedings of STOC, pp.210-217 (1987).

[FiS]   A.Fiat and A.Shamir, "How to Prove Yourself: Practical Solutions to Identification and Signature Problems", Proceedings of CRYPTO '86, LNCS 263, Springer-Verlag, pp.186–194 (1987).

[GGM]   O.Goldreich, S.Goldwasser, and S.Micali, "How to Construct Random Functions," Journal of the ACM, Vol.33, No.4 (1986).

[GK]   O.Goldreich and H.Krawczyk "On the Composition of Zero-Knowledge Proof Systems," Proceedings of ICALP, LNCS 443, Springer-Verlag, pp.268-282 (1990).

[GMRa]   S.Goldwasser, S.Micali and C.Rackoff, "The Knowledge Complexity of Interactive Proofs," SIAM J. Comput., 18, 1, pp.186-208 (1989).

[GMRi]   S.Goldwasser, S.Micali and R.Rivest, "A Digital Signature Scheme Secure Against Adaptive Chosen-Message Attacks," SIAM J. Comput., 17, 2, pp.281–308 (1988).

[GQ]   L.S.Guillou, and J.J.Quisquater, "A Practical Zero-Knowledge Protocol Fitted to Security Microprocessors Minimizing both Transmission and Memory," Proceedings of Eurocrypt '88, LNCS 330, Springer-Verlag, pp.123-128 (1988).

[HMV]   G.Harper, A.J.Menezes, S.A.Vanstone, "Public-Key Cryptosystems with Very Small Key Length", to appear in the Proceedings of Eurocrypt'92.

[Kob1] N.Koblitz, *A Course in Number Theory and Cryptography*, Berlin: Springer-Verlag, (1987).

[Kob2] N.Koblitz, "CM-Curves with Good Cryptographic Properties," Proceedings of Crypto '91 (1992).

[Kun]   D.E.Knuth, *The Art of Computer Programming*, Vol.2, 2nd Ed. Addison-Wesley (1981).

[Mil]   V.Miller, "Uses of Elliptic Curves in Cryptography," Proceedings of Crypto '85, LNCS 218, Springer-Verlag, pp.417-426 (1986).

[Miy]   A.Miyaji, " On Ordinary Elliptic Curve Cryptosystems," to appear in the Proceedings of Asiacrypt'91, LNCS, Springer-Verlag.

[Mon]   P.L.Montgomery, "Modular Multiplication without Trial Division," Math. of Computation, Vol.44, pp.519–521 (1985).

[MOV]   A.J.Menezes, T.Okamoto, S.A.Vanstone, "Reducing Elliptic Curve Logarithms to Logarithms in a Finite Field", Proceedings of STOC, pp.80-89 (1991).

[OhO1] K.Ohta, and T.Okamoto, "A Modification of the Fiat-Shamir Scheme," Proceedings of Crypto '88, LNCS 403, Springer-Verlag, pp.232-243 (1990).

[OhO2] K.Ohta, and T.Okamoto, "A Digital Multisignature Scheme Based on the Fiat-Shamir Scheme," to appear in the Proceedings of Asiacrypt'91.

[Oka]   T.Okamoto, "A Single Public-Key Authentication Scheme for Multiple Users," *Systems and Computers in Japan*, 18, 10, pp.14-24 (1987), Previous version, Technical Report of IECE Japan, IN83-92 (1984).

[OkO]   T.Okamoto, and K.Ohta, "Divertible Zero-Knowledge Interactive Proofs and Commutative Random Self-Reducible," Proceedings of Eurocrypt '89, LNCS 434, Springer-Verlng, pp.134–149 (1990).

[PH]   S.C.Pohlig, and M.E.Hellman, "An Improved Algorithm for Computing Logarithmsover GF(p) and Its Cryptographic Significance," IEEE Trans. Inform. Theory, 24, pp.106–110 (1978)

[RSA]   R.Rivest, A.Shamir and L.Adleman, "A Method for Obtaining Digital Signatures and Public-Key Cryptosystems", Communications of the ACM, Vol.21, No.2, pp.120-126 (1978).

[Sch]   C.P. Schnorr, "Efficient Signature Generation by Smart Cards," Journal of Cryptology, Vol.4, No.3, pp.161-174 (1991).

[Sha]   A.Shamir, "Identity-Based Cryptosystems and Signature Scheme," Proceedings of Crypto '84, LNCS 196, Springer-Verlag, pp.47–53 (1986).

[SI]   K.Sakurai, and T.Itoh, "On the Discrepancy between Serial and Parallel of Zero-Knowledge Protocols," These proceedings.

[TW]   M.Tompa and H.Woll, "Random Self-Reducibility and Zero Knowledge Interactive Proofs of Possession of Information," Proceedings of FOCS, pp.472-482 (1987).

Appendix A

In this appendix, we introduce a variant of "no useful information transfer" [FFS] given by Ohta and Okamoto [OhO1], called "no transferable information with (sharp threshold) security level".

Definition 21. An identification scheme (A, B) is *secure with security level ρ if*

1. $(\overline{A}, \overline{B})$ succeeds with overwhelming probability.
2. There is no coalition of $\widetilde{A}, \widetilde{B}$ with the property that, after a polynomial number of executions of $(\overline{A}, \overline{B})$ and relaying a transcript of the communication to $\widetilde{A}$, it is possible to execute $(\widetilde{A}, \overline{B})$ with $c \cdot \rho$ probability of success, where $c = (1 + 1/|n|^d)$ and d is an arbitrary constant. The probability is taken over the distribution of the public key and the secret key as well as the coin tosses of $\overline{A}, \widetilde{B}, \widetilde{A}$, and $\overline{B}$, up to the time of the attempted impersonation.

Definition 22. An identification scheme (A, B) is *secure with sharp threshold security level ρ* if

1. (A, B) is secure with security level ρ.
2. There exists $\widetilde{A}$ such that it is possible to execute $(\widetilde{A}, \overline{B})$ with ρ probability of success.

Appendix B

In this appendix, we introduce the identity-based variant and blind signature scheheme of the Schnorr scheme.

B.1 Identity-Based Variant of the Schnorr scheme

A trusted center T (or key authentication center) generates a public key (p, q, g, t, v_T) and its secret key s_T, and publishes the public key as a system parameter. T generates T's digital signature, (e_A, y_A), of A's identity, Id_A. T gives A A's secret key s_A and e_A, where $s_A = q - y_A$. Then A generates A's public key $v_A = g^{-s_A} \bmod p$ from the secrete key given by T.

In this identity-based identification protocol, A first sends (Id_A, v_A, e_A) to verifier B along with x B checks the validity of Id_A and v_A by checking whether $e_A = h((v_A v_T^{e_A} \bmod p), Id_A)$ holds or not. If the check passes, the remainig protocol is the same as the Schnorr scheme.

B.2 Blind Signature of the Schnorr scheme

Alice's public key is (p, q, g, t, v) and her secret key is s.

Protocol: Blind signature based on the Schnorr scheme

Step 1 Alice (blind signer) picks random number $r \in Z_q$, computes $x = g^r \bmod p$, and sends x to Bob (client).

Step 2 Bob picks random numbers $d, u \in Z_q$, computes

$$x^* = g^u v^{-d} x \bmod p, \quad e^* = h(x^*, m), \quad e = e^* + d \bmod q.$$

Bob sends e to Alice. Here, m is a message to be signed.

Step 3 Alice computes y such that $y = r + es \bmod q$, and sends y to Bob.

Step 4 Bob computes $y^* = y + u \bmod q$.

(e^*, y^*) is Alice's signature of message m.

An Efficient Digital Signature Scheme Based on an Elliptic Curve over the Ring Z_n

Tatsuaki Okamoto   Atsushi Fujioka   Eiichiro Fujisaki

NTT Laboratories
Nippon Telegraph and Telephone Corporation
1-2356, Take, Yokosuka-shi, Kanagawa-ken, 238-03 Japan

Abstract. We propose a practical digital signature scheme based on the elliptic curve modulo n, where $n = p^2 q$ such that p and q are large secret primes. The signature generation speed of our scheme is more than 10 times faster than that of the RSA scheme. Moreover, a pre-processing technique can significantly increase the signature generation speed.

1   Introduction

The use of *Digital signatures* is being increasingly demanded to ensure the integrity and authenticity of digital messages and documents. Applications include electronic mail, office automation, and electronic funds transfer.

Many digital signature schemes have been developed since Diffie and Hellman's seminal paper on public key cryptosystems [DH] was presented in 1976. Among these schemes, the RSA scheme [RSA] appears to be very promising from the practical viewpoint. However, the RSA scheme has the disadvantage of low processing speed, and is somewhat insecure against low multiplier attacks [Ha] and attacks using the homomorphic property [EH]. Although effective countermeasures are known against these attacks, the existence of these attacks may imply some implicit weaknesses in the RSA scheme.

The security of the RSA scheme can be increased with the scheme based on an elliptic curve over a ring Z_n [KMOV]. This variant (the KMOV scheme) seems to be more secure than the original RSA scheme against some attacks such as low multiplier attacks, although it is less efficient.

In this paper, we propose a new digital signature scheme based on an elliptic curve over a ring Z_n, that is more efficient than the RSA scheme as well as the KMOV scheme. We construct the new scheme on an elliptic curve over a ring using the idea of Okamoto's scheme [Ok]. The new scheme seems to be more secure than Okamoto's scheme against low degree attacks (or lattice attacks) and seems to be more secure than the RSA scheme against the homomorphic attacks. That is, our scheme with parameter $k = 2$, the double version, seems to be secure, while Okamoto's scheme with $k = 2$, the quadratic version, has been broken [BD, VGT]. Our scheme has no homomorphic property since the relationship between a message and its signature is randomized (or our signature is verified by an inequality not by an equation), so no homomorphic attack seems to apply to our scheme. This implies a possibility that our scheme may still be secure even if security weaknesses in Okamoto's or RSA scheme are found in the future.

The pre-processing technique (off-line processing) is possible with our scheme, as is true for Okamoto's scheme and DSA proposed by NIST as DSS (the Digital Signature Standard) [NIST]. This dramatically increases the signature generation (on-line processing) speed of our scheme. Thus, signature generation with our scheme is effectively instantaneous even if implemented on a smart card.

2   Notations

Z_n denotes the set of numbers between 0 and $n - 1$, and Z_n^* denotes the set of numbers between 0 and $n - 1$ which are relatively prime to n. $\lceil M \rceil$ denotes the least integer which is larger than or equal to M. $x \equiv y \pmod{n}$ denotes that n divides $x - y$. $f(x) \bmod n$ denotes an integer such that n divides $f(x) - (f(x) \bmod n)$ and $f(x) \bmod n \in Z_n$. $x/y \bmod n$ denotes an integer such that n divides $x - y(x/y \bmod n)$ and $x/y \bmod n \in Z_n$. $|X|$ denotes $\lfloor \log_2 X \rfloor + 1$, or the bit size of X.

3   Elliptic Curves over a Field and a Ring

Assume that K is the finite prime field $GF(p)$ with $p \neq 2, 3$. An elliptic curve over K (in affine coordinates), denoted by C_p, is the set of all solutions $(x, y) \in K \times K$ to the equation

$$C_p \; : \; y^2 \equiv x^3 + ax + b \quad \bmod p, \tag{1}$$

where $a, b \in K$, and $4a^3 + 27b^2 \not\equiv 0 \bmod p$, together with a special point O, called the point at infinity. Here, the group law operation [Ko1] (usually we call it the addition, and use the notation $+$) is defined over the points on C_p, $P(x_1, y_1), Q(x_2, y_2)$, and $R(x_3, y_3)$ as follows:

- $P(x_1, y_1) + Q(x_2, y_2) = R(x_3, y_3)$

$$\begin{cases} x_3 = \left(\dfrac{y_2 - y_1}{x_2 - x_1} \right)^2 - x_1 - x_2 \\ y_3 = -y_1 + \left(\dfrac{y_2 - y_1}{x_2 - x_1} \right)(x_1 - x_3) \end{cases} \quad \text{over } Z_p \tag{2}$$

- $2P(x_1, y_1) = R(x_3, y_3)$

$$\begin{cases} x_3 = \left(\dfrac{3x_1^2 + a}{2y_1} \right)^2 - 2x_1 \\ y_3 = -y_1 + \left(\dfrac{3x_1^2 + a}{2y_1} \right)(x_1 - x_3) \end{cases} \quad \text{over } Z_p \tag{3}$$

Let p and q be primes and $n = p^2 q$. Consider an elliptic curve modulo n: C_n. The addition operation on C_n is analogous to the usual one over $GF(p)$, although C_n is not a group.

4 Okamoto's Digital Signature Scheme

4.1 Procedures

- *Keys:*
  - *Secret key:* large prime numbers p, q $(p > q)$.
  - *Public key:* a positive integer $n = p^2 q$.
- *Signature generation:*
  - The signature s of a message m is computed by the originator as follows:
    * Pick a random number $t \in Z_{pq}^*$.
    * Compute s such that

$$w = \left\lceil \frac{h(m) - (t^k \bmod n)}{pq} \right\rceil,$$

$$u = w/(kt^{k-1}) \bmod p,$$

$$s = t + upq,$$

  where h is a one-way hash function ($h(m) \in Z_n$ for any positive integer m), k is an integer ($4 \le k$).
- *Signature verification:*
  - The signature message (s, m) is considered valid if the following verification inequality holds.

$$h(m) \le s^k \bmod n < h(m) + 2^{2 \cdot |n|/3}.$$

5 Proposed Digital Signature Scheme Based on Elliptic Curves over a Ring

Before describing our new proposed scheme, we introduce two extensions of Okamoto's scheme. The first one is the extension of the function type; from the *polynomial function* to the *rational function*. The other extension is the number of variables; from the *one variable function* to the *multi-variable function*.

5.1 Mathematical Preparations

The Taylor series expansion and the generalized Taylor series expansion for a multi-variable function are essential to prove the correctness of our schemes.

Proposition 1. (The Taylor expansion)
When function f is a one-variable infinitely differentiable function,

$$f(a + x) = f(a) + f^{(1)}(a)x + \frac{f^{(2)}(a)}{2}x^2 + \cdots + \frac{f^{(l)}(a)}{l!}x^l + \cdots,$$

where a is not a singular point, x is less than the convergence radii, and $f^{(l)}$ denotes l-th derived function of f.

Proposition 2. (The generalized Taylor expansion)
When function f is a t-variable infinitely differentiable function,

$$f(a_1 + x_1, a_2 + x_2, \ldots, a_t + x_t) =$$

$$f(a_1, a_2, \ldots, a_t) + (x_1 \frac{\partial}{\partial x_1} + \cdots + x_t \frac{\partial}{\partial x_t}) f(a_1, a_2, \ldots, a_t) +$$

$$\cdots + \frac{1}{l!} (x_1 \frac{\partial}{\partial x_1} + \cdots + x_t \frac{\partial}{\partial x_t})^l f(a_1, a_2, \ldots, a_t) + \cdots,$$

where $(a_1, a_2, \ldots, a_t)$ is not a singular point, $(x_1, x_2, \ldots, x_t)$ is less than the convergence radii, and $(x_1 \frac{\partial}{\partial x_1} + \cdots + x_t \frac{\partial}{\partial x_t})^l f(a_1, a_2, \ldots, a_t)$ denotes the value at $(a_1, a_2, \ldots, a_t)$ of $(x_1 \frac{\partial}{\partial x_1} + \cdots + x_t \frac{\partial}{\partial x_t})^l f(x_1, x_2, \ldots, x_t)$.

5.2 Extension Using a Rational Function

In this section, we show an extension of Okamoto's scheme, in which a *rational* function f is used in place of the polynomial function.

5.2.1 Procedures

- *Keys:*
  - *Secret key:* large prime numbers p, q $(p > q)$.
  - *Public key:* a positive integer $n = p^2 q$.
    a *rational* function f.
- *Signature generation:*
  - The signature s of a message m is computed by an originator as follows:
    * Pick a random number $t \in Z_{pq}^*$. If one of the following cases occurs, pick another random number $t \in Z_{pq}^*$: (1) $f(t) \bmod p = \infty$, (2) $f(t) \bmod q = \infty$, (3) $f(t) \bmod p = 0$, (4) $f(t) \bmod q = 0$, (5) $f'(t) \bmod p = 0$. Here, f is a rational function, or there exist polynomial functions, a and b, satisfying $f = a/b$. f' is the derived function of f, or $f'(x) = \frac{df(x)}{dx}$. Note that this check is not necessary in practice, since these cases occur with negligible probability.
    * Compute s such that

$$w = \left\lceil \frac{h(m) - (f(t) \bmod n)}{pq} \right\rceil,$$

$$u = w / f'(t) \bmod p,$$

$$s = t + upq.$$

Here, h is a one-way hash function ($h(m) \in Z_n$ for any positive integer m). Functions h and f can be fixed in the system.

- *Signature verification:*
  - The signature message (s, m) is considered valid if the following verification inequality holds.

$$h(m) \leq f(s) \bmod n < h(m) + 2^{2|n|/3}.$$

5.2.2 Correctness

Theorem 3. *Let* $0 \le h(m) < n - pq$, *and* s *be the signature of* m, *which is generated through the above-described procedure. Then,*

$$h(m) \le f(s) \bmod n < h(m) + pq.$$

Proof. First, let $\tilde{f}(x) \equiv f(x) \pmod{n}$ for all $x \in Z_n$ and all singular points of $\tilde{f}(x)$ do not lie in the interval $[0, n)$. For any rational function $f(x)$, $\tilde{f}(x)$ always exists. This is because: Let $a_i \in [0, n)$ $(i = 1, \ldots, k)$ be the singular points of $f(x)$. Then $f(x) = \frac{c(x)}{(x-a_1)(x-a_2)\cdots(x-a_k)b(x)}$. Let $\tilde{f}(x) = \frac{c(x)}{(x-\tilde{a}_1)(x-\tilde{a}_2)\cdots(x-\tilde{a}_k)b(x)}$, where $\tilde{a}_i = a_i + n$ $(i = 1, \ldots, k)$. Then, $\tilde{f}(x)$ satisfies the above conditions.

Since $\tilde{f}(x)$ is an analytic function and there exists no singular point in interval $[0, n)$, the Taylor expansion of $\tilde{f}(t + v)$ around t converges for any $t \in [0, n)$ and $t + v \in [0, n)$. That is,

$$\tilde{f}(t + v) = \tilde{f}(t) + \tilde{f}^{(1)}(t)v + \frac{\tilde{f}^{(2)}(t)}{2}v^2 + \cdots + \frac{\tilde{f}^{(l)}(t)}{l!}v^l + \cdots,$$

for any $t \in [0, n)$ and $t + v \in [0, n)$. Hence,

$$\tilde{f}(t + upq) \bmod n = \tilde{f}(t) + \tilde{f}^{(1)}(t)upq + (upq)^2(\frac{\tilde{f}^{(2)}(t)}{2} + \cdots) \bmod n$$
$$= \tilde{f}(t) + \tilde{f}^{(1)}(t)upq \bmod n,$$

for any $t \in Z_n$ and $t + upq \in Z_n$. From the definition of $\tilde{f}(x)$, $\tilde{f}(t+upq) \bmod n = f(t + upq) \bmod n$. Therefore,

$$f(t + upq) \bmod n = f(t) + f^{(1)}(t)upq \bmod n.$$

Furthermore from the equation $w = f^{(1)}(t)u \bmod p$, we have

$$f(t + upq) \bmod n = f(t) + wpq \bmod n.$$

On the other hand, from the definition $w = \left\lceil \frac{h(m)-(f(t)\bmod n)}{pq} \right\rceil$, we obtain

$$wpq = h(m) - (f(t) \bmod n) + \gamma,$$

where $0 \le \gamma < pq$. Therefore we have the following equation:

$$f(t+upq) \bmod n = f(t) + h(m) - (f(t) \bmod n) + \gamma \bmod n = h(m) + \gamma \bmod n.$$

Since $0 \le h(m) < n - pq$,

$$h(m) \le h(m) + \gamma \bmod n = h(m) + \gamma < h(m) + pq.$$

Hence we obtain

$$h(m) \le f(s) \bmod n < h(m) + pq,$$

where $s = t + upq$. $\qquad\qquad\square$

5.3 Extension Using a Multi-Variable Function

In this section, we present another extension of Okamoto's scheme, in which a *multi-variable* rational function f takes the place of the single-variable function.

5.3.1 Procedures

Let f_j $(j = 1, \ldots, J)$ be an I-variable rational function and $\boldsymbol{f}$ denote $(f_1, \ldots, f_J)$. Let $\boldsymbol{x} = (x_1, \ldots, x_I)$, $\boldsymbol{y} = (y_1, \ldots, y_J)$, where $x_i \in \boldsymbol{Z}_n$ $(i = 1, \ldots, I)$, and $y_j \in \boldsymbol{Z}_n$ $(j = 1, \ldots, J)$. We write $\boldsymbol{y} = \boldsymbol{f}(\boldsymbol{x})$ as $y_j = f_j(x_1, \ldots, x_I)$ $(j = 1, \ldots, J)$.

In this subsection, we show a signature scheme that uses $\boldsymbol{f}$ only once. However, by repeating the following procedure, we can easily construct a signature scheme based on a more complicated multi-variable rational function. In the next section, we will show an example in which the basic procedure is repeatedly executed.

For explanation simplicity, we suppose that $I = J$.

- *Keys:*
  - *Secret key:* large prime numbers p, q $(p > q)$.
  - *Public key:* a positive integer $n = p^2 q$.
    a *multi-variable* function $\boldsymbol{f}$.
- *Signature generation:*
  - The signature $\boldsymbol{s} = (s_1, \ldots, s_I)$ $(s_i \in \boldsymbol{Z}_n^*; i = 1, \ldots, I)$ of a message m is computed by originator A as follows:
    * Pick a random number vector $\boldsymbol{t} = (t_1, \ldots, t_I)$ $(t_i \in \boldsymbol{Z}_{pq}^*; i = 1, \ldots, I)$. If one of the following cases occurs, pick another random number vector $\boldsymbol{t}$: for $j \in \{1, \ldots, I\}$, (1) $f_j(\boldsymbol{t}) \bmod p = \infty$, (2) $f_j(\boldsymbol{t}) \bmod q = \infty$, (3) $f_j(\boldsymbol{t}) \bmod p = 0$, (4) $f_j(\boldsymbol{t}) \bmod q = 0$, (5) $I \times I$ matrix $\Delta \boldsymbol{f}(\boldsymbol{t}) \bmod p$ is not regular. Here, f_j is a I-variable rational function, and

$$\Delta \boldsymbol{f}(\boldsymbol{t}) \bmod p = \begin{pmatrix} \frac{\partial f_1(\boldsymbol{t})}{\partial x_1} & \cdots & \frac{\partial f_1(\boldsymbol{t})}{\partial x_I} \\ \vdots & \ddots & \vdots \\ \frac{\partial f_I(\boldsymbol{t})}{\partial x_1} & \cdots & \frac{\partial f_I(\boldsymbol{t})}{\partial x_I} \end{pmatrix} \bmod p.$$

Note that this check is not necessary in practice, since these cases occur with negligible probability.
    * Compute $\boldsymbol{s}$ such that

$$(m_1, \ldots, m_I) = h(m), \quad (m_j \in \boldsymbol{Z}_n; j = 1, \ldots, I)$$

$$w_j = \left\lceil \frac{m_j - (f_j(\boldsymbol{t}) \bmod n)}{pq} \right\rceil, \quad (j = 1, \ldots, I)$$

$$\begin{pmatrix} u_1 \\ \vdots \\ u_I \end{pmatrix} = \begin{pmatrix} \frac{\partial f_1(\boldsymbol{t})}{\partial x_1} & \cdots & \frac{\partial f_1(\boldsymbol{t})}{\partial x_I} \\ \vdots & \ddots & \vdots \\ \frac{\partial f_I(\boldsymbol{t})}{\partial x_1} & \cdots & \frac{\partial f_I(\boldsymbol{t})}{\partial x_I} \end{pmatrix}^{-1} \begin{pmatrix} w_1 \\ \vdots \\ w_I \end{pmatrix} \bmod p$$

$$s_i = t_i + u_i pq \quad (i = 1, \ldots, I)$$

$$s = (s_1, \ldots, s_I)$$

Here, h is a one-way hash function. Functions h and f can be fixed in the system.

- *Signature verification:*
  - ∘ The signature message (s, m) is considered valid if the following verification inequality holds for all $j = 1, \ldots, I$,

$$m_j \leq f_j(s) \bmod n < m_j + 2^{2|n|/3},$$

where $(m_1, \ldots, m_I) = h(m)$.

5.3.2 Correctness

Theorem 4. *Let* $0 \leq m_j < n - pq$ *for all* $j = 1, \ldots, I$, *and* s *be the signature of* m, *which is generated through the above-described procedure. Then,*

$$m_j \leq f_j(s) \bmod n < m_j + pq.$$

This theorem can be proven in a manner similar to Theorem 3, using the generalized Taylor expansion.

5.4 A New scheme Based on Elliptic Curve over Z_n

This section introduces our new scheme based on an elliptic curve over Z_n. The correctness of the scheme is given as a combined specific example of two previous extensions of Okamoto's scheme; the new scheme is the *two-variable rational* function version.

5.4.1 Elliptic Curve and Some Definitions

We consider an elliptic curve C_n:

$$y^2 = x^3 + ax + b \quad \text{over } Z_n.$$

As described in Section 3, the addition operation is defined over the points on C_n, $P(x_1, y_1)$, $Q(x_2, y_2)$, and $R(x_3, y_3)$, by equations (2) and (3) over Z_n.

Here, let $f = (f_x, f_y)$, $g = (g_x, g_y)$ such that

$$\begin{cases} f_x(x_1, y_1) = \left(\frac{3x_1^2 + a}{2y_1}\right)^2 - 2x_1 \\ f_y(x_1, y_1) = -y_1 + \left(\frac{3x_1^2 + a}{2y_1}\right)(x_1 - f_x(x, y)) \end{cases}$$

$$\begin{cases} g_x(x_1, y_1, x_2, y_2) = \left(\frac{y_2 - y_1}{x_2 - x_1}\right)^2 - x_1 - x_2 \\ g_y(x_1, y_1, x_2, y_2) = -y_1 + \left(\frac{y_2 - y_1}{x_2 - x_1}\right)(x_1 - g_x(x_1, y_1, x_2, y_2)) \end{cases}$$

Then we can express

$$2P = f(P) \bmod n,$$

$$P + Q = g(P, Q) \bmod n.$$

Therefore, for an integer k, we can calculate $R = kP$ over Z_n by using an addition chain corresponding to k, where P and R are points on C_n.

Let

$$\Delta f = \begin{pmatrix} \frac{\partial f_x}{\partial x} & \frac{\partial f_x}{\partial y} \\ \frac{\partial f_y}{\partial x} & \frac{\partial f_y}{\partial y} \end{pmatrix},$$

$$\Delta g = \begin{pmatrix} \frac{\partial g_x}{\partial x_1} & \frac{\partial g_x}{\partial y_1} & \frac{\partial g_x}{\partial x_2} & \frac{\partial g_x}{\partial y_2} \\ \frac{\partial g_y}{\partial x_1} & \frac{\partial g_y}{\partial y_1} & \frac{\partial g_y}{\partial x_2} & \frac{\partial g_y}{\partial y_2} \end{pmatrix}.$$

Next, let $A = (A_1, A_2)$, $B = (B_1, B_2)$, $C = (C_1, C_2)$ such that

$$A_1 = (a_x, a_y), \quad B_1 = (b_x, b_y), \quad C_1 = (c_x, c_y),$$

$$A_2 = \begin{pmatrix} a_{11} & a_{12} \\ a_{21} & a_{22} \end{pmatrix}, \quad B_2 = \begin{pmatrix} b_{11} & b_{12} \\ b_{21} & b_{22} \end{pmatrix}, \quad C_2 = \begin{pmatrix} c_{11} & c_{12} \\ c_{21} & c_{22} \end{pmatrix},$$

where $a_x, a_y, b_x, b_y, c_x, c_y \in Z_n^*$, and $a_{ij}, b_{ij}, c_{ij} \in Z_p^*$ ($i, j \in \{1, 2\}$).

Definition 5. (Functions F and G)
Let F be a function such that $F(A) = (C_1, C_2)$, where

$$C_1 = f(A_1) \bmod n, \text{ and } C_2 = \Delta f(A_1) \cdot A_2 \bmod p.$$

Let G be another function such that $G(A, B) = (C_1, C_2)$, where

$$C_1 = g(A_1, B_1) \bmod n, \text{ and } C_2 = \Delta g(A_1, B_1) \cdot [A_2, B_2] \bmod p,$$

where $[A_2, B_2]$ denotes the 4×2 matrix in which i-th (1st and 2nd) row of A_2 is the i-th (1st and 2nd) row and the i-th (1st and 2nd) row of B_2 is the $(i+2)$-nd (3rd and 4th) row.

5.4.2 Procedures

- *Keys:*
  - o *Secret key:*   large prime numbers p, q ($p > q$).
  - o *Public key:*   a positive integer $n = p^2 q$.
             a parameter (of the curve) a.
- *Signature generation:*
  - o Signature $S = (s_x, s_y)$ of a message m is computed by the originator as follows:

* Pick a random number vector $T = (t_x, t_y)$ $(t_x, t_y \in Z_{pq}^*)$. If one of the following cases occurs during executing the following signature generation procedure, return to this stage and pick another random number vector t: for $i \in \{x, y\}$, (1) $(kT$ over $Z_n)_i$ mod $p = \infty$, (2) $(kT$ over $Z_n)_i$ mod $q = \infty$, (3) $(kT$ over $Z_n)_i$ mod $p = 0$, (4) $(kT$ over $Z_n)_i$ mod$q = 0$, (5) 2×2 matrix $D(T)$ mod p is not regular, where kT over Z_n means k times point of T by the addition formula on C_n, and $(\cdot)_x$ (or $(\cdot)_y$) means the x-coordinate (or y-coordinate) of point $(\cdot)$. (Note that the calculation, kT over Z_n, here is formally executed by the addition formula, and that T is not necessary to be on C_n.) Note that this check is not necessary in practice, since these cases occur with negligible probability.
* Compute S such that

$$M = (m_x, m_y) = h(m), \quad (m_x, m_y \in Z_n)$$

where, h is a one-way hash function$(m_x, m_y \in Z_n)$.

$$w_x = \left\lceil \frac{m_x - (kT \text{ over } Z_n)_x}{pq} \right\rceil,$$

$$w_y = \left\lceil \frac{m_y - (kT \text{ over } Z_n)_y}{pq} \right\rceil.$$

Next 2×2 matrix $D(T, k)$ is computed from T and k by Algorithm D below. Then

$$\binom{u_x}{u_y} = D(T, k)^{-1} \binom{w_x}{w_y} \text{ mod } p$$

$$s_i = t_i + u_i pq \quad (i = x, y)$$

$$S = (s_x, s_y)$$

Integer k and functions h can be fixed in the system. Note that the parameter a in the public key can be fixed in the system. Therefore the real public key for each user is considered to be an only n.
* Note that kT over Z_n and $D(T, k)^{-1}$ can be computed as pre-processing works since they are independent of a message m.

Algorithm D

Input: T, k
Output: 2×2 matrix $D(T, k)$, whose element is in Z_p^*.
Step 1:   Set $A = (A_1, A_2)$ such that

$$A_1 \leftarrow T, \quad A_2 \leftarrow \begin{pmatrix} 1 & 0 \\ 0 & 1 \end{pmatrix}.$$

Set $l \leftarrow 1$ and $t \leftarrow 0$.
The bit expression of k is "$b_L b_{L-1} \cdots b_1$."
(The initial setting for $B = (B_1, B_2)$ is not necessary, since the value of B is set in Step 2.)

63

Step 2: If $b_l = 1$ and $t = 0$, then $B \leftarrow A$ and $t \leftarrow 1$.
If $b_l = 1$ and $t = 1$, then $B \leftarrow G(A, B)$.

Step 3: If $l = L$, then output B_2 as $D(T, k)$.
Otherwise $l \leftarrow l + 1$, and $A \leftarrow F(A)$.
Return to Step 2.

Note that the value of B_1 that corresponds to the output value of B_2 or $D(T, k)$ is equivalent to kT over $\boldsymbol{Z}_n$.

- *Signature verification:*
  - The signature message (S, m) is considered valid if the following verification inequalities hold

$$m_x \leq (kS \text{ over } \boldsymbol{Z}_n)_x < m_x + 2^{2|n|/3},$$

$$m_y \leq (kS \text{ over } \boldsymbol{Z}_n)_y < m_y + 2^{2|n|/3},$$

where $(m_x, m_y) = h(m)$. Note that the first parameter a of $\mathcal{C}_n$ is fixed and given for the calculation kS over $\boldsymbol{Z}_n$, but that the other parameter b of $\mathcal{C}_n$ is not necessary for the calculation and is determined by the value $S = (s_x, s_y)$ such that

$$b = s_y^2 - s_x^3 - as_x \text{ over } \boldsymbol{Z}_n.$$

6 Security Consideration

The security of our scheme depends on the difficulty of factoring $n = p^2 q$. Although it has not been proven that our scheme is as secure as factoring, our scheme seems to be more secure than Okamoto's scheme, against which no attack is known so far when its degree is greater than three. The quadratic version of Okamoto's scheme was broken by Brickell et.al. [BD], and this attack was generalized by Vallée et.al. [VGT] using the lattice algorithm. Their attacks essentially use and generalize the approximation property that $\left[\sqrt{N}\right]^2 - N = O(N^{1/2}) < O(N^{2/3})$. However, this approximation technique does not appear applicable to our scheme even if it is the double version $(k = 2)$, since the rational function mod n is essentially used in our scheme. Although it is not clear that factoring $n = p^2 q$ is as hard as factoring $n = pq$, no attack has been reported so far, that is specifically effective for a number with the square of a prime.

7 Performance

We have estimated the amount of work needed to generate a signature with our scheme and compare it with that of the RSA scheme. We assume that $n(= p^2 q)$ is 96 bytes and $k = 2$ for our scheme, and $n'(= p'q')$ is 64 bytes for RSA.

Signature generation with the new scheme requires 4 modulo-n multiplications, 1 modulo-n division, 17 modulo-p multiplications, and 1 modulo-p division.

So, in total, it is almost equivalent to $(4 + 17/9) + (1 + 1/9)c$ modulo-n multiplications, which is less than $(6 + 1.2c)$ modulo-n multiplications. Here, c is the ratio of the amount of work for modulo-n division to that for modulo-n multiplication, and is considered to be less than 10 from our implementation data based on algorithm L (p.329) in [Kn]. The RSA scheme requires 750 modulo-n' multiplications.

As the computational complexity of one modulo-n multiplication is almost equivalent to that of 2.25 $(=1.5^2)$ modulo-n' multiplications, signature generation with our scheme is considered to require less than 40 modulo-n' multiplications.

The signature generation speed of our new scheme is more than 10 times faster than that of the RSA scheme. If the Chinese Remainder Theorem technique is applied to the RSA scheme, the amount of work is theoretically reduced by 75%, while the work of our scheme is reduced by about 50%. In addition, the m-ary exponentiation and Montgomery arithmetic techniques can reduce the amount of work needed by the RSA scheme, however, they can also applied to our scheme. Therefore our new scheme is still at least several times faster than the RSA scheme.

Moreover, the pre-processing technique (off-line processing) is possible with our scheme, as is true for Okamoto's scheme [FOM] and DSA. In the pre-processing phase, some computations that do not depend on the message are executed. This dramatically increases the signature generation (on-line processing) speed of our scheme. Thus, signature generation with our scheme is effectively instantaneous even if implemented on a smart card, since the amount of work needed for signature generation is less than one modulo-n multiplication.

8   Conclusion

We have proposed a new practical digital signature scheme based on elliptic curves over a ring. To construct this scheme, we introduced two extensions of Okamoto's scheme. The signature generation speed of our scheme is more than 10 times faster than that of the RSA scheme. Moreover, a pre-processing technique can significantly increase the signature generation speed.

References

[BD]   E. Brickell and J. DeLaurentis, "An Attack on a Signature Scheme Proposed by Okamoto and Shiraishi", *Advances in Cryptology — CRYPTO'85*, Lecture Notes in Computer Science No.218, Springer-Verlag, pp.28–32 (1986).

[DH]   W. Diffie and M. E. Hellman, "New Directions in Cryptography", *IEEE Transactions on Information Theory*, Vol.IT-22, No.6, pp.644–654 (Nov., 1976).

[EH]   J. Evertse and E. van Heyst, "Which New RSA Signatures can be Computed from some Given RSA Signatures?", *Advances in Cryptology — EUROCRYPT'90*, Lecture Notes in Computer Science No.473, Springer-Verlag, pp.83–97 (1991).

[FOM] A. Fujioka, T. Okamoto, and S. Miyaguchi, "ESIGN: An Efficient Digital Signature Implementation for Smart Cards", *Advances in Cryptology — EUROCRYPT'91*, Lecture Notes in Computer Science No.547, Springer-Verlag, pp.446–457 (1991).

[Ha] J. Hastad, "On Using RSA with Low Exponent in a Public Key Network", *Advances in Cryptology — CRYPTO'85*, Lecture Notes in Computer Science No.218, Springer-Verlag, pp.403–408 (1985).

[Ka] B. S. Kaliski, Jr., "A Pseudo-Random Bit Generator Based on Elliptic Logarithms", *Advances in Cryptology — CRYPTO'86*, Lecture Notes in Computer Science No.263, Springer-Verlag, pp.84–103 (1986).

[KMOV] K. Koyama, U. Maurer, T. Okamoto, and S. A. Vanstone, "New Public-Key Schemes Based on Elliptic Curves over the Ring Z_n", *Advances in Cryptology — CRYPTO'91*, Lecture Notes in Computer Science No.576, Springer-Verlag, pp.252–266 (1992).

[Kn] D. E. Knuth, *The Art of Computer Programming*, 2nd Edition, Addison-Wesley Publishing Company (1981).

[Ko1] N. Koblitz, *A Course in Number Theory and Cryptography*, Springer-Verlag (1987).

[Ko2] N. Koblitz, "Constructing Elliptic Curve Cryptosystems in Characteristics 2", *Advances in Cryptology — CRYPTO'90*, Lecture Notes in Computer Science No.537, Springer-Verlag, pp.156–167 (1991).

[NIST] National Institute for Standards and Technology, "Specifications for a Digital Signature Standard", *Federal Information Processing Standard Publication XX*, draft (Aug., 1991).

[Ok] T. Okamoto, "A Fast Signature Scheme Based on Congruential Polynomial Operations", *IEEE Transactions on Information Theory*, Vol.IT-36, No.1, pp.47–53 (Jan., 1990).

[RSA] R. Rivest, A. Shamir, and L. Adleman, "A Method for Obtaining Digital Signatures and Public-Key Cryptosystems", *Communications of the ACM*, Vol.21, No.2, pp.120–126 (Feb., 1978).

[VGT] B. Vallée, M. Girault, and P. Toffin, "How to Break Okamoto's Cryptosystem by Reducing Lattice Bases", *Advances in Cryptology — Eurocrypt'88*, Lecture Notes in Computer Science No.330, Springer-Verlag, pp.281–292 (1988).

Designing and Detecting Trapdoors for Discrete Log Cryptosystems

Daniel M. Gordon*

Department of Computer Science, University of Georgia, Athens, GA 30602

Abstract. Using a number field sieve, discrete logarithms modulo primes of special forms can be found faster than standard primes. This has raised concerns about trapdoors in discrete log cryptosystems, such as the Digital Signature Standard. This paper discusses the practical impact of these trapdoors, and how to avoid them.

1 Introduction

The National Institute of Standards and Technology (NIST) recently announced a proposal for a federal digital signature standard, DSS [21]. This proposal gives an algorithm for electronically signing documents, to guarantee the integrity of the message and the identity of the sender. The Digital Signature Algorithm (DSA) given in the proposal is based on the difficulty of solving discrete logarithms modulo large primes. It has already excited a great deal of discussion regarding its efficiency and security.

In the DSA, the public key consists of a prime p of 512 bits, a prime q dividing $p-1$ of 160 bits, and a number g which is a $((p-1)/q)$th power mod p. The private key is a number x, and $y = g^x$ is also made public. Then to sign a message m, the sender calculates

$$r \equiv (g^k \bmod p) \bmod q$$

and

$$s \equiv (k^{-1}(H(m) + xr) \bmod q.$$

Here H is any one-way hash function, m is the message, and k is a random number less than q. To authenticate a message, a recipient computes:

$$w = s^{-1} \bmod q,$$

$$u_1 = (H(m)w) \bmod q,$$

$$u_2 = (rw) \bmod q,$$

* Current address: Center for Communications Research, 4320 Westerra Court, San Diego, CA 92121

$$v = (g^{u_1} y^{u_2} \bmod p) \bmod q.$$

The signature is correct if $v = r$.

The only known way to break this system is to find x from g and y (i.e. find the discrete logarithm $\log_g y \bmod p$). Several other schemes (e.g. [3], [6], [18]) also depend on the difficulty of discrete logarithms. Subexponential algorithms are known for finding discrete logarithms modulo large primes, but the largest prime for which the problem has been solved is 224 bits in length, by LaMacchia and Odlyzko [9], using the Gaussian integer method of Coppersmith, Odlyzko and Schroppel [5].

In [7], an algorithm is given for finding discrete logarithms using a number field sieve, which is asymptotically faster than other known methods. The general number field sieve is impractical, but a variant of the algorithm for primes of special forms is practical. The idea of using the number field sieve to make trapdoor primes is mentioned in [1], page 50.

In Sect. 2, we give a brief description of how the special number field sieve for discrete logarithms works. Estimates for the time to break the DSS with regular versus various trapdoor primes are given in Sect. 3. The rest of the paper deals with how to detect trapdoors, how to construct trapdoors to avoid detection, and how one or more people can choose primes for which the probability of a trapdoor existing is negligible.

2 The Number Field Sieve

Here we give a short presentation of the special number field sieve. For a more complete description of the algorithm, and the heuristic assumptions involved, see [7].

Let p be a prime and f be an irreducible monic polynomial of degree k with reasonably small coefficients, such that for some integers X and Y near $p^{1/k}$ we have $Y^k f(X/Y) \equiv 0 \pmod{p}$. Let $\alpha \in \mathbb{C}$ denote a root of f, and $K = \mathbb{Q}(\alpha)$. For constructing trapdoor primes, it is convenient to pick f so that $\mathcal{O}_K = \mathbb{Z}[\alpha]$ is a unique factorization domain.

We may define a homomorphism φ from $\mathbb{Z}[\alpha]$ to $\mathbb{Z}/p\mathbb{Z}$ by sending α to $X/Y \bmod p$, so that for any integers c and d,

$$cY + dX = Y(c + dX/Y) \equiv Y\varphi(c + d\alpha) \pmod{p}.$$

The factor base $\mathcal{B}$ will consist of rational primes less than a bound B $(\mathcal{B}_{\mathbb{Q}})$, first-degree primes in $\mathcal{O}_K$ with norm less than B $(\mathcal{B}_K)$, a fundamental set of units in $\mathcal{O}_K$, and Y. Calculating the primes and units for the field is not difficult when f is, say, $x^5 - 2$ (see [14]), but will be more difficult for polynomials with larger coefficients. We will discuss this problem in the next section.

Call a rational or algebraic integer *smooth* if its prime factors are all in the factor base. We will need to find many pairs of coprime integers c, d such that $cY + dX$ and $c + d\alpha$ are both smooth. This can be accomplished efficiently by sieving $cY + dX$ and the norm

$$|N(c + d\alpha)| = |(-d)^k f(-c/d)|$$

for fixed c and large range of d. The smoothness of $c + d\alpha$ and $N(c + d\alpha)$ are related by the following (see [7], Proposition 2):

Theorem 1. *If c and d are relatively prime and $r^l \parallel N(c + d\alpha)$ for a prime r, then $(r, \alpha - c_r)^l \parallel (c + d\alpha)$ in $\mathcal{O}_K$, for $c_r \equiv -c/d \pmod{r}$.*

We will choose g, the base for the discrete logarithm to be smooth and a primitive root modulo p. Note that this cannot be the same as the base g for the DSA, since that g is a $(p-1)/q$th power. Thus, the first step in breaking the DSS would be to find the log of its base.

The precomputation step involves sieving through small c and d, looking for pairs with $cY + dX$ and $N(c + d\alpha)$ both smooth. Each hit gives us an equation involving logarithms of the factor base. Suppose that we find a c and d for which both are smooth, say

$$cY + dX = \prod_{s \in \mathcal{B}_{\mathbb{Q}}} s^{w_s(c,d)},$$

and

$$|N(c + d\alpha)| = \prod_{s \in \mathcal{B}_{\mathbb{Q}}} s^{v_s(c,d)},$$

for $v_s, w_s \in \mathbb{Z}_{\geq 0}$. Then

$$(c + d\alpha) = \prod_{s \in \mathcal{B}_K} s^{v_s(c,d)}$$

by Theorem 1. Since $\mathcal{O}_K$ is a UFD, this equation involving ideals can be replaced with one involving algebraic integers, by replacing each s in the above equation by a generator for the ideal. Then $c + d\alpha$ divided by the generators is a unit, which can be explicitly computed in terms of a fundamental set of units, using Theorem 5 of [7].

From this, we obtain:

$$(cY + dX)\varphi(c + d\alpha)^{-1} \equiv \prod_{s \in \mathcal{B}} s^{u_s(c,d)} \equiv 1 \pmod{p},$$

which gives us an equation for the logs of the factor base:

$$\sum_{s \in \mathcal{B}} u_s(c, d) \log_g s \equiv 0 \pmod{p-1}.$$

Once we have more than $|\mathcal{B}|$ hits, we solve the resulting matrix equation over $\mathbb{Z}/(p-1)\mathbb{Z}$ using structured Gaussian elimination to reduce the size of the matrix, and then solving a smaller, dense matrix using the conjugate gradient method or Wiedemann's algorithm (see [10]). This completes the precomputation.

To find an individual logarithm, we reduce the problem to finding the logs of medium-sized primes. Choose random values of s and attempt to factor $g^s y \pmod{p}$ using the elliptic curve method (ECM) until one is found for which

$$g^s y \equiv q_1 q_2 \cdots q_r \pmod{p},$$

with each q_i is less than a bound Q. (This can be improved as in [9], by finding $z_1, z_2 = O(\sqrt{p})$ such that $g^s y \equiv z_1/z_2 \pmod{p}$, and testing whether z_1 and z_2 are both Q-smooth.)

For each q_i, we will sieve c and d for which $q_i|(cY + dX)$, say fixing d and taking $c = c_0 + eq_i$, to find one value for which $(cY + dX)/q_i$ and $N(c + d\alpha)$ are both smooth. Once this happens we are done, since from the precomputation we know the logs of the whole factor base.

The choices for the size of the factor base and q_i's depend on how time is to be divided between the two stages. Enlarging the factor base reduces the time needed to find individual logarithms, but at the cost of increasing the precomputation time. Let

$$L_n[v; c] = \exp\{(c + o(1))(\log n)^v (\log\log n)^{1-v}\},$$

for $n \to \infty$. Assuming some reasonable heuristics (see [7]), the optimal choice of parameters is

$$k = \left\lceil 10^{1/5} \left(\frac{\log p}{\log\log p} \right)^{1/5} \right\rceil,$$

$$B = L_p[2/5; (4/125)^{1/5}],$$

and

$$Q = L_p[3/5; (1/100)^{1/5}],$$

which results in both the precomputation and individual logarithms taking expected time

$$L_p[2/5; \left(\frac{128}{125} \right)^{1/5}] \approx L_p[2/5; 1.00475].$$

If many instances are to be done for one p, more time could be spent on the precomputation by taking a larger factor base. For $\mu \geq (128/125)^{1/5}$, if we spend $L_p[2/5; \mu]$ time on the precomputation, each logarithm can be found in time

$$L_p\left[2/5; \left(\frac{128}{125\mu^2} \right)^{1/3}\right].$$

The Gaussian integer method is a special number field sieve with $k = 2$ and K a complex quadratic field. For any $c \geq 1$, the Gaussian integer method can find logarithms in time $L_p[1/2; 1/(2c)]$ if $L_p[1/2; c]$ is spent on the precomputation. Even for fairly small primes with good polynomials, the special number field sieve is faster than the Gaussian integer method.

For primes which cannot be represented by good polynomials, a similar procedure called the general number field sieve can be done. The difference is that the polynomial f will have large coefficients, so operations in the resulting field will be impractical. To avoid them, the equations must be solved over the rationals instead of modulo $p - 1$, to eliminate ideals and units.

The better asymptotic time for the general number field sieve comes from using different fields for finding individual logarithms. Instead of sieving through

c and d such that $q_i|(cY+dX)$, we search through polynomials for which $q_i|\varphi(\alpha)$. This allows us to take Q as big as X and Y, which asymptotically speeds up the algorithm. The time for the general number field sieve is

$$L_p[1/3; 3^{2/3}] \approx L_p[1/3; 2.08].$$

Oliver Schirokauer [17] has developed a method to avoid solving equations over the rationals, so that the time can be improved to $L_p[1/3, 1.902]$. The larger constant and $o(1)$ terms make the general number field sieve impractical for numbers we are interested in.

3   Complexity Estimates

There are four parts of the algorithm which dominate the timing estimates. For the precomputation, there is the sieve to gather equations, and then the linear algebra modulo $p-1$ to solve the equations. For finding individual logarithms, the medium-sized primes are found by repeated trials of the ECM, and then another sieve must be done for each q_i.

How much time is devoted to each part depends on the choice of parameters: the degree k of the polynomial f, the polynomial chosen (and the resulting field), the size of the factor base, and the size of a medium-sized prime.

For $k = 2$, we can take $f = x^2 + r$, for r a small positive integer for which $-r$ is a quadratic residue modulo p. Then the resulting field is just a complex quadratic field $\mathbb{Q}(\sqrt{-r})$, and we have the Gaussian integer method. This can be applied to any prime, but is impractical for 512-bit primes. Breaking the DSS using the Gaussian integer method using $B = 50,000,000$ would require sieving 10^{20} numbers. Even if this could be accomplished, the resulting matrix would have over $5,000,000$ columns, and the linear algebra problem would be a major hurdle.

The numbers in Table 1 show the difficulty of finding discrete logarithms for 512-bit primes using the special number field sieve with polynomials of degree 2–5 with small coefficients. They assume that the large prime variation described in [14] is being used. They are intended as rough estimates only, but serve to give an idea of the time required. For comparison, the factorization of F_9 required sieving about 10^{14} numbers, and solving a matrix with 199,203 columns modulo 2 [14]. For larger k, X and Y are smaller, so a smaller factor base can be used, speeding the precomputation. But then for individual logarithms $N(c+d\alpha) \approx Q^k$ is larger, so we need to take Q smaller and do more ECM trials.

Table 1 indicates that the ideal polynomial for a trapdoor would have degree four. Its coefficients should be small, to keep down the size of $N(c + d\alpha)$. The field generated by a root of the polynomial should have small discriminant and regulator, class number one and index one, so that field operations can be done efficiently. If the polynomial has four complex roots, then the unit group will have rank one.

For example, the polynomial $x^4 + x + 1$ satisfies all the above conditions. The problem is that the polynomial could only be used with primes p for which there

Table 1. Statistics for 512-bit primes with good polynomials.

k	2	3	4	5
B	5×10^7	5×10^6	3×10^6	2×10^6
sieve range	10^{20}	2×10^{16}	2×10^{14}	10^{14}
matrix size	$5,600,000$	$650,000$	$400,000$	$280,000$
Q	10^{20}	10^{19}	10^{15}	10^{13}
# ECM trials	$29,000$	$78,000$	2×10^7	2×10^9
second sieve	2.5×10^{14}	2.4×10^{15}	1.5×10^{14}	3×10^{14}

exist $X, Y \approx p^{1/4}$ such that $X^4 + XY^3 + Y^4 \equiv 0 \pmod{p}$. This is a thin set of primes, which can easily be detected (see the next section).

For polynomials with larger coefficients, the special number field sieve is more complicated. The sieving stage takes slightly longer, since the norms being tested for smoothness are larger. For polynomials with coefficients of, say, up to 100 in absolute value, the sieving range must be increased by roughly a factor of ten.

Another difficulty is dealing with a field of larger discriminant. The problem is finding generators for the unit group and prime ideals in the factor base. In [13], these are found by searching through algebraic integers of the form $\sum_{i=0}^{4} h_i \alpha^i$, for α a generator of K and small values of h_i. For fields generated by polynomials with larger coefficients, this will be impractical.

There have been several papers on efficient algorithms to find units and algebraic integers of given norms in general number fields, (see [4], [16]). The computations are involved, but they only need to be done once for a given f.

The matrix equation resulting from the sieving may be solved using intelligent Gaussian elimination to greatly reduce the size of the matrix, and then the conjugate gradient algorithm to solve the reduced equation. In [10] these methods were used to solve matrices with up to 96,321 columns.

4 Trapdoor Primes and Polynomials

From Table 1, we see that some 512-bit primes may not be safe, but general ones (at least for the moment) are. We want to ensure that for a given prime p there is a no polynomial f which can be used for the special number field sieve. Currently, the only way to check for this is to check one polynomial at a time.

Let p be a 512-bit prime and f be a polynomial of degree $k \geq 3$. We will say that X and Y are a trapdoor for p and f if they are both less than (say) $1,000\, p^{1/k}$ in absolute value, and $Y^k f(X/Y) \equiv 0 \pmod{p}$.

Theorem 2. *If a trapdoor X, Y exists for p and f, then for a root c_p of f mod p, (X, Y) is a short vector in the lattice $\mathcal{L} = \langle (p, 0), (c_p, 1) \rangle$.*

Proof. Let c_p be the root of f mod p congruent to X/Y mod p. The lattice $\mathcal{L}$ contains (X, Y), since $(c_p, 1)Y = (c_p Y, Y) \equiv (X, Y) \pmod{p}$.

The shortest vector in $\mathcal{L}$ has length at most $O(\sqrt{p})$, and for most choices of p, f and c_p, the short vector will be $\Theta(\sqrt{p})$. If such an X and Y do exist, (X, Y) has length $< \sqrt{2}\, 1,000\, p^{1/k}$.

Conversely, all vectors $(X, Y) \in \mathcal{L}$ satisfy $Y^k f(X/Y) \equiv 0 \pmod{p}$, so any such short vector is a trapdoor for f and p. □

This gives an efficient algorithm for testing whether a trapdoor exists for a given f and p. One may find linear factors of f mod p efficiently by eliminating square factors (dividing by the greatest common divisor of f and f' mod p), and then taking the gcd of $(x^p - x)$ and f mod p (see [8]). Then X and Y, if they exist, can be found using lattice reduction.

The main problem with this is that every polynomial f needs to be considered separately, so a limited range of polynomials can be searched. On a Sparcstation 1, one fourth-degree polynomial can be tested for a 512-bit prime in about a minute. On a parallel machine many polynomials could be searched at once, and a fairly large range of polynomials could be tested. With this test, a trapdoor with a good polynomial could be found. This forces an adversary to choose a polynomial from a set too large to be exhaustively searched, say an f of degree fourth with coefficients chosen randomly between -100 and 100.

Note that in the special case $Y = 1$, where $p = f(X)$, the polynomial can be found much faster. In this case $(p/a_k)^{1/k}$ is close to X, where a_k is the leading coefficient of f. All polynomials with a given a_k could be tested at once, very efficiently, so such a trapdoor would be much easier to discover.

Similar techniques can be used to construct a trapdoor prime. Suppose we wish to compute q and p for the DSA such that for a given polynomial $f(x) = x^3 + bx^2 + cx + d$ and some $X, Y \approx p^{1/3}$, $p = Y^3 f(X/Y)$. Begin by finding a 160-bit prime q, and choosing any $Y_0 \approx 2^{170}$. Let $g(x) = Y_0^3 f(x/Y_0)$. Then we may find an a mod q such that $g(a) \equiv 1 \pmod{q}$, by looking for linear factors of $g(x) - 1 \pmod{q}$. If none exists, then another Y_0 may be tried.

For any $X \equiv a \pmod{q}$ and $Y \equiv Y_0 \pmod{q}$, we have $Y^3 f(X/Y) \equiv 1 \pmod{q}$. Taking $X = a + l_1 q$ and $Y = Y_0 + l_2 q$, with l_1 and l_2 chosen so that $Y^3 f(X/Y) \approx 2^{512}$. we expect to soon find a pair for which $p = Y^3 f(X/Y)$ is prime. This p and q could be used as a trapdoor for the DSS.

This is not an ideal trapdoor, since from Table 1, a degree four polynomial would work better. The problem with constructing a better trapdoor using the above method is that a is usually a 160-bit number, which is bigger than $p^{1/4}$, so X would be too large. The revised DSS will allow primes p up to 1024 bits [19]. For primes with 640 or more bits, the above method can be used to make a trapdoor with a degree four polynomial. For primes with 800 or more bits, a degree five polynomial can be used.

Another way to generate a trapdoor would be to choose a polynomial f, and try random values of X and Y until $p = Y^k f(X/Y)$ is prime and divisible by a 160-bit prime q. To find such a value, one could sieve by small primes or use the ECM factoring method to find an X, Y pair for which $p - 1$ is smooth except for one 160-bit prime factor. This has the drawback that $(p-1)/q$ would be smooth, which while it is not known to weaken the system, does seem undesirable.

5   Protocols for Choosing a Prime

The ideal way to avoid worries about a trapdoor would be to come up with a way of generating primes for which one can guarantee that no such polynomial exists. An alternative is to use a random prime, which is almost certain to be safe. Call a prime p *unsafe* if an f exists with $Y^k f(X/Y) \equiv 0 \pmod{p}$, where k is between 3 and 10, X and Y are less than $1,000\, p^{1/k}$, and the absolute values of the coefficients of f are less than 500. Then the fraction of 512-bit primes which are unsafe is at most

$$\frac{1}{\pi(2^{512}) - \pi(2^{511})} \sum_{k=3}^{10} \left(1,000 \cdot 2^{512/k}\right)^2 \cdot 1000^{k+1} < 2^{-100}.$$

Suppose two people wish to agree on a safe key for the DSS. They can choose a random seed for the random number generator, using a protocol due to Blum [2]. From this they can use the method of Appendix 2 of [21] to create a key which is as likely to be safe as any random key.

On the other hand, a central authority might want to announce a key for general use, so that everyone is convinced there is no trapdoor. To do this, the authority must have a pseudo-random number generator and algorithm for constructing keys so that

1. Any user can verify that a key was generated using the approved method.
2. Keys produced by this method should be no more likely than random keys to contain trapdoors.
3. The choice of seed used for the random number generator should not allow the authority to create a particular key.

With a few modifications, the random number generator mentioned in Appendix 3 of [21] can be made to satisfy the above criteria. That method uses DES with a 64-bit seed, DES key and 64-bit date/time-stamp. To satisfy the above conditions, the DES key used should be fixed as part of the algorithm, the seed should be made public with the DSS key, and the time-stamp format should be specified.

It could be argued that the 64-bit seed gives too much freedom, putting the third condition at risk. This can be remedied by restricting the choice of seeds, or eliminating the seed entirely and just using the time-stamp.

For an example of a "trustable" key, consider:

$$q = 1147860701762054730346201299935827782113538756127$$

and

$$p = 7156194764397802049278787791933618087377339058379247 6383$$
$$4406258190286105951717150792702081842023182021408216 9894$$
$$3733340787353141262972727789275248126274 11$$

These numbers were generated using the binary expansions of π and e. The prime q is the smallest prime greater than $\left[2^{158}\pi\right]$, and $(p-1)/q$ is the smallest number larger than $\left[2^{350}e\right]$ for which p is prime. There is no reason to suspect this number of being any more likely than a random number to have a trapdoor, and tests of p by many polynomials have not found any.

6   Conclusion

In this paper, we have tried to quantify the threat of trapdoors for discrete logarithm-based cryptosystems, in particular DSS. While trapdoors do give a definite advantage over standard keys, with a few easy precautions in the choosing of p and q it is possible to prevent them, and they do not seem to pose a major problem for such systems.

In [15], Maurer and Yacobi present a public key distribution system, based on computing discrete logarithms modulo a composite number n. The factorization of n is a trapdoor which allows a trusted authority to compute secret keys. Unlike the DSS, their system relies on the trapdoor, and they ask if a similar trapdoor can be made for primes. The special number field sieve does provide a trapdoor which could be used to construct a similar system with a prime modulus, but such a system would be impractical.

References

1. T. Beth, M. Frisch and G.J. Simmons, eds., *Public-key Cryptography, State of the Art and Future Directions*, LNCS #578, Springer, 1992.
2. M. Blum, Coin flipping by telephone: A protocol for solving impossible problems, (*Proceedings of the 24th IEEE Computer Conference*, 1982, pp. 133-137.
3. E.F. Brickell and K.S. McCurley, An Interactive Identification Scheme Based on Discrete Logarithms and Factoring, *Journal of Cryptology*, to appear.
4. J. Buchmann and A. Pethö, Computation of independent units in number fields by Dirichlet's method, *Math. Comp.*, **52** (1989), pp. 149-159.
5. D. Coppersmith, A.M. Odlyzko and R. Schroeppel, Discrete logarithms in GF(p), *Algorithmica*, **1** (1986), pp. 1-15.
6. W. Diffie and M.E. Hellman, New directions in cryptography, *IEEE Trans. Info. Theory*, **22** (1976), pp. 644-654.
7. D. Gordon, Discrete logarithms in $GF(p)$ using the number field sieve, *SIAM Journal on Discrete Math.*, to appear.
8. D.E. Knuth, *The Art of Computer Programming*, Vol. 2, *Seminumerical Algorithms*, Second Edition, Addison-Wesley, Massachusetts, 1981.
9. B. LaMacchia and A.M. Odlyzko, Computation of discrete logarithms in prime fields, *Designs, Codes and Cryptography*, **1** (1991), pp. 47-62.
10. B. LaMacchia and A.M. Odlyzko, Solving large sparse linear systems over finite fields, Advances in Cryptology: Proceedings of Crypto '90, (A. Menezes, S. Vanstone, eds.), *Lecture Notes in Computer Science*, Springer-Verlag, New York, 1991.
11. H.W. Lenstra, Jr., Factoring integers with elliptic curves, *Ann. of Math.*, **126** (1987), pp. 649-673.

12. A.K. Lenstra, H.W. Lenstra, Jr., and L. Lovász, Factoring polynomials with rational coefficients, *Math. Ann.* **261** (1982), pp. 515-534.

13. A.K. Lenstra, H.W. Lenstra, Jr., M.S. Manasse and J.M. Pollard, The number field sieve, *Proc. 22nd ACM Symposium on Theory of Computing* (1990) pp. 564-572.

14. A.K. Lenstra, H.W. Lenstra, Jr., M.S. Manasse and J.M. Pollard, The factorization of the ninth Fermat number, preprint, 1991.

15. U.M. Maurer and Y. Yacobi, A non-interactive public-key distribution system, Advances in Cryptology: Proceedings of Eurocrypt '91, (D.W. Davies, ed.), *Lecture Notes in Computer Science*, Springer-Verlag, New York, 1991, pp. 498-507.

16. M. Pohst and H. Zassenhaus, *Algorithmic Algebraic Number Theory*, Cambridge University Press, Cambridge, 1989.

17. O. Schirokauer, *On pro-finite groups and on discrete logarithms,* Ph.D. thesis, University of California, Berkeley, May 1992.

18. C.P. Schnorr, Efficient signature generation by smart cards, *Journal of Cryptology*, to appear.

19. M.E. Smid and D.K. Branstad, Response to comments on the NIST Proposed Digital Signature Standard, *Advances in Cryptology: Proceedings of Crypto '92*, to appear.

20. D.H. Wiedemann, Solving sparse linear equations over finite fields, *IEEE Trans. Info. Theory*, **32** (1986), pp. 54-62.

21. Specifications for a digital signature standard, National Institute for Standards and Technology, *Federal Information Processing Standard Publication XX*, draft, August 1991.

Response to Comments on the NIST Proposed

Digital Signature Standard

Miles E. Smid
Dennis K. Branstad

National Institute of Standards and Technology[1]

Abstract. NIST received comments from 109 separate government agencies, companies, and private individuals concerning the proposed Digital Signature Standard. Both positive and negative comments were received. However the number of negative comments was significantly larger than normally received for a proposed Federal Information Processing Standard (FIPS). This paper summarizes the major comments, both positive and negative, and provides responses where appropriate. The paper highlights the anticipated significant modifications to the proposed standard and concludes by discussing the future milestones that need to be accomplished before the proposed DSS becomes a FIPS.

1. Introduction

1.1 History of the DSA

In August, 1991 [FRDSS], the National Institute of Standards and Technology (NIST) proposed a Digital Signature Algorithm (DSA) for use in computing and verifying digital signatures in government applications. The DSA was proposed in a draft Digital Signature Standard (DSS) [DFIPSXX] as the initial step of a process leading to a Federal Information Processing Standard.

The goal was to provide a standard for government organizations to use for applications in which a digital signature is required. Private and commercial organizations are encouraged to adopt and use the DSS as well. This paper discusses the primary issues that were raised during the public comment period on the DSS.

The Digital Signature Algorithm is used for mathematically computing and verifying a digital signature. The algorithm explicitly defines the parameters (name, type, size but not value) and specifies the computations for signature generation and verification. A digital signature is simply a number that depends upon the contents of the message and the private key of the message signer. The signature is normally transmitted with the message. A verifier, who has possession of the message, the signature, and the public key of the signer, can determine that the signature was generated by the signer and was not modified, either accidentally or intentionally. In addition, the verifier can provide the message, the digital signature, and the signer's public key as evidence to a third party that the message was, in fact, signed by the claimed signer. Given the evidence, the third party can also verify the signature. This capability is called "nonrepudiation". Of course, one can sign data other than messages, for example, electronic contracts, computer programs, and any valuable electronic information.

1.2 Factors Considered

In selecting the Digital Signature Algorithm for the proposed DSS, the following factors were considered important:

---

[1] U.S. Government contribution not subject to copyright.

the level of security provided, the applicability of patents, the ease of export from the U.S., the impact on national security and law enforcement, and the efficiency in a number of government and commercial applications. A number of techniques were reviewed and deemed appropriate for providing adequate protection in Federal systems. Among these, NIST placed primary emphasis on selecting the technique that best assures appropriate security for Federal information and does not require payment of royalties by U.S. private or commercial interests. All proposals were coordinated with the national security and law enforcement communities.

A Digital Signature Algorithm should have several technical characteristics. First, it must compute a signature which depends on the contents of the message and the private key of the person that originated it. Second, the private key used for signature generation should not be computable knowing the public key used for signature verification. Third, the efficiency of generating keys, signing messages and verifying messages should have an acceptable impact on performance in various implementations and applications. Fourth, a digital signature algorithm should be useful in many different applications and provide a level of security commensurate with the value or sensitivity of the data being protected.

Several digital signature algorithms have been proposed in the technical literature. Each exhibits the above characteristics to a greater or lesser degree. NIST proposed an algorithm which satisfies the desired technical characteristics in addition to the established non-technical criteria. This paper summarizes the comments received during the first public solicitation for comments on the proposed standard, provides responses to the comments, and discusses planned revisions to the proposed DSS.

1.3 GAO Decision B-245714

Government agencies have often raised questions concerning the legality of using a digital rather than a written signature. A "catch 22" condition existed. Agencies would not use digital signature technology because the regulations appeared to require written signatures, and the regulations were not changed or clarified because agencies were not using the new technology. In order to help clarify the issue, NIST requested a formal decision from the General Accounting Office (GAO) [NLET]. Based on its analysis of an agency's financial system and operating procedures, the GAO often grants relief against financial loss. If funds are lost as the result of a weakness in the system or the operational procedures, the loss will come out of general revenues rather than the funds of the agency.

NIST asked the GAO whether NIST standards for electronic signatures could be used to record obligations in government Electronic Data Interchange (EDI) payments. The GAO decision [GAO91] established the criteria for government use of electronic signatures for EDI technologies consistent with 31 U.S.C. Section 1501. Electronic signatures had to be unique and they had to provide a verifiable binding of the individual to the transaction. In particular, the GAO stated that "EDI systems using message authentication codes which follow NIST's Computer Data Authentication Standard (Federal Information Processing Standard (FIPS) 113) or digital signatures following NIST's Digital Signature Standard, as currently proposed, can produce a form of evidence that is acceptable under section 1501."

2. Overview of Comments

NIST received comments form 109 separate government agencies, companies, and private individuals concerning the proposed DSS. Both positive and negative comments were received. While government agencies tended to support the proposed standard, the number of negative comments was significantly larger than normally received for a proposed FIPS. The comments are public and copies are available for inspection at the Central Reference and Records Inspection Facility, room 6020, Herbert C. Hoover Building, 14th Street between Pennsylvania and Constitution Avenues, N.W., Washington, DC 20230.

3. Sample of Positive Comments

Many responders to the NIST solicitation for comments stated their belief that a digital signature capability will be necessary in electronic funds transfer, electronic data interchange, payroll, and administrative systems. Several responders supported the government's goal of having a standard that was free of patent impediments and expressed their desire that there be a federal standard for digital signatures which would provide for interoperability and a common level of security. Many government agencies supported the proposed standard. A sample of some positive comments is provided below:

1. The DSA will be especially useful to the financial services industry

2. The DSS is the key to robust and secure transfer of funds between individuals, financial institutions, governments and corporations

3. There will be minimal cost impact if the proposed standard is implemented

4. Generating keys for the DSA is a relatively efficient operation

5. The DSA is the only signature algorithm that has been publicly proposed by any government

6. We recommend that the algorithm be adopted as a FIPS

7. The Department applauds NIST's work in developing a DSS that will help to meet the needs of Federal departments and agencies....

4. Response to Negative Comments

Like the Data Encryption Standard (DES) proposed fifteen years earlier as a Federal Information Processing Standard, the DSS received many negative comments, but the comments generally fell into one of several categories. Some responders believed that since the selection process of the proposed DSA had not been public, the usual standards making process was not followed. Other people thought the solicitation for comments was the end of the standards process rather than just the beginning and therefore did not believe sufficient time was being provided for evaluation of the proposal. Many noted that the proposal was an alternative to the Rivest, Shamir and Adleman (RSA) algorithm [RIVEST] that has achieved a high degree of public acceptance. Selecting an alternative to the RSA was felt to have a negative impact by those that had a financial interest and a positive impact by some that had alternative financial interests. Finally, several technical concerns were expressed regarding the security and efficiency of the proposed algorithm. These concerns and responses are summarized below.

4.1 The DSA selection process was not public

Response:

The early discussions leading to the proposal of the DSA algorithm were not public. The Computer Security Act of 1987 states that NIST "shall draw upon computer system technical security guidelines developed by the National Security Agency ...." [CSA87]. NIST followed its normal standards development procedures, the provisions of the act, and the memorandum of understanding established with the National Security Agency (NSA). Several alternatives were considered before the DSA was selected. The cooperation between NIST and NSA was publicly known. NIST advised the appropriate ANSI accredited standards committees, as well as others, of the joint effort.

In the normal standards development process, NIST identifies the need for a standard, produces technical specifications of a standard using inputs from different sources and then solicits government and public comment on the proposal. After the comment period, the comments are analyzed, appropriate changes are made and a revised standard issued (or further comment is solicited if the revisions are substantial). This public process is being followed. NIST made the specification of the algorithm public and then solicited comments on the proposed algorithm. NIST personnel have given talks on the DSS to Accredited Standards Committee (ASC) X9, Working Group X9F1, Interop '92, the First International Symposium on Cryptographic Security, the Federal Computer

Security Program Managers' Forum, and the NIST Computer Security and Privacy Advisory Board. Working Group X9F1, which makes financial standards related to public key cryptography, is now developing a standard that is equivalent to the DSA [DANSIX9].

4.2 Sufficient time for analysis has not been provided

Several parties felt that the three month comment period did not provide sufficient time for analysis of the algorithm. In response to a formal request, NIST extended the comment period for another three months. Few new comments were provided after the initial three month period.

Response:

NIST considered the initial three month comment period to be only part of the total DSS evaluation process. The security of the DSA is believed to be equivalent to the difficulty of solving the discrete logarithm problem which has been studied for several years. The ElGamal technique, upon which the DSA is based, has been studied since 1984 and remains basically sound. The DSA does have some new features. In particular, r is calculated by computing $(g^k \bmod p) \bmod q$. However, the new features as well as the entire algorithm were evaluated by the NSA and underwent the same analysis used by NSA to evaluate classified cryptographic systems. In fact, the DSA may be used to sign unclassified data processed by "Warner Amendment" systems (10 U.S.C. 2315 and 44 U.S.C. 3502(2)) as well as classified data in selected applications [FRDSS].

It is now almost a year since the algorithm was publicly proposed and no cryptographic shortcut attacks have been found. NIST will continue to evaluate the merits of any proposed attack and will formally review the DSS at five year intervals. However, to be sure that there is no additional, currently unknown information about the algorithm or its revision (see Section 5.2 below), NIST has stated there will be a second public comment period on a revised DSS proposal before it is published as a standard.

4.3 The DSA may infringe on other patents

Response:

One of the selection criteria for the DSA was that it be free of patent impediments to the maximum extent possible. An agreement to grant non-exclusive, royalty free licenses had been made by the International Business Machines Corporation in 1975 prior to adopting the DES, which was covered by IBM patents, as a Federal Information Processing Standard. A similar status was desired for the DSS. Some alternative algorithms were considered less desirable because of known patent impediments. The DSS was designed by the government specifically to meet the selection criteria, including the patent criteria. However, two claims of infringement (by Public Key Partners and Professor Claus P. Schnorr) were received during the comment period. In addition other comments expressed a concern that the DSS infringed the patents held by these entities.

A major criterion for the invention and selection of the DSS by the government was to avoid patented technology that could result in payment of royalties for government, commercial and private use. This was stated in Congressional testimony in June, 1991, shortly before the DSS was issued for comment. A patent application was filed for the DSA on behalf of the government with the intent of making the DSS available on a non-exclusive, royalty-free basis. The patent claims were recently allowed by the U.S. patent office. The patents that are claimed to be infringed were directly or indirectly referenced in the DSA patent application.

Based on its initial analysis of existing patents, NIST believed the DSA did not infringe on any known patents. As a result of the claims of infringement, NIST is attempting to clarify the patent issue (see Section 5.1). The judgment of infringement is a complex legal issue and outside the scope of this paper.

4.4 The DSA does not provide for secret key distribution

Response:

The DSA does not provide for secret key distribution because the DSA is not intended for secret key distribution. In many applications a digital signature capability for integrity and nonrepudiation is sufficient and secret key distribution is not necessary. NIST does recognize the need for secret key distribution in other applications (e.g., where encryption is used). However, NIST and NSA have not yet selected such a method. NIST decided that it would be better to provide a public key based signature system immediately than to wait for both a signature system and secret key distribution system at some later time.

In addition, there are certain advantages to having separate algorithms for signature and key distribution. First, cryptographic algorithms that do not encipher data clearly come under the Department of Commerce export rules whereas export of encryption algorithms is controlled by the Department of State procedures which tend to be more restrictive [NBUL]. Secondly, certain countries readily permit the use of signature algorithms within their borders, but they restrict the use of encryption algorithms.

4.5 The DSA is incomplete because no hash algorithm is specified

Response:

On January 30, 1992 [FRSHS], NIST proposed a Secure Hash Standard (SHS) [DFIPSYY] which specifies a Secure Hash Algorithm (SHA) that is required for use with the DSA and whenever a secure hash algorithm is needed for federal applications. Copies of the SHS may be obtained by writing to the Standards Processing Coordinator (ADP), National Institute of Standards and Technology, Technology Building, room B-64, Gaithersburg, MD 20899. The SHA produces a 160-bit message digest on any data string up to 2^{64}-1 bits. The SHS comment period ended on April 30. Comments were received from twenty-four separate government agencies, companies, and private individuals. The vast majority of the comments were favorable, and no technical flaws in the algorithm were found. NIST now plans to proceed with the process of making the proposed SHS a FIPS.

Table 1 shows sample SHA processing rates obtained for C code implementations of the SHA on three different computers. Other implementors may obtain differing rates based upon the degree to which the code has been optimized, the compiler used, and other factors. The rates appear adequate for many data security applications.

Machine	Rate (bytes/second)
AT	2,523
486 (33 MHz)	28,169
SUN SPARC	222,233

Table 1: *Sample SHA Processing Rates*

4.6 The DSA is not compatible with IS 9796

International Standard 9796 [IS9796] is a standard for digital signatures with message recovery. According to this standard the message must be half the block size of a reversible public key encryption algorithm. The message is then redundantly padded to fill the entire block size and then "encrypted" with the user's private key to form the signature. An n-bit message results in a 2n-bit signature. Any verifier of the signature can use the public key of the signer to recover the redundantly encoded message. Rather than having the signer send the message as well as the signature, IS 9796 permits the recovery of the message from the signature itself.

Response:

IS 9796 specifies a digital signature scheme which provides message recovery from the signature. It is inefficient for signing moderate or long messages one half block at a time. The standard does allow for signing a message digest instead of a message, but then one would have to transmit the message along with the signature and the reversibility of the algorithm would provide no apparent advantage.

Since the DSA is not reversible, it could not meet the requirements of IS 9796 for a reversible algorithm. However, producing a 2n-bit signature from an n-bit message (as with IS 9796) is inefficient and causes unnecessary data expansion. When the DSA is used with the SHA algorithm an n-bit message will result in a 320-bit signature, and only n+320 bits need be transmitted. Thus, messages longer than 320 bits, or shorter than the block size minus 320 bits, will have less data transmission requirements if signed using the DSA.

In addition, there have been proposals for an alternative international signature standard, called "Digital Signature with Appendix". This alternative standard would permit the use of nonreversible algorithms for digital signatures and would not require that a n-bit message produce a 2n-bit signature. NIST will propose that the DSA algorithm be one of the algorithms that may be used in conjunction with the proposed alternative standard.

4.7 The modulus is fixed at 512 bits

Some parties responding to the request for comments believed that the DSA was insecure because the modulus was fixed at 512 bits. Others felt that although 512 bits provided adequate security for most of today's applications, it was not adequate for public key certificates and long term security.

Response:

The security of the DSA is based on the difficulty of solving the discrete log problem. Most security experts consider the discrete log problem to be at least as difficult as factoring (i.e., solving $y = g^x$ mod p for x is as difficult as solving $n = a * b$ for a and b when p is the same size as n). Therefore, the 512-bit DSA is at least as secure as many products, whose security is based on factoring, that are currently on the market today. One responder estimated that today it would take over eight million dollars (2.1 million MIPS[2]-years @ \$4 per MIPS-year) to break the DSA but recommends allowing a modulus size of at least 710 bits.

Currently, smart card systems have limited computational capabilities which would be heavily utilized in implementing a 512-bit public key algorithm. Smart card implementations of larger modulus sizes are not yet practical. However, implementing a 512-bit algorithm in a smart card where the private key never needs to leave the card may offer much greater overall security than implementing a larger size modulus in a shared PC.

In response to the comments that a larger modulus size is required for certificates and long term security, modulus sizes of up to 1024 bits will be allowed. The revised standard will allow modulus sizes of 512, 576, 640, 704, 768, 832, 896, 960, and 1024 bits. This array of sizes should be sufficient for protecting sensitive unclassified data for the foreseeable future.

4.8 The 160-bit size of q is too small

Response:

Some parties claimed that the 160-bit size of q is too small but no analytical justification for this claim was provided. The 160-bit q provides a work factor of 2^{80} which is consistent with the 160-bit message digest provided by the SHA. (Note that the 160-bit SHA message digest is already 32 bits longer than most other accepted message

---

[2] MIPS = Million Instructions Per Second.

digests.) Assuming 32×10^{12} operations per MIPS-year and a cost of $4 per MIPS-year, one would expect to spend at least $[(2^{80} \text{operations})/(32 \times 10^{12} \text{operations/MIPS-year})] \times (\$4/\text{MIPS-year}) = \$151,000,000,000$ to recover a single key, x. It has been estimated by Andrew Odlyzko that this is roughly the same effort that would be required to break a discrete log system with a 1024-bit modulus using the number field sieve. Therefore, the 160-bit q appears to be sufficient even when a 1024-bit p is used.

4.9 Compromise of k would compromise the private key

Response:

Compromise of k would compromise the private key x. However it has not been shown that compromising k is any easier than compromising x itself. Both x and k are randomly or pseudorandomly generated; both x and k are kept in the most secure area of the cryptographic module; and neither x nor k need be known to any human being. If an adversary can gain physical access to k, then the adversary could also gain physical access to x. The DSA is designed so that neither x nor k can be determined from the signature.

NIST will suggest techniques for generating the x, k, and other values in an appendix of the DSS. In addition, the authors highly recommend the use of smart cards to protect private keys and any other secret parameters used by public key algorithms.

4.10 Weak values of p could be selected by a dishonest CA

A claim was made that a dishonest Certification Authority (CA) could purposely select a value of p for its own users which would permit the CA to recover the private keys of the users.

Response:

The proposed DSS specifies a Digital Signature Algorithm. It does not discuss all the ways the algorithm may be used or misused. The qualifications section of the DSS Announcement states that "The responsible authority in each agency or department shall assure that an overall implementation provides an acceptable level of security." The proposed DSS specifically states that, "Systems for certifying credentials and distributing certificates are beyond the scope of this standard." Therefore, one would not expect an algorithm specification standard to cover the case of a dishonest certification authority.

The DSS allows users to generate their own primes, p and q. The DSS also allows the user to use primes generated by a trusted party or a certification authority. If primes are known to be randomly generated, the user can even accept primes generated by a distrusted party. One can construct special primes that are considered weak. If they were used the private keys of the users might be recovered. (Note that many other algorithms have similar weak values.) However, the probability of generating a weak prime at random is infinitesimally small. (The probability of generating a weak p at random has been estimated to be less than 10^{-90}.) Two parties pointed out that the use of a one-way function, such as the SHA, in the process that generates p and q could ensure that weak values occur only randomly. By making publicly known the input to the SHA, the resulting p, the resulting q, and the process, the user would be able to verify that weak primes were not purposely constructed. A technique which makes use of the SHA in the generation of DSA primes is proposed in Appendix A of this paper.

The claim that a trapdoor was purposely placed in the DSA was the subject of a panel session at Eurocrypt '92. No evidence of an intent to put a trapdoor in the DSA was presented and by the end of the session the claim was substantially discredited.

Warning! As with all systems using a certification authority, the certification authority must be trusted to correctly establish the binding between the user's identity and the user's public key.

4.11 The DSA is less efficient for verification

Response:

Some of the comments provided inaccurate estimates of the computation time required for the DSA. Obviously one would like a signature algorithm to be as efficient as possible while still providing adequate security. The real issue is whether the DSA verification speed is sufficient. On a 386 personal computer[3], the DSA can validate a signature in less than one second and the same computation can be done in milliseconds in hardware. These times are adequate for most applications.

In order to fully understand the computational differences between the DSA and RSA one must consider five different computations: global computations, key generation computations, pre-computations, signature computations and verification computations.

Global computations may be performed once for a set of users and need not be recomputed for a long period of time. Therefore, these computations do not normally impose a severe penalty on the operational system. For the DSA, the computation of p, q, and g could be considered global computations. The RSA does not have a similar computation.

Key generation computations are performed in generating the public and private keys. For DSA one must generate x and y as the private and public keys. For RSA, primes p and q must be generated and e and d computed. (Note that when using the Chinese remainder theorem, d mod (p-1) and d mod (q-1) are generated instead of d.)

The *pre-computations* for the DSA are performed for each message to be signed. However, these computations may be performed before any message is selected to be signed. These pre-computations involve generating k^{-1} and r as inputs to the signature generation computation. RSA has no similar computation.

For DSA the *signature generation computations* involve generation of the message digest, H(m), and the s portion of the signature. For RSA signature generation, one must compute $s = (H(m))^d \bmod n$.

When performing the *signature verification computations* the DSA computes a putative r from the received message m, the received r, and the received s. If the computed value of r equals the received value of r the signature is verified. Otherwise the signature is rejected. Using the RSA one computes $s^e \bmod n$ and compares it to the message digest of the received message.

Table 2 indicates some sample computation times for the DSA and RSA algorithms performed either in a Hitachi H8-310 smart card processor or in a host personal computer. Efficient smart card implementations of public key cryptography are difficult to achieve because of the limited capabilities of current 8-bit smart card processors. On faster computers or special purpose smart card processors, the differences in computation times between the DSA and RSA algorithms become less significant to the human observer.

The DSS offers an advantage with regard to its extremely efficient computation of the private and public keys. The private key is any randomly generated 160-bit value called x and the public key is y where $y = g^x \bmod p$. Since both computations are efficient the private and public keys can be easily generated on a smart card. While the public key can be read from the smart card at any time, the private key never needs to leave the protection of the card. Observed DSA key generation computations are 40-80 times faster than RSA key generation computations.

---

[3] Products are mentioned in this paper for informational purposes only and do not constitute an endorsement.

In addition, the DSS has the capability of performing most of the signature computations before the actual message to be signed has been selected. This is done by pre-computing k, k^{-1}, and r. In fact several k, k^{-1}, and r values may be precomputed in a fashion that is transparent to the user. Then, when the user selects the message or data to be signed, the signature will be computed in a fraction of a second. This feature is especially useful in today's smart card systems where the card will perform the necessary pre-computations while the user is selecting and forming the message to be signed. Therefore, the signature process appears very efficient to the user.

Algorithm	DSA	RSA	DSA Common p,q,g Estimated
Global Computation	Off Card (P)	NA	Off Card (P)
Key Generation	14	Off Card (S)	4
Pre-computation	14	NA	4
Signature	.03	15	.03
Verification	16 On Card 1-5 Off Card (P)	1.5	10 On Card 1-3 Off Card (P)

Table 2: *Smart Card DSA & RSA Computation Times* (All times are given in seconds. Off card computations performed on a 386, 33 MHz, personal computer. (P) indicates public parameters off card and (S) indicates secret parameters off card. Both algorithms use a 512-bit modulus.)

The signature verification computations for the DSA require more computations than signature generation and 10-15 times more than for the RSA algorithm. However, verification involves only public keys and can therefore be implemented in personal computers or in some other medium where more computational capability exists. This is an important distinction in smart card systems where signature generation would be performed in the secure card having a modest computational capability while verification could be performed elsewhere.

The DSA parameters p, q, and g can be selected by individual users or be common to a group of users. If individually selected, they must be passed along with the user's public key to anyone desiring to verify that user's signature. If common values are selected by a group of users, they need not be transmitted with each message or each user's public key. Efficiency is improved by reducing the number of parameters that have to be transmitted and by permitting the one-time computation of certain intermediate results.

When common or preestablished public values are employed, a technique due to Brickell, Gordon, and McCurley [BRICKELL] can be used to reduce the DSA pre-computation and verify times. NIST estimates that the pre-computation time can be reduced to approximately 1/4 the un-optimized time and the verify time to 1/2 the un-optimized time in a smart card implementation. Computer programs which make use of this work are now being developed at NIST and Sandia Laboratories.

In summary, the DSA validation computation appears to be adequate for many government and commercial applications. The DSA generates keys very efficiently and provides a pre-computation feature that can make the signature computation transparent to the user. Verification, although less efficient, is adequate for nearly all applications. These features may make the DSA highly desirable for many applications involving smart cards.

4.12 The DSS is "buggy"

One responder claimed that the DSS is "buggy" because if s = 0 then the computation of s^{-1} at signature verification would "blow up". In addition if s = 0, then the user's private key x could be recovered.

Response:

The computation of s^{-1} would not "blow up" on the verification calculation because the standard clearly states that the signature is rejected for any received s' outside of the range $0 < s' < q$. As far as the security issue is concerned, it is true that if $s = 0$ then x could be recovered. However, there is no need to check for a condition which occurs with probability 2^{-160}. The proposed DSS allows implementors to either check for $s = 0$ upon signature generation or to ignore the unlikely event depending on their own preferences.

5. Future Efforts

The following set of activities are presently planned by NIST in adopting a DSS as a FIPS:

1. Complete analysis and summary of comments;

2. Analyze and attempt to resolve patent issues;

3. Develop and evaluate alternative signature certification authority infrastructures;

4. Propose technical enhancements to DSA;

5. Issue second solicitation of comments on revised DSA;

6. Hold a symposium on the applications of the DSA;

7. Investigate the economic interests involving the DSS;

8. Coordinate and harmonize the revised DSS with ANSI and ISO standards activities;

9. Conduct final coordination of DSS within government;

10. Recommend Secretary of Commerce approval of DSS;

11. Publish revised DSS after Secretary of Commerce approval.

A brief discussion of some of the major activities are presented below.

5.1 Resolution of Patent Issues

NIST is presently attempting to resolve the patent issues in accordance with its desire to make the manufacture, sale and use of devices and systems implementing the DSA free of royalties for patents. The U.S. government already has rights to use patented techniques assigned to Public Key Partners because the government sponsored some of the research leading to the patents. However, private users presently do not enjoy such rights. Neither the U.S. government nor private users presently enjoy rights to the Schnorr patent. Alternative solutions to potential problems are being reviewed.

5.2 Second Federal Register Solicitation of Comments

As currently envisioned, the DSS will be revised to allow the use of a larger modulus, to add a new method for generating p and q, to add a method for pseudorandom generation of k values, and to correct or clarify minor editorial and technical issues. In order to assure an adequate opportunity for review of the revised proposed DSS, NIST is planning to publish the proposal for a second comment period.

5.3 Applications Symposium

NIST plans to host a Symposium on the Applications of Digital Signature Technology. The purpose of the symposium is to provide a forum for discussion of common problems, goals, and issues pertaining to the application of the DSA. Further information will be provided as plans develop.

5.4 International Infrastructure

NIST is studying the legal and technical issues related to development and operation of an international digital signature infrastructure. The infrastructure would be a system of organizations, people and computers used for distributing certificates to individuals, government agencies and private companies. The study will examine the legal and regulatory requirements which must be addressed, propose a certification authority architecture, and attempt to clarify the roles that various government agencies wish to perform. Several U.S. government agencies are participating in and financing the study.

NIST perceives a great need for such an infrastructure. Electronic filing of corporate and personal tax returns could be made more efficient and more secure if such a structure were available. Federal payments to contractors, vendors and social security recipients could be fully automated if the integrity and authenticity of electronic payments were assured. An international infrastructure is needed to provide security for worldwide business communications. NIST is presently working with the federal organizations responsible for such large scale applications. NIST intends to hold workshops with potential users and knowledgeable technical people in order to develop an infrastructure that will meet these anticipated needs.

It is intended that the infrastructure will utilize existing concepts and systems. International Standard X.509 (a security part of the Directory standard) describes a tree structure for certifying digital signatures. A digital signature certificate distribution system has been designed in conjunction with the Privacy Enhanced Mail project. NIST plans to build on these efforts to produce a recommendation for consideration by federal organizations planning to use digital signatures. Results of the present study are anticipated in the middle of 1993.

6. Conclusion

Several milestones have been met and several still need to be accomplished. NIST will continue the work required for adoption of the proposed Digital Signature Standard as an approved Federal Information Processing Standard. NIST also believes that an international infrastructure is required in order for digital signatures to be widely used throughout the U.S. government and the world.

References

[BRICKELL] E. Brickell, D. M. Gordon, K. S. McCurley, D. Wilson, Fast Exponentiation with Precomputation, Eurocrypt '92 Extended Abstracts, p 193-201.

[CSA87] Computer Security Act of 1987, June 11, 1987, Sec. 3.

[DANSIX9] Working Draft American National Standard X9.30-199X, Public Key Cryptography Using Irreversible Algorithms for the Financial Services Industry: Part 1: The Digital Signature Algorithm (DSA), American Bankers Association, Washington, DC.

[DFIPSXX] Draft Federal Information Processing Standards Publication XX, Announcement and Specifications for a Digital Signature Standard (DSS), August 19, 1991.

[DFIPSYY] Draft Federal Information Processing Standards Publication YY, Announcement and Specifications for a Secure Hash Standard (SHS), January 22, 1992.

[FRDSS]   A Proposed Federal Information Processing Standard for Digital Signature Standard (DSS), Federal Register Announcement, August 30, 1991, p 42980-41982.

[FRSHS]   A Proposed Federal Information Processing Standard for Secure Hash Standard, Federal Register Announcement, January 31, 1992, p 3747-3749.

[GAO91]   Comptroller General of the United States Decision, Matter of: National Institute of Standards and Technology--Use of Electronic Data Interchange Technology to Create Valid Obligations, file B-245714, December 13, 1991.

[IS9796]   Information technology - Security techniques - Digital signature scheme giving message recovery, IS 9796, International Organization for Standardization, Geneva, Switzerland.

[NBUL]   NCSL Bulletin, Data Encryption Standard, Exportability of DES Devices and Software Products, June 1990, p 3-4.

[NLET] Letter to General Counsel, U.S. General Accounting Office, from James H. Burrows, Director of NIST Computer Systems Laboratory, September 13, 1990.

[RIVEST]   R. Rivest, A. Shamir, and L. Adleman, A Method for Obtaining Digital Signatures and Public-Key Cryptosystems, Communications of the ACM, No. 2, p 120-126, 1978.

Appendix A: Generation of Primes p and q

The Digital Signature Standard requires two primes, p and q, satisfying the following three conditions:

a) $2^{159} < q < 2^{160}$

b) $2^{L-1} < p < 2^L$ for a specified L, where L = 512 + 64j for some $0 \leq j \leq 8$

c) q divides p-1.

This prime generation scheme starts by using the SHA and a user supplied SEED to construct a prime, q, in the range $2^{159} < q < 2^{160}$. Once this is accomplished, the same SEED value is used to construct an X in the range $2^{L-1} < X < 2^L$. The prime, p, is then formed by rounding X to a number congruent to 1 mod 2q as described below.

An integer x in the range $0 \leq x < 2^g$ may be converted to a g-long sequence of bits by using its binary expansion as shown below:

$$x = x_1 * 2^{g-1} + x_2 * 2^{g-2} + ... + x_{g-1} * 2 + x_g \rightarrow \{ x_1,...,x_g \}.$$

Conversely, a g-long sequence of bits $\{ x_1,...,x_g \}$ is converted to an integer by the rule

$$\{ x_1,...,x_g \} \rightarrow x_1 * 2^{g-1} + x_2 * 2^{g-2} + ... + x_{g-1} * 2 + x_g.$$

Note that the first bit of a sequence corresponds to the most significant bit of the corresponding integer and the last bit to the least significant bit.

Let L - 1 = n*160 + b, where both b and n are integers and $0 \leq b < 160$.

Step 1.   Choose an arbitrary sequence of at least 160 bits and call it SEED. Let g be the length of SEED in bits.

Step 2.   Compute

$$U = SHA[SEED] \text{ XOR } SHA[(SEED+1) \bmod 2^g].$$

Step 3.   Form q from U by setting the most significant bit (the 2^{159} bit) and the least significant bit to 1. In terms of boolean operations, q = U OR 2^{159} OR 1. Note that $2^{159} < q < 2^{160}$.

Step 4.   Use a robust primality testing algorithm to test whether q is prime[1].

Step 5.   If q is not prime, go to step 1.

Step 6.   Let counter = 0 and offset = 2.

Step 7.   For k = 0,...,n let

$$V_k = SHA[(SEED + offset + k) \bmod 2^g].$$

Step 8.   Let W be the integer

$$W = V_0 + V_1*2^{160} + ... + V_{n-1}*2^{(n-1)*160} + (V_n \bmod 2^b)*2^{n*160}$$

and let $X = W + 2^{L-1}$. Note that $0 \leq W < 2^{L-1}$ and hence $2^{L-1} \leq X < 2^L$.

Step 9.   Let c = X mod 2q and set p = X - (c-1). Note that p is congruent to 1 mod 2q.

Step 10.   If $p < 2^{L-1}$, then go to step 13.

Step 11.   Perform a robust primality test on p.

Step 12.   If p passes the test performed in step 11, go to step 15.

Step 13.   Let counter = counter+1 and offset = offset + n + 1.

Step 14.   If counter $\geq 2^{12}$ = 4096 go to step 1, otherwise (i.e., if counter < 4096) go to step 7.

Step 15.   Save the value of SEED and the value of counter for use in certifying the proper generation of p and q.

---

[1] A robust primality test is one where the probability of a non-prime number passing the test is at most 2^{-80}.

Wallet Databases with Observers

(Extended Abstract)

David Chaum
CWI
The Netherlands

Torben Pryds Pedersen*
Aarhus University
Denmark

Abstract

Previously there have been essentially only two models for computers that people can use to handle ordinary consumer transactions: (1) the tamper-proof module, such as a smart card, that the person cannot modify or probe; and (2) the personal workstation whose inner working is totally under control of the individual. The first part of this article argues that a particular combination of these two kinds of mechanism can overcome the limitations of each alone, providing both security and correctness for organizations as well as privacy and even anonymity for individuals.

Then it is shown how this combined device, called a wallet, can carry a database containing personal information. The construction presented ensures that no single part of the device (i.e. neither the tamper-proof part nor the workstation) can learn the contents of the database — this information can only be recovered by the two parts together.

1 Introduction

In this paper we shall be concerned with a general system consisting of a number of individuals and organizations. Each individual has a small database with (personal) information (for example credentials), and the purpose of a transaction is to either update this database (obtain a new credential) or read some information in it (show a credential). For such a system it is important that the data in the database are correct:

- The organizations want to be sure that the contents of each database corresponds to what they have written in it.

- The individuals want to be sure that the organizations only store correct information in the database, and that they can only read and update those parts of the database that they are entitled to.

---

*Research partly done while visiting CWI

Another basic requirement is that it should be possible for the individuals to participate anonymously in certain transactions. If, for example, the database contains medical information, which is needed in an investigation of a particular disease, the owner might require anonymity in order to participate in this investigation. There are, however, many other kinds of transactions, such as financial transactions, in which the issue of privacy is essential as well.

Section 2 argues that the *electronic wallet* is well suited for this scenario. An electronic wallet consists of two parts:

- A small, hand-held computer controlled by the user—denoted by C, for "computer"; and

- A tamper-proof module issued by the organizations—denoted by T, for "tamper-proof".

These two parts are arranged in such a way that T can only talk with C and not the outside world. This might be achieved by embedding T inside C. All communication with organizations is via C. It is essential that there is no "alternative way" that T can send messages to or receive messages from the outside world.

In the second part of the paper practical protocols are presented. First a new blind signature technique is presented in Section 3. Then Section 4 shows how, using the blind signatures, T can get a certified public key, which it can use to sign messages and thereby authenticate the actions taken by C. Then Section 5 presents the database protocols. In particular, it is shown how T can validate the information sent from the wallet without even knowing the contents of the database.

2 Possible Settings

This section discusses the advantages and disadvantages of different devices for use by individuals in a system including users and organizations, as described above. We shall primarily be concerned with how well the various alternatives support the requirements of correctness and privacy.

2.1 Correctness and Privacy

Correctness basically means that the data stored in a person's database can only be read or updated by the organizations/individuals that have permission to do so (according to some initially agreed rules). Note that these rules could say that a person is not allowed to change (parts of) his own database, and they could even (in extreme situations) specify that the user may not read parts of the database.

The terms *positive credential* and *negative credential* will be used to denote information in the database, which is to the advantage and disadvantage of the person, respectively. A bad criminal record is an example of a negative credential. The user may want to delete negative credential in the database, but this should, of course, be infeasible.

By *one-show credential* we mean a credential that the individual is allowed to show only once. Electronic money, which may be spent only once, is a typical example of a one-show credential.

While correctness is the most important requirement for organizations, privacy might be the important issue for individuals (at least in some situations), and it is essential for general acceptance of the system. We distinguish three levels of privacy:

- *Pure trust*:
  Information about the individual may be revealed during a transaction—the individual cannot do anything to enhance his privacy, but must trust the organization to maintain it.

- *Computational privacy*:
  If the individual follows the prescribed protocols, the organization cannot learn anything about him unless it can make a computation assumed to be infeasible.

- *Unconditional privacy*:
  If the individual follows the prescribed protocols, even an all powerful organization cannot learn anything extra about him.

2.2 Possible Approaches

We now analyze how two very different devices meet the demands of correctness and privacy outlined above. We first consider a device trusted completely by the individuals and then a tamper-proof device issued by the organizations (or an issuing center trusted by the organizations). This analysis then leads to the definition of electronic wallets.

Computer alone

First consider the situation where the user just has a computer, which he controls completely. In particular, he can delete or change any part of the memory, and he determines all messages which the device sends to the outside world.

Using the techniques of [Cha84] and [CFN90] it is possible to obtain unconditional privacy in this scenario in an efficient way. However, this setting makes it very difficult for the organizations to prevent users from deleting negative credentials or using one-show credentials more than once.

For example, in the case of an off-line electronic payments system, it is only known how to catch cheaters, who spend copies of the same electronic coin more than once, "after the fact". This method furthermore requires a large central database in which all valid coins are collected and compared (see [CFN90]), but this only has to be done periodically.

In short, this setting can give unconditional privacy, whereas no really efficient method for correctness is known.

Tamper-proof only

In this setting each individual has a tamper-proof module (packaged as a smart-card, for instance) issued by the organizations. Hence the organizations trust the correctness of the messages sent by the card, whereas the user does not even know which messages are being sent.

This approach gives correctness quite easily, because the tamper-proof part has to be broken in order to compromise the system. Furthermore, if cryptographic techniques are added it is sometimes possible to make systems in which cheating requires breaking the tamper resistant part *and* the cryptographic methods. Off-line electronic cash is an example of such a system.

If a tamper-proof unit is used to store negative credentials, the owner can delete these by destroying or throwing away the card. However, in this way he will also delete the positive credentials and, furthermore, the organizations will detect it the next time they need the card. In order to recover from such intentional as well as accidental losses of credentials, the system can have a back up facility for recovering such lost credentials. Hence, this approach can provide a very high degree of correctness.

The disadvantage of using the tamper-proof unit alone is that it only provides a low level of privacy, as the user has no control over the messages sent from the card. Therefore the card can (in principle) send any message that it likes during a transaction (e.g. the identity of the user). Hence, it can only give pure trust.

Electronic Wallets

The above analysis of two extreme settings shows that neither a user controlled computer nor a tamper-proof device alone can give sufficiently efficient and secure solutions. Electronic wallets can be thought of as a way to obtain the benefits of both approaches by a suitable combination.

Since no device that allows a tamper-proof device to communicate directly with the organizations can give a higher level of privacy than pure trust, the device must be constructed in such a way that the tamper-proof device cannot send messages to the organizations.

Thus the device should consist of a user controlled computer, C, with a tamper-proof unit, T, (sometimes called an *observer*), which on behalf of the organizations ensures that C cannot deviate from the prescribed protocols or change any information in its database. The electronic wallet is the simplest such device as it only has a single such observer (T). It might be useful (for example in order to make fault recovery easier) to have more than one observer, but such an approach does not seem to add significantly more power to the wallet.

Note, that C can freely communicate with the outside world without the knowledge of T, but the honest organizations will only accept messages which are approved by T.

The rest of this paper presents protocols, which show how T can control the actions of C. The fact that T may not communicate directly with organizations means that these protocols must be secure against

- *Inflow:*
  No matter how T and the organization deviate from the prescribed protocol, if C follows the protocol, the organization cannot send any extra (subliminal) information to T.

- *Outflow:*

  No matter how T and the organization deviate from the prescribed protocol, if C follows the protocol, T cannot send any extra (subliminal) information to the organization.

This means that even if the organization places a malicious observer in the wallet, there is no way that it can send back any information about the owner.

If all protocols are secure against outflow, then the security against inflow is not that significant, because T cannot tell other organizations what it learns. However, if it is important that T does not reveal any secrets in case it is returned to the organizations, the protocols must be secure against inflow as well.

3 The Signature Scheme

This section presents the signature scheme which will be used in this paper. The notation is introduced, the basic signature scheme is described, and it is shown how it can be used in wallets. Then it is shown how to make blind signatures.

3.1 Notation

Let q be a prime. The protocols to be presented work for any group, G_q of order q. As an example of such a group we consider another prime, p, such that q divides $p-1$, and define G_q as the unique subgroup of $\mathbb{Z}_p^*$ of order q. The element $g \in G_q$ will always be a generator of G_q. It will be assumed that all parties know p, q and g.

The discrete logarithm of $h \in G_q$ with respect to g is denoted by $\log_g h$, and the number of bits of an integer, x, will be denoted $|x|$.

3.2 The Basic Scheme

This subsection presents the signature scheme which will be used in the following protocols.

The public key of the scheme is

$$(p, q, g, h),$$

where $h \in G_q \setminus \{1\}$ and the corresponding secret key is $x = \log_g h$.

Let $m \in G_q$ be a message. The signature on m consists of $z = m^x$ plus a proof that

$$\log_g h = \log_m z.$$

Given m and z, consider the following protocol:

1. The prover chooses $s \in \mathbb{Z}_q$ at random and computes $(a, b) = (g^s, m^s)$. This pair is sent to the verifier.

2. The verifier chooses a random challenge $c \in \mathbb{Z}_q$ and sends it to the prover.

3. The prover sends back $r = s + cx$.

4. The verifier accepts the proof if

$$g^r = ah^c \qquad \text{and} \qquad m^r = bz^c.$$

If the prover can send correct responses r_1 and r_2 to two different challenges, c_1 and c_2 then

$$g^{r_1 - r_2} = h^{c_1 - c_2} \qquad \text{and} \qquad m^{r_1 - r_2} = z^{c_1 - c_2},$$

and hence

$$\log_g h = \log_m z = \frac{c_1 - c_2}{r_1 - r_2}$$

since $c_1 \neq c_2 \bmod q$ implies that $r_1 \neq r_2 \bmod q$. Now let H be a one-way hash function (as in the Fiat-Shamir scheme, see [FS87]). Given this function and the above protocol the signature on m is

$$\sigma(m) = (z, a, b, r).$$

It is correct if $c = H(m, z, a, b)$ and

$$g^r = ah^c \qquad \text{and} \qquad m^r = bz^c.$$

Hence, a signature on a $|q|$ bits message is $|q| + 3|p|$ bits long.

Now consider attempts to forge signatures given only the public key. If H has the property that it is as difficult to convince a verifier, who chooses $c := H(m, z, a, b)$, as a verifier who chooses the challenge at random (H is like a random oracle), it is not feasible to make signatures without knowing x.

Furthermore, it does not seem to help a forger to execute the proof that $\log_g h = \log_m z$ with the signer for the following reason. Consider the modification of the proof system in which the challenge, c, is chosen from a subset $A \subseteq \mathbb{Z}_q$ instead of $\mathbb{Z}_q$. For any such subset an execution of this modified scheme can be simulated perfectly in expected time $O(|A|)$. In particular this simulation is feasible if $|A|$ is polynomial in $|q|$. It is an open question to prove that executions of the protocol are secure, when A equals $\mathbb{Z}_q$, but we conjecture that no matter which $c \in \mathbb{Z}_q$ is chosen as challenge, the signer reveals no other information than the fact that $\log_g h$ equals $\log_m z$.

Finally remains the possibility that a forger can construct a false signature by combining various given signatures (m_i, σ_i), where the forger has chosen m_i adaptively (see [GMR88]). If $z_i = m_i^x$ then

$$z_1 z_2 = (m_1 m_2)^x.$$

Hence there is a multiplicative relation which might be useful for a forger. However, the use of H should prevent the forger from combining different signatures into a new signature.

3.3 Signatures by T

This section shows how the above signature scheme can be used by the tamper-proof device T in a wallet. The problem, that we have to deal with, is that T cannot be allowed to choose a and b alone, as it can encode some information in these two numbers. We therefore generate these two numbers using a coin-flipping protocol. If T has a public key (p, q, g, h_T) and a corresponding secret key $x_T = \log_g h_T$ it can sign a message $m \in G_q$ as follows:

1. C chooses $s_0 \in \mathbb{Z}_q$ and $t_0 \in \mathbb{Z}_q$ at random and sends $\alpha := g^{s_0} h_T^{t_0}$ to T (a commitment to s_0).

2. T chooses $s_1 \in \mathbb{Z}_q$ at random and sends $a_1 := g^{s_1}$ and $b_1 := m^{s_1}$ to C.

3. C sends (s_0, t_0) to T and computes $a := a_1 g^{s_0}$ and $b := b_1 m^{s_0}$.

4. T verifies that α equals $g^{s_0} h_T^{t_0}$ and computes $(a, b) := (a_1 g^{s_0}, b_1 m^{s_0})$.

5. T computes $c := H(m, m^{x_T}, a, b)$ and $r := s_0 + s_1 + c x_T \bmod q$.

The signature on m is (m^{x_T}, a, b, r).

It is not hard to see that if C follows the protocol then a and b are uniformly distributed in G_q. Furthermore, C can only open α as some $s_0' \neq s_0$ if it can find x_T. Hence, if T follows the protocol and C does not know x_T, then a and b are random elements of G_q.

Proposition 3.1
The above protocol for making signatures has the following two properties:

1. If C follows the protocol, then the signature is randomly distributed among the signatures on m — even if a cheating T has unlimited computing power.

2. If T follows the protocol, then a polynomially bounded cheating C learns no more than a random signature on m.

Proof
Both claims follow from the fact that the coin-flipping protocol in Step 1– 4 above has the following two properties:

1. (a, b) is uniformly distributed among the possible pairs, if C follows the protocol — even if a cheating T has unlimited computing power (because α contains no Shannon information about s_0).

2. A polynomially bounded C can only open α in two different ways if it knows $\log_g h_T$. ∎

3.4 Blind Signatures

To get a blind signature on the message m in the above scheme one chooses a random $t \in \mathbb{Z}_q^*$ and asks the signer to sign $m_0 = m^t$. Let $z_0 = m_0^x$. Then the signer proves that $\log_g h = \log_{m_0} z_0$ in such a way that the messages are blinded:

1. The signer chooses $s \in \mathbb{Z}_q$ at random and computes $(a_0, b_0) = (g^s, m_0^s)$. This pair is sent to the verifier.

2. The verifier chooses $u \in \mathbb{Z}_q^*$ and $v \in \mathbb{Z}_q$ at random and computes

$$a = (a_0 g^v)^u \quad \text{and} \quad b = (b_0^{1/t} m^v)^u.$$

(If both parties follow the protocol $a = (g^{s+v})^u$ and $b = (m^{s+v})^u$.) Then the verifier computes $z = z_0^{1/t}$, the challenge $c = H(m, z, a, b)$ and the blinded challenge $c_0 = c/u \bmod q$. The verifier sends c_0 to the signer.

3. The signer sends back $r_0 = s + c_0 x$.

4. The verifier accepts if

$$g^{r_0} = a_0 h^{c_0} \quad \text{and} \quad m_0^{r_0} = b_0 z_0^{c_0}.$$

The verifier computes $r = (r_0 + v)u \bmod q$ and

$$\sigma = (z, a, b, r).$$

Proposition 3.2
σ is a correct signature on m, if the verifier accepts in the above protocol.

Proof
Let $c = H(m, z, a, b)$. We have to prove that

$$g^r = ah^c \quad \text{and} \quad m^r = bz^c.$$

The first equality follows from

$$g^r = (g^{r_0} g^v)^u = (a_0 h^{c_0} g^v)^u = (a_0 g^v)^u h^{c_0 u} = ah^c$$

and the second from

$$
\begin{aligned}
m^r &= m^{ur_0} m^{uv} \\
&= (m_0^{1/t})^{ur_0} m^{vu} \\
&= (m_0^{r_0})^{u/t} m^{vu} \\
&= (b_0 z_0^{c_0})^{u/t} m^{vu} \\
&= (b_0^{1/t} m^v)^u z_0^{uc_0/t} \\
&= bz^c.
\end{aligned}
$$

∎

Proposition 3.3
The signer gets no information about m and σ if the receiver follows the protocol.

Proof

We will show that for all m, z, a, b and r such that

$$g^r = ah^c$$
$$m^r = bz^c$$
$$c = H(m, z, a, b)$$

and for all m_0, z_0, a_0, b_0, c_0 and r_0 such that

$$g^{r_0} = a_0 h^{c_0}$$
$$m_0^{r_0} = b_0 z_0^{c_0}$$

there is exactly one set of values of t, u and v such that the signer sees $(m_0, z_0, a_0, b_0, c_0, r_0)$, when making the signature σ on m. In other words, that there is exactly one set of values of t, u and v such that

$$m = m_0^{1/t}$$
$$a = (a_0 g^v)^u$$
$$b = (b_0^{1/t} m^v)^u$$
$$c = c_0 u$$
$$r = (r_0 + v)u.$$

First, m and m_0 determine t as

$$m_0 = m^t \iff t = \log_m m_0.$$

Secondly, u and v are determined by c, c_0, r and r_0 as

$$u = \frac{c}{c_0} \quad \text{and} \quad v = \frac{c_0}{c} r - r_0.$$

Thus we just need to show that these values of t, u and v satisfy

$$a = (a_0 g^v)^u \quad \text{and} \quad b = (b_0^{1/t} m^v)^u.$$

In doing this it can be assumed that $z_0 = m_0^x$ and $z = m^x$, because the signer actually proves that z_0 equals m_0^x when making a blind signature. Hence $m_0 = m^t$ implies that

$$z_0 = z^t.$$

The first equality is proven as follows

$$a = g^r h^{-c} = g^{(r_0+v)u} h^{-uc_0} = (g^{r_0} g^v h^{-c_0})^u = (a_0 g^v)^u.$$

The second equality follows by similar rewritings:

$$\begin{aligned}
b &= m^r z^{-c} \\
&= m^{(r_0+v)u} z^{-c_0 u} \\
&= (m^{r_0} m^v z^{-c_0})^u \\
&= ((m_0^{1/t})^{r_0} m^v (z_0^{1/t})^{-c_0})^u \\
&= ((m_0^{r_0} z_0^{-c_0})^{1/t} m^v)^u \\
&= (b_0^{1/t} m^v)^u.
\end{aligned}$$

This completes the proof. ∎

Hence, this signature scheme allows the receiver to obtain blind signatures. In particular it is possible for the receiver to get a signature on any message that he chooses. In order to avoid this problem in the application to wallets, the organization only signs a blinded message if the challenge is signed by T. The resulting scheme is presented in the next subsection.

3.5 Blind Signatures in Wallets

We assume that a center Z is the signer. The public key of Z is h_Z and the secret key is $x_Z = \log_g h_Z$.

1. C chooses the blinding factor $t \in \mathbb{Z}_q^*$ at random and sends $m_0 := m^t$ to Z.

2. Z and C choose a_0 and b_0 using a coin-flipping (as in Section 3.3) protocol, such that only Z knows $s = \log_g a_0 = \log_{m_0} b_0$.

3. Z computes $z_0 := m_0^{x_Z}$ and sends it to C.

4. C computes $z := z_0^{1/t}$ and chooses u and v at random. Then it sends (a_0, b_0, z, u, v, t) to T.

5. Both T and C can then compute $a := (a_0 g^v)^u$, $b := (b_0^{1/t} m^v)^u$, $c := H(m, z, a, b)$ and $c_0 := c/u$. T signs c_0 and sends it to C.

6. C verifies the signature before sending the challenge and the signature to Z.

7. From now on the protocol for constructing and verifying blind signatures is followed. Hence Z computes the response, r_0, and sends it to C. C verifies this response before forwarding it to T. Finally T unblinds r_0 and verifies the signature.

Theorem 3.4
If C follows the protocol then

1. Z gets no information about the signature on m.

2. T sends no information to Z except a random signature on c_0.

3. Z sends no information to T except z_0.

Proof
Assume that C follows the protocol.

1. Z sees messages with the same distribution as in the original protocol for making blind signatures — except that Z cannot choose (a_0, b_0) freely anymore. But this pair is chosen at random. Hence this property follows from Proposition 3.3.

2. The only information, which originates from T is the signature on c_0. However, Proposition 3.1 implies that this signature is randomly chosen among the possible signatures.

3. T sees the following messages from Z:

$$(a_0, b_0), z_0^{1/t} \text{ and } r_0,$$

and T receives u, v and t from C. Here (a_0, b_0) is uniformly distributed (by the same argument as in the proof of Proposition 3.1), and r_0 is uniquely determined. Hence, Z can only send information to T via z.

■

Note that if Z does not compute z_0 as $m_0^{x_Z}$ then C will discover it. Thus, it is impossible for Z to send information to T without being detected. However, as we shall see in the next section even this possibility of inflow is eliminated in our application of the protocol.

We now look at the security of the protocol and assume that T and Z both follow the protocol. It will be argued that if the basic signature scheme is secure, and if T's signatures cannot be faked, then no matter what a polynomially bounded $\tilde{C}$ does, it learns no more than a random signature on m.

As $\tilde{C}$ cannot forge T's signatures, it can be assumed that c_0 is computed as $c_0 := H(m, \tilde{z}, \tilde{a}, \tilde{b})$, where $\tilde{C}$ can choose $\tilde{z}$, $\tilde{a}$ and $\tilde{b}$, but not m. By the assumption about H this means that $\tilde{C}$ cannot control the value of c_0 ($\tilde{C}$ cannot force c_0 to be any particular value, except by trying different values for $\tilde{z}$, $\tilde{a}$ and $\tilde{b}$ and hoping they will give a "good" value of c_0). Thus $\tilde{C}$ does not seem be better off in this situation than when it just gets a "normal" signature from the signer.

4 Obtaining a Pseudonym

This section shows how the wallet can get a public key, which is signed by a key authentication center. The signature on the public key will be called a *validator*. This protocol has the property that neither the center nor any other unlimited powerful organization can link the identity of the user to the public key (or its validator).

Combining this result with Section 3.3 gives a method for T to sign messages without revealing any information at all about the owner of the wallet. This provides a method for T to validate the messages, which C sends to the outside world, without revealing anything about the identity of the user; these messages are only accepted by the organizations if they are signed properly (by T).

We now show how T can generate a secret key $x \in \mathbb{Z}_q^*$ and obtain a certificate on the corresponding public key $h = g^x \bmod p$. In order to get started, it is assumed that each T is born with a secret key, x_T, and a corresponding public key, h_T, to the signature scheme described in Section 3.3. These signatures can be traced to T (and hence to the individual), and they are therefore only used in an initial step where T gets a validated key from a key authentication center (Z). The center issues validators using the blind signature scheme from the previous section with secret key x_Z and public key h_Z.

The basic idea of the protocol for issuing validators is that C and T first execute a coin-flipping protocol in order to choose a secret key, x, which only T learns. The corresponding public key is denoted by h. Then C chooses a blinding factor, $t \in \mathbb{Z}_q^*$, and C signs the blinded public key $(h_1 = h^t)$. Note, that in the process of making the blind signature, T has to sign a challenge computed as $H(h, h^{xz}, a, b)$. This signature guarantees to Z that it validates a public key which is accepted by T. There is no need that T signs h_1 before Z starts making the blind signature, because before Z computes the response, it only produces random messages, which a cheating C could have produced by itself. In more detail the protocol goes like this:

1. C chooses $y_0 \in \mathbb{Z}_q^*$ at random and sends a commitment to y_0 to T.

2. T chooses $y_1 \in \mathbb{Z}_q^*$ at random and sends $h_0 := g^{y_1}$ to C.

3. C opens the commitment and sends y_0 to T.

4. T and C compute $h := h_0^{y_0}$, and T computes the secret key $x := y_0 y_1 \bmod q$.

5. T computes $z := h_Z^x$ and sends it to C.

6. C chooses $t \in \mathbb{Z}_q$ at random and sends $h_1 := h^t$ to Z.

7. Z makes a blind signature on h by signing h_1 as follows:

   (a) Z computes $z_0 := h_1^{xz}$. Then Z and C choose $(a_0, b_0) := (g^{s_0}, h_1^{s_0})$ at random such that only Z knows s_0, whereas both know a_0 and b_0. Z sends z_0 to C.

   (b) C first verifies that $z_0 = z^t$, and then it chooses $u \in \mathbb{Z}_q^*$ and $v \in \mathbb{Z}_q$ at random and computes
   $$a := (a_0 g^v)^u \quad \text{and} \quad b := (b_0^{1/t} h^v)^u.$$
   C then sends u, v, t and (a_0, b_0) to T.

   (c) T computes the pair (a, b) just as C did, the challenge $c := H(h, z, a, b)$, and $c_0 := c/u \bmod q$. Then it signs c_0 using x_T (with help from C) and sends the signature to C.

   (d) C computes $c := H(h, z, a, b)$, $c_0 := c/u$, and verifies the signature. C then forwards c_0 and the signature to Z.

   (e) Z verifies the signature on c_0 and computes $r_0 := s_0 + c_0 s_Z \bmod q$.

   (f) C verifies that
   $$g^{r_0} = a_0 h_Z^{c_0} \quad \text{and} \quad h_1^{r_0} = b_0 z_0^{c_0}$$
   and computes $r = (r_0 + v)u \bmod q$. Then C forwards r_0 to T.

   (g) T computes $r := (r_0 + v)u \bmod q$ and verifies that:
   $$g^r = a h_Z^c \quad \text{and} \quad h^r = b z^c.$$

Theorem 4.1
This protocol satisfies:

1. If T, C and Z follow the protocol, then T gets Z's signature on h.

2. If C follows the protocol then Z gets no information about h or σ. This is true even if T and Z have unlimited computing power.

3. If C follows the protocol then Z can construct all messages with the same distribution in expected polynomial time except the signature on c_0.

4. If C follows the protocol, then T can simulate all messages that it receives — except r_0.

5. If the blind signature scheme is secure then a polynomially bounded $\tilde{C}$ cannot get a validated public key for which he knows the corresponding secret key.

Proof
The first three properties are straightforward to prove, and the fourth follows from Theorem 3.4 and the fact that T can compute z by itself. As for the last property, note that the security of the blind signature scheme means that $\tilde{C}$ can only get a signature on h, but $\tilde{C}$ cannot find the secret key corresponding to h (i.e. $\log_g h$) unless it can compute discrete logarithms in G_q. ∎

As C can make sure that the signature on c_0 is random among all possible signatures, this theorem shows that the protocol for issuing a validated public key has no outflow. Furthermore, as r_0 is uniquely determined from the other messages the protocol protects against inflow.

5 An Application to Databases

This section first describes how a very simple database offering unconditional privacy as well as correctness can be constructed, and then it is shown how a database in which the information is kept secret from both T and C can be constructed. By similar techniques, it is also possible to construct databases in which

1. The data is known by T, but kept secret from C; and

2. The data is known by C, but kept secret from T.

Whenever T signs a message (anonymously) with respect to a public key, which is validated by the key authentication center, the signature will be referred to as a certified signature.

5.1 A Simple Database

The wallet can be used to store the personal database described in the introduction as follows:

- All information in the database is stored by T and C.

- Whenever an organization updates a field in the database, it sends a signed message to the wallet. C verifies the signature before it updates the database and forwards the new information plus the signature to T. Finally T verifies the signature and updates the database.

- When an organization wants to read a field in the database (or a function-value of several fields), a certified signature on the value is sent to the organization.

5.2 Database with Hidden Information

The implementation presented above has the property that both T and C know all information in the database. This could be a little dangerous for the user, because T could leak all information, in case it is captured by another person, who is able to break the tamper-resistance. On the other hand, there might be certain very sensitive data in the database, which the user should not know either (or does not want to be stored in his computer).

In the following it is therefore shown how the above database can be modified such that neither T nor C knows the data, but T is still able to control that C does not change anything in the database. We shall, however, only give protocols which allow the organization to read or write a single bit in the database. The following scheme for probabilistic encryption is an important ingredient in these protocols.

Probabilistic Encryption

Let $n = pq$, where p and q are primes both equivalent to 3 modulo 4. In order to encrypt a bit b, the committer chooses $r \in \mathbb{Z}_n^*$ at random and computes

$$BC(n, b, r) := (-1)^b r^2 \bmod n.$$

A person knowing p and q can decipher a given ciphertext by determining whether it is a quadratic residue or not. However, for a person not knowing p and q this is presumably infeasible.

Let n_1 and n_2 be two different moduli as above, and let $\beta_1 = (-1)^b r_1^2 \bmod n_1$ and $\beta_2 = (-1)^b r_2^2 \bmod n_2$ be probabilistic encryptions of the same bit $b \in \{0, 1\}$.

Theorem 5.1

There exists a four-round protocol with security parameter k in which a person, P, knowing r_1 and r_2 can prove to another person, V, that β_1 and β_2 are in fact encryptions of the same bit. More precisely this protocol satisfies:

1. If P and V follow the protocol, then V will accept, if $a = b$.

2. If V follows the protocol and $a \neq b$, then V will reject the proof with probability at least 2^{-k} no matter what an unlimited powerful prover does.

3. It is a proof of knowledge of r_1 and r_2.

4. It is (computational) witness hiding (see [FS90]).

Proof

The protocol uses the cut-and-choose technique. The details are omitted here. ∎

The Protocols

It is assumed that each organization, W, has a modulus, n_W, as above, and that W can make digital signatures. Prior to the execution of the read and write protocols to be described, the following start-up protocol is executed:

1. W constructs a request of the form $(n_W, op, name, time)$, where $op \in \{read, write\}$, $name$ identifies the bit which W wants to read or write, and $time$ is a time-stamp. This request is signed and sent to the wallet together with certificates, which show that n_W is a valid modulus and that the public key of W (for the signature scheme) is valid.

2. C verifies the request and certificates, and if they are legal, C forwards them to T. In particular, C verifies that $time$ is constructed correctly so that W has not encoded any information in it.

3. T verifies the request and the certificates.

Whenever T and C sign a message in the certified signature scheme op, $name$, $time$ and n_W are included in the message. This prevents obvious frauds by C in which signatures from previous executions of the same or different protocols are reused.

Furthermore, each write protocol must be immediately followed by a protocol in which T sends a signed message to W (through C) in which it confirms having received the required messages.

For each bit b in the database, T has given C a commitment $\beta_T = BC(n_0, b_T, r_T)$ to a bit b_T, and C has given T a commitment $\beta_C = BC(n_0, b_C, r_C)$ to a bit b_C such that $b = b_T \oplus b_C$. The modulus n_0 is the modulus of the organization which wrote b. An organization, W, with public modulus n_W can read b as follows

1. T chooses $s_T \in \mathbb{Z}_{n_W}^*$ at random and sends $\alpha_T := (-1)^{b_T} s_T^2 \bmod n_W$ to C.
   T proves to C that α_T and β_T are encryptions of the same bit.

2. C chooses $s_C \in \mathbb{Z}_{n_W}^*$ at random and sends $\alpha_C := (-1)^{b_C} s_C^2 \bmod n_W$ to T.
   C proves to T that α_C and β_C encrypt the same bit.

3. T and C sign $\alpha := \alpha_T \alpha_C$ using the certified signature scheme.
   This signature (and α) is sent to W (through C).

4. W verifies the signature and finds the encrypted bit by deciphering α.

This protocol has the following properties:

- If C follows the protocol: No matter what (an unlimited powerful) T does, α is a random encryption of b. Furthermore, the signature on α does not contain any information other than the fact that a legal T produced it.

- If T follows the protocol, then α is an encryption of b as long as C cannot fake T's signatures (or break the tamper-proofness).

- It does not make it easier for T and/or C to find b unless W tells them how to distinguish encryptions of 0 from encryptions of 1 modulo n_W.

The proofs of these properties are quite straightforward, and they are omitted from this extended abstract. The organization, W, can write a bit, b, in a given field in the database as follows:

1. T chooses $a_T \in \{0,1\}$ and $r_T \in \mathbb{Z}_{n_W}^*$ at random and sends $\alpha_T := (-1)^{a_T} r_T^2 \bmod n_W$ to C.

2. C chooses $a_C \in \{0,1\}$ and $s_C \in \mathbb{Z}_{n_W}^*$ at random and sends $\alpha_C := (-1)^{a_C} s_C^2 \bmod n_W$ to T.

3. T and C sign $\alpha := \alpha_T \alpha_C$ in the certified signature scheme. This signature (and α) is sent to W (through C).

4. W verifies the signature and finds the bit a by deciphering α.

5. W and C choose $r \in \mathbb{Z}_{n_W}$ at random using a coin-flipping protocol.

6. W then computes $b' = a \oplus b$ and $\alpha_W := (-1)^{b'} \alpha r^2$, which it subsequently signs. W sends the signature (σ_W) to C.

7. C computes α_W, verifies σ_W and computes $\beta_C := \alpha_W \alpha_T^{-1}$ and $\beta_T := \alpha_T$ and $r_C := r s_C$ and

$$
b_C := \left\{ \begin{array}{ll} a_C & \text{if } \alpha_W = \alpha r^2 \\ a_C \oplus 1 & \text{if } \alpha_W = -\alpha r^2. \end{array} \right.
$$

C then forwards α_W and σ_W to T.

8. T verifies the signature and computes $\beta_C := \alpha_W \alpha_T^{-1}$ and $\beta_T := \alpha_T$ and lets $b_T = a_T$.

This protocol satisfies (again the proofs are omitted):

- If T, C and W follow the protocol then after the execution the following holds:

  1. $\beta_T = BC(n_W, b_T, r_T)$;
  2. $\beta_C = BC(n_W, b_C, r_C)$;
  3. $b = b_T \oplus b_C$.

- If C cannot fake T's or W's signatures then $b \oplus b_T$ equals the plaintext corresponding to β_C.

- After the execution $b \oplus b_C$ equals the plaintext corresponding to β_T no matter what an unlimited powerful T does.

- C and/or T can only find b if they can distinguish quadratic residues from quadratic non-residues modulo n_W.

- If C follows the protocol, then W just gets a signature on a random encryption of a random bit. Similarly, T just gets a random encryption of a random bit chosen by W.

In the above two protocols the amount of inflow and outflow is very limited. Note, that W could have told T the factorization of n_W in advance. Hence, T learns the bit. However, this does seem to be a serious problem as W already knows this bit.

6 Conclusion and Future Work

We have argued that the electronic wallets presented here are an excellent way to store personal databases. And we have shown protocols that allow T to control and validate all messages from the user to the outside world. These protocols allow C to ensure that the privacy of the person is not compromised. They provide organizations with security against abuse by individuals that relies on the assumption that the tamper-proofness cannot be broken and that the signatures cannot be forged.

The protocols presented do, however, have a limited kind of inflow because T and W see the same random values (such as those used to form the signatures). In case T gets captured, these values would let organizations who could read out the contents of a captured T link it to specific protocol instances. Forthcoming joint work with Stefan Brands, Ronald Cramer and Niels Ferguson shows how the need for observers and organizations to share such information can be avoided altogether.

References

[CFN90] D. Chaum, A. Fiat, and M. Naor. Untraceable electronic cash. In *Advances in Cryptology - proceedings of CRYPTO 88*, Lecture Notes in Computer Science, pages 319 – 327. Springer-Verlag, 1990.

[Cha84] D. Chaum. Blind signature systems. In *Advances in Cryptology - proceedings of CRYPTO 83*, 1984.

[FS87] A. Fiat and A. Shamir. How to prove yourself: Practical solutions to identification and signature problems. In *Advances in Cryptology - proceedings of EUROCRYPT 86*, Lecture Notes in Computer Science, pages 186 – 194. Springer-Verlag, 1987.

[FS90] U. Feige and A. Shamir. Witness indistinguishable and witness hiding protocols. In *Proceedings of the 22nd Annual ACM Symposium on the Theory of Computing*, pages 416 – 426, 1990.

[GMR88] S. Goldwasser, S. Micali, and R. L. Rivest. A digital signature scheme secure against adaptive chosen message attack. *SIAM Journal on Computing*, 17(2):281 – 308, April 1988.

Making Electronic Refunds Safer

Rafael Hirschfeld

Laboratory For Computer Science
Massachusetts Institute of Technology
Cambridge, MA 02139

Abstract. We show how to break an electronic cash protocol due to van Antwerpen (a refinement of the system proposed by Chaum, Fiat, and Naor), and give an alternative protocol that fixes the problem.

1 Introduction

There has been much recent interest in electronic money—ways to perform monetary transactions by computer, telephone, fax machine, etc. Most proposed electronic money schemes rely on cryptography for their security, in particular, on digital signatures [7].

In response to privacy concerns, electronic money systems in which payments are untraceable without the cooperation of the payer have been developed. This untraceable electronic money is called **electronic cash**. The cryptographic mechanism used to provide untraceability is that of blind signatures [3].

Several different forms of electronic cash have been proposed. In addition to electronic coins [4], which have a fixed value, there are electronic checks [4] [2], which can be used for any amount up to a maximum value and then returned for a refund of the unused portion, and divisible electronic cash [5], which can be broken into smaller pieces that can be spent separately.

We concentrate on a recent electronic check scheme [1] that is based upon earlier check schemes, but with great improvement in efficiency. We show how a weakness in the refund mechanism of the earlier systems becomes a fatal flaw in the newer system, and how that flaw can be exploited to cheat undetectably. We propose a revised protocol to correct the flaw.

2 Transactions

We call the participants in an electronic cash system the **bank**, the **user**, and the **shop**. The bank is the issuer of the electronic cash. The user obtains electronic cash from the bank in a **withdrawal** transaction, spends it at a shop in a **payment** transaction, and the shop then redeems it at the bank in a **deposit** transaction. In addition, for electronic check systems, the unused portion of a check is returned by the user to the bank in a **refund** transaction.

We distinguish the user and the shop only to emphasize their roles in a payment; in fact, users can act as shops and shops can act as users. The user and the shop are also called the **payer** and the **payee**, respectively.

3   Checks

Untraceable electronic checks were introduced by Chaum, Fiat and Naor [4], and refined by den Boer, Chaum, van Heyst, Mjølsnes, and Steenbeek [2], with a significant improvement in efficiency. Checks are made up of three kinds of elements: challenge terms, denomination terms, and refund terms.

Challenge terms are used to prevent double-spending of a check. Each challenge term c_i contains a random a_i chosen by the user, as well as $a_i \oplus u$, where u is the user's identity (bank account number). During payment, each challenge term is opened to reveal either a_i or $a_i \oplus u$, depending on the corresponding bit of a challenge chosen by the shop. If a check is spent with two different challenges, then (for some i) both a_i and $a_i \oplus u$ will be revealed, from which u can be obtained.

Denomination terms are used to represent the value of the check. Each denomination term d_i contains a unique coin number b_i, and corresponds to a different power-of-two denomination. The denomination terms are either ordered or are signed with different roots to indicate which denomination they represent. During payment, only those terms corresponding to denominations used are actually opened. In order to keep the user from mixing terms from different checks (which would expose the protocol to a simple attack), the challenge and denomination terms are tied to a check number. The shop (and the bank) can easily verify that all the terms presented for payment (or deposit) belong to the same check.

Refund terms are used to obtain refunds for unspent denomination terms in a check. Unlike the other terms, they are not tied to the check number. If they were, the bank could link deposits to their corresponding refunds, defeating the untraceability of the scheme. Instead, the refund terms r_i contain the same coin numbers b_i as the denomination terms. The bank keeps track of spent and refunded coin numbers to ensure that a denomination is not both spent and refunded.

The coin numbers and identity numbers are built into the terms by the user. Because the terms are blinded for untraceability, the bank doesn't actually see these numbers at the time of withdrawal. To ensure that the terms are correctly formed, the protocol uses the cut-and-choose methodology introduced by Rabin [6]: the bank asks for more candidate terms than it actually needs, chooses a random subset of them and asks the user to demonstrate that they are well-formed, and uses the remaining unopened candidates only if all of the opened candidates were legitimate. We will assume that the bank asks for twice as many candidates as it needs, so that the probability of the user getting caught attempting to slip in a single bogus term is $1/2$.

This is the source of the weakness in these check systems. With probability $1/2$, the user can slip a bogus candidate past the bank. If there are enough challenge terms, a single bogus challenge term is unlikely to be of much use to a cheater, because the probability is still overwhelming that two challenges will differ at some other position. A user who tries to slip enough bad challenge terms into a check to cheat effectively will with high probability be caught.

For the other terms, though, this presents a problem. If the user can slip in a denomination term/refund term pair for which the coin numbers b_i don't match, then that denomination can be both spent and refunded, undetectably.

To address this problem, Chaum, Fiat, and Naor suggest penalizing detected cheating attempts so that the net expected effect favors the bank. This solution is somewhat unsatisfying. At the time of the detection—withdrawal—the user has not yet cheated (by spending and refunding the same term). When a bad term is detected, the user may claim it is due to a data error, or may even refuse to open a bad term, claiming that the data has been lost. Since data errors and losses do occur, it would be difficult for the bank to fully justify imposing the penalty.

Another possibility is to require more than one pair of terms for each denomination; Chaum, Fiat, and Naor suggest using two, as an alternative to imposing penalties upon detection. This would lower the probability of cheating because in order to escape detection, both pairs would have to be bad. The more pairs make up a denomination, the lower the chance of successfully cheating, but the bigger and costlier to handle the check becomes.

As we will see, this troublesome problem proves to be fatal in the latest incarnation of the system.

4 "Improved" Electronic Cash

A new protocol by van Antwerpen [1] further refines this electronic cash system. This is a sophisticated protocol with order-of-magnitude improvements in both the size of the checks (and consequently the amount of storage required for them) and the amount of communication required, at the expense of some extra computation that can be done in the background. For efficiency reasons, several checks are grouped into a single pack; during withdrawal the user obtains an entire pack from the bank, but then spends them one at a time.

For a complete description of the protocol, the reader is referred to van Antwerpen's paper, but we will give a brief synopsis. A pack of k checks is made up of $2k$ *pseudochecks* and $2k$ *pseudo-refund-parts*, each of which is a single RSA-sized number. The terms that make up each pseudocheck (pseudo-refund-part) are multiplied together in such a way that they can later be separated. The pseudochecks contain the denomination and challenge terms, and the pseudo-refund-parts contain the refund terms. The terms that make up an *actual* check (refund part) are distributed among the pseudochecks (pseudo-refund-parts) by permutations chosen by the bank. The refund terms in the pseudo-refund-parts are also permuted by the user so that they won't line up with the corresponding denomination terms in the pseudochecks. This is to prevent the bank from gleaning information that might be used to link by by amount, *i.e.*, to link a deposit to a corresponding refund by checking that they are for complementary amounts.

In addition, the cut-and-choose process has been modified. The bank still chooses a random subset of the terms, and the user still provides opening information for those terms, but the bank, rather than verifying that they are correct (which would be difficult because of the way the pseudochecks are formed), instead multiplies the resulting pseudocheck by a *protection factor* that renders it useless if the user lied about any of the opened terms. Veugen [8] has proved that the security of this technique is equal to that of the original, in the sense that if the user attempts to cheat, the probability that she will be unable to use the resulting checks is the same as the probability that she would be caught by the original cut-and-choose process.

But this is precisely where the new protocol is flawed. Although there is still no problem with the challenge terms, recall that the solution (of imposing penalties) to the problem with the refund terms relied upon detection by the bank of cheating attempts. With this new mechanism, cheating attempts are *never* detected.

Let b_i be the coin numbers in the denomination terms and b_i' be the coin numbers in the refund terms. The user could make $b_i \neq b_i'$ for *all i*. Then when asked to open some terms, the user provides the b_i, so that she will be able to remove the protection factors. The opened terms are divided out of the check, but the user will be able to both spend *and* refund all of the denominations that actually make up the check, undetectably. Needless to say, this is not a desirable property of an electronic cash system.

5 Attempted Fixes

There are several ways that one might attempt to patch the protocol. One possibility is to put protection factors on the pseudo-refund-parts in addition to the ones on the pseudochecks. That way if the user were asked to open the ith terms, where $b_i \neq b_i'$, then if she produced b_i she wouldn't be able to refund the term, and if she produced b_i' she wouldn't be able to spend it. So in the previous scenario (with $b_i \neq b_i'$ for all i), she would still be able to spend all of the denomination terms, but not be able to refund any of the refund terms, and so would gain nothing.

But this isn't good enough. The user need not make all of the denomination/refund pairs bad. If only some are bad, it is possible that the bank won't select any of them during the cut-and-choose. The fewer bad terms there are, the more likely that they will all slip by. Whenever the bank picks a bad term to be opened, the user won't gain anything, because she'll only be able to spend the check and not refund any of it. But she won't lose anything either. Since the attempts are undetectable, she can just keep trying until she succeeds.

To work around this, we could add an extra refund term that has no corresponding denomination term, so that the user must refund it or else lose money. But this isn't quite good enough either, because the user could reveal the coin numbers b_i' from the refund terms instead of the coin numbers b_i from the denomination terms, and when "caught," she would just refund the whole check and not spend any of it. To prevent this, we would need also to add an extra denomination term that has no corresponding refund term, forcing the user to spend as well as to refund. This works after a fashion, provided that the values of these extra terms are large enough to ensure negative expectation from cheating attempts, but it is cumbersome and makes the checks inconvenient to use. We will present a much cleaner solution.

6   A Modified Protocol

Our problems stem from the difficulty of ensuring that two coin numbers that are provided separately are in fact the same. The solution is actually quite simple— just provide a single coin number and use it for both terms! With the earlier systems, it is difficult to see how to do this without compromising the unlinkability of the terms, but in the newer system it is fairly straightforward.

In van Antwerpen's system, candidates provided by the user to the bank during withdrawal look like

$$X = R^{PQ} \prod_i F(a_i)^{P_i Q} \prod_i G(b_i)^{P Q_i} \tag{1}$$

$$Y = T^r S^Q \prod_i G(b_i)^{Q_i} \tag{2}$$

where X is a pseudocheck candidate, Y is a pseudo-refund-part candidate, R, S, and T are blinding factors, the P's and Q's are exponents related to the prime roots used for signatures, the F's are challenge terms with random numbers a_i (as well as $a_i \oplus u$) incorporated, and the G's are denomination or refund terms, with the random coin numbers b_i incorporated. There are multiple X and Y for a given check pack; not shown here is a collection of permutations θ_i chosen by the user, the ith refund term of the jth pseudo-refund-part corresponds to the ith denomination term of not the j'th pseudocheck, but rather the $\theta_i(j)$th pseudocheck. These permutations are to prevent the bank from linking by amount, as previously mentioned.

For now we will ignore these permutations as well as the blinding factors. Note that the basic form of X and Y are

$$X = \prod_i F(a_i)^{P_i Q} \prod_i G(b_i)^{P Q_i} \tag{3}$$

$$Y = \prod_i G(b_i)^{Q_i} \tag{4}$$

The b_i are given in both X and Y, and it is difficult to ensure that they in fact match. But instead of giving X and Y to the bank, the user can instead give

$$X_1 = \prod_i F(a_i)^{P_i} \tag{5}$$

$$X_2 = \prod_i G(b_i)^{Q_i} \tag{6}$$

and now the bank can compute

$$X = X_1^Q X_2^P \tag{7}$$
$$Y = X_2 \tag{8}$$

and *voila!*, the denomination and refund terms are now *guaranteed* to contain the same coin numbers.

This is the primary protocol change. Except for the refund transaction, the rest of the protocol is the same as before.

We ignored the blinding factors and the user permutations. The blinding factors are not a problem; we could just put blinding factors as before on X_1 and X_2. The resulting X would have a different blinding factor from the original protocol because it would be the product of the two blinding factors, but that has no detrimental effects on either the privacy or the security of the system.

The permutations are another matter. Because the denomination and refund terms are now provided together as a single unit, there is no way to put them in separate places as before. This may not be as serious as it seems, because we need not refund the pseudo-refund-parts in the same order in which we withdrew them, so linking by amount can be made very difficult. Still, it would be preferable for the new protocol to leak no more information than the old. In the next section we will show how to recapture the full degree of unlinkability by slightly altering the structure of the blinding factors.

7 Regaining Lost Privacy

Notice that the pseudo-refund-part candidate has *two* blinding factors, T^r and S^Q. The blinding factor T^r is used to blind during withdrawal (r is the prime root used to sign the pseudo-refund-part, so the user can divide by T afterwards). The blinding factor S^Q is used during refund; terms that have been spent are moved into this blinding factor so that they are not revealed when refunding the unspent terms.

If the user could separate out and refund each unspent term individually, she could rearrange them into any order she wanted, getting the same privacy effect as from the permutations θ_i. Although she could do this just by moving unspent terms along with spent terms into the blinding factor, she would have to do it once for each unspent term in the pseudo-refund-part, and the resulting blinding factors would be correlated because they would contain some of the same factors. The bank could use this fact to link terms that come from the same pseudo-refund-part, undoing the privacy gain we thought we had achieved.

If we change the refund blinding factor, however, to be of the form S^{Qr}, then we can make this work. After the bank takes the rth root, the user is left with S^Q rather than $S^{Q/r}$, which means that she can change S. She couldn't do this before because she couldn't extract rth roots. By changing the S for each separated unspent refund term, the user can make them unlinkable from each other.

By making this change to the blinding factors we can recapture all of the privacy lost by eliminating the permutations θ_i. The rest of the protocol is also simplified by the elimination of the permutations. The cost is in efficiency: each refund term from the same pseudo-refund-part must be separated and refunded separately. But the blowup is only by the average number of unspent denominations per pseudocheck, and can be traded off against a slight increase in traceability if desired.

Acknowledgements

I would like to thank David Chaum for his guidance and support for this work, and Thijs Veugen for his assistance in working out the method for regaining privacy.

References

1. C. J. van Antwerpen, "Electronic cash," master's thesis, Eindhoven University of Technology (1990).
2. B. den Boer, D. Chaum, E. van Heyst, S. Mjølsnes, and A. Steenbeek, "Efficient off-line electronic checks," Proceedings of Eurocrypt '89, 294–301.
3. D. Chaum, "Blind signatures for untraceable payments," Proceedings of Crypto '82, 199–203.
4. D. Chaum, A. Fiat, and M. Naor, "Untraceable electronic cash," Proceedings of Crypto '88, 319–327.
5. T. Okamoto and K. Ohta, "Universal electronic cash," Proceedings of Crypto '91.
6. M. O. Rabin, "Digitalized signatures," Foundations of Secure Computation, Academic Press, NY (1978).
7. R. Rivest, A. Shamir, and L. Adleman, "A method for obtaining digital signatures and public-key cryptosystems," CACM 21, 2 (February 1978).
8. T. Veugen, "Some mathematical and computational aspects of electronic cash," master's thesis, Eindhoven University of Technology (1991).

Fair   Public-Key   Cryptosystems

by

Silvio Micali

Laboratory for Computer Science
Massachusetts Institute of Technology
545 Technology Square, Cambridge, MA 02139

(Rough Draft)

Abstract. We show how to construct public-key cryptosystems that are *fair*, that is, strike a good balance, in a democratic country, between the needs of the Government and those of the Citizens. Fair public-key cryptosystems guarantee that: (1) the system cannot be misused by criminal organizations and (2) the Citizens mantain exactly the same rights to privacy they currently have under the law.

We actually show how to transform any public-key cryptosystem into a fair one. The transformed systems preserve the security and efficiency of the original ones. Thus one can still use whatever system he believes to be more secure, and enjoy the additional properties of fairness. Moreover, for today's best known cryptosystems, we show that the transformation to fair ones is particularly efficient and convenient.

As we shall explain, our solution compares favorably with the Clipper Chip, the encryption proposal more recently put forward by the Clinton Administration for solving similar problems.

Note For The Reader. Since privacy and law enforcement interest most of society, and since we would welcome an informed debate before making crucial policy decisions in this area, we have made a sincere attempt to reach a broad audience. We thus hope that at least the goals and the properties of our approach will be understandable by the Government official and the Citizen who do not have any familiarity with cryptography. Further, the basic technical ideas of our solution --which are quite simple to begin with-- are presented at a very intuitive level, so as to be enjoyable for the reader generally familiar with the field of cryptography, though not necessarily an expert in secure protocol design. Such an expert will not have great difficulty in filling in the formalization and the occasionally subtle technical details that have been omitted in this draft. (We actually hope to have given her sufficient indications to make her journey through this draft as short as possible.)

We apologize for not having the time to write different versions of this paper for different audiences.

1.   Introduction

A wrong debate

Currently, Court-authorized line tapping is an effective method for securing criminals to justice. More importantly, in our opinion, it also prevents the further spread of crime by deterring the use of ordinary communication networks for unlawful purposes. Thus, there is a legitimate concern that wide-spread use of public-key cryptography may be a big boost for criminal and terrorist organizations. Indeed, many bills propose that a proper governmental agency, under the circumstances allowed by the law, be able to obtain the clear text of any communication over a public network. At the present time, this requirement would translate into coercing citizens into either (1) *using weak cryptosystems* --i.e., cryptosystems that the proper authorities (but also everybody else!) could crack with a moderate effort-- or (2) *surrendering, a priori, their secret key* to the authority. It is not surprising that such alternatives have legitimately alarmed many concerned citizens, generating the feeling that privacy should come before national security and law enforcement.

It is our opinion that this debate is wrong. It is wrong because it is a "one-bit debate," that is, it envisages either unconstrained privacy or no privacy at all. Extreme positions are more likely to be unjust and, indeed, having to choose only between the above alternatives is quite uncomfortable. Fortunately, we are not bound to choose only among what is currently available. It is indeed the goal of Science to understand reality and to change it to our advantage, so as to enlarge our options.

Broadening the debate
In this paper we show how cryptographic protocols can be successfully and efficiently used to build cryptosystems that are fairer, that is, that strike a better balance, in a democratic country, between the needs of society and those of the individual. More precisely, we show a *simple* and *general* methodology for transforming any public-key cryptosystem into a *fair* one, that is, one enjoying the following properties:

1   *(Unabusing)* The privacy of the law-obeying user *cannot* be compromised, while

2   *(Unabusable)* Unlawful users *will not* enjoy any privacy.

Our transformation preserves the original security of the underlying cryptosystem and its efficiency. Since we believe that public-key cryptosystems are best suited for adoption in a large nation, in this paper we solely focus on making fair this type of cryptosystems.

2.   Public-Key Cryptosystems

A conventional cryptosystem allows two users X and Y, who have previously agreed on a common secret key (e.g., by meeting in a secure physical location) to exchange private messages over a public network. The usefulness of such systems is quite limited. While there is plenty of need for private communication, agreeing on a common secret key without the help of a modern communication network is quite cumbersome. In the case of the military it may not be too inconvenient, since in this application it may be clearer beforehand with whom one will need to exchange private messages. But in other cases, as in business applications, it is very hard to know a priori with whom one will need to talk in private and thus establish a common secret key in advance. The type of cryptosystem best suited for these latter settings is a *public-key cryptosystem* (PKC for short) as introduced by Diffie and Hellman in [DiHe]. While in a conventional cryptosystem each secret key was used both for encrypting and decrypting, in a PKC the encryption and decryption processes are governed by pairs of *matching* keys, which are generated together so to satisfy the following three properties: letting (E,D) be one such pair of matching encryption/decryption keys,

1   Any message can be encrypted using E.

2   Knowledge of D enables one to read any message encrypted with E; on the contrary, ignoring D it is practically impossible to understand messages encrypted with E.

3   Knowing E does not enable one to compute its corresponding decryption key D.

PKCs thus dismiss the need for agreeing beforehand on a common secret key, by using instead a bit of initial interaction. Assume that a user X generates a pair of matching encryption/decryption keys (E_X, D_X), and that a user Y wants for the first time to send him a private message and tells him so. Then X sends E_X to Y over the phone; Y easily encrypts her message to X with E_X because of Property 1; X easily decrypts it because of Property 2; and, because of Properties 2 and 3, no one else can understand the message so exchanged. Interaction (like in the case of electronic mail) is not however always available, and PKCs are thus most useful by having stipulating what *de facto* is a "*social agreement*" between users and a *key-management center*. Each user X comes up with a pair of matching encryption and decryption keys (E_X, D_X). After generating a (E_X, D_X) pair, the user keeps D_X for himself and gives E_X to the key-management center. The center is responsible (and is trusted!) for *updating* and *publicizing* a directory of *correct* encryption keys, one for each user --i.e., a list of entries of the type (X, E_X) which, for example, may be publicized in a "phone-book format" or via a "411-like service." If, as in the latter example, this distribution occurs over a public network, a digital authentication that E_X comes from the center must be provided, for instance by using one of the existing digital signature schemes. Clearly the users must trust the center, as an untrustworthy center may enable a user Y to read the messages intended for user X by falsely claiming that E_Y is X's encryption key. Thus, in ultimate analysis, the security of a PKC depends on the key-management center. Since setting up such a center on a grand scale requires a great deal of effort by society, the precise protocols the center must follow (and thus its properties) must be properly chosen.

Every advantage has a drawback, and public-key cryptography is no exception. Here a main disadvantage is that any such system can be abused; for example, by terrorists and criminal organizations who can now conduct their illegal business with great secrecy and yet with extreme convenience. Very often scientists have jumped into new technical ventures without giving much thought to the consequences of their actions. Developing nuclear plants without solving first their associated nuclear waste problems is a notable example of the social blindness of Science in this century. Certainly, all of us envisage good uses for public-key cryptography, but the risk exists that the main fruits of this development may be harvested by criminal organizations, and it is thus our responsibility to give a more thorough thought to the matter. *Fair Public-Key Cryptosystems* (Fair PKCs for short) are our proposal to enjoy public-key cryptography while protecting society from the problems arising from its blind utilization. We hope that our proposal will start a fruitful scientific debate, and other scientific solutions will be sought to this important problem in order to avoid further plaguing a crime-ridden world.

3. Fair PKCs

3.1 The Informal Notion of a Fair PKC

Let S be a public-key cryptosystem. Informally speaking, we say that

S is a Fair PKC if it guarantees a special agreed-upon party --and solely this party!-- under the proper circumstances envisaged by the law --and solely under these

circumstances!-- to understand all messages encrypted using S, even without the users' consent and/or knowledge.

That is, the philosophy behind a Fair PKC is *improving* the security of the existing communication systems while *keeping the legal procedures* already holding and accepted by the society. The following proposition immediately follows from the above definition.

Proposition: Let C be a ciphertext exchanged by two users in a Fair PKC S. Then, under the proper circumstances envisaged by the law, the proper third party will either

1) find the *cleartext* of C relative to S (whenever C was obtained by encrypting a message according to S) or

2) obtain a (court-presentable) *proof* that the two users were not using S for their secret communication.

Of course, if using any other type of public-key cryptosystem were to be made *illegal,* Fair PKCs would be most effective in guaranteing both private communication to law-obeying citizens and law enforcement. (In fact, if a criminal uses a phone utilizing a Fair PKC to plan a crime, he can still be secured to justice by court-authorized line tapping. If he, instead, illegally uses another cryptosystem, the content of his conversations will never be revealed even after a court authorization for tapping his lines, but, at least, he will be convicted for something else: his use of an unlawful cryptosystem.) Nonetheless, as we shall discuss in section 4, Fair PKCs are quite useful even without such a law.

3.2 An Abstract Way for Constructing Fair PKCs

We shall now present, in a very *abstract* way, our prefered method for constructing Fair PKCs. We shall see in section 5 that this very abstract and almost paradoxical method can not only be concretly implemented, but actually be implemented in a most efficient way.

Below, for concreteness of presentation, we shall use the *Government* for the special agreed-upon party, a *court order* for the circumstances contemplated by the law for monitoring a user's messages, and the *telephone system* for the underlying method of communication. We also assume the existence of a key-distribution center as in an ordinary PKC.

In a Fair PKC there are a fixed number of predesignated *trustees* and an arbitrary number of users. The trustees may be federal judges (as well as different entities, such as the Government, Congress, the Judiciary, a civil rights group, etc.) or computers controlled by them and especially set up for this purpose. Even if efforts have been made to choose *trustworthy* trustees, a Fair PKC does not blindly rely on their being honest. The trustees, together with the individual users and the key-distribution center, play a crucial role in deciding which encryption keys will be publicized in the system. Here is how.

For concreteness of exposition, assume that there are 5 trustees. Each user independently chooses his own public and private keys according to a given double-key system. Since the user himself has chosen both keys, he can be sure of their "quality" and of the privacy of his decryption key. He then breaks his private decryption key into five *special* "pieces" (computing from his decryption key 5 special strings/numbers) possessing the following properties:

1)   The private key can be reconstructed given knowledge of all five special pieces;

2)   The private key cannot be reconstructed if one only knows (any) 4, or less, of special pieces;

3)   For $i=1,...,5$, the i-th special piece can be *individually* verified to be *correct*.

Comment. Of course, given all 5 special pieces, one can verify that they are correct by checking that they indeed yield the private decryption key. The difficulty and power of property 3 consists of the fact that each special piece can be verified to be correct (i.e., that together with the other 4 special pieces yields the private key) individually; that is, without knowing the secret key at all, and without knowing the value of any of the other special pieces! (How these special pieces can be generated is explained in the full paper. Below we will show how they can be used.)

The user then privately (e.g., in encrypted form) gives trustee i his own public key and the i-th piece of its associated private key. Each trustee individually inspects his received piece, and, if it is correct, *approves* the public key (e.g., signs it) and safely *stores* the piece relative to it. These approvals are given to the key-management center, either directly by the trustees, or (possibly in a single message) by the individual user who collects them from the trustees. The center, which may or may not coincide with the Government, itself approves (e.g., it itself signs) any public key which *is approved by all trustees*. These center-approved keys are the public keys of the Fair PKC and they are distributed and used for private communication as in an ordinary PKC.

Since the special pieces of each decryption key are privately given to the trustees, an adversary who taps a user's communication line possesses the same information as in the underlying, ordinary PKC. Thus if this is secure, so is the Fair PKC. Moreover, even if the adversary were one of the trustees himself, or even a cooperating collection of any 4 out of five of the trustees, due to property 2, he would still have the same information as in the underlying ordinary PKC. Since the possibility that an adversary corrupts 5 out of 5 federal judges is absolutely remote, the security of the resulting Fair PKC is the same as in the underlying, ordinary one.

When presented with a court order, and only in this case, the trustees will reveal to the Government the pieces of a given decryption key in their possession. This enables the Government to reconstruct the given key. Recall that, by property 3, each trustee has already verified that he was given a correct piece of the decryption key in question. Thus, the Government is *guaranteed* that, *in case of a court order*, it will be given all correct pieces of any given decryption key. By property 1, it follows that the Government will be able to reconstruct any given decryption key if necessary.

4. Basic Questions About Fair PKCs

Before addresing the real technical question of how Fair PKCs can be concretly constructed, let us consider some legitimate and broader questions.

Q: *Are Fair PKCs less secure?*

A: No. Unless an adversary corrupts 5 out of 5 trustees --a rather unlikely event-- they provably provide just the same security as the underlying, ordinary PKC. (Only the Government, and in case of a court order, may have the cooperation of all 5 trustees.)

Q: *Are Fair PKCs less efficient?*

A: No. Communication is exactly as efficient as in an ordinary PKC. The only differences are (1) when a public-key is registered, and (2) when a private key is, in a lawful manner, retrieved by the Government. Each user validates his public key only once. Thus only once does he need to give pieces of his private key to the trustees. Moreover, as we have seen in section 4, this step can be implemented by sending 5 short messages, one to each trustee. Second, the lawful reconstruction of a private key by the Government is essentially instantaneous once the five special pieces are obtained from the trustees. Collecting these five pieces electronically is no more cumbersome than issuing or checking a court order as it is needed in a lawful procedure. (As we have seen in section 4, private-key reconstruction may just consist of receiving 5 short messages and one addition.)

Q: *In a totalitarian system, what confidence can we have in a Fair PKC?*

A: Most probably, in a totalitarian system the trustees will be selected with rather different criteria. It is thus conceivable that all of them (whether individuals or organizations) may routinely conspire so as to reconstruct all private keys, destroying all confidence in the privacy of a Fair PKC. On the other hand, believing that ordinary PKCs may be the way to guarantee individual privacy during a dictatorship is quite *naive*. Outlawing any form of PKC will be among the first measures taken by any dictator. Indeed, public use of cryptography is a gift of democracy (and it is important that this gift cannot be turned against it). In fact, Fair PKCs are close in spirit to Democracy itself, in that power is not trusted to any chosen individual (read "trustee") but to a multiplicity of delegated individuals.

Q: *Aren't Fair PKCs the same as ordinary PKCs in which users are obliged to give the Government the private key corresponding to every public key?*

A: No. This deprives the individual of his right to privacy *a priori* and without any just cause. Someone who has not committed (nor is suspected to have committed) a crime should not be required to surrender his right to private communication to anybody, not even to the Government. And this is exactly what he would be obliged to do by revealing his own private key at the time of registering his public one with the key-management authority.

       People consent that their right to privacy may be taken away under special circumstances, but do not agree to lose it in an automatic manner. Fair PKCs guarantee the users that they will keep exactly the same rights they currently have in a phone network, and with greater security. (In fact, due to technological advances or collusions with phone operators, eavesdropping ordinary phone conversations will become easier and easier for unauthorized parties.)

Q: *What is the difference between a Fair PKC and a PKC with a "hidden trapdoor" chosen by the Government?*

A: There are three main differences:

       1) A PKC with a hidden trapdoor is very dangerous: if an enemy finds it, the security of the entire system is compromised.

       By contrast, in a Fair PKC, each user chooses his key independently. Thus even if a single user's key is compromised, this does not affect other users at all.

2) Society may never consent to using a PKC with a hidden trapdoor, since this is equivalent to asking the citizen to surrender their right to privacy even before being suspected of any wrong doing! (On the other hand, should a government maliciously ask its citizens to use a special type of PKC concealing the presence of a master secret key, things may get quite unpleasant if the existence of such a key is later discovered!)

3) PKCs with a hidden trapdoor may be weaker than ordinary PKCs, since in the former case the public and private keys must be chosen in a constrained way. In fact, enforcing the existence of a single master secret key for all public keys in the system is a very severe constraint in choosing the individual users' keys. Indeed, it is easy to speak of a system with a single master key, but it is also quite conceivable that any such cryptosystem may be easy to break.

By contrast, a Fair PKC, unless all trustees unlawfully collaborate, offers *the same* security of the underlying PKC. Even if 4 out of 5 trustees are traitors, the time that an adversary should invest for understanding anything about a message encrypted in a Fair PKC *provably equals* the time he needs to invest when the same message has been encrypted in the underlying ordinary PKC.

Q: *Granted that Fair Cryptosystems protect Society and the individual. But what is their advantage if criminals do not use them for their communications?*

A: We must distingush two settings: First, when the use of any PKC which is not Fair is made illegal. Second, when all commercially available PKCs are Fair (e.g., because thay are the only ones to be standardized), even though non-Fair PKC are not illegal.

Setting 1 has a short answer: a criminal who uses a non-Fair PKC could be brough to justice at least on this charge (recal that Al Capone was convicted for tax evasion).

Let us now consider setting 2. First, note that this is the current setting: anyone in the U.S.A. can use any cryptosystem he or she chooses (though the market for encryption product has not yet reached its full potential). Still, if Society ensures, via standardization, that all *easily available* PKCs are Fair, there are big advantages to be gained.

1) Criminals will have difficulty in distributing their own keys.

In fact, they could not enjoy the convenience of a well-kept and well-publicized public file; that is, they could not call up anyone they want and have a secret conversation with her. They thus would need alternative, cumbersome, and secretive methods to exchange their own keys.

In other words, it is one thing that criminals go out of their way to avoid being controlled by the Government in presenc of a court order, and a *very different thing* that the Government goes out of their way to provide criminals with this capability by setting up an ordinary PKC on a grand scale!

2) Besides difficulty in key distribution, criminals will have no convenient access to "alternative" cryptographic *products* which use their keys.

In fact, most products whose usefulness may be greatly enhanced by public-key cryptography --such as "secure" phones, "secure" faxes, etc.-- could become reasonably available, economic, reliable, and compatible, only if *mass produced;* that is, only after intensive engineering effort and big initial investments. Thus, if essentially only the criminals were to use non-fair cryptography, industry would not have sufficient interest in developing products incorporating such technology. (Else, the "criminal market" should have grown so much that we would have nothing more to worry about: civil society as we know it would have already ceased to exist.) Also, big and reputable companies would refrain anyway from manufacturing "questionable" products. Finally, even if a company were willing to manufacture products utilizing non-Fair PKCs, the list of its customers or any record of its sales would be excellent tips for the Police.

3) In an ordinary PKC, the Government is in a difficult position. Since it cannot understand any conversation at all, it has no way to distinguish even potential criminals from non-criminals (setting aside what criminals are saying). In a Fair PKC, instead, the Government can at least make this distinction. Assume that a Fair PKC is standardized, X is one of its users, and a court order authorizes the Government to listen to all messages addressed to X. If the Government is still unable to understand these calls, it means that X really uses a different cryptosystem, and thus intends not to be understood by the Government even in case of a court order. This may be crucial information, and information not available in an ordinary PKC.

4) If all commercially available cryptographic products (e.g., "secure" phones) were based on Fair-PKCs, there would be several advantages. True: a powerful criminal organization could succeed in having designed and produced phones made secure by a non-Fair PKC. This would, however, be less easy for isolated criminals; moreover, it would be most inconvenient for two or three people to get hold of "alternative" products just to discuss their FIRST crime. *At least,* Fair PKC-based products prevent their initially (but no longer) honest buyers from conveniently and undetectably shift to illegal communications.

5) In any case, punishing *abuse* is secondary with respect to enabling *legitimate use.*

Q: *Fair PKCs may strike a good balance between the needs of the Government and those of the citizens in a democratic country, but: is there any use of Fair PKCs for "less democratic" settings?*

A: Yes. Consider the case of a large organization, say a private company, where there is a need for privacy, there is an established "superior" --say, a president,-- but not all employees can be trusted since there are too many of them. The need for privacy requires the use of encryption. Since not all employees can be trusted, using a single encryption key for the whole company is unthinkable. So is using lots of single-key cryptosystems, since this would generate enormous key-distribution problems. Having each employee use his own double-key system is also dangerous, since he might conspire against the company with great secrecy, impunity, and convenience. Obliging every employee to surrender his decryption key to the president is certainly more possible than in the public sector, since a private company need not to be too democratic an organization. But *it may not be a good idea* for many reasons, two of which are the following. First, the identity of the president may change, and change quite often, but an employee should not change his keys for every new president. Second, a storage device containing all or many of the decryption keys would require to be overwhelmingly guarded.

Even in this context Fair PKCs may be of help. Again, key distribution will not be a problem. Each employee will be in charge of choosing his own keys, which makes the system more distributed and agile. While enjoying the advantages of a more distributed procedure, the company will retain an absolute control, since the president is guaranteed to be able to decrypt every employee's communications when necessary. There is no need to change keys when the president does, since the trustees need not to be changed. The trustees' storage places need less surveillance, since only compromising all of them will give an adversary any advantage.

Finally, Fair PKCs can be used as better secret sharing, since one has the guarantee that the secret will be reconstructed if all pieces (or the majority of them, depending on the implementation) will be made available.

5. A Concrete But Impractical Construction of Fair PKCs

We now show that any ordinary PKC can actually be made fair along the lines of the abstract construction of Section 3. The construction below, though concrete, is however too general for being practical, and thus more direct solutions are described in the next two sections for making fair the most popular, ordinary PKCs. The practically-oriented reader may thus prefer to procede directly to those sections.

5.1 A Sketch For The Expert

The expert in secure protocol theory may be satisfied with the following sketch.
Cuttng corners, each user should (1) come up with a pair of matching public and private keys and give the trustees his chosen public key, (2) encrypt (by a different cryptosystem, even one based on a one-way function) his chosen private key, (3) give the trustees the just computed ciphertext and a zero-knowledge proof that the corresponding "decryption" really consists of the private key corresponding to the given public key, and (4) give the trustees shares of this decryption by means of a proper Verifiable Secret Sharing protocol.

5.2 A More Informative Discussion

In expanding the above sketch for the non-expert in protocol design, we feel important to illustrate both similarities and differences between Fair PKCs and other related prior notions.

SECRET SHARING
As independently put forward by Shamir [Sh] and Blakley [Bl], secret sharing (with parameters n,T,t) is a cryptographic scheme consisting of two phases: in phase 1, a secret value chosen by a distinguished person, the *dealer*, is put in "safe storage" with n people or computers, the *trustees,* by giving each one of them a piece of information, a *share,* of the secret value. In phase 2, when the trustees pool together the information in their possession, the secret is recovered . In a secret sharing, this storage is *safe* only in two senses:

1   *Redundancy.*
     Not all trustees need to reveal their shares in phase 2: it is enough that T of them do. (Thus the system tolerates that some of the trustees "die" or accidentally destroy the shares in their possession)

2   *Privacy.*
   If less than t of the trustees accidentally or even intentionally divulge the
   information in their possession to each other or to an outside party, the secret
   remains unpredictable until phase 2 occurs.

Secret sharing suffers, though, of a main problem: *Assumed honesty*; namely,

   Secret sharing presupposes that the dealer gives the trustees correct "shares" (pieces
   of information) about his secret value. This is so because each trustee cannot verify
   that he has received a meaningful share of anything. A dishonest dealer may thus
   give "junk" shares in phase 1, so that, when in phase 2 the trustees pool together
   the shares in their possession, there is no secret to be reconstructed.

EXAMPLE (Shamir)
The following is a secret sharing scheme with parameters n=2t+1 and T=t+1.

   Let p be a prime >n, and let S belong to the interval [0,p-1]. Choose a polynomial
   P(x) of degree t by choosing at random each of its coefficients in [0,p-1], except for
   the last one which is taken to be equal to S , that is, P(0)=S. Then the n shares are
   so computed: S1=P(1),...,Sn=P(n). *Redundancy* holds since the polynomial P(x)
   can be interpolated from its value at any t+1 distinct points. (This, in turn, allows
   the computation of P(0) and thus of the secret.) *Privacy* holds since P(0) is totally
   undetermined by the value of P at any t points X1 ... Xt different from 0 (in fact,
   any value v for P(0), together with the value of P at points X1 ... Xt uniquely
   determines a polynomial).

As it can be easily seen, if the dealer is dishonest, he may give each trustee a random
number mod p. If this is the case, then (a) each trustee cannot tell that he has a junk share,
and (b) in phase 2 there will be no secret to reconstruct. The consequence of this is that
secret sharing is more useful in those occasions in which the dealer is certainly honest, for
instance, because being honest is *in his own interest.* (A user that encrypts his own files
with a secret key has a big interest in properly secret sharing his key with, say, a group of
colleagues: if he accidentally looses it, he needs to reconstruct it!) Secret sharing alone,
instead, cannot be too useful for building Fair Cryptosystems: we cannot expect that a
criminal give proper shares of his secret key to some federal judges when the only purpose
of his doing this is allowing the authorities, under a court order, to understand his
communications!

VERIFIABLE SECRET SHARING
A closer connection exists between Fair PKCs and verifiable secret sharing (VSS)
protocols. While the two concepts are not identical, a special type of VSS can be used to
build Fair PKCs. As put forward by Awerbuch, Chor, Goldwasser, and Micali [CGMA],
a verifiable secret sharing (VSS) scheme is a scheme that, while guaranteeing both the
redundancy and the privacy property, overcomes the "honesty problem." In fact, in a VSS
scheme each trustee can *verify* that the share given to him is genuine *without knowing at all
the shares of other trustees or the secret itself.* That is, he can verify that, if T verified
shares are revealed in phase 2, the original secret will be reconstructed, no matter what the
dealer or dishonest trustees might do.

EXAMPLE (Goldreich, Micali, and Wigderson [GMW1])

Assume that a PKC is in place and let Ei be the public encryption function of trustee i. Then, as in Shamir's scheme, the dealer selects a random polynomial P of degree t such that P(0)=the secret, and gives each trustee the n-vector of encryptions E1(P(1)) E2(P(2))...En(P(n)). Trustee i will therefore properly decode P(i), but has no idea about the value of the other shares, and, consequently, whether these shares "define" a unique t-degree polynomial passing through them. The dealer thus proves to each trustee that the following sentence is true "*if you were so lucky to guess all decryption keys, you could easily verify that there exists a unique t-degree polynomial interpolating the encrypted shares.*" Since easily verifying something after a lucky guess corresponds to NP, the above is an "NP sentence." Since, further, the whole of NP is in zero-knowledge [GMW1], the dealer proves the correctness of the sentence, in zero knowledge, to every trustee. This guarantees each trustee that he has a legitimate share of the secret, since he has a legitimate share of P, but does not enable him (or him and other t-1 trustees) to guess what the secret is before phase 2.

VSS AND FAIR PKCs

Assume that each user chooses a secret/public key pair, and then VSS shares his secret key with some federal judges. Does this constitute a Fair PKC? Not necessarily. In a VSS scheme, in fact, the secret may be *unstructured*. That is, each trustee can only verify that he got a genuine share of some secret value, but this value can be "anything." For instance, if the dealer promises that his secret value is a prime number, in an unstructured VSS a trustee can verify that he got a genuine share of some number, but has no assurances that this number is prime.

Unstructured VSS is not enough for Fair PKCs. In fact, the trustees should not stop at verifying that they possess a legitimate share of a "generic" secret number: they should verify that the number they have a share of actually is the decryption key of a given public key! The GMW scheme, as described above, is an unstructured VSS, and thus unsuitable for directly building Fair PKCs. The same is true for other VSS schemes (e.g. the ones of Ben-Or, Goldwasser and Wigderson [BeGoWi]; of Chaum, Crepeau and Damgard [ChCrDa}; and of Rabin and Ben-Or [RaBe], just to mention a few).

Some VSS schemes are *structured*, that is each trustee can further verify that the secret value of which he possesses a genuine share satisfies some additional property. What this property is depends on the VSS scheme used. For instance, Feldman proposes a VSS in which, given an RSA modulus N and an RSA ciphertext $E(m)= m^e$ mod N (of some cleartext message m), the trustees can verify that they do possess genuine shares of the decryption of E(m) (i.e., of m). This scheme is attractive in that it is "non-interactive," but *cannot* be used to hand out in a verifiable way shares of the decryption key of a given public key. In fact,

the trustees have no guarantee that the decryption of E(m) actually consists of N's factorization.

In other words, the trustees can verify that they have genuine shares of the decryption (m) of a ciphertext E(m), but m is *unstructured* (with respect to N's factorization and anything else).

CONSTRUCTING FAIR PKCs WITH A GENERIC VSS

Can a generic VSS scheme be transformed so as to yield Fair PKCs? The answer is YES, but at a formidable cost. All of the above mentioned VSS protocols can be "structured" so that the extra property verifiable by the trustees is that the dealer's secret actually is the decryption key of a given public key. In fact, this can be achieved as an instance of *secure function evaluation* between many parties as introduced by Goldreich, Micali, and Wigderson in a second paper [GoMiWib]. Such secure evaluation protocols are possible, though, more in theory than in practice in light of the complexity of the particular functions involved. In the case of the GMW VSS scheme, since the encryption of all the shares is publicly known, the transformation can actually be achieved by a simpler machinery: an additional zero-knowledge proof. But even in this case the computational effort involved is formidable. Essentially, one has to encode the right statement (i.e., the secret, whose proper shares are the decodings of these public ciphertexts, is the decryption key of this given public key) as a VERY BIG graph, 3-colorable if and only if the statement is true, and then prove, in zero-knowledge, that indeed the graph is 3-colorable. Not only are these transformations of a generic VSS to one with the right property computationally expensive, but they require INTERACTION (on top, if any, of the interaction required by the VSS scheme itself)! All these considerations may rule out constructing Fair PKCs this way in practice. Thus CUSTOM-TAILORED methods should be sought, whenever possible, to transform ordinary PKCs to Fair ones. This is our next goal.

6. Making Fair the Diffie-Hellman Scheme

Let us now exhibit concrete and efficient methods for turning two popular PKCs into Fair ones. We start by making Fair the scheme of Diffie and Hellman, since this is the simplest of the two.

Recall that, a bit differently than in other systems, in Diffie-Hellman's scheme each pair of users X and Y succeeds, without any interaction, in agreeing upon a common, secret key S_{xy} to be used as a conventional single-key cryptosystem. Here is how.

The Ordinary Diffie-Hellman PKC

There are a *prime p* and a *generator* (or high-order element) g common to all users. User X *secretly* selects a random integer Sx in the interval *[1,p-1]* as his private key and publicly announces the integer $Px=g^{Sx} \bmod p$ as his public key. Another user, Y, will similarly select Sy as his private key and announce $Py=g^{Sy} \bmod p$ as his public key. The value of this key is determined as $S_{xy}=g^{Sx.Sy} \bmod p$. User X computes Sxy by raising Y's public key to his secret key mod p; user Y by raising X's public key to his secret key mod p. In fact

$$(g^{Sx})^{Sy}=g^{Sx.Sy}=S_{xy}=g^{Sy.Sx}=(g^{Sy})^{Sx}= \bmod p.$$

While it is easy, given g, p, and x, to compute $y=g^x \bmod p$, no efficient algorithm is known for computing, given y and p, x such that $g^x=y \bmod p$ when g has high enough order. This is, in fact, the famous *discrete logarithm problem*. This problem has been used as the basis of security in many cryptosystems, and in the recently proposed U.S. standard for digital signatures. We now transform Diffie and Hellman's PKC into a fair one. Again, to keep things as simple as possible we imagine that there are 5 trustees and that ALL of them should cooperate to reconstruct a secret key, that is, that ALL shares are needed to reconstruct a secret key. Relaxing this condition involves another idea and will be dealt with in section 5.

A Fair Diffie-Hellman Scheme
(All-Shares Case)

Instructions for the users
Each user X randomly chooses 5 integers $Sx1,...,Sx5$ in the interval *[1,p-1]* and lets Sx be their sum *mod p*. From here on, it will be understood that all operations are modulo *p*. He then computes the numbers

$$t1 = g^{Sx1}, ..., t5 = g^{Sx5} \text{ and } Px = g^{Sx}.$$

Px will be user X's public key and *Sx* his private key. The *ti*'s will be referred to as the *public pieces* of *Px*, and the *Sxi*'s as its *private pieces*. Notice that the product of the public pieces equals the public key Px. In fact,

$$t1 \cdot ... \cdot t5 = g^{Sx1} \cdot ... \cdot g^{Sx5} = g^{(Sx1 + ... + Sx5)} = g^{Sx}.$$

Let T1,...,T5 be the five trustees. User X now gives *Px* and pieces *t1* and *Sx1* to trustee T1, *t2* and *Sx2* to T2, and so on. It is important that piece *Sxi* be privately given to trustee T_i.

Instructions for the trustees
Upon receiving public and private pieces *ti* and *Sxi*, trustee Ti verifies whether $g^{Sxi} = ti$. If so, it stores the pair *(Px,Sxi)*, signs the pair *(Px,ti)*, and gives the signed pair to the key-management center. (Or to user X, who will then give all of the signed public pieces at once to the key-management center.)
Instructions for the key-management center
Upon receiving all the signed public pieces, *t1...t5*, relative to a given public key *Px*, the center verifies that the product of the public pieces indeed equals *Px*. If so, it approves *Px* as a public key, and distributes it as in the original scheme (e.g., signs it and gives it to user X.)

This ends the instructions relative to the keys of the Fair PKC. The encryption and decryption instructions for any pair of users X and Y are exactly as in the Diffie and Hellman scheme (i.e., with common, secret key *Sxy*). It should be noticed that, like the ordinary Diffie-Hellman, the Fair Diffie-Hellman scheme does not require any special hardware and is actually easily to implement in software.

Why does this work?
First, the privacy of communication offered by the system is the same as in the Diffie and Hellman scheme. In fact, the validation of a public key *does not compromise at all* the corresponding private key. Each trustee Ti receives, as a special piece, the discrete logarithm, *Sxi*, of a *random number*, ti. This information is clearly irrelevant for computing the discrete logarithm of *Px*! The same is actually true for any 4 of the trustees taken together, since any four special pieces are independent of the private decryption key *Sx*. Also the key-management center does not possess any information relevant to the private key; that is, the discrete logarithm of *Px*. All it has are the public pieces signed by the trustees. (The public pieces simply are *5* random numbers whose product is *Px*. This type of information is irrelevant for computing the discrete logarithm of *Px;* in fact, any one could choose four integers at random and set the fifth to be *Px* divided by the product of the first four[1]. As for a trustee's signature, this just represents the promise that *someone else* has a secret piece. As a matter of fact, even the information in the hands of the center together with any four of the trustees is irrelevant for computing the private key *Sx*.) Thus, not only is the user guaranteed that the validation procedure will not betray his private key, but he also knows that this procedure has been properly followed because he himself has computed his own keys and the pieces of his private one!

Second, if the key-management center validates the public key *Px*, then the corresponding private key is guaranteed to be reconstructible by the Government in case of a court order.

---

[1] The result would be integral because division is modulo p.

In fact, the center receives all 5 public pieces of *Px*, each signed by the proper trustee. These signatures testify that trustee Ti possesses the discrete logarithm of public piece *ti*. Since the center verifies that the product of the public pieces equals *Px*, it also knows that the sum of the secret pieces in storage with the trustees equals the discrete logarithm of *Px;* that is, user X's private key. Thus the center knows that, if a court order is issued requesting the private key of X, by summing the values received by the trustees, *the Government is guaranteed* to obtain the needed private key.

It should be noticed that, for efficiency considerations, we split the verification of the structure of the secret among trustees and key-management center. In fact a trustee verifying that Sxi is the discrete log of ti cannot possibly verify that Sxi is a share of the secret key of public key Px, since he has never seen Px! (If we wanted we could have defined the public key to consist of Px t1 t2 t3 t4 t5. In this case giving trustee Ti the entire public key and the private piece (share) Sxi, we would have enabled him to verify the structure of the secret as well.)

7. Making Fair the RSA Scheme

Let us now just OUTLINE a custom-tailored method to make the RSA Fair. We will be more precise in the final paper. Our method, while simple algorithmically, does require some more knowledge of number theory. (We wish to note that our effort could be consirerably simplified if we were willing to make Fair not the basic RSA scheme, but some variants of its that essentially exhibit its same security.)

In the basic RSA PKC, the public key consists of an integer N product of two primes and one exponent e (relatively prime with f(N), where f is Euler's totient function). No matter what the exponent, the private key may always be chosen to be N's factorization. Before we show how to make a Fair PKC out of RSA we need to recall some facts from number theory.

Fact 1. Let Z_N^* denote the multiplicative group of the integers between 1 and N which are relatively prime with N. If N is the product of two primes N=pq (or two prime powers: $N=p^a p^b$), then

* a number s in Z_N^* is a square mod N if and only if it has four distinct square-roots mod N: x, -x mod N, y, and -y mod N. (That is, $x^2=y^2=s$ mod N.) Moreover, from the greatest common divisor of +-x+-y and N, one easily computes the factorization of N. Also,

* one in four of the numbers in Z_N^* is a square mod N.

Fact 2. Among the integers in Z_N^* is defined a function easy to evaluate, the Jacobi symbol, that evaluates to either 1 or -1. The Jacobi symbol of x is denoted by (x/N). The Jacobi symbol is multiplicative; that is, (x/N)(y/N)=(xy/N). If N is the product of two primes N=pq (or two prime powers: $N=p^a p^b$), and p and q are congruent to 3 mod 4, then, letting x, -x, y, and -y mod N be the four square roots of a square mod n, (x/N)=(-x/N)=+1 and (y/N)=(-y/N)=-1. Thus, because of fact 1, if one is given a Jacobi symbol 1 root and a Jacobi symbol -1 root of any square, he can easily factor N.

We are now ready to describe how the RSA cryptosystem can be made fair in a simple way. For simplicity we again assume that we have 5 trustees and that *all* of them must collaborate to reconstruct a secret key, while no 4 of them can even predict it.

A Fair RSA Scheme
(All-Shares Case)

Instructions for the user
A user chooses P and Q primes and congruent to 3 mod 4 as his private key, and N=PQ as his public key. Then he chooses 5 Jacobi 1 integers X_1 X_2 X_3 X_4 and X_5 at random in Z_N^* and computes their product, X, and X_i^2 mod N for all i=1,...,5. The product of these 5 squares, Z, is itself a square. One square root of Z mod N is X, which has Jacobi symbol equal to 1 (since the Jacobi symbol is multiplicative). The user thus computes Y one of the Jacobi -1 roots mod N. X_1...X_5 will be the public pieces of public key N, and the X_is its private pieces. The user gives trustee T_i private piece X_i (and possibly the public piece).

Instructions for the trustees
Trustee Ti checks that X_i has Jacobi symbol 1 mod N, then he squares X_i mod N, gives the key-management center his signature of X_i^2 mod N, and stores X_i and X_i^2 (or X_i and N).

Instructions for the key-management center
The center first checks that (-1/N)=1, that is, that for all x: (x/N)=(-x/N); which is partial evidence that N is of the right form. Upon receiving the valid signature of the public pieces of N and the Jacobi -1 value Y from the user, the center checks whether, mod N, the square of Y equals the product of the 5 public pieces. If so, the center is now guaranteed that it has a *split* of N. To make sure that it actually has the *complete factorization* of N, it must now perform the *missing procedure* (i.e., a procedure whose description we temporarily postpone) to check that N is the product of two prime powers. If this is the case, it *approves* N.

Again, it should be noticed that the Fair RSA scheme can be conveniently implemented in software.

Why does this work?
The reasoning behind the scheme is the following. The trustees' signatures of the X_i^2's (mod N) guarantee the center that every trustee Ti has stored a Jacobi symbol 1 root of X_i^2 mod N. Thus, in case of a court order, all these Jacobi symbol 1 roots can be retrieved. Their product mod N will also have Jacobi symbol 1, since this function is multiplicative, and will be a root of X^2 mod N. But since the center has verified that $Y^2 = X^2$ mod N, one would have two roots X and Y of a common square mod N; moreover, Y is different from X since it has a different Jacobi symbol, and is also different from -x, since (-x/N)=(x/N); in fact: (a) (-1/N) has been checked to be 1 and (b) the Jacobi symbol is multiplicative. Possession of such square roots, by Facts 1 and 2, is equivalent to having the factorization of N, *provided that N is a product of at most two prime powers*. That's why this last property has also been checked by the center before it approved N.

The reason that 4 (or less) trustees cannot factor N with the information in their possession is similar to the one of the discrete log scheme. Namely, the information in their possession solely consists of 4 random squares and their square roots mod N. This cannot be of any help in factoring N, since anybody could randomly choose 4 integers in Z_N^* and square them mod N.

The missing procedure

The center can easily verify that N is not prime. It can also easily verify that N is not a prime power by checking that N is not of the form x^y, for x and y positive integers, y>1. In fact, for each fixed y one can perform a binary search for x, and there are at most $\log_2(N)$ y's to check, since x must be at least 2 if N>1. It is thus now sufficient to check that N is the product of at most 2 prime powers. Since no efficient algorithm is known for this task when N's factorization is not known, any such check must involve the user who chose N, since he will be the only one to know N's factorization. In the spirit of what we have done so far, we seek a verification method that is (1) *simple*, (2) *non-interactive*, and (3) *provably safe*. The key to this is the older idea of Goldwasser and Micali of counting the number of prime divisors of N by estimating the number of quadratic residues in $Z_N{}^*$. In fact, if N is the product of no more than two prime powers, at least one number in four is a square mod N, otherwise at most 1 in 8 is. Thus the user can demonstrate that N has at most two different prime divisors by computing and sending to the center a square root mod N for at least, say, 3/16 of the elements of a prescribed list of numbers that are guaranteed to be randomly chosen. This list may be taken to be part of the system. Requiring the user to give the square roots of those numbers in such a random sequence that are squares mod N does not enable the center --or anybody else for that matter-- to easily factor N. To make this idea viable one would need some additional details. For instance, the trustees may be involved in choosing this public sequence so as to guarantee to all users the randomness of their elements; also the sequence should be quite long, else a user may "shop around" for a number N' that, though product of --say-- 3 prime powers, is such that at least 3/16 of the numbers in the sequence are squares modulo it; and so on. In "practice" this idea can be put to work quite efficiently by one-way hashing the user's chosen N to a small "random" number H(N), where H is a publicly known one-way hash function, and then generating a sufficiently long sequence of integers S(N) by giving H(m) as a seed to a reasonable pseudo-random number generator. This way, the number sequence may be assumed to be random enough by everybody, since the user cannot really control the seed of the generator. Moreover, the sequence changes with N, and thus a dishonest user cannot shop around for a tricky N as he might when the sequence is chosen before hand. Thus, the sequence chosen may be much shorter than before. If a dishonest user has chosen his N to be the product of three or more prime powers, then it would be foolish for him to hope that roughly 1/4 of the integers in the sequence are squares mod N. The scheme is of course non-interactive, since the user can compute on his own H(N), the number sequence S(N), and the square roots mod N of those elements in S(N) that are quadratic residues, and then sends the center only N and the computed square roots. Given N, the center will compute on its own the same value H(N) and thus the same sequence S(N). Then, without involving the user at all, it will check that, by squaring mod N the received square roots, it obtains a sufficiently high number of elements in S(N).

8. Basic Variants of the Basic Notion

Independent of the underlying PKC, several variants of the notion of a Fair PKC are possible, each, of course, possessing its own advantages and disadvantages, either in efficiency or fairness. Here, let us briefly discuss two important variants and then just mention a few others.

8.1 Relying on Fewer Shares

The schemes developed so far are robust only in the sense that some trustees, accidentally or maliciously, may reveal the shares in their possession without compromising the security of the system. However, our schemes so far rely on the fact that the trustees will collaborate during the recovering stage. In fact, we insisted that all of the shares should be needed for recovering a secret key. This may be disadvantageous, either because some trustees may after all be untrustworthy and refuse to give the Government the key in their possession, or because, despite all file back-ups, they may have genuinely lost the information in their possession. Whatever the reason, in this circumstance the reconstruction of a secret key will be prevented. Since VSS protocols exist (such as the GMW one) which tolerate any minorities of trustees to be bad, this problem can, in principle, be solved. However, the cost to be paid would be very very high, *independently of whether or not the number of trustees is small*. Thus, once again, one should resort to direct constructions. The ones discussed below have been selected because of their *simplicity*, their being quite practical whenever the number of trustees is small (in particualr they continue to be non-interactive), and their sufficient generality (though they will be illustrated only in the context of a single PKC). Slicker solutions can be obtained, but at the expense of greater complications. (One such method has been recently developed by Sidney based on a previous construction of Feldman [Fe87].)

THE SUBSET METHOD.

Each Fair PKC described so far is based on a (properly structured, non-interactive) VSS scheme with parameters n=5, T=5 and t=4. It may be preferable to have different values for our parameters; for instance, n=5, T=3, and t=2. That is, any majority of the trustees can recover a secret key, while no minority of trustees can predict it at all. This is achieved as follows (and it is easily generalized to any desired values of n,T and t in which T>t). We confine ourselves to exemplifying our method in conjunction with the Diffie-Hellman scheme. The same method essentially works for the RSA case as well.

The Subset Method for the Diffie-Hellman scheme

After choosing a secret key Sx in [1,p-1], user X computes his public key $Px=g^{Sx}$ mod p. (All computations from now on will be mod p.) User X now considers all triplets of numbers between 1 and 5: (1,2,3), (2,3,4), etc.
For each triplet (a,b,c), he randomly chooses *3* integers *S1abc,...,S3abc* in the interval *[1,p-1]* so that their sum *mod p* equals Sx. Then he computes the 3 numbers

$$t1abc=g^{S1abc}, \quad t2abc=g^{S2abc}, \quad t3abc=g^{S3abc}$$

The *tiabc'*s will be referred to as *public pieces* of *Px*, and the *Sxiabc'*s as *private pieces*. Again, the product of the public pieces equals the public key Px. In fact,

$$t1abc \cdot t2abc \cdot t3abc = g^{S1abc} \cdot g^{S2abc} \cdot g^{S3abc} =$$
$$= g^{(S1abc+ S2abc +S3abc)} = g^{Sx} = Px$$

User X then gives trustee Ta *t1abc* and *S1abc*, trustee Tb *t2abc* and *S2abc*, and trustee Tc *t3abc* and *S3abc*, always specifying the triplet in question.

Upon receiving these quantities, trustee Ta (all other trustees do something similar) verifies that $t1abc=g^{S1abc}$, signs the value (Px,t1abc,(a,b,c)) and gives the signature to the key management center.

The key-management center, for each triple (a,b,c), retrieves the values *t1abc t2abc* and *t3abc* from the signed information received from trustees Ta, Tb and Tb. If the product of these three values equals Px and the signatures are valid, it approves Px as a public key.

The reason the scheme works, assuming that at most 2 trustees are bad, is that all secret pieces of a triple are needed for computing (or predicting) a secret key. Thus no secret key in the system can be retrieved by any 2 trustees. On the other hand, when after a court order, at least 3 trustees reveal all the secret pieces in their possession about a given public key, the Government has all the necessary secret pieces for at least one triple, and thus can compute easily the desired secret key.

THE SHARE REPLICATION METHOD.

In this solution, each of the 5 trustees is replaced by a group of new trustees. For instance, instead of a single trustee T_1, there may be 3 trustees, $T_1{}^1 \; T_2{}^1 \; T_3{}^1$; each of these trustees will receive and check the same share of trustee T_1. Thus, it is going to be very unlikely that all 3 trustees will refuse to surrender their copy of the first share. This scheme is a bit "trustee-wasteful" since it requires 15 trustees while it is enough that an adversary corrupts 5 of them to defeat the scheme. (However, one should appreciate that defeating the share-replication scheme is not as easy as corrupting any 5 trustees out of 15, since it must be true that a trustee is corrupted in each group.) The scheme has, nonetheless, two strong advantages: (1) *Scalability*: denoting by n the number of trustee groups, the computational effort of the scheme grows polynomially in n, no matter what the group size is, and thus -- if desired-- one can choose a large value for n; (2) *Repetitiveness*: if there are n trustee groups of size k each, one should only perform n "operations," in fact, each member of a trustee group gets a "xerox copy" of the same computation.

In the final paper we shall demonstrate that both methods can be optimized, but here let us instead move on to consider a far more important problem than efficiency.

8.2 Making Trustees Oblivious

There is another point that requires attention. Namely, a trustee requested by a court order to surrender his share of a given secret key may alert the owner of that key that his communications are going to be monitored. This serious problem can be attacked by a general-purpose machinery, yielding a purely theoretical solution. But, here, let us outline a simple and practical one, available when the cryptosystem used by the trustees possesses a nice algebraic property (essentially, *random self-reducibility* as introduced by Blum and Micali [BlMi]). This practical strategy is exemplified below by making oblivious (and Fair) the Diffie-Hellman scheme for the "all-shares" case, but also works for the RSA scheme and for fewer shares.

Oblivious and Fair Diffie-Hellman Scheme
(All-Shares Case)

The trustees' encryption algorithms
Since RSA itself possesses a sufficient algebraic property, let us assume that all trustees use *deterministic* RSA for receiving private messages. Thus, let Ni be the public RSA modulus of trustee Ti and ei his encryption exponent (i.e., to send Ti a message m in encrypted form, one would send m^{ei}mod Ni.)

Instructions for user U
User U prepares his public and secret key, respectively Px and Sx (thus $Px = g^{Sx} \bmod p$), as well as his public and secret pieces of the secret key, respectively ti and Sxi's (thus Px= $t1 \cdot t2 \cdot \ldots \cdot t5 \bmod p$ and $ti = g^{Sxi} \bmod p$ for all i). Then he gives to the key-management center Px, all of the ti's and the n values Ui=$(Sxi)^3 \bmod Ni$; that is, he encrypts the i-th share with the public key of trustee Ti.

(Comment: Since the center does not know the factorization of the Ni's this is no useful information to predict Sx, nor can it verify that the decryption of the n ciphertexts are proper shares of Sx. For this, the center will seek the cooperation of the n trustees, but without informing them of the identity of the user.)

Instructions for the center/trustees
The center stores the values tj's and Uj's relative to user U and then forwards Ui and ti to trustee Ti. If every trustee Ti responds to have verified that the decryption of Ui is a proper private piece relative to ti, the center approves Px.

Instructions in case of a court order
To lawfully reconstruct secret key Sx without leaking to a trustee the identity of the suspected user U, a judge (or another authorized representative) randomly selects a number $Ri \bmod Ni$ and computes $yi = Ri^{ei} \bmod Ni$. Then, he sends trustee Ti the value $zi = Ui \cdot yi \bmod Ni$, asking with a court order to compute and send back wi, the *ei*-th root of zi mod Ni. Since zi is a random number mod Ni, no matter what the value of Ui is, trustee Ti cannot guess the identity of the user U in question. Moreover, since zi is the product of Ui and yi mod Ni, the ei-th root of zi is the product mod Ni of the ei-th root of Ui (i.e., Sxi) and the ei-th root of yi (i.e., Ri). Thus, upon receiving wi, the judge divides it by yi mod Ni, thereby computing the desired Sxi. The product of these Sxi's equals the desired Sx.

8.3 Time-Bounded Court-Authorized Eavesdropping

At present the Citizens have no guarantees that an illegal wiretapping will be initiated, or that a legitimate eavesdropping will be stopped at the prescribed date --indeed, courts usually authorize line-tapping for a bounded length of time only.

Fair PKCs are preferable to the *status quo* : the users are guaranteed that no illegal wire-tapping will be initiated, because without the help of the trustees their cryptosystems are impenetrable. Fair PKCs, however, are just as "bad" as the current system with respect to the time-bound issue. In fact, once the private key of the user of a Fair PKC erroneously suspected of unlawful activities is reconstructed, thanks to the collaboration of the trustees in response to a legitimate court order, it would be very easy for the agent monitoring her conversations (say, the Police) to exceed its mandate and keep on tapping (or allow someone else to tap) her line for a longer period of time.

Because it is our goal to strike a better balance between the needs of the Government and those of the Citizens in a modern democracy, we have developed various strategies for improving on the *status quo* and removing this weakness altogether.

8.3.1 Multiple Public-Keys

A very simple way to ensure time-bounded court-authorized line tapping consists of having each user choose a sufficient amount of matching public and secret keys, say one per month. Each public key will then be publicized specifying the month to which it refers. Someone who wants to send user X a private message in March, will then encrypt it with X's public March key. If this level of granularity is acceptable, the court may then ask the trustees to reveal X's secret keys for a prescribed set of months.

The disadvantage of this approach is that it requires a rather large "total public key," and it may be totally impractical if a fine granularity is desired.

8.3.2   Tamper-Proof Chips

One simple method to ensure time-bounded court-authorized eavesdropping makes use of *secure* chips; these are special chips that cannot be "read" from the outside, and cannot be tampered with. Thus, in particular, upon receiving an input they produce a specific output, but effectively hide all intermediate results. (Such chips are central to the Clipper Chip proposal.)

Time-bounded legal eavesdropping can be achieved by having the Police use secure chips possessing an internal and thus untamperable clock, the *Polchips*, in order to monitor the communications of a suspected user. Assume that a proper court order is issued to tap the line of user X from February to April. Then, each trustee will send the Polchip a digitally signed message consisting of his own share of user X's private key (encrypted so that only the Polchip will understand it). The Polchip can now easily compute X's secret key. Thus, if the Court sends to the Polchip a signed message consisting of, say, "decode, X, February-April"[1], since the Polchip has an internal clock, it can easily decrypt all messages relative to X for the prescribed time period. Then, it will destroy X's secret key, and, in order to allow further line tapping, a new court order will be required.

A main advantage of this approach is its simplicity; it does, however, require some additional amount of trust. In fact, the citizens cannot check, but must believe, that each Polchip is manufactured so as to work as specified above.

8.3.3 Algorithmically-Chosen Session Keys

In the multiple public-key method described above, each user selected and properly shared with the Trustees a number of secret keys of a PKC equal to the number of possible transmission "dates" (in the above example, each possible month). Within each specified date, the same public-secret key pair was used for directly encrypting and decrypting any message sent or received by any user. Time-bounded Fair PKCs, however, can be more efficiently achieved by using public keys only to encrypt session keys, and session keys to encrypt real messages (by means of a conventional single-key system). This is, in fact, the most common and efficient way to proceed.

Session keys are usually unique to each pair of users and date of transmission. Indeed, if each minute or second is considered a different date, there may be a different session key for every transmission between two users. Abstractly, the date may just be any progressive number identifying the transmission, but not necessarily related to physical time.

To achieve time-bounded court-authorized line tapping, we suggest to choose session keys *algorithmically* (so that the Trustees can compute each desired session key from information received when users enter the system), but *unpredictably* (so that, though some session keys may become known --e.g., because of a given court order-- the other session keys remain unknown).

The particular mechanics to exploit this approach is, however, important, because not all schemes based on algorithmically selected session keys yield equally convenient time-bounded Fair PKCs.[2]

---

[1] Alternatively, the time interval can be specified in the message of the trustees, since they learned it from the Court anyway.

[2] For instance, a time-bounded FAIR PKC that required the Police to contact the Trustees specifying the triplet (X,Y,D) in order to understand X's communication to Y at time D (belonging to the court-authorized time interval), might be deemed inpractical. A better scheme may allow the Police to contact the Trustees only once, specifying only X, Y, and D1 and D2,

An effective method is described below, basic properties first and technical details later.

The high-level mechanics of our Suggestion
In presence of a court order to tap X's lines beween dates D1 and D2, no matter how many dates there may be between D1 and D2, our method allows the Trustees to easily compute and give the Police a small amount of information, i=i(X,D1,D2), that makes it easy to tap X's lines in the specified time interval. The method consists of using a Fair PKC **F** together with a special additional step for selecting session keys for a conventional single-key cryptosystem **C**. In our suggested method, call it the *(F,C) method*, for any users X and Y, and any date D, there is a session key SXDY for enabling X to send a private message to Y at time D. Each user X is asked to provide the trustees not only with proper shares of his secret key in **F**, but also with *additional pieces of information* that enable them, should they receive a legitimate court order for tapping X between dates D1 and D2, to compute easily i(X,D1,D2) and hand it to the Police.

While the trustees can verify that they possess correct shares of X's secret key in **F**, we do not insist that the same holds for X's session keys. This decreased amount of verifiability is not crucial in this context for the following reasons. Assume in fact that the Police, after receiving i(X,D1,D2) from the Trustees in response to a legitimate court order, is unable to reconstruct a session key of X during the given time interval. This inability proves that X did not originally give the Trustees the proper additional pieces of information about his session keys. If so, the protocol will then ask the cooperation of the Trustees so as to reconstruct X's secret key in **F** (which is guaranteed possible since the trustees could verified to have legitimate shares of that key). Consequently, from that point on, all messages sent to X will cease to be private. Moreover, the adoption of a proper "hand-shaking protocol" will ensure the Police to understand all messages sent by X to any user who replies to him in the **(F,C)** system.[1]

In sum, therefore, malicious users who want to hide their conversations from law-enforcement agents even in presence of a court order, cannot do so by taking advantage of

---

in order to understand all the communications between X and Y at any date D in the time interval (D1,D2). Since, however, there may be quite many users Y to which the suspected user X talks to, also this scheme may be considered impractical.

[1] Of course, one may object that nothing is guaranteed about conversations between two users that are both malicious, since they may be using their own, altogether-different cryptosystems. Once more, however, we should remember that this is impossible to prevent, unless use of non-government-approved cryptosystems is made illegal. It is instead important to realize that, though all good citizens can enjoy a nation-wide PKC, the Government is at least guaranteed to have done NOTHING to facilitate private communications between malicious users. In fact, they cannot use **F** to exchange session keys for the recommended conventional cryptosystem **C**, since after reconstructing the relevant secret keys of **F** the Government could reconstruct such session keys and understand what any two malicious users would be saying to each other via **C**. Nor can all malicious users use **F** for exchanging secret keys relative to a special conventional cryptosystem **C'** that is known to criminals but unknown to the Government. In fact, any conventional cryptosystem that is used by a sufficiently large group of people will eventually become known to the Government. On the other hand, if each pair of malicious users X and Y were to use a dedicated conventional cryptosystem **Cxy** to talk to each other, they would have no convenience to gain from using the society-provided public-key cryptosystem **F**! In fact, if they could establish beforehand (i.e., without using **F**) a common and secret cryptosystem **Cxy**, they might as well exchange (without using **F**) a common secret key Kxy to be used with any conventional cryptosystem.

the convenience of a nation-wide $(\mathbf{F},\mathbf{C})$ system. They must go back to the cumbersome practice of exchanging common secret keys before hand, outside any major communication network. It is my firm opinion that the amount of illegal business privately conducted in this cumbersome way should be estimated minuscule with the respect to the one that might be conducted via a nation-wide *ordinary* PKC.

The Specifics Of Our Suggestion

The hand-shaking protocol of our suggested $(\mathbf{F},\mathbf{C})$ cryptosystem is the following. When X wants to initiate a secret conversation with Y at date D, she computes a secret session key SXDY and sends it to Y using the Fair PKC $\mathbf{F}$ (i.e., encrypts it with Y's public key in $\mathbf{F}$). User Y then computes his secret session key SYDX and sends it to X after encrypting it with the received secret key SXDY (by means of the agreed-upon conventional cryptosystem $\mathbf{C}$). User X then sends SYDX to Y by encrypting it with SXDY. Throughout the session, X sends messages to Y conventionally encrypted with SXDY, and Y sends messages to X via SYDX. (If anyone spots that the other disobeys the protocol the communication is automatically terminated, and an alarm signal may be generated.) Thus in our example, though X and Y will understand each other perfectly, they will not be using a common, conventional key. Notice that, if the Police knows SXDY (respectively, SYDX), it will also know SYDX (respectively, SXDY).

Assume now that the Court authorizes tapping the lines of user X from date D1 to date D2, and that a conversation occurs at a time D in the time interval [D1,D2] between X and Y. The idea is to make SXDY available to the Police in a convenient manner, because knowledge of this quantity will enable the Police to understand X's out-going and in-coming messages, if the hand-shaking has been performed, independently of whether X or Y initiated the call. To make SXDY conveniently available to the Police, we make sure that it is easily computable on input SXD, a master secret key that X uses for computing his own session key at date D with every other user. For instance, $SXDY = H(SXD,Y)$, where H is a one-way (possibly hashing) function.

Since there may be many dates D in the desired interval, however, we make sure that SXD is easily computable from a short string, $i(X,D1,D2)$, immediately computable by the Police from the information it receives from the Trustees when they are presented with the court order "tap X from D1 to D2." For instance, in a 3-out-of-3 case, if we denote by $i_j(X,D1,D2)$ the information received by the Police from Trustee j in response to the court order, we may set

$$i(X,D1,D2)= H(i_1(X,D1,D2), i_2(X,D1,D2), i_3(X,D1,D2)),$$

where H is a one-way (preferably hashing) function. Now, we must specify one last thing: what should $i_j(X,D1,D2)$ consist of? Letting X_j be the value originally given to Trustee j by user X when she entered the system (i.e., X gives X_j to Trustee j together with the j-th piece of her own secret key in the FAIR PKC $\mathbf{F}$), we wish that $i_j(X,D1,D2)$ easily depend on X_j. Let us thus describe effective choices for X_j, $i_j(X,D1,D2)$, and SXD. Assume that there are 2^d possible dates. Imagine a binary tree with 2^d leaves, whose nodes have n-bit identifiers --where n=0,...,d. Quantity $i_j(X,D1,D2)$ is computed from X_j by storing a value at each of the nodes of our tree. The value stored at the root, node Ne (where e is the empty word), is X_j. Then a *secure* function G is evaluated on input X_j so as to yield two values, X_j0 and X_j1. The effect of G is that the value X_j is unpredictable given X_j0 and X_j1. (For instance, X_j is a random k-bit value and G is a secure pseudo-random number generator that, using X_j as a seed, outputs 2k bits: the first k will constitute value X_j0, the second k value X_j1.) Value X_j0 is then stored in the left child of the root (i.e., it is stored in node N0) and value X_j1 is stored in the right child of the root (node N1). The values of below nodes in the tree are computed using G and the value stored in their ancestor in a similar way. Let SX_iD be the value stored in leaf D (where D is a n-bit date) and $SXD=H(SX_1D,SX_2D,SX_3D)$. If D1 < D2 are n-bit dates, say that a node N *controls* the interval [D1,D2] if every leaf in the tree that is a descendent of N belongs to [D1,D2], while no proper ancestor of N has this property. Then, if $i_j(X,D1,D2)$ consists of the (ordered) sequence of values stored in the nodes that control [D1,D2], then

I. $i_j(X,D1,D2)$ is quite short (with respect to the interval [D1,D2]), and

II.   For each date D in the interval [D1,D2], the value SX_jD stored in leaf D is easily computable from $i_j(X,D1,D2)$, and

III.   The value stored at any leaf not belonging to [D1,D2] is not easily predictable from $i_j(X,D1,D2)$.

Thus if each user X chooses her X_j values (sufficiently) randomly and (sufficiently) independently, the scheme has all the desired properties. In particular,

1.   user X computes SXD very efficiently for every value of D.

2.   When presented with a court order to tap the line of user X between dates D1 and D2, each Trustee j quickly computes $i_j(X,D1,D2)$. (In fact, he does not need to compute all values in the 2^n-node tree, but only those of the nodes that control [D1,D2].)

3.   Having received $i_j(X,D1,D2)$ from every trustee j, the Police can, *very quickly* and *without further interaction with the Trustees,* compute
(3.1)SX_jD from $i_j(X,D1,D2)$ for every date D in the specified interval (in fact, its job is even easier since the SXiD's are computed in order and intermediate results can be stored)
(3.2) the master secret-session key SXD from the SX_jD's, and
(3.3) the session key SXDY from SXD from any user Y talking to X in the specified time interval.

Note, however, that no message sent or received before or after the time-interval specified by the court order will be intelligible to the Police (unless a new proper court order is issued).

9. Fair PKCs vs. the Clipper Chip

9.1 A Quick Review of the Clipper Chip

Also the Clipper Chip proposal is based on the notion of a set of trustees, but it is primarily aimed at conventional cryptosystems. Under the new proposal, users encrypt messages by means of secure chips (as defined in subsection 8.2). All these chips contain in their protected memory a common classified encryption algorithm E and possess a unique identifier. To "initialize" chip x, two Trustees A and B independently choose a secret number (call ax the secret choice of Trustee A and bx that of Trustee B), and remember which secret choice they have made relative to x. These two numbers are then given (somehow) to a chip factory that computes their exclusive-or, cx, and stores it into the protected memory of the chip. This ends the initialization of chip x. Thus after being initialized, each clipper chip possesses a secret key, whose value is at this point only known to the chip itself, though shares of it are stored with the two trustees. Since the chip is assumed to be tamper-proof, it can be handled and sold without any further precautions after being initialized. Assume now that user X has bought chip x, that user Y has bought an analogous chip y, and that the two users have *somehow* exchanged a common secret key Kxy. To privately send a message m to Y, X inputs m to chip x, which will then use the classified algorithm to (1) encrypt Kxy with key cx, and (2) encrypt message m with key Kxy, and then send both ciphertexts to Y. Y ignores the first ciphertext, but decodes the second one with the same key Kxy so as to obtain m. In case of a court order for monitoring X's conversations, the two trustees will retrieve their respective secret numbers ax and bx, and reveal them to the Police, which will then xor them so as to compute cx, decode the first ciphertext with cx so as to compute Kxy, and finally decode the second ciphertext with Kxy so as to compute m.

9.2 A Potential Weakness of the Clipper Chip

Before making any comparison with Fair PKCs, it should be noted that, in absence of a properly specified protocol, the step of having the trustees send their secret shares of the (future) secret key cx to the factory is a dangerous one. In fact, this step introduces a special party, the factory, that "single-handedly knows" the chip's secret (thus nullifying the very notion of a set of trustees), and is therefore single-handedly capable of tapping X's conversations independently of any court order. Worse, while we can hope that trustees will be chosen so as to be considered trustworthy by most people, the same trust will not presumably be enjoyed by a "factory party."

Though more inconvenient, it would thus be preferable to have trustee A itself first insert secret ax in the protected memory of chip x, and then ship chip x to trustee B so that it can directly insert its own secret bx, and then have the chip itself compute cx.

9.3 Comparison with Fair PKCs.

Though they share a common approach, we believe Fair PKCs to be superior to the Clipper Chip proposal in a variety of ways; in particular,

1.   *Software versus Hardware*

      While Fair PKCs can be implemented in hardware or software, the Clipper Chip requires the use of secure hardware, and thus will drive up the cost of any devise using encryption.[1]

2.   *Citizen Control*

      While in the Clipper Chip the user does not choose all keys on which her privacy depends, in a Fair PKC the user chooses all of her keys (and algorithms for that matter).

---

[3] It should be noted that even though in the particular implementation of time-bounded Fair PKCs of subsection 8.2 we recommend the use of secure hardware, this hardware is used by the legitimate monitoring agent, and thus it does not constitute a direct cost of the users. Moreover there will be much less monitoring agents than users.

On the other hand, the Government has at least as much control as in the Clipper Chip proposal. In either case, in fact, the Trustees have pieces that are guaranteed to be right.

3.   *Flexibility*

Since in a Fair PKC the user chooses and knows all of her keys, it is easy to have the system satisfy convenient additional properties; for instance, relying on fewer shares (in the sense of section 8.1) could be a feature of crucial importance for the Government. As for another example, users may find it advantageous to use the same keys in different contexts (e.g., for their phones at work or at home) even if each of these different contexts has a different set of Trustees. This is not a problem for Fair PKCs; in fact, users, knowing all of their secret keys, can break them into a different set of proper shares, and give different set of shares to different sets of trustees, each time easily proving that they hold legitimate shares. (It should be noticed that, unless an enemy has all the shares of one set of trustees, having some of the shares of both sets is useless.).

4.   *Public-Key*

If the Clipper Chip proposal wants to control crime in an effective manner, it should properly address the public-key scenario. In fact, once a nation-wide public-key distribution center is created[1] --with or without the help of the Government-- it will be easier for criminals to bypass the protection of the Clipper Chip. In fact, having one's encryption key properly publicized (e.g., by a nation-wide 411-like mechanism) may be more crucial and difficult a goal to achieve than entering in possession of a conventional cryptosystem chip. If not specifically forbidden, there will certainly be widely available "alternative" conventional-cryptosystem chips for use in conjunction with the publicly-available PKC. It is thus crucial for law-enforcement, in my opinion, to make sure that any public encryption key of a national PKC cannot be used to encrypt messages in a way that avoids court-authorized line tapping. This is the best way to extend to the field of encryption the proper system of "checks-and-balances" necessary in a democracy.

10. Final Thoughts

Fair PKCs are a new technical tool possessing the potential to improve on the *status quo*. Society must though decide which is the best way to use such a tool. Who should the Trustees be? How many should they be? For how long should line-tapping be authorized? We believe that answering questions like these requires a debate as public and wide as possible.

Acknowledgments

I wish to thank Dennis Branstad, Michael Dertouzos, John Deutch, Shafi Goldwasser, Oded Goldreich, Carl Keysen, and my wife Daniela for their comments and encouragement.

---

[4] A not unlikely event since it provides the most convenient way to achieve private communication.

References

[AwChGoMi] B. Awerbuch, B. Chor, S. Goldwasser and S. Micali. Verifiable Secret Sharing and Achieving Simultaneity in the Presence of Faults. In *Proceedings of the 26th Annual IEEE Symposium of Foundations of Computer Science.* IEEE, New York, 1986, pp. 383-395.

[Be] J. Benaloh. Secret Sharing Homomorphisms: Keeping Shares of a Secret Secret. Advances in Cryptology --Proceedings of Crypto '86. Springer Verlag, 1986.

[BeGoWi] M. Ben-Or, S. Goldwasser, and A. Wigderson. Completeness Theorems for Fault-Tolerant Distributed Computing. In *Proceedings of the 20th ACM Symposium of Theory of Computing.* ACM, New York, 1988, pp. 1-10.

[Bl] G. Blakley. Safeguarding Cryptographic Keys. In *AFIPS - Conference Proceedings.* NCC, New Jersey, 1979, Vol. 48 (June), pp. 313-317.

[BlMi] M. Blum and S. Micali. How to Generate Cryptographically Strong Sequences of Pseudo-Random Bits. *Siam Journal on Computing,* 1984, vol. 13 (Novenber), pp. 850-863.
Proceeding Version: FOCS 1982

[ChCrDa] D. Chaum, C. Crepeau, and I. Damgard. Multi-party Unconditionally Secure Protocols. In *Proceedings of the 20th ACM Symposium of Theory of Computing.* ACM, New York, 1988, pp. 11-19.

[DiHe] W. Diffie and M. Hellman. New Directions in Cryptography. *IEEE Trans. Inform. Theory.* IT-22, 6 (Nov. 1976), IEEE, New York, pp. 644-654.

[Fe87] P. Feldman. A Practical Scheme for Non-Interactive verifiable Secret Sharing. In *Proceedings of the 28th Annual IEEE Symposium of Foundations of Computer Science.* IEEE, New York,1987, pp. 427-438.

[GoMi] S. Goldwasser and S. Micali. Probabilistic Encryption. *Journal of Computer Systems Science.* Academic Press, New York, Vol. 28 No. 2 (1984), pp. 270-299.

[GoMiWia] O. Goldreich, S. Micali, and A. Wigderson. Proofs that yield Nothing but their Validity and a Methodology of Cryptographic Protocol Design. In *Proceedings of the 27th Annual IEEE Symposium of Foundations of Computer Science.* IEEE, New York,1986, pp. 174-187.

[GoMiWib] O. Goldreich, S. Micali, and A. Wigderson. How To Play ANY Mental Game or A Completeness Theorem for Protocols with Honest Majority. In *Proceedings of the 19th Annual ACM Symposium of Theory of Computing.* ACM, New York, 1987, pp. 218-229.

[RaBe] T. Rabin and M. Ben-Or. Verifiable Secret Sharing and Multiparty Protocols with Honest Majority. In *Proceedings of the 21st ACM Symposium of Theory of Computing.* ACM, New York, 1989, pp. 73-85.

[RSA] R. Rivest, A. Shamir, and L. Adleman. A Method for Obtaining Digital Signatures and Public-Key Cryptosystens. Comm. ACM 21, 2 (Feb. 1978), pp. 120-126.

[Sh] A. Shamir. How to Share a Secret. *Communications of the ACM.* ACM, New York, 1979, Vol. 22, No. 11 (Nov.), pp. 612-613.

Pricing via Processing
or
Combatting Junk Mail

Cynthia Dwork and Moni Naor

IBM Almaden Research Center
650 Harry Road
San Jose, CA 95120

Abstract. We present a computational technique for combatting junk mail in particular and controlling access to a shared resource in general. The main idea is to require a user to compute a moderately hard, but not intractable, function in order to gain access to the resource, thus preventing frivolous use. To this end we suggest several *pricing functions*, based on, respectively, extracting square roots modulo a prime, the Fiat-Shamir signature scheme, and the Ong-Schnorr-Shamir (cracked) signature scheme.

1   Introduction

Recently, one of us returned from a brief vacation, only to find 241 messages in our reader. While junk mail has long been a nuisance in hard (snail) mail, we believe that electronic junk mail presents a much greater problem. In particular, the ease and low cost of sending electronic mail, and in particular the simplicity of sending the same message to many parties, all but invite abuse. In this paper we suggest a computational approach to combatting the proliferation of electronic mail.[1] More generally, we have designed an *access control mechanism* that can be used whenever it is desirable to restrain, but not prohibit, access to a resource.

Two general approaches have been used for limiting access to a resource: legislation and usage fees. For example, it has been suggested that sending an unsolicited FAX message should be a misdemeanor. This approach encounters obvious definitional problems. Usage fees may be a deterrent; however, we do not want a system in which to send a letter or note between friends should have a cost similar to that of a postage stamp; similarly we do not wish to charge a high fee to transmit long files between collaborators. Such an approach could lead to underutilization of the electronic medium.

Since we believe the real cost of using the medium will not serve as a deterrent to junk mail, we propose a system that imposes another type of cost on transmissions. These costs will deter junk mail but will not interfere with other uses of the system. The main idea is for the mail system to require the

---

[1] A simple solution, due to Blum and Micali [1], is simply not to read one's mail. We have another solution.

sender to compute some moderately expensive, but not intractable, function of the message and some additional information. Such a function is called a *pricing function*.

In the more general setting, in which we have an arbitrary resource and a resource manager, a user desiring access to the resource would compute a moderately hard function of the *request id*. (The request id could be composed of the user's identifier together with, say the date and time of the request.)

The pricing function may be chosen to have something like a trap door: given some additional information the computation would be considerably less expensive. We call this a *shortcut*. The shortcut may be used by the resource manager to allocate cheap access to the resource, as the manager sees fit, by bypassing the control mechanism. For example, in the case of electronic mail the shortcut permits the post office to grant bulk mailings at a price chosen by the post office, circumventing the cost of directly evaluating the pricing function for each recipient.

We believe our approach to be of practical interest. It also raises the point that, unlike the situation with one-way functions, there is virtually no complexity theory of moderately hard functions, and therefore yields excellent motivation for the development of such a theory.

The rest of this paper is organized as follows. Section 2 contains a description of the properties we require of pricing functions. Section 3 focusses on combatting junk mail. Section 4 describes three possible candidates for pricing functions. We require a family of hash functions satisfying certain properties. Potentially suitable hash functions are discussed in more detail in Section 5. Section 6 contains conclusions and open problems.

2   Definitions and Properties

We must distinguish between several grades of difficulty of computation. Rather than describe the hardness of computing a function in terms of asymptotic growth, or in terms of times on a particular machine, we focus on the *relative* difficulty of certain computational tasks.

We require three classes of difficulty: *easy*, *moderate*, and *hard*. The term *moderate* can be viewed in two different ways. As an upper bound, it means that computation should be at *most* moderately hard (as opposed to hard); as a lower bound it means that computation should be at *least* moderately easy (as opposed to easy). The precise definition of easy and moderate and hard will depend on the particular implementation. However, there must be some significant gap between easy and moderately easy. As usual, *hard* means intractable in reasonable time, such as factoring a 1024-bit product of two large primes.

The functions we consider for implementing our scheme have a *difference parameter* that serves a role analogous to that of a *security parameter* in a cryptosystem. A larger difference parameter stretches the difference between easy and moderate. Thus, if it is desired that, on a given machine, checking that a function has been correctly evaluated should require only, say, .01 seconds of

CPU time, while evaluating the function directly, without access to the shortcut information, should require 10 seconds, the difference parameter can be chosen appropriately.

A function f is a *pricing function* if

1. f is moderately easy to compute;
2. f is not amenable to amortization: given ℓ values $m_1, \ldots m_\ell$, computing $f(m_1), \ldots, f(m_\ell)$ has amortized cost comparable to computing $f(m_i)$ for any $1 \le i \le \ell$;
3. given x and y it is easy to determine if $y = f(x)$.

We use the term "function" loosely: sometimes f will be a relation.

$F = \{f_s\}$ is a *family* of pricing functions indexed by $s \in S \subseteq \{0,1\}^*$, such that S is not hard to sample.

$\mathcal{F} = \{F_k\}$ is a collection of families of pricing functions indexed by a difference parameter k.

It is important not to choose a function that after some preprocessing can be computed very efficiently. Consider the following family of pricing functions F, based on subset sum. The index s is a set of ℓ numbers $a_1, a_2, \ldots a_\ell$, $1 \le a_i \le 2^\ell$, such that 2^ℓ is moderately large. For a given request x, $f_s(x)$ is a subset of $a_1, a_2, \ldots a_\ell$ that sums to x. Computing f_s seems to require time proportional to 2^ℓ. As was shown by Schroepel and Shamir [17], after preprocessing, using only a moderate amount of storage, such problems can be solved much more efficiently. Thus, there could be large difference between the time spent evaluating f_s on a large number k of different inputs, such as would be necessary for sending bulk mail, and k individual computations of f_s from scratch. This is clearly undesirable.

We now introduce the notion of a *shortcut*, similar in spirit to a trapdoor. A pricing function with a shortcut is easy to evaluate given the shortcut. In particular, the shortcut is used for bypassing the access control mechanism, at the discretion of the resource manager.

A collection of families of pricing functions is said to have the *shortcut property* if

1. there exists a polynomial time algorithm A that generates a pair s, c;
2. f_s is a function in $\mathcal{F}$;
3. c is a *shortcut*: computing f_s is easy given c.

Note that since f_s is a pricing function, it is not amenable to amortization. Thus, given s, finding c or an equivalent shortcut, should be hard.

Remark. The consequences of a "broken" function are not severe. For example, if a cheating sender actually sends few messages, then little harm is done; if it sends many messages then the cheating will be suspected, if not actually detected, and the pricing function or its key can be changed.

In the context of junk mail we use hash functions so that we never apply the pricing function to a message, which may be long, but only to its hash value.

Ideally, the hash function should be very easy to compute. However, given m, h, and m', it should not be easy to find m'' closely related to m' such that $h(m'') = h(m)$. For example, if Macy's sends an announcement m of a sale, and later wishes to send an announcement m' of another sale, it should not be easy to find a suffix z such that $h(m' \cdot z) = h(m)$.

Suitable hash functions could be based on DES, subset sum, MD4, and Snefru. We briefly discuss each of these in Section 5.

3   Junk Mail

The primary motivation for our work is combatting electronic junk mail. We envision an environment in which people have computers that are connected to a communication network. The computers may be used for various anticipated activities, such as, for example, updating one's personal database (learning that a check has cleared), subscribing to a news service, and so on. This communication requires no human participation. This is different from the situation when one receives a personal letter, or an advertisement, which clearly require one's attention. Our interest is in controlling mail of this second kind.

The system requires a single pricing function f_s, with shortcut c, and a hash function h. There is a pricing authority who controls the selection of the pricing function and the setting of usage fees. All users agree to obey the authority. There can be any number of trusted agents that receive the shortcut information from the pricing authority. The functions h and f_s are known to all users, but only the pricing authority and its trusted agents know c.

To send a message m at time t to destination d, the sender computes $y = f_s(h(m \cdot t \cdot d))$ and sends y, m, t, d to d. The recipient's mail program verifies that $y = f_s(h(m \cdot t \cdot d))$. If the verification fails, or if t is significantly different from the current time, then the message is discarded and (optionally) the sender is notified that transmission failed. If the verification succeeds and the message is timely, then the message is routed to the reader.

Suppose the pricing function f has no short-cut. In this case, if one wants to write a personal letter, the computation of f_s may take time proportional to the time taken to compose the letter. For typical private use that may be acceptable. In contrast, the computational cost of a bulk mailing, even a "desirable" (not junk) mailing, would be prohibitive, defeating the whole point of high bandwidth communication.

In our approach bulk mail, such as notification of acceptance or rejection from a conference, is sent using the shortcut c, which necessarily requires the participation of the system manager. The sender pays a fee and prepares a set of letters, and one of the trusted agents evaluates the pricing function as needed for all the letters, using the shortcut. Since the fee is chosen to deter junk mail, and not to cover the actual costs of the mailing, it can simply be turned over to the recipients of the message.[2]

---

[2] Another possible scenario would be that in order to send a user a letter, some compu-

Finally, each user can have a *frequent correspondent list* of senders from whom messages are accepted without verification. Thus, friends and relatives could circumvent the system entirely. Moreover, one could join a mailing list by adding the name of the distributor to one's list of frequent correspondents.[3] The list, which is maintained locally by the recipient, can be changed as needed. Thus, when submitting a paper to a conference, an author can add the name of the conference to the list of frequent corresponders. In this way the conference is spared the fees of bulk mailing.

4 Pricing Functions

In this section we list three candidate families of pricing functions. The first one is the simplest, but has no shortcut.

4.1 Extracting Square Roots

The simplest implementation of our idea is to base the difficulty of sending on the difficulty (but not infeasibility) of extracting square roots modulo a prime p. Again, there is no known shortcut for this function.

- **Index:** A prime p of length depending on the difference parameter; a reasonable length would be 1024 bits.
- **Definition of f_p:** The domain of f_p is Z_p. $f_p(x) = \sqrt{x} \bmod p$.
- **Verification:** Given x, y, check that $y^2 \equiv x \bmod p$.

The checking step requires only one multiplication. In contrast, no method of extracting square roots $\bmod p$ is known that requires fewer than about $\log p$ multiplications. Thus, the larger we take the length of p, the larger the difference between the time needed to evaluate f_p and the time needed for verification.

4.2 A Fiat-Shamir Based Scheme

This implementation is based on the signature scheme of Fiat and Shamir [6].

- **Index:** Let $N = pq$, where p and q are primes of sufficient length to make factoring N infeasible (currently 512 bits suffice). Let $y_1 = x_1^2, \ldots, y_k = x_k^2$ be k squares modulo N, where k depends on the difference parameter. Finally, let h be a hash function whose domain is $Z_N^* \times Z_N^*$, and whose range is $\{0, 1\}^k$. h can be obtained from any of the hash functions described in Section 5 by taking the k least significant bits of the output. The index s is the $(k + 2)$-tuple $(N, y_1, \ldots, y_k, h)$.
- **Shortcut:** The square roots $x_1, \ldots, x_k$.

---

tation that is *useful* to the recipient must be done. We currently have no candidates for such useful computation.

[3] Similarly, one could have a list of senders to whom access is categorically denied.

- **Definition of f_s:** The domain of f_s is Z_N^*. Below, we describe a moderately easy algorithm for finding z and r^2 satisfying the following conditions. Let us write $h(x, r^2) = b_1 \ldots b_k$, where each b_i is a single bit. Then z and r^2 must satisfy

$$z^2 = r^2 x^2 \prod_{i=1}^{k} y_i b_i \bmod N.$$

  $f_s(x) = (z, r^2)$ (note that f_s is a relation).
- **Verification:** Given x, z, r^2, compute $b_1 \ldots b_k = h(x, r^2)$ and check that

$$z^2 = r^2 x^2 \prod y_i b_i \bmod N.$$

- **To Evaluate f_s with Shortcut Information:** Choose an r at random, compute $h(x, r^2) = b_1 \ldots b_k$, and set $z = rx \prod x_i b_i$. $f_s(x) = (z, r^2)$.

$f_s(x) = (z, r^2)$ can be computed as follows.

Guess $b_1 \ldots b_k \in \{0, 1\}^k$.
Compute $B = \prod_{i=1}^{k} y_i b_i \bmod N$.
Repeat:
    Choose random $z \in Z_N^*$
    Define r^2 to be $r^2 = (z^2 / Bx^2) \bmod N$
Until $h(x, r^2) = b_1 \ldots b_k$.

The expected number of iterations is 2^k, which, based on the intuition driving the Fiat-Shamir signature scheme, seems to be the best one can hope for. In particular, if h is random, then one can do no better. In particular, retrieving the shortcut $x_1, \ldots, x_k$ is as hard as factoring [15]. In contrast, the verification procedure involves about $2k$ multiplications and one evaluation of the hash function. Similarly, given the shortcut the function can be evaluated using about k multiplications and one evaluation of the hash function. Thus, k is the difference parameter. A reasonable choice is $k = 10$.

4.3 An Ong-Schnorr-Shamir Based Scheme or Recycling Broken Signature Schemes

A source of suggestions for pricing functions with short cuts is signature schemes that have been broken. The "right" type of breaking applicable for our purposes is one that does not retrieve the private signature key (analogous to factoring N in the previous subsection), but nevertheless allows forging signatures by some moderately easy algorithm.

In this section we describe an implementation based on the proposed signature scheme of Ong, Schnorr and Shamir and the Pollard algorithm for breaking it. In [12, 13] Ong, Schnorr, and Shamir suggested a very efficient signature scheme based on quadratic equations modulo a composite: the public key is a modulus N (whose factorization remains secret) and an element $k \in Z_N^*$. The private key is u such that $u^2 = -k^{-1} \bmod N$, (i.e a square root of the inverse of

$-k$ modulo N). A signature for a message m (which we assume is in the range $0 \ldots N-1$) is a solution (x_1, x_2) of the equation $x_1^2 + k \cdot x_2^2 = m \mod N$. There is an efficient signing algorithm, requiring knowledge of the private key:

- choose random $r_1, r_2 \in Z_n^*$ such that $r_1 \cdot r_2 = m \mod N$
- set $x_1 = \frac{1}{2} \cdot (r_1 + r_2) \mod N$ and $x_2 = \frac{1}{2} \cdot u \cdot (r_1 - r_2) \mod N$.

Note that verifying a signature is extremely easy, requiring only 3 modular multiplication.

Pollard (reported in [14]) suggested a method of solving the equation without prior knowledge of the private key (finding the private key itself is hard – equivalent to factoring [15]). The method requires roughly $\log N$ iterations, and thus can be considered moderately hard, as compared with the verification and signing algorithms, which require only a constant number of multiplications and inversions. For excellent descriptions of Pollard's method and related work see [4, 9].

We now describe how to use the Ong-Schnorr-Shamir signature scheme as a pricing function.

- **Index:** Let $N = pq$ where p and q are primes let $k \in Z_n^*$. Then $s = (N, k)$.
- **Shortcut:** u such that $u^2 = k^{-1} \mod N$
- **Definition of f_s:** The domain of f_s is Z_N^*. Then $f_s(x) = (x_1, x_2)$, where $x_1^2 + kx_2^2 = x \mod N$. f_s is computed using Pollard's algorithm, as described above.
- **Verification:** Given x_1, x_2, x, verify that $x = x_1^2 + kx_2^2$.
- **To Evaluate f_s with Shortcut Information:** Use the Ong-Schnorr-Shamir algorithm for signing.

5 Hash Functions

Recall that we need hash functions for two purposes. First, in the context of junk mail, we hash messages down to some reasonable length, say 512 bits, and apply the pricing function to the hashed value of the message. In addition, we need hashing in the pricing function based on the signature scheme of Fiat-Shamir.

We briefly discuss four candidate hash functions. Each of these can be computed very quickly.

- **DES:** Several methods have been suggested for creating a one-way hash function based on DES (*e.g.* [10] and the references contained therein). Since DES is implemented in VLSI, and such a chip might become widely used for other purposes, this approach would be very efficient. Note that various attacks based on the "birthday paradox" [5] are not really relevant to our application since the effort needed to carry out such attacks is moderately hard.
- **MD4:** MD4 is a candidate one-way hash function proposed by Rivest [16]. It was designed explicitly to have high speed in *software*. The length of the output is either 128 or 256 bits. Although a simplified version of MD4 has been successfully attacked [3], we know of no attack on the full MD4.

- **Subset Sum:** Impagliazzo and Naor [8] have proposed using "high density" subset sum problems as one-way hash functions. They showed that finding colliding pairs is as hard as solving the subset sum problem for this density. Although this approach is probably less efficient than the others mentioned here, the function enjoys many useful statistical properties (*viz.* [8]). Moreover, it is parameterized and therefore flexible.
- **Snefru:** Snefru was proposed by Merkle [11] as a one-way hash function suitable for software, and was broken by Biham and Shamir [2]. However, the Biham and Shamir attack still requires about 2^{24} operations to find a partner of a given message. Thus, it may still be viable for our purposes.

6   Discussion and Open Problems

Of the three pricing functions described in Section 4, the Fiat-Shamir is the most flexible and enjoys the greatest difference function: changing k by 1 doubles the difference. The disadvantage is that this function, like the Fiat-Shamir scheme, requires the "extra" hash function.

As mentioned in the Introduction, there is no theory of moderately hard functions. The most obvious theoretical open question is to develop such a theory, analogous, perhaps, to the theory of one-way functions. Another area of research is to find additional candidates for pricing functions. Fortunately, a trial and error approach here is not so risky as in cryptography, since as discussed earlier, the consequences of a "broken" pricing function are not severe. If someone tries to make money from having found cheaper ways of evaluating the pricing function, then he or she underprices the pricing authority. Either few people will know about this, in which case the damage is slight, or it will become public.

Finally, the evaluation of the pricing function serves no useful purpose, except serving as a deterrent. It would be exciting to come up with a scheme in which evaluating the pricing function serves some additional purpose.

References

1. M. Blum and S. Micali, *personal communication.*
2. E. Biham and A. Shamir, *Differential Cryptanalysis of Snefru, Khafre, REDOC-II, LOKI, and Lucifer,* Crypto '91 abstracts.
3. B. den Boer and A. Bosselaers, *An attack on the last two rounds of MD4,* Crypto '91 abstracts.
4. E. F. Brickell and A. M. Odlyzko, *Cryptanalysis: A Survey of Recent Results,* Proceedings of the IEEE, vol. 76, pp. 578-593, May 1988.
5. D. Coppersmith, *Another Birthday Attack,* Proc. CRYPTO '85, Springer Verlag, LNCS, Vol. 218, pp. 369-378.
6. A. Fiat and A. Shamir, *How to prove yourself,* Proc. of Crypto 86, pp. 641-654.
7. B. A. Huberman, **The Ecology of Computing**, Studies in Computer Science and Artificial Intelligence 2, North Holland, Amsterdam, 1988.
8. R. Impagliazzo and M. Naor, *Cryptographic schemes provably secure as subset sum,* Proc. of the 30th FOCS, 1989.

9. K. McCurley, *Odd and ends from cryptology and computational number theory*, in **Crypttoloy and computational number theory**, edited by C. Pomerance, AMS short course, 1990, pp. 145-166.

10. R. C. Merkle, *One Way Functions and DES*, Proc. of Crypto'89, pp. 428–446.

11. R. C. Merkle, *Fast Software One-Way Hash Function*, J. of Cryptology Vol 3, No. 1, pp. 43–58, 1990.

12. H. Ong, C. P. Schnorr and A. Shamir, *An efficient signature scheme based on quadratic equations*, Proc 16th STOC, 1984, pp. 208-216.

13. H. Ong, C. P. Schnorr and A. Shamir, *Efficient signature scheme based on polynomial equations*, Proc of Crypto 84, pp. 37-46.

14. J. M. Pollard and C. P. Schnorr, *Solution of $X^2 + ky^2 = m \bmod n$*, IEEE Trans. on Information Theory., 1988.

15. M. O. Rabin, *Digital Signatures and Public Key Functions as Intractable as Factoring* Technical Memo TM-212, Lab. for Computer Science, MIT, 1979.

16. R. L. Rivest, *The MD4 Message Digest Algorithm*, Proc of Crypto'90, pp. 303–311.

17. R. Schroepel and A. Shamir, *A $T = O(2^{n/2})$, $S = O(2^{n/4})$ algorithm for certain NP-complete problems*. SIAM J. Computing, 10 (1981), pp. 456-464.

On the Information Rate
of Secret Sharing Schemes*

Extended Abstract

C. Blundo, A. De Santis, L. Gargano, U. Vaccaro

Dipartimento di Informatica ed Applicazioni, Università di Salerno
84081 Baronissi (SA), Italy

Abstract. We derive new limitations on the information rate and the average information rate of secret sharing schemes for access structure represented by graphs. We give the first proof of the existence of access structures with optimal information rate and optimal average information rate less that $1/2 + \epsilon$, where ϵ is an arbitrary positive constant. We also provide several general lower bounds on information rate and average information rate of graphs. In particular, we show that any graph with n vertices admits a secret sharing scheme with information rate $\Omega((\log n)/n)$.

1 Introduction

A secret sharing scheme is a technique to distribute a secret S among a set of participants P in such a way that only qualified subsets of P can reconstruct the value of S whereas any other subset of P, non-qualified to know S, cannot determine anything about the value of the secret. We briefly recall the results on secret sharing schemes that are more closely related to the topics of this paper.

Shamir [19] and Blackley [2] were the first to consider the problem of secret sharing and gave secret sharing schemes where each subset A of P of size $|A| \geq k$ can reconstruct the secret, and any subset A of participants of size $|A| < k$ have absolutely no information on the secret. These schemes are known as (n, k) *threshold schemes*; the value k is the threshold of the scheme and n is the size of P.

Ito, Saito and Nishizeki [15] considered a more general framework and showed how to realize a secret sharing scheme for any access structure. An access structure is a family of all subsets of P which are qualified to recover the secret. Their technique requires that the size of set where the shares are taken be very large compared to the size of the set where the secret is chosen. Benaloh and Leichter [1] proposed a technique to realize a secret sharing scheme for any access structure more efficient than Ito, Saito and Nishizeki's methodology. It should be pointed out that threshold schemes are insufficient to realize a secret sharing

* Partially supported by Italian Ministry of University and Research (M.U.R.S.T.) and by National Council for Research (C.N.R.) under grant 91.02326.CT12.

scheme for general access structures $\mathcal{A}$ [1]. Moreover, Benaloh and Leichter also showed that there exist access structures for which any secret sharing scheme must give to some participant a share which is from a domain strictly larger than that of the secret.

Brickell and Davenport [5] analyzed ideal secret sharing schemes in terms of matroids. An ideal secret sharing scheme is a scheme for which the the shares are taken has the same size of the set where the secret is chosen. In particular, they proved that an ideal secret sharing scheme exists for a graph G, if and only if G is a complete multipartite graph. Equivalently, if we define the information rate as the ratio between the size of the secret and that of the biggest share given to any participant, Brickell and Davenport's result can be stated saying that a graph has information rate 1 if and only if it is a complete multipartite graph. Brickell and Stinson [6] gave several upper and lower bounds on the information rate of access structures based on graphs.

Capocelli, De Santis, Gargano, and Vaccaro [7] gave the first example of access structures with information rate bounded away from 1.

Blundo, De Santis, Stinson, and Vaccaro [4] analyzed the information rate and the average information rate of secret sharing schemes based on graphs. The average information rate is the ratio between the secret size and the arithmetic mean of the size of the shares for such schemes. They proved the existence of a gap in the values of information rates of graphs, more precisely they proved that if a graph G with n vertices is not a complete multipartite graph then any secret sharing scheme for it has information rate not greater than 2/3 and average information rate not greater than $n/(n+1)$. These upper bounds arise by applying entropy argument due to Capocelli, De Santis, Gargano, and Vaccaro [7].

The recent survey by Stinson [21] contains an unified description of recent results in the area of secret sharing schemes. For different approaches to the study of secret sharing schemes, for schemes with "extended capabilities" as disenrollment, fault-tolerance, and pre-positioning and for a complete bibliography we recommend the survey article by Simmons [20].

In this paper we derive new limitations on the information rate and the average information rate for access structures represented by graphs. In the first part we prove new upper bounds on the information rate and the average information rate. These bounds are obtained by using the entropy approach by [7] and are the best possible for the considered structures since we exhibit secret sharing schemes that meet the bounds. In particular, we give the first proof of the existence of access structures with information rate and average information rate strictly less that 2/3. This solves a problem of [4]. In the second part we consider the problem of finding good lower bounds on the information rate and the average information rate and we give several general lower bounds that improve on previously known results.

2 Preliminaries

In this section we review the basic concepts of Information Theory we shall use. For a complete treatment of the subject the reader is advised to consult [8] and [11]. We shall also recall some basic terminology from graph theory.

Given a probability distribution $\{p(x)\}_{x\epsilon X}$ on a set X, we define the *entropy* of X, $H(X)$, as

$$H(X) = -\sum_{x\epsilon X} p(x)\log p(x)^2.$$

The entropy $H(X)$ is a measure of the average uncertainty one has about which element of the set X has been chosen when the choices of the elements from X are made according to the probability distribution $\{p(x)\}_{x\epsilon X}$. The entropy enjoys the following property

$$0 \leq H(X) \leq \log|X|, \tag{1}$$

where $H(X) = 0$ if and only if there exists $x_0 \in X$ such that $p(x_0) = 1$; $H(X) = \log|X|$ if and only if $p(x) = 1/|X|$, for all $x \in X$.

Given two sets X and Y and a joint probability distribution $\{p(x,y)\}_{x\epsilon X,y\epsilon Y}$ on their Cartesian product, the *conditional entropy* $H(X|Y)$, also called the equivocation of X given Y, is defined as

$$H(X|Y) = -\sum_{y\epsilon Y}\sum_{x\epsilon X} p(y)p(x|y)\log p(x|y).$$

The conditional entropy can be written as

$$H(X|Y) = \sum_{y\epsilon Y} p(y)H(X|Y=y)$$

where $H(X|Y=y) = -\sum_{x\epsilon X} p(x|y)\log p(x|y)$. From the definition of conditional entropy it is easy to see that

$$H(X|Y) \geq 0. \tag{2}$$

If we have $n+1$ sets $X_1,\ldots,X_n,Y$, the entropy of $X_1\ldots X_n$ given Y can be expressed as

$$H(X_1\ldots X_n|Y) = H(X_1|Y) + H(X_2|X_1Y) + \cdots + H(X_n|X_1\ldots X_{n-1}Y) \tag{3}$$

The *mutual information* between X and Y is defined by

$$I(X;Y) = H(X) - H(X|Y) \tag{4}$$

and enjoys the following properties:

$$I(X;Y) = I(Y;X), \tag{5}$$

---

[2] All logarithms in this paper are of base 2

and

$$I(X;Y) \geq 0,$$

from which one gets

$$H(X) \geq H(X|Y). \tag{6}$$

Given $n + 2$ sets $X, Y, Z_1, \ldots, Z_n$ and a joint probability distribution on their Cartesian product, the *conditional mutual information* between X and Y given $Z_1, \ldots, Z_n$ can be written as

$$I(XY|Z_1, \ldots, Z_n) = H(X|Z_1, \ldots, Z_n) - H(X|Z_1, \ldots, Z_nY). \tag{7}$$

Since the conditional mutual information is always non negative we get

$$H(X|Z_1, \ldots, Z_n) \geq H(X|Z_1, \ldots, Z_nY). \tag{8}$$

We now present some basic terminology from graph theory. A graph, $G = (V(G), E(G))$ consists of a finite non empty set of vertices $V(G)$ and a set of edges $E(G) \subseteq V(G) \times V(G)$. Graphs do not have loops or multiple edges. We consider only undirected graphs. In an undirected graph the pair of vertices representing any edge is unordered. Thus, the pairs (X, Y) and (Y, X) represent the same edge. To avoid overburdening the notation we often describe a graph G by the list of all edges $E(G)$. We will use reciprocally (X, Y) and XY to denote the edge joining the vertices X and Y. G is *connected* if any two vertices are joined by a path. The *complete graph* K_n is the graph on n vertices in which any two vertices are joined by an edge. The *complete multipartite graph* $K_{n_1, n_2, \ldots, n_t}$ is a graph on $\sum_{i=1}^{t} n_i$ vertices, in which the vertex set is partitioned into subsets of size n_i $(1 \leq i \leq t)$ called *parts*, such that vw is an edge if and only if v and w are in different parts.

Suppose G is a graph and $G_1, \ldots, G_t$ are subgraphs of G, such that each edge of G occurs in at least one of the G_i's. We say that $\Pi = \{G_1, \ldots, G_t\}$ is a covering of G and if each G_i, $i = 1, \ldots, t$ is a complete multipartite graph then we say that Π is a *complete multipartite covering* (CMC) of G.

3 Secret Sharing Schemes

A secret sharing scheme permits a secret to be shared among n participants in such a way that only qualified subsets of them can recover the secret, but any non-qualified subset has absolutely no information on the secret. An access structure $\mathcal{A}$ is the set of all subsets of P that can recover the secret.

Definition 1. Let P be a set of participants, a monotone access structure $\mathcal{A}$ on P is a subset $\mathcal{A} \subseteq 2^P$, such that

$$A \in \mathcal{A}, A \subseteq A' \subseteq P \Rightarrow A' \in \mathcal{A}.$$

Definition 2. Let P a set of participants and $A \subseteq 2^P$. The closure of A, $cl(A)$, is the set

$$cl(A) = \{C | B \in A \text{ and } B \subseteq C \subseteq P\}.$$

For a monotone access structure $\mathcal{A}$ we have $\mathcal{A} = cl(\mathcal{A})$.

A secret sharing scheme for secrets $s \in S$ and a probability distribution $\{p(s)\}_{s \in S}$ naturally induce a probability distribution on the joint space defined by the shares given to participants. This specifies the probability that participants receive given shares.

In terms of the probability distribution on the secret and on the shares given to participants, we say that a secret sharing scheme is a *perfect* secret sharing scheme, or simply a secret sharing scheme, for the monotone access structure $\mathcal{A} \subseteq 2^P$ if

1. *Any subset $A \subseteq P$ of participants not enabled to recover the secret have no information on the secret value.*[3]
   If $A \notin \mathcal{A}$ then for all $s \in S$ and for all $a \in A$ it holds $p(s|a) = p(s)$.
2. *Any subset $A \subseteq P$ of participants enabled to recover the secret can compute the secret:*
   If $A \in \mathcal{A}$ then for all $a \in A$ a unique secret $s \in S$ exists such that $p(s|a) = 1$.

Notice that the property 1. means that the probability that the secret is equal to s given that the shares held by $A \notin \mathcal{A}$ are a, is the same of the *a priori* probability that the secret is s. Therefore, no amount of knowledge of shares of participants not enabled to reconstruct the secret enables a Bayesian opponent to modify an *a priori* guess regarding which the secret is. Property 2. means that the value of the shares held by $A \in \mathcal{A}$ univocally determines the secret $s \in S$.

Let P be a set of participants, and $\mathcal{A}$ be a monotone access structure on P. Following the approach of [13], [14], and [7] we can restate above conditions 1. and 2. using the information measures introduced in the previous section. Therefore, we say that a secret sharing scheme is a sharing of the secret S among participants in P such that

1′. *Any qualified subset can reconstruct the secret.*
   Formally, for all $A \in \mathcal{A}$, it holds $H(S|A) = 0$.
2′. *Any non-qualified subset has absolutely no information on the secret.*
   Formally, for all $A \notin \mathcal{A}$, it holds $H(S|A) = H(S)$.

Notice that $H(S|A) = 0$ means that each set of values of the shares in A corresponds to a unique value of the secret. In fact, by definition, $H(S|A) = 0$ is equivalent to the fact that for all $a \in A$ with $p(a) \neq 0$ exists $s \in S$ such that $p(s|a) = 1$. Moreover, $H(S|A) = H(S)$ is equivalent to state that S and A are statistically independent, i.e., for all $a \in A$ for all $s \in S$, $p(s|a) = p(s)$ and therefore the knowledge of a gives no information about the secret. Notice that the condition $H(S|A) = H(S)$ is equivalent to say that for all $a \in A$ it holds $H(S|A = a) = H(S)$.

---

[3] To maintain notation simpler, we denote with the same symbol (sets of) participant(s) and the set(s) from which their shares are taken.

3.1 The Size of the Shares

One of the basic problems in the field of secret sharing schemes is to derive bounds on the amount of information that must be kept secret. This is important from the practical point of view since the security of any system degrades as the amount of secret information increases.

Let P be a set of n participants and $\mathcal{A} \subseteq 2^P$ be an access structure on P. We denote by $X \in P$ either the participant X or the random variable defined by the value of his share. Different measures of the amount of secret information that must be distributed in a secret sharing scheme are possible. If we are interested in limiting the maximum size of shares for each participant (i.e., the maximum quantity of secret information that must be given to any participant), then a worst-case measure of the maximum of $H(X)$ over all $X \in P$ naturally arises. To analyze such cases we use the *information rate* of $\mathcal{A}$ defined as

$$\rho(\mathcal{A}, \mathcal{P}_S) = \frac{H(S)}{\max_{X \in P} H(X)},$$

for a given secret sharing scheme and non-trivial probability distribution $\mathcal{P}_S$ on the secret. This measure was introduced by Brickell and Stinson [6] when the probability distributions over the secret and the shares are uniform. In such a case the definition becomes $\rho(\mathcal{A}) = \log|S|/\max_{X \in P} \log|X|$. The optimal information rate is then defined as:

$$\rho^*(\mathcal{A}) = \sup_{\mathcal{T}, \mathcal{Q}} \frac{H(S)}{\max_{X \in P} H(X)},$$

where $\mathcal{T}$ is the space of all secret sharing schemes for the access structure $\mathcal{A}$ and $\mathcal{Q}$ is the space of all non-trivial probability distributions $\mathcal{P}_S$.

In many cases it is preferable to limit the sum of the size of shares given to all participants. In such a case the arithmetic mean of the $H(X)$, $X \in P$, is a more appropriate measure. We define the *average information rate* as follows

$$\tilde{\rho}(\mathcal{A}, \mathcal{P}_S) = \frac{H(S)}{\sum_{X \in P} H(X)/|P|},$$

for a given secret sharing scheme and non-trivial probability distribution $\mathcal{P}_S$ on the secret. This measure was introduced in [3], [16], and [17] when an uniform probability distribution on the set of secrets is assumed. Blundo, De Santis, Stinson, and Vaccaro [4] analyzed secret sharing schemes by means of this measure, when the probability distributions over the secret and the shares are uniform. If the secret and the shares are chosen under a uniform probability distribution, considering previous measure is equivalent to consider the "average size" of the shares assigned to each participant to realize a secret sharing scheme. The optimal average information rate is then defined as:

$$\tilde{\rho}^*(\mathcal{A}) = \sup_{\mathcal{T}, \mathcal{Q}} \frac{H(S)}{\sum_{X \in P} H(X)/|P|}.$$

It is clear that, for the same secret sharing scheme and non-trivial probability distribution $\mathcal{P}_S$ on the secret, the information rate is no greater than the average information rate, that is $\tilde{\rho} \geq \rho$ and $\tilde{\rho} = \rho$ if and only if all $H(X)$, $X \in P$, have the same value. As done in [4] we denote, for a graph G, the optimal information rate with $\rho^*(G)$ and the average information rate with $\tilde{\rho}^*(G)$.

3.2 Auxiliary Results

In this section we recall some auxiliary results. We will improve some of them in the next sections and we will use others in our constructions.

Brickell and Stinson [6] proved the following lower bound on the information rate for any graph of maximum degree d.

Theorem 3. *Let G be a graph with maximum degree d, then*

$$\rho^*(G) \geq \frac{1}{\lceil d/2 \rceil + 1}.$$

In Section 4 we will show how to improve on it for odd d. Blundo, De Santis, Stinson, and Vaccaro [4] proved the following results for acyclic graphs

Lemma 4. *Let G be a tree, then a secret sharing scheme for G exists with information rate equal to $1/2$. Thus $\rho^*(G) \geq 1/2$.*

In Section 4 we will show how to improve this bound for any tree.

The following result, proved in [4] will be used to obtain good secret sharing schemes for graphs with maximum degree 3.

Theorem 5. *Let P_n be a path of length n, $n \geq 3$. A secret sharing scheme for P_n exists with optimal information rate $2/3$.*

The following lemmas have been proved by Capocelli, De Santis, Gargano, and Vaccaro [7]; we will use them to find new upper bounds on the information rate of access structures. Since their proofs are simple, we report them for reader's convenience.

Lemma 6. *Let $\mathcal{A}$ be an access structures on a set P of participants and $X, Y \subset P$. Let $Y \notin \mathcal{A}$ and $X \cup Y \in \mathcal{A}$. Then $H(X|Y) = H(S) + H(X|YS)$.*

Proof. The conditional mutual information $I(X; S|Y)$ can be written either as $H(X|Y) - H(X|YS)$ or as $H(S|Y) - H(S|XY)$. Hence, $H(X|Y) = H(X|YS) + H(S|Y) - H(S|XY)$. Because of $H(S|XY) = 0$ for $X \cup Y \in \mathcal{A}$ and $H(S|Y) = H(S)$ for $Y \notin \mathcal{A}$, we have $H(X|Y) = H(S) + H(X|YS)$. □

Lemma 7. *Let $\mathcal{A}$ an access structures on a set P of participants and $X, Y \subset P$. If $X \cup Y \notin \mathcal{A}$ then $H(Y|X) = H(Y|XS)$.*

Proof. The conditional mutual information $I(Y, S|X)$ X can be written either as $H(Y|X) - H(Y|XS)$ or as $H(S|X) - H(S|XY)$. Hence, $H(Y|X) = H(Y|XS) + H(S|X) - H(S|XY)$. Because of $H(S|XY) = H(S|X) = H(S)$, for $X \cup Y \notin \mathcal{A}$, we have $H(Y|X) = H(Y|XS)$. □

Finally, we briefly recall a technique introduced in [4] to obtain lower bounds on the information rate of a graph G.

Suppose G is a graph and $G_1, \ldots, G_n$ are subgraphs of G, such that each edge of G occurs in at least one of the G_i's. Suppose also that each G_i is a complete multipartite graph. Then we say that $\Pi = \{G_1, \ldots, G_t\}$ is a *complete multipartite covering* (or CMC) of G. Let $\Pi_j = \{G_{j1}, \ldots, G_{jn_j}\}$, $j = 1, \ldots L$, comprise a complete enumeration of the minimal CMCs of G. For every vertex v and for $j = 1, \ldots L$ define $R_{jv} = |\{i : v \in G_{ji}\}|$ and consider the following optimization problem $\mathcal{O}(G)$:

Minimize T subject to:

$$a_j \geq 0, \ 1 \leq j \leq L$$

$$\sum_{j=1}^{L} a_j = 1$$

$$T \geq \sum_{j=1}^{L} a_j R_{jv}, \ v \in V(G)$$

In citeBlDeStVa it is proved that if T^* is the optimal solution to $\mathcal{O}(G)$ then $\rho^*(G) \geq 1/T^*$.

4 Upper Bounds on the Information Rate and Average Information Rate

In this section we will exhibit an access structure having information rate less than $2/3$. This solves an open problem in [4]. The result is obtained using the entropy approach of [7].

Consider the graph $\mathcal{AS}_k = (V(\mathcal{AS}_k), E(\mathcal{AS}_k))$, $k \geq 1$, where

$$V(\mathcal{AS}_k) = \{Y_0, X_0, X_1, \ldots, X_k, X_{k+1}, \ldots, X_{2k}\}$$

and

$$E(\mathcal{AS}_k) = \{(Y_0, X_0), (X_0, X_1), \ldots, (X_0, X_k), (X_1, X_{k+1}), \ldots, (X_k, X_{2k})\}.$$

As an example, the graph $\mathcal{AS}_k$ for $k = 3$ is depicted in Figure 1(a).

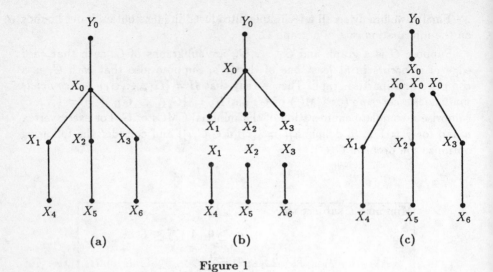

Figure 1

Theorem 8. *The optimal information rate of the graph $\mathcal{AS}_k$, $k \geq 1$, is*

$$\rho^*(\mathcal{AS}_k) = \frac{1}{2} + \frac{1}{4k+2},$$

and the optimal average information rate is

$$\tilde{\rho}^*(\mathcal{AS}_k) = \frac{2}{3} + \frac{2}{9k+6}.$$

Proof: Consider the conditional entropy $H(X_1 \ldots X_k|Y_0)$. We have

$$
\begin{aligned}
H(X_1 \ldots X_k|Y_0) &= H(X_1|Y_0) + H(X_2|X_1Y_0) + \cdots + H(X_k|X_1 \ldots X_{k-1}Y_0) \quad \text{(from (3))} \\
&\geq H(X_1|Y_0X_{k+1}) + H(X_2|X_1Y_0X_{k+2}) + \\
&\quad\quad H(X_3|X_1X_2Y_0X_{k+3}) + \cdots + H(X_k|X_1 \ldots X_{k-1}Y_0X_{2k}) \quad \text{(from (8))} \\
&\geq kH(S) \quad \text{(from Lemma 6 and (2))}.
\end{aligned}
$$

On the other hand, we have also

$$
\begin{aligned}
H(X_1 \ldots X_k|Y_0) &= H(X_1 \ldots X_k|Y_0S) \quad \text{(from Lemma 7)} \\
&\leq H(X_0X_1 \ldots X_k|Y_0S) \quad \text{(from (3) and (2))} \\
&\leq H(X_0|Y_0S) + H(X_1|X_0S) + \cdots + H(X_k|X_0S) \quad \text{(from (3) and (8))} \\
&= H(X_0|Y_0) - H(S) + \cdots + H(X_k|X_0) - H(S) \quad \text{(from Lemma 6)} \\
&\leq H(X_0) + \cdots + H(X_k) - (k+1)H(S) \quad \text{(from (6))}.
\end{aligned}
$$

Therefore, we get

$$H(X_0) + H(X_1) + \ldots + H(X_k) \geq (2k+1)H(S). \tag{9}$$

From (9) it follows that there exists $i \in \{0, 1, \ldots, k\}$ such that

$$H(X_i) \geq \frac{2k+1}{k+1} H(S).$$

Therefore, the optimal information rate of $\mathcal{AS}_k$ $\rho^*(\mathcal{AS}_k)$ is upper bounded by

$$\rho^*(\mathcal{AS}_k) = \frac{H(S)}{\max H(X)} \leq \frac{k+1}{2k+1} = \frac{1}{2} + \frac{1}{4k+2}.$$

From (9) and from Lemma 6 it follows that

$$H(Y_0) + \sum_{i=0}^{2k} H(X_i) \geq (3k+2)H(S).$$

Therefore, the optimal average information rate of $\mathcal{AS}_k$ is upper bounded by

$$\frac{2k+2}{3k+2} = \frac{2}{3} + \frac{2}{9k+6}.$$

Actually, $1/2 + 1/(4k+2)$ is the true value of the optimal information rate. This value can be attained by using the CMC technique presented in [4] as solution of the following linear programming problem.

Consider the following two minimal complete multipartite coverings of $\mathcal{AS}_k$

$$\Pi_1 = \Big\{ \{Y_0X_0, X_0X_1, \cdots, X_0X_k\}, \{X_1X_{k+1}, \cdots, X_kX_{2k}\} \Big\}$$

$$\Pi_2 = \Big\{ \{Y_0X_0\}, \{X_0X_1, X_1X_{k+1}\}, \cdots, \{X_0X_k, X_kX_{2k}\} \Big\}.$$

An example of these two covering of $\mathcal{AS}_k$ are depicted in Figure 1(b) and 1(c) for $k = 3$. The matrix of entries R_{jv} is

$$\begin{pmatrix} 1 & 1 & \overbrace{2 \cdots 2}^{k} & \overbrace{1 \cdots 1}^{k} \\ 1 & k+1 & \underbrace{1 \cdots\cdots\cdots 1}_{2k} \end{pmatrix}.$$

Hence the linear programming problem to be solved is the following:

Minimize T subject to

$$a_j \geq 0, \; j = 1, 2$$
$$a_1 + a_2 = 1$$
$$T \geq a_1 + (k+1)a_2$$
$$T \geq 2a_1 + a_2$$

The optimal solution is

$$(a_1, a_2, T) = \left(\frac{k}{k+1}, \frac{1}{k+1}, \frac{2k+1}{k+1} \right).$$

Hence, $\rho_C^*(\mathcal{AS}_k) = (2k+1)/(k+1)$, and this rate can be attained by taking k copies of Π_1, and one copy of Π_2. Thus, the optimal information rate of $\mathcal{AS}_k$ is $1/2 + 1/(4k+2)$. The optimal average information rate equal to $2/3 + 2/(9k+6)$ can be attained by either Π_1 or Π_2. $\quad\square$

Suppose that $p(s) = 1/|S|$, for any $s \in S$. Above result and inequality (1) imply that any perfect secret sharing scheme for $\mathcal{AS}_k$ must give to at least a participant a share of size greater than $2 - 1/(k+1)$ times the size of the secret.

Theorem 8 is a generalization of Theorem 4.1 of [7]. In fact if we choose $k = 1$ the access structure $\mathcal{AS}_k$ is the closure of the edge-set of P_3, the path on four vertices.

In Appendix A are depicted all graphs on six vertices that have $\mathcal{AS}_2$ as induced subgraph and, therefore, have optimal information rate less than $3/5$. It turns out that the optimal information rate for all those graphs is equal to $3/5$, and all but one have also an optimal average information rate equal to $3/4$.

Using the previous theorem we can show the existence of access structures having *average information* rate less than $2/3$, which represented the best upper bound known so far [7]. Consider the graph $\mathcal{M}_k$, where $V(\mathcal{M}_k) = \{X_1, X_2, \ldots, X_{2k+3}, X_{2k+4}\}$ and

$$E(\mathcal{M}_k) = \{X_1 X_2\} \bigcup \{X_2 X_i, X_i X_{k+i}, X_{k+i} X_{2k+3} | 3 \leq i \leq k+2\} \bigcup \{X_{2k+3} X_{2k+4}\}.$$

The graph $\mathcal{M}_3$ is depicted in Figure 2. The following theorem holds.

Theorem 9. *The optimal average information rate for $\mathcal{M}_k$, $k \geq 1$, is*

$$\widetilde{\rho}^*(\mathcal{M}_k) = \frac{k+2}{2k+2}.$$

Proof : From Lemma 6 we get $H(X_1) \geq H(S)$ and $H(X_{2k+4}) \geq H(S)$, whereas from Theorem 8 we have

$$\sum_{i=2}^{k+2} H(X_i) \geq 2k + 1$$

and

$$\sum_{i=k+3}^{2k+3} H(X_i) \geq 2k + 1.$$

Thus,

$$\sum_{i=1}^{2k+4} H(X_i) \geq 4k + 4.$$

Hence,

$$\widetilde{\rho}^*(\mathcal{M}_k) \leq \frac{k+2}{2k+2}.$$

It is easy to see that the following complete multipartite covering Π of the graph $\mathcal{M}_k$ meets this bound.

$$\Pi = \Big\{ \{X_1 X_2, X_2 X_3, \ldots, X_2 X_{k+2}\},$$
$$\{X_3 X_{k+3}, X_{k+3} X_{2k+3}\},$$
$$\vdots$$
$$\{X_{k+2} X_{2k+2}, X_{2k+2} X_{2k+3}\},$$
$$\{X_{2k+3} X_{2k+4}\} \Big\}.$$

□

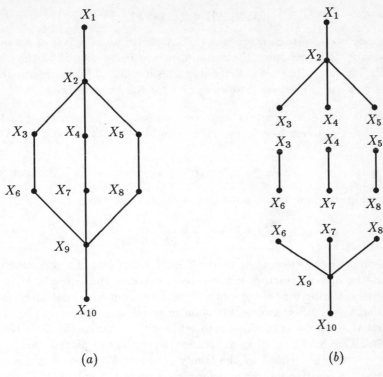

(a) (b)

Figure 2

4.1 A NP-completeness result

A close look to the proof of the upper bound in Theorem 8 shows that it can be applied also to any access structure $\mathcal{A}$ on $2k+2$ participants, $Y_0, X_0, X_1, \ldots, X_{2k}$, such that the set $\mathcal{A}$-*allowed* defined as

$$\mathcal{A}\text{-}allowed = \{Y_0 X_0\} \bigcup \{X_0 X_i, X_i X_{k+i} | 1 \le i \le k\}$$

is in the access structure, i.e. $\mathcal{A}$-*allowed* $\subseteq \mathcal{A}$, but the set $\mathcal{A}$-*forbidden* defined as

$$\mathcal{A}\text{-}forbidden = \{X_1 X_2 \ldots X_k Y_0\} \bigcup \{Y_0 X_{k+1}\} \bigcup \{X_1 \ldots X_i Y_0 X_{k+i+1} | 1 \le i \le k-1\}$$

has no intersection with the access structure, i.e. $\mathcal{A}$-*forbidden* $\bigcap \mathcal{A} = \emptyset$. Let $\mathcal{B}_k$ be the set of all access structures which satisfy the above requirements. The sequence $(X_1, X_2, \ldots, X_k)$ is called the *children list* of access structure $\mathcal{A}$ (the name is inspired by the fact that the set $\mathcal{A}$-*allowed* has the form of a tree). To maintain simpler notation we denote a set $\{a_1, a_2, \ldots, a_n\}$ by the sequence $a_1 a_2 \ldots a_n$. In case the access structure is the closure of a graph, the set $\mathcal{A}$-*forbidden* can be written as

$$\mathcal{A}\text{-}forbidden\text{-}edges = \{Y_0 X_i | 1 \le i \le 2k\} \bigcup \{X_i X_j | 1 \le i < j \le k\}$$

$$\bigcup\{X_i X_{k+j} | 1 \le i < j \le k\}.$$

Let $\mathcal{A}$ be an access structure on a set P of participants. Given a subset of participants $P' \subseteq P$, we define the access structure *induced by P'* as the family of sets $\mathcal{A}[P'] = \{x \in \mathcal{A} | x \subseteq P'\}$. Extending Theorem 3.3 of [6] to general access structures and using Theorem 8 we can prove the following theorem.

Theorem 10. *Let $\mathcal{A}$ be an access structure on a set P of participants and $P' \subseteq P$. If $\mathcal{A}[P'] \in \mathcal{B}_k$, where $k \ge 1$, then the optimal information rates for $\mathcal{A}$ and $\mathcal{A}[P']$ satisfy*

$$\rho^*(\mathcal{A}) \le \rho^*(\mathcal{A}[P']) \le \frac{1}{2} + \frac{1}{4k+2},$$

and optimal average information rate for $\mathcal{A}[P']$ satisfies

$$\widetilde{\rho}^*(\mathcal{A}[P']) \le \frac{2}{3} + \frac{2}{9k+6}.$$

Above theorem gives an upper bound on the information rate of access structures given that the access structure induced by a subset of participants is in $\mathcal{B}_k$. Unfortunately, testing for this property is an hard computational problem, as we show that this is NP–complete. Let $\mathcal{A}$ be an access structure, a set $C \in \mathcal{A}$ is a *minimal* set of $\mathcal{A}$ if it does not contain any set in $\mathcal{A} \setminus \{C\}$. Define the BOUNDED–INFORMATION–RATE problem as follows: Given a set of participants P and an access structure $\mathcal{A}$ defined by the family of minimal sets which can recover the secret and a positive integer k, determine if there is a subset $P' \subseteq P$ such that the induced access structure $\mathcal{A}[P']$ is in $\mathcal{B}_k$.

Theorem 11. *BOUNDED–INFORMATION–RATE is NP–complete.*

Proof. The proof will be given in the final version of the paper. □

4.2 Upper bounds for more general access structures

A general technique to upper bound the average information rate $\widetilde{\rho}^*(G)$, of graphs G who have one or more induced subgraphs of a given form is given below.

If G is a graph and $V_1 \subseteq V(G)$, then we define the *induced* graph $G[V_1]$ to have vertex set V_1 and edge set $\{XY \in E(G) : X, Y \in V_1\}$.

Let G be a graph. We define a subgraph F_G of G, that we will call the *foundation* of G, in the following manner. This is an extension of the notion of foundation presented in [4]. Let $X \in V(G)$. Let k be the maximum integer such that there is a set V' of $2k + 1$ vertices $Y_0, X_1, \ldots, X_{2k} \in V(G)$ such that the induced subgraph $G[V' \cup \{X\}]$ is in $\mathcal{B}_k$; that is, $E(G[V' \cup \{X\}])$ contains the set $\mathcal{A}$-allowed but does not contain any edge in the set $\mathcal{A}$-forbidden-edges. Clearly $k < deg(X)$, where $deg(X)$ is the degree of vertex X. A set V' satisfying above properties is called a X-set of vertex X, with size k. Denote by $f_{X;X_1,\ldots X_k}$ the set of edges XX_i, $i = 1, \ldots, k$. We call $f_{X;X_1,\ldots X_k}$ the *local foundation* of vertex X and X-set V' and we call the vertices $X_1, \ldots, X_k$ *descendants* of X in $f_{X;X_1,\ldots X_k}$. Let $\{V_1, \ldots, V_{m_X}\}$ be the family of all X-sets of vertex $X \in V(G)$, and $\{f_X^1, \ldots, f_X^{m_X}\}$ be the family of the corresponding local foundations. Observe that this approach might not be feasible for large values of m, since m might be exponentially large in the worst case. Now we can define the *foundation F_G* of a graph G as follows

$$F_G = \{f_X^1, \ldots, f_X^{m_X} \mid X \in V(G)\}.$$

If $f_{X_0}^i$ is in F_G, the foundation of a graph G, and $X_1, \ldots, X_k$ are descendants of X_0 in $f_{X_0}^i$, then by Theorem 8, we have $\sum_{i=0}^{i=k} H(X_i) \geq (2k+1)H(S)$ for any secret sharing scheme with access structure $cl(E(G))$. Consider the following linear programming problem $\mathcal{A}(G)$:

Minimize $\qquad C = \sum_{X \in V(G)} a_X$

subject to:

$$a_X \geq 0, \quad X \in V(G)$$

$$a_{X_0} + \cdots + a_{X_k} \geq k, \quad X_0 \in V(G), \quad f_{X_0}^i \in F_G, \text{ and}$$

$$X_1, \ldots, X_k \text{ descendants of } X_0 \text{ in } f_{X_0}^i$$

The following upper bound on the average information rate holds.

Theorem 12. *Let G be a graph with foundation G_1. Let C^* be the optimal solution to the problem $\mathcal{A}(G)$. Then*

$$\tilde{\rho}^*(G) \leq \frac{|V(G)|}{C^* + |V(G)|}.$$

Proof. The proof will be given in the final version of the paper. $\qquad \square$

5 Lower Bounds on Information Rate and Average Information Rate

In this section we will give several general lower bounds on the information rate and on the average information rate of access structures represented by graphs.

We first improve on the bound of Theorem 3 for graphs with n vertices and odd maximum degree d.

Lemma 13. *Let G be a graph of n vertices and maximum degree d, d odd. Then*

$$\rho^*(G) \geq \frac{1}{\lceil d/2 \rceil + 1 - \lceil d/2 \rceil / n}.$$

Proof. Let $Adj(X)$, $Inc(X)$, $degree\_one(X)$ be the following sets :

- $Adj(X) = \{Y : (X, Y) \in E\}$
- $Inc(X) = \{(X, Y) : (X, Y) \in E\}$
- $degree\_one(X) = \{Y \in Adj(X) : |Inc(Y)| = 1\}$

Let $X \in V(G)$ and G_x be a subgraph of G such that $V(G_x) = \{X\} \bigcup Adj(X)$ and $E(G_x) = Inc(X)$. It is well known a secret sharing scheme for G_x exists with information rate equal to 1 (G_x is a complete multipartite graph). Consider the graph G' where $V(G') = V(G) - \{X\} \bigcup degree\_one(X)$ and $E(G') = E(G) - Inc(X)$. We realize a secret sharing scheme for G', for a secret of one bit, using the technique showed in Theorem 3.8 of [BrSt]. Each vertex in $Adj(X) \bigcap V(G')$ gets at most $\lceil (d-1)/2 \rceil + 1$ bits while other vertices get at most $\lceil d/2 \rceil + 1$ bits. A secret sharing scheme for G can be realized joining the scheme for G_x and the scheme for G'. In this scheme the vertex X will receive one bit, the vertices in $Adj(X) \bigcap V(G')$ will receive at most $\lceil (d-1)/2 \rceil + 2$ bits, while other vertices will get at most $\lceil d/2 \rceil + 1$ bits. Since $\lceil (d-1)/2 \rceil + 2 = \lceil d/2 \rceil + 1$, if d is odd, there is a secret sharing schemes for G, for a secret consisting of a single bit, that gives to each vertex in G at most $\lceil d/2 \rceil + 1$ bits while a predeterminated vertex gets only one bit. If we consider n of these secret sharing schemes, one for each vertex in V, and then we compose them, we can realize a secret sharing scheme, for a secret of n bits, giving to each vertex at most $1 + (n-1)(\lceil d/2 \rceil + 1)$ bits, so we can realize a secret sharing scheme with an information rate equal to

$$\frac{1}{\lceil d/2 \rceil + 1 - \lceil d/2 \rceil / n},$$

and the lemma follows. □

For a graph G of maximum degree 3, the bound of [6] gives $\rho^*(G) \geq 1/3$ while the bound of lemma 13 gives $\rho^*(G) \geq 1/(3 - 2/n)$. The following lemma gives an improved bound.

Lemma 14. *Let G be a graph of maximum degree 3. Then, $\rho^*(G) \geq 2/5$.*

Proof. Consider a covering $\mathcal{C}$ of G consisting of maximal length paths $P_1, \ldots, P_m$. It is well know a secret sharing scheme for a path exists with an optimal information rate equal to $2/3$ (see Theorem 5), this scheme, for a secret of two bits, gives two bits to terminal vertices in the path while other vertices gets three bits. We can realize a secret sharing scheme for G, for a secret of two bits, using secret sharing schemes, with optimal information rate, for the paths belonging to $\mathcal{C}$. A vertex of G of degree one can only be a terminal vertex of a path so it receive two bits. If a vertex has degree two then it belongs to only one path and it receives three bits, it cannot be a terminal vertex of two different paths since we consider a covering of maximal length paths. If a vertex has degree three then it can't belong to three different paths since we consider a covering of maximal length paths so it belongs to two paths, it is a terminal vertex of a path and it is a central vertex of another path and it gets totally five bits. Thus we can construct a secret sharing scheme for G, giving to each vertex at most five bits for a secret of two bits obtaining a secret sharing scheme with information rate equal to 2/5. □

If we know the number of vertices in the graph G then we can improve previous bound as stated by next lemma.

Lemma 15. *Let G a graph of maximum degree 3 with n vertices. Then,*

$$\rho^*(G) \geq \frac{2}{5 - 3/n}.$$

Proof. Let G_X, with $X \in V(G)$, be the graph defined in Lemma 13. Consider the graph G' where $V(G') = V(G) - \{X\} \bigcup degree\_one(X)$ and $E(G') = E(G) - Inc(X)$. We realize a secret sharing scheme for G', for a secret of two bit, using the technique showed in Lemma 14. Each vertex $Y \in Adj(X) \bigcap V(G')$ gets at most 3 bits, since $|Inc(Y)| \leq 2$, while the other vertices get at most 5 bits. A secret sharing scheme for G can be realized joining the scheme for G_X and the scheme for G'. Thus we can realize a secret sharing scheme for G, for a secret consisting of two bits, giving two bits to a predeterminated vertex while other vertices get at most five bits. If we consider n of these schemes, one for each vertex, and then we compose them we obtain a secret sharing scheme for a secret of $2n$ bits giving to each vertex at most $2 + 5(n-1) = 5n - 3$ bits so the information rate for this scheme is $2/(5 - 3/n)$. □

Applying the same reasoning of Lemma 14 to graphs of odd degree d leads to the bound $\rho^*(G) \geq 1/(\lfloor d/2 \rfloor 1.5 + 1)$ which is worse than previous bounds.

Regardless of the degree, it is possible to obtain better bounds for trees. We recall that an internal node is a vertex of degree greater than one.

Lemma 16. *Let G be a tree with n internal vertices. Then*

$$\rho^*(G) \geq \frac{n}{2n - 1}.$$

Proof. In [4] was showed how to obtain a secret sharing scheme for any tree with information rate equal to 1/2. This scheme, for a secret consisting of a single bit, gives one bit to a predeterminated vertex $X \in V(G)$ and to all non-internal vertices, whereas each other vertex gets two bits. We will use this construction as basic construction. If we consider n of these schemes, one for each internal vertex, and we compose them then it is possible to realize a secret haring scheme for G, for a secret of n bits, giving to each vertex at most $2(n-1) + 1 = 2n - 1$ bits. Thus

$$\rho^*(G) \geq \frac{n}{2n - 1}.$$

□

If only the number of vertices are known, what can we say on the information rate of a graph G? The maximum degree of G can be as bad as $n-1$. Thus, the bound of [6] gives $\rho^*(G) \geq 1/(\lceil(n-1)/2\rceil + 1)$, while the bound of Lemma 13 gives $\rho^*(G) \geq 1/(\lceil(n-1)/2\rceil + 1 - \lceil(n-1)/2\rceil/n)$, if n is even.

In this last part of the paper we present general lower bounds on the information rate and average information rate for *any* graph G with n vertices. The lower bounds are obtained by using known results on the covering of the edges of a graphs by means of complete bipartite graphs. We first recall that Brickell and Davenport [5] proved that a graph G has information rate 1 if and only if G is complete multipartite graph.

Tuza [22] proved that the edge-set of an arbitrary graph G can be covered by complete bipartite subgraphs such that the sum of the number of the vertices of such subgraphs is less than $3n^2/2\log n + o(n^2/\log n)$. Using the above quoted result by Brickell and Davenport we get that the optimal average information rate for any graph G with n vertices is greater than n times the inverse of $3n^2/2\log n + f(n)$, where $|f(n)| < \epsilon n^2/\log n$, for all $\epsilon > 0$ and sufficiently large n. Therefore, the average information rate is greater than $2\log n/3n + g(n)$, where $|g(n)| \leq (2\epsilon/3(\epsilon + 3/2))\log n/n$, if $|f(n)| < \epsilon n^2/\log n$.

Feder and Motwani [10] proved that the problem of partitioning the edges of a graph G into complete bipartite graphs such that the sum of the cardinalities of their vertex sets is minimized is NP–complete. However, they proved that the edge set of a graph $G = (V, E)$, with $|V| = n$ and $|E| = m$ can be partitioned into complete bipartite graphs with sum of the cardinalities of their vertex sets $O(\frac{m\log\frac{n^2}{m}}{\log n})$, and presented an efficient algorithm to compute such a partition. Using their result, it follows that there is a secret sharing scheme with average information rate at least $\Omega(\frac{n\log n}{m\log\frac{n^2}{m}})$.

Finally, we recall a result of Erdös and Pyber [9] (see also [18]) which states that edges of a graph G with n vertices can be partitioned into complete bipartite graphs such that each vertex of G is contained by at most $O(n/\log n)$ complete bipartite graphs. This result directly implies that the optimal information rate of G is $\rho^*(G) = \Omega\left(\frac{\log n}{n}\right)$.

These results can be summarized in the following theorem.

Theorem 17. *Let G be a graph with n vertices and m edges. Then, the optimal average information rate for G satisfies*

$$\widetilde{\rho}^*(G) > \frac{2\log n}{3n} + o\left(\frac{\log n}{n}\right),$$

and

$$\widetilde{\rho}^*(G) = \Omega\left(\frac{n\log n}{m\log\frac{n^2}{m}}\right).$$

The optimal information rate for G satisfies

$$\rho^*(G) = \Omega\left(\frac{\log n}{n}\right).$$

It is worth pointing out that if G is a sparse graph, i.e., $m = \alpha n$, where α is a constant, then above theorem implies that $\hat{\rho}^*(G)$ is limited from below by a constant. This result describes a wide class of graphs having average information rate that does not go to zero as the number of participants increases.

Acknowledgments

We are indebted with professor Capocelli for his constant encouragement and support. We would like to dedicate this paper to his memory as a sign of appreciation and love.

We would like thank L. Pyber for providing us reference [18] and A. Marchetti–Spaccamela and E. Feuerstein for bringing to our attention reference [10].

References

1. J. C. Benaloh and J. Leichter, *Generalized Secret Sharing and Monotone Functions,* in "Advances in Cryptology - CRYPTO 88", Ed. S. Goldwasser, vol. 403 of "Lecture Notes in Computer Science", Springer-Verlag, pp. 27–35.

2. G. R. Blakley, *Safeguarding Cryptographic Keys,* Proceedings AFIPS 1979 National Computer Conference, pp.313–317, June 1979.

3. C. Blundo *Secret Sharing Schemes for Access Structures based on Graphs,* Tesi di Laurea, University of Salerno, Italy, 1991, (in Italian).

4. C. Blundo, A. De Santis, D. R. Stinson, and U. Vaccaro, *Graph Decomposition and Secret Sharing Schemes,* Eurocrypt 1992, Hungary.

5. E. F. Brickell and D. M. Davenport, *On the classification of ideal secret sharing schemes,* J. Cryptology, 4:123–134, 1991.

6. E. F. Brickell and D. R. Stinson, *Some Improved Bounds on the Information Rate of Perfect Secret Sharing Schemes,* Lecture Notes in Computer Science, 537:242–252, 1991. To appear in J. Cryptology.

7. R. M. Capocelli, A. De Santis, L. Gargano, and U. Vaccaro, *On the Size of Shares for Secret Sharing Schemes,* in "Advances in Cryptology - CRYPTO 91", Ed. J. Feigenbaum, vol. 576 of "Lecture Notes in Computer Science", Springer-Verlag, pp. 101–113. To appear in J. Cryptology.

8. I. Csiszár and J. Körner, *Information Theory. Coding theorems for discrete memoryless systems,* Academic Press, 1981.

9. P. Erdös and L. Pyber, unpublished.

10. T. Feder and R. Motwani, *Clique Partition, Graph Compression and Speeding-up Algorithms,* Proceedings of the 23rd Annual ACM Symposium on Theory of Computing, New Orleans, 1991, pp. 123–133.

11. R. G. Gallager, *Information Theory and Reliable Communications,* John Wiley & Sons, New York, NY, 1968.

12. M. Garey and D. Johnson, *Computers and Intractability: a Guide to the Theory of NP-Completeness,* W. H. Freeman & Co., New York, 1979.

13. E. D. Karnin, J. W. Greene, and M. E. Hellman, *On Secret Sharing Systems,* IEEE Trans. on Inform. Theory, vol. IT-29, no. 1, Jan. 1983, pp. 35–41.

14. S. C. Kothari, *Generalized Linear Threshold Schemes,* in "Advances in Cryptology - CRYPTO 84", G. R. Blakley and D. Chaum Eds., vol 196 of "Lecture Notes in Computer Science", Springer-Verlag, pp. 231–241.

15. M. Ito, A. Saito, and T. Nishizeki, *Secret Sharing Scheme Realizing General Access Structure*, Proc. IEEE Global Telecommunications Conf., Globecom 87, Tokyo, Japan, 1987.

16. K. M. Martin, *Discrete Structures in the Theory of Secret Sharing*, PhD Thesis, University of London, 1991.

17. K. M. Martin, *New secret sharing schemes from old*, submitted to Journal of Combin. Math. and Combin. Comput..

18. L. Pyber, *Covering the Edges of a Graph by ...*, in Sets, Graphs and Numbers, Colloquia Mathematica Soc. János Bolyai, L. Lovász, D. Miklós, T. Szönyi, Eds., (to appear).

19. A. Shamir, *How to Share a Secret*, Communications of the ACM, vol. 22, n. 11, pp. 612–613, Nov. 1979.

20. G. J. Simmons, *An Introduction to Shared Secret and/or Shared Control Schemes and Their Application*, Contemporary Cryptology, IEEE Press, pp. 441–497, 1991.

21. D. R. Stinson, *An Explication of Secret Sharing Schemes*, Technical Report UNL-CSE-92-004, Department of Computer Science and Engineering, University of Nebraska, February 1992.

22. Z. Tuza, *Covering of Graphs by Complete Bipartite Subgraphs; Complexity of 0-1 matrices*, Combinatorica, vol. 4, n. 1, pp. 111–116, 1984.

Appendix A

In this appendix we analyze all graphs who have optimal information rate less than 2/3 accordingly to Theorem 10. The schemes for these graphs are obtained by using the Multiple Construction Technique [4] based on complete multipartite coverings of the graph. The optimal information rate is not greater than 3/5 and the optimal average information rate is less than or equal to 3/4 for all graphs from Theorem 10. All these results are summarized in Table 1, and the first CMC of each graph gives the scheme with average information rate showed in Table 1. Below are depicted some of the minimal CMCs for 5 graphs on 6 vertices.

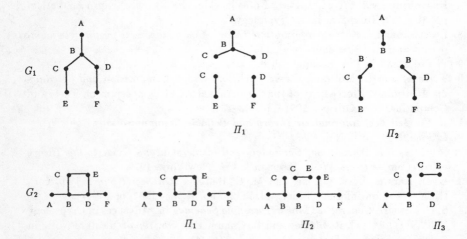

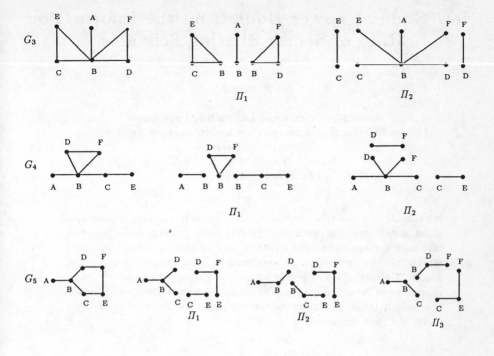

Table 1. Information Rate and Average Information Rate

Graph	Information Rate	Average information Rate
G_1, G_2, G_3, G_4	$\rho^* = 3/5$	$\widetilde{\rho}^* = 3/4$
G_5	$\rho^* = 3/5$	$2/3 \leq \widetilde{\rho}^* \leq 3/4$

New General Lower Bounds on the Information Rate of Secret Sharing Schemes

D. R. Stinson

Computer Science and Engineering Department
and Center for Communication and Information Science
University of Nebraska
Lincoln, NE 68588-0115, U.S.A.
stinson@bibd.unl.edu

Abstract. We use two combinatorial techniques to apply a decomposition construction in obtaining general lower bounds on information rate and average information rate of certain general classes of access structures. The first technique uses combinatorial designs (in particular, Steiner systems $S(t, k, v)$). The second technique uses equitable edge-colourings of bipartite graphs. For uniform access structures of rank t, this second technique improves the best previous general bounds by a factor of t (asymptotically).

1 Introduction and Terminology

Informally, a secret sharing scheme is a method of sharing a secret key K among a finite set of participants in such a way that certain specified subsets of participants can compute the secret key K. The value K is chosen by a special participant called the *dealer*.

We will use the following notation. Let $\mathcal{P} = \{P_i : 1 \leq i \leq w\}$ be the set of participants. The dealer is denoted by D and we assume $D \notin \mathcal{P}$. $\mathcal{K}$ is *key set* (i.e. the set of all possible keys) and $\mathcal{S}$ is the *share set* (i.e. the set of all possible shares). Let Γ be a set of subsets of $\mathcal{P}$; this is denoted mathematically by the notation $\Gamma \subseteq 2^{\mathcal{P}}$. The subsets in Γ are those subsets of participants that should be able to compute the secret. Γ is called an *access structure* and the subsets in Γ are called *authorized subsets*.

When a dealer D wants to share a secret $K \in \mathcal{K}$, he will give each participant a share from $\mathcal{S}$. The shares should be distributed secretly, so no participant knows the share given to another participant. At a later time, a subset of participants will attempt to determine K from the shares they collectively hold. We will say that a scheme is a *perfect secret sharing scheme realizing* the access structure Γ provided the following two properties are satisfied:

1. If an authorized subset of participants $B \subseteq \mathcal{P}$ pool their shares, then they can determine the value of K.
2. If an unauthorized subset of participants $B \subseteq \mathcal{P}$ pool their shares, then they can determine nothing about the value of K.

The security of such a scheme is unconditional, since we do not place any limit on the amount of computation that can be performed by a subset of participants.

Suppose that $B \in \Gamma$, $B \subseteq C \subseteq \mathcal{P}$ and the subset C wants to determine K. Since B is an authorized subset, it can already determine K. Hence, the subset C can determine K by ignoring the shares of the participants in $C \backslash B$. Stated another way, a superset of an authorized set is again an authorized set. What this says is that the access structure should satisfy the *monotone* property:

$$\text{if } B \in \Gamma \text{ and } B \subseteq C \subseteq \mathcal{P}, \text{ then } C \in \Gamma.$$

If Γ is an access structure, then $B \in \Gamma$ is a *minimal* authorized subset if $A \notin \Gamma$ whenever $A \subseteq B$, $A \neq B$. The set of minimal authorized subsets of Γ is denoted Γ_0 and is called the *basis* of Γ. Since Γ consists of all subsets of $\mathcal{P}$ that are supersets of a subset in the basis Γ_0, Γ is determined uniquely as a function of Γ_0. Expressed mathematically, we have

$$\Gamma = \{C \subseteq \mathcal{P} : B \subseteq C, B \in \Gamma_0\}.$$

We say that Γ is the *closure* of Γ_0 and write $\Gamma = cl(\Gamma_0)$.

We define the *rank* of an access structure Γ to be the maximum cardinality of a minimal authorized subset. An access structure is *uniform* if every minimal authorized subset has the same cardinality. Observe that the rank of Γ is two if and only if $\Gamma = cl(E(G))$, where $E(G)$ denotes the edge set of a graph G.

We now briefly describe a general mathematical model for secret sharing and discuss the concept of security. In this model, we represent a secret sharing scheme by a set $\mathcal{F}$ of *distribution rules*. A distribution rule is a function

$$f : \mathcal{P} \cup \{D\} \to \mathcal{K} \cup \mathcal{S}$$

which satisfies the conditions $f(D) \in \mathcal{K}$, and $f(P_i) \in \mathcal{S}$ for $1 \leq i \leq w$. A distribution rule f represents a possible distribution of shares to the participants, where $f(D)$ is the secret key being shared, and $f(P_i)$ is the share given to P_i.

If $\mathcal{F}$ is a set of distribution rules and $K \in \mathcal{K}$, denote

$$\mathcal{F}_K = \{f \in \mathcal{F} : f(D) = K\}.$$

If $K \in \mathcal{K}$ is the value of the secret that D wishes to share, then D will choose a random distribution rule $f \in \mathcal{F}_K$, and use it to distribute shares.

Suppose Γ is an access structure and $\mathcal{F}$ is a set of distribution rules. Suppose the following two properties are satisfied:

(*) Let $B \in \Gamma$, and suppose $f, g \in \mathcal{F}$. If $f(P_i) = g(P_i)$ for all $P_i \in B$, then $f(D) = g(D)$.

(**) Let $B \notin \Gamma$ and suppose $f : B \to \mathcal{S}$. Then there exists a non-negative integer $\lambda(f, B)$ such that, for every $K \in \mathcal{K}$,

$$|\{g \in \mathcal{F}_K : g(P_i) = f(P_i) \forall P_i \in B\}| = \lambda(f, B).$$

Then $\mathcal{F}$ is a perfect secret sharing scheme that realizes the access structure Γ. The property (*) is relatively straightforward: it says that the shares given to an authorized subset uniquely determine the value of the secret. The property (**) guarantees that the shares given to an unauthorized subset give no information as to the value of the secret. The list of shares $(f(P_i) : P_i \in B)$ given to an unauthorized subset B will restrict the possible distribution rules to some subset of $\mathcal{F}$. However, the remaining possible rules will be equally divided among the possible keys. More precisely, for any assignment of shares f to B, there will remain $\lambda(f, B)$ possible rules corresponding to each value of the secret. The formal security proof uses probability distributions; it can be found in [9].

As an example, in Figure 1 we present a perfect secret sharing scheme from [9] for the access structure having basis

$$C_6 = \{\{A, B\}, \{B, C\}, \{C, D\}, \{D, E\}, \{E, F\}, \{F, A\}\}.$$

(C_6 is the graph which is a cycle of length six.)

Fig. 1. A Secret Sharing Scheme For C_6

	D	A	B	C	D	E	F
f_1	0	0	0	1	1	2	2
f_2	0	0	0	2	2	1	1
f_3	0	1	1	2	2	0	0
f_4	0	1	1	0	0	2	2
f_5	0	2	2	0	0	1	1
f_6	0	2	2	1	1	0	0
f_7	1	0	1	1	2	2	0
f_8	1	0	2	2	1	1	0
f_9	1	1	2	2	0	0	1
f_{10}	1	1	0	0	2	2	1
f_{11}	1	2	0	0	1	1	2
f_{12}	1	2	1	1	0	0	2

The construction of secret sharing schemes for arbitrary access structures has been studied by several researchers. General construction methods are described in [14, 1, 21, 20].

2 Information Rate

We measure the efficiency of a secret sharing scheme by the information rate. Suppose $\mathcal{F}$ is a set of distribution rules for a secret sharing scheme. For $1 \leq i \leq w$, define

$$\mathcal{S}_i = \{f(P_i) : f \in \mathcal{F}\}.$$

S_i represents the set of possible shares that P_i might receive; of course $S_i \subseteq S$. Now, since the secret key K comes from a finite set $\mathcal{K}$, we can think of K as being represented by a bit-string of length $\log_2 |\mathcal{K}|$, by using a binary encoding, for example. In a similar way, a share given to P_i can be represented by a bit-string of length $\log_2 |S_i|$. Intuitively, P_i receives $\log_2 |S_i|$ bits of information (in his or her share), but the information content of the secret is $\log_2 |\mathcal{K}|$ bits. The information rate for P_i is the ratio

$$\rho_i = \frac{\log_2 |\mathcal{K}|}{\log_2 |S_i|}.$$

The *information rate* [9] of the scheme is denoted by ρ and is defined as

$$\rho = \min\{\rho_i : 1 \leq i \leq w\}.$$

The *average information rate* [3, 17], denoted by $\tilde{\rho}$, is the harmonic mean of the ρ_i's:

$$\tilde{\rho} = \frac{w}{\sum_{i=1}^{w} \frac{1}{\rho_i}} = \frac{w \log_2 |\mathcal{K}|}{\sum_{i=1}^{w} \log_2 |S_i|}.$$

The scheme of Figure 1 has $\rho = \tilde{\rho} = \log_2 2 / \log_2 3 \approx .63$. (This is not optimal: the optimal scheme has rate $2/3$ [4].)

It is easy to prove that $\rho \leq \tilde{\rho} \leq 1$ in any scheme, and that $\rho = 1$ if and only if $\tilde{\rho} = 1$. Since $\rho = \tilde{\rho} = 1$ is the optimal situation, we refer to such a scheme an *ideal* scheme. Ideal schemes have been studied extensively; see for example [7, 8, 17, 15, 18]. In the cases where ideal schemes do not exist, the objective is to construct a scheme with (average) information rate as close to one as possible. Research in this direction can be found in [9, 10, 4, 22, 16].

3 A Decomposition Construction

Our main recursive construction uses small schemes as building blocks in the construction of larger schemes. We call this the decomposition construction. Note that various versions of this construction have been described in several papers, such as [9, 4, 22, 17, 16].

We will use the notation $PS(\Gamma, \rho, q)$ to denote a perfect secret sharing scheme with access structure $cl(\Gamma)$ and information rate at least ρ for a set of q keys. Analogously, a perfect secret sharing scheme with access structure $cl(\Gamma)$ and average information rate at least $\tilde{\rho}$ for a set of q keys will be denoted by $\overline{PS}(\Gamma, \tilde{\rho}, q)$.

Suppose Γ is an access structure having basis Γ_0. A *decomposition* of Γ_0 consists of a set $\{\Gamma_1, \ldots, \Gamma_n\}$ such that the following properties are satisfied:

1. $\Gamma_k \subseteq \Gamma_0$ for $1 \leq k \leq n$
2. $\cup_{k=1}^{n} \Gamma_k = \Gamma_0$

Often, $\{\Gamma_1, \ldots, \Gamma_n\}$ will form a partition of Γ_0, but this is not a requirement. For $1 \leq k \leq n$, define $\mathcal{P}_k = \cup_{B \in \Gamma_k} B$; $\mathcal{P}_k$ denotes the set of participants in a scheme with access structure $cl(\Gamma_k)$.

We present the following two results, both of which use the same construction.

Theorem 1. *Let Γ be an access structure on w participants having basis Γ_0 and suppose that $\{\Gamma_1, \ldots, \Gamma_n\}$ is a decomposition of Γ_0. Let q be an integer and for $1 \leq k \leq n$, suppose there exists a $PS(\Gamma_k, \rho_k, q)$. For $1 \leq i \leq w$, let*

$$R_i = \frac{1}{\sum_{\{k : P_i \in \mathcal{P}_k\}} \frac{1}{\rho_k}}.$$

Then there exists a $PS(\Gamma, \rho, q)$, where

$$\rho = \min\{R_i : 1 \leq i \leq w\}.$$

Theorem 2. *Let Γ be an access structure on w participants having basis Γ_0 and suppose that $\{\Gamma_1, \ldots, \Gamma_n\}$ is a decomposition of Γ_0. Let q be an integer and for $1 \leq k \leq n$, suppose there exists a $\widetilde{PS}(\Gamma_k, \widetilde{\rho}_k, q)$. Then there exists a $\widetilde{PS}(\Gamma, \widetilde{\rho}, q)$, where*

$$\widetilde{\rho} = \frac{w}{\sum_{k=1}^{n} \frac{|\mathcal{P}_k|}{\rho_k}}.$$

Remark. If we define

$$\widetilde{R}_i = \frac{1}{\sum_{\{k : P_i \in \mathcal{P}_k\}} \frac{1}{\rho_k}}$$

for $1 \leq i \leq w$, then

$$\widetilde{\rho} = \frac{w}{\sum_{i=1}^{w} \frac{1}{R_i}}.$$

Proof. Let $\mathcal{K}$ be a fixed set of q keys. For $1 \leq k \leq n$, let $\mathcal{F}^k$ denote the distribution rules in a $PS(\Gamma_k, \rho_k, q)$ with key set $\mathcal{K}$. For any $K \in \mathcal{K}$, and for $1 \leq k \leq n$, we have

$$\mathcal{F}^k = \bigcup_{K \in \mathcal{K}} \mathcal{F}_K^k,$$

where $\mathcal{F}_K^k$ consists of the distribution rules in $\mathcal{F}^k$ for which the key value is K. For $1 \leq k \leq n$, suppose $f_K^k \in \mathcal{F}_K^k$. Define a distribution function $f_K^1 \times f_K^2 \times \ldots \times f_K^n$ which gives to each participant P_j the list of shares

$$(f_K^k(P_j) : P_j \in \mathcal{P}_k).$$

We construct a $PS(\Gamma, \rho, q)$ in which $\mathcal{F} = \cup_{K \in \mathcal{K}} \mathcal{F}_K$, where

$$\mathcal{F}_K = \{f_K^1 \times f_K^2 \times \ldots \times f_K^n : f_K^k \in \mathcal{F}_K^k, 1 \leq k \leq n\}.$$

The verifications and the computation of the information rate are straightforward. $\qquad \square$

Let us look at an example to illustrate these constructions. Consider the access structure having basis

$$\Gamma_0 = \{\{A, B\}, \{A, C\}, \{B, C\}, \{C, D\}, \{C, E\}, \{D, E\}, \{E, F\}, \{E, A\}, \{F, A\}\}.$$

Consider the decomposition

$$\Gamma_1 = \{\{A, B\}, \{B, C\}, \{C, D\}, \{D, E\}, \{E, F\}, \{F, A\}\}$$
$$\Gamma_2 = \{\{A, C\}, \{C, E\}, \{E, A\}\}.$$

We have already seen in Fig. 1 that there is a $PS(\Gamma_1, 2, \log 2/\log 3)$. For all $q \geq 3$, a $PS(\Gamma_2, q, 1)$ exists from [9]. However, in order to apply the decomposition construction, we need schemes with the same number of keys. This creates no problem, as it follows from [9] that a $PS(\Gamma_1, 2, \log 2/\log 3)$ implies the existence of a $PS(\Gamma_1, 2^j, \log 2/\log 3)$ for all $j \geq 1$. So we can take $q = 2^j$, $j \geq 2$. From Theorem 1 we get a $PS(\Gamma, 2, \rho)$ where $\rho = \log 2/\log 6 \approx .38$, and Theorem 2 yields a $\widetilde{PS}(\Gamma, 2, \widetilde{\rho})$ where $\widetilde{\rho} = \log 4/\log 18 \approx .47$.

However, if we use a different decomposition, we can do better. Define

$$\Gamma_3 = \{\{A, B\}, \{B, C\}, \{A, C\}\}$$
$$\Gamma_4 = \{\{C, D\}, \{D, E\}, \{C, E\}\}$$
$$\Gamma_5 = \{\{E, F\}, \{F, A\}, \{E, A\}\}.$$

For any $q \geq 3$, there exists a $PS(\Gamma_i, q, 1)$ for $i = 3, 4, 5$, and we obtain a $PS(\Gamma, q, 1/2)$ and a $\widetilde{PS}(\Gamma, q, 2/3)$.

This scheme could be implemented as follows: Suppose $q \geq 3$ is prime and let $\mathcal{K} = GF(q)$. Then $\mathcal{F}_K = \{f_{r_1, r_2, r_3, K} : r_1, r_2, r_3 \in GF(q)\}$, where

$$f_{r_1, r_2, r_3, K}(A) = (r_3, 2K + r_5)$$
$$f_{r_1, r_2, r_3, K}(B) = K + r_3$$
$$f_{r_1, r_2, r_3, K}(C) = (r_4, 2K + r_3)$$
$$f_{r_1, r_2, r_3, K}(D) = K + r_4$$
$$f_{r_1, r_2, r_3, K}(E) = (r_5, 2K + r_4)$$
$$f_{r_1, r_2, r_3, K}(F) = K + r_5.$$

In the remaining sections of this paper, we use two combinatorial techniques to apply the decomposition construction in obtaining general lower bounds on information rate and average information rate of certain general classes of access structures. The first technique uses combinatorial designs (in particular, Steiner systems $S(t, k, v)$). (Due to a lack of knowledge of infinite classes of Steiner systems for $t > 3$, this technique is applicable primarily to access structures of ranks two and three.) The second technique uses equitable edge-colourings of bipartite graphs. We first give a new proof of a result proved by Brickell and Stinson [9] which applies to access structures of rank two. Then we describe some generalizations to access structures of higher rank which improve the best previous general bounds by a factor of t (asymptotically).

4 Applications Using Steiner Systems

4.1 Two Corollaries of the Decomposition Construction

In this section we discuss applications of the decomposition construction using combinatorial designs. A *Steiner system* $S(t, k, w)$ is a pair $(X, \mathcal{A})$, where X is a set of w elements (called *points*) and $\mathcal{A}$ is a set of k–subsets of X (called *blocks*), such that every t–subset of points occurs in exactly one block. An $S(t, k, w)$ is said to be *non-trivial* if $t < k < w$. We note that no non-trivial Steiner systems are known to exist for $t > 5$, and very few are known to exist for $t > 3$. For general information on the existence of Steiner systems, we refer to [2].

Suppose Γ is an access structure of rank t on w participants, having basis Γ_0. Suppose also that $(X, \mathcal{A})$ is an $S(t, k, w)$. We can use $(X, \mathcal{A})$ to construct a decomposition of Γ_0, as follows: For every block $A \in \mathcal{A}$, define

$$\Gamma_A = \{B \in \Gamma_0 : B \subseteq A\}.$$

Then $\{\Gamma_A : A \in \mathcal{A}\}$ is a decomposition of Γ_0 (observe that it is a partition if and only if Γ is uniform).

Now suppose that we compute values $\pi_{k,t}$ and $q_{k,t}$ such that there exists a $PS(\Gamma', \pi_{k,t}, q_{k,t})$ for *any* access structure Γ' of rank $\leq t$ on k participants. Now, in the Steiner system, elementary counting shows that each point occurs in exactly $\binom{w-1}{t-1}/\binom{k-1}{t-1}$ blocks. Hence, when we apply Theorem 1, we get

$$R_i = \frac{\pi_{k,t}\binom{k-1}{t-1}}{\binom{w-1}{t-1}}$$

for every point i. The resulting scheme is a $PS(\Gamma, \rho, q_{k,t})$ for $\rho = \pi_{k,t}\binom{k-1}{t-1}/\binom{w-1}{t-1}$.

Summarizing, we have the following result.

Theorem 3. *Suppose Γ is an access structure of rank t on w participants, and suppose that an $S(t, k, w)$ exists. Suppose there exists a $PS(\Gamma', \pi_{k,t}, q_{k,t})$ for any access structure Γ' of rank $\leq t$ on k participants. Then there exists a $PS(\Gamma, \rho, q_{k,t})$ for $\rho = \pi_{k,t}\binom{k-1}{t-1}/\binom{w-1}{t-1}$.*

For average information rate, we get the following similar result by applying Theorem 2.

Theorem 4. *Suppose Γ is an access structure of rank t on w participants, and suppose that an $S(t, k, w)$ exists. Suppose there exists a $PS(\Gamma', \widetilde{\pi}_{k,t}, \widetilde{q}_{k,t})$ for any access structure Γ' of rank $\leq t$ on k participants. Then there exists a $\widetilde{PS}(\Gamma, \widetilde{\rho}, \widetilde{q}_{k,t})$ for $\widetilde{\rho} = \widetilde{\pi}_{k,t}\binom{k-1}{t-1}/\binom{w-1}{t-1}$.*

4.2 Graph Access Structures

The situation that has been studied the most is when the basis consists of the edges of a graph (i.e. the access structure has rank two); see [9, 4, 10], for example. If G is a graph, then we will denote the vertex set of G by $V(G)$, the edge set by $E(G)$, and a $PS(cl(E(G)), \rho, q)$ by $PS(G, \rho, q)$.

Considerable attention has been paid to the graphs on at most five vertices. Lower bounds on the (average) information rate have been obtained in [4] by applying various versions of the decomposition construction. The following result updates the bounds of [4]:

Theorem 5. *1. If G is a graph with $|V(G)| \leq 3$, then there is a $PS(G, 1, q)$ for any prime power $q \geq 3$.*
2. If G is a graph with $|V(G)| = 4$, then there is a $PS(G, 2/3, q^2)$ and a $\widetilde{PS}(G, 4/5, q)$ for any prime power $q \geq 4$.
3. If G is a graph with $|V(G)| = 5$, then there is a $PS(G, 2/3, q^2)$ and a $\widetilde{PS}(G, 5/7, q)$ for any prime power $q \geq 5$.

Proof. The only cases left unresolved in [4] concern the following four graphs on five vertices:

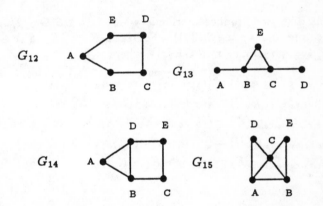

For G_{12}, we produce a scheme which is simultaneously a $PS(G_{12}, 2/3, q^2)$ and a $\widetilde{PS}(G_{12}, 2/3), q^2)$, where $q \geq 5$ is a prime and $\mathcal{K} = (GF(q))^2$. For each $K = (K_1, K_2) \in \mathcal{K}$, $\mathcal{F}_K = \{f_{r_1, r_2, r_3, r_4, r_5, K} : r_1, r_2, r_3, r_4, r_5 \in GF(q)\}$, where

$$f_{r_1, r_2, r_3, r_4, r_5, K}(A) = (r_1, r_4, r_5 + K_1 + 3K_2)$$
$$f_{r_1, r_2, r_3, r_4, r_5, K}(B) = (r_2, r_5, r_1 + K_1)$$
$$f_{r_1, r_2, r_3, r_4, r_5, K}(C) = (r_3, r_1, r_2 + K_2)$$
$$f_{r_1, r_2, r_3, r_4, r_5, K}(D) = (r_4, r_2, r_3 + K_1 + K_2)$$
$$f_{r_1, r_2, r_3, r_4, r_5, K}(E) = (r_5, r_3, r_4 + K_1 + 2K_2).$$

For G_{13}, we exhibit a scheme (constructed by Dean Hoffman) which is simultaneously a $PS(G_{13}, 2/3, q^2)$ and a $\widetilde{PS}(G_{12}, 10/13), q^2)$, where $q \geq 3$ is a prime and $\mathcal{K} = (GF(q))^2$. For each $K = (K_1, K_2) \in \mathcal{K}$, $\mathcal{F}_K = \{f_{r_1, r_2, r_3, r_4}K :$ $r_1, r_2, r_3, r_4 \in GF(q)\}$, where

$$f_{r_1, r_2, r_3, r_4, K}(A) = (r_1 + K_1, r_2)$$
$$f_{r_1, r_2, r_3, r_4, K}(B) = (r_1, r_2 + K_2, r_3)$$
$$f_{r_1, r_2, r_3, r_4, K}(C) = (r_2, r_3 + K_1, r_4)$$
$$f_{r_1, r_2, r_3, r_4, K}(D) = (r_3, r_4 + K_2)$$
$$f_{r_1, r_2, r_3, r_4, K}(E) = (r_1 + r_4 + K_1 + K_2, r_4 + K_2, r_2 - r_3).$$

For G_{14}, we produce a scheme which is a $PS(G_{12}, 2/3, q^2)$, where $q \geq 5$ is a prime and $\mathcal{K} = (GF(q))^2$. Ror each $K = (K_1, K_2) \in \mathcal{K}$, $\mathcal{F}_K = \{f_{r_1, r_2, r_3, r_4, r_5, K} :$ $r_1, r_2, r_3, r_4, r_5 \in GF(q)\}$, where

$$f_{r_1, r_2, r_3, r_4, r_5, K}(A) = (r_1 + K_1, r_2 + K_2, r_4)$$
$$f_{r_1, r_2, r_3, r_4, r_5, K}(B) = (r_1, r_4 + K_1 + 2K_2, r_5)$$
$$f_{r_1, r_2, r_3, r_4, r_5, K}(C) = (r_1 + K_1, r_3 + K_1 + K_2, r_5 + 2K_1 + K_2)$$
$$f_{r_1, r_2, r_3, r_4, r_5, K}(D) = (r_2, r_4 + 2K_1 + 4K_2, r_5 + 2K_1 + K_2)$$
$$f_{r_1, r_2, r_3, r_4, r_5, K}(E) = (r_2 + K_2, r_3, r_5).$$

Finally, for G_{15}, we produce a scheme which is a $PS(G_{12}, 2/3, q^2)$, where $q \geq 5$ is a prime and $\mathcal{K} = (GF(q))^2$. For each $K = (K_1, K_2) \in \mathcal{K}$, $\mathcal{F}_K = \{f_{r_1, r_2, r_3, r_4, r_5, K} : r_1, r_2, r_3, r_4, r_5 \in GF(q)\}$, where

$$f_{r_1, r_2, r_3, r_4, r_5, K}(A) = (r_1 + K_1, r_4, r_5 + K_1 + 2K_2)$$
$$f_{r_1, r_2, r_3, r_4, r_5, K}(B) = (r_2 + K_2, r_4 + 2K_1 + K_2, r_5)$$
$$f_{r_1, r_2, r_3, r_4, r_5, K}(C) = (r_3, r_4 + 4K_1 + 2K_2, r_5 + 2K_1 + 4K_2)$$
$$f_{r_1, r_2, r_3, r_4, r_5, K}(D) = (r_1, r_3 + K_1 + K_2, r_4 + 2K_1 + K_2)$$
$$f_{r_1, r_2, r_3, r_4, r_5, K}(E) = (r_2, r_3 + K_1 + K_2, r_5 + K_1 + 2K_2).$$

$\square$

Remark. With the schemes presented above, the optimal value of the information rate and average information rate is now determined for all graph access structures on at most five vertices. In each case, the upper bound presented in [4] turns out to be the correct value. Also, the constructions for G_{12}, G_{14} and G_{15} are based on a new generalization of the decomposition that we will present in a forthcoming paper. Finally, we remark that minor modifications of the above constructions will produce schemes where the number of keys is a prime power.

Using the notation of Section 4.1, we can take $\pi_{3,2} = 1$, $\pi_{4,2} = 2/3$, and $\pi_{5,2} = 2/3$; $\tilde{\pi}_{3,2} = 1$, $\tilde{\pi}_{4,2} = 4/5$, and $\tilde{\pi}_{5,2} = 5/7$.

In order to apply Theorems 3 and 4, we need information about Steiner systems $S(2, k, w)$ for $k = 3, 4, 5$. This information is summarized in the following theorem:

Theorem 6. *[13] Suppose $3 \leq k \leq 5$. Then there exists an $S(2, k, w)$ if and only if $w \equiv 1, k \pmod{k(k-1)}$.*

We obtain lower bounds on the (average) information rate of any graph on w vertices that are presented in Table 1. For example, we see that there is a $PS(G, 1/3, q)$ for any graph G having seven vertices, where $q \geq 3$ is a prime power.

Table 1. Bounds on the Information Rate for Access Structures of Rank Two

k	w	lower bound on ρ or $\widetilde{\rho}$	number of keys
3	$w \equiv 1, 3 \pmod 6$	$\rho \geq \frac{2}{w-1}$	q, where $q \geq 3$ is a prime power
4	$w \equiv 1, 4 \pmod{12}$	$\rho \geq \frac{2}{w-1}$	q^2, where $q \geq 3$ is a prime power
4	$w \equiv 1, 4 \pmod{12}$	$\widetilde{\rho} \geq \frac{12}{5(w-1)}$	q, where $q \geq 4$ is a prime power
5	$w \equiv 1, 5 \pmod{20}$	$\rho \geq \frac{8}{3(w-1)}$	q^2, where $q \geq 5$ is a prime power
5	$w \equiv 1, 5 \pmod{20}$	$\widetilde{\rho} \geq \frac{20}{7(w-1)}$	q, where $q \geq 5$ is a prime power

It is interesting to observe how the bounds improve as we use designs with larger block size. Also, note that if there does not exist an $S(2, k, w)$, then we can take the smallest integer $w_0 > w$ such that there does exist an $S(2, k, w_0)$, and delete $w_0 - w$ points from the Steiner system, thereby constructing a pairwise balanced design [2]. Then apply Theorem 1 or 2 to obtain a scheme where the information rate is computed by replacing w by w_0 in Table 1.

4.3 Rank Three Access Structures

We can apply the same techniques to access structures of rank three, using the following results concerning access structures on four participants, proved in [22, 17].

Theorem 7. *1. If Γ is a rank three access structure on four participants, then there is a $PS(\Gamma, 2/3, q^2)$ and a $\widetilde{PS}(\Gamma, 4/5, q)$ for any prime power $q \geq 4$.*
2. If Γ is a uniform rank three access structure on four participants, then there is a $PS(\Gamma, 1, q)$ for any prime power $q \geq 4$.

Using the notation of Section 4.1, we can let $\pi_{4,3} = 2/3$ and $\widetilde{\pi}_{4,3} = 4/5$. The relevant Steiner systems $S(3, 4, w)$ exist as follows:

Theorem 8. *[12] There exists an $S(3,4,w)$ if and only if $w \equiv 2,4 \pmod 6$.*

Application of Theorems 3 and 4 yield the bounds for access structures of rank three presented in Table 2.

Table 2. Bounds on the Information Rate for Access Structures of Rank Three

w	lower bound on ρ or $\widetilde{\rho}$	number of keys
$w \equiv 2,4 \pmod 6$	$\rho \geq \frac{4}{(w-1)(w-2)}$	q^2, where $q \geq 4$ is a prime power
$w \equiv 2,4 \pmod 6$	$\widetilde{\rho} \geq \frac{24}{5(w-1)(w-2)}$	q, where $q \geq 4$ is a prime power
$w \equiv 2,4 \pmod 6$	$\rho \geq \frac{6}{(w-1)(w-2)}$ if Γ is uniform	q, where $q \geq 4$ is a prime power

5 Applications Using Edge-colourings of Bipartite Graphs

The following result was proved in [9].

Theorem 9. *Suppose G is a graph in which the maximum vertex degree is d. Then there exists a $PS(G, 1/(\lceil \frac{d}{2}\rceil + 1), q)$ for any prime power $q \geq 2$.*

Remark. For the case of odd d, an improved bound is given in [5].

Theorem 9 is proved by decomposing G into complete bipartite graphs $K_{1,m}$ (called *stars*) in such a way that any vertex of G is in at most $\lceil \frac{d}{2}\rceil + 1$ of the stars. It has been shown in [8] that there is a $PS(K_{1,m}, 1, q)$ for any prime power $q \geq 2$. Hence, the result follows from Theorem 1.

The star decomposition was obtained in [9] by first constructing an eulerian tour in a multigraph related to G. We will present an alternative proof of Theorem 9 which appears to be more easily generalizable. This proof makes use of a result concerning edge-colourings of bipartite graphs.

For a graph G, denote the degree of a vertex x by $d_G(x)$. Suppose ℓ is an integer. An $\ell-edge\ colouring$ of G is a function $f : E(G) \to \{1, \ldots, \ell\}$. f induces a partition $E(G) = \cup_{i=1}^{\ell} E_i(G)$, where $E_i(G) = f^{-1}(i)$, $1 \leq i \leq \ell$ (that is, $E_i(G)$ consists of the edges of G receiving colour i). An $\ell-edge$ colouring is said to be *equitable* if, for every vertex $x \in V(G)$ and for every colour i ($1 \leq i \leq \ell$), the number of edges in $E_i(G)$ incident with vertex x is either $\lfloor d(x)/\ell \rfloor$ or $\lceil d(x)/\ell \rceil$.

The following theorem of de Werra [11] (see also [6, pp. 62-63]) is of use to us:

Theorem 10. *If G is a bipartite graph, then there exists an equitable ℓ—edge colouring of G for any positive integer ℓ.*

Here now is an alternate proof of Theorem 9:

Proof of Theorem 9. Construct a bipartite graph H with bipartition $(V(G), E(G))$ having edge set

$$E(H) = \{xe : x \in V(G), e \in E(G), x \in e\}.$$

By Theorem 10, there is an equitable 2—edge colouring of H. Each vertex $x \in V(G)$ has degree $d_G(x)$ in H and each vertex $e \in E(G)$ has degree 2 in H. Hence, every vertex $e \in E(G)$ is incident with one edge of $E_1(H)$ and every vertex $x \in V(G)$ is incident with $\lfloor d(x)/2 \rfloor$ or $\lceil d(x)/2 \rceil$ edges of $E_1(H)$.

For every vertex $x \in V(G)$, define a subgraph $G_x = \{e \in E(G) : xe \in E_1(H)\}$. It is not difficult to see that $\{G_x : x \in V(G)\}$ forms the desired star decomposition. □

Let's consider how to generalize this result to uniform access structures of higher rank. As our "building blocks" we use a class of access structures that we call *generalized stars*. Let $t \geq 2$ and $m \geq t - 1$. Define a basis on $m + t - 1$ participants as follows:

$$\Gamma_0^*(t, m) = \{\{P_1, \ldots, P_{t-1}, P_j\} : t \leq j \leq m + t - 1\}.$$

(In the case $t = 2$, $\Gamma_0^*(t, m)$ consists of the edges of a star graph $K_{1,m}$.) Define the *centre* of a generalized star to be the intersection of the basis subsets (i.e. $\{P_1, \ldots, P_{t-1}\}$ in the above example). Any access structure $\Gamma^*(t, m)$ is easily seen to be ideal. In fact, there exists a $PS(\Gamma^*(t, m), 1, q)$ for any prime power $q \geq t$ by a simple modification of a Shamir (t, t)—threshold scheme [19].

Now, suppose Γ_0 is the basis of a uniform access structure of rank t. Construct a bipartite graph H as follows: The bipartition is (X, Y), where $Y = \Gamma_0$ and

$$X = \{A : A \subseteq B \in \Gamma_0, |A| = t - 1\};$$

and the edges in H are

$$E(H) = \{AB : A \in X, B \in Y, A \subseteq B\}.$$

(In the case $t = 2$, the graph H is the same as the one constructed earlier.) Note that every vertex $A \in X$ has degree t in H. Now, apply Theorem 10 to obtain an equitable t—edge colouring of H. For every vertex $A \in X$, define

$$\Gamma_A = \{B \in Y : AB \in E_1(H)\}.$$

Then each Γ_A is a $\Gamma_0^*(t, m)$ where $m = \lceil d_H(A)/t \rceil$ or $m = \lfloor d_H(A)/t \rfloor$. $\{\Gamma_A : A \in X\}$ is a decomposition of Γ_0, and for every $A \in X$, there is a $PS(\Gamma_A, 1, q)$ for any prime power $q \geq t$.

It remains to compute bounds on the R_i's. Define d_i (the *degree* of P_i) to be the number of t–subsets in Γ_0 which contain P_i. Then

$$d_i = \frac{1}{t-1} \sum_{\{A:P_i \in A \in X\}} d_H(A).$$

Now, P_i is in the centre of $|\{A \in X : P_i \in A\}|$ of the Γ_A's. Since we used an equitable colouring to construct the Γ_A's, this accounts for at least

$$\sum_{\{A:P_i \in A \in X\}} \left\lfloor \frac{d_H(A)}{t} \right\rfloor$$

of the d_i t–subsets in Γ_0 that contain P_i. Hence, the number of Γ_A's that contain P_i is at most

$$|\{A \in X : P_i \in A\}| + \sum_{\{A:P_i \in A \in X\}} \left(\frac{d_H(A)}{t-1} - \left\lfloor \frac{d_H(A)}{t} \right\rfloor \right)$$

$$\leq |\{A \in X : P_i \in A\}| + \sum_{\{A:P_i \in A \in X\}} \left(\frac{d_H(A)}{t-1} - \frac{d_H(A) - t + 1}{t} \right)$$

$$= \frac{d_i}{t} + \frac{2t-1}{t} |\{A \in X : P_i \in A\}|$$

It is easy to see that

$$|\{A \in X : P_i \in A\}| \leq \binom{w-1}{t-2};$$

hence,

$$R_i \geq \frac{t}{(2t-1)\binom{w-1}{t-2} + d_i}$$

for $1 \leq i \leq w$.

Now ρ is just the minimum of the R_i's. To compute a bound on $\tilde{\rho}$, we use the remark following Theorem 2. We calculate:

$$\tilde{\rho} = \frac{w}{\sum_{i=1}^{w} \frac{1}{R_i}}$$

$$\geq \frac{wt}{\sum_{i=1}^{w} \left((2t-1)\binom{w-1}{t-2} + d_i \right)}$$

$$= \frac{wt}{w(2t-1)\binom{w-1}{t-2} + t|\Gamma_0|}.$$

Summarizing, we have the following generalization of Theorem 9:

Theorem 11. *Let Γ be a uniform access structure of rank t on w participants, and denote by d the maximum degree of any participant. Then there exists a $PS(\Gamma, \frac{t}{(2t-1)\binom{w-1}{t-2}+d}, q)$ and a $\widetilde{PS}(\Gamma, \frac{wt}{w(2t-1)\binom{w-1}{t-2}+t|\Gamma_0|}, q)$ for any prime power $q \geq t$.*

Asymptotically, the bound on ρ represents an improvement by a factor of t to the rate that would be obtained from the Benaloh-Leichter construction [1] using a disjunctive normal form boolean circuit.

Finally, note that if Γ is a non-uniform access structure of rank t, we can first partition the basis as $\Gamma_0 = \cup_{i=1}^{t}\Gamma_i$, where each Γ_i is uniform of rank i, and then apply the techniques of this section to each Γ_i.

References

1. J. Benaloh and J. Leichter. Generalized secret sharing and monotone functions. Lecture Notes in Computer Science **403** (1990) 27–35.
2. T. Beth, D. Jungnickel, and H. Lenz. Design Theory. Bibliographisches Institut, Zurich, 1985.
3. C. Blundo. Secret Sharing Schemes for Access Structures based on Graphs. Tesi di Laurea, University of Salerno, 1991.
4. C. Blundo, A. De Santis, D. R. Stinson, and U. Vaccaro. Graph decompositions and secret sharing schemes. Presented at EUROCRYPT '92, submitted to Journal of Cryptology.
5. C. Blundo, A. De Santis, L. Gargano, and U. Vaccaro. On the information rate of secret sharing schemes. Presented at CRYPTO '92.
6. B. Bollobás. Graph Theory – An Introductory Course. Springer-Verlag, 1979.
7. E. F. Brickell. Some ideal secret sharing schemes. J. Combin. Math. and Combin. Comput. **9** (1989) 105–113.
8. E. F. Brickell and D. M. Davenport. On the classification of ideal secret sharing schemes. J. Cryptology **4** (1991), 123–134.
9. E. F. Brickell and D. R. Stinson. Some improved bounds on the information rate of perfect secret sharing schemes. J. Cryptology (to appear), preliminary version appeared in Lecture Notes in Computer Science **537** (1991) 242–252.
10. R. M. Capocelli, A. De Santis, L. Gargano, and U. Vaccaro. On the size of shares for secret sharing schemes. Submitted to Journal of Cryptology, preliminary version appeared in Lecture Notes in Computer Science **576** (1992) 101–113.
11. D. de Werra. Equitable colorations of graphs. Rev. Franc. Automat. Informat. Rech. Operat. Sér. Rouge **3** (1971) 3–8.
12. H. Hanani. On quadruple systems. Canad. J. Math. **12** (1960) 145–157
13. H. Hanani. Balanced incomplete block designs and related designs. Discrete Math. **11** (1975) 255–369.
14. M. Ito, A. Saito, and T. Nishizeki. Secret sharing scheme realizing general access structure. Proc. IEEE Globecom '87 (1987) 99–102.
15. W.-A. Jackson and K. M. Martin. On ideal secret sharing schemes. Submitted to J. Cryptology.
16. K. M. Martin. New secret sharing schemes from old. Submitted to J. Comb. Math. Comb. Comp.

17. K. M. Martin. Discrete Structures in the Theory of Secret Sharing. PhD thesis, University of London, 1991.
18. P. D. Seymour. On secret-sharing matroids. Journal of Combin. Theory B (to appear).
19. A. Shamir. How to share a secret. Commun. of the ACM **22** (1979) 612–613.
20. G. J. Simmons. An introduction to shared secret and/or shared control schemes and their application. In G. J. Simmons, editor, Contemporary Cryptology, The Science of Information Integrity, IEEE Press, 1991, pp. 441–447.
21. G. J. Simmons, W. Jackson, and K. Martin. The geometry of shared secret schemes. Bulletin of the ICA **1** (1991) 71–88.
22. D. R. Stinson. An explication of secret sharing schemes. Designs, Codes and Cryptography (to appear).

Universally Ideal Secret Sharing Schemes (Preliminary Version)

Amos Beimel* and Benny Chor **

Department of Computer Science
Technion, Haifa 32000, Israel

"I weep for you," the Walrus said,
"I deeply sympathize."
With sobs and tears he sorted out
Those of the largest size,
Holding his pocket-handkerchief
Before his streaming eyes.

"O Oysters," said the Carpenter.
"You've had a pleasant run!
Shall we be trotting home again?"
But answer came there none –
And this scarcely odd, because
They'd eaten every one.

from "Through the looking Glass" by Lewis Caroll

Abstract. Given a set of parties $\{1, \ldots, n\}$, an access structure is a monotone collection of subsets of the parties. For a certain domain of secrets, a secret sharing scheme for an access structure is a method for a dealer to distribute shares to the parties, such that only subsets in the access structure can reconstruct the secret.

A secret sharing scheme is *ideal* if the domains of the shares are the same as the domain of the secrets. An access structure is *universally ideal* if there is an ideal secret sharing scheme for it over every finite domain of secrets. An obvious necessary condition for an access structure to be universally ideal is to be ideal over the binary and ternary domains of secrets. In this work, we prove that this condition is also sufficient. In addition, we give an exact characterization for each of these two conditions, and show that each condition by itself is not sufficient for universally ideal access structures.

1 Introduction

A secret sharing scheme involves a dealer who has a secret, a finite set of n parties, and a collection $\mathcal{A}$ of subsets of the parties called the access structure. A secret-sharing scheme for $\mathcal{A}$ is a method by which the dealer distributes shares to the parties such that any subset in $\mathcal{A}$ can reconstruct the secret from its shares, and any subset not in $\mathcal{A}$ cannot reveal any partial information about the secret (in the information theoretic sense). A secret sharing scheme can only exist for monotone access structures, i.e. if a subset A can reconstruct the secret, then every superset of A can also reconstruct the secret. If the subsets that can reconstruct the secret are all the sets whose cardinality is at least a certain

* email: beimel@cs.technion.ac.il
** Supported by the Fund for Promotion of Research at the Technion. email: benny@cs.technion.ac.il

threshold t, then the scheme is called t out of n threshold secret sharing scheme. Threshold secret sharing schemes were first introduced by Blakley [Bla79] and by Shamir [Sha79]. Secret sharing schemes for general access structures were first defined by Ito, Saito and Nishizeki in [ISN87]. Given any monotone access structure, they show how to realize a secret sharing scheme for the access structure. Benaloh and Leichter [BL88] describe a more efficient way to realize such secret sharing schemes.

Even with the more efficient scheme of [BL88], most access structures require shares of exponential size: Even if the domain of the secret is binary, the shares are strings of length $2^{\Theta(n)}$, where n is the number of participants. The question of lower bounds on the size of shares for some (explicit or random) access structures is still open. On the other hand, certain access structures give rise to very economical secret sharing schemes. A secret sharing scheme is called *ideal* if the shares are taken from the same domain as the secrets. An access structure is called m−ideal if there is an ideal secret sharing scheme which realizes the access structure over a domain of secrets of size m.

Brickell [Bri89] was the first to introduce the notion of m−ideal access structures. Brickell and Davenport [BD91] have shown that such structures are closely related to matroids over a set containing the participants plus the dealer. They give a necessary condition for an access structure to be m−ideal (being a matroid) and a somewhat stronger sufficient condition (the matroid should be representable over a field or algebra of size m). Certain access structures, such as the threshold ones, are m−ideal for m that is at least n. However, for domains of secrets which contain m elements where m is smaller then n, the threshold access structures are *not* m−ideal (for threshold t such that $2 \leq t \leq n - 1$), as proved by Karnin, Greene and Hellman [KGH83]. This qualitative result was improved by Kilian and Nisan [KN90], who showed that the t out of n threshold secret sharing scheme over a binary domain of secrets requires shares from a domain that is at least of size $n - t + 2$ (for $2 \leq t \leq n - 1$).

We say that an access structure is *universally* ideal if for every positive integer m, it is m−ideal. Universally ideal access structures are particularly convenient to work with because they are very efficient no matter what the domain of secrets is. A simple example of a universally ideal access structure is the n out of n threshold access structure. In this work we give a complete characterization of universally ideal access structures. Our work builds upon results of Brickell and Davenport which relate ideal access structures to matroids, as well as some known results from matroid theory. An obvious necessary condition for an access structure to be universally ideal is to be both 2−ideal and 3−ideal. Interestingly, our main result states that this condition is also sufficient. We give examples which demonstrate that just one of these two requirements is not a sufficient condition to be universally ideal.

The remaining of this paper is organized as following. In section 2 we give formal definitions and quote the results of Brickell and Davenport. Section 3 states our main theorem, and details its proof. Section 4 illustrates some clarifying examples.

2   Definitions and Related Results

This section contains formal definitions and known related results, that will be used in the rest of this paper.

2.1   Secret Sharing Schemes

The definition of secret sharing schemes is based on [CK89].

Definition 1. Let $S = \{0, \ldots, m-1\}$ be a finite set of secrets. Let $\mathcal{A} \subseteq 2^{\{1,\ldots,n\}}$ be a monotone set (such that $\emptyset \notin \mathcal{A}$) called the *access structure*. We say that a *secret-sharing scheme* Π realizes an access structure $\mathcal{A}$ with domain of secrets S if Π is a mapping $\Pi : S \times R \to S_1 \times S_2 \times \ldots \times S_n$ from the cross product of secrets and a set of random inputs to a set of n-tuples (the shares) such that the following two requirements hold:

1. The secret s can be reconstructed by any subset in $\mathcal{A}$. That is, for any subset $A \in \mathcal{A}$ ($A = \{i_1, \ldots, i_{|A|}\}$), there exists a function $h_A : S_{i_1} \times \ldots \times S_{i_{|A|}} \to S$ such that for every random inputs r it holds that if $\Pi(s, r) = \{s_1, s_2, \ldots, s_n\}$ then $h_A(\{s_i\}_{i \in A}) = s$.

2. Every subset not in $\mathcal{A}$ can not reveal any partial information about the secret (in the information theoretic sense). Formally, for any subset $A \notin \mathcal{A}$, for every two secrets $a, b \in S$, and for every possible shares $\{s_i\}_{i \in A}$:

$$\Pr_r[\{s_i\}_{i \in A} \mid a] = \Pr_r[\{s_i\}_{i \in A} \mid b]$$

We denote the shares of party i by $\Pi_i(s, r)$.

Given a collection $\Gamma \subseteq 2^{\{1,\ldots,n\}}$ the closure of Γ, denoted by $\mathrm{cl}(\Gamma)$, is the minimum collection that contains Γ and is monotone (if $B \in \mathrm{cl}(\Gamma)$ and $B \subseteq C$ then $C \in \mathrm{cl}(\Gamma)$). Given an access structure $\mathcal{A}$, we denote $\mathcal{A}_m$ to be the collection of minimal sets of $\mathcal{A}$, that is $B \in \mathcal{A}_m$ if $B \in \mathcal{A}$ and for every $C \subsetneq B$ it holds that $C \notin \mathcal{A}$. If $\mathcal{A} = \{A : |A| \geq t\}$, then a secret sharing for $\mathcal{A}$ is called a t out of n threshold secret sharing scheme, and the access structure $\mathcal{A}$ is called the t out of n threshold access structure.

Definition 2. A secret sharing scheme $\Pi : S \times R \to S_1 \times \ldots \times S_n$ is $m-ideal$ if $|S_1| = |S_2| = \ldots = |S_n| = |S| = m$, that is the domain of the shares of each party has the same size as the domain of the secrets, and this domain contains m elements. An *access structure* $\mathcal{A}$ *is* $m-ideal$ if there exists a $m-$ideal secret sharing scheme that realizes $\mathcal{A}$. An access structure $\mathcal{A}$ is *universally ideal* if for every positive integer m the access structure $\mathcal{A}$ is $m-$ideal.

2.2 Matroids

Before we continue, we recall the definition of matroids . Matroids are well studied combinatorial objects (see for example Welsh [Wel76]). A matroid is an axiomatic abstraction of linear independence. We give here one of the equivalent axiom systems that define matroids. A matroid $T = (V, \mathcal{I})$ is a finite set V and a collection $\mathcal{I}$ of subsets of V such that **(I1)** through **(I3)** are satisfied.

(I1) $\emptyset \in \mathcal{I}$.
(I2) If $X \in \mathcal{I}$ and $Y \subseteq X$ then $Y \in \mathcal{I}$.
(I3) If X, Y are members of $\mathcal{I}$ with $|X| = |Y| + 1$ there exists $x \in X \backslash Y$ such that $Y \cup \{x\} \in \mathcal{I}$.

For example every finite vector space is a matroid, in which V is the set of vectors and $\mathcal{I}$ is the collection of the independent sets of vectors. The elements of V are called the *points* of the matroid and the sets in $\mathcal{I}$ are called *independent sets*. A *dependent set* of a matroid is any subset of V that is not independent. The minimal dependent sets are called *circuits*. A matroid is said to be *connected* if for any two elements in V, there is a circuit containing both of them. The maximal independent sets are called *bases*. In every matroid, all bases have the same cardinality, which is defined as the *rank* of a matroid. A matroid is *representable* over a field $\mathcal{F}$ if there exists a dependence preserving mapping from the points of the matroid into the set of vectors of a vector space over the field. In other words, there exist k and a mapping $\phi : V \to \mathcal{F}^k$ that satisfies:

$$A \subseteq V \text{ is a dependent set of the matroid iff } \phi(A) \text{ is linearly dependent.}$$

2.3 Relation between Secret Sharing Schemes and Matroids

The next definition relates access structures and matroids.

Definition 3. Let $\mathcal{A}$ be an access structure with n parties $\{1, \ldots, n\}$ and let $T = (V, \mathcal{I})$ be a connected matroid. We say that the matroid T is *appropriate* for the access structure $\mathcal{A}$ if $V = \{0, \ldots, n\}$ and

$$\mathcal{A} = \mathrm{cl}(\{C \backslash \{0\} : 0 \in C \text{ and } C \text{ is a minimal dependent set of } T\})$$

That is, the minimal sets of the access structure $\mathcal{A}$ correspond to the minimal dependent sets in the matroid which contain 0. Intuitively, 0 is added to the set $\{1, \ldots, n\}$ to "play the role" of the dealer.

There are various properties which the collection of minimal dependent sets in a matroid must satisfy, and these properties do not necessarily hold for an arbitrary access structure. Not every access structure has an appropriate matroid. But if a connected matroid is appropriate for an access structure, then it is the only matroid with this property (see [Wel76], Theorem 5.4.1). Brickell and Davenport [BD91] have found relations between the two notions when $\mathcal{A}$ is an ideal access structure. The next two theorems almost characterize m−ideal access structures.

Theorem 4 (necessary condition) [BD91]. *If a non-degenerate access structure A is m-ideal for some positive integer m, then there exists a connected matroid T that is appropriate for A.*

Theorem 5 (sufficient condition) [BD91]. [3] *Let q be a prime power, and A be a non-degenerate access structure. Suppose that there is a connected matroid T that is appropriate for A. If T is representable over the field $\mathrm{GF}(q)$, then A is q-ideal.*

3 The Characterization Theorem

The two theorems of Brickell and Davenport almost characterize q-ideal access structures for q a prime power. However, If there is a connected matroid T that is appropriate for A but is not representable over the field $\mathrm{GF}(q)$, then the theorems do not determine whether or not A is q-ideal. While we do not close the remaining gap for q-ideal access structures, we do give a complete characterization for universally ideal ones. We recall that an access structure A is universally ideal if it is q-ideal for any finite domain of secrets. Our main result is:

Theorem 6. *The access structure A is universally ideal if and only if A is binary-ideal (2-ideal) and ternary-ideal (3-ideal).*

The proof of the theorem proceeds along the following lines: We strengthen Theorem 4 of Brickell and Davenport for the binary and ternary domains of secrets. We show that over these domains, every reconstruction function can be expressed as a linear combination of the shares of the parties. This enables us to show that if an access structure A is binary ideal, then there is a matroid T that is appropriate for A and is representable over the binary field. The same result is proved for the ternary field. Then, using a known result from matroid theory, we conclude that if an access structure A is binary and ternary ideal, then there is a matroid T appropriate for A which is representable over *any* field. Thus, by Theorem 5 of Brickell and Davenport, the access structure is q-ideal for any prime power q. Using the Chinese remainder Theorem, A is m-ideal over any finite domain, namely is universally ideal, as desired.

Definition 7. Let Π be a secret sharing scheme for n parties $\{1, \ldots, n\}$, and the dealer which we denote by 0. The secret will be considered as the share of party 0 – the dealer. Let $A \subseteq \{0, \ldots, n\}$ and $i \in \{0, \ldots, n\}$. The parties in A *cannot reveal any information* about the share of i if for every distribution on the secrets, every possible shares $\{s_a\}_{a \in A}$, and every possible shares s_i, s_i'

$$\Pr_{s,r}[\ \Pi_i(s,r) = s_i \mid \{s_a\}_{a \in A}\ ] = \Pr_{s,r}[\ \Pi_i(s,r) = s_i' \mid \{s_a\}_{a \in A}\ ]$$

We also say that i is independent of A with respect to Π.

---

[3] The Theorem in [BD91] had a slightly weaker condition, which we omit for simplicity.

Definition 1 implies that if $A \subseteq \{1, \ldots, n\}$ and $A \notin \mathcal{A}$, then in every secret sharing scheme realizing $\mathcal{A}$ the *secret* (i.e. the share of the dealer) is independent of the shares of the parties in A.

Definition 8. Let Π be a secret sharing scheme. We say that a subset $A \subseteq \{0, 1, \ldots, n\}$ is *dependent* with respect to Π if there exists an $i \in A$ such that the parties in $A \setminus \{i\}$ can reconstruct the share of i (in the sense of definition 1). A subset $A \subseteq \{0, \ldots, n\}$ is *independent* if for every $i \in A$, i is independent of $A \setminus \{i\}$ with respect to Π.

Notice that the notions of dependent and independent set with respect to a given secret sharing schemes are *not* complementary. There could be a subset A of parties which could neither reconstruct the share of any of its members (and thus A in not dependent), yet could reveal some information on the share of one of its members (and thus A is not independent). However, for *ideal* secret sharing scheme, the following theorem of Brickell and Davenport [BD91] establishes the desired relation between the two notions.

Theorem 9 [BD91]. *Let Π be an ideal secret sharing scheme realizing a non-degenerate access structure $\mathcal{A}$ with n parties $\{1, \ldots, n\}$ over some domain of secrets S. Let $A \subseteq \{0, \ldots, n\}$. Then*

1. *The subset A is either dependent or independent with respect to Π.*
2. *The subset A is independent with respect to Π if and only if A is an independent set in a matroid $\mathcal{T}$ which is appropriate for $\mathcal{A}$.*

Definition 10. Let q be a prime power, and Π a q-ideal secret sharing scheme. We say that Π is *linear* if for every set that is dependent with respect to Π, the reconstruction function is linear. That is, for every $A \subseteq \{0, \ldots, n\}$ and every $0 \leq i \leq n$ such that $i \notin A$ and i depends on A with respect to Π, there are constants $\{\alpha_j\}_{j \in A}$, σ (all in $\mathrm{GF}(q)$) such that for every secret $s \in \mathrm{GF}(q)$ and choice of random inputs $r \in R$

$$\Pi_i(s, r) = \sigma + \sum_{j \in A} \alpha_j \Pi_j(s, r)$$

where the sum is mod q.

We remark that the secret sharing scheme of Shamir [Sha79] is linear. The secret (or any other share) is reconstructed from the shares by substitution in the interpolating polynomial. The sufficient condition of Brickell and Davenport [BD91] (theorem 5) states that if an access structure $\mathcal{A}$ has an appropriate matroid which is representable over $\mathrm{GF}(q)$, then $\mathcal{A}$ is q-ideal. Their scheme, using our terminology, is a linear q-ideal secret sharing scheme. Our next lemma states the reverse direction.

Lemma 11. *If an access structure $\mathcal{A}$ has a linear q-ideal secret sharing scheme, then $\mathcal{A}$ has an appropriate matroid which is representable over $\mathrm{GF}(q)$.*

Proof (sketch). By Theorem 4 there is a matroid which is appropriate for $\mathcal{A}$. Let Π be a linear q-ideal secret sharing scheme for the access structure $\mathcal{A}$. Using Π, we will construct a dependence preserving mapping ϕ from the set of points of the matroid, $\{0, \ldots, n\}$, into a vector space over $\mathrm{GF}(q)$.

The mapping ϕ will be constructed in two stages. In the first stage we will map $V = \{0, \ldots, n\}$ to $\mathrm{GF}(q)^{q \times |R|}$, where R is the source of randomness used in Π. For every $a \in V$ we define

$$\phi_1(a) = (\ \Pi_a(s_1, r_1), \Pi_a(s_1, r_2), \ldots, \Pi_a(s_q, r_{|R|})\ )$$

intuitively $\phi_1(a)$ describes the shares of party a with every secret and every random input. In the second stage we construct a mapping ϕ_2 which fixes some remaining technicalities. We leave the details to the final version of this paper. These two mappings ϕ_1 and ϕ_2 have the property that $A \subseteq V$ is dependent in $\mathcal{T}$ if and only if $\phi_2 \circ \phi_1(A)$ is linearly dependent in $\mathrm{GF}(q)^t$. Thus $\phi = \phi_2 \circ \phi_1$ is a dependence preserving mapping, and by definition the appropriate matroid $\mathcal{T}$ is representable over $\mathrm{GF}(q)$. □

Definition 12. We say that a function $f : S^t \to S$ is *component sensitive* if for every $1 \le i \le t$, every $s_1, \ldots, s_{i-1}, s_i, s_i', s_{i+1}, \ldots, s_t \in S$ $(s_i' \ne s_i)$:

$$f(s_1, \ldots, s_{i-1}, s_i, s_{i+1}, \ldots, s_t) \ne f(s_1, \ldots, s_{i-1}, s_i', s_{i+1}, \ldots, s_t).$$

In other words, every change of the value of one variable of f, changes the value of f.

Lemma 13. *Let Π be a q-ideal secret sharing scheme. Let $i \in \{0, \ldots, n\}$, and $A \subseteq \{0, \ldots, n\}$ be a minimal subset such that i depends on A and $i \notin A$. Let $f : S^{|A|} \to S$ be the reconstruction function of the i-th share from the shares of the parties in A. Then f is component sensitive.*

Proof. Omitted from this preliminary version.

We now show that the only component sensitive functions for the binary and for the ternary domains are linear. We start with the binary case.

Lemma 14. *Let $f : \mathrm{GF}(2)^t \to \mathrm{GF}(2)$ be a component sensitive function. Then f can be expressed as a linear function with non-zero coefficients over $\mathrm{GF}(2)$:*

$$f(x_1, \ldots, x_t) = \sigma + \sum_{i=1}^{t} \alpha_i x_i \quad (\alpha_i \ne 0 \text{ for all } i).$$

Proof. Omitted from this preliminary version.

We use Lemma 14 to give an exact characterization of binary-ideal access structures.

Corollary 15. *An access structure $\mathcal{A}$ is binary-ideal if and only if there is a matroid which is representable over $\mathrm{GF}(2)$ and is appropriate for $\mathcal{A}$.*

Proof. Let Π be a binary-ideal secret sharing scheme that realizes the access structure $\mathcal{A}$. By lemma 13 the reconstruction function of every dependent set is component sensitive. Therefore by lemma 14 every reconstruction function is linear over $GF(2)$, or in other words Π is a linear scheme. By lemma 11, We conclude that if $\mathcal{A}$ is binary-ideal then $\mathcal{A}$ has an appropriate matroid that is representable over $GF(2)$. The other direction is implied by the sufficient condition of Brickell and Davenport [BD91] (theorem 5). □

The next lemma paralles Lemma 14, this time for the ternary case.

Lemma 16. *Let $f : GF(3)^t \to GF(3)$ be a component sensitive function. Then f can be expressed as a linear function with non-zero coefficients over $GF(3)$:*

$$f(x_1, \ldots, x_t) = \sigma + \sum_{i=1}^{t} \alpha_i x_i \quad (\alpha_i \neq 0 \text{ for all } i).$$

Proof (sketch). The proof relies on the observation that any partial assignment to the variables of a component sensitive function results in a new component sensitive function (of the remaining variables). In addition, a component sensitive function of *one* variable is a permutation of its domain.

For any finite field $GF(q)$, any function which maps $GF(q)^t$ into $GF(q)$ can be expressed as a multivariable polynomial over the field, in which every monomial of f contains variables whose powers do not exceed $q - 1$ (since $x^q \equiv x$). In our case the power will not exceed 2.

We first show that no term in the polynomial f contains a variable of degree 2. Suppose, without loss of generality, that x_1^2 appears in some monomial. The polynomial f will have the form:

$$x_1^2 \cdot p_1(x_2, \ldots, x_n) + x_1 \cdot p_2(x_2, \ldots, x_n) + p_3(x_2, \ldots, x_n)$$

where the polynomial p_1 is not identically zero, and p_2, p_3 are arbitary polynomials. Hence there exists a substitution to the variables $x_2, \ldots, x_n$ such that the value of p_1 after the substitution is not zero. This substitution to f yeilds a polynomial in x_1, of the form $ax_1^2 + bx_1 + c$. The coefficient of x_1, a, is non–zero. By the observation mentioned above, the resulting function of x_1 should also be component sensitive. It is not hard to check that any degree 2 polynomial over $GF(3)$ is not a permutation[4], and therefore is not component sensitive. Thus f contains no variable of degree 2, so all its monomials are multilinear.

We still have to show that f contains no monomial with two variables. We leave the details to the final version of the paper. □

We remark that $GF(3)$ is the largest field where every component sensitive function is linear. Already for $GF(4)$, there are $4! = 24$ component sensitive functions of one variable (permutations), but only $3 \cdot 4 = 12$ non-constant linear

---

[4] Every polynomial of the form $a \cdot x_1 + b$ where $a \neq 0$ is a permutation. There are 6 such polynomials and there are 6 permutations over $GF(3)$, therefore every degree 2 polynomial cannot be a permutation.

functions. Now using the same arguments as in the proof of Corollary 15 (for the binary case), we conclude with the following charcterization of ternary-ideal access structures.

Corollary 17. *An access structure* $\mathcal{A}$ *is ternary-ideal, if and only if there is a matroid which is representable over* $GF(3)$ *and is appropriate for* $\mathcal{A}$.

We saw that representation over $GF(2)$ determines if an access structure is binary-ideal, and representation over $GF(3)$ determines if an access structure is ternary-ideal. Therefore, if an access structure is both binary-ideal and ternary-ideal, then it has an appropriate matroid that is representable over $GF(2)$ and over $GF(3)$. The next proposition from [Wel76] states strong implications of the representatability over the two finite fields. It will be used to complete the proof of our main theorem.

Proposition 18. *A matroid* $\mathcal{T}$ *is representable over* $GF(2)$ *and over* $GF(3)$ *if and only if* $\mathcal{T}$ *is representable over any field.*

Using this proposition we get:

Corollary 19. *If an access structure* $\mathcal{A}$ *is binary-ideal and ternary-ideal then for every* q *such that* q *is a prime power,* $\mathcal{A}$ *is* $q-ideal$.

Proof. If an access structure $\mathcal{A}$ is binary-ideal and ternary-ideal, then by corollaries 15 and 17 the access structure $\mathcal{A}$ has an appropriate matroid $\mathcal{T}$ that is representable over $GF(2)$ and over $GF(3)$ (remember that there can be only one appropriate matroid for $\mathcal{A}$). Hence proposition 18 implies that $\mathcal{T}$ is representable over any field. From Theorem 4 we conclude that the access structure $\mathcal{A}$ is ideal over any finite field, i.e. $\mathcal{A}$ is $q-$ideal for every prime-power q. □

Corollary 20. *If an access structure* $\mathcal{A}$ *is binary-ideal and ternary-ideal then for every positive integer* m, *the access structure* $\mathcal{A}$ *is* $m-ideal$.

Proof. Let S be a finite domain of secrets of size m. Let $m = p_1^{i_1} \cdot p_2^{i_2} \cdot \ldots \cdot p_t^{i_t}$ where p_j are distinct primes. Given a secret $s \in S$ for every $1 \le j \le t$, independently, we use the ideal secret sharing scheme to share $s \bmod p_j^{i_j}$. Every subset of parties $A \in \mathcal{A}$ can reconstruct $s \bmod p_j^{i_j}$, therefore using the Chinese remainder Theorem, they can reconstruct the secret. Since for each j the secret $s \bmod p_j^{i_j}$ is shared independently, then every subset $A \notin \mathcal{A}$ does not know anything about the secret s. □

This last corollary is a restatement of Theorem 6, and it completes the arguments in the proof of our main result.

4 Examples

In this section we formulate several known constructions from matroid theory as ideal access structures. Our first two examples show that the condition of

Theorem 6 cannot be relaxed: Being either just 2–ideal or just 3–ideal is not sufficient for being universally ideal. Then, we demonstrate how graphic and cographic matroids give rise to interesting classes of universally ideal access schemes.

Example 1 (the 2 out of 3 access structure) . We recall that the 2 out of 3 access structure is the access structure with 3 parties in which every two parties together can reconstruct the secret, and every party by itself does not know anything about the secret. The appropriate matroid for this access structure is the matroid with $V = \{0, 1, 2, 3\}$ and $\mathcal{I} = \{A : |A| \leq 2\}$. It is not difficult to verify that this matroid is not representable over GF(2), hence the 2 out of 3 access structure is not 2-ideal. But this access structure is 3-ideal, as the following scheme demonstrates:

Let $s \in \{0, 1, 2\}$ be the secret. The dealer chooses at random a number $r \in \{0, 1, 2\}$. the share of party 1 is r, the share of party 2 is $r + s$, and the share of party 3 is $r + 2s$. This access structure demonstrates that being 3–ideal does not suffice to guarantee that an access scheme is universally ideal.

Example 2. Consider the following access structure $\mathcal{F}$ (see Fig. 1). The set of parties is $\{1, 2, 3, 4, 5, 6\}$. The Access structure is the closure of the set

$$\mathcal{F}_m = \{\{1, 4\}, \{2, 5\}, \{3, 6\}, \{1, 2, 6\}, \{1, 3, 5\}, \{2, 3, 4\}, \{4, 5, 6\}\}.$$

The matroid that is appropriate for this access structure is the Fano matroid [Wel76], which is representable only over fields of characteristic 2. Hence $\mathcal{F}$ is 2–ideal, and is not 3–ideal. The 2–ideal secret sharing scheme for $\mathcal{F}$ uses two random bits r_0, r_1 which are chosen independently with uniform distribution. The scheme is described in Fig. 2. This access structure demonstrates that being 2–ideal does not suffice to guarantee that an access scheme is universally ideal.

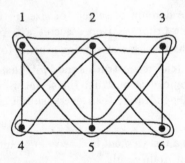

Fig. 1. The minimal sets of the access structure $\mathcal{F}$

The access structure $\mathcal{F}' = \mathrm{cl}(\mathcal{F}_m \cup \{3, 4, 5\})$ has a appropriate matroid that is representable over GF(3) but not over GF(2) [Wel76]. Actually, the 3–ideal

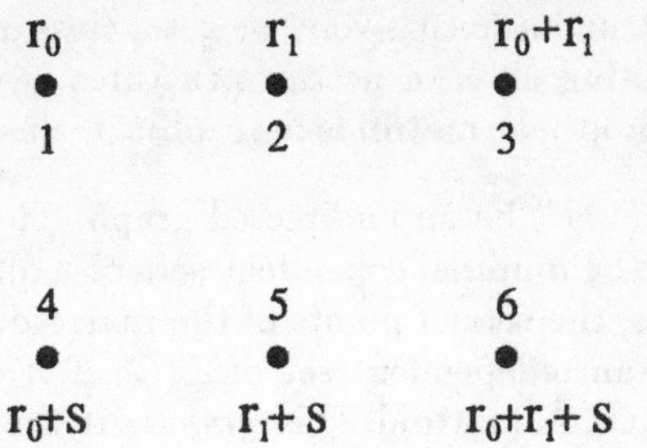

$$\begin{array}{ccc} \mathbf{r_0} & \mathbf{r_1} & \mathbf{r_0 + r_1} \\ \bullet & \bullet & \bullet \\ 1 & 2 & 3 \\ \\ 4 & 5 & 6 \\ \bullet & \bullet & \bullet \\ \mathbf{r_0 + S} & \mathbf{r_1 + S} & \mathbf{r_0 + r_1 + S} \end{array}$$

Fig. 2. An ideal scheme for $\mathcal{F}$ with secret s and random independent inputs r_0, r_1.

secret sharing scheme for $\mathcal{F}'$ is the same as the binary scheme for $\mathcal{F}$, except here r_0, r_1 are chosen uniformly and independently from $\{0, 1, 2\}$. Notice that the parties $\{3, 4, 5\}$ can reconstruct $2s$ over the two fields, which is useless over $GF(2)$, but enables to reconstruct the secret over $GF(3)$. This access structure demonstrates again that being 3–ideal does not suffice to guarantee that an access scheme is universally ideal.

Example 3. Here we give a method for combining two ideal access structures for n and ℓ parties into a new ideal access structure for $n + \ell - 1$ parties. Let $\mathcal{A}$ be a non-degenerate access structure with parties $\{1, \ldots, n\}$, and let $\mathcal{A}_1$ be an access structure with parties $\{n + 1, \ldots, n + \ell\}$. We denote by $\mathcal{A}' = \mathcal{A}(i, \mathcal{A}_1)$ the access structure with $n + \ell - 1$ parties $\{1, \ldots, i - 1, i + 1, \ldots, n, n + 1, \ldots, n + \ell\}$, and reconstructing sets

$$\mathcal{A}' = \{e : e \in \mathcal{A} \text{ and } i \notin e\} \cup \left\{ (e \setminus \{i\}) \bigcup e_1 : e \in \mathcal{A}, i \in e, \text{ and } e_1 \in \mathcal{A}_1 \right\}.$$

That is, the sets that can reconstruct the secret in the new access structure are:

- The sets from $\mathcal{A}$ that do not contain party i.
- The sets from $\mathcal{A}$ that do contain party i, in which we replace the party i with each set of $\mathcal{A}_1$.

Let $\mathcal{A}$ be a non-degenerate access structure, let i be a party in $\mathcal{A}$, and let $\mathcal{A}_1$ be an access structure. We will show that if $\mathcal{A}$ and $\mathcal{A}_1$ are universally ideal then $\mathcal{A}' = \mathcal{A}(i, \mathcal{A}_1)$ is universally ideal, by describing (for every m) an m–ideal secret sharing scheme for $\mathcal{A}'$. Given a secret s use an m–ideal scheme to generate shares for the parties in $\mathcal{A}$. Let a be the random variable that denotes the share of party i in the scheme for $\mathcal{A}$. Now use an m–ideal scheme for $\mathcal{A}_1$ with secret a to generate shares for the parties in $\mathcal{A}_1$.

It is easy to see that the 1 out of 2 threshold access structure is universally ideal (give the secret to the two parties). The 2 out of 2 threshold access structure is also universally ideal (give the first party a random input r, and to the second party $s + r \bmod m$). Using these two access structures as building blocks, and using the above construction recursively, we get a class of universally ideal access structures. The resulting class of access structures is a special case of access structures whose appropriate matroids is graphic, a class which we discuss next.

Example 4. Let $G = (V, E)$ be an undirected graph. The cycles of G (as defined in graph theory) are the minimal dependent sets of a matroid $T(G)$ on the edge set E. In other words, the sets of points of the matroid $T(G)$ is the set of *edges* of G, and $A \subseteq E$ is an independent set of $T(G)$ if A does not contain cycles, i.e. A is a forest in G. A matroid T is *graphic* if there exists some graph G such that T is isomorphic to the cycle matroid $T(G)$. Every graphic matroid is representable over any field [Wel76]. Therefore if an access structure $\mathcal{A}$ has a graphic appropriate matroid, then $\mathcal{A}$ is universally ideal. To be more precise, let $G = (V, E)$ where $V = \{0, 1, \ldots, n\}$, $E \subseteq V \times V$, and $e_0 = (0, 1) \in E$ be a special edge which corresponds to the dealer. Let

$$\mathcal{A}(G) = \text{cl}(\{C \setminus \{e_0\} : C \subseteq E \text{ is a minimal cycle that contains } e_0\})$$

Then $\mathcal{A}(G)$ is universally ideal. The scheme Π for graphic matroids is actually quite simple. Let $r = < r_1, r_2, \ldots, r_{|V|-1} >$ be the random input ($|V| - 1$ independent values). Then for every $(i, j) \in E$ ($i \leq j$)

$$\Pi_{(i,j)}(s, r) = \begin{cases} r_i - r_j & i \neq 0 \\ r_1 + s - r_j & i = 0 \end{cases}$$

For every simple path which starts at node 1, and ends at node 0, it is possible to assign ± 1 weights to the shares along the path, such that the weighted sum is equel to the secret s. This scheme was found previously (not in the context of graphic matroids) by Benaloh and Rudich [BR89].

We demonstrate this construction on a specific graph G_0, shown in Fig. 3. The cycles in the graph are:

$$\{e_0, e_2, e_3\}, \{e_0, e_1, e_2, e_4\}, \{e_1, e_3, e_4\},$$

and these sets are the minimal dependent sets of $T(G_0)$. The access structure $\mathcal{A}(G_0)$ is the closure of $\{\{e_2, e_3\}, \{e_1, e_2, e_4\}\}$. The dealer is the edge e_0. The shares of the parties e_2 and e_3 are $r_1 - r_2$ and $r_1 + s - r_2$ respectably and they can reconstruct the secret by substructing their shares.

Example 5. Let $G = (V, E)$ be an undirected graph. A cut in G is a collection of edges such that deleting them from G, increases the number of connected components in the remaining graph. The cuts of G are the minimal dependent sets of a matroid $T^*(G)$ on the edge set E. A matroid T is cographic if there exists some graph G such that T is isomorphic to the cut matroid $T^*(G)$. Every cographic matroid is representable over any field [Wel76]. Therefore if an access structure $\mathcal{A}$ has a cographic appropriate matroid, then $\mathcal{A}$ is universally ideal. To be more precise, let $G = (V, E)$ where $V = \{0, 1, \ldots, n\}$, $E \subseteq V \times V$, and $e_0 = (0, 1) \in E$ be a special edge which coresponds to the dealer. Let

$$\mathcal{A}^*(G) = \text{cl}(\{C \setminus \{e_0\} : C \subseteq E \text{ is a minimal cut that contains } e_0\})$$

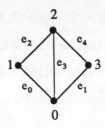

Fig. 3. The graph G_0.

Then $\mathcal{A}^\star(G)$ is universally ideal. We again demonstrate this example on the graph G_0 shown in Fig. 3. The cuts of G_0 are

$$\{e_0, e_1, e_3\}, \{e_0, e_2\}, \{e_0, e_3, e_4\}, \{e_1, e_2, e_3\}, \{e_1, e_4\}, \{e_2, e_3, e_4\},$$

and these are the minimal dependent sets of the matroid $T^\star(G_0)$.

Acknolidgments. We would like to thank Guy Even, Oded Goldreich, and Eyal Kushilevitz for thier useful comments and suggestions.

References

[BD91]   E.F. Brickell and D.M. Davenport. On the classification of ideal secret sharing schemes. *Journal of Cryptology*, 4(73):123–134, 1991.

[BL88]   J.C. Benaloh and J. Leichter. Generalized secret sharing and monotone functions. In S. Goldwasser, editor, *Advanced in Cryptology - CRYPTO '88 proceeding*, volume 403 of *Lecture notes in computer Science*, pages 27–35. Springer-Verlag, 1988.

[Bla79]   G.R. Blakley. Safeguarding cryptographic keys. In *Proc. AFIPS 1979 NCC, vol. 48*, pages 313–317, June 1979.

[BR89]   J.C. Benaloh and S. Rudich. Private communication, 1989.

[Bri89]   E.F. Brickell. Some ideal secret sharing schemes. *Journal of Combin. Math. and Combin. Comput.*, 6:105–113, 1989.

[CK89]   B. Chor and E. Kushilevitz. Secret sharing over infinite domains. In G. Brassard, editor, *Advanced in Cryptology - CRYPTO '89 proceeding*, volume 435 of *Lecture notes in computer Science*, pages 299–306. Springer-Verlag, 1989. To appear in Jour. of Cryptology.

[ISN87]   M. Ito, A. Saito, and T. Nishizeki. Secret sharing schemes realizing general access structure. In *Proc. IEEE Global Telecommunication Conf., Globecom 87*, pages 99–102, 1987.

[KGH83]   E.D. Karnin, J.W. Greene, and M.E. Hellman. On secret sharing systems. *IEEE Trans. on Inform. Theory*, IT-29 no. 1:35–41, 1983.

[KN90]   J. Kilian and N. Nisan. Private communication, 1990.

[Sha79]   A. Shamir. How to share a secret. *Communications of the ACM*, 22:612–613, November 1979.

[Wel76]   D.J.A. Welsh. *Matroid theory*. Academic press, London, 1976.

Perfect Zero-Knowledge Arguments for *NP* Can Be Based on General Complexity Assumptions

(EXTENDED ABSTRACT)

Moni Naor[1], Rafail Ostrovsky[2]*, Ramarathnam Venkatesan[3], Moti Yung[4]

[1] IBM Research Division, Almaden Research Center, San Jose, CA
[2] International Computer Science Institute at Berkeley and
University of California at Berkeley
[3] Bellcore, 445 South Street, Morristown, N.J. 07960.
[4] IBM Research, T.J. Watson Research Center, Yorktown Heights, NY 10598.

Abstract. "Zero-knowledge arguments" is a fundamental cryptographic primitive which allows one polynomial-time player to convince another polynomial-time player of the validity of an NP statement, without revealing any additional information in the information-theoretic sense. Despite their practical and theoretical importance, it was only known how to implement zero-knowledge arguments based on specific algebraic assumptions; basing them on a general complexity assumption was open since their introduction in 1986 [BCC, BC, CH]. In this paper, we finally show a general construction, which can be based on *any* one-way permutation.

We stress that our scheme is *efficient*: both players can execute only polynomial-time programs during the protocol. Moreover, the security achieved is *on-line*: in order to cheat and validate a false theorem, the prover must break a cryptographic assumption on-line *during the conversation*, while the verifier can not find (ever!) any information unconditionally (in the information theoretic sense).

---

* Part of this work was done while visiting Bellcore, and part at IBM T.J. Watson Research Center.

1 Introduction

Reducing complexity assumptions for basic cryptographic primitives is a major current research program in cryptography. Characterizing the necessary and sufficient complexity conditions needed for primitives helps us develop the theoretical foundations of cryptography, and further, reducing requirements for a primitive may imply more concrete underlying functions for its practical implementations.

Here we study the problem of secure transfer of the proof of "validity of an NP assertion" in this perspective. We note that the ability to convey proofs for NP in a secure way (i.e., in *zero-knowledge* (ZK) fashion, as defined by [GMR]) has a large variety of applications in cryptography and distributed computing.

Informally, proving some fact in zero-knowledge is a way for one player (called "prover") to convince another player (called "verifier") that certain fact is true, while not revealing any additional information. In our setting, we assume that both players are polynomially bounded (thus NP proofs where the prover has a witness, are the natural setting). We must make complexity assumptions for implementing the above task since in our setting these protocols imply existence of a one-way function. The assumptions could be used in two different ways:

1. Zero-knowledge *proofs* [GMR, GMW]: The prover can not convince the verifier to accept a false theorem, even if he gets help from an infinitely powerful computation; while the verifier (or anyone overhearing the protocol), if he ever breaks the assumption (say, after 100 years), *can* extract additional information about the proof (thus, the security is only ensured computationally).

2. Zero-knowledge *arguments* [CH, BC, BCC]: The verifier can not extract additional information even if he is given infinite time (i.e., security is perfect); however, the prover (assumed to be polynomial-time) can cheat in his proof only if he manages to break the assumption *on-line during the execution of the protocol*. This is the reason to call it an "argument" rather than a "proof".

In many practical settings, ZK-arguments may be preferable to ZK- proofs: the verifier must only be sure that the prover did not break the assumption *during their interaction* (which lasted, say, ten seconds or minutes). Notice that while assuring that the assumption can *never* be broken is unreasonable, the assumption that something can not be broken *during the next ten minutes* can be based on the current state of the art. On the other hand, the prover has absolute (i.e. information-theoretic) guarantee that no additional information is released, even if the verifier spends as much time as it desires trying (off-line) to extract it. (Thus, the notion of zero-knowledge arguments is useful if there is a need to maintain the secrecy for very long time independent of the possible future advance of cryptanalysis).

So far the complexity assumptions needed for perfect-zero-knowledge arguments were too strong — they required specific algebraic assumptions. This is in contrast with zero-knowledge interactive proofs, which can be based on any one-way function. In this work we finally dispose of specific algebraic assumptions for zero-knowledge arguments:

Main result: If one-way permutations exist, then it is possible for polynomial-time players to perform a perfect zero-knowledge arguments for all of $\mathcal{NP}$

In our proof, we construct an information-theoretically secure bit-commitment scheme, which has additional applications like information-theoretically secure coin-flipping. We can implement the scheme (with almost-perfect security) based on k-regular one-way functions. One practical implication of our result is that secure arguments can now be based on functions which are DES-like ciphers.

1.1 Background and organization

Past successes in establishing basic cryptographic primitives on general assumptions (initiated in [Y82]) have shown that various primitives, which were originally based on specific algebraic functions, can be based on the existence of general one-way functions or permutations. For example, Naor [N] showed that computationally secure bit commitments (i.e., bit commitments which *can be* broken off-line given sufficient resources) can be constructed from a pseudo-random generators (a notion originated and first implemented based on a discrete logarithm assumption in [BM]). The later, in turn (after a long sequence of papers) can now be based on any one-way function [ILL, H]. Another primitive that can now be based on any one-way function as well is digital-signature [NY, Ro]. Furthermore these primitives (and primitives derived from them, e.g. identification) were shown to imply a one-way function (thus they are equivalent) [IL]. On the other hand, basing the primitive of oblivious transfer on a general one-way permutation which is not a trapdoor[5] was shown to be "a seemingly

---

[5] a trapdoor implies that there is an information which enables easy inversion

hard task" [IR] – when based on black box reductions, it will separate P and NP (on the positive side, a trapdoor permutation is sufficient).

Concerning secure proofs, Goldreich, Micali and Wigderson showed that zero-knowledge *proofs* for $\mathcal{NP}$ can be done and require secure encryption functions (the results of [N, ILL, H] give such functions under any one-way function); this applies to general $\mathcal{IP}$ proofs as well [IY]. Further, zero-knowledge proofs and zero-knowledge arguments for non-trivial languages as well as non-interactive zero-knowledge proofs of [BFM, BDMP] imply the existence of one-way functions [OW].

In contrast to computational zero-knowledge *proofs*, the primitive of perfect zero-knowledge *arguments* for NP was much inferior in this respect: their constructions were known only under specific algebraic assumptions [BCC, BKK, IY, BY, IN]. Our result gives the first general reduction: zero-knowledge NP-arguments can be constructed given any one-way permutation.

Our construction has two stages. First, we show how to design an information-theoretically secure bit commitment between two polynomial-time parties based on any one-way permutation (we employ a technique that can be called "interactive-hashing" introduced initially in a different model involving an all-powerful party [OVY1]). Moreover, we do it in such a way that the conversations in the commitment protocol are *simulatable* (i.e. by an expected polynomial time algorithm). Then, we apply the reduction of "perfectly-secure simulatable bit commitment" to "perfect ZK-argument". (A general scheme connecting various commitments to various ZK-systems was given in e.g. [IY] and can be used).

We note that this work differs from [OVY1] in that there the sender must be able to invert one-way functions, whereas here the sender is efficient (this is the traditional cryptographic model). In [OVY1] we deal with oblivious transfer and any technique succeeding in allowing a weak sender there, would be quite significant since it would implement oblivious transfer between polynomial time parties using one-way permutations (see [IR]).

1.2 Relation to recent work on bit-commitment

Recently, models in which parties may have power beyond polynomial-time were investigated; it is worth while pointing out the differences between the current work and the recent one. By "From Strong to Weak BC", we denote Bit-commitments (BC) protocols, in which even an infinitely-powerful "Commiter" can not cheat, (i.e. change the value of the committed bit) except with negligible probability, but the polynomial-time "Receiver" can "see" the commitment, if he breaks the assumption. The result of [N] imply that under any one-way function, there is a (Strong-to-Weak) BC from a polynomial-time Commiter to a polynomial-time Receiver (that is, it is an efficient protocol and the underlying assumption in this case is optimal [IL]).

The work in [OVY2] investigated commitments between a strong and a polynomial-time players where the strong player actually uses its non-polynomial-time power. Thus, the main issue in that paper is how cryptographic assumptions changes and can be relaxed when the power of players differs (rather than being polynomial-time for both players, as needed in practical applications). It is shown that unless Distributional-NP=RP there is a (Strong-to-Weak) BC from a Commiter with an (NP union co-NP) power to a polynomial-time Receiver. Similarly, unless Distributional-PSPACE=RP, there is a (Strong-to-Weak) BC from a (PSPACE) Commiter to a polynomial-time Receiver. Distributional-NP is defined by Levin in the theory of average-case NP, whereas Distributional-PSPACE is a complete (in Levin's sense) problem for PSPACE under a uniform distribution. Thus, when allowing the commiter to use non-polynomial power this theoretical result relaxes the assumptions in [N].

By "from Weak to Strong BC" we denote BC in which even an infinitely-powerful "receiver" can not "see" the commitment, but the polynomial-time commiter can not change the value of the commitment if a complexity assumption holds. In [OVY2] it is also shown, based on an oblivious transfer protocols among unequal-power players introduced in [OVY1] (where interactive hashing was presented), that given any one-way function, there is a (Weak-to-Strong) BC from a polynomial-time Commiter to a (PSPACE) Receiver (and if the receiver is NP, the same holds under a one-way permutation).

The main results in [OVY1] yield oblivious transfer under one-way function when players have unequal power. The cryptographic application of [OVY1] (when both parties are polynomial time), is basing two-party secure computation with one party having information theoretic security under general *trapdoor* permutation assumption (whereas previously known under specific algebraic trapdoor functions). This is done by applying the results for one-way permutation but by adding a trapdoor property to be useful in cryptographic scenarios (so that computations are in polynomial-time).

In the current paper, we assume polynomial-time parties and do not use non-polynomial-time computations. We stress again that this is the model for cryptographic applications. Further, we make no use of trapdoor properties, as

BC's and secure interactive proofs do not need decryptions, but rather displaying of pre-images (for decommitals). Our result here for BC can be stated as: given any one-way permutation, there is an efficient (Weak-to-Strong) BC protocol from a polynomial-time Commiter to a polynomial-time Receiver (which may be stronger); the BC is simulatable and is a commitment of knowledge.

1.3 Organization of the paper

In section 2, we give the model, the formal definitions of the problem, and the assumptions. (Specifically, we present the model of interactive machines, the definitions of perfect zero-knowledge arguments, the notion of commitment, and the definition of one-way functions and permutations). In Section 3, we present the new method for basing a perfectly-secure bit commitment on a one-way permutation, and discuss its reduction to zero-knowledge arguments. In section 4 we present additional applications of our methods.

2   Model and Definitions

Let *Alice* (the prover) and *Bob* (the verifier) be interacting Turing machine [GMR, B] which share an access to a security parameter n, and a common communication tapes. Each has a private input and output tapes and a private random tape. When *Alice* and *Bob*'s programs are both polynomial time, we say that the protocol is "efficient" (we will assume this throughout), *Alice* usually has a private tape in which a "witness" to the correctness of the common input is written. We may consider *Bob* to be infinitely-powerful when he wishes to extract information from a protocol conversation, although he needs only poly time computations to execute the protocol. Both parties share an input tape of size k and and two "communication tapes": tapes for *Alice* to write in and *Bob* to read and vice versa. *Bob* has a private history tape h.

2.1   Perfect Zero-Knowledge Arguments

An NP-proof protocol with polynomial-time prover is a protocol between two polynomial time parties: a prover *Alice* and a verifier *Bob*. The parties take turns being "active", that is, reading the tapes and performing the computation, outputting a "message" on the corresponding communication tape. Both parties are probabilistic machines, (i.e., they have a read-only infinite tape of truly random bits which is private and read left-to-right). *Alice* also has a private input with a witness to the input. (Without lose of generality, we can assume that the input is a legal satisfiability (SAT) statement, since otherwise any NP statement can be translated first to SAT, and *Alice* can translate the witness to a witness to the SAT-statement). At the end of the protocol *Bob* moves to one of two states: ACCEPT or REJECT.

Definition 1 *An NP-proof protocol with polynomial-time prover is called an* **argument** *if:*

1. *There exists a polynomial-time program (in the statement size which is a security parameter) for Alice such that given any statement in NP, Alice can always convince polynomial-time Bob (that is make Bob move to ACCEPT at the end of the interaction).*
2. *No polynomial-time Alice* interacting with Bob can convince Bob to AC-CEPT, when the input is not true, except with negligible small probability (that is for a polynomial p for large enough input the error becomes smaller than $1/p(n)$.*

For an input I and history h let $CONV_{Bob^*}(I, h)$ be the random variable (depending on the parties' random tapes), which Bob^* produces throughout an interaction with *Alice*.

We note that similarly an argument can be prove "a possession of knowledge" in the sense that one formally shows that a machine employing the prover can extract a witness to the claimed NP statement [FFS, TW, BG]. (In the next version we describe this as well).

We say that two distributions μ_1 and μ_2 on $\{0,1\}^n$ are almost identical if for all polynomials $p(n)$, large enough n and for all $A \subset \{0,1\}^n$, $|\mu_1(A) - \mu_2(A)| < 1/p(n)$.

Definition 2 *An argument is* **perfectly zero-knowledge** *if: for all verifier Bob*, there is a simulator which is a probabilistic expected polynomial-time machine M_{Bob^*}, such that for any input I , it produces a random variable $SIM_{Bob^*}(I, h)$ so that the distribution of $SIM_{Bob^*}(I, h)$ is identical to that of $CONV_{Bob^*}(I, h)$.*

2.2 Commitment

Definition 3 *A* **bit commitment** *protocol consists of two stages:*

− *The* commit *stage: Alice has a bit b on her input tape, to which she wishes to commit to Bob. She and Bob exchange messages. At the end of the stage Bob has some information that represents b written on its output tape.*

- *The* reveal (opening) stage: *Alice and Bob exchange messages (where their output tapes from the commit stage are serving as input tapes for this stage). At the end of the exchange Bob writes on its output tape b.*

Definition 4 *To be* **perfectly-secure commitment,** *the protocol must obey the following: for all Turing machines Bob, for all probabilistic polynomial time Alice, for all polynomials p and for large enough security parameter n*

1. *(Security property:) After the commit stage, when Alice follows the protocol Bob cannot guess b with probability greater than $\frac{1}{2} + \frac{1}{p(n)}$ (even if Bob is given unbounded computational resources).*
2. *(Binding property:) After the commit stage in which Bob follows the protocol, with probability at least $1 - \frac{1}{p(n)}$ the polynomial-time Alice can reveal only one possible value.*

Note that the security property does not rely on *Bob* being polynomial time. In addition, if *Bob*'s algorithm can be performed in polynomial-time, we say that the bit commitment is "efficient"– we concentrate on this case.

We say that a commitment scheme is *polynomial-time simulatable* (with respect to the receiver) if given a polynomial-time receiver Bob^*, its history of conversations is a probability space simulatable by having Bob^* taking part in a computation with an expected polynomial time machine S (as in the definition of zero-knowledge).

We call a commitment a *commitment of knowledge* if there is a polynomial-time machine X (extractor) interacting with the sender performing the commit stage, such that the probability that X outputs a bit b is close to the probability that the reveal stage outputs same bit b (assuming reveal ended successfully). (A formal definition, is postponed to the full version).

In defining the properties that a bit commitment protocol must obey, we have assumed a scenario where *Bob* cannot guess b with probability greater than $\frac{1}{2}$ prior to the execution of the commit protocol. In the more general case, *Bob* has some auxiliary input that might allow him to guess b with probability $q > \frac{1}{2}$. The definition for this case is that as a result the commit stage the advantage that *Bob* gains in guessing b is less than $\frac{1}{p(n)}$. All the results of this paper hold for this more general case as well.

2.3 One-way functions and permutations

We define the underlying cryptographic operations we assume.

Let f be a length preserving function $f : \{0,1\}^* \to \{0,1\}^*$ computable in polynomial time.

Definition 5 [One-way function.] f *is one-way if for every probabilistic polynomial time algorithm A, for all polynomials p and all sufficiently large n,*

$$Pr[f(x) = f(A(f(x))) \mid x \in_R \{0,1\}^n] < 1/p(n).$$

The above definition is of a *strong one-way function*. Its existence is equivalent to the existence of the weaker *somewhat one-way function* using Yao's amplification technique [Y82] or the more efficient method of [GILVZ] (which is applicable only to permutations or regular functions). (A somewhat one-way function has the same definition as above, but the hardness of inversion is smaller, i.e. its probability is inverse polynomially away from 1.)

If in addition f is 1-1 then we say the f is a **One-Way Permutation**. For the construction outlined in Section 3 we require a one-way permutation f. (We note that we can also employ k-regular one-way functions in our protocol, since they can be converted into an "almost a permutation" [GKL]).

3 Perfectly-Secure Simulatable Bit Commitment

We present a perfectly-secure scheme and its proof of security. The polynomial commiter generates a bit encryption which comes from two possible distributions. The commiter will be able to open the encryption only as a member of one distribution (even though the distribution are identical).

3.1 The Scheme based on any one-way permutation

Let f be a strong one-way permutation f on $\{0,1\}^n$. Let S denote the sender *Alice* (as defined in 2.1) and R the receiver *Bob* (as defined). In the beginning of the protocol, S is given a secret input bit b. $B(x,y)$ denotes the dot-product mod 2 of x and y.

Commit Stage.

Commit to a bit b.

1. The sender S selects $x \in_R \{0,1\}^n$ at random and computes $y \leftarrow f(x)$. S keeps both x and y secret from R.

2. The receiver R selects $h_1, h_2, \ldots h_{n-1} \in \{0,1\}^n$ such that each h_i is a random vector over $GF[2]$ of the form $0^{i-1}1\{0,1\}^{n-i}$ (i.e. $i-1$ 0's followed by a 1 followed by an arbitrary choice for the last $n-i$ positions). Note that $h_1, h_2, \ldots h_{n-1}$ are linearly independent over $GF[2]$

3. For j from 1 to $n-1$
   - R sends h_j to S.
   - S sends $c_j \leftarrow B(h_j, y)$ to R.

4. At this point there are exactly two vectors $y_0, y_1 \in \{0,1\}^n$ such that for $i \in \{0,1\}$, $c_j = B(y_i, h_j)$ for all $1 \leq j \leq n-1$. y_0 is defined to be the lexicographically smaller of the two vectors. Both S and R compute y_0 and y_1. Let

$$c = \begin{cases} 0 \text{ if } y = y_b \\ 1 \text{ if } y = y_{1-b} \end{cases}$$

5. S computes c and sends it to R.

Reveal Stage.

1. S sends b and x to R.
2. R verifies that $y = f(x)$ obeys $c_j = B(h_j, y)$ for all $1 \leq j \leq n-1$ and verifies that if $c = 0$, then $y = y_b$ and if $c = 1$, then $y = y_{1-b}$.

end-commit-protocol

It is clear that the protocol described above can be executed in polynomial time by both parties. In the next subsection we will see that it is indeed a perfectly secure bit commitment protocol.

3.2 Proof of security

Theorem 1. *If f is a one-way permutations exist, then the scheme presented in Section 3.1 is a perfectly-secure computationally-binding bit commitment scheme.*

Theorem 1 follows from the two theorems below, the security theorem and the binding theorem, respectively.

Theorem 2. *For any receiver R', after the commit stage the bit b is hidden information-theoretically.*

Proof : We can prove inductively on j, that for any choice of $h_1, h_2, \ldots h_j$ the conditional distribution of y given $h_1, h_2, \ldots h_j\ c_1, c_2, \ldots c_j$ is uniform in the subspace defined by $h_1, h_2, \ldots h_j$ and $c_1, c_2, \ldots c_j$. Thus, at step 4 the probability that $y = y_0$ is exactly $\frac{1}{2}$. Therefore giving away c yields nothing about b. $\square$

Theorem 3. *Assume there exists a probabilistic polynomial time $S'(n)$ that following the commit stage can reveal to a honest receiver two different values for b with non-negligible probability (over its coin-flips) $\varepsilon = \varepsilon(n)$. Then there exists a probabilistic polynomial time algorithm A that inverts f on non-negligible fraction of the y's in $\{0,1\}^n$.*

Proof : Using such an S' we now construct the algorithm A to invert f. A has a fixed polynomial time bound and it aborts if its runtime exceeds the bound. By assumption, there exists a set Ω of $\varepsilon(n)$ fraction of strings such that if the tape of S' is initialized with $\omega \in \Omega$, S' succeeds in revealing two different values for b after the commit stage of $n-1$ rounds. We may fix such an ω and view S' as deterministic. This is true, since one can repeatedly run A with the random tape of S' initialized with $\omega_i, i := 1, \ldots, m = 1/\varepsilon^2$ and with probability $1 - e^{-\sqrt{m}}$ some $\omega_i \in \Omega$. We treat S' as a deterministic algorithm from now on.

The responses c_i of S' to the queries h_i sent by R define a rooted tree T whose edges are labeled in $\{0,1\}$. A path from the root to a leaf is defined by an assignment to $h_1, h_2, \ldots h_{n-1}$ and it is labeled with $c_1, c_2, \ldots c_{n-1}$. A node U at level i corresponds to a state of S' after $i-1$ stages. It defined by $h_1, \ldots, h_{i-1}$ and $c_1, \ldots, c_{i-1}$. The outgoing edges of U correspond to R's 2^{n-i} possible queries. These edges are labeled with the responses of S'. Note that since S' may be cheating, his answers need not be consistent and that on the same query S' may give different answers depending on the previous queries.

For a leaf u, let $\{y_0(u), y_1(u)\}$ be the set consistent with S's answers; we say u is *good* if given that R's queries define u, then S' succeeds in opening the bit committed in two different ways: i.e. S' inverts on both $y_0(u)$ and $y_1(u)$.

Description of A: A gets as an input a random image y in $\{0,1\}^n$ and it attempts to invert y. In order to compute $f^{-1}(y)$, A tries to find a good leaf u such that $y \in \{y_0(u), y_1(u)\}$. Starting at the root, A develops node by node a path consistent with y. Fix j to be $n - 8(\log n/\epsilon + 1)$. For j rounds A does as follows: for $1 \leq i < j$ at the i round the path so far is defined by $h_1, h_2, \ldots h_{i-1}$ and the labels are $c_1, c_2, \ldots c_{i-1}$ such that $c_i = B(h_i, y)$. Now, a random h of the $0^{i-1}1\{0,1\}^{n-i}$ is chosen (note that h is linearly independent from $h_k, k < i$ is chosen. If the edge h is labeled with $B(h, y)$, then $h_i \leftarrow h$ and the path is expanded by the new node. Otherwise, S' is reset to the state before its reply, and a new candidate for h_i is chosen. This is repeated until either a success or until there are no more candidates left, in which case A aborts. If A reaches the jth level, it guesses the remaining $n-j$ queries $h_j, h_{j+1}, \ldots h_{n-1}$ and checks whether the path to the leaf is labeled consistently with $B(y, h_i)$. If it is and the leaf reached is good, then A has succeeded in inverting y.

The rest of this proof is devoted for showing that $\mathcal{A}$ as defined above has probability at least $\varepsilon^{10}/8e^3n^8$ for inverting y. Note that $\mathcal{A}$ as described above does not necessarily halt after a polynomial number of steps. However, as we shall see at the end of the proof, we can limit the total number of unsuccessful attempts at finding a consistent h to $8n$ without decreasing significantly the probaiblity that $\mathcal{A}$ succeeds in inverting y.

Before we continue we introduce some notation. Since we are dealing with several types of vectors of length n over $GF[2]$ we will distinguish them by calling those vectors that are sent by R as *queries* and those vectors which may be the image that y attempts to invert as *images*. Let U be a node at the ith of the tree defined by $h_1, h_2, \ldots h_{i-1}$ and $c_1, c_2, \ldots c_{i-1}$. We say that $y \in \{0,1\}^n$ is an image in U if $B(h_k, y) = c_k$ for all $1 \le k < i$. We denote the set of images of U by $\mathcal{I}(U)$. We know that $|\mathcal{I}(U)| = 2^{n-i+1}$. We say that $h \in \{0,1\}^n$ is a query of U if it is of the form $0^{i-1}1\{0,1\}^{n-i})$.

Let $A(U, y) = |\{h | h$ is a query of U and $B(h, y)$ agrees with the label h of $U\}|$

An image y is *balanced* in U_i, a node of the ith level if

$$1 - \frac{1}{n} \le \frac{A(U_i, y)}{2^{n-i-1}} \le 1 + \frac{1}{n}$$

An image y is *fully balanced* in U, a node of the jth level, if it is balanced in all the ancestors of U. Define $\mathcal{F}(U)$ as the set of all $y \in \mathcal{I}(U)$ **and** are fully balanced in U. For a set of queries H at a node U and an image y of U the *discrepancy* of y at H is the absolute difference between $|H|/2$ and the number of queries in H that agree with y. Finally, recall that $j = n - 8(\log n/\varepsilon + 1)$.

Lemma 4. *For any node U of level j at least $2^{n-j}(1 - \beta)$ for $\beta = 2^{-3/4(n-j)}$ of the images of U have the property that $2^{n-j} - 2^{7/8(n-j)} \le A(U,y) \le 2^{n-j} + 2^{7/8(n-j)}$*

Proof : First note that any pair of queries h', h'' of U has the property that h'' is linearly independent of $h', h_1, h_2, \ldots h_{j-1}$. Now suppose that an image y of U is chosen at random and consider the indicator a_h which is 1 whenever $B(h, y)$ is equal to U's response on h. For any h we have that $Prob[a_h = 1] = 1/2$ and for every pair h', h'' the events $a_{h'}$ and $a_{h''}$ are pairwise independent. We are essentially interested in

$$Prob\left(|\sum_{h \ query \ of \ U} a_h - E[\sum_{h \ query \ of \ U} a_h]| \ge 2^{7/8(n-j)}\right) \qquad (1)$$

By Chebyschev's inequality

$$Prob\left(|\sum_{h \text{ query of } U} a_h - E[\sum_{h \text{ query of } U} a_h]| \geq \lambda\sqrt{VAR[\sum a_h]}\right) \leq \frac{1}{\lambda^2}$$

$Var[\sum_h a_h]$ is 2^{n-j} and hence (1) is at most $2^{-3/4(n-j)}$.

Lemma 5. *For any node U of level j and random image y of U the probability that y is fully balanced in U is at least $1 - \gamma$ for $\gamma = n2^{-5/8(n-j)}$*

Proof : Let $U_1, U_2, \ldots U_j = U$ be the nodes on the path to U. For any $1 \leq i \leq j$ we can partition the 2^{n-i} queries of U_i into 2^{j-i} subsets $H_1, H_2, \ldots H_{2^{j-i}}$ of size 2^{n-j} each such that for any $1 \leq \ell \leq 2^{j-i}$ and $h', h'' \in H_\ell$ we have that h' is linearly independent of $h_{i+1}, \ldots h_j, h''$. Therefore, similar to Lemma 4, we have that $Prob[|\sum_{h \in H_\ell} -E[\sum_{h \in H_\ell} a_h]| > 2^{7/8(n-j)}] \leq 2^{-3/4(n-j)}$. Therefore by Markov's inequality the probability that more than $2^{-1/8(n-j)}$ fraction of the H_ℓ's have a discrepancy larger than $2^{7/8(n-j)}$ is at most $2^{-5/8(n-j)}$. Therefore with probability at least $1 - 2^{-5/8(n-j)}$ the total discrepancy at node U_i is at most

$$2^{-1/8(n-j)}2^{n-j}2^{j-i} + (1 - 2^{-1/8(n-j)})2^{7/8(n-j)}2^{j-i} \leq 2 \cdot 2^{7/8n+1/8j-i} \qquad (2)$$

and hence with the probability at least $1 - 2^{-5/8(n-j)}$ we have

$$1 - \frac{1}{n} \leq 1 - 2^{-1/8(n-j)+1} \leq \frac{A(U_i, y)}{2^{n-i-1}} \leq 1 + 2^{-1/8(n-j)+1} \leq 1 + \frac{1}{n}$$

The probability that y is balanced in all the levels is therefore at least $1 - n2^{-5/8(n-j)} = 1\gamma$.

Lemma 6. *The probability that a node U of the jth level is reached by an execution of A is at least $\frac{1-\gamma}{e}$ of the probability that it is reached by an execution of S'*

Proof : Let $U_1, U_2, \ldots U_j = U$ be the nodes on the path to U's from the root. For any node U_i the probability that U_i is reached in S' is $\prod_{i=1}^{j-1} \frac{1}{2^{n-i}}$. On the other hand

$$Prob[U \text{ is reached by } A] = \sum_{y \in \mathcal{I}(U)} Prob[y \text{ is chosen and } U \text{ is reached}] \geq$$

$$\sum_{y \in \mathcal{F}(U)} Prob[y \text{ is chosen and } U \text{ is reached}] = \sum_{y \in \mathcal{F}(U)} 1/2^n \prod_{i=1}^{j-1} \frac{1}{A(U_i, y)} \geq$$

$$\sum_{y \in \mathcal{F}(y,U)} 1/2^n \prod_{i=1}^{j-1} \frac{1}{(1+1/n)2^{n-i-1}} \geq \sum_{y \in \mathcal{F}(U)} 1/2^n \prod_{i=1}^{j-1} \frac{1}{(1+1/n)2^{n-i-1}} \geq$$

$$\frac{2^{n-j+1}(1-\gamma)}{2^n} \cdot \frac{1}{(1+1/n)^n} \prod_{i=1}^{j-1} \frac{1}{2^{n-i-1}} \geq \frac{(1-\gamma)}{e} \prod_{i=1}^{j-1} \frac{1}{2^{n-i}}$$

Lemma 7. *The probability that the image $\mathcal{A}$ is trying to invert is fully balanced at the jth level is at least $\frac{(1-\gamma)^2}{e}$*

Proof : For every node of the jth level and every fully balanced image y of U we have that $Prob[y \text{ is chosen and } U \text{ is reached}] \geq \frac{(1-\gamma)}{2^n e} \prod_{i=1}^{j-1} \frac{1}{2^{n-i-1}}$. Hence,

$$Prob[U \text{ is reached with a fully balanced } y] \geq$$

$$2^{n-j}(1-\gamma) \cdot \frac{1-\gamma}{e2^n} \prod_{i=1}^{j-1} \frac{1}{2^{n-i-1}} = \frac{(1-\gamma)^2}{e} \prod_{i=1}^{j-1} \frac{1}{2^{n-i}}$$

The number of nodes at the jth level is $\prod_{i=1}^{j-1} 2^{n-i}$ and therefore the probability that the image chosen is fully balanced at the jth level is at least $\frac{(1-\gamma)^2}{e}$.

Call a node *good* if at least ε of the leaves at the subtree rooted at U have the property that S' succeeds in cheating, i.e., inverting both images. By assumption, the fraction of good nodes U is at least ε. Hence, by Lemma 6 the probability that $\mathcal{A}$ reaches a good U at level j is at least $\frac{1-\gamma}{e}\varepsilon$.

Lemma 8. *In any good node U of level j the fraction of the good leaves that have at least one image that is in $\mathcal{F}(U)$ is at least $\varepsilon/2$.*

Proof : Any pair of images $y_1 \neq y_2$ in $\mathcal{I}(U)$ can be together in at most $1/2^{n-j}$ of the leaves: in any node U'' along the way from U to the leaves and for random query h of U' we have $Prob[B(h, y_1) = B(h, y_2)] = 1/2$. Since there are at most $\gamma 2^{n-j+1}$ images that are not fully balanced in U, then at most

$$\binom{\gamma 2^{n-j+1}}{2} /2^{n-j-1} \leq 2\gamma^2 2^{n-j} \leq n^2 2^{-1/4(n-j)+1} = n^2 2^{-2(\log n/\varepsilon + 1)+1} \leq \frac{\varepsilon^2}{2}$$

of the leaves have both of their images from the unbalanced. Therefore at least $\varepsilon - \frac{\varepsilon^2}{2} \geq \varepsilon/2$ of the leaves are both good and have at least one image which is fully balanced at U.

Lemma 9. *For any good node U of level j and $z \in \mathcal{F}(U)$, given that U was reached with a fully balanced y, the probability that $y = z$ is at least $\frac{1}{e^2 2^{n-j+1}}$*

Proof : We would like to bound from below

$$\frac{Prob[z \text{ is chosen and } U \text{ is reached}]}{Prob[U \text{ is reached and the image is fully balanced}]} \tag{3}$$

We know that $Prob[U \text{ is reached and the image is fully balanced}] =$

$$\sum_{z \in \mathcal{F}(U)} Prob[z \text{ is chosen and } U \text{ is reached}] = \sum_{y \in \mathcal{F}(U)} 1/2^n \prod_{i=1}^{j-1} \frac{1}{A(U_i, y)} \le$$

$$\sum_{y \in \mathcal{F}(U)} 1/2^n \prod_{i=1}^{j-1} \frac{1}{(1 - 1/n)2^{n-i-1}} \le \sum_{y \in \mathcal{F}(U)} e/2^n \prod_{i=1}^{j-1} \frac{1}{2^{n-i-1}} \le$$

$$\frac{2^{n-j}}{2^n} \cdot e \prod_{i=1}^{j-1} \frac{1}{2^{n-i-1}} \le e \prod_{i=1}^{j-1} \frac{1}{2^{n-i}}$$

As can be seen from the proof of Lemma 6 for any $z \in \mathcal{F}(U)$ we have that

$$Prob[z \text{ is chosen and } U \text{ is reached}] \ge \frac{1}{e2^{n-j+1}} \prod_{i=1}^{j-1} \frac{1}{2^{n-i}}$$

Therefore (3) is at least $\frac{1}{e^2 2^{n-j+1}}$

Lemma 10. *The probability that $\mathcal{A}$ is successful is at least $\frac{\varepsilon^{10}}{4e^3 n^8}$*

Proof : Suppose that (a) $\mathcal{A}$ reaches a good node U at level j and the y is fully balanced and (b) that $h_j, h_{j+1}, \dots h_{n-1}$ define a path to a good leaf that has at least one image in $\mathcal{F}(U)$. Call this image z. Then by Lemma 9 we know that the probability that $y = z$ is at least $\frac{1}{e^2 2^{n-j}}$. The probability that (a) occurs is at least $\varepsilon \frac{(1-\gamma)^2}{e}$ by Lemma 7 and that (b) occurs given (a) is at least $\varepsilon/2$ by Lemma 8. Therefore the probability of success is at least $\varepsilon^2 \frac{(1-\gamma)^2}{2^{n-j} e^3} \ge \varepsilon^{10}/4e^3 n^8$

Note that we have only considered $\mathcal{A}$ successes when y was fully balanced at level j. However, given that y is fully balanced at level j, the probability that $\mathcal{A}$ had many unsuccessful candidates until he reached the jth level is small: we know that y is balanced at U_i for all $1 \le i < j$ and therefore $A(U, y)/2^{n-i} > 1/4$. Therefore the probability that $\mathcal{A}$ had to try more than $8n$ candidate for the h_i's until reaching level j is exponentially small in n and we have that even if we bound the run time of $\mathcal{A}$ by $8n^2$ the probability of success is still at least $\varepsilon^{10}/8e^3 n^8$. If ε is non negligible, then this is non negligible as well. This concludes the Proof of Theorem 3.

For our applications we need a simulatable bit commitment and commitment of knowledge (to be defined in the full version along the lines of [BG]).

Theorem 11. There is a perfectly-secure commitment scheme which is simulatable, and is commitment of knowledge.

Proof sketch: All actions of S are in polynomial time, so simulatability (generating the same distribution in polynomial time) is given.

To achieve simulatable commitment of knowledge, one has to modify the basic protocol described above as follows. The protocol's steps 1,2, and 3 will be first performed twice. At this point R asks S to open the chosen x which is the pre-image of y of one of the instances and continue the protocol with the other instance. Obviously, the security and binding properties are maintained.

To get a commitment of knowledge, we have an extraction algorithm X which plays the steps 1,2, and 3 twice. Then, it decides on which instance to continue, it asks to open it and gets y, then the simulation is backtracked and the other instance is asked to be opened, and the actual commitment is done using the (by now known) y in step 4 and 5 (given the input bit b to the machine X). The probability that the commitment will be different is negligible assuming the hardness assumption as was shown above. □

Next, we can state the following known "reduction theorems" present in the works on computational (perfect) zero-knowledge proofs (arguments) [GMW, BCC, IY].

Theorem 12. If there is a (perfectly-secure commitment) [commitment] scheme which is simulatable by an expected probabilistic polynomial-time machine ("interacting" with the receiver) and the receiver is polynomial-time, then there is a (perfect zero-knowledge argument) [computationally zero-knowledge proof] for any statement in NP.

The perfectly-secure simulatable bit-commitment protocol can be used in the general scheme above. In addition, the general proof system scheme can also be shown to give a "proof of possession of a witness" (i.e., proof of knowledge) as was formalized [FFS, TW, BG]. Thus, combining the above, gives our main result:

Theorem 13. If any one-way permutation exists, then there exist perfect zero-knowledge arguments for proving language-membership as well as for proving knowledge-of-witness.

4 Discussion

There are various other applications to information-theoretically secure bit commitment. For example, another application of the bit commitment above is a "coin-flipping protocol" (introduced by Blum [B]), with perfect security, and assuming only a one-way permutations.

For practical purposes consider the data encryption standard (DES) [Kon]. Given a k-regular [GKL] one-way function (i.e. the number of pre-images of a point is $\leq k$ and is k on a significant fraction), one can transform it into a one-way function which is 1-1 almost everywhere [GILVZ]. We apply this to the function $DES(k, m) = y$ ($k = $ key, $m = $ message) where (actual used parameters are) $k \in \{0, 1\}^{56}$, $m, y \in \{0, 1\}^{64}$. Assuming that DES is not breakable on-line (say in 10 seconds), then it is a good candidate for our scheme. We explore this further in the full version of the paper. The security of the commitment is not perfect but rather almost-perfect (guessing the commitment is not exactly $1/2$, but it is close to $1/2$). We note that DES is available in many machines and usually on an optimized hardware circuit.

It is an interesting question whether a general one-way function with no additional property suffices for zero-knowledge arguments. Reducing the rounds (by more than the achievable logarithmic factor) is interesting as well.

Acknowledgments

We would like to thank Dalit Naor for helpful advice.

References

[BDMP] Blum M., A. DeSantis, S. Micali and G. Persiano, "Non-Interactive Zero-Knowledge" *SIAM J.Comp. 91*

[BM] Blum M., and S. Micali "How to Generate Cryptographically Strong Sequences Of Pseudo-Random Bits" *SIAM J. on Computing,* Vol 13, 1984, pp. 850-864, FOCS 82.

[BFM] Blum M., P. Feldman, and S. Micali "Non-Interactive Zero-Knowledge Proof Systems and Applications" *STOC 88.*

[BMO] Bellare, M., S. Micali and R. Ostrovsky, "The (True) Complexity of Statistical Zero Knowledge" STOC 90.

[B] Blum, M., "Coin Flipping over the Telephone," IEEE COMPCON 1982, pp. 133-137.

[BKK]   J. Boyar, S. Kurtz, and M. Krental *A Discrete Logarithm Implementation of Perfect Zero-Knowledge Blobs*, Journal of Cryptology, V. 2 N. 2, 1990, pp. 63-76, Springer International.

[BG]   Bellare M., and O. Goldreich, "On Defining Proof of Knowledge," *CRYPTO 92* (this proceedings).

[BC]   G. Brassard, C. Crépeau, *"Non-Transitive Transfer of Confidence: A Perfect Zero-Knowledge Interactive Protocol for SAT and Beyond"*, FOCS 86 pp. 188-195.

[BCC]   G. Brassard, D. Chaum and C. Crépeau, *Minimum Disclosure Proofs of Knowledge*, JCSS, v. 37, pp 156-189.

[BCY]   Brassard G., C. Crépeau, and M. Yung, "Everything in NP can be proven in Perfect Zero Knowledge in a bounded number of rounds," *ICALP* 89. (Also in TCS).

[BY]   Brassard G., and M. Yung, "One-Way Group Action," *CRYPTO* 90.

[CH]   Chaum, D., "Demonstrating that a public predicate can be satisfied without revealing any information about how", Crypto 86.

[CDG]   D. Chaum, I. Damgård and J. van de Graaf, *Multiparty Computations Ensuring Secrecy of each Party's Input and Correctness of the Output*, Proc. of Crypto 87, pp. 462.

[FFS]   U. Feige, A. Fiat and A. Shamir, *Zero-Knowledge Proofs of Identity*, STOC 87, pp. 210-217.

[GILVZ] O. Goldreich, R. Impagliazzo, L. Levin, R. Venkatesan, and D. Zuckerman, *Security Preserving Amplification of Hardness*, FOCS 90.

[GKL]   O. Goldreich, H. Krawczyk, and M. Luby, *On the Existence of Pseudo-Random Generators* , FOCS 88.

[GMW]   S. Goldreich, S. Micali and A. Wigderson, *Proofs that Yields Nothing But their Validity*, FOCS 86, pp. 174-187. (also in JACM).

[GMR]   S. Goldwasser, S. Micali and C. Rackoff, *The Knowledge Complexity of Interactive Proof-Systems*, STOC 85, pp. 291-304. (also in SIAM J. COMP.)

[H]   Håstad, J., "Pseudo-Random Generators under Uniform Assumptions", *STOC 90*.

[ILL]   I. Impagliazzo, L. Levin and M. Luby, *Pseudo-random generation from one-way functions*, Proc. 21st Symposium on Theory of Computing, 1989, pp. 12-24.

[IL]   R. Impagliazzo and M. Luby, *One-way Functions are Essential for Complexity-Based Cryptography* FOCS 89, pp. 230-235.

[IN]   R. Impagliazzo and M. Naor, *Efficient Cryptographic Schemes Provably as Secure as Subset-Sum*, Proc. of FOCS 89, pp. 236-241.

[IR]   R. Impagliazzo and S. Rudich, *On the Limitations of certain One-Way Permutations* , Proc. of STOC 89, pp. 44-61.

[IY]   R. Impagliazzo and M. Yung, *Direct Minimum-Knowledge Computations* , Proc. of Crypto 87, Springer Verlag.

[Kon]   A. G. Konheim, *Cryptography: a primer* , Wiley, New York, 1981.

[N]   M. Naor *"Bit Commitment Using Pseudo-Randomness"* Crypto-89 pp.123-132.

[NY]   M. Naor and M. Yung, *Universal One-Way Hash Functions and their Cryptographic Applications*, STOC 89.

[OVY1]   R. Ostrovsky, R. Venkatesan, M. Yung, *Fair Games Against an All-Powerful Adversary, Sequences 91*, (to appear).

[OVY2]   R. Ostrovsky, R. Venkatesan, M. Yung, *Secure Commitment Against A Powerful Adversary, STACS 92*, Springer Verlag LNCS, 1992.

[OW]   R. Ostrovsky and A. Wigderson, *One-Way Functions are Essential For Non-Trivial Zero-Knowledge Proofs*, (preliminary manuscript).

[Ro]   J. Rompel "One-way Functions are Necessary and Sufficient for Secure Signatures" STOC 90.

[TW]   M. Tompa and H. Woll, *Random Self-Reducibility and Zero-Knoweldge Interactive-Proofs of Possession of Information*, Proc. of FOCS 1987.

[Y82]   A. C. Yao, *Theory and Applications of Trapdoor functions*, Proceedings of the 23th Symposium on the Foundation of Computer Science, 1982, pp 80-91.

Low communication 2-prover zero-knowledge proofs for NP

(Preliminary Version)

Cynthia Dwork[1] Uri Feige[2] Joe Kilian[3] Moni Naor[1] Muli Safra[1]

[1] IBM Research Division, Almaden Research Center
[2] IBM Research Division, T. J. Watson Research Center
[3] NEC Research

Abstract. We exhibit a two-prover perfect zero-knowledge proof system for 3-SAT. In this protocol, the verifier asks a single message to each prover, whose size grows logarithmically in the size of the 3-SAT formula. Each prover's answer consists of only a constant number of bits. The verifier will always accept correct proofs. Given an unsatisfiable formula S the verifier will reject with probability at least $\Omega((|S|-\mathbf{max\text{-}sat}(S))/|S|$, where $\mathbf{max\text{-}sat}(S)$ denotes the maximum number of clauses of S that may be simultaneously satisfied, and $|S|$ denotes the total number of clauses of S. Using a recent result by Arora et al [2], we can construct for any language in NP a protocol with the property that any non-member of the language be rejected with constant probability.

1 Introduction

In a multiple-prover interactive proof system, several provers, $P_1, P_2, \ldots$ try to convince a verifier V that a common input x belongs to a language L. The verification proceeds in rounds; in each round, the verifier sends to each prover a private message (query) and receives an answer. Each prover sees only the queries addressed to it, and cannot communicate with the other provers (at least until the end of the round). When the protocol ends, the verifier decides, based on the input string and the messages received, whether or not to accept.

Multi-prover proof systems were introduced by Ben-Or, Goldwasser, Kilian and Wigderson [7] in order to obtain zero knowledge proofs without relying on complexity assumptions such as the existence of one-way functions. In this paper we show another advantage of multi-prover proof systems by exhibiting a low communication two-prover perfect zero-knowledge proof system for 3-SAT (and thus for every language in NP). In contrast, no such low communication zero knowledge protocol is possible in a *single* prover proof system, unless $NP \subset BPP$.

Kilian [16] has provided additional motivation for striving for low communication in the two prover setting: he suggests enforcing the separation of the two provers by keeping them (say the two provers are implemented on a smart card) at some distance from each other. If the distance is long enough and the communication complexity is low, then the two provers do not have enough time to communicate during the execution of the protocol.

In the protocol we present, the verifier sends to each of the two provers a query whose length is logarithmic in the length of the input string, and receives back answers whose length is constant. If the input string is not in the language, then the verifier detects cheating with some fixed probability $\alpha > 0$. The protocol is perfect zero-knowledge, i.e. there is a polynomial time machine, called the simulator, that produces for every possible (possibly cheating) verifier the same distribution of conversations as the verifier would have had with two "real" provers.

To reduce the probability of error to 2^{-k} (rather than $1 - \alpha$), the protocol can be executed $O(k)$ times sequentially. Lapidot and Shamir [18] have provided an elegant zero-knowledge two prover protocol which is parallelizable, i.e. running copies of it in parallel decreases the probability of error exponentially in the number of copies. However, it is not known whether this is true for general protocols. Feige and Lovasz [13] (continuing [19]) have provided a method that can be applied to any protocol in order to obtain a parallelizable protocol, however the method does not preserve zero-knowledge. Finding such a method that preserves zero-knowledge is an open question.

In our protocol the two provers share a common random string of only logarithmic length. Thus, even if we consider the shared random string to be part of the communication complexity of the protocol, then it is still logarithmic. The existence of a shared random string is necessary, since we show that for low communication zero-knowledge protocols, the only languages that do not require the two provers to share a common random string are exactly those in BPP.

Our protocol is constructive in the sense that once two provers know a satisfying assignment to the formula, all they are required to do is some polynomial time computation.

1.1 Definitions

Definition 1 We say that a language L has a *two prover interactive proof system* if there exists an interactive probabilistic polynomial time machine (called the verifier) V and two interactive machines P_1, P_2 called Prover 1 and Prover 2 respectively, satisfying the following conditions. All three machines have a common input x which may or may not be in L. The two provers once and for all agree on a common strategy. Moreover, prior to each execution of the protocol, they may interact in order to share random bits. Once the protocol begins, they are assumed to be isolated from each other. The three machines follow a prescribed protocol consisting of several rounds; in each round, the verifier sends to each prover in private a message (query) and receives an answer. When the protocol ends, the verifier decides whether or not to accept, based on the input string and the messages received. The protocol must satisfy

- $\forall x \in L$ there exist machines P_1 and P_2 such that V accepts with probability 1 (completeness);
- there is a fixed constant $\alpha > 0$ such that $\forall x \notin L$ and $\forall P_1, P_2$ the probability that V accepts on input x is at most $1 - \alpha$.

Note that this definition is not standard in that α is not required to be say 2/3. However, by running the protocol *sequentially* several times (as a function of α) one can get arbitrary small probability of accepting erroneously. Showing that the probability goes down when the protocols are run in parallel is a major open problem in this area.

Part of the strategy that the two provers agree on may simply be a common random string. This is used to obtain the zero-knowledge property defined below.

Definition 2 For a given verifier V, provers P_1 and P_2, and input x, we define $View_{V,P_1,P_2}(x)$ be the distribution over the interaction between verifier V and provers P_1 and P_2. This distribution is over V's coin tosses and the random choices made by P_1 and P_2

Definition 3 A two prover interactive protocol V, P_1, P_2 is *perfect zero knowledge for V* if there exists a probabilistic polynomial time machine S that on input x outputs a string whose distribution is $View_{V,P_1,P_2}(x)$. A language L is said to have a perfect zero-knowledge protocol if it has a two-prover interactive proof system V, P_1, P_2 such that for every V' the protocol V', P_1, P_2 is perfect zero-knowledge for V'.

The communication complexity of a protocol is composed of three parts:

1. the total length of the queries sent by the verifiers;
2. the total length of the answers given by the provers;
3. the length of the random string shared by the two provers.

The term *low communication* will mean that the sum of these three components is logarithmic in the length of the input string.

1.2 Background

Multi-prover proof systems have inspired much research in Complexity Theory [5, 8, 9, 10, 11, 13, 14, 19]. In particular, Babai, Fortnow and Lund have shown that the class of languages that are recognized by multiple prover proof system where the verifier is a polynomial time machine and the communication is restricted to be of polynomial length is exactly NEXP-Time. This was scaled down to the NP setting [3, 4, 12], culminating in the result of Arora, Lund, Motwani, Sudan and Szegedy [2] showing a two prover proof system for NP in which the length of the queries that the verifier sends to the provers is logarithmic in the length of the input string, and the answers are of constant length. From this they derive:

Theorem 1 *[2] There is a $\beta > 0$ such that for any language $L \in NP$ there is a polynomial time reduction R from L to 3-CNF formulas such that for $x \in L$ $R(x)$ is a satisfiable 3-CNF and for all $x \notin L$, a fraction of at most $1 - \beta$ of the clauses of $R(x)$ can be satisfied simultaneously. The proof is constructive in the sense that given a witness for x's membership in L, there is a polynomial time procedure that yields a satisfying assignment to $R(x)$.*

We will apply this theorem to get our protocols. This theorem (or actually its precursor [4]) was already used by Kilian [17] to lower the communication complexity of single prover zero knowledge arguments and proof systems. However, by a simple observation, the only languages that have a single prover proof system with logarithmic communication are those in BPP. Thus, if we are aiming at logarithmic communication we must have two provers.

We further observe that the two provers must share a random string in order for a low-communication protocol to be zero-knowledge; for if not, by running the simulator enough times we can get the response on any query to each prover, and thus can simulate each prover *on-line*. If the two provers do not share a random string, then their responses are independent polynomial time samplable distributions and thus there is a probabilistic polynomial time machine that can compute the probability that the verifier accepts, whence L is in BPP. We do not know whether it is possible for the two provers to share fewer than the logarithmically many random bits required by our protocol. However, in Section 3 we show that $\Omega(\log \log n)$ random bits are essential.

2 The Interactive Proof System

We construct the interactive proof system in two steps. In the first step we use the result of Barrington [6] to reduce checking that an assignment satisfies F to checking that an assignment to variables in the permutation group S_5 satisfies certain equations (over S_5). More precisely, each clause of F gives rise to one equation over S_5. We also provide a way for the verifier to check consistency among distinct occurrences of each literal in F. In the second step we use the randomizing tableaux of Kilian [15] to construct for each equation a 2-prover interactive proof system for an assertion about a product.

The entire proof system is therefore as follows. All parties apply Barrington's result to obtain the set of equations over variables in S_5. The Verifier then randomly chooses either to check consistency or to check that a randomly chosen clause is satisfied. We now describe each of these steps and checks.

2.1 Reduction to Equations over S_5

For reasons of zero-knowledge we first make F a little more "robust" by expressing each variable $y_a \in F$ as the exclusive or of three new sub-variables x_{a1}, x_{a2}, x_{a3}. Note that information about up to three variables in the robust formula gives no information about any variable in F. From now on we simply assume that F is in this robust form.

Following the exposition in [6], a *permutation branching program* of width 5 and depth d is a level graph. Each level is labeled with one of n input variables $x_1, \ldots, x_n$, and contains 5 vertices. Associated with each level ℓ is a pair of permutations $\pi_0^\ell, \pi_1^\ell \in S_5$. Given a setting of the input variables, the level *yields* the permutation π_j^ℓ if the variable associated with level ℓ has value $j \in \{0, 1\}$ in the assignment. On input setting $\mathbf{x}$ the branching program yields the product of the permutations yielded by each of the levels. For level ℓ we let g_ℓ denote the variable over S_5 that has value either π_0^ℓ or π_1^ℓ according to the value of the (Boolean) input variable associated with level ℓ.

A permutation branching program B is said to *5-cycle recognize* a set $A \subseteq \{0,1\}^n$ if there exists a five-cycle $\sigma \in S_5 \backslash e$ (called the *output*) such that $B(\mathbf{x}) = \sigma$ if $\mathbf{x} \in A$ and $B(\mathbf{x}) = e$ if $\mathbf{x} \notin A$, where e is the identity permutation.

Theorem 2 *(Barrington [6]): Let A be recognized by a depth d fan-in 2 Boolean circuit. Then A is five-cycle recognized by a permutation branching program of depth 4^d.*

We will apply Barrington's result to a very specific type of circuit: one that checks that the clause

$$(y_{i_1 1} \oplus y_{i_1 2} \oplus y_{i_1 3}) \vee (y_{i_2 1} \oplus y_{i_2 2} \oplus y_{i_2 3}) \vee (y_{i_3 1} \oplus y_{i_3 2} \oplus y_{i_3 3})$$

is satisfied by the input. The clause has at most 9 distinct variables.

We assume that the robust F is a conjunction of clauses of the type just described (that is, F is in a sort of *robust 3-CNF*), so each clause has constant size. For each clause c_i having variables $x_{i1}, x_{i2}, \ldots, x_{i9}$ all three parties create a constant-depth Boolean circuit C_i, which, given an assignment $\mathbf{x_i}$ to the x_{ij}'s, checks that $\mathbf{x_i}$ satisfies c_i. Letting A_i be the set of assignments to $x_{i1}, x_{i2}, \ldots, x_{i9}$ satisfying C_i, the parties then apply Barrington's result to obtain a permutation branching program B_i that five-cycle recognizes A_i. Let $\sigma_i \in S_5$ be the output of B_i. Letting d be the depth of B_i, the construction yields an equation $g_{i1} \ldots g_{id} = \sigma_i$. Here, g_{ij} is associated with the (Boolean) variable that labels level j in B_i, taking on $\pi_0^{i,j}$ or $\pi_1^{i,j}$ according to the value of the associated Boolean variable. Thus, F is satisfiable if and only if (1) for all $1 \leq i \leq m$, the equations $g_{i1} \ldots g_{id} = \sigma_i$ are satisfiable (over the $\pi^{i,j}$'s), and (2) for all ℓ, p, j, q such that the same variable is associated with level ℓ of B_p and level j of B_q, $g_{\ell p} = \pi_0^{p,\ell}$ iff $g_{jq} = \pi_0^{q,j}$ in this satisfying assignment to the g's.

2.2 Checking an Equation

Consider the ith equation $g_{i1} g_{i2} \ldots g_{id} = \sigma_i$. Let us suppress the subscript i for ease of notation, so that we get $g_1 g_2 \ldots g_d = \sigma$. Let $\mathbf{h} = \{h_j | 1 \leq j \leq d, h_j \in \{\pi_0^j, \pi_1^j\}\}$ be an assignment to the g's satisfying the equation. We use a slight modification of the *randomizing tableaux* of Kilian [15] to allow the Provers to convince the verifier of the existence of $\mathbf{h}$.

Let T be the following array with 3 rows and d columns. $T[1, j] = h_j$ for all $1 \leq j \leq d$. Note that $\prod_{1 \leq j \leq d} T[1, j] = \sigma$. Let $r_{1,1}, \ldots, r_{1,d-1}$ be elements of S_5 chosen independently and uniformly at random. Then $T[2, 1] = h_1 r_{1,1}$, $T[2, d] = r_{1,d-1}^{-1} h_d$, and for all $1 < j < d$, $T[2, j] = r_{1,j-1}^{-1} h_j r_{1,j}$. Note that again $\prod_{1 \leq j \leq d} T[2, j] = \sigma$. Finally, we randomize again, choosing $d - 1$ new random elements $r_{2,1} \ldots r_{2,d-1} \in S_5$, and setting $T[3, 1] = T[2, 1] r_{2,1}$, $T[3, d] = r_{2,d-1}^{-1} T[2, d]$, and for all $1 < j < d$, $T[3, j] = r_{2,j-1}^{-1} T[2, j] r_{2,j}$. Once again $\prod_{1 \leq j \leq d} T[3, j] = \sigma$. Moreover, neither the second nor the third row of T contains any information about the assignment $\mathbf{h}$.

For any i, j such that $i \in \{1, 2\}$ and $j \in \{1, \ldots d\}$, let the i, j *rectangle* be the two entries $T[i, j], T[i + 1, j]$. Given the i, j rectangle and the random elements $r_{i+1,j-1}, r_{i+1,j}$ (if $j = 1$ or $j = d$ then only one of these is defined), it is easy to check that $r_{i+1,j-1}^{-1} T[i, j] r_{i+1,j} = T[i + 1, j]$. In addition, if T is not a randomizing tableau for $\mathbf{h}, \sigma$ then some rectangle will fail this test [15]. This suggests the following 2-prover interactive proof system.

The Verifier interacts with each prover once. In each interaction it may make the following requests. From P_1 it can request to see *one* of: (1) the third row of the tableau ($T[3, j], 1 \leq j \leq d$); (2) the i, j rectangle, for some $1 \leq i \leq 2$ and $1 \leq j \leq d$.

From P_2 the Verifier can request to see *one* of: (1) an element from the second and third rows of the of the tableau; (2) all the random elements $r_{1,j}, 1 \leq j \leq d$; (3) all the random elements $r_{2,j}, 1 \leq j \leq d$; (4) the assignment $\mathbf{x}_j$, where x_j labels one of the levels in the branching program.

The Verifier chooses either to check that the equation is satisfied or that the tableau is correctly constructed. To check that the equation is satisfied, the Verifier requests the third row from P_1 (option (1)) and an element of the top row from P_2 (option (1)). To check that the tableau is correctly constructed, the verifier has three possible options. In all three, it requests an i, j rectangle from P_1.

If $i = 1$: (a) The Verifier can request the assignment to the (Boolean) variable associated with level j. This checks consistency with P_1 and that the h_j's are chosen from the right sets (the π's). (b) The Verifier can request the randomizers for row 2. This checks that row 2 is formed correctly from row 1. (c) The Verifier can request the element from $T[2, j]$. This checks consistency with P_1.

If $i = 2$: (a) The Verifier can request the randomizers for row 3, checking that row 3 is formed correctly from row 2. (b) The Verifier can request an element from the rectangle, checking consistency with P_1. This completes the description of the protocol.

Intuitively, the most information a cheating verifier can possibly obtain about the bottom row (the assignment $\mathbf{h}$) is the assignment to two of the permutations g_j. Since each of these is associated with only one variable of the *robust* form of the Boolean formula F, and since the values of any two variables in the robust form yield no information about the value of any Boolean variable in the satisfying assignment to the original F, the procedure is truly zero-knowledge. Finally, since the randomizing tableau is for a single clause, it is of constant size. Thus any error in the construction of the tableau is detected with constant probability.

Remark: Checking Consistency

Let x_a be a variable in the robust form of F. Clearly, x_a may appear several times, and it must have the same assignment each time it appears. Let x_a appear in clauses p and q (p and q may be equal). Then for some j, k, x_a is the variable associated with level j of B_p and level k of B_q. Letting $\pi_0^{p,j}$ and $\pi_1^{p,j}$ be the two permutations at level j of B_p, and making analogous definitions for level k of B_q, the verifier must check that $\mathbf{h}_{pj} = \pi_0^{p,j} \Leftrightarrow \mathbf{h}_{qk} = \pi_0^{q,k}$. To check this, the Verifier asks P_1 for the $1, j$ rectangle from the tableau for B_p, and asks P_2 for the assignment $\mathbf{x}_a$ *without* disclosing to P_2 the name of the clause (p or q) that it is examining. This is covered by Case $i = 1(a)$ above.

Remark: Reducing the Number of Shared Random Bits In the description above it was assumed that the random bits used by the provers were completely independent. However, a closer examination reveals that since the verifier never sees more than a constant number of bits, they can be chosen to be c-wise independent for some constant c. Thus, the size of the probability space that generates them can be $O(\log n)$ bits (see e.g. [1]).

2.3 Putting it All Together

Without communicating, the Provers and Verifiers construct the robust form of F and the 5-cycle permutation branching programs for each of the m clauses of the robust form of F. Using their shared random bits, the Provers construct randomizing tableaux for all clauses consistent with a fixed satisfying assignment to the robust form of F. The Verifier randomly chooses a clause and one of the six legal pairs of questions described in the previous subsection, and proceeds accordingly. Note that the Verifier must tell P_1 which clause it has chosen, while it does not tell P_2 the chosen clause when it requests from P_2 the value of an assignment.

We now sketch proofs that our proof system is complete, partially sound and secure.

Theorem 3 (completeness) *If* $\mathbf{x}$*, the assignment known to* P_1 *and* P_2*, satisfies the robust form of* F*, then* V *will always accept.*

Proof. (Sketch) By construction of the randomizing tableaux, a simple case analysis shows that any constraints that V chooses to check will be satisfied.

Theorem 4 (soundness) *There exists a constant* $c > 0$ *such that* V *will reject with probability at least*

$$\frac{c(|S| - \mathbf{max\text{-}sat}(S))}{|S|},$$

where $\mathbf{max\text{-}sat}(S)$ *denotes the maximum number of clauses of* S *that may be simultaneously satisfied, and* $|S|$ *denotes the total number of clauses of* S*. This theorem holds regardless of the strategies of the provers,* $\hat{P}_1$ *and* $\hat{P}_2$*.*

Proof. (Sketch) First, by a standard lemma [7], there exist optimal *deterministic* provers, $\hat{P}_1$ and $\hat{P}_2$, that cause V to accept with the highest possible probability. It suffices to show that even with these provers, V will reject sufficiently often.

$\hat{P}_2$'s responses to queries about $\mathbf{x}$ constitute an assignment. Its responses to queries about rows 2 and 3 of the tableaux define these rows, just as its responses to queries about the randomizers define these objects as well. Let x_i be associated with some level ℓ of B_q, for some clause c_q.

Let c_p be chosen at random. Then with probability at least

$$\frac{c(|S| - \mathbf{max\text{-}sat}(S))}{|S|},$$

c_p is not satisfied by $\mathbf{x}$. It suffices to show that when this happens V will reject with some constant probability, regardless of what $\hat{P}_1$. does. In this case, $B_p(\mathbf{x}) = e$, so either the product of the elements of the top row of the randomizing tableau for B_p equals e, or the tableau is badly formed. Because the tableau is of constant size the error will be detected with constant probability.

Theorem 5 *The proof system achieves perfect zero-knowledge.*

Proof. (sketch) In order to prove this theorem, we construct a simulator M such that for any satisfiable 3-SAT formula F, any verifier $\hat{V}$ will obtain the same view by interacting with M as by interacting with P_1 and P_2. Recall that in the first step of the interactive proof system, before any communication begins, F is made "robust" by replacing every variable in F with 3 new variables. Let $\mathbf{x}_i$ denote the provers' assignment to x_i in the original formula. Then the provers may choose any random assignment to the sub-variables $x_{i1} \ldots x_{i3}$ so that the exclusive-or of these is $\mathbf{x}_i$.

The verifier makes one of 2 kinds of queries to P_1 and 4 kinds of queries to P_2 for a total of 8 kinds of pairs. The analysis is straightforward; we discuss only the case in which V requests a rectangle from P_1 and an assignment to some x_i from P_2.

Let the (possibly faulty) Verifier request rectangle i, j in the randomizing tableau for B_p from P_1. If $i = 2$ then the rectangle contains two independent randomly chosen elements of S_5, so simulating P_1's response is trivial. If $i = 1$ then since the variable associated with level j of B_p is from the robust form of F, both possible assignments to this variable are equally likely. Thus, either element of $\{\pi_0^{p,j}, \pi_1^{p,j}\}$ is equally likely, so the simulator can choose $T[i,j]$ from this set, and $T[i+1, j]$ from S_5. Finally, the response from P_2 needs only to be consistent with the response from P_1.

In the final version of the paper we will show how the number of bits that the provers send can be reduced to three - two by one prover and a single bit by the other. Note that this is the best possible, unless $P = NP$, since the existence of a two bit proof system can be translated to a 2-SAT problem.

3 Lower Bound on the Number of Shared Random Bits

In this section we show that the two provers must share $\Omega(\log\log n)$ random bits.

Let r be the number of shared random bits. We make several simplifying assumptions: let the total number of possible queries to each prover be polynomial in n; let the protocol be one round, i.e. the verifier sends the queries to the provers and they respond; let the provers responses be limited to $c \le 2^r$ possibilities (in this section we do not require c to be constant); let the two provers have no random bits other than the r shared random bits. Some of the above assumptions can be relaxed (see remarks at the end of this section).

We will show that if 3-SAT is recognized by a 2-prover zero-knowledge interactive proof system obeying these constraints and $r \in o(\log\log n)$, then 3-SAT $\in BPP$. The main idea is to first show that a small number of random bits implies that the two provers have only a small number of different strategies for answering the queries. We then show that this implies that on inputs of 3-SAT of any length n, in polynomial time it is possible, using the simulator whose existence is guaranteed by the zero-knowledge property, to reduce the problem to an instance of 3-SAT of size strictly less than n. By repeating this at most n times (a more careful analysis shows that $\log\log n$ times suffice), we can therefore, in polynomial time, reduce the problem to one that can be efficiently solved by brute force.

Fix a satisfiable input formula F of n variables for the remainder of the discussion.

Let $u_1, \ldots, u_m$ ($v_1, \ldots, v_m$) be all the possible queries, over all random choices of the verifier, that the verifier could send to P_1 (P_2). The first step in the reduction is to split the u's (v's) into a (relatively) small number of equivalence classes. We describe the procedure for splitting the u's. The v's are handled similarly.

Intuitively, u_1 and u_2 will be in the same class if P_1 does not distinguish between them. However, for any random string s shared by the two provers, even using the simulator, there is no way to compare the behavior of P_1 on query u_1 with its behavior on query u_2, since each invocation of the simulator queries P_1 exactly once and on different invocations of the simulator the simulated P_1 may have different random strings. We must therefore define the equivalence classes in a slightly more roundabout fashion, so that we can compute them using the simulator.

Let (u, v) be an arbitrary pair of queries to P_1 and P_2, respectively. Let Answers$((u, v), s)$ denote the pair of responses on this pair of queries when the provers share s. Let Pairs$(u, v) = \{\text{Answers}((u, v), s) | s \in \{0, 1\}^r\}$. Then

$$u_1 \sim u_2 \Leftrightarrow \forall v(\text{Pairs}(u_1, v) = \text{Pairs}(u_2, v)).$$

Intuitively, although the verifier might distinguish between similar queries, P_1 does not.

At a high level, we will proceed as follows. To reduce the size of the problem we use the simulator to compute the equivalence classes, arguing that there are not too many of them. The entire strategy of the two provers can then be described by the number of pairs of classes times the number of pairs of responses (c^2, assuming each prover sends one of only c possible answers on each query). But the description of the strategy is just a string, so we have reduced the problem to one of finding a string of at most this length that causes the verifier to accept. We now give more details.

By assumption, the number of u's is at most polynomial in n. We now show that, using the simulator, we can compute Pairs(u, v) for all pairs of queries u, v, in BPP. For each u we proceed as follows. For each v_i run the simulator many times with a verifier that asks the pair of queries (u, v_i), to obtain the set Pairs(u, v_i). (It may be that the honest verifier never asks this particular pair of queries. However, some cheating verifier must do so.) Note that as long as the number of shared random bits is at most $O(\log n)$ every element of this set will be discovered with arbitrarily high probability in polynomial time. The sets Pairs(u, v) are then used to determine the equivalence classes.

Note that for every query u there is a vector of possible replies, each an element in $\{1, \ldots, c\}$, and indexed by the shared random string s. Let this vector of reply be the *color* of the query. There are only c^{2^r} possible colors. Moreover, if two queries have the same color then they are in the same equivalence class (an equivalence class may include queries of different colors). Thus, the number of equivalence classes is at most c^{2^r}. If r is sufficiently small, then we can obtain a representative from each equivalence class on the u's and on the v's. Using the simulator with the real verifier we can obtain, with arbitrarily high probability, for all pairs of representatives (u, v) such that on some execution the verifier actually asks this pair of queries, the set Pairs(u, v). Call this set a *constraint*. Note that $|\text{Pairs}(u, v)| \le c^2$.

To reduce the size of the problem, we make the following definitions. Let $u_1, \ldots, u_\ell$ be representatives of the classes of queries to P_1, and let $v_1, \ldots, v_k$ be representatives of the classes of queries to P_2; note that $\ell, k \le c^{2^r}$. Let S_1 be a function from the representatives u_i to $\{1, \ldots, c\}$, and let S_2 be a function from the representatives v_j to $\{1, \ldots, c\}$. The problem now reduces to finding S_1 and S_2 satisfying the following condition. For all pairs of representatives u_i, v_j such that in some execution of the interactive proof system, V sends a member of the class represented by u_i to P_1 and a member of the class represented by v_j to P_2,

$$(S_1(u_i), S_2(v_j)) \in \text{Pairs}(u_i, v_j).$$

Thus, the problem of proving that $F \in 3 - SAT$ can be reduced to the problem of finding a strategy for the provers that satisfies these constraints. It follows that the question of whether a strategy exists can be defined by a string that is at most the square of the number of classes times the square of the number of possible responses. That is, the length of the description of the constraints that the strategy must satisfy is at most $(c^{2^r})^2 \cdot c^2 = 2^{2^r + O(1)}$. Since this question is clearly in NP, it follows from the Cook-Levin Theorem that there exists a polynomial p such that a string x is such a strategy if and only if some (effectively computable) formula F_x of length $p(|x|)$ is satisfiable. Thus, if $p(2^{2^{r+O(1)}}) < n$ then the original problem of size n can be reduced, in BPP, to a problem of strictly smaller size. This happens when $r = o(\log\log n)$. We therefore have the following theorem.

Theorem 6 *Let L be an NP-complete language recognizable by a perfectly complete perfect zero-knowledge two prover interactive proof system in which the verifier poses a single query to each prover, the reply from each prover is restricted to a single element from a set of size $c \leq 2^r$, and the provers have no random bits other than the shared random bits. Then if the number of shared random bits is $o(\log\log n)$ then $L \in BPP$.*

Note that we have not used the fact that the probability of acceptance in case the formula is not satisfiable is less than $\alpha < 1$ and that the provers are polynomial time machines (with access to a satisfying assignment).

Remarks:
(1) Virtually the same proof shows that the provers must share $\Omega(\log\log n)$ random bits also in *statistical* zero knowledge proofs for NP.
(2) If the provers do not use private random bits, we can assume that the range of possibilities of the provers' replies (denoted by c) is at most of size 2^r. Given that there are only r shared random bits and no private random bits, then on every possible query there are at most 2^r answers that the prover may give. Using the simulator these answers can be enumerated. The protocol can then be changed with the prover giving a pointer of r bits into this list, instead of sending the full answer. The resulting protocol would be only statistical zero knowledge, and would not have perfect completeness. Nevertheless, the proof of the lower bound would still hold with minor modifications.
(3) If the number of possible queries is not polynomial in n then it is still possible, in polynomial time, to find all the equivalence classes that are "likely" to be asked and all the constraints that are likely to influence. The construction proceeds the same way, only we simply ignore "unlikely" queries.
(4) The protocol may contain several rounds instead of one round. The concatenation of a prover's answers plays the role of the prover's answer in the single round case. The only difficulty is in implementing remark 2. However this can be solved by making c no larger than 2^{2^r}, which does not affect the lower bound.
(5) We can allow the provers to have an arbitrary number r of private random bits, provided $\log c + r \in o(\log\log n)$. The main difference is in the definition of the color of a query. In the new definition, each entry in the vector is replaced by a list of possible responses, which vary according to the private random bits of the prover.
(6) Under most assumptions, the lower bound on the number of random bits shared by the provers can be pushed up to $\log\log n - 3$.
(7) While the lower bound shows that $\Omega(\log\log n)$ shared random bits are necessary, the proof relies on the fact that the protocol must be zero-knowledge for *all* verifiers. Indeed, our protocol can be easily modified to use only $O(1)$ shared random bits if zero-knowledge is only required against the honest Verifier.

References

1. N. Alon, J. H. Spencer, **The Probabilistic Method**, John Wiley & Sons, New-York, 1992.

2. S. Arora, C. Lund, R. Motwani, M. Szegedy and M. Sudan, *Proof Verification and the Hardness of Approximations*, Proc. 33rd IEEE Symp. on Foundation of Computer Science, 1992, to appear.

3. S. Arora and M. Safra *Probabilistic Checking of Proofs* Proc. 33rd IEEE Symp. on Foundation of Computer Science, 1992, to appear.

4. L. Babai, L. Fortnow, L. Levin and M. Szegedy *Checking Computations in Poly-logarithmic Time* Proc. 23rd ACM Symposium on Theory of Computing, 1991, pp. 21–31.

5. L. Babai, L. Fortnow, C. Lund, *Non-Deterministic. Exponential Time has Two-Prover Interactive Protocols*, Proc. 31st IEEE Symp. on Foundation of Computer Science, 1990, pp. 16-25.

6. D. A. Barrington, *Bounded-Width Polynomial-Size Branching Programs Recognize Exactly Those Languages in NC^1*, JCSS (38), 1989, pp. 150-164.

7. M. Ben-or, S. Goldwasser, J. Kilian, A. Wigderson, *Multi Prover Interactive Proofs: How to Remove Intractability*, Proc. 20th ACM Symposium on Theory of Computing, 1988, pp. 113-131.

8. J. Cai, A. Condon, R. Lipton, *On Bounded Round Multi-Prover Interactive Proof Systems*, Proc. of Structure in Complexity, 1990, pp. 45-54.

9. J. Cai, A. Condon, R. Lipton, *Playing Games of Incomplete Information, STACS 1990*.

10. J. Cai, A. Condon, R. Lipton, *PSPACE is Provable by Two Provers in One Round*, Proc. Structure in Complexity, 1991, pp. 110-115.

11. U. Feige, *On the Success Probability of the Two Provers in One Round Proof Systems*, Proc. Structure in Complexity, 1991, pp. 116-123.

12. U. Feige, S. Goldwasser, L. Lovasz, M. Safra, M. Szegedy, "Approximating Clique is Almost NP-Complete", Proc. 32nd IEEE Symp. on Foundation of Computer Science, 1991, pp. 2-12.

13. U. Feige and L. Lovasz, *Two-Provers One Round Proof Systems: Their Power and Their Problems*, Proc. 24th ACM Symposium on Theory of Computing, 1992,

14. L. Fortnow, J. Rompel, M.Sipser, *On the Power of Multi-Prover Interactive Proto-cols*, Proc. of Structure in Complexity 1988, pp. 156-161. Erratum in Proc. Structure in Complexity, 1990, pp. 318-319.

15. J. Kilian, **Use of Randomness in Algorithms and Protocols**, MIT Press, 1990.

16. J. Kilian, *Strong Separation Models of Multi Prover Interactive Proofs*, DIMACS Workshop on Cryptography, October 1990.

17. J. Kilian, *A Note on Efficient Zero-Knowledge Proofs and Arguments*, Proc. 24th ACM Symposium on Theory of Computing, 1992,

18. D. Lapidot, A. Shamir, *A One-Round, Two-Prover, Zero-Knowledge Protocol for NP*, Crypto'91 abstracts.
19. D. Lapidot, A. Shamir, *Fully Parallelized Multi Prover Protocols for NEXP-time* Proc. 32nd IEEE Symp. on Foundation of Computer Science, 1991, pp. 13-18.

Invariant Signatures and Non-Interactive Zero-Knowledge Proofs are Equivalent

(EXTENDED ABSTRACT)

Shafi Goldwasser*   and   Rafail Ostrovsky**

Abstract. The standard definition of digital signatures allows a document to have many valid signatures. In this paper, we consider a subclass of digital signatures, called *invariant* signatures, in which all legal signatures of a document must be identical according to some polynomial-time computable function (of a signature) which is hard to predict given an unsigned document. We formalize this notion and show its equivalence to non-interactive zero-knowledge proofs.

* MIT. This research was supported in part by NSF-FAW CCR-9023313, NSF-PYI CCR-865727, Darpa N0014-89-J-1988, BSF 89-00312.
** International Computer Science Institute at Berkeley and University of California at Berkeley. Supported by NSF Postdoctoral Fellowship. Parts of this work were done at MIT, Bellcore and IBM T.J. Watson Research Center.

1   Introduction

Currently, due to the lack of proven non-trivial lower bounds on NP problems, the theory of cryptography is primarily based on unproven assumptions such as the difficulty of particular computational problems such as integer factorization, or more generally the existence of one-way and trapdoor functions. It is thus naturally desirable to establish minimal complexity assumptions for basic cryptographic primitives, and to establish connections among these primitives. Indeed, it has been an active and in many cases successful area of research. For example, pseudo-random generators [BM] were shown to be equivalent to the existence of any one-way function [ILL, H]. On the other hand, several other primitives, such as secret-key exchange seem to require the trapdoor [IR] property.

Digital signatures have been an especially interesting case in point. Originally introduced by Diffie and Hellman [DH], the first implementation was based on the RSA trapdoor function [RSA] which yields a deterministic signature scheme where each document has a unique valid signature. Later, the notion of digital signatures which are secure against chosen message attack[3] was formally defined by [GoMiRi] and proved to exist under a sequence of decreasingly weaker assumptions: the existence of claw-free permutations [GoMiRi] (e.g. factoring), the existence of trapdoor permutations [BeMi], the existence of one-way permutations by [NY], and finally the existence of one-way functions by [Ro]. In all of these schemes, each document may have many valid signatures.

The fact that digital signatures can be implemented if one-way functions exist without the need for a trapdoor [NY, Ro] is somewhat remarkable, as by definition a digital signature seems to posses the essential flavor of a trapdoor function: namely, it should be easy for everyone to verify the correctness of a signature, while it should be hard for everyone except a privileged user (with access to the private file) to sign. In this paper, we study which aspects of digital signatures allows for this dichotomy and whether digital signatures can in some cases be used in cryptographic protocols instead of trapdoor functions.

We show that the issue of having many different valid signatures of the same document plays a role in the above question. That is, on the positive side, we

---

[3] Note that RSA does not satisfy security against adaptive chosen message attack as there do exist messages for which the signature can be forged.

show that digital signatures can sometimes be used instead of trapdoor functions, provided that all valid signatures of the same document have an *invariant* property which is unpredictable from the document itself. On the negative side, we show that this *invariant* property for a signature scheme may require a trapdoor for its implementation (unless non-interactive zero-knowledge proofs among polynomial-time participants can also be implemented without a trapdoor).

Invariant signatures are interesting in their own right, as they capture the flavor of having a unique valid signature per document as in the case of RSA, and yet can be proven secure against adaptive chosen message attack as in the case of [GoMiRi, BeMi, NY, Ro]. Achieving these two aspects simultaneously may prove valuable in applications.

1.1 Invariant Signatures

Let us recall the definition of digital signatures as defined in [GoMiRi]. Informally, the setting is as follows: in a network, every user can generate (using a polynomial-time algorithm) a pair of keys: the public key and the corresponding secret key. In addition to the generation algorithm, the signature scheme is provided with two probabilistic polynomial-time algorithms: one for signing and one for verifying. Given an arbitrary document, a user applies his signing algorithm to the document, his public key, and his secret key. Given a signature of a document, any other user can verify the validity of the signature by applying the polynomial time verification algorithm to the signature, document, and the public key of the signer. No adversary can forge a signature for a new document, even after asking for arbitrary signature samples in an adaptive fashion.

The additional constraint we put on digital signatures so as to make them *invariant*, is (informally) that there exists a deterministic poly-time computable function g computed on signatures such that with high probability (1) for any document D and for any two legitimate signatures $\sigma_1(D)$ and $\sigma_2(D)$, $g(\sigma_1(D)) = g(\sigma_2(D))$ and (2) given D, $g(\sigma(D))$ is pseudo-random. If the above conditions hold we say that the signature scheme is invariant under g.

Although not the subject of this paper, we suggest that our definition of invariant signatures might serve as a good definition for what we may want from a *finger print* of a document: hard to predict for any document even in an adaptive setting, dependent perhaps on the time of inquiry, and yet unique.

1.2 Non-Interactive Zero-Knowledge Proofs and Digital Signatures

We investigate the comparative difficulty of non-interactive zero-knowledge proofs ($\mathcal{NIZK}$) [BFM] and digital signatures ($\mathcal{DS}$) [GoMiRi]. These seemingly different primitives were shown to be connected in a paper by Bellare and Goldwasser [BG], where it was shown that the existence of one-way functions and non-interactive zero-knowledge proofs implies the existence of digital signatures (secure against adaptive chosen-message attacks). We remark that the known constructions of non-interactive zero-knowledge proofs with polynomial-time participants use the *trapdoor* permutations assumption [FLS], while digital signatures can be implemented based on *any* one-way function [Ro].

We show that the existence of invariant digital signatures is equivalent to the existence of non-interactive zero-knowledge proofs. That is, we show that while a signature scheme in which a document can be signed in an unconstrained plurality of ways requires the existence of any one-way function, a signature scheme in which each document has unique or at least "similar signatures" (according to any "nontrivial" poly-time computable function — this is the invariant property!) requires the same assumptions as non-interactive zero-knowledge proofs (i.e. currently the trapdoor assumption is necessary).

More precisely, we consider non-interactive zero-knowledge proofs in the random string model, where users in the system can read a pre-existing common (polynomial size) random string set up by the system (a model defined by [BFM]). We prove that in this common random string model, the existence of invariant digital signatures is equivalent to the existence of non-interactive zero-knowledge proofs for any hard to predict NP language (see definition in 2.2). To prove this theorem we must define invariant signatures in the common random string model.

1.3 A simple example: using digital signatures to achieve asymmetry

Suppose two probabilistic polynomial-time players (Alice and Bob) wish to agree on a boolean predicate $B(\cdot)$, so that when later given a randomly chosen x as a common input, Bob can *not* predict $B(x)$ with probability (over x and Bob's coin tosses) bounded away from half, but Alice can compute $B(x)$ and convince Bob of the value of $B(x)$. Under what assumptions can we implement such a protocol?

Before we examine the above question, let us recall definitions of a one-way function and a trapdoor function. Informally, a poly-time computable function

f is one-way if when we pick x uniformly at random and compute $y \leftarrow f(x)$, it is infeasible for any polynomial time machine to find x' in $f^{-1}(y)$ for a non-negligible fraction of the instances. Again informally, a trapdoor function, is a one-way function with an additional secret key, the knowledge of which makes inversion easy.

Assuming the existence of one-way trapdoor permutations, Alice and Bob can achieve the above task. In particular, they can agree on a trapdoor one-way permutation (f, f^{-1}), so that Alice knows (f, f^{-1}) and Bob knows only f. In addition, they agree on a hard-core [GL] bit $B(\cdot)$ for f. (Notice that Alice and Bob must make sure that f is really a permutation for $B(\cdot)$ to be well defined.) Subsequently, when x is given, Alice can invert f and compute a hard-core bit, while Bob can not.

Can we achieve the above task using one-way functions which are not trapdoor? Let us examine if digital signatures (which do not need trapdoor in their implementation) might be useful.

At first glance, to implement a simple protocol specified above could be done using digital signatures as follows: Alice prepares a public and a secret key (of a signature scheme), gives her public key to Bob and convinces him that her public key is produced using an appropriate key-generation algorithm. Moreover, they agree on a hard-core bit B of a signature for any document x'. Notice that given x and a public key of Alice, the signature of x is hard to find for any polynomial-time player, and thus Bob can not predict the hard-core bit of a signature of x, while Alice can easily compute it. Since we can implement signatures based on one-way functions (without the trapdoor) it seems that we can implement the above protocol without the trapdoor... What is wrong in this argument?

The problem, is that this bit is *not* well defined. That is, the specification of digital signatures allows for many legal signatures of x. However, if we put an additional constraint on the digital signature scheme, then the above argument will go through. The additional constraint is to have an *invariant* signature scheme (as above). Then, to implement the above game, Alice can use a hard-core bit of $g(\sigma(D))$ (where all signatures of D are invariant under g) and the bit is well-defined. Thus, notice that *invariant* digital signatures can be used in the above setting instead of a trapdoor function.

2 Model and Definitions

2.1 Negligible, noticeable and infeasible functions

We use the usual O, o and $1/o(1)$ (asymptotically tending to ∞) notation. We fix some function $s(n) = n^{1/o(1)}$ and call it *infeasible*. We call $\epsilon(n) = 1/s^{O(1)}(n)$ *negligible* and $\delta(n) = 1/O(n^c), c > 0$ *noticeable*. In this case, n is a security parameter, which we omit when clear from the context. We use standard definitions of one-way functions and computationally indistinguishable distributions (see, for example,[GL, ILL, H]). If S is a probability space then $x \leftarrow S$ denotes the algorithm which assigns to x an element randomly selected according to S. For probability spaces $S, T, \ldots$, the notation $\Pr(p(x, y, \cdots) : x \leftarrow S; y \leftarrow T; \cdots)$ denotes the probability that the predicate $p(x, y, \cdots)$ is true after the (ordered) execution of the algorithms $x \leftarrow S$, $y \leftarrow T$, etc. The notation $\{f(x, y, \cdots) : x \leftarrow S; y \leftarrow T; \cdots\}$ denotes the probability space which to the string σ assigns the probability $\Pr(\sigma = f(x, y, \cdots) : x \leftarrow S; y \leftarrow T; \cdots)$, f being some function. If S is a finite set we will identify it with the probability space which assigns to each element of S the uniform probability $\frac{1}{|S|}$. (Then $x \leftarrow S$ denotes the operation of selecting an element of S uniformly at random).

2.2 Non-Interactive Zero-Knowledge ($\mathcal{NIZK}$) Proofs in the Common Random String Model

Non-interactive zero-knowledge proofs were introduced in [BFM]. We note that this is where the "common random string model" was introduced as well.

Common random string model: at the time of the system set-up a string of a fixed (polynomial in the security parameter) length is chosen uniformly at random and published by a trusted center for everyone in the system (provers, verifiers, users etc.) such that it can be read but not modified.

Informally, a $\mathcal{NIZK}$ proof of an $\mathcal{NP}$ statement in a common random string model is a way for any polynomial-time user to convince other users that some statement is true without revealing anything else. That is, given a common random string, and a witness to an $\mathcal{NP}$ statement, there should be a probabilistic poly-time algorithm (for the prover) which constructs a proof of that statement, and a probabilistic poly-time algorithm (for the verifiers) to check that the proof is correct. Moreover, such proof should not reveal anything about the witness.

Formally, the following definition is essentially taken from [BDMP].

Definition 1. We fix an $\mathcal{NP}$ language L (with poly-time relation $\rho(\cdot, \cdot)$ and constant d such that $x \in L$ iff $\exists w, |w| < |x|^d, \rho(x, w) = 1$.) We say that two probabilistic polynomial-time algorithms $(prover(\cdot, \cdot, \cdot), verifier(\cdot, \cdot, \cdot))$ constitute **bounded** $\mathcal{NIZK}$ for language L if the following conditions are satisfied: there exist a polynomial l such that

Completeness: For all $x \in L$, $|x| = n$, sufficiently large n, and ϵ negligible, where w is such that $|w| < n^d$ and $\rho(x, w) = 1$, the $\Pr(verifier(x, w, c) = accept : c \leftarrow \{0, 1\}^{l(n)}; y \leftarrow prover(x, w, c)) > 1 - \epsilon(n)$.

(Here, c is the "common random string", w is the NP witness, and y is the output of the prover which is computed non-interactively. The probability is taken over the choice of c and the prover's coin tosses).

Soundness: For all probabilistic polynomial-time players $prover'$, $x \notin L$, $|x| = n$, for sufficiently large n, and negligible ϵ, the $\Pr(verifier(x, w, c) = accept : c \leftarrow \{0, 1\}^{l(n)}; y \leftarrow prover'(x, c)) < \epsilon(n)$.

(Here, the probability is taken over the choice of c and prover's coin tosses).

Zero-Knowledge: There exists a probabilistic expected polynomial-time algorithm $S(\cdot, \cdot)$ such that for all $x \in L$, $|x| = n$, and w such that $|w| < n^d$ and $\rho(x, w) = 1$, for all probabilistic polynomial time algorithms D, for all sufficiently large n, the

$$| \Pr(D(c, x, y) = 1 : c \leftarrow \{0, 1\}^{l(n)}; y \leftarrow prover(x, w, c)) -$$
$$\Pr(D(c, x, S(x, c)) - 1 : c \leftarrow \{0, 1\}^{l(n)})| < \epsilon(n)$$

In the above c is called the "common random string"., and l the length of the common random string.

REMARKS:

- One difference from above definition to [BDMP] is that we impose the soundness condition only on probabilistic polynomial-time $prover'$s. This is not actually necessary as known constructions achieve soundness against all $prover'$s. However, as in the context of this paper we show equivalence to a digital signatures in which a reasonable forger to consider is probabilistic polynomial time, we relax the soundness requirement here as well.

- The above definition is specified for a single theorem of a fixed polynomial size. This **bounded** $\mathcal{NIZK}$ definition can be extended to polynomially-many theorems each of polynomial length and to many users in the roles of both provers and verifier. This is the notion of $\mathcal{NIZK}$ we adopt here. To modify the above definition to accommodate this extension, we must require (as in [BDMP]) the existence of many pairs of $prover_i, verifier_i$ for which completeness and soundness are true, and change the zero-knowledge condition as follows.

  [Zero-Knowledge':] There exists a probabilistic expected polynomial time algorithm S such that for all $x_1, x_2, \ldots \in L \cap \{0,1\}^n$, where $|w_1|, |w_2|, \ldots < n^d$ and $\rho(x_1, w_1) = 1$, $\rho(x_2, w_2) = 1, \ldots$, for all probabilistic polynomial time algorithm D, for all sufficiently large n, for all negligible ϵ,

  $|\Pr(D(c, (x_1, y_1), (x_2, y_2), \ldots) = 1 : c \leftarrow \{0,1\}^{l(n)}; y_1 \leftarrow prover_1(x_1, w_1, c);$
  $y_2 \leftarrow prover_2(x_2, w_2, c); \ldots) -$
  $\Pr(D(c, (x_1, S(x_1, c)), (x_2, S(x_2, c)), \ldots) = 1 : c \leftarrow \{0,1\}^{l(n)})| < \epsilon(n).$

- Another aspect of $\mathcal{NIZK}$ is a preservance of zero-knowledge in an adaptive setting, which means that even after requesting polynomially-many proofs one by one, the probability for polynomial-time Adv (over its coin-flips) of being able to distinguish an NIZK proof of a new theorem from the run of the simulator is negligible. Notice that if $\mathcal{NIZK}$ proofs remains Zero-Knowledge even in an adaptive setting, then the statements may be *dependent* on the previous proofs and on the common random string. From now on, when we refer to $\mathcal{NIZK}$, we refer to $\mathcal{NIZK}$ which is secure in an adaptive setting. To modify the above definition to accommodate this extension we further refine the zero knowledge condition as follows.

  [Zero-Knowledge'':] There exists a probabilistic expected polynomial time algorithm S such that for all polynomial time Adv, for all probabilistic polynomial time D, for all sufficiently large n, for all negligible ϵ,

  $|\Pr(D(c, (x_1, y_1), (x_2, y_2), \ldots) = 1 : c \leftarrow \{0,1\}^{l(n)}; x_1 \leftarrow Adv(c);$
  $y_1 \leftarrow prover_1(x_1, w_1, c); x_2 \leftarrow Adv(c, x_1, y_1); y_2 \leftarrow prover_2(x_2, w_2, c); \ldots) -$
  $\Pr(D(c, (x_1, S(x_1, c)), (x_2, S(x_2, c)), \ldots) = 1 : c \leftarrow \{0,1\}^{l(n)}; x_1 \leftarrow Adv(c);$
  $y_1 \leftarrow S(x_1, c); x_2 \leftarrow Adv(c, x_1, S(x_1, c)); y_2 \leftarrow S(x_2, c); \ldots)| < \epsilon(n).$

- We note that in our setting, provers are polynomial-time machines.

- An additional property of $\mathcal{NIZK}$ that we must stress is of being *publically verifiable* $\mathcal{NIZK}$ proof system, which means that the proof can be verified by any polynomial-time machine which has access to a common random string.

In [BFM, DMP1, BDMP] it was shown how $\mathcal{NIZK}$ could be implemented, based on algebraic assumptions. In [DMP2, KMO] the $\mathcal{NIZK}$ was implemented based on the general complexity assumptions and without a common random string, but at a price of a small pre-processing stage, which was interactive. Finally, in [FLS] it was shown how $\mathcal{NIZK}$ could be implemented without pre-processing, based on (verifiable) trapdoor one-way permutations. (In [BY], they show how verifiability requirement could be implemented based on trapdoor one-way permutations). Moreover, in [FLS] it was shown how to convert $\mathcal{NIZK}$ into publically-verifiable and adaptively secure (see remarks above) $\mathcal{NIZK}$ proof system. Again, we mention that it is not known how the assumptions (of one-way trapdoor permutations) could be reduced further.

Definition 2. We say that a language L is *hard to predict* if there exist a probabilistic polynomial time algorithm $S(1^n)$ (which samples $X \in \{0,1\}^n$) such that for every probabilistic polynomial-time algorithm Adv, for all sufficiently large n and for all negligible ϵ, the probability (over S and Adv coin tosses) that Adv can correctly decide if $X \in L$ is less then $\frac{1}{2} + \epsilon(n)$.

REMARK: The above definition can be modified as follows: we say that a language L is *sometimes hard to predict* if there exist a probabilistic polynomial time algorithm $S(1^n)$ (which samples $X \in \{0,1\}^n$) such that on a noticeable fraction H of $S(1^n)$, for every probabilistic polynomial-time algorithm Adv, for all sufficiently large n and for all negligible ϵ, the probability (over S and Adv coin tosses) that Adv on X in H can correctly decide if $X \in L$ is less then $\frac{1}{2} + \epsilon(n)$.

Definition 3. We say that nontrivial $\mathcal{NIZK}$ exists, if there exists a (sometimes) hard to predict $L \in \mathcal{NP}$ which possesses an $\mathcal{NIZK}$ proof system.

We note that the existence of $\mathcal{NIZK}$ proofs for (sometimes) hard to predict L implies the existence of one-way functions [OW].

2.3 Invariant Digital Signatures ($\mathcal{INV} - \mathcal{DS}$)

The formulation of the digital signatures of [GoMiRi] allows any document to have many valid signatures (i.e. accepted by the signature verification algorithm as valid) of the same document. For invariant signatures we make the additional requirement that all valid signatures of the same document be "similar", that is, there exists an easy to compute function defined on signatures which yields the same value for all signatures of the same document. This function should be hard to compute from the document itself with access to the public key (but without access to the secret key).

In the following definition we incorporate the possibility that a common random string c was published by a trusted center at the time of a system set up for everyone in the system (signers and verifiers) to read but not to modify. This is similar to the set up of NIZK (see previous section). The definition of an invariant digital signature scheme can be made in the standard model as well (without the presumption of the existence of c), but as in this paper we show the equivalence of invariant signatures and NIZK in the common random string model, we present the definition of invariant digital signatures in this model. The polynomial $l(n)$ will denote the length of the common random string with security parameter n.

Definition 4. An **invariant signature scheme** π is a quadruple (G, S, V, g) such that the following conditions hold: let l be a polynomial function

G: is a probabilistic poly-time computable algorithm (the "key generation" algorithm) which on input 1^n (the security parameter), $c \in \{0,1\}^{l(n)}$ (the common random string) outputs a pair of strings (*secret-key, public-key*). We let the random variables $G_1(1^n)$ denote the first output and $G_2(1^n)$ the second output. (Wlog we let $|G_1(1^n)| = |G_2(1^n)| = n$. The probability is over $c \leftarrow \{0,1\}^{l(n)}$ and G's coin tosses.)

S: is a probabilistic poly-time computable algorithm (the "signing" algorithm) which on input strings 1^n, $c \in \{0,1\}^{l(n)}$ (the common random string), $D \in \{0,1\}^*$ of length polynomial in n (the document), and a pair of strings $\{secret-key, public-key\}$ in the range of $G(1^n, c)$ outputs a string. The output is referred to as the "signature" of D (with respect to $public - key$ and c). When the context is clear we let $\sigma(D)$ denote an output of $S(1^n, D, G(1^n, c), c)$.

V: is a probabilistic poly-time computable algorithm (the "verification" algorithm) which receives as inputs the strings 1^n (the security parameter), $D \in \{0,1\}^*$ of length polynomial in n (the document), s (the presumed signature of D), $c \in \{0,1\}^{l(n)}$ and $public - key \in G_2(1^n)$, and outputs either true or false. We require that for all D in n, the $\Pr(V(1^n, D, s, public - key, c) = true : c \leftarrow \{0,1\}^{l(n)};$ $\{secret - key, public - key\} \leftarrow G(1^n, c); s \leftarrow S(1^n, D, \{secret - key, public - key\}, c)) = 1$

(Namely, signatures produced by the signing algorithm S are always accepted by the verifying algorithm V for any pair of public and private keys produced by key generation algorithm G).

If $V(1^n, D, s, public\text{-}key, c) = true$ then we say that s is a "valid" signature of D (with respect to $public - key$ and c).

security: Let F be a probabilistic poly time forging algorithm which receives as input the strings 1^n, $c \in \{0, 1\}^{l(n)}$, and $public - key \in G_2(1^n)$; can request and receive signatures with respect to $public - key$ and c of polynomially-many adaptively chosen documents $\{D_i\}$ and finally outputs a pair of strings (D, s). Then, for all such F, for all sufficiently large n, for all negligible functions ϵ, the probability that F outputs (D, s) where $D \notin \{D_i\}$, and s is a valid signature of D with respect to $public - key$ and c is less than $\epsilon(n)$.

(The probability is taken over the outcome of G, signatures of D_i, and the coin tosses of F).

invariant function $g(\cdot, \cdot)$: is a polynomial time computable function which takes as input strings 1^n and s (when clear we use notation $g(s)$ for $g(1^n, s)$) and produces as output a string $t \in \{0, 1\}^{r(n)}$ where r is a fixed polynomial, such that:

invariance Let Adv be a probabilistic polynomial-time algorithm which receives as input strings 1^n, $c \in \{0, 1\}^{l(n)}$, and produces as output the tuple $(public - key, D, \sigma_1(D), \sigma_2(D))$ where $public - key \in \{0, 1\}^n$, and $\sigma_1(D)$ and $\sigma_2(D)$ are both valid signatures of D with respect to $public - key$ and c. Then, for all such Adv, for all $public - key \in \{0, 1\}^n$, for any negligible ϵ, and for sufficiently large n, the probability that $g(\sigma_1(D)) \neq g(\sigma_2(D))$ is less than $\epsilon(n)$.

(Here the probability is taken over $c \leftarrow \{0, 1\}^{l(n)}$, and the coin tosses of Adv.) (Note, that the definition implies that even the honest signer who has access to the secret key can not produce two signatures of the same document for which g is not the same with non-negligible probability.)

pseudo-randomness Let Adv be a probabilistic polynomial time algorithm which operates in two stages on input strings 1^n, $c \in \{0, 1\}^{l(n)}$, and $public - key \in G_2(1^n)$. In the first stage Adv can request and receive signatures with respect to $public - key$ and c of polynomially-many (in n) adaptively chosen documents $\{D_i\}$.

At the end of the first stage, Adv outputs a polynomial length string D not in $\{D_i\}$. In the second stage, Adv is presented with a string t on which it outputs 0 or 1 (we let $Adv(t)$ denote the output bit). Let $\alpha = \Pr(Adv(t) = 1 : c \leftarrow \{0,1\}^{l(n)}; \{secret-key, public-key\} \leftarrow G(1^n, c); s \leftarrow S(1^n, D, \{secret-key, public-key\}, c); t \leftarrow g(s))$ and let $\beta = \Pr(Adv(t) = 1 : t \leftarrow \{0,1\}^{r(n)})$. Then, for all Adv, for all negligible ϵ, for all sufficiently large n, $|\alpha - \beta| < \epsilon(n)$.

We call g the *invariant function* of the signature scheme, and l the length of the invariant function.

REMARK: We note that in the above definition the invariant property holds for *any* public file *public − key*, and not just over $G_2(1^n)$. This requirement ensures that *invariant* property holds for any public key, even a maliciously chosen one, and avoids problem pointed out in [BY] of lack of certification in [FLS].

The most important aspect of invariant signature scheme for our application is:

Lemma: If $\pi = (G, S, V, g)$ is an invariant signature scheme, then there exists a polynomial time computable Boolean predicate P which on input 1^n and s, outputs 0 or 1 such that the following conditions hold:

1. "P is invariant for all signatures of a document ": Let Adv be a probabilistic polynomial-time algorithm which takes as input strings $c \in \{0,1\}^{l(n)}$, and produces as output $(public - key, D, \sigma_1(D), \sigma_2(D))$ where $public - key \in \{0,1\}^n$, $\sigma_1(D), \sigma_2(D)$ are valid signatures of D with respect to $public-key$ and c. Them for all Adv, for all $public - key$, for all negligible ϵ, for all sufficiently large n, the probability that $P(\sigma_1(D)) \neq P(\sigma_2(D)$ is less than $\epsilon(n)$. (The probability is taken over $c \leftarrow \{0,1\}^{l(n)}$ and coin-tosses of Adv.)

2. "P is unpredictable from D": Let Adv be a probabilistic polynomial time algorithm which receives as input strings $1^n, c \in \{0,1\}^{l(n)}, public - key \in G_2(1^n)$; can request and receive signatures with respect to $public-key$ and c of polynomially-many (in n) adaptively chosen documents $\{D_i\}$; and finally outputs a polynomially length string D not in $\{D_i\}$ and a bit b. Let $\alpha = \Pr(b = P(t) : c \leftarrow \{0,1\}^{l(n)}; \{secret - key, public - key\} \leftarrow G(1^n, c); s \leftarrow S(1^n, D, \{secret - key, public - key\}, c); t \leftarrow g(s))$. Then, for every Adv, for every negligible ϵ, and for all sufficiently large n, $|\alpha - \frac{1}{2}| \leq \epsilon(n)$.

We refer to the predicate P as, the *invariant property* of π.

This lemma follows immediately from the definition of invariant signature scheme.

REMARK: We must stress that digital signatures of [GoMiRi, BeMi, NY, Ro] are *not* known to be invariant in the above sense. In fact, while honest signer can sign in some predetermined (in fact, deterministic [G]) way, there exists many valid signatures for the same document which bear no similarity to each other. In contrast, invariant signatures require all valid signatures of a document to be "similar" according to some polynomial time computable function which is unpredictable from the document itself.

3 Preliminaries

Before we show the equivalence between the existence of $\mathcal{NIZK}$ and $\mathcal{INV} - \mathcal{DS}$, we review necessary ingredients of [FLS] and [BG] scheme.

3.1 Where Feige-Lapidot-Shamir use Trapdoor?

The [FLS] solution for $\mathcal{NIZK}$ for $\mathcal{NP}$ when the participants are polynomial-time requires the assumption that trapdoor permutations exist. This assumption is not necessary throughout their construction. In fact, the only place where the trapdoor property is used is to construct a "hidden random string". In particular, they show how to use a common random string in order to get a "hidden random string" as follows:

- prover picks a trapdoor one-way permutation (f, f^{-1}) and sends to the verifier the code of f. In addition, let B be a hard-core predicate associated with f [GL].
- A common random tape can be interpreted as a sequence of $(y_1, y_2, \ldots, y_m)$, with each $|y_i|$ of length n (a security parameter of f). Then *hidden random string* is defined as: $(B(f^{-1}(y_1)), B(f^{-1}(y_2)), \ldots, B(f^{-1}(y_m)))$, where $B(\cdot)$ is a hard-core bit [GL]. Notice that since f is a permutation, the hidden random string is well-defined[4]. Notice that since f is a *trapdoor* permutation, the polynomial time prover can compute $f^{-1}(y_i)$.

---

[4] In [FLS] it is assumed that f is a *verifiable* permutation. That is, verifier can check that it is a permutation by inspecting the code of f. In [BY], this is extended to arbitrary trapdoor one-way permutations.

Using different f's the prover can construct new hidden random bits (for each new theorem). Thus, they show how assuming a common fixed (polynomial length) random string and the existence of a trapdoor one-way permutations, a $\mathcal{NIZK}$ which is publically verifiable and Zero-Knowledge (in an adaptive setting) can be constructed for $\mathcal{NP}$.

3.2 Bellare-Goldwasser Signature Scheme

In [BG], it is shown how assuming publically verifiable non-interactive zero-knowledge proofs and pseudo-random functions of [GGM], a signature scheme can be constructed. (As was shown by [GGM], pseudo-random functions can be based on any one-way function.)

We outline their scheme below:

Step1: The signer chooses at random a seed s for a pseudo random function $F_s(\cdot)$ and publishes an encryption $E(s)$ along with the public information necessary to verify $\mathcal{NIZK}$ proofs (i.e., the random string etc.) as his public key, and keeps s as his secret key.

Step2: The signature of a document D is the value $v = F_s(D)$ together with an $\mathcal{NIZK}$ proof that indeed v was computed correctly.

We remark that in their construction the public-key contains the random string which is necessary for the signer for producing $\mathcal{NIZK}$ proofs. Since the signer serves here in the role of the prover, and it is to his advantage to chose the random string truly with uniform probability (else the chance of a successful forgery increases) the random string is made part of the signers public key rather than part of the systems choice.

In what follows, we will use a similar scheme except that the random string needed by the $\mathcal{NIZK}$ proof system will be specified by the system as a common random string.

4 The Equivalence of NIZK and INV-DS

Recall that when we say that nontrivial $\mathcal{NIZK}$ exist, we mean that $\mathcal{NIZK}$ proof system exist for some hard to predict language L. First, we state our main result:

MAIN THEOREM: $\mathcal{INV} - \mathcal{DS}$ exist if and only if nontrivial $\mathcal{NIZK}$ exist.

Proof outline: We prove our main result in two parts: (1) $\mathcal{INV} - \mathcal{DS}$ imply the existence of nontrivial $\mathcal{NIZK}$; (2) nontrivial $\mathcal{NIZK}$ imply the existence of $\mathcal{INV} - \mathcal{DS}$;

First, we prove (1). We claim that digital signatures (and, hence, $\mathcal{INV} - \mathcal{DS}$) already imply the existence of a one-way function [Ro]. Thus, it remains to show that based on $\mathcal{INV} - \mathcal{DS}$ and the existence of one-way functions we can construct $\mathcal{NIZK}$ for some hard language. Assuming that one-way function f exist, we can construct a hard language in a straight-forward fashion. For example, let $L_f \triangleq \{x | B(f^{-1}(x)) = 1\}$, where B is hard core bit for f [GL]). We now give intuition for the fact that $\mathcal{INV} - \mathcal{DS}$ imply $\mathcal{NIZK}$ in the common random string model.

Let us first consider the case of one theorem $\mathcal{NIZK}$, with the common random string $R = (y_1, \ldots, y_m)$. To specify a hidden random string $H = (b_1, \ldots, b_m)$, instead of using a trapdoor function (i.e, $b_i = B(f^{-1}(y_i))$ where B is a hard core bit as in section 3.1) the intuition is to use digital signatures (i.e., $b_i = P(\sigma(y_i))$ where P is some boolean function of the digital signature of y_i). Clearly, this intuition is correct if indeed for every y_i there exists a unique fixed boolean value b_i computable from any valid signature of y_i. Unfortunately, this is not the case for digital signatures in general [GoMiRi, BeMi, NY, Ro]. We remark that if it were true, then we could have implemented $\mathcal{NIZK}$ based on *any* one-way function instead of one-way trapdoor permutations as currently known. However, the above intuition is true for *invariant* digital signatures with high probability. That is, for *invariant* signatures it is the case that for all y_i there exist some invariant function g defined over the signatures $\sigma(y_i)$, and therefore an invariant Boolean predicate P defined over the signatures $\sigma(y_i)$.

Now, let us consider the case for many theorems. In this case, we need different hidden random strings for each new theorem. Thus, how do we extend the above intuition to obtain many hidden random strings for different theorems? (Recall that the solution of [FLS] was to pick a *new* trapdoor 1-way permutation f for each new proof so that a common random string $(y_1, y_2, \ldots, y_m)$, defines a hidden random string $(B(f^{-1}(y_1)), B(f^{-1}(y_2)), \ldots, B(f^{-1}(y_m)))$. The solution here is simple: when proving the i'th theorem T_i we use as a common hidden random string the sequence: $(P(\sigma(y_1 + i)), P(\sigma(y_2 + i)), \ldots, P(\sigma(y_m + i)))$. By adding a new i when proving each new theorem, we note that each new hidden random string is unpredictable even when given proofs of all the previous theorems. This is so, since the definition of $\mathcal{INV} - \mathcal{DS}$ requires that the hard bit

$P(\sigma(y + i))$ be unpredictable in the adaptive setting (i.e., even if for all $j < i$, $P(\sigma(y + j))$ is given.)

We are now ready to outline how to use an $\mathcal{INV} - \mathcal{DS}$ to construct a $\mathcal{NIZK}$ for an $\mathcal{NP}$ language. Let n be a security parameter and nm is a length of a common random string (where m is as specified in [FLS]). (1) Run key-generation algorithm for $\mathcal{INV} - \mathcal{DS}$ m times and publish all m public keys as a "common" public key; (2) Keep a counter i (initialized to 0) of the number of theorems proven so far. (3) to prove theorem T_i utilize $(P(\sigma(y_1 + i)), P(\sigma(y_2 + i)), \ldots, P(\sigma(y_m + i)))$ as a hidden random sequence of the [FLS] construction.

Note that the completeness follows from that fact that both P and σ are efficiently computable, and the rest of the protocol is analogous to [FLS]. The soundness holds since the signature scheme we are using is *invariant*, and hence any particular choice of i with high probability specifies uniquely a hidden random string. Thus, for a sufficiently long random string, even if prover picks an arbitrary (but polynomially-bounded) i the conditions that at least one "block" has a property required by [FLS] proof do hold with very high probability (over common random string chosen with uniform distribution). The Zero-Knowledge property holds due to the fact that if the adversary can distinguish the $\mathcal{NIZK}$ and the simulator then [FLS] show that such a distinguisher can be turned into a good predictor of a hidden random bit. (The idea there is to use witness-indistinguishable proof that either the graph is Hamiltonian or that the first $2n$ random bits of the common random string a pseudo-random and are produced from a seed of length n. Exploring properties of witness-indistinguishability [FLS] show that the distinguisher of the simulator can be turned into a distinguisher for a pseudo-random generator or into a predictor of a hidden random bits.) In our construction, predicting a hidden random bit provides us with predictor of the *invariant property*, which by definition enables us to to forge a $\mathcal{INV} - \mathcal{DS}$ for some new D'. Since our signature scheme is secure against existential adaptive chosen-message attacks, we get a contradiction.

In order to show (2), we first note that $\mathcal{NIZK}$ for hard to predict L imply the existence of one-way functions [OW]. Hence, we must show that assuming one-way functions and $\mathcal{NIZK}$ proofs is sufficient to construct $\mathcal{INV} - \mathcal{DS}$. This, however, is essentially established for us by [BG] with the following modification of their construction. The idea is to make sure that $E(s)$ (of [BG] Step 1) uniquely specifies s, i.e., is a commitment to s. If this is the case, then for any document D, $F_s(D)$ (of [BG] Step 2) is uniquely defined, and the invariant function will be simply $F_s(D)$. Any bit of $F_s(D)$ can be used as a hard-core bit for the invariant predicate P (as discussed in the section on the definition of invariant signatures).

Now, we specify how to perform step 1 of [BG], based on any one-way function. In order to commit to a seed s, consisting of bits $s_1, s_2, \ldots, s_n$, the player commits *to each bit* s_i separately using a modification of Naor's scheme [N]. (The scheme of [N] is interactive, in which the player who receives committed bits (called Bob) chooses a random string during the conversation). In our protocol, the challenges of Bob are substituted by a (dedicated for this purpose) portion of the common random string. Following through an argument analogous to [N] shows that this scheme uniquely determines s with overwhelming probability (over uniformly distributed common random string), and hence we can use the proof of security presented in [BG] here as well. Hence we are done with (2). □

References

[BM]   Blum M., and S. Micali "How to Generate Cryptographically Strong Sequences Of Pseudo-Random Bits" *SIAM J. on Computing,* Vol 13, 1984, pp. 850-864, FOCS 1982.

[BeMi]   Bellare, M., and S. Micali "How to Sign Given Any Trapdoor Function" *STOC* 88.

[BFM]   Blum M., P. Feldman, and S. Micali, "Non-interactive zero-knowledge proofs and their applications," *Proceedings of the* 20th STOC, ACM, 1988.

[BDMP]   Blum M., A. DeSantis, S. Micali and G. Persiano, "Non-Interactive Zero-Knowledge" *SIAM J.Comp. 91*

[BG]   M. Bellare, S. Goldwasser "New Paradigms for digital signatures and Message Authentication based on Non-Interactive Zero Knowledge Proofs" Crypto 89 proceedings, pp. 194 -211

[BY]   Bellare, Yung, "Certifying Cryptographic Tools: The Case of Trapdoor Permutations" CRYPTO-92 proceedings.

[DMP1]   De Santis, A., S. Micali and G. Persiano, "Non-Interactive Zero Knowledge Proof Systems," *CRYPTO-87*

[DMP2]   De Santis, A., S. Micali and G. Persiano, "Bounded-Interaction Zero-Knowledge proofs," *CRYPTO-88*.

[DH]   W. Diffie, M. Hellman, "New directions in cryptography", *IEEE Trans. on Inf. Theory,* IT-22, pp. 644–654, 1976.

[EGM]   Even S., O. Goldreich and S. Micali "On-line/Off-line Digital Signatures" *CRYPTO* 89.

[FLS]   Feige, U., D. Lapidot and A. Shamir, "Multiple Non-Interactive Zero-Knowledge Proofs Based on a Single Random String", Proc. IEEE Symp. on Foundations of Computer Science, 1990.

[G]   Goldreich O., "Two remarks Concerning the GMR Signature Scheme" MIT Tech. Report 715, 1986.

[GGM]   Goldreich O., S. Goldwasser and S. Micali "How to Construct Random Functions" *JASM* V. 33 No 4. (October 1986) pp. 792-807.

[GL]   Goldreich, O., and L. Levin "A Hard-Core Predicate for all One-Way Functions" Proc. 21st STOC, 1989, pp.25-32.

[GMR]   S. Goldwasser, S. Micali and C. Rackoff, "The Knowledge Complexity of Interactive Proof-Systems", *SIAM J. Comput.* 18 (1989), pp. 186-208; (also in STOC 85, pp. 291-304.)

[GoMiRi] Goldwasser, S., S. Micali and R. Rivest "A Digital Signature Scheme Secure Against Adaptive Chosen-Message Attacks" *SIAM Journal of Computing* vol 17, No 2, (April 1988), pp. 281-308.

[GMY]   Goldwasser S., S. Micali and A. Yao, "Strong Signature Schemes" *STOC* 83, pp.431-439.

[H]   Hastad, J., "Pseudo-Random Generators under Uniform Assumptions", STOC 90.

[IL]   R. Impagliazzo and M. Luby, "One-way Functions are Essential for Complexity-Based Cryptography" FOCS 89.

[IR]   R. Impagliazzo and S. Rudich, "On the Limitations of certain One-Way Permutations", Proc. ACM Symp. on Theory of Computing, pp 44-61, 1989.

[ILL]   R. Impagliazzo, R., L. Levin, and M. Luby "Pseudo-Random Generation from One-Way Functions," *STOC* 89.

[KMO]   J. Kilian, S. Micali, R. Ostrovsky "Minimum Resource Zero-Knowledge Proofs", FOCS-89.

[N]   M. Naor "Bit Commitment Using Pseudo-Randomness", Crypto-89.

[NY]   M. Naor and M. Yung, "Universal One-Way Hash Functions and their Cryptographic Applications", STOC 89.

[OW]   R. Ostrovsky, A. Wigdeson, "One-Way Functions are Essential for Non-Trivial Zero-Knowledge", preliminary draft.

[RSA]   Rivest, R.L., Shamir, A. and Adleman, L., "A Method for Obtaining Digital Signatures and Public Key Cryptosystems" Comm. ACM, Vol 21, No 2, 1978.

[Ro]   J. Rompel "One-way Functions are Necessary and Sufficient for Secure Signatures" STOC 90.

On the Discrepancy between
Serial and Parallel of Zero-Knowledge Protocols

(Extended Abstract)

Kouichi Sakurai

Computer &
Information Systems Laboratory,
Mitsubishi Electric Corporation,
5-1-1 Ofuna, Kamakura 247, Japan.
sakurai@isl.melco.co.jp

Toshiya Itoh

Department of Information Processing,
Interdisciplinary Graduate School
of Science and Engineering,
Tokyo Institute of Technology,
4259 Nagatsuta, Midori-ku,
Yokohama 227, Japan.
titoh@ip.titech.ac.jp

Abstract

In this paper, we investigate the discrepancy between a serial version and a parallel version of zero-knowledge protocols, and clarify the information "leaked" in the parallel version, which is not zero-knowledge unlike the case of the serial version. We consider two sides: one negative and the other positive in the parallel version of zero-knowledge protocols, especially of the Fiat-Shamir scheme.

1 Introduction and motivation

The notions of interactive proofs and zero knowledge were introduced by Goldwasser, Micali and Rackoff [GMR]. Fiat and Shamir [FiS] exhibited a practical identification scheme, which is zero-knowledge, based on the intractability of the factorization.

A common weakness in such zero-knowledge protocols is that the protocols require many iterations of a basic (three move) protocol, then such zero-knowledge protocols are not efficient.

The straightforward parallelization of the basic protocol decreases the round complexity of the protocols. However, a problem on the straightforward parallelization of zero-knowledge protocols is that a technique of the proof of zero-knowledge in the serial version, so called *resettable simulation*, fails in the parallel version.

Feige, Fiat and Shamir [FSS] showed that the parallel version of the Fiat-Shamir identification scheme releases no "useful" knowledge that could help the verifier to impersonate the prover within the identification system.

On the other hand, Goldreich and Krawczyk [GKr] observed that non zero-knowledgeness is an intrinsic property of the three move protocols, and showed that the parallel version of the Fiat-Shamir scheme is not zero-knowledge unless the factorization is tractable.

Our motivation of this study is derived from these contradictive results on the security of the parallel version of the Fiat-Shamir scheme (generally, the three move protocols).

Some researchers characterize the security of the parallel execution of the Fiat-Shamir type identification scheme [FSS, FeS, OhOk'88, BM]. However, none has investigated what kind of information is leaked by the parallel version or how useful these knowledge is for the verifier.

In this paper, we investigate the essential discrepancy between the serial version and the parallel version of the Fiat-Shamir scheme (more generally, zero-knowledge protocols), and clarify properties which the parallel version has but the serial version does not have.

Our main observation is that the information "leaked" in the parallel version of the Fiat-Shamir identification scheme is closely related to a digital signature which is a modification of the Fiat-Shamir identification scheme, and the parallel version of zero-knowledge protocols leave a trace.

Furthermore, we consider two sides of the discrepancy, one negative and the other positive.

Organization of this paper

In section 2, we give the definitions and overview the Fiat-Shamir scheme. In section 3, we consider the reason why straightforward parallelization fail to be zero-knowledge. In section 4, we point out abuses of the parallel version. In section 5, we positively apply the parallel version. Finally, we conclude with future topics.

2 Preliminaries

In this section, we give some definitions on zero-knowledge [GMR] and overview of the Fiat-Shamir scheme [FiS, FSS]. The reader who is familiar with these topics may skip this section.

2.1 Notation and Definitions

Our model of computation is the interactive probabilistic Turing machines (both for the prover P and for the verifier V) with an auxiliary input. The common input is denoted by x and, and its length is denoted by $|x| = n$. We use $\nu(n)$ to denote any function vanishing faster than the inverse of any polynomial in n. More formally,

$$\forall k \in \mathbf{N} \ \exists n_0 \ s.t. \ \forall n > n_0 \ 0 \leq \nu(n) < \frac{1}{n^k}.$$

We define *negligible* probability to be the probability behaving as $\nu(n)$, and *overwhelming* probability to be the probability behaving as $1 - \nu(n)$.

Let $A(x)$ denote the output of a probabilistic algorithm A on input x. This is a random variable. When we want to make the coin tosses of A explicit, for any $\rho \in \{0,1\}^*$ we write $A[\rho]$ for the algorithm A with ρ as its random tape. Let $V_P(x)$ denote V's output after interaction with P on common input x, and let $M(x; A)$ (where A may be either P or V) denote the output of the algorithm M on input x, where M may use the algorithm A as a (blackbox) subroutine. Each call M makes to A is counted as a single computation step for M.

Definition 2.1 [GMR]: *An interactive proof for membership of the language L is a pair of interactive probabilistic Turing machines (P, V) satisfying:*

Membership Completeness: *If x belongs to L, V accepts P's proof with overwhelming probability. Formally:*

$$\forall x \in L \ \ Prob(V_{P(x)}(x) \, accepts) > 1 - \nu(|x|),$$

where the probability is taken over all of the possible coin tosses of P and V.

Membership Soundness: *If x does not belong to L and P^* may act in any way, V accepts P^*'s proof with negligible probability. Formally:*

$$\forall x \notin L \, \forall P^* \ \ Prob(V_{P^*(x)}(x) \, accepts) < \nu(|x|),$$

where the probability is taken over all of the possible coin tosses of P^* and V.

It should be noted that P's resource is computationally unbounded, while V's resource is bounded by probabilistic polynomial time in $|x|$.

Definition 2.2: Let R be a relation $\{(x, w)\}$ testable in $\mathcal{BPP}$. Namely, given x and w, checking whether $(x, w) \in R$ is computed in probabilistic polynomial time. For any x, its witness set $w(x)$ is the set of w such that $(x, w) \in R$.

Definition 2.3 [FSS]: An interactive proof of knowledge for the relation R is a pair of interactive probabilistic Turing machines (P, V) satisfying:

Knowledge Completeness: For any $(x, w) \in R$, V accepts P's proof with overwhelming probability. Formally:

$$\forall (x, w) \in R \quad Prob(V_{P(x,w)}(x) \, accepts) > 1 - \nu(|x|),$$

where the probability is taken over all of the possible coin tosses of P and V.

Knowledge Soundness: For any x, for any P^*, P^* can convince V to accept only if he actually "knows" a witness for $x \in dom \, R$. An expected polynomial time knowledge extractor M is used in order to demonstrate P^*'s ability to compute a witness. Formally:

$$\forall a \, \exists M \, \forall P^* \, \forall x \, \forall w' \, \forall \rho$$
$$Prob(V_{P^*[\rho](x,w')}(x) \, accepts) > 1/|x|^a \Rightarrow$$
$$Prob(M(x; P^*[\rho](x, w')) \in w(x)) > 1 - \nu(|x|),$$

where the probability is taken over all of the possible coin tosses of M and V. P^* is assumed not to toss coins, since his favorable coin tosses can be incorporated into the auxiliary input w'. The knowledge extractor M is allowed to use P^* as a blackbox subroutine and runs in expected polynomial time. Each message that P^* sends M costs a single computation step for M.

Note that both P's and V's resource are bounded by probabilistic polynomial time in $|x|$.

We recall that the *view* of the verifier is everything he sees during an interaction with the prover, that is, his own coin tosses and the conversation between himself and the prover.

Definition 2.4 [GMR]: Let (P, V) be an interactive protocol and let $x \in \{0,1\}^*$. The view of V' on input x is the probability space

$$VIEW_{(P,V')}(x) = \{(R, C) : R \leftarrow \{0,1\}^{p(|x|)} ; \ C \leftarrow (P \leftrightarrow V'[R])(x)\},$$

where p is a polynomial bounding the running time of V', and $(P \leftrightarrow V'[R])(x)$ denotes the probability space of conversations between P and $V'[R]$ on input x (the probability is taken over all of the possible coin tosses of P).

Denote by $Time_P^{V'}(x)$ the running time of machine V' when interacting with P on input x.

Definition 2.5 [GO]: An interactive proof system (P, V) of knowledge for the relation R is blackbox simulation perfect zero knowledge if there exists a universal simulator S_u which runs in expected polynomial time, such that for every polynomial Q and any pair (x, y, V') such that $(x, y) \in R$ and $Time_{P(y)}^{V'}(x) \leq Q(|x|)$, $S_u(x; V'(x))$ is exactly identical to $VIEW_{(P(y),V')}(x)$.

Formally:

$$\exists S_u \, \forall Q \, \forall x \, \forall y \, \forall V' \ \ s.t. \ \ (x,y) \in R \ \& \ Time^{V'}_{P(y)}(x) \le Q(|x|),$$
$$VIEW_{(P(y),V')}(x) = S_u(x; V'(x)).$$

Blackbox simulation zero knowledge represents the strongest notion of zero knowledge among the types of the simulation (e.g. auxiliary input zero-knowledge [GO]) although all known concrete zero knowledge protocols are in fact blackbox simulation zero knowledge. Thus these definitions above are reasonable and never too restrictive.

Throughout this paper, we use a term "zero knowledge" in the sense of *blackbox simulation zero knowledge*.

$\overline{A}$ (resp. $\overline{B}$) represents the real prover (resp. verifier) who follows its designated protocol. $\tilde{A}$ represents a polynomial time cheater who does not possess the witness (or secret) but can derive from the protocol in an arbitrary way. $\tilde{B}$ represents an arbitrary polynomial time verifier who tries to extract additional information from $\overline{A}$.

Definition 2.6 [FSS]: *The protocol (A,B) releases no transferable information if:*

1. *It succeeds with overwhelming probability.*

2. *There is no coalition of $\tilde{A}, \tilde{B}$ with the property that, after a polynomially many number of executions of $(\overline{A}, \tilde{B})$ it is possible to execute $(\tilde{A}, \overline{B})$ with a non negligible probability of success.*

Ohta and Okamoto [OhOk'88] defined rigorous notions on "revealing no transferable information".

For more precise definition of *no transferable* that is suitable for the identification system, see the journal version of the reference [FSS].

2.2 The Fiat-Shamir scheme

Fiat and Shamir [FiS] exhibited a practical identification scheme and a signature scheme that are provably secure if factoring is difficult. We overview their scheme.

Fiat-Shamir identification scheme (FSIS)

1. PREPROCESSING STAGE BETWEEN THE TRUSTED CENTER AND EACH USER
   The unique trusted center's secret key in the system is (p,q), and the public key is N, where p, q are distinct large primes, $N = p \times q$. The center generates user A's secret key s_A, where $1/s_A = \sqrt{I_A} \pmod{N}$. I_A is the identity of user A and is published to other users.

2. IDENTIFICATION STAGE BETWEEN USER A AND USER B
   Repeat step (a) to (d) t times.

   (a) The user A picks $r \in_R Z_N^*$, and sends $x \equiv r^2 \pmod{N}$ to a user B.

   (b) The user B generates $e \in_R \{0,1\}$, and sends e to the user A.

   (c) The user A sends $y \equiv s_A^e r \pmod{N}$ to the user B.

   (d) The user B checks that $x \equiv y^2 I_A^e \pmod{N}$. If the check is not valid, the user B quits the procedure.

   The user B accepts A's proof of identity only if all t round checks are successful.

Remark 2.7: In the parallel version of the protocol above, A sends B all the x_i $(i = 1, \ldots, t)$ simultaneously, then B sends A all the e_i $(i = 1, \ldots, t)$, and finally A sends all the y_i $(i = 1, \ldots, t)$ to B.

Furthermore, Fiat and Shamir modified the identification scheme above into a non-interactive digital signature scheme by replacing the verifier B's role by the prover with a pseudo-random function f.

Fiat-Shamir digital signature scheme (FSDS)

1. PREPROCESSING STAGE BETWEEN THE TRUSTED CENTER AND EACH USER
   Same as the preprocessing stage in FSIS.

2. TO SIGN A MESSAGE M:
   The user A picks $r_i \in_R Z_N^*$ $(i = 1, \ldots, t)$, and calculates $x_i \equiv r_i^2 \pmod{N}$ $(i = 1, \ldots, t)$, $f(M, x_1, \ldots, x_t)$ and sets its first t bits to e_i $(i = 1, \ldots, t)$. Furthermore, the user A computes $y_i \equiv s^{e_i} r_i \pmod{N}$ $(i = 1, \ldots, t)$ and sends M, e_i, y_i $(i = 1, \ldots, t)$ to the user B.

3. TO VERIFY A'S SIGNATURE ON M:
   The user B calculates $z_i = y_i^2 I_A^{e_i} \pmod{N}$ $(i = 1, \ldots, t)$, $f(M, z_1, \ldots, z_t)$, and checks that its first t bits are equal to e_i $(i = 1, \ldots, t)$. If the checks are valid, the user B recognizes that M is A's valid message.

2.3 Known properties of the Fiat-Shamir scheme

Feige, Fiat and Shamir [FSS] showed that FSIS is provably secure. Namely,

Proposition 2.8 [FSS]: *The serial version of FSIS, where $t = O(|N|)$, is a zero-knowledge proof of knowledge.*

Although Feige, Fiat and Shamir [FSS] did not show that the parallel version of FSIS is zero knowledge, they did show that the parallel version of FSIS releases no "useful" knowledge that could help the verifier to impersonate the prover within the identification system. Namely,

Proposition 2.9 [FSS]: *If factoring is difficult, the parallel version of FSIS releases no transferable information.*

Note that Proposition 2.9 does not imply that the parallel version of FSIS releases no "useful" knowledge that could help the verifier to cheat *outside* the identification system.

Goldreich and Krawczyk [GKr] observed that non-zero-knowledgeness is an intrinsic property of the parallel version of the FSIS protocol.

Proposition 2.10 [GKr]: *If factoring is difficult, the parallel version of FSIS is not (black-box simulation) zero knowledge.*

Although the straightforward parallel version of FSIS is not zero-knowledge, Bellare, Micali, and Ostrovsky [BMO] proposed how to parallelize FSIS with preserving zero-knowledgeness. Their scheme is not three move and needs some additional interactions between the prover and the verifier.

In this paper, we use a term "parallel" version of protocols in the sense of the (three move) straightforward parallelization as in **Remark 2.7**.

With respect to the security of FSDS, Fiat and Shamir showed

Proposition 2.11 [FiS]: *When f is a truly random function, FSDS is existentially unforgeable under an adaptive chosen message attack unless factoring is easy.*

Remark 2.12: A variant of the Fiat-Shamir scheme has proposed [GQ1] and the security as in Proposition 2.9 has been considered [OhOk'88]. Brickell and McCurley [BM] proposed a modified Schnorr's identification scheme [Sch] based on a special discrete logarithm problem, and gave a formal proof on the security. Probably secure three move identification scheme based on the general problems is proposed by Okamoto [Oka].

3   Why does straightforward parallelization fail to be zero-knowledge ?

Feige, Fiat and Shamir's result in Proposition 2.9 guarantees a security of the parallel version of FSIS. On the other hands, Goldreich and Krawczyk's statement in Proposition 2.10 implies the parallel version of FSIS is not (blackbox simulation) zero knowledge. Many researchers [FSS, BC] remarked that the parallel version of FSIS could leak some "partial" information on the prover's secret.

Our first question is :

Question A: What information is released in the parallel version of FSIS ?

To prove a protocol to be zero knowledge, a main technique is to reset a (cheating) verifier, so called *resettable simulation* [GMR].

Many researchers [BC, BMO] have observed that the resettable simulation may not be applied to the following cheating verifier in the parallel version of FSIS.

> After receiving the prover's the message x_i $(i = 1, \ldots, t)$, the (cheating) verifier sends back bits e_i $(i = 1, \ldots, t)$ which are computed with dependence on x_i $(i = 1, \ldots, t)$, for example, $(e_1, \ldots, e_t) = g(x_1, \ldots, x_t)$ for a random hash function g.

In fact, Goldreich and Krawczyk's proof on the non-zero-knowledgeness of the parallel version of FSIS (generally, on the triviality of three move protocols) is based on a careful analysis of the cheating verifier with random hash function.

In the cheating strategy above, the verifier learns $(x_1, \ldots, x_t, y_1, \ldots, y_t)$ satisfying the conditions that $(e_1, \ldots, e_t) = g(x_1, \ldots, x_t)$ and $y_i \equiv s^{e_i} r_i \pmod{N}$ $(i = 1, \ldots, t)$. The (polynomial-time bounded) verifier without the secret s seems not to be able to generate such information by himself. Thus, we regard the information above as knowledge leaked in the parallel version of FSIS.

Our second question is as follows.

Question B: How useful is this information for the verifier ?

To clarify the role of this information above, we consider a verifier who acts as below.

> After receiving the prover's message x_i $(i = 1, \ldots, t)$, the verifier selects a message M and sends back bits e_i $(i = 1, \ldots, t)$ which are computed as $g(M, x_1, \ldots, x_t)$ for a *one-way* hash function g.

In the cheating method, the verifier learns $(x_1, \ldots, x_t, y_1, \ldots, y_t)$ satisfying the conditions that $(e_1, \ldots, e_t) = g(M, x_1, \ldots, x_t)$ and $y_i \equiv s^{e_i} r_i \pmod{N}$ $(i = 1, \ldots, t)$ for the message M selected by him. If g is a *one-way* hash function, we can regard $(e_1, \ldots, e_t, y_1, \ldots, y_t)$ as the prover's digital signature for the message M in FSDS with respect to the function g.

Our observation above implies that in the parallel version of FSIS a cheating verifier, who makes an access to the true prover in the parallel version of FSIS, gets the prover's digital signature of FSDS for any message M. Note that in the *serial* version of FSIS, even if a cheating verifier acts as the same as the above, the verifier cannot get any digital signature of FSDS.

4   Abuses of the parallel version

In this section we point out abuses of the parallel Fiat-Shamir scheme based on our remarks in the previous section.

4.1   Non-transferable information helps to forge secure digital signatures

We consider a practical system which consists of FSIS and FSDS.

Suppose a prover uses only one secret key s for his public information I in the system. Namely, the prover shows his identity via the *serial* version of FSIS using the secret s, and the prover signs messages via FSDS using the same secret s. This system is convenient for the prover because he keeps only one secret information.

However, if the prover shows his identity via the parallel FSIS, not via the *serial* one, this system is not secure for the prover. As we noted in the previous section, in the parallel version of FSIS a cheating verifier can get the prover's digital signature of FSDS for any message M while the verifier interacts with the prover in FSIS. In this system, FSDS(or FSIS) is not secure.

Note that "releasing no-transferable information" by Feige, Fiat, and Shamir [FSS] guarantees the security of the case only when the prover's secret information is used in the identification systems.

Remark 4.1:   We may prevent the verifier's cheating above by using a different security parameter t in the signature stage and in the identification stage. However, such temporary protection never implies the provable security of the system.

4.2   Message authentication based on the public key

The message authentication is used as a data integrity mechanism to detect whether data have been altered in an unauthorized manner. An implementation of message authentication based on the conventional secret key cipher (e.g. DES) is Message Authentication Codes (MACs) [ISO]. The *public-key based message authentication* is defined as:

> **Validity:**   In the authentication stage, *only* the user A can prove
> the validity of a message to any user B by using A's public key.

The authentication stage based on the public key needs an interaction between the prover and the verifier, while MACs is *non-interactively* verified by the only receiver who knows the same secret key as the sender has. Note that the digital signature [DH] is verified by anybody *without interaction* using only the signer's public key.

Desmedt [Des] and Guillou-Quisquater [GQ2] applied FSIS to the public-key based message authentication. Guillou and Quisquater modified the (extended) Fiat-Shamir identification scheme into a message authentication by using a one-way hash function. The one-way hash function is used to mix the message into the communication for the identification.

Guillou-Quisquater's Message Authentication
based on the (extended) Fiat-Shamir scheme

1. PREPROCESSING STAGE BETWEEN THE TRUSTED CENTER AND EACH USER

   In this system, the center's secret key is p, q (distinct large primes) and the public key is $N = pq$ and L. The center generates prover A's secret key s_A satisfying $1/s_A = (I_A)^{1/L} \pmod{N}$, where I_A is the identity of user A and is published to other users. Furthermore, a one-way hash function g is published to each user.

2. AUTHENTICATION STAGE BETWEEN THE USER A AND THE USER B

(a) The user A sends his message M with his identify I_A to the user B.

Repeat step (b) to (e) t times.

(b) The user A picks $r \in_R Z_N^*$, and computes $x \equiv r^L \pmod{N}$ and $u = g(M, x)$. The user A sends x and u to the user B.

(c) The user B sends $d \in_R Z_L$ to the user A.

(d) The user A sends y such that $y \equiv s_A^d r \pmod{N}$ to the user B.

(e) The user B checks that $u = g(M, y^L I_A^d \pmod{N})$. If the check is not valid, the user B quits the procedure.

The user B recognizes that M is A's valid message only if all t round checks are successful.

The serial version of the protocol above (when $t = O(|N|)$ and $L = O(1)$) is zero-knowledge, and the security of parallel versions, which are not zero-knowledge, is studied by Ohta and Okamoto [OhOk'88].

However, no discrepancy between the serial and the parallel of the message authentication based on the (extended) Fiat-Shamir scheme has known. We clarify the discrepancy.

Desmedt [Des] considered the *one-time-validity* of the message authentication and Okamoto and Ohta [OkOh'90] called the same notion *non-transitive signature*:

> **Validity:** *Only* the user A can prove the validity of a message M to any user B by A's public key.
>
> **Non-transitivity:** The user B cannot transfer the proof of A's origin of the message M to another user C.

We should notice that the ordinary (transitive) digital signature [DH] does not satisfy the condition of non-transitivity, i.e, in the digital signature any user B can transfer the proof of A's origin of the message M to another user C and the user C can check the correctness of the proof of A's origin of the message M using only A's public key.

Okamoto and Ohta implemented message authentication based on the modification of the prover's randomness in the (extended) Fiat-Shamir scheme.

Desmedt [Des] mentioned that the serial version of his message authentication is non-transitive (one-time-valid), however nothing was mentioned in the case of the parallel version. Note that the serial version of Guillou-Quisquater's message authentication is non-transitive. Okamoto-Ohta [OkOh'90] claimed, without formal discussion, that both the serial and the parallel version of the message authentication are non-transitive. But, our claim is as follows.

> **Claim:** The parallel version of the Guillou-Quisquater, Okamoto-Ohta, and Desmedt's message authentication are *not* non-transitive.

A cheating method for a verifier in the Guillou-Quisquater message authentication is as follows. (This cheating is applied to other message authentication like as the Desmedt and Okamoto-Ohta's one.)

A (cheating) verifier manages to record the history of the communication with the prover. After receiving prover's message $M, x_1, \ldots, x_t$ and $u = g(M, x_1, \ldots, x_t)$ and the verifier sends back $d_i(i = 1, \ldots, t)$ which is computed as $(d_1, \ldots, d_t) = h(x_1, \ldots, x_t)$ by a one-way hash function h. After receiving the prover's answer $y_i(i = 1, \ldots, t)$ for $d_i(i = 1, \ldots, t)$, the verifier records $H = (M, x_1, \ldots, x_t, h, d_1, \ldots, d_t, y_1, \ldots, y_t)$ as the history of the communication with prover A. Once the verifier publishes the history H, anyone can check the validity and the origin of message M by calculating $u = g(M, y_1^L I_A^{d_1} \pmod{N}, \ldots, y_t^L I_A^{d_t} \pmod{N})$, and $(d_1, \ldots, d_t) = h(x_1, \ldots, x_t)$.

Remark 4.2: The same kind of abuse as above cannot be applied to the scheme based on the *serial* version of the extended Fiat-Shamir scheme.

5 Positive applications of the parallel version

In this section, we consider positive applications of the parallel version.

Okamoto and Ohta [OkOh'89] proposed a blind signature scheme, which was introduced by Chaum [Ch'82], based on a combination of the parallel version of FSIS and FSDS. This is the first positive application of the parallel version of Fiat-Shamir scheme although Okamoto and Ohta did not clarify the distinction between the parallel version and the serial one of the Fiat-Shamir scheme. The technique used in Okamoto-Ohta scheme is more sophisticated than one observed in subsection 4.1, however Okamoto and Ohta's technique is applied to a special class of problems which satisfy a condition, so called *random self reducibility* [TW], and seems not to be applied to the parallel version of more general zero-knowledge protocols (e.g. the references [GMW, BCC]).

We consider positive applications of the parallel version of the Fiat-Shamir scheme, which can be applied to the parallel version of the more general protocols.

5.1 The parallel version of the Fiat-Shamir scheme leaves a trace

Our observations in the previous sections suggest that the parallel version of FSIS leaves some trace, unlike the case of the serial version of zero-knowledge FSIS. We positively apply the trace to message authentication with the proof of the origin and to a protection of divertibility of interactive protocols.

5.2 Testifiable message authentication

As we pointed out in the previous section, the message authentication based on the parallel FSIS does not satisfy the non-transitivity. We positively apply the transitive trace of authentication stage in the parallel version of FSIS.

In the message authentication based on the serial FSIS, the sender (signer) can deny the fact that the signer has shown authentication, because there are no evidence of the prover's proving stage. Okamoto and Ohta [OkOh'90] remarked this property as a merit to show the distinction between non-transitive signatures and Chaum's undeniable signature [CA]. Occasionally, however, we needs an evidence to avoid prover's denying the fact of his authentication on the message. The trace in the parallel version is useful for the evidence.

Suppose that user A sends a message M to user B. A *testifiable* message-authentication has the following properties.

> **Validity:** In the authentication stage, *only* the user A can prove the validity of a message M to any user B by A's public key.
>
> **Testifiability:** Any user C can check the fact that the user A has given the proof of A's origin on the message M by A's public-key *without interaction with A*.

It must be noted that the digital signatures [DH] satisfy the condition of testifiability, however, the digital signatures do not have the authentication stage where A can prove to B that he is A.

We propose a message-authentication which is a modification of the verifier's randomness in the parallel version of the message authentication using the Guillou and Quisquater's idea.

Proposed testifiable message authentication

1. PREPROCESSING STAGE BETWEEN THE TRUSTED CENTER AND EACH USER
   Same as the preprocessing stage in FSIS. Furthermore, two one-way hash function g and h are published to all users.

2. AUTHENTICATION STAGE BETWEEN THE USER A AND THE USER B

   (a) The user A sends his identity I_A and a message M_A to user B.

   (b) The user A picks $r_i \in_R Z_N^*$ $(i = 1, \ldots, t)$, and computes $x_i \equiv r_i^2 \pmod{N}$ $(i = 1, \ldots, t)$, and $u = g(M_A, x_1, \ldots, x_t)$. The user A sends $x_1, \ldots, x_t$ and u to the user B.

   (c) The user B selects a message R_B at random, calculates $h(R_B, x_1, \ldots, x_t)$. The user B sets its first t bits to e_i $(i = 1, \ldots, t)$ and sends e_i $(i = 1, \ldots, t)$ and R_B to the user A.

   (d) The user A computes $h(R_B, x_1, \ldots, x_t)$ and checks if the first t bits of $h(R_B, x_1, \ldots, x_t)$ are e_i $(i = 1, \ldots, t)$. If the check is not valid, the user A quits the procedure. Otherwise, the user A sends to B $y_i \equiv s^{e_i} r_i \pmod{N}$ $(i = 1, \ldots, t)$.

   (e) The user B checks that $u = g(M_A, x_1, \ldots, x_t)$ and $x_i \equiv y_i^2 I_A^{e_i} \pmod{N}$ $(i = 1, \ldots, t)$. If the check is not valid, the user B quits the procedure.

   After all procedures are passed, the user B accepts that M_A is A's valid message.

3. PUBLICATION AND VERIFICATION OF THE EVIDENCE OF THE AUTHENTICATION
   If the prover denies his authentication on the message M_A, the verifier shows
   $H = (I_A, M_A, u, x_1, \ldots, x_t, R_B, y_1, \ldots, y_t)$ as an evidence of the A's authentication on M_A. Anyone can accepts the A's authentication on M_A only if H satisfies the conditions that $u = g(M_A, x_1, \ldots, x_t)$, $(e_1, \ldots, e_t) = h(R_B, x_1, \ldots, x_t)$, and $x_i \equiv y_i^2 I_A^{e_i} \pmod{N}$ $(i = 1, \ldots, t)$.

The authors [SI] applied the proposed testifiable message authentication to a digital credit card system, where both the identification and the digital signature are required.

5.3 Protection against divertibility

Desmedt et al. [DGB] pointed out an abuse of FSIS, so called *Mafia fraud problem*, where an intermediate verifier B can masquerade as the genuine prover A to another (victimized) verifier C while A proves his identity to B; and B cancels any evidence which shows that B is assisted by A. This concept was formulated as divertibility of (zero-knowledge) protocols by Okamoto and Ohta [OkOh'89]. They proposed some types of measure to protect against such an abuse.

We propose a simple technique to protect against the abuse of divertibility of the parallel version of the Fiat-Shamir scheme, which cannot be applied to the serial one. Figure 1 describes the technical details on the divertibility of the parallel version of the Fiat-Shamir scheme. The divertibility is arisen from the property that there are no evidence which distinguishes two communication data,
$((x_1, \ldots, x_t), (\tilde{e}_1, \ldots, \tilde{e}_t), (y_1, \ldots, y_t))$ and $((\tilde{x}_1, \ldots, \tilde{x}_t), (e_1, \ldots, e_t), (\tilde{y}_1, \ldots, \tilde{y}_t))$.

Proposed countermeasure

The technique used in our proposed testifiable message authentication is useful to create an evidence which distinguishes the data. Consider the following modified protocol:

After receiving the prover's first message $(x_1, \ldots, x_t)$, the verifier selects a random message R_V, computes $h(R_V, x_1, \ldots, x_t)$ and sets its first t bits to $e_1, \ldots, e_t$. Then the verifier sends the random message R_V to the prover instead of sending $e_1, \ldots, e_t$. The prover sends back the verifier $y_i \equiv u_i(y_i)^{e_i} (i = 1, \ldots, t)$, where $(e_1, \ldots, e_t) = h(R_V, x_1, \ldots, x_t)$ as the ordinary parallel Fiat-Shamir scheme. The verifier accepts the prover only if the checks $(e_1, \ldots, e_t) = h(R_V, x_1, \ldots, x_t)$, and $x_i \equiv y_i^2 (I_A)^{e_i} (i = 1, \ldots, t)$ are passed.

Prover	Verifier	Victimized Verifier

$r_i \in_R Z_n^*$

$x_i \equiv r_i^2$

$$\xrightarrow{\quad (x_1, \ldots, x_t) \quad}$$

$b_i \in_R \{0,1\}$, $u_i \in_R Z_n^*$

$\tilde{x}_i \equiv u_i^2 x_i^{1-2b_i} I_A^{b_i}$

$$\xrightarrow{\quad (\tilde{x}_1, \ldots, \tilde{x}_t) \quad}$$

$e_i \in_R \{0,1\}$

$$\xleftarrow{\quad (e_1, \ldots, e_t) \quad}$$

$\tilde{e}_i = e_i \oplus b_i$

$$\xleftarrow{\quad (\tilde{e}_1, \ldots, \tilde{e}_t) \quad}$$

$y_i \equiv s^{\tilde{e}_i} r_i$

$$\xrightarrow{\quad (y_1, \ldots, y_t) \quad}$$

$\tilde{y}_i \equiv u_i (y_i)^{1-2b_i}$

$$\xrightarrow{\quad (\tilde{y}_1, \ldots, \tilde{y}_t) \quad}$$

$\tilde{y}_i^2 I_A^{e_i} \equiv \tilde{x}_i$

Figure 1: Divertible ZK on the parallel Fiat-Shamir scheme

In this modified protocol, the way of the verifier's generating the challenge bits $(e_1, \ldots, e_t)$ is restricted and the verifier's computation in the original divertible protocol (Figure 1) cannot be apply to the modified protocol.

The proof on the correctness of our countermeasure is obtained from the same argument as the proof of the security of FSDS (Proposition 2.11). The protection is rather practical than theoretical because it is assumed in a way similar to Proposition 2.11 that the function h is a (blackbox) truly random function.

Remark 5.1: Ohta, Okamoto, and Fujioka [OOF] proposed how to protect the divertibility by using a bit commitment function. Their countermeasure is useful for both the serial version and the parallel version. However, our proposed countermeasure is applied to only the *parallel* version.

6 Concluding remarks

In this paper, we clarify the discrepancy between the serial version and the parallel version of zero-knowledge protocols, especially point out the relation between the "information" leaked in the parallel version of the Fiat-Shamir identification scheme and the Fiat-Shamir digital signature scheme. Furthermore, we consider the merit and demerit of the parallel version with comparing to the serial one. Note that our observation is applied to general zero-knowledge protocols, which is a sequential iteration of a three move protocol.

The security of the straightforward parallel execution of the Fiat-Shamir type identification scheme is characterized by some researchers [FSS, OhOk'88, FeS, BM, Oka]. However, their results heavily depend on the structure of the underlying problems (e.g. factorization, or discrete logarithm), and the technique of the proofs fails in the case of the straightforward parallel execution of the zero-knowledge protocol for general problems like as Graph-3-Colourability [GMW, BCC]. The security of these protocols are still unclear.

The security of three move protocols [FSS, OhOk'88, FeS, BM, Oka], which are based on some algebraic problems, guarantees only the case within the identification system, and nothing is mentioned outside the identification system.

The security of an identification and a signature is one of the central topics in modern cryptography, and many results are known. However, the aspect of these researches on the security is irrelevant to each other. We must study the security of the combination of the different objects.

Acknowledgments

The authors wish to thank Dr. Tatsuaki Okamoto of NTT Laboratories for his advice on digital signatures. The authors also would like to thank Dr. Hiroki Shizuya of Tohoku University. His observation on a weakness of zero-knowledge proofs in electronic fund systems inspired us to study this work. The authors are grateful to Prof. James L. Massey of Swiss Federal Institute of Technology Zurich for affording the first author an opportunity to present this research at ETH and for his encouragement of this work.

References

[BC] Brassard, G. and Crépeau, C., "Sorting out zero-knowledge," Advances in Cryptology – Eurocrypto'89, Lecture Notes in Computer Science 434, *Springer-Verlag*, Berlin, pp.181-191 (1990).

[BCC] Brassard, G., Chaum, D., and Crépeau, C., "Minimum Disclosure Proofs of Knowledge," *Journal of Computer and System Sciences*, Vol.37, No.2, pp.156-189 (October 1988).

[BM] Brickell, E. F. and McCurley, K.S "An Interactive Identification Scheme Based on Discrete Logarithms and Factoring," *Journal of Cryptology*, Vol.5, pp.29-40 (1992).

[BMO] Bellare, M., Micali, S., and Ostrovsky, R., "Perfect Zero-Knowledge in Constant Rounds," *ACM Annual Symposium on Theory of Computing*, pp.482-493 (May 1990).

[CA] Chaum,D. and van Antwerpen, H "Undeniable Signatures," Advances in Cryptology – Crypto'89, Lecture Notes in Computer Science 435, *Springer-Verlag*, Berlin, pp.212-216 (1989).

[Ch'82] Chaum,D., "Blind signature for Untraceable Payments," Advances in Cryptology – Crypto'82, *Plenum Press*, New York, pp.199-203 (1983).

[Des] Desmedt, Y., "Major security problems with the "unforgeable" (Feige-)Fiat-Shamir proofs of identity and how to overcome them,"In *Securicom 88, 6th worldwide congress on computer and communications security and protection*, pp.147-159, (March 1988).

[DGB] Desmedt, Y., Goutier, C. and Bengio,S.: "Special Uses and abuses of the Fiat-Shamir Passport Protocol,"Advances in Cryptology – Crypto'87, Lecture Notes in Computer Science 293, *Springer-Verlag*, Berlin, pp.21-39 (1988).

[DH] Diffie, W., and Hellman, M. "New Directions in Cryptology", IEEE Trans. on Info. Technology, vol. IT-22, 6(1976) pp.644-654 (1976).

[FeS] Feige, U. and Shamir, A., "Witness Indistinguishable and Witness Hiding Protocols," *ACM Annual Symposium on Theory of Computing*, pp.416-426 (May 1990).

[FiS] Fiat, A. and Shamir, A., "How to Prove Yourself," Advances in Cryptology – Crypto'86, Lecture Notes in Computer Science 263, *Springer-Verlag*, Berlin, pp.186-199 (1987).

[FSS] Feige, U., Fiat, A., and Shamir, A., "Zero-Knowledge Proofs of Identity," *ACM Annual Symposium on Theory of Computing*, pp.210-217 (May 1988), the final version: *Journal of Cryptology*, Vol.1, pp.179-194 (1988). v

[GKr] Goldreich, O. and Krawczyk, H., "On the Composition of Zero-Knowledge Proof Systems," ICALP'90, Lecture Notes in Computer Science 443, *Springer-Verlag*, Berlin, pp.268-282 (1990).

[GMR] Goldwasser, S., Micali, S., and Rackoff, C., "The Knowledge Complexity of Interactive Proof Systems," *SIAM Journal of Computing*, Vol.18, No.1, pp.186-208 (February 1989).

[GMW] Goldreich, O., Micali, S., and Wigderson, A., "Proofs that Yield Nothing But Their Validity and a Methodology of Cryptographic Protocol Design," *IEEE Annual Symposium on Foundations of Computer Science*, pp.174-187 (October 1986).

[GO] Goldreich, O. and Oren, Y., "Definitions and Properties of Zero-Knowledge Proof Systems," *Technical Report #610*, Technion – Israel Institute of Technology, Department of Computer Science, Haifa, Israel (February 1990).

[GQ1] Guillou,L.C., and Quisquater,J.J, "A Practical Zero-Knowledge Protocol Fitted to Security Microprocessors Minimizing Both Transmission and Memory," Advances in Cryptology – Eurocrypt'88, Lecture Notes in Computer Science 330, *Springer-Verlag*, Berlin, pp.123-128 (1988).

[GQ2] L.C.Guillou, and J.J.Quisquater, "A "Paradoxical" Identity-Based Signature Scheme Resulting from Zero-Knowledge," Advances in Cryptology – Crypto'88, Lecture Notes in Computer Science 403, *Springer-Verlag*, Berlin, pp.216-231 (1990).

[ISO] International Standard, "Banking – Approved algorithm for message authentication – ," *ISO 8731-1* (1987).

[Oka] Okamoto,T., "Provably Secure and Practical Identification Schemes and Corresponding Signature Schemes," Entenxed Abstract for CRYPTO'92 (1992).

[OhOk'88] Ohta,K., and Okamoto,T., "A Modification of the Fiat-Shamir Scheme," Advances in Cryptology – Crypto'88, Lecture Notes in Computer Science 403, *Springer-Verlag*, Berlin, pp.232-243 (1990).

[OkOh'89] Okamoto,T., and Ohta,K., "Divertible Zero-Knowledge Interactive Proofs and Commutative Random Self-Reducibility," Advances in Cryptology – Eurocrypt'89, Lecture Notes in Computer Science 434, *Springer-Verlag*, Berlin, pp.134-149 (1989).

[OkOh'90] Okamoto,T., and Ohta,K., "How to utilize the randomness of Zero-Knowledge Proofs," Advances in Cryptology – Crypto'90, Lecture Notes in Computer Science 537, *Springer-Verlag*, Berlin, pp.456-475 (1991).

[OOF] Ohta,K., Okamoto,T., and Fujioka,A., "Secure bit commitment function against divertibility," *EUROCRYPTO'92 Extended Abstracts*, (May 1992).

[SI] Sakurai,K. and Itoh,T "Testifiable Identification and Its application to a Digital Credit Card," Proc. of the 1992 Symposium on Cryptography and Information Security, 1D, Japan (April 1992).

[Sch] Schnorr, C. P., "Efficient identification and signatures for smart cards," Advances in Cryptology – Crypto'89, Lecture Notes in Computer Science 435, *Springer-Verlag*, Berlin, pp.239-252 (1990).

[TW] Tompa, M. and Woll, H., "Random Self-Reducibility and Zero-Knowledge Interactive Proofs of Possession of Information," *IEEE Annual Symposium on Foundations of Computer Science*, pp.472-482 (October 1987).

On the Design of SP Networks from an Information Theoretic Point of View

M. Sivabalan, S. E. Tavares and L. E. Peppard

Department of Electrical Engineering

Queen's University at Kingston

Kingston, Ontario, Canada, K7L 3N6

e-mail   : tavares@ee.queensu.ca

Abstract : The cryptographic strength of an SP network depends crucially on the strength of its substitution boxes (S-boxes). In this paper we use the concept of information leakage to evaluate the strength of S-boxes and SP networks. We define an equivalence class on n×n S-boxes that is invariant in information leakage. Simulation results for a 16×16 SP network suggest that after a sufficient number of rounds the distribution of the output XOR in the SP network looks random. We further present simulation results to show that the information leakage for an SP network diminishes more rapidly with the number of rounds when the S-boxes are cryptographically strong.

1. Introduction

The concept of "confusion" and "diffusion", which led to the design of Substitution-Permutation Network (SPN) cryptosystems (e.g., DES [1]), was first introduced by Shannon [2] and was elaborated on in concrete and practical ways by Feistel [3] and Feistel, Notz and Smith [4]. The strength of an SP network depends highly on the strength of the substitution boxes (S-boxes). Work on the design and analysis of S-boxes has been presented in [5][6][7][8][9][10].

Kam and Davida [11] presented an approach to the design of S-boxes and SP networks which is guaranteed to satisfy *completeness*, a property which requires that each output bit depends on every input bit. Since then, very little work has been done on the design and analysis of a general SP network [12][13], even though many fully designed cryptosystems have been published [14][15].

In this work we review some previously proposed evaluation criteria based on information leakage and extend them for an n×n bijective S-box. We then define an equivalence class on S-boxes which will enable one to create cryptographically strong S-boxes more efficiently. We also present simulation results to show that cryptographically strong S-boxes improve the performance of an SP network.

2. Evaluation Criteria for a Cryptographically Strong S-box

Forré [9] presented a set of cryptographic properties of S-boxes based on information theory. Dawson & Tavares [10] extended Forré's ideas to define an expanded set of design criteria for cryptographically strong S-boxes. The authors viewed an S-box in two different ways : *static view*, which models an S-box when the inputs are steady and *dynamic view*, which models an S-box when the inputs change. Forré's criteria, however, apply to the static model only. In the Dawson & Tavares' design framework both an S-box and its inverse were designed to have low information leakage. The expanded set of design criteria was developed at a "single" bit level, where information leakage between a single output bit and the input bits or between a single output bit and the rest of the output bits were computed. We extend the design criteria to a "multiple" bit level, where information leakage between one or more output bits and the input bits or between one or more output bits and the rest of the output bits are considered. We further show that some of the new design criteria defined in [10] are redundant. We also introduce a useful information theoretic property, which we call "XOR Information Leakage" (XL[I;O]) for an S-box. The attractive feature of this property is that it uses a "single quantity" to compare the XOR distributions of S-boxes and SP networks. The n×n S-box S considered in this section is a bijective S-box with an n-bit input $X = \{x_1, x_2, ..., x_n\}$ and an n-bit output $Y = \{y_1, y_2, ..., y_n\}$; where x_i and y_i ; $1 \leq i \leq$ n are binary variables.

2.1. Static Input-Output Information Leakage (SL[I;O])

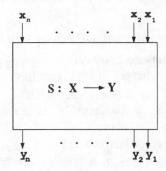

Figure 1.   Static view of an n×n S-box

The input-output mapping of an S-box is assumed to be known, i.e., the output is assumed to be known when the input is completely known (or vice versa). In an ideal S-box, however, partial information about the input bits should not reduce the uncertainty in the unknown output bits (or vice versa).

The static view of the S-box is shown in Figure 1. If $X_k = \{x_{j_1}, x_{j_2}, ..., x_{j_k}\}$; where $1 \leq k \leq n-1$; $1 \leq j_1, j_2, ..., j_k \leq n$, is a subset of the input bits and $Y_t = \{y_{l_1}, y_{l_2}, ..., y_{l_t}\}$; where $1 \leq t \leq n-1$; $1 \leq l_1, l_2, ..., l_t \leq n$, is a subset of the output bits, then the Static Input-Output Information Leakage is the mutual information between $\mathbf{Y}_t$ and $\mathbf{X}_k$ which is given by :

$$SL[I; O] = I(\mathbf{Y}_t; \mathbf{X}_k) = H(\mathbf{Y}_t) - H(\mathbf{Y}_t | \mathbf{X}_k).$$

The averaged[*] SL[I;O] matrices of a 4×4 DES S-box and one of the S-boxes found by Dawson & Tavares are given in Table 1. The detailed SL[I;O] matrices for these two S-boxes are given in Tables 2 and 3. In these tables, the information leakage is given in bits/input. The S-boxes considered in the example are as follows :

DES S-box   : { 0,15,7,4,14,2,13,1,10,6,12,11,9,5,3,8 }
Dawson & Tavares S-box   : { 7,9,1,10,12,14,0,5,4,13,11,6,2,3,15,8 }.

k	DES S-box			Dawson & Tavares S-box		
	t			t		
	1	2	3	1	2	3
1	0.0228	0.1060	0.3750	0.0114	0.0786	0.3750
2	0.0865	0.4271	1.0938	0.0786	0.4284	1.1250
3	0.3594	1.0885	2.0000[†]	0.3750	1.1250	2.0391

Table 1. Averaged SL[I;O] matrices for the DES and the Dawson & Tavares S-box

2.2. Dynamic Input-Output Information Leakage (DL[I;O])

In an ideal S-box information about any changes in the input bits should not reduce the uncertainty in the changes in the output bits.

The dynamic view of an S-box (delta S-box) is shown in Figure 2 in which the steady state value of the input $\mathbf{X}_c$ is assumed to be unknown. If $\Delta\mathbf{X}_k = \{\Delta x_{j_1}, \Delta x_{j_2}, ..., \Delta x_{j_k}\}$; where $1 \leq k \leq n$; $1 \leq j_1, j_2, ..., j_k \leq n$, is a set of changes in the input bits and $\Delta\mathbf{Y}_t = \{\Delta y_{l_1}, \Delta y_{l_2}, ..., \Delta y_{l_t}\}$; where $1 \leq t \leq n$; $1 \leq l_1, l_2, ..., l_t \leq n$, is a set of changes in the output bits then the Dynamic Input-Output Information Leakage is the mutual information between $\Delta\mathbf{Y}_t$ and $\Delta\mathbf{X}_k$ which is given by :

$$DL[I; O] = I(\Delta\mathbf{Y}_t; \Delta\mathbf{X}_k) = H(\Delta\mathbf{Y}_t) - H(\Delta\mathbf{Y}_t | \Delta\mathbf{X}_k).$$

---

[*]   averaged means that for any k and t, the leakage is averaged over all the choices of $\mathbf{Y}_t$ and $\mathbf{X}_k$.
[†]   In all the DES S-boxes, when t=k=3 the static input-output information leakage is 2 bits/input which is the minimum possible value for $I(\mathbf{Y}_3, \mathbf{X}_3)$ in a 4×4 S-box.

Input Bit(s)	Output Bit(s)													
	y_1	y_2	y_3	y_4	y_1y_2	y_1y_3	y_1y_4	y_2y_3	y_2y_4	y_3y_4	$y_1y_2y_3$	$y_1y_2y_4$	$y_1y_3y_4$	$y_2y_3y_4$
x_1	0.0000	0.0000	0.0000	0.0456	0.0000	0.0000	0.0944	0.0000	0.0944	0.0944	0.0000	0.7500	0.2500	0.2500
x_2	0.0456	0.0456	0.0000	0.0000	0.3444	0.0944	0.0944	0.0944	0.0944	0.0000	0.5000	0.5000	0.2500	0.2500
x_3	0.0456	0.0456	0.0456	0.0000	0.3444	0.0944	0.0944	0.0944	0.0944	0.0944	0.5000	0.5000	0.2500	0.5000
x_4	0.0000	0.0000	0.0456	0.0456	0.0000	0.0944	0.0944	0.0944	0.0944	0.3444	0.2500	0.2500	0.5000	0.5000
x_1x_2	0.0944	0.0944	0.1887	0.0944	0.5000	0.5000	0.3750	0.5000	0.6250	0.5000	1.0000	1.3750	1.0000	1.1250
x_1x_3	0.0944	0.0944	0.0944	0.0944	0.5000	0.5000	0.6250	0.2500	0.3750	0.3750	1.0000	1.3750	1.1250	1.0000
x_1x_4	0.0000	0.0000	0.0944	0.0944	0.0000	0.2500	0.2500	0.2500	0.2500	0.6250	1.0000	1.0000	1.0000	1.0000
x_2x_3	0.0944	0.0944	0.0944	0.0000	0.7500	0.3750	0.2500	0.3750	0.2500	0.2500	1.1250	1.0000	1.1250	1.0000
x_2x_4	0.0944	0.0944	0.0944	0.0944	0.7500	0.3750	0.3750	0.3750	0.3750	0.7500	1.1250	1.1250	1.1250	1.1250
x_3x_4	0.0944	0.0944	0.0944	0.0944	0.7500	0.3750	0.3750	0.3750	0.3750	0.6250	1.1250	1.1250	1.0000	1.2500
$x_1x_2x_3$	0.2500	0.2500	0.5000	0.2500	1.0000	1.1250	1.0000	1.1250	1.0000	1.0000	2.0000	2.0000	2.0000	2.0000
$x_1x_2x_4$	0.2500	0.2500	0.5000	0.5000	1.0000	1.1250	1.0000	1.0000	1.1250	1.3750	2.0000	2.0000	2.0000	2.0000
$x_1x_3x_4$	0.2500	0.2500	0.5000	0.5000	1.0000	1.1250	1.1250	1.0000	1.0000	1.2500	2.0000	2.0000	2.0000	2.0000
$x_2x_3x_4$	0.5000	0.2500	0.2500	0.2500	1.5000	1.0000	1.0000	1.0000	1.0000	1.1250	2.0000	2.0000	2.0000	2.0000‡

Table 2. Detailed SL[I;O] matrix for the DES S-box

‡ Minimum value of SL[I;O] when t=k=3

Input Bit(s)	Output Bit(s)													
	y_1	y_2	y_3	y_4	y_1y_2	y_1y_3	y_1y_4	y_2y_3	y_2y_4	y_3y_4	$y_1y_2y_3$	$y_1y_2y_4$	$y_1y_3y_4$	$y_2y_3y_4$
x_1	0.0000	0.0000	0.0000	0.0456	0.1887	0.0000	0.0944	0.0000	0.0944	0.0944	0.5000	0.5000	0.2500	0.2500
x_2	0.0000	0.0000	0.0456	0.0000	0.0000	0.0944	0.0000	0.0944	0.1887	0.0944	0.2500	0.5000	0.2500	0.5000
x_3	0.0456	0.0000	0.0000	0.0000	0.0944	0.0944	0.0944	0.0000	0.0000	0.1887	0.2500	0.2500	0.5000	0.5000
x_4	0.0000	0.0456	0.0000	0.0000	0.0944	0.0000	0.1887	0.0944	0.0944	0.0944	0.2500	0.5000	0.5000	0.2500
x_1x_2	0.1887	0.0000	0.0944	0.0944	0.5000	0.5000	0.5000	0.2500	0.5000	0.6250	1.1250	1.2500	1.1250	1.1250
x_1x_3	0.0944	0.0000	0.0000	0.0944	0.5000	0.2500	0.6250	0.0000	0.2500	0.5000	1.0000	1.1250	1.1250	1.0000
x_1x_4	0.0000	0.0944	0.0000	0.0944	0.5000	0.0000	0.5000	0.2500	0.6250	0.2500	1.0000	1.2500	1.0000	1.1250
x_2x_3	0.0944	0.1887	0.0944	0.0000	0.5000	0.6250	0.2500	0.5000	0.5000	0.5000	1.1250	1.1250	1.1250	1.1250
x_2x_4	0.0000	0.0944	0.0944	0.1887	0.2500	0.2500	0.5000	0.6250	0.6722	0.5000	1.0000	1.2500	1.1250	1.2500
x_3x_4	0.0944	0.0944	0.1887	0.0000	0.3750	0.5000	0.5000	0.5000	0.2500	0.5000	1.1250	1.0000	1.2500	1.1250
$x_1x_2x_3$	0.5000	0.5000	0.2500	0.2500	1.2500	1.1250	1.1250	1.0000	1.1250	1.1250	2.0000	2.1250	2.0000	2.0000
$x_1x_2x_4$	0.5000	0.2500	0.2500	0.5000	1.1250	1.0000	1.2500	1.1250	1.2500	1.1250	2.0000	2.1250	2.0000	2.1250
$x_1x_3x_4$	0.2500	0.2500	0.5000	0.2500	1.1250	1.0000	1.1250	1.0000	1.0000	1.1250	2.0000	2.0000	2.0000	2.0000
$x_2x_3x_4$	0.2500	0.5000	0.5000	0.5000	1.0000	1.1250	1.1250	1.2500	1.2500	1.2500	2.0000	2.0000	2.1250	2.1250

Table 3. Detailed SL[I;O] matrix for the Dawson & Tavares S-box

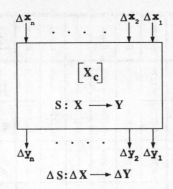

Figure 2. An n × n delta S-box

The averaged DL[I;O] matrices of the 4×4 DES S-box and the Dawson & Tavares S-box of the above example are given in Table 4. The detailed DL[I;O] matrix for the DES S-box is given in Table 5. In these tables, the information leakage is given in bits/input change.

k	DES S-box				Dawson & Tavares S-box			
	t				t			
	1	2	3	4	1	2	3	4
1	0.0014	0.0102	0.0462	0.1725	0.0007	0.0104	0.0437	0.1333
2	0.0066	0.0371	0.1586	0.4958	0.0104	0.0476	0.1484	0.3558
3	0.0317	0.1286	0.4362	1.0202	0.0437	0.1484	0.3756	0.7741
4	0.1333	0.4220	0.9866	1.7541	0.1333	0.3558	0.7741	1.4024

Table 4. Averaged DL[I;O] matrices for the DES and the Dawson & Tavares S-box

In [10] Output-Input Information Leakage, which is the same as the Input-Output Information Leakage, except that the input and the output are interchanged, has been defined as a separate property in both the static and the dynamic cases. But due to the symmetry in mutual information, i.e., $I(A; B) = I(B; A)$, the output-input information leakage matrix is simply the transposition of the input-output information leakage matrix in both the static and the dynamic cases for any bijective S-box. Therefore, the output-input information leakage is a redundant criterion for both the static and the dynamic conditions for any bijective S-box.

Input Change															Output Change
	y_1	y_2	y_3	y_4	y_1y_2	y_1y_3	y_1y_4	y_2y_3	y_2y_4	y_3y_4	$y_1y_2y_3$	$y_1y_2y_4$	$y_1y_3y_4$	$y_2y_3y_4$	$y_1y_2y_3y_4$
x_1	0.0000	0.0000	0.0000	0.0028	0.0000	0.0000	0.0057	0.0000	0.0057	0.0057	0.0000	0.2557	0.0114	0.0114	0.2901
x_2	0.0028	0.0028	0.0000	0.0000	0.0516	0.0057	0.0057	0.0057	0.0057	0.0000	0.0578	0.0578	0.0114	0.0114	0.1333
x_3	0.0028	0.0028	0.0028	0.0000	0.0516	0.0057	0.0057	0.0057	0.0057	0.0057	0.0578	0.0578	0.0114	0.0578	0.1333
x_4	0.0000	0.0000	0.0028	0.0028	0.0000	0.0057	0.0057	0.0057	0.0057	0.0516	0.0114	0.0114	0.0578	0.0578	0.1333
x_1x_2	0.0057	0.0057	0.0456	0.0057	0.0578	0.0578	0.0164	0.0578	0.0618	0.0578	0.1333	0.3830	0.0776	0.1279	0.5186
x_1x_3	0.0057	0.0057	0.0057	0.0057	0.0578	0.0578	0.0618	0.0114	0.0164	0.0164	0.1333	0.3830	0.1279	0.0776	0.5186
x_1x_4	0.0000	0.0000	0.0057	0.0057	0.0000	0.0114	0.0114	0.0114	0.0114	0.0618	0.2901	0.2901	0.0776	0.0776	0.6676
x_2x_3	0.0057	0.0057	0.0057	0.0000	0.1099	0.0164	0.0114	0.0164	0.0114	0.0114	0.1279	0.1333	0.2778	0.0776	0.5186
x_2x_4	0.0057	0.0057	0.0057	0.0057	0.1099	0.0164	0.0164	0.0164	0.0164	0.1099	0.1279	0.1279	0.1279	0.1279	0.3603
x_3x_4	0.0057	0.0057	0.0057	0.0057	0.1099	0.0164	0.0164	0.0164	0.0164	0.0618	0.1279	0.1279	0.0776	0.1659	0.3909
$x_1x_2x_3$	0.0114	0.0114	0.0578	0.0114	0.1333	0.1279	0.0776	0.3423	0.0776	0.0776	0.5186	0.5590	0.4591	0.5022	1.0990
$x_1x_2x_4$	0.0114	0.0114	0.0578	0.0578	0.1333	0.1279	0.0776	0.0776	0.1279	0.2093	0.5186	0.5186	0.3249	0.3249	1.0319
$x_1x_3x_4$	0.0114	0.0114	0.0578	0.0578	0.1333	0.1279	0.1279	0.0776	0.0776	0.1659	0.5186	0.5186	0.3444	0.3295	1.0472
$x_2x_3x_4$	0.0578	0.0578	0.0114	0.0114	0.3444	0.0776	0.0776	0.0776	0.0776	0.1320	0.3909	0.3909	0.4591	0.3013	0.9027
$x_1x_2x_3x_4$	0.1333	0.1333	0.1333	0.1333	0.5000	0.3909	0.3608	0.5590	0.3608	0.3608	1.1173	0.9847	0.9222	0.9222	1.7541

Table 5. Detailed DL[I;O] matrix for the DES S-box

2.3. Dynamic Ouput-Output Information Leakage (DL[O;O])

For any given change ΔX at the input, if $\Delta Y_k = \{\Delta y_{j_1}, \Delta y_{j_2}, ..., \Delta y_{j_k}\}$; where $1 \leq k \leq n - 1$; $1 \leq j_1, j_2, ..., j_k \leq n$, is a set of changes in the output bits and $\Delta Y_t = \{\Delta y_{l_1}, \Delta y_{l_2}, ..., \Delta y_{l_t}\}$; where $1 \leq t \leq n - 1$; $1 \leq l_1, l_2, ..., l_t \leq n$, is another set of changes in the output bits such that $\Delta Y_k \cap \Delta Y_t = \{\emptyset\}$, then the Dynamic Output-Output Information Leakage (with respect to ΔX) is the mutual information between ΔY_k and ΔY_t which is given by :

$$DL[O; O] = I(\Delta Y_t; \Delta Y_k) = H(\Delta Y_t) - H(\Delta Y_t \mid \Delta Y_k).$$

In any bijective S-box, under the static condition, for any given subset of output bits Y_k, each of the 2^t combinations of the bits from another subset of output bits Y_t (such that $Y_k \cap Y_t = \{\emptyset\}$) occurs with equal probability over all the possible static states. Therefore, the mutual information between Y_k and Y_t must be zero. This may not be true under the dynamic condition where the correlation in the output bits could be exploited to gain information about the unknown changes in the output bits. Thus, this information theoretic property is cryptographically meaningful only under the dynamic condition for a bijective S-box.

The averaged DL[O;O] matrices of the 4×4 DES S-box and the Dawson & Tavares S-box of the above examples are given in Table 6. The detailed DL[O;O] matrix for the DES S-box is given in Table 7. In these tables, the information leakage is given in bits/output change.

k	DES S-box			Dawson & Tavares S-box		
	t			t		
	1	2	3	1	2	3
1	0.1659	0.4600	0.6766	0.0952	0.3040	0.5280
2	0.4600	0.9707	-§	0.3040	0.7368	-
3	0.6766	-	-	0.5280	-	-

Table 6. The averaged DL[O;O] matrices for the DES and the Dawson & Tavares S-box

In fact, there is an averaged DL[O;O] matrix for each value of ΔX (i.e., for each pattern of input change). In this paper, however, the average values of DL[O;O] for each value of ΔX (from a single bit change to four bit change) are calculated and averaged again to form a single matrix. Note that due to the symmetry in mutual information the element a_{ij} is equal to the element a_{ji} in the DL[O;O] matrix.

---

§   " - " means $\Delta Y_k \cap \Delta Y_t \neq \{\emptyset\}$

Change in Output (Subset 2)	Change in Output (subset 1)													
	y_1	y_2	y_3	y_4	y_1y_2	y_1y_3	y_1y_4	y_2y_3	y_2y_4	y_3y_4	$y_1y_2y_3$	$y_1y_2y_4$	$y_1y_3y_4$	$y_2y_3y_4$
y_1	-	0.2490	0.1327	0.1005	-	-	-	0.4534	0.5234	0.4567	-	-	-	0.7453
y_2	0.2490	-	0.3119	0.1005	-	0.6327	0.5234	-	-	0.4567	-	-	0.7453	-
y_3	0.1327	0.3119	-	0.1005	0.5163	-	0.4567	-	0.4567	-	-	0.6786	-	-
y_4	0.1005	0.1005	0.1005	-	0.3748	0.4245	-	0.2453	-	-	0.5371	-	-	-
y_1y_2	-	-	0.5163	0.3748	-	-	-	-	-	0.9529	-	-	-	-
y_1y_3	-	0.6327	-	0.4245	-	-	-	-	1.0693	-	-	-	-	-
y_1y_4	-	0.5234	0.4567	-	-	-	-	0.8900	-	-	-	-	-	-
y_2y_3	0.4534	-	-	0.2453	-	-	0.8900	-	-	-	-	-	-	-
y_2y_4	0.5234	-	0.4567	-	-	1.0693	-	-	-	-	-	-	-	-
y_3y_4	0.4567	0.4567	-	-	0.9529	-	-	-	-	-	-	-	-	-
$y_1y_2y_3$	-	-	-	0.5371	-	-	-	-	-	-	-	-	-	-
$y_1y_2y_4$	-	-	0.6786	-	-	-	-	-	-	-	-	-	-	-
$y_1y_3y_4$	-	0.7453	-	-	-	-	-	-	-	-	-	-	-	-
$y_2y_3y_4$	0.7453	-	-	-	-	-	-	-	-	-	-	-	-	-

Table 7. Detailed DL[O;O] matrix for the DES S-box

$= \Delta Y_t \cap \Delta Y_k \neq \{\emptyset\}$

2.4. XOR Input-Output Information Leakage (XL[I;O])

The XOR distribution gives the probability distribution of the input XOR and the output XOR for an S-box. Biham & Shamir [16] first used the XOR distribution for their differential attack on DES-like cryptosystems. The XOR distribution of the DES S-box of the above example is given in Table 8.

If ΔX is the input XOR and ΔY is the output XOR, then the XOR Input-Output Information Leakage is the mutual information between ΔX and ΔY and is given by :

$$XL[I; O] = I(\Delta Y; \Delta X) = H(\Delta Y) - H(\Delta Y \mid \Delta X).$$

In an $n \times n$ S-box, for any given input XOR, if each output XOR occurs with equal probability, the XOR distribution must have all identical entries. Such an XOR distribution is called a "uniform" or "flat" distribution. The "differential probability" corresponding to an entry in the XOR distribution is obtained by dividing that entry by 2^n, where n is the block size of the S-box (or SP-network). Thus, in a uniform XOR distribution the highest differential probability is $1/2^n$. For a uniform XOR distribution XL[I;O] is zero. However, due to the nature of the XOR operation, each output XOR either occurs an even number of times or does not occur at all. Further, in an S-box when $\Delta X = 0$, $\Delta Y = 0$. Thus, a zero input XOR and the corresponding output XORs are trivial.

In the XOR distribution for an $n \times n$ S-box the sum of the entries in a row is 2^n, and if the S-box is bijective the sum of the entries in a column is also 2^n. Therefore, in the "best possible distribution" for an $n \times n$ S-box, an entry corresponding to a non-zero input XOR can be either 0 or 2. Thus, for a non-zero input XOR half of the possible output XORs do not occur. For a 4×4 S-box with the best possible distribution, XL[I;O] will be 1.1875 bits/input XOR which is the minimum value of XL[I;O] for any 4×4 S-box. However, Adams [17] showed that an $n \times n$ S-box with the best possible XOR distribution cannot be bijective when n is even.

It should be noted that in an $n \times n$ S-box, XL[I;O] is the same as DL[I;O] when $t = k = n$. Thus, XL[I:O] is not an independent evaluation criterion. However, XL[I;O] is useful in measuring how far an XOR distribution deviates from a uniform XOR distribution, using a single "quantity". Further, due to the symmetry in mutual information, the XOR output-input information leakage is the same as the XOR input-output information leakage, i.e., XL[O;I]=XL[I;O].

XL[I;O] for the DES S-box and the Dawson & Tavares S-box of the above examples are 1.7541 bits/input XOR and 1.4024 bits/ input XOR respectively (note that these values correspond to DL[I;O] in Table 4 when $t = k = 4$). The highest XL[I;O] among the 4×4 DES S-boxes is 1.8438 bits/input XOR. We note that using S-boxes with uniform XOR distribution does not necessarily increase the immunity of an SPN cryptosystem against a differential attack [18]. In order to develop resistance to a differential attack, other design criteria must also be taken into consideration.

XL[I;O] = 1.7541 bits / input XOR

Input XOR	Output XOR															
	0000	0001	0010	0100	1000	0011	0101	1001	0110	1010	1100	0111	1011	1101	1110	1111
0000	16	0	0	0	0	0	0	0	0	0	0	0	0	0	0	0
0001	0	0	0	0	0	2	0	0	0	0	8	2	2	0	0	2
0010	0	0	0	0	0	4	0	0	2	2	0	2	2	4	0	0
0100	0	0	0	0	0	6	2	0	0	2	0	0	0	2	2	2
1000	0	0	0	0	0	0	0	4	0	2	0	4	2	0	2	2
0011	0	4	0	2	2	0	0	0	2	2	0	0	0	0	0	4
0101	0	2	2	2	2	0	0	2	2	0	0	0	0	0	0	4
1001	0	0	2	0	2	0	4	0	2	0	2	0	4	0	0	0
0110	0	0	0	6	0	0	2	4	2	2	0	0	0	2	6	0
1010	0	0	2	6	4	0	0	0	0	4	2	0	0	4	0	0
1100	0	2	0	6	4	0	0	2	0	2	2	0	0	0	0	0
0111	0	2	6	0	0	0	4	2	0	2	0	0	0	0	0	0
1011	0	4	0	0	4	2	0	0	2	0	0	0	2	0	2	0
1101	0	0	0	0	0	0	2	0	4	0	0	2	0	2	0	2
1110	0	2	2	0	0	2	0	2	0	0	0	2	0	2	2	2
1111	0	0	2	0	2	0	2	0	0	0	2	0	4	2	2	0

Table 8. XOR distribution for the DES S-box

3. An Equivalence Class on S-boxes

Consider the n×n S-box S' created by XORing $\mathbf{X_r}$ and $\mathbf{Y_s}$ with the input and the output respectively of the n×n S-box S as shown in Figure 3. $\mathbf{X_r}$ and $\mathbf{Y_s}$ are arbitrary fixed n-bit binary vectors.

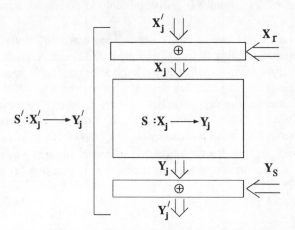

Figure 3. Equivalent S-boxes S and S' with invariant information leakage

Since

$$Prob(\mathbf{X_j}) = Prob(\mathbf{X_j'} \oplus \mathbf{X_r}) = Prob(\mathbf{X_j'})$$

and

$$Prob(\mathbf{Y_j}) = Prob(\mathbf{Y_j'} \oplus \mathbf{Y_s}) = Prob(\mathbf{Y_j'})$$

the SL[I;O] of S is the same as that of S'. Also, since the properties related to the changes in the input and the output bits are invariant to the XOR operations at the input and the output, all the dynamic information leakages (DL[I;O], DL[O;O] and XL[I;O]) will be the same for both S and S'. Therefore, in this fashion, we can generate 2^{2n} equivalent n×n S-boxes with invariant information leakage and with different input-output mapping.

A new S-box S'' can be generated by permuting the input and/or output bits of the original S-box S. S'' will have similar cryptographic properties to S. However, due to the bit permutation, the entries of the leakage matrices and the XOR distribution of S'' may be located differently. Starting with S'', a new class of S-boxes can be generated using the above procedure. Hence, if a single S-box with low information leakage is found (possibly through a computer search), a large number of S-boxes with similar information leakage can be created easily.

4. Differential Attack on SP Networks

The differential attack developed by Biham & Shamir is a statistical chosen plaintext attack on DES-like block ciphers. If a pair of distinct plaintexts with known XOR difference ΔX produces a pair of $(r-1)^{th}$ round ciphertexts $\mathbf{Y_{r-1}}$ and $\mathbf{Y_{r-1}'}$ such that $Y_{r-1} \oplus Y'_{r-1} = \Delta Y_{r-1}$; then an r round cipher is vulnerable to the differential attack if and only if the following conditions hold [19] :

I. There exists a pair of $(r-1)^{th}$ round outputs Y_{r-1} and Y'_{r-1} such that $Prob(\Delta Y_{r-1} \mid \Delta X)$ is greater than $1/2^m$, where m is the cipher block size.

II. Given some pairs of Y_{r-1} and Y'_{r-1} it is possible to determine some key bits in the rth round.

The effectiveness of this line of attack depends on how confidently the $(r-1)^{th}$ round XOR values (corresponding to the chosen input XOR) can be predicted in the SP network. In the cryptosystem, if XL[I;O] is zero after the $(r-1)^{th}$ round then the maximum differential probability reaches the ideal value which is $1/2^m$ (m is the cipher block size). Hence, the first condition will be satisfied. However, due to the nature of the XOR operation, an SP network with even the best possible XOR distribution will have a differential probability of $1/2^{m-1}$ (i.e., $2/2^m$).

In an SP network, keying can be introduced in one of the two ways shown in Figure 4. DES uses a combination of these two methods. In Figure 4 (a), the $(r-1)^{th}$

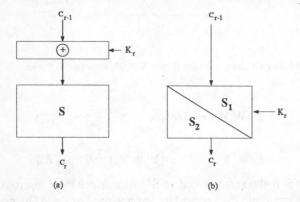

Figure 4.   Two possible methods of keying

round ciphertext C_{r-1} is XORed with the rth round key K_r to form the actual input to the S-box. A given $(\Delta C_{r-1}, \Delta C_r)$ pair, where $\Delta C_{r-1} \neq 0$, restricts the possible values for the actual input to the S-box. Using the actual input and the value of C_{r-1} (if known) the uncertainty in K_r can be reduced. However, in an SPN cryptosystem using this keying arrangement, if the value of C_{r-1} is not available (note that this condition is not satisfied in DES-like systems), a differential cryptanalyst cannot learn about the key using the knowledge of the input XOR and the output XOR of the S-box.

In Figure 4 (b), one bit in K_r is used to select one of the two S-boxes : S_1 and S_2. In this arrangement, K_r is not mixed with C_{r-1} to form the actual input to the S-box in the rth round. Since in this illustration only two S-boxes are used, a single key bit is sufficient to select an S-box. This arrangement is vulnerable to a differential attack if the two S-boxes do not have identical XOR distributions [13]. It has been pointed out by Heys [20] that a differential attack is possible, even if the S-boxes have identical XOR distributions (i.e., if S_1 and S_2 are chosen from an equivalence class). This can be explained with the help of Figure 5.

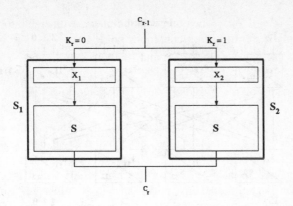

Figure 5. Differential attack on the arrangement shown in Figure 4 (b)

Assume that in Figure 5 the equivalent S-boxes S_1 and S_2 are derived by XORing the vectors X_1 and X_2 respectively at the input of the S-box S. The knowledge of a $(\Delta C_{r-1}, \Delta C_r)$ pair, where $\Delta C_{r-1} \neq 0$, would suggest the actual values of the input of S. Using these suggested values and the values of C_{r-1} (assumed to be known) we can obtain the possible values of the vectors X_1 and X_2, and compare them with the known values of X_1 and X_2 to get the keying information. Since only two S-boxes were used in this example, the described attack does not seem efficient. However, if a large number of S-boxes are used in this fashion, the differential attack would become more efficient.

Therefore, an SP network using one or a combination of the above keying techniques should be designed to minimize the maximum entry in the XOR distribution (i.e., maximum differential probability), in order to increase the immunity against differential attack.

5. Analysis of a 16–bit SP Network

In an r round SPN cryptosystem the substitution-permutation function is iterated r times so that the final product (ciphertext) is cryptographically stronger than the intermediate products. The number of rounds required depends strongly on the strength of the individual layers. If the individual layers are strong, the number of rounds required can be smaller which means that higher data encryption/decryption rates can be achieved. In order to study the influence of the S-boxes on the cryptographic properties of an SP network, a 16×16 SP network (which is tractable) shown in Figure 6 was evaluated with respect to various criteria explained above. The DES, Dawson & Tavares and some randomly selected S-boxes were used for this analysis. We found that some of the DES S-boxes are relatively stronger than the others with respect to information leakage. Therefore, under each evaluation criterion, the DES S-boxes with relatively low information leakage (DES-L) and relatively high information leakage (DES-H) were analyzed separately.

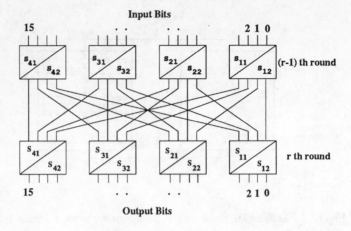

Figure 6.   A 16 bit SP network

We first studied how the maximum differential probability of the SP network varies with the number of rounds (for the purpose of this test DES S-boxes were ranked according to their XL[I;O]). We know that even in the best case a non-zero minimum entry in the XOR distribution of the SP network is 2. Hence, for any non-zero input XOR, at least 50% of the output XORs do not occur. Since the S-boxes used are bijective, the 16–bit SP network is also bijective. As in the case of an S-box, a bijective SP network with XOR distribution containing only 0's and 2's is not realizable when the block size is even, which is true for the 16–bit SP network. Therefore, we can expect some entries in the XOR distribution which are greater than 2. Figure 7 shows the variation of the maximum entry in the XOR distribution (for 100 randomly selected non-zero input XORs) with the number of rounds. For all the S-boxes used, the highest entry in the XOR distribution converged to 14 (i.e., the maximum differential probability is $14/2^{16}$) after 5 rounds. Further, after 3 rounds there was not much difference in the maximum differential probability regardless of the selection of S-boxes. However, the S-boxes with low XL[I;O] led to faster convergence. In addition, we noted that for all the S-boxes the percentage of 0's (in a row) was 60.7% of the number of possible output XORs, once the convergence was achieved. We and Heys [20] observed that the distribution of entries in a given row in the XOR distribution, after a sufficient number of rounds, behaves like a random placement of $n/2$ balls in n bins, where each ball has a value of 2. The maximum entry in a row corresponding to the random placements was observed to be less than or equal to 14.

A well designed SP network can be regarded as a large strong S-box. Hence an ideal SP network should satisfy all the cryptographic properties of an ideal S-box. We then examined the 16–bit SP network on a round-by-round basis with respect to the four types of information leakages. For selected input and output bits the system was tested exhaustively, where feasible, or using a large number of randomly chosen inputs. The simulation results are shown in Figures 8 through 11.

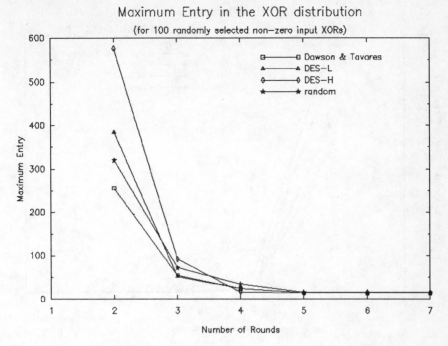

Figure 7. Variation of Maximum Entry in the XOR
distribution with number of rounds for the 16–bit SP network

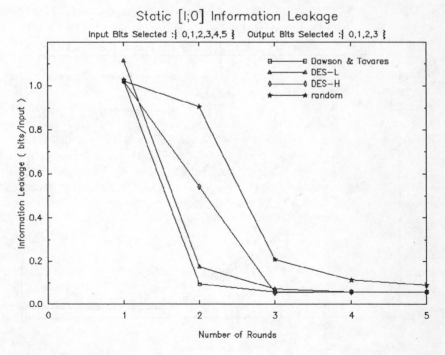

Figure 8. Variation of SL[I:O] with number of rounds for the 16–bit SP network

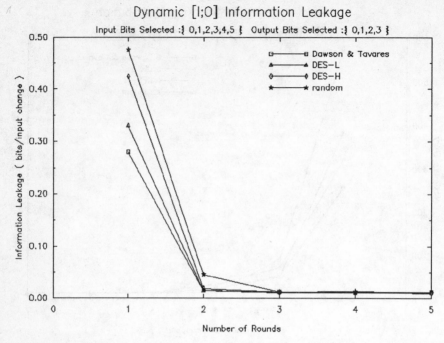

Figure 9. Variation of DL[I:O] with number of rounds for the 16–bit SP network

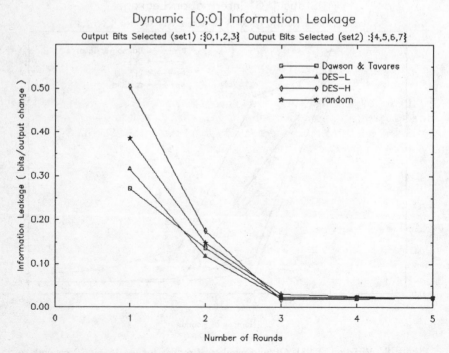

Figure 10. Variation of DL[O:O] with number of rounds for the 16–bit SP network

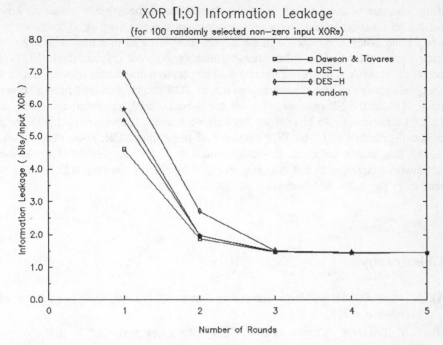

Figure 11. Variation of XL[I:O] with number of rounds for the 16–bit SP network

7. Conclusions

We reviewed evaluation criteria for n×n bijective S-boxes based on information leakage and introduced the concept of XOR Information Leakage (XL[I;O]), which is useful in comparing the XOR distributions of S-boxes. We then defined an equivalence class on n×n S-boxes which have invariant information leakage. The equivalence classes will reduce the search space for the design of cryptographically strong S-boxes with low information leakage. We also found that not all the DES S-boxes are equally strong with respect to information leakage.

We studied the impact of the choice of S-boxes on the cryptographic properties of a 16×16 SP network using various S-boxes. Sample S-boxes were chosen from the DES, Dawson & Tavares, and randomly constructed ones. The variation of the maximum entry in the XOR distribution with the number of rounds is shown in Figure 7. This experimental XOR distribution (corresponding to 100 randomly selected non-zero input XORs) closely approximates a random distribution of the output XORs after 5 rounds. After 3 rounds there is not much difference in the maximum entry in the XOR distribution regardless of the selection of S-boxes. However, the S-boxes with low XL[I;O] lead to faster convergence to the random XOR distribution.

We finally studied the influence of the S-boxes on the information leakage of the SP network. The simulation results are shown in Figures 8 through 11. Here four types of information leakages are plotted against the number of rounds. After 3 rounds there is not much difference in the information leakage of any kind, regardless of the selection of the S-boxes. However, the choice of the S-boxes influences how fast the information leakage achieves the minimum value. Using the S-boxes which produce the fastest convergence in the SP network will lead to a more efficient and faster implementation of a substitution-permutation network cryptosystem. XL[I;O] for the SP network is of special interest with respect to a differential attack because it is a good measure of how confidently an output XOR can be predicted from a known input XOR in the SP network. For all the S-boxes used, the minimum value of XL[I;O] achieved is 1.45 bits / input XOR after 5 rounds. The value of XL[I;O] for a random distribution of 16–bit XORs is also 1.45 bits / input XOR. These observations suggest that, after a sufficient number of rounds, the XOR distribution of the 16–bit SP network converges to a distribution obtained by placing the output XOR pairs at random in the XOR distribution.

Bibliography

[1] N. B. of Standards, "Data Encryption Standard," No. 46, U.S. Department of Commerce, 1977.

[2] C. E. Shannon, "Communication theory of secrecy systems," in *Bell Systems Technical Journals*, vol. 28, pp. 656–715, 1949.

[3] H. Feistel, "Cryptography and Computer privacy," in *Scientific American*, vol. 228, pp. 15–23, 1973.

[4] H. Feistel, W. Notz, and J. L. Smith, "Some Cryptographic Techniques for Machine-to-Machine Data Comminications," in *Proc. of the IEEE*, vol. 63, pp. 1545–1554, 1975.

[5] A. F. Webster and S. E. Tavares, "On the Design of S-boxes," in *Advances in Cryptology, Proc. of CRYPTO '85*, pp. 523–534, Springer-Verlag, New York, 1986.

[6] E. F. Brickell, J. H. Moore, and M. R. Purtill, "Structure in the S-boxes of the DES (extended abstract)," in *Advances in Cryptology : Proc. of CRYPTO '86*, pp. 3–8, Springer-Verlag, New York, 1987.

[7] J. Pieprzyk and G. Finkelstein, "Towards effective nonlinear cryptosystem design," in *IEE Proceedings, Part E : Computers and Digital Techniques*, vol. 135, pp. 325–335, 1988.

[8] C. A. Adams and S. Tavares, "The structured design of cryptographically good S-boxes," in *Journal of Cryptology*, vol. 3, pp. 27–41, 1990.

[9] R. Forré, "Methods and Instruments for designing S-boxes," in *Journal of Cryptology*, vol. 2, pp. 115–130, 1990.

[10] M. H. Dawson and S. E. Tavares, "An Expanded Set of S-box Design Criteria Based on Information Theory and its Relation to Differential-Like Attacks," in *Advances in Cryptology, Proc. of EUROCRYPT' 91*, pp. 352–367, Springer-Verlag, New York, 1992.

[11] J. B. Kam and G. I. Davida, "Structured Design of Substitution-Permutation Encryption Networks," in *IEEE Transaction on Computers*, C-28, pp. 747–753, 1979.

[12] A. F. Webster, "Plaintext/Ciphertext Bit Dependence in Cryptographic Systems," Master's thesis, Queen's University at Kingston, Canada, 1985.

[13] L. O'Connor, "A Differential-like cryptanalysis of SP-networks," tech. rep., Department of Computer Science, University of Waterloo, Canada, 1992. , (submitted for publication).

[14] A. Shimizu and S. Miyaguchi, "Fast data encryption algorith FEAL," in *Advances in Cryptology, EUROCRYPT '87*, pp. 267–278, 1988.

[15] L. Brown, J. Pieprzyk, and J. Seberry, "LOKI - cryptographic primitive for authentication and secrecy applications," in *Advances in Cryptology, Proc. of AUSCRYPT '90*, pp. 229–236, Springer-Verlag, New York, 1990.

[16] E. Biham and A. Shamir, "Differential cryptanalysis of DES-like cryptosystem," in *Journal of Cryptology*, vol. 4, pp. 3–72, 1991.

[17] C. A. Adams, "On immunity against Biham and Shamir's "differential cryptanalysis"," in *Information Processing Letters*, vol. 41, pp. 77–80, 1992.

[18] E. Biham, *Differential Cryptanalyst of Iterative Cryptosystem*. PhD thesis, Weizmann Institute of Science, Rehovolt, Israel, 1992.

[19] X. Lai and J. Massey, "Markov Ciphers and Differential Cryptanalysis," in *Advances in Cryptology, Proc. of EUROCRYPT' 91*, pp. 17–38, Springer-Verlag, New York, 1992.

[20] H. Heys , Department of Electrical Engineering, Queen's University, Kingston, Canada, (personal communication).

Partially-bent functions

C. Carlet[1]

INRIA, Domaine de Voluceau, Rocquencourt,
Bat 10, BP 105, 78153 Le Chesnay Cedex, FRANCE

Abstract

We study a conjecture stated in [6] about the numbers of non-zeros of, respectively, the auto-correlation function and the Walsh transform of the function $(-1)^{f(x)}$, where $f(x)$ is any boolean function on $\{0,1\}^n$. The result that we obtain leads us to introduce the class of partially-bent functions. We study within these functions the propagation criterion. We characterize those partially-bent functions which are balanced and prove a relation between their number (which is unknown) and the number of non-balanced partially-bent functions on $\{0,1\}^{n-1}$. Eventually, we study their correlation immunity .

1 Introduction

The study of the properties of the substitution transformations of DES has resulted in nonlinearity criteria for boolean functions. Perfect nonlinear boolean functions, also called *bent functions*, are defined to be at maximum Hamming distance from affine functions. Those functions, of great importance in cryptography, seem to be rare, and very few are known. They are neither balanced nor correlation-immune. So, it seems useful to define a larger class of boolean functions, containing balanced functions, and preserving a high level of nonlinearity. That is what this paper obtains through the proof of a conjecture stated in [6]. The class of functions that we obtain is also a superclass of the class of quadratic functions. It shares with this class all its nice properties relative to the propagation criterion, the balancedness and the correlation immunity.

n is a positive integer, $G = \{0,1\}^n$.

The dot product on G is defined by :

$$\forall x = (x_1, \ldots, x_n), s = (s_1, \ldots, s_n) \in G \quad x \cdot s = x_1 s_1 + \ldots + x_n s_n \in \{0,1\}$$

where the operations on $\{0,1\}$ are the usual operations on $GF(2)$.

---

[1] Université de Picardie, France

Let f be a real-valued function on G. The Walsh (or Hadamard) transform of $f(x)$ is the function on G :

$$\hat{f}(s) = \sum_{x \in G} f(x)(-1)^{x \cdot s}.$$

Let f be a boolean function on G. We will denote by $\hat{F}$ the Walsh transform of the real-valued function $F(x) = (-1)^{f(x)}$:

$$\hat{F}(s) = \sum_{x \in G} (-1)^{f(x) + x \cdot s}.$$

It satisfies the Parseval's relation (cf.[5], p.416, corollary 3 or the lemma below) :

$$\sum_{s \in G} (\hat{F}(s))^2 = 2^{2n}.$$

f is kth-order correlation-immune if (cf.[1], [9]) :
$\hat{F}(s) = 0 \quad 1 \le w(s) \le k$ (where $w(s)$ denotes the Hamming weight of s).
The auto-correlation function of F is defined by :

$$\hat{r}(s) = \sum_{x \in G} (-1)^{f(x) + f(x+s)}.$$

f satisfies the propagation criterion $PC(k)$ of degree k $\quad (1 \le k \le n)$ if :

$$\hat{r}(s) = 0 \quad 1 \le w(s) \le k.$$

There exists functions satisfying $PC(n)$ if and only if n is even (cf.[4]). In that case, any boolean function f satisfies $PC(n)$ if and only if, for any element s of G, the number $\hat{F}(s)$ is equal to : $\pm 2^{n/2}$ (cf.[4] or the lemma below). Such functions are called bent. According to Parseval's relation, the bent functions are those functions which are at maximum Hamming distance from affine functions.

The definition of bent functions is invariant under any linear isomorphism, and we may define the bent functions on any $GF(2)$-space E of even dimension as the functions satisfying :

$$\sum_{x \in E} (-1)^{f(x) + f(x+s)} = 0, \forall s \in E, s \ne 0 \text{ or equivalently :}$$

$$\sum_{x \in E} (-1)^{f(x) + x \cdot s} = \pm \sqrt{|E|} \quad \forall s \in E.$$

In [6], the authors conjecture that the numbers of zeros $N_{\hat{r}}$ and $N_{\hat{F}}$ of the functions $\hat{r}$ and $\hat{F}$ associated with any boolean function satisfy :

$$(2^n - N_{\hat{r}})(2^n - N_{\hat{F}}) \geq 2^n$$

and that equality holds only for functions of order 2 (that are functions whose algebraic normal forms have degrees at most 2 : we will call them quadratic) or satisfying $PC(n)$ or $PC(n-1)$. At Las Vegas Conference on Finite Fields, they changed the second part of their conjecture in : "equality holds only for functions of order 2 or satisfying $PC(n)$ (n even) or such that $N_{\hat{r}} = 2^n - 2$ (n odd)".

In section 2, we prove that the first part of that conjecture : $(2^n - N_{\hat{r}})(2^n - N_{\hat{F}}) \geq 2^n$ is true. We characterize those functions for which equality holds. We call these functions partially-bent for they are related to bent functions (cf. the theorem below). Any quadratic function is partially-bent.

In section 3, we study those partially-bent functions which satisfy $PC(k)$, those which are balanced, kth-order correlation-immune (we deduce that both versions of the second part of the conjecture are false). We prove that the number of partially-bent balanced functions on G is equal to the number of partially-bent non-balanced functions on $\{0,1\}^{n-1}$, times $(2^n - 1)$. All the results of that section hold for quadratic functions, and we deduce that there are more balanced quadratic functions than non-balanced quadratic functions on G if and only if n is odd.

2 Partially-bent functions

Let f be any boolean function on G, let us recall that the functions $\hat{r}$ and $\hat{F}$ defined in section 1 are related to each other the following way :

Lemma 2.1 *The Walsh transform of the function $\hat{r}$ is equal to the function $\hat{F}^2$:*

$$\forall t \in G, \sum_{s \in G} \hat{r}(s)(-1)^{s \cdot t} = (\hat{F}(t))^2.$$

Proof: According to the definition of the autocorrelation function, we have :

$$\forall t \in G, \sum_{s \in G} \hat{r}(s)(-1)^{t \cdot s} = \sum_{s \in G}\left(\sum_{x \in G}(-1)^{f(x)+f(x+s)+t \cdot s}\right) = \sum_{x \in G}\left(\sum_{s \in G}(-1)^{f(x)+f(x+s)+t \cdot s}\right).$$

Since G is invariant under any translation, we may replace s by $x + s$ in the second sum. We obtain :

$$\sum_{s \in G} \hat{r}(s)(-1)^{t \cdot s} = \sum_{x \in G} \left(\sum_{s \in G} (-1)^{f(x)+f(s)+t.(x+s))} \right) = \left(\sum_{x \in G} (-1)^{f(x)+t \cdot x} \right) \left(\sum_{s \in G} (-1)^{f(s)+t \cdot s} \right)$$

$$= \left(\sum_{x \in G} (-1)^{f(x)+t \cdot x} \right)^2 = (\hat{F}(t))^2.$$
□

We now prove the first part of the conjecture stated in [6] and characterize those functions for which equality holds :

Theorem 2.1 *Any boolean function f on G satisfies $(2^n - N_{\hat{f}})(2^n - N_{\hat{F}}) \geq 2^n$. Equality holds if and only if :*

(i) there exists an element t in G such that for any s in G, $\hat{r}(s)$ is equal to 0 or to $(-1)^{t \cdot s} 2^n$ that is if and only if :

(ii) there exists a linear form $x \to t \cdot x$ on G, two subspaces E and E' in G (E' of even dimension), such that :

- G is the direct sum of E and E'

- the restriction of f to E' is bent

- for all x in E, and all y in E', $f(x + y)$ is equal to : $f(y) + t \cdot x$.

Proof: - Since the values of the function $\hat{r}$ all are at most equal to 2^n, we have :

$$2^n - N_{\hat{f}} \geq 2^{-n} \sum_{s \in G} \hat{r}(s) = 2^{-n}(\hat{F}(0))^2.$$

The number $N_{\hat{f}}$ clearly does not change when we replace the function $f(x)$ by any of the functions $f(x) + x \cdot t$ $(t \in G)$. Replacing $f(x)$ by $f(x) + x \cdot t$, we change $\hat{F}(0)$ in $\hat{F}(t)$. Thus :

$$2^n - N_{\hat{f}} \geq 2^{-n}(\hat{F}(t))^2 \quad \forall t \in G \tag{1}$$

We also have :

$$2^n - N_{\hat{F}} \geq \frac{\sum_{s \in G}(\hat{F}(t))^2}{\sup(\hat{F}(t))^2} = \frac{2^{2n}}{\sup(\hat{F}(t))^2} \tag{2}$$

Multiplying these two inequalities, we obtain :

$$(2^n - N_{\hat{f}})(2^n - N_{\hat{F}}) \geq 2^n.$$

We now shall prove that if equality holds then (i) is true, if (i) is satisfied then so is (ii) and if (ii) is true then equality holds.

- If equality holds then, according to (1) and (2) :

$$2^n - N_{\tilde{r}} = 2^{-n} \sup(\hat{F}(t))^2 \quad (\text{and} \quad 2^n - N_{\hat{F}} = \frac{2^{2n}}{\sup(\hat{F}(t))^2}).$$

Let $\tilde{r}$ be the auto-correlation function associated with the function $f(x) + x \cdot t$ where $(\hat{F}(t))^2$ is maximal. By applying the previous lemma to the function $f(x) + x \cdot t$, we obtain :

$$\sum_{s \in G} \tilde{r}(s) = (\hat{F}(t))^2 \text{ and therefore} : \sum_{s \in G} \tilde{r}(s) = 2^n(2^n - N_{\tilde{r}}) = \sum_{s \in G/\tilde{r}(s) \neq 0} 2^n.$$

Thus : $\forall s \in G, \tilde{r}(s) = 0$ or 2^n. We have : $\forall s \in G, \tilde{r}(s) = (-1)^{t \cdot s}\hat{r}(s)$, and (i) is true.

- If (i) is true, then let E be the set of all the elements x of G such that :
$\hat{r}(x) = (-1)^{t \cdot x}2^n$ that is $\forall s \in G, f(x+s) = f(s) + x \cdot t$.
E is clearly a subspace of G. Let E' be any subspace of G such that G is the direct sum of E and E'. Then :

$$\forall v \in E', v \neq 0 \Rightarrow v \notin E \Rightarrow \hat{r}(v) = 0 \Rightarrow$$

$$\sum_{x \in E} \sum_{y \in E'} (-1)^{f(y)+f(y+v)} = |E| \sum_{y \in E'} (-1)^{f(y)+f(y+v)} = 0.$$

Thus (ii) is satisfied.

- Suppose (ii) is true. We may without loss of generality suppose that $t = 0$ since changing the value of t does not change $N_{\tilde{r}}$ or $N_{\hat{F}}$. Then, the value of $f(x+y)$ $(x \in E, y \in E')$ does not depend on x, and we have :

$$\forall s = u + v, u \in E, v \in E', \hat{r}(s) = |E| \sum_{y \in E'} (-1)^{f(y)+f(y+v)} = \begin{cases} 0 & \text{if } s \notin E \quad (v \neq o) \\ |E||E'| = 2^n & \text{otherwise} \end{cases}$$

$$2^n - N_{\tilde{r}} = |E| \text{ and } \forall s \in G, \hat{F}^2(s) = \sum_{x \in G} \hat{r}(x)(-1)^{s \cdot x} = \sum_{x \in E} 2^n(-1)^{s \cdot x} = \begin{cases} 0 & \text{if } s \notin E^\perp \\ 2^n|E| & \text{otherwise} \end{cases}$$

where $E^\perp = \{s \in G/\forall x \in E, s \cdot x = 0\}$.
So, $2^n - N_{\hat{F}} = |E^\perp|$ and $(2^n - N_{\tilde{r}})(2^n - N_{\hat{F}}) = 2^n$. $\square$
REMARK :
1) We have in fact : $\forall x \in E, \forall y \in G, f(x+y) = f(y) + t \cdot x$

2) We have proved :

$$\frac{2^n - N_{\hat{f}}}{2^n} \geq \sup_t \left(\frac{\hat{F}(t)}{2^n}\right)^2 \geq \frac{1}{2^n - N_{\hat{F}}},$$

which shows the trade-offs between the highest correlation to linear functions (in the middle), a certain measure of correlation immunity (on the right) and the non-vanishing of the auto-correlation function.

Definition 2.1 *A function f which satisfies the equality $(2^n - N_{\hat{f}})(2^n - N_{\hat{F}}) = 2^n$ is called partially-bent.*

Let f be a partially-bent function, E and E' two linear subspaces of G such that G is the direct sum of E and E', f is bent on E' and $f(x + y) = f(y) + t.x$, $x \in E, y \in E'$. Let φ_f be the function defined on $G \times G$ by : $\varphi_f(u, v) = f(0) + f(u) + f(v) + f(u + v)$. Then : $\forall x, x' \in E, \forall y, y' \in E', \varphi_f(x + y, x' + y') = \varphi_f(y, y')$. Since $f|_{E'}$ is bent , the restriction of φ_f to $E' \times E'$ is non-degenerate, and :

$$(\varphi_f(x + y, v) = 0 \quad \forall v \in G) \Leftrightarrow (y = 0).$$

E is the set of all the elements u of G such that $\varphi_f(u, v) = 0 \quad \forall v \in G$.
Thus E is unique.
Clearly, E' is not.
If E has dimension $n - 2h$, then t may take 2^{2h} values since the values of the linear form $x \to t \cdot x$ are fixed only on E.

Definition 2.2 *Let f be a partially-bent function, φ_f be the function defined on $G \times G$ by :*

$$\varphi_f(u, v) = f(0) + f(u) + f(v) + f(u + v).$$

The linear space $E = \{u \in G / \varphi_f(u, v) = 0 \quad \forall v \in G\}$ is called the kernel associated with f.

Any quadratic function is partially-bent (cf [6]) and the kernel associated with f is the kernel of its associated symplectic form φ_f.

REMARK :
1) The definition and the linearity of the set E are valid for any boolean function
2) Since the degree of any bent function on a linear space of dimension $2p$ is at most p, the degree of a partially-bent function is at most the half of the codimension of its kernel.

3) the set of partially-bent functions on G is not a linear space : for instance, if $n = 6$, the non-quadratic partially-bent functions are the non-quadratic bent functions which all are known (cf [7]) and it is easy to find two bent functions whose sum is neither bent nor quadratic.

4) The number of partially-bent functions seems to be difficult to obtain : it depends on the number of bent functions which is unknown (except for small values of n).

5) Let f be a boolean quadratic function on G and 1 an affine boolean function on the same space, then the following boolean function on $\{0,1\}^{n+1}$:

$$(x_1, \ldots, x_n, x_{n+1}) \in \{0,1\}^{n+1} \rightarrow f(x_1, \ldots, x_n) + x_{n+1}1(x_1, \ldots, x_n)$$

is quadratic and any quadratic function on $\{0,1\}^{n+1}$ is of that type (thus, the number of quadratic functions on $\{0,1\}^{n+1}$ equals the number of quadratic functions on $\{0,1\}^n$, times 2^{n+1}). *That is no more true if we replace "quadratic" by "partially-bent".*

3 Properties of partially-bent functions

Since the authors conjecture in [6] that, if n is even, the non-quadratic partially-bent functions satisfy $PC(n)$, let us begin with the propagation criterion :

Proposition 3.1 *A partially-bent function f on G satisfies $PC(k)$ $(k = 1, \ldots, n)$ if and only if its associated kernel E only contains elements of Hamming weight $> k$, or equal to 0.*

Proof: The proof is straightforward : $\hat{r}(x) = 0$ if and only if $x \notin E$. □

Thus, the second parts of the conjectures stated by B. Preneel in [6] and at Las Vegas Conference on Finite Fields (which characterize the functions for which equality holds) are false :

if n is even, suppose that E contains an element of weight 1, then f does not satisfy $PC(1)$,

if n is odd, $2^n - N_{\hat{f}} = |E|$ may be any odd power of 2, and if the codimension of E is at least 6, then f may be non-quadratic.

REMARK :
The number of partially-bent functions satisfying $PC(k)$ seems to be even more difficult to obtain than that of the partially-bent functions : it depends on the number of linear spaces of minimum weights greater than k, which is unknown except for small values of n.

The weight of a boolean function on G is the size of its support. A function $f(x)$ is called balanced if its weight is 2^{n-1}, that is if $\hat{F}(0) = 0$.

Proposition 3.2 *A partially-bent function f on G is balanced if and only if its restriction to its associated kernel is non-constant, that is if and only if there exists an element u in G such that :*

$$\forall x \in G, f(x + u) = f(x) + 1.$$

Otherwise, its weight is equal to $2^{n-1} \pm 2^{n-1-h}(h \in \mathbb{N}, h \leq n/2).$

Proof: Let f be a partially-bent function, E its associated kernel, and E' a subspace such that G is the direct sum of E and E'.
$\hat{F}(0)$ is equal to : $\sum_{u \in G}(-1)^{f(u)} = \sum_{x \in E}(-1)^{t \cdot x}\sum_{y \in E'}(-1)^{f(y)}$ and these two last sums satisfy :

$$\sum_{y \in E'}(-1)^{f(y)} = \pm\sqrt{|E'|} \neq 0 \text{ since } f|_{E'} \text{ is bent, and} : \sum_{x \in E}(-1)^{t \cdot x} = \begin{cases} 0 & \text{if } t \notin E^{\perp} \\ |E| & \text{otherwise.} \end{cases}$$

Thus, f is balanced if and only if t does not belong to $E^{\perp}$, that is if and only if f is non-constant on E.
In that case, let u be any element in $E \backslash t^{\perp}$, where $t^{\perp} = \{x \in G / x \cdot t = 0\}$. We have :

$$\forall x \in G, f(x + u) = f(x) + t.u = f(x) + 1.$$

Conversely, if u satisfies that property, then f is non-constant on E.

If f is non-balanced, suppose E has dimension $n - 2h$, then the sum $\sum_{u \in G}(-1)^{f(u)}$, which is equal to $2^n - 2w(f)$, is also equal to : $\pm|E|\sqrt{|E'|} = \pm 2^{n-2h}2^h = \pm 2^{n-h}$.
So, $w(f) = 2^{n-1} \pm 2^{n-h-1}$. $\quad\square$

Proposition 3.3 *The number λ_n of partially-bent balanced functions on $G = \{0,1\}^n$ is equal to $(2^n - 1)$ times the number λ'_{n-1} of partially-bent non-balanced functions on $\{0,1\}^{n-1}$ $(n \geq 2)$.*

Proof: Let f be a partially-bent balanced function on G and E its associated kernel (since f is balanced, E is not the trivial space $\{0\}$).
Let E' be a subspace of G such that G is the direct sum of E and E', t any element of G such that $f(x + y) = f(y) + t.x$, $x \in E$, $y \in E'$ ($t \neq 0$ since f is balanced) and H the linear hyperplane $t^{\perp} = \{x \in G / x \cdot t = 0\}$.
Let $\phi : \{0,1\}^{n-1} \to H$ be a linear isomorphism. Then the boolean function $g = f \circ \phi$

is clearly partially-bent of associated kernel $\phi^{-1}(E)$. According to proposition 3.2, it is non-balanced since :

$$\forall x \in \phi^{-1}(E), \forall y \in \phi^{-1}(E'), g(x+y) = g(y).$$

Let us now calculate the number of (H, ϕ, g) so associated with f :
suppose E has dimension $n - 2h(2h < n)$, then the set of all the possible values of t is an affine set of direction $E^{\perp}$, so its size (which is the number of possible H) is 2^{2h}.
H being chosen, there exists (cf [5]) $(2^{n-1} - 1)(2^{n-1} - 2) \ldots (2^{n-1} - 2^{n-2})$ isomorphisms ϕ from $\{0,1\}^{n-1}$ to H, and if H and ϕ are chosen, then g is unique. So the number of (H, ϕ, g) associated with f is $2^{2h}(2^{n-1} - 1) \ldots (2^{n-1} - 2^{n-2})$.
Notice that the dimension of the associated kernel $\phi^{-1}(E)$ of g is $n - 1 - 2h \geq 0$.

Let now g be any partially-bent non-balanced function on $\{0,1\}^{n-1}$, suppose its associated kernel E'' has dimension $n - 1 - 2h$, and let H be a linear hyperplane of $\{0,1\}^n$ and ϕ an isomorphism from $\{0,1\}^{n-1}$ onto H. Let us calculate the number of partially-bent balanced functions f on G such that $g = f \circ \phi$.
The associated kernel E of f necessarily contains $\phi(E'')$, has dimension $n - 2h$, and is not contained in H. So, it is equal to a linear space of the type : $\{u + v, u \in \phi(E''), v \in \{0, s\}\}$ where s is any element outside H. The number of such E is equal to the number of such elements s in $\{0,1\}^n \backslash H$, divided by the size of $\phi(E'')$, since two elements s and s' define the same set E if and only if $s + s'$ belongs to $\phi(E'')$. The number of kernels E is therefore 2^{2h}.
E being chosen, f is unique since the value of f on $E \backslash \phi(E'')$ must be equal to $f(0) + 1$.
So the number of partially-bent balanced functions f on G corresponding to (H, ϕ, g) is 2^{2h} and the number λ_n of partially-bent balanced functions on G equals the number of ordered pairs (H, g) where H is any linear hyperplane and g any partially-bent non-balanced function on $\{0,1\}^{n-1}$. The number of linear hyperplanes being $2^n - 1$, we have
$\lambda_n = (2^n - 1)\lambda'_{n-1}.$ □

REMARK :
1) The previous proof is valid when we restrict ourselves to the quadratic functions since f is quadratic if and only if g is quadratic.
Therefore, the number μ_n of balanced quadratic functions on G is equal to $(2^n - 1)$ times the number μ'_{n-1} of non-balanced quadratic functions on $\{0,1\}^{n-1}$ $(n \geq 2)$.
This result can be recovered by another way : the number mn is known (cf[5]) :

$$\mu_n = 2^{\binom{n}{2}+n+1} - 2 - 2\sum_{h=1}^{\left[\frac{n}{2}\right]} 2^{h(h+1)} \frac{(2^n - 1)\ldots(2^{n-2h+1} - 1)}{(2^{2h} - 1)(2^{2h-2} - 1)\ldots(2^2 - 1)}$$

(where [ ] denotes the integer part), and therefore equality $\mu_n = (2^n - 1)\mu'_{n-1}$ is equivalent with :

$$2^{\binom{n}{2}+n+1} - 2 - 2\sum_{h=1}^{[\frac{n}{2}]} 2^{h(h+1)} \frac{(2^n - 1)\ldots(2^{n-2h+1} - 1)}{(2^{2h} - 1)(2^{2h-2} - 1)\ldots(2^2 - 1)} =$$

$$(2^n - 1)\left(2 + 2\sum_{h=1}^{[\frac{n-1}{2}]} 2^{h(h+1)} \frac{(2^{n-1} - 1)\ldots(2^{n-2h} - 1)}{(2^{2h} - 1)\ldots(2^2 - 1)}\right).$$

That last equality is checked in [2].

2) Proposition 3 would give us a chance to evaluate the number of partially-bent balanced functions if the number of partially-bent functions was known.

That is not the case, but we have :

Proposition 3.4 *The number of balanced quadratic functions on G is greater than that of the quadratic non-balanced functions when n is odd and smaller when n is even.*

Proof: Let μ_n (respectively μ'_n) be the number of quadratic balanced (respectively non-balanced) functions on G.

Since the number of quadratic functions is $2^{\binom{n+1}{2}+1}$ (cf [5]),we have :

$$\forall n \geq 2, \mu'_n = 2^{\binom{n+1}{2}+1} - (2^n - 1)\mu'_{n-1}.$$

Let us prove by induction on n that :

$\mu_n < \mu'_n$, that is $\mu'_n > 2^{\binom{n+1}{2}}$ if n is even, $n \geq 2$

$\mu_n > \mu'_n$, that is $\mu'_n < 2^{\binom{n+1}{2}}$ if n is odd, $n \geq 3$.

That is true for $n = 2, 3$ since $\mu_2 = 6$, $\mu'_n = 10$, $\mu_3 = 70$, $\mu'_3 = 58$.

Suppose it is true for odd $n > 2$, then :

$$\mu'_n < 2^{\binom{n+1}{2}} \Rightarrow \mu'_{n+1} > 2^{\binom{n+2}{2}+1} - (2^{n+1} - 1)2^{\binom{n+1}{2}}$$

$$= 2^{\binom{n+2}{2}} + 2^{\binom{n+1}{2}}$$

$$\Rightarrow \mu'_{n+2} < 2^{\binom{n+3}{2}+1} - (2^{n+2} - 1)\left(2^{\binom{n+2}{2}} + 2^{\binom{n+1}{2}}\right)$$

$$= 2^{\binom{n+3}{2}} - 2^{\binom{n+2}{2}} + 2^{\binom{n+1}{2}}$$

$$< 2^{\binom{n+3}{2}}.$$

And the proof is complete. ◻

Proposition 3.5 *A partially-bent function defined by : $\forall x \in E, \forall y \in E', f(x + y) = f(y) + t.x$ is kth-order correlation-immune (respectively kth-order correlation-immune and balanced) if and only if $t + E^\perp$ only contains elements of Hamming weight greater than k or equal to 0 (respectively greater than k).*

Proof: We have :

$$\hat{F}(s) = \sum_{x \in E, y \in E'} (-1)^{x \cdot t + x \cdot s + f(y) + y \cdot s} = \sum_{x \in E} (-1)^{x \cdot (t+s)} \sum_{y \in E'} (-1)^{f(y) + y \cdot s}.$$

Since f is bent on E', the sum $\displaystyle\sum_{y \in E'} (-1)^{f(y) + y \cdot s}$ is different from 0.

Therefore : $\hat{F}(s) \neq 0 \Leftrightarrow s + t \in E^\perp$. $\qquad\qquad$ □

REMARK :

If f is non-balanced, then we may take $t = 0$ and the condition becomes :

$E^\perp \backslash \{0\}$ only contains elements of Hamming weight greater than k.

According to the singleton bound (cf [5]), we then have : $\dim E^\perp \leq n - k$ and since the degree of the restriction of f to a subspace E' of G is bounded by $\dim E'/2$ (cf [4]), the degree of f is bounded by $(n - k)/2$.

So, there does not exist any function which would be partially-bent and kth-order correlation immune of maximal degree : the maximal degree of the kth-order correlation immune functions is $n - k$ (cf [9]).

On the contrary, there does exist partially-bent balanced k-th order correlation-immune functions of maximal degree (that degree is $n - k - 1$) : see [1] or [2] for the case $k = n - 3$.

4 Conclusion

The main interest of the class of quadratic functions is in its nice properties : we know the weights of the functions and we can characterize the functions which satisfy $PC(k)$, those which are balanced, kth-order correlation-immune. But the quadratic functions are of a poor interest from a cryptographic point of view since they are too simple.

The class of partially-bent functions shares the same qualities since all the properties of the quadratic functions can be generalized to the partially-bent functions (with three exceptions : it is not a linear space, we are not able to calculate its size or to give the general algebraic normal form of these functions).

The interest of this class of functions is greater from a cryptographic point of view because the partially-bent functions involve bent functions whose complexity may be great (clearly, a partially-bent function will have a high level of nonlinearity if its associated kernel is small).

Acknowledgement

The author is grateful to Kaisa Nyberg who made an interesting remark on what the theorem proves (see the remark which follows it).

References

[1] P. CAMION, C. CARLET, P. CHARPIN AND N. SENDRIER *On correlation-immune functions*, Crypto'91, Advances in Cryptology, Lecture Notes on Computer Science, Springer Verlag n° 576.

[2] C. CARLET *Codes de Reed-Muller, codes de Kerdock et de Preparata, thèse*, publication du LITP, Institut Blaise Pascal, Université Paris 6, n° 90.59 (1990).

[3] C. CARLET *A transformation on boolean functions, its consequences on some problems related to Reed-Muller codes*, Eurocode 90, Lecture notes in computer science 514 (1991).

[4] J. F. DILLON *Elementary Hadamard Difference sets*, Ph. D. Thesis, Univ. of Maryland (1974).

[5] F. J. MAC WILLIAMS & N. J. A. SLOANE *The theory of error correcting codes*, North Holland 1977.

[6] B. PRENEEL, R. GOVAERTS AND J. VANDEWALLE *Boolean Functions Satisfying Higher Order Propagation Critereia* Eurocrypt'91, Lecture Notes in Computer Science 547 p 141-152 also presented at Las Vegas Int. Conf. on Finite Fields and Adv. in Com. and Comp. 1991.

[7] O. S. ROTHAUS *On bent functions*, J. Comb. Theory, 20A (1976) 300- 305.

[8] T. SIEGENTHALER *Correlation-immunity of nonlinear combining functions for cryptographic applications*, IEEE on Inf. Theory, vol IT-30, n° 5, Sept. 84.

[9] G.-Z. XIAO & J. L. MASSEY *A spectral characterization of correlation-immune combining functions*, IEEE, Vol 34, n° 3, May 88.

Practical Approaches to Attaining Security against Adaptively Chosen Ciphertext Attacks (Extended Abstract)

Yuliang Zheng and Jennifer Seberry *

The Centre for Computer Security Research
Department of Computer Science
University of Wollongong
Locked Bag 8844, Wollongong, NSW 2521
AUSTRALIA
E-mail: {yuliang,jennie}@cs.uow.edu.au

Abstract. This paper presents three methods for strengthening public key cryptosystems in such a way that they become secure against *adaptively* chosen ciphertext attacks. In an adaptively chosen ciphertext attack, an attacker can query the deciphering algorithm with any ciphertexts, *except* for the exact object ciphertext to be cryptanalyzed. The first strengthening method is based on the use of one-way hash functions, the second on the use of universal hash functions and the third on the use of digital signature schemes. Each method is illustrated by an example of a public key cryptosystem based on the intractability of computing discrete logarithms in finite fields. Two other issues, namely applications of the methods to public key cryptosystems based on other intractable problems and enhancement of information authentication capability to the cryptosystems, are also discussed.

1 Introduction

A considerable amount of research has been done in recent years, both from the theoretical [BFM88, NY90, DDN91, RS92] and practical [Dam92] points of view, in the pursuit of the construction of public key cryptosystems secure against chosen ciphertext attacks. In such an attack, the attacker (cryptanalyst) has access to the deciphering algorithm of a cryptosystem. The attacker can query the deciphering algorithm with any ciphertexts, obtain the matching plaintexts and use the attained knowledge in the cryptanalysis of an object ciphertext.

The theoretical results are appealing in that the schemes which embody them are provably secure under certain assumptions. However, most of these schemes are impractical due to the large expansion of the resulting ciphertext. The recent and notable schemes by Damgård overcome the problem of impracticality, but they are totally insecure against *adaptively* chosen ciphertext attacks in which

* Supported in part by the Australian Research Council under the reference numbers A49030136, A49130102 and A49131885.

an attacker has access to the deciphering algorithm even after he or she is given an object ciphertext to be cryptanalyzed. The attacker is allowed to query the deciphering algorithm with any ciphertext, *except* for the exact object ciphertext.

Adaptively chosen ciphertext attacks would impose serious problems on many services provided by modern information technology. To illustrate the possible attacks, consider the case of a security-enhanced electronic mail system where a public key cryptosystem is used to encipher messages passed among users. Nowadays it is common practice for an electronic mail user to include the original message he or she received into a reply to the message. For instance, a reply to a message may be as follows

```
(original message)
> ......
> Hi, is Yum-Cha still on tonight ?
> ......

(reply to the message)
......
Yes, it's still on.  I've already made the bookings.
......
```

this practice provides an avenue for chosen ciphertext attacks, as an attacker can send a ciphertext to a target user and expect the user to send back the corresponding plaintext as part of the reply. Now suppose that a user Alice is in the process of negotiating, through the electronic mail system, with two other users Bob and Cathy who are rivals of each other in a business. Let c be a ciphertext from Bob to Alice. Naturally, Cathy would like to know the contents of the communications between Alice and Bob. Cathy can obtain the ciphertext c by eavesdropping. However, it would be infeasible for her to extract its contents immediately. Instead, Cathy might try to discover *implicitly* the contents of c through discussions with Alice using the electronic mail. The problem facing Cathy is that she can not simply pass c to Alice with the hope that Alice would include the contents of c into her reply, as Alice would detect that c is actually a ciphertext created by Bob but not by Cathy. Nevertheless, if the cryptosystem is insecure against adaptively chosen ciphertext attacks, Cathy might still be able to obtain *indirectly* what she wants in the following way

1. Send Alice ciphertexts $c_1, c_2, \ldots, c_n$, none of which is the same as the object ciphertext c.
2. Receive the matching plaintext messages (hopefully) and
3. Extract the contents of c by the use of information obtained from the n plaintext-ciphertext pairs.

In this paper we present three pragmatic methods for immunizing public key cryptosystems against adaptively chosen ciphertext attacks. The first method is based on the use of one-way hash functions, the second on the use of universal hash functions and the third on the use of digital signature schemes. Each method is illustrated by an example of a public key cryptosystem based on the intractability of computing discrete logarithms in finite fields. Security of the three cryptosystems against adaptively chosen ciphertext attacks is formally proved under reasonable assumptions.

In Section 2, we introduce notion and notations that are needed, and summarize various types of possible attack to cryptosystems. In Section 3 previous proposals together with their problems are reviewed. Our immunization methods are illustrated in Section 4, by three public key cryptosystems based on the intractability of computing discrete logarithms in finite fields. Section 5 is concerned with two other issues, namely applications of the immunization methods to public key cryptosystems based on other intractable problems, such as the problem of factoring large composite numbers, and the addition of information authentication capability to the three cryptosystems. Finally Section 6 presents some concluding remarks.

The reader is directed to [ZS93] where the three cryptosystems are formally proved to be secure against adaptively chosen ciphertext attacks.

2 Notion and Notations

We will be concerned with the alphabet $\Sigma = \{0, 1\}$. The length of a string x over Σ is denoted by $|x|$, and the concatenation of two strings x and y is denoted by $x||y$. The bit-wise exclusive-or of two strings x and y of the same length is denoted by $x \oplus y$. The i-th bit of x is denoted by x_i and the substring of x from x_i to x_j, where $i \leq j$, is denoted by $x_{[i \cdots j]}$. $\#S$ indicates the number of elements in a set S, and $x \in_R S$ means choosing randomly and uniformly an element x from the set S. The Cartesian product of two sets S and T is denoted by $S \times T$.

Denote by $\mathbb{N}$ the set of all positive integers, and by n a security parameter which determines the length of messages, the length of ciphertexts, the security of cryptosystems etc. As in the Diffie-Hellman/ElGamal's public key scheme [DH76, ElG85], p is an n-bit prime and g is a generator for the multiplicative group $GF(p)^*$ of the finite field $GF(p)$. Both p and g are public. To guarantee the security of cryptosystems based on the discrete logarithm problem, the length n of p should be large enough, preferably $n > 512$, and $p - 1$ should contain a large prime factor [PH78, LO91]. Unless otherwise specified, all exponentiation operations appearing in the remaining part of this paper are assumed to be over the underlying groups.

Note that there is a natural one-to-one correspondence between strings in Σ^n and elements in the finite field $GF(2^n)$. Similarly, there is a natural one-to-one correspondence between strings in Σ^n and integers in $[0, 2^n - 1]$. Therefore, we will not distinguish among strings in Σ^n, elements in $GF(2^n)$ and integers in $[0, 2^n - 1]$.

A *public key cryptosystem*, invented by Diffie and Hellman [DH76], consists of three polynomial time algorithms (C, E, D). C is called a *key-generation algorithm* which, on input n, generates probabilistically a pair (pk, sk) of public and secret keys. Following the tradition in the field, when a security parameter n is used as input to an algorithm, it will be represented by the all-1 string of n bits which is denoted by 1^n. E is called an *enciphering algorithm* which, on input a public key pk and a plaintext message m, outputs a ciphertext c. Here m is chosen from a message space M_n. D is called a *deciphering algorithm* which, on input a secret key sk and a ciphertext c, outputs a message m or a special symbol $\emptyset$ meaning "no plaintext output". E and D satisfy the following unique decipherability condition, namely $D(sk, E(pk, m)) = m$.

There are four common types of attack to a cryptosystem, namely *ciphertext only attacks*, *known plaintext attacks*, *chosen plaintext attacks* and *chosen ciphertext attacks* [Riv90]. Related attacks against digital signatures are fully discussed in [GMR88].

In a ciphertext only attack, which is the least severe among the four types of attack, an attacker is given an object ciphertext and tries to find the plaintext which is hidden in the object ciphertext.

In a known plaintext attack, an attacker has a collection of plaintext-ciphertext pairs besides an object ciphertext. The attacker may use the knowledge gained from the pairs of plaintexts and ciphertexts in the cryptanalysis of the object ciphertext.

In a chosen plaintext attack, an attacker has access to the enciphering algorithm. During the cryptanalysis of an object ciphertext, the attacker can choose whatever plaintexts he or she desires, feed the enciphering algorithm with the desired plaintexts and obtain the corresponding ciphertexts. Note that this type of attack is always applicable to a public key cryptosystem, since the attacker always has access to the public enciphering algorithm.

In a chosen ciphertext attack, which is the most severe among the four types of attack, an attacker has access to the deciphering algorithm. The attacker can query the deciphering algorithm with any ciphertexts and obtain the corresponding plaintexts. Then the attacker can use the knowledge obtained in the query and answer process to extract the plaintext of an object ciphertext.

Researchers further distinguish two forms of chosen ciphertext attack: *indifferently chosen ciphertext attacks* and *adaptively chosen ciphertext attacks*. An indifferently chosen ciphertext attack is also called a *lunchtime attack* or a *midnight attack* [NY90]. In such an attack the ciphertexts fed into the deciphering algorithm are chosen without being related to the object ciphertext. However the ciphertexts fed into the deciphering algorithm may be correlated with one another. This form of attack models the situation where the attacker has access to the deciphering algorithm *before* he or she is actually given the object ciphertext.

In adaptively chosen ciphertext attacks all ciphertexts fed into the deciphering algorithm can be correlated to the object ciphertext. This form of attack is more severe than the indifferently chosen ciphertext attacks and it models the situation where the attacker has access to the deciphering algorithm even *after* he or she is given the object ciphertext. The attacker is thus permitted to give the deciphering algorithm any available ciphertexts, *except for* the exact object ciphertext, and obtain the matching plaintexts. See the Introduction for a practical application where adaptively chosen ciphertext attacks would be a considerable threat.

3 Problems with Previous Proposals

Rabin pioneered the research of constructing provably secure public key cryptosystems by designing a public key cryptosystem with the property that extracting the complete plaintext of an object ciphertext is computationally equivalent to factoring large numbers [Rab79]. Goldwasser and Micali invented the first public key cryptosystem that hides all partial information [GM84]. The cryptosystem is a probabilistic one and it enciphers a plaintext in a bit-by-bit manner. A common drawback of these and many other cryptosystems is that, although secure against chosen plaintext attacks, they are easily compromised by chosen ciphertext attackers. On the other hand, much progress has been made in recent years in the construction of public key cryptosystems secure against chosen ciphertext attacks. We will review this development, and point out problems and weakness of the proposed schemes.

3.1 Theoretical Results

Theoretical study into the construction of public key cryptosystems secure against chosen ciphertext attacks was initiated by Blum, Feldman and Micali [BFM88], who suggested the potential applicability of non-interactive zero-knowledge proofs to the subject. Naor and Yung carried further the study and gave the first concrete public key cryptosystem that is (semantically) secure against indifferently chosen ciphertext attacks [NY90]. Rackoff and Simon considered a more severe type of attack, namely adaptively chosen ciphertext attacks, and gave a concrete construction for public key cryptosystems withstanding the attacks [RS92]. In [DDN91] Dolev, Dwork and Naor proposed a non-malleable (against chosen plaintext attacks) public key cryptosystem and proved that the cryptosystem is also secure against adaptively chosen ciphertext attacks.

All of these cryptosystems are provably secure under certain assumptions. However since they rely heavily on non-interactive zero-knowledge proofs, the resulting ciphertexts are in general much longer than original plaintexts. This disadvantage makes the cryptosystems highly impractical and difficult to realize in practice.

3.2 Damgård's Schemes

In [Dam92], Damgård took a pragmatic approach to the subject. He proposed two simple public key cryptosystems that appear to be secure against indifferently chosen ciphertext attacks. The first is based on deterministic public key cryptosystems. Let (E_0, D_0) be the pair of enciphering and deciphering algorithms of a deterministic public key cryptosystem. Let (pk_1, sk_1) and (pk_2, sk_2) be two pairs of public and secret keys and h be an invertible one-to-one length-preserving function. The enciphering algorithm of Damgård's first cryptosystem operates in the following way:

$$E(pk_1, pk_2, m) = (E_0(pk_1, r), \; E_0(pk_2, h(r)) \oplus m) = (c_1, c_2)$$

where $m \in \Sigma^n$ is a plaintext message and $r \in_R \Sigma^n$ is a random string. The corresponding deciphering algorithm is as follows:

$$D(sk_1, pk_2, c_1, c_2) = E_0(pk_2, h(D_0(sk_1, c_1))) \oplus c_2$$

Damgård's second scheme is based on the Diffie-Hellman/ElGamal public key cryptosystem [DH76, ElG85], whose security relies on the intractability of computing discrete logarithms in finite fields. A user Alice's secret key is a pair (x_{A1}, x_{A2}) of elements chosen independently at random from $[1, p-1]$. Her public key is (y_{A1}, y_{A2}), where $y_{A1} = g^{x_{A1}}$ and $y_{A2} = g^{x_{A2}}$. When a user Bob wants to send an n-bit message m in secret to Alice, he sends her the following enciphered message

$$E(y_{A1}, y_{A2}, p, g, m) = (g^r, \ y_{A1}^r, \ y_{A2}^r \oplus m) = (c_1, c_2, c_3)$$

where $r \in_R [1, p-1]$. Note that here n is the length of the prime p. The deciphering algorithm for Alice, who possesses the secret key (x_{A1}, x_{A2}), is as follows

$$D(x_{A1}, x_{A2}, p, g, c_1, c_2, c_3) = \begin{cases} c_1^{x_{A2}} \oplus c_3 \text{ if } c_1^{x_{A1}} = c_2 \\ \\ \emptyset \qquad \qquad \text{otherwise} \end{cases}$$

Here $\emptyset$ is a special symbol meaning "no plaintext output".

Although Damgård's schemes are very simple and seem to be secure against indifferently chosen ciphertext attacks, they are *insecure* against adaptively chosen ciphertext attacks. Given an object ciphertext c ($c = (c_1, c_2)$ for the first scheme, and $c = (c_1, c_2, c_3)$ for the second scheme), an attacker can choose a random message m_r from Σ^n, calculate the bit-wise exclusive-or of m_r and the last part of the ciphertext c, and feed the deciphering algorithm with the modified ciphertext c'. The attacker will get $m' = m \oplus m_r$ as an answer, and obtain the desired message [2] m by computing $m' \oplus m_r$. Our cryptosystems to be described below share the same simplicity possessed by Damgård's cryptosystems, yet they attain a higher level of security, namely security against adaptively chosen ciphertext attacks.

4 Strengthening Public Key Cryptosystems

This section presents three simple methods for immunizing public key cryptosystems against chosen ciphertext attacks. The nature of the three immunization

---

[2] One might argue that, since at least half bits in the original ciphertext c remain untouched in the modified ciphertext c', adding a checking step to the deciphering algorithms would effectively thwart the attack. This countermeasure, however, does *not* work in general, as the deciphering algorithms may *not* know c. Even if the deciphering algorithms have a list of ciphertexts containing c, a more sophisticated attacker might still succeed in extracting m by generating c' in such a way that it passes the checking step.

methods is the same — they all immunize a public key cryptosystem by appending to each ciphertext a tag that is correlated to the message to be enciphered. This is also the main technical difference between our proposals and Damgård's schemes. The three methods differ in the ways in which tags are generated. In the first method tags are generated by the use of a one-way hash function, in the second method by the use of a function chosen from a universal class of hash functions, and in the third method by the use of a digital signature scheme. The second immunization method is superior to the other two immunization methods in that no one-way hash functions are needed. This property is particularly attractive given the current state of research, whereby many one-way hash functions exist, few are efficient, and even fewer are provably secure.

We will illustrate our immunization methods with cryptosystems based on the Diffie-Hellman/ElGamal public key scheme. In Section 5, applications of the immunization methods to cryptosystems based on other intractable problems will be discussed. Denote by G the cryptographically strong pseudo-random string generator based on the difficulty of computing discrete logarithms in finite fields [BM84, LW88, Per85]. G stretches an n-bit input string into an output string whose length can be an arbitrary polynomial in n. This generator produces $O(\log n)$ bits output at each exponentiation. In the authors' opinion, for practical applications the generator could produce more than $\frac{3n}{4}$ bits at each exponentiation, without sacrificing security. Recently Micali and Schnorr discovered a very efficient pseudo-random string generator based on polynomials in the finite field $GF(p)$ (see Section 4 of [MS91]). The generator can produce, for example, $\frac{n}{2}$ bits with 1.25 multiplications in $GF(p)$. The efficiency of our cryptosystems to be described below can be further improved if Micali and Schnorr's pseudo-random string generator is employed.

A user Alice's secret key is an element x_A chosen randomly from $[1, p-1]$, and her public key is $y_A = g^{x_A}$. It is assumed that all messages to be enciphered are chosen from the set Σ^P, where $P = P(n)$ is an arbitrary polynomial with $P(n) \geq n$. Padding can be applied to messages whose lengths are less than n bits. In addition, let $\ell = \ell(n)$ be a polynomial which specifies the length of tags. It is recommended that ℓ should be at least 64 for the sake of security.

4.1 Immunizing with One-Way Hash Functions

Assume that h is a one-way hash function compressing input strings into ℓ-bit output strings. A user Bob can use the following enciphering algorithm to send in secret a P-bit message m to Alice.

> **Algorithm 1** $E_{owh}(y_A, p, g, m)$
> 1. $x \in_R [1, p-1]$.
> 2. $z = G(y_A^x)_{[1 \cdots (P+\ell)]}$.
> 3. $t = h(m)$.
> 4. $c_1 = g^x$.
> 5. $c_2 = z \oplus (m \| t)$.
> 6. output (c_1, c_2).
>
> **end**

The deciphering algorithm for Alice, who possesses the secret key x_A, is as follows:

Algorithm 2 $D_{owh}(x_A, p, g, c_1, c_2)$
    1. $z' = G(c_1^{x_A})_{[1 \cdots (P+\ell)]}$.
    2. $w = z' \oplus c_2$.
    3. $m' = w_{[1 \cdots P]}$.
    4. $t' = w_{[(P+1) \cdots (P+\ell)]}$.
    5. if $h(m') = t'$ then
         output (m')
      else
         output $(\emptyset)$.
end

When messages are of n bits, i.e. $P = n$, instead of the one-way hash function h the exponentiation function can be used to generate the tag t. In this case, the enciphering algorithm can be modified as follows: (a) Change the step 2 to "$z = G(y_A^x)_{[1 \cdots 2n]}$." (b) Change the step 3 to "$t = g^m$." The deciphering algorithm can be modified accordingly.

4.2 Immunizing with Universal Hash Functions

A class H of functions from Σ^P to Σ^ℓ is called a *(strongly) universal class of hash functions* [CW79, WC81] mapping P-bit input into ℓ-bit output strings if for every $x_1 \neq x_2 \in \Sigma^P$ and every $y_1, y_2 \in \Sigma^\ell$, the number of functions in H taking x_1 to y_1 and x_2 to y_2 is $\#H/2^{2\ell}$. An equivalent definition is that when h is chosen uniformly at random from H, the concatenation of the two strings $h(x_1)$ and $h(x_2)$ is distributed randomly and uniformly over the Cartesian product $\Sigma^\ell \times \Sigma^\ell$. Wegman and Carter found a nice application of universal classes of hash functions to unconditionally secure authentication codes [WC81].

Now assume that H is a universal class of hash functions which map P-bit input into ℓ-bit output strings. Also assume that $Q = Q(n)$ is a polynomial and that each function in H is specified by a string of exactly Q bits. Denote by h_s the function in H that is specified by a string $s \in \Sigma^Q$. The enciphering algorithm for Bob who wants to send in secret a P-bit message m to Alice is the following:

Algorithm 3 $E_{uhf}(y_A, p, g, m)$
    1. $x \in_R [1, p-1]$.
    2. $r = y_A^x$.
    3. $z = G(r)_{[1 \cdots P]}$.
    4. $s = G(r)_{[(P+1) \cdots (P+Q)]}$.
    5. $c_1 = g^x$.
    6. $c_2 = h_s(m)$.
    7. $c_3 = z \oplus m$.
    8. output (c_1, c_2, c_3).
end

The deciphering algorithm for Alice, who possesses the secret key x_A, is as follows:

Algorithm 4 $D_{uhf}(x_A, p, g, c_1, c_2, c_3)$

    1. $r' = c_1^{x_A}$.
    2. $z' = G(r')_{[1 \cdots P]}$.
    3. $s' = G(r')_{[(P+1) \cdots (P+Q)]}$.
    4. $m' = z' \oplus c_3$.
    5. if $h_{s'}(m') = c_2$ then
        output (m')
    else
        output $(\emptyset)$.

end

Note that the second part $c_2 = h_s(m)$ in the ciphertext can be obscured in the same way as Algorithm 1. This would improve practical security of the cryptosystem, at the expense of more computation time spent in generating pseudo-random bits.

The following is a simple universal class of hash functions which is originated from linear congruential generators in finite fields. (See also Propositions 7 and 8 of [CW79].) Let k be an integer. For $k + 1$ elements $a_1, a_2, \ldots, a_k, b \in GF(2^\ell)$, let s be their concatenation, i.e., $s = a_1 || a_2 || \cdots || a_k || b$, and let h_s be the function defined by $h_s(x_1, x_2, \ldots, x_k) = \sum_{i=1}^{k} a_i x_i + b$ where $x_1, x_2, \ldots, x_k$ are variables in $GF(2^\ell)$. Then the collection H of the functions h_s defined by all $k+1$ elements from $GF(2^\ell)$ is a universal class of hash functions. Functions in H compress $k\ell$-bit input into ℓ-bit output strings. By padding to input strings, these functions can be applied to input strings whose lengths are not exactly $k\ell$. In particular, when $k = \lceil \frac{P}{\ell} \rceil$, they can be used to compress P-bit input into ℓ-bit output strings. In this case, a function in H can be specified by a string of $Q = P + (1+\alpha)\ell$ bits, where $0 \leq \alpha = \frac{P \bmod \ell}{\ell} < 1$. This universal class of hash functions is particularly suited to the case where the length P of messages to be enciphered is much larger than the length ℓ of tags. We refer the reader to [WC81, Sti90] for other universal classes of hash functions.

4.3 Immunizing with Digital Signature Schemes

Assume that h is a one-way hash function compressing input strings into n-bit output strings. Also assume that Bob wants to send in secret a P-bit message m to Alice. The enciphering algorithm employed by Bob is the following:

Algorithm 5 $E_{sig}(y_A, p, g, m)$

    1. $x \in_R [1, p-1]$.

    2. $k \in_R [1, p-1]$ such that $\gcd(k, p-1) = 1$.

    3. $r = y_A^{x+k}$.

    4. $z = G(r)_{[1 \cdots P]}$.

    5. $c_1 = g^x$.

    6. $c_2 = g^k$.

    7. $c_3 = (h(m) - xr)/k \bmod (p-1)$.

    8. $c_4 = z \oplus m$.

    9. output (c_1, c_2, c_3, c_4).

end

The corresponding deciphering algorithm for Alice, who possesses the secret key x_A, is as follows:

Algorithm 6 $D_{sig}(x_A, p, g, c_1, c_2, c_3, c_4)$

    1. $r' = (c_1 c_2)^{x_A}$.

    2. $z' = G(r')_{[1 \cdots P]}$.

    3. $m' = z' \oplus c_4$.

    4. if $g^{h(m')} = c_1^{r'} c_2^{c_3}$ then

        output (m')

    else

        output $(\emptyset)$.

end

Similar to the cryptosystem based on the use of universal hash functions described in Section 4.2, security of the cryptosystem can also be improved by hiding the third part $c_3 = (h(m) - xr)/k \bmod (p-1)$ with extra pseudo-random bits produced by the pseudo-random string generator G. In addition, when messages to be enciphered are of n bits, neither the one-way hash function h nor the pseudo-random string generator G is necessary. The enciphering algorithm for this case can be simplified by changing the step 4 of the above enciphering algorithm to "$z = r$." and the step 7 into "$c_3 = (m - xr)/k \bmod (p-1)$." The deciphering algorithm can be simplified accordingly.

The first three parts (c_1, c_2, c_3) of the ciphertext represents an adaptation of the ElGamal's digital signature. However, since everyone can generate these parts, they do *not* really form the digital signature of m. This immunization method was first proposed in [ZHS91], where other ways for generating the third part c_3 in the ciphertext were also suggested.

In [ZS93] it is proved that, under reasonable assumptions, all the three cryptosystems are secure against adaptively chosen ciphertext attacks. We introduce in the paper an interesting notion called *sole-samplability*, and apply the notion in the proofs of security.

5 Extensions of the Cryptosystems

We have focused our attention on cryptosystems based on the discrete logarithm problem in finite fields. The cryptosystems can also be based on discrete logarithms over other kinds of finite abelian groups, such as those on elliptic or hyper-elliptic curves defined over finite fields [Kob87, Kob89]. Another variant of the cryptosystems is to have a different large prime for each user. This variant can greatly improve practical security of the cryptosystems when a large number of users are involved.

Our first two methods for immunization, namely immunization with one-way hash functions and immunization with universal hash functions, can be applied to public key cryptosystems based on other intractable problems. For example, the methods can be used to immunize the probabilistic public key cryptosystem proposed in [BG85], which is based on the intractability of factoring large composite numbers. The methods might be extended further in such a way that allows us to construct from *any* trap-door one-way function a public key cryptosystem secure against adaptively chosen ciphertext attacks.

Authentication is another important aspect of information security. In many situations, the receiver of a message needs to be assured that the received message is truly originated from its sender and that it has not been tampered with during its transmission. Researchers have proposed many, unconditionally or computationally, secure methods for information authentication [Sim88]. We take the second cryptosystem which uses universal has functions as an example to show that our cryptosystems can be easily added with information authentication capability.

To do so, it is required that the sender Bob also has a pair (y_B, x_B) of public and secret keys. Information authentication is achieved by letting Bob's secret key x_B be involved in the creation of a ciphertext. More specifically, we change the step 2 of the enciphering Algorithm 3 to "$r = y_A^{x_B + x}$." and the step 1 of the corresponding deciphering Algorithm 4 to "$r' = (y_B c_1)^{x_A}$." Although ciphertexts from Alice to Bob are indistinguishable from those from Bob to Alice, it is infeasible for a user differing from Alice and Bob to create a "legal" ciphertext from Alice to Bob or from Bob to Alice. This property ensures information authentication capability of the cryptosystem. It is not hard to see that computing $g^{x_1(x_2 + x_3)}$ from g^{x_1}, g^{x_2} and g^{x_3}, and computing $g^{x_1 x_2}$ from g^{x_1} and g^{x_2}, are equally difficult. Therefore the authentication-enhanced cryptosystem is as secure as the original one.

The first cryptosystem which is based on the use of a one-way hash function can be enhanced with information authentication capability in a similar way. For the third cryptosystem, the capability can be added by simply replacing x, a random string chosen from $[1, p-1]$, with Bob's secret key x_B.

6 Conclusions

We have presented three methods for immunizing public key cryptosystems against chosen ciphertext attacks, among which the second immunization method based on the use of universal hash functions is particularly attractive in that no one-way hash functions are needed. Each immunization method is illustrated by an example of a public key cryptosystem based on the intractability of computing discrete logarithms in finite fields. The generality of our immunization methods is shown by their applicability to public key cryptosystems based on other intractable problems, such as that of factoring large composite numbers. An enhancement of information authentication capability to the example cryptosystems has also been suggested.

Acknowledgments

We would like to thank Thomas Hardjono for fruitful discussions, and to anonymous referees for helpful comments.

References

[BFM88] M. Blum, P. Feldman, and S. Micali. Non-interactive zero-knowledge proof systems and applications. In *Proceedings of the 20-th Annual ACM Symposium on Theory of Computing*, pages 103–112, 1988.

[BG85] M. Blum and S. Goldwasser. An efficient probabilistic public key encryption scheme which hides all partial information. In G. R. Blakeley and D. Chaum, editors, *Advances in Cryptology - Proceedings of Crypto'84*, Lecture Notes in Computer Science, Vol. 196, pages 289–299. Springer-Verlag, 1985.

[BM84] M. Blum and S. Micali. How to generate cryptographically strong sequences of pseudo-random bits. *SIAM Journal on Computing*, 13(4):850–864, 1984.

[CW79] J. Carter and M. Wegman. Universal classes of hash functions. *Journal of Computer and System Sciences*, 18:143–154, 1979.

[Dam92] I. Damgård. Towards practical public key systems secure against chosen ciphertext attacks. In J. Feigenbaum, editor, *Advances in Cryptology - Proceedings of Crypto'91*, Lecture Notes in Computer Science, Vol.576, pages 445–456. Springer-Verlag, 1992.

[DDN91] D. Dolev, C. Dwork, and M. Naor. Non-malleable cryptography. In *Proceedings of the 23-rd Annual ACM Symposium on Theory of Computing*, 1991.

[DH76] W. Diffie and M. Hellman. New directions in cryptography. *IEEE Transactions on Information Theory*, IT-22(6):472–492, 1976.

[ElG85] T. ElGamal. A public key cryptosystem and a signature scheme based on discrete logarithms. *IEEE Transactions on Information Theory*, IT-31(4):469–472, 1985.

[GM84] S. Goldwasser and S. Micali. Probabilistic encryption. *Journal of Computer and System Sciences*, 28(2):270–299, 1984.

[GMR88] S. Goldwasser, S. Micali, and R. Rivest. A digital signature scheme secure against adaptively chosen message attacks. *SIAM Journal on Computing*, 17(2):281–308, 1988.

[Kob87] N. Koblitz. Elliptic curve cryptosystems. *Mathematics of Computation*, 48:203–209, 1987.

[Kob89] N. Koblitz. Hyperelliptic cryptosystems. *Journal of Cryptology*, 1(3):139–150, 1989.

[LO91] B. A. LaMacchia and A. M. Odlyzko. Computation of discrete logarithms in prime fields. *Designs, Codes and Cryptography*, 1:47–62, 1991.

[LW88] D. L. Long and A. Wigderson. The discrete logarithm hides $O(\log n)$ bits. *SIAM Journal on Computing*, 17(2):363–372, 1988.

[MS91]   S. Micali and C. P. Schnorr. Efficient, perfect polynomial random number generators. *Journal of Cryptology*, 3(3):157–172, 1991.

[NY90]   M. Naor and M. Yung. Public-key cryptosystems provably secure against chosen ciphertext attacks. In *Proceedings of the 22-nd Annual ACM Symposium on Theory of Computing*, pages 427–437, 1990.

[Per85]   R. Peralta. Simultaneous security of bits in the discrete log. In Franz Pichler, editor, *Advances in Cryptology - Proceedings of EuroCrypt'85*, Lecture Notes in Computer Science, Vol. 219, pages 62–72. Springer-Verlag, 1985.

[PH78]   S. C. Pohlig and M. E. Hellman. An improved algorithm for computing logarithms over $GF(p)$ and its cryptographic significance. *IEEE Transactions on Information Theory*, IT-24(1):106–110, 1978.

[Rab79]   M. Rabin. Digitalized signatures as intractable as factorization. Technical Report MIT/LCS/TR-212, MIT, Laboratory for Computer Science, 1979.

[Riv90]   R. Rivest. Cryptography. In J. van Leeuwen, editor, *Handbook of Theoretical Computer Science, Volume A, Algorithms and Complexity*, chapter 13, pages 717–755. The MIT Press, Cambridge, Massachusetts, 1990.

[RS92]   C. Rackoff and D. Simon. Non-interactive zero-knowledge proof of knowledge and chosen-ciphertext attacks. In J. Feigenbaum, editor, *Advances in Cryptology - Proceedings of Crypto'91*, Lecture Notes in Computer Science, Vol.576, pages 433–444. Springer-Verlag, 1992.

[Sim88]   G. J. Simmons. A survey of information authentication. *Proceedings of IEEE*, 76:603–620, 1988.

[Sti90]   D. R. Stinson. Combinatorial techniques for universal hashing. Report Series #127, Department of Computer Science, University of Nebraska, Lincoln, November 1990. (Also submitted to *Journal of Computer and System Sciences*).

[WC81]   M. Wegman and J. Carter. New hash functions and their use in authentication and set equality. *Journal of Computer and System Sciences*, 22:265–279, 1981.

[ZHS91]   Y. Zheng, T. Hardjono, and J. Seberry. A practical non-malleable public key cryptosystem. Technical Report CS91/28, Department of Computer Science, University College, University of New South Wales, 1991.

[ZS93]   Y. Zheng and J. Seberry. Immunizing public key cryptosystems against chosen ciphertext attacks. *Special Issue on Secure Communications, IEEE Journal on Selected Areas on Communications*, 1993. (to appear).

On the Security of the Permuted Kernel Identification Scheme

Thierry BARITAUD Mireille CAMPANA Pascal CHAUVAUD Henri GILBERT

Centre National d'Etudes des Télécommunications

PAA/TSA/SRC

38-40 Rue du Général Leclerc

92131 Issy les Moulineaux

FRANCE

Abstract

A zero-knowledge identification scheme built upon the so-called Permuted
Kernel Problem (PKP) was proposed by Adi Shamir in 1989 [1].
In this paper, we present a time-memory trade-off leading to a reduction of
the computation time for solving the PKP problem, as compared with the
best known attack mentioned in [1].

1. Introduction

In 1989, Adi Shamir proposed a new identification scheme, based on the intractability of the Permuted Kernel Problem (PKP)[1]. This scheme requires quite limited time, memory and communication resources, and is well suited for smart card implementation.

This paper investigates the security of the new scheme for some of the parameter values suggested in [1] as possible choices, subject to a further analysis of their security. Our main conclusion is that the smallest parameter values mentioned in [1] (n=32; m=16; p=251) are not recommended, at least for applications with strong security requirements. As a matter of fact, we show that for these values there exists a very simple time-memory trade-off leading to a faster solution of the PKP problem than the best known attack mentioned in [1]. For larger parameter values, this time-memory trade-off does not endanger the practical security of the PKP scheme, while the time, space and communication complexities of the PKP scheme stay within acceptable limits.

2. The Permuted Kernel Problem

We are using, as far as possible, the notations of [1].

The PKP problem is the following :

Given :

> a prime number p;
> a $m \times n$ matrix $A = (a_{ij})_{i=1..m;j=1..n}$ over Z/pZ;
> a n-vector $V = (V_j)_{j=1..n}$ over Z/pZ

Find : a permutation π over $\{1,..,n\}$ such that $A.V_\pi = 0$, where $V_\pi = (V_{\pi(j)})_{j=1..n}$.

In the PKP identification scheme, each prover uses a public instance (p,A,V) of the above problem. The values p, A, V are generated as follows :

- p and A are fixed values agreed by the users (and can be used by several provers). We will here assume that A is of rank m and is generated under the form [A',I], where A' is a fixed $m \times (n-m)$ matrix and I is the $m \times m$ identity matrix. (As mentioned in [1], this is not restrictive because both the prover and the verifier can apply Gaussian elimination to any $m \times n$ matrix without changing the kernel).

- V (the public key of a prover) is generated from a random permutation π and a random vector of the kernel of A (which serve as his secret key) in such a way that $A.V_\pi = 0$.

To convince a verifier of his identity, a prover gives him evidence of his knowledge of a solution π to the (p,A,V) instance by using a zero-knowledge protocol. The detail of this protocol, which is the main subject of [1], is outside the scope of this paper, since we are merely interested in the computational difficulty for an attacker of solving the (p,A,V) instance.

3. A time-memory trade-off for the PKP problem

Let (p, A, V) be an instance of the PKP problem generated as explained in Section 2. We are trying to find a permutation π such that $A.V_\pi = 0$.

Using the notations introduced in Section 2, we can rewrite $A.V_\pi = 0$ as :

$$\begin{pmatrix} a'_{1,1} \cdots a'_{1,n-m} & 1 & \\ \vdots & & \vdots \\ a'_{m,1} \cdots a'_{m,n-m} & & 1 \end{pmatrix} \begin{pmatrix} V\pi(1) \\ \cdot \\ \cdot \\ \cdot \\ \cdot \\ V\pi(n) \end{pmatrix} = \begin{pmatrix} 0 \\ \vdots \\ 0 \end{pmatrix}$$

We denote by (1) to (m) the relations provided by the rows 1 to m of the above matrix.

Note : in the sequel we are sometimes using the notation $A=(a_{ij})\ _{i=1..m;j=1..n}$ instead of the above 'reduced' notation $A=[A',I]; A'=(a'_{ij})\ _{i=1..m;j=1..n-m}$ to denote the elements of the A matrix.

We first try to solve the equations (1) to (k), where k is a parameter of our algorithm ($0 \le k \le m$). Because of the structure of the A matrix, these k relations involve only the n-m+k unknown values $V_{\pi(1)}, ..., V_{\pi(n-m+k)}$.

We introduce an additional parameter k' (such that $0 \le k' \le n-m+k$), which determines the amount of storage to be performed in the precomputation phase.

There are two main steps in the proposed searching method :

Step1 : precomputation

For each of the $\frac{n!}{(n-k')!}$ possible values for the $(V_{\pi(1)}, ..., V_{\pi(k')})$ k'-uple, we calculate the corresponding contributions :

$$b_1 = \sum_{j=1}^{k'} a_{1,j}.V\pi(j)\ ;$$

$$\cdot\ \cdot$$

$$b_k = \sum_{j=1}^{k'} a_{k,j}.V\pi(j)\ .$$

to the relations (1) to (k).

We store these values and the obtained results $b_1, ..., b_k$ in such a way that for each of the p^k possible $(b_1, ..., b_k)$ values, the list of the corresponding $(V_{\pi(1)}, ..., V_{\pi(k')})$ k'-uples can be accessed in very few elementary operations.

The cost of this precomputation step is $\frac{n!}{(n-k')!}$ matrix-vector products. The storage required is about $\frac{n!}{(n-k')!}$ k'-uples. The average number of k'-uples corresponding to a $(b_1, ...,b_k)$ value is $\frac{n!}{(n-k')!}$ p^{-k}.

We also introduce the convention that k'=0 means : no precomputation.

Step 2 : exhaustive trial

We perform an exhaustive trial of the $\frac{n!}{(m+k'-k)!}$ possible values for the $(V_{\pi(k'+1)}, ..., V_{\pi(n-m+k)})$ vector.

For each tried value, we calculate the corresponding contributions to the relations (1) to (k) :

$$c_1 = \sum_{j=k'+1}^{n-m+k} a_{1,j}.V\pi(j) ;$$

..

$$c_k = \sum_{j=k'+1}^{n-m+k} a_{k,j}.V\pi(j) .$$

We can now use the precomputations of Step 1 to obtain a list of possible $(V_{\pi(1)}, ..., V_{\pi(k')})$ k'-uples. As a matter of fact, the relations (1) to (k) can be rewritten :

$$b_1+c_1 = 0;$$

..

$$b_k+c_k = 0.$$

so that the $(V_{\pi(1)}, ..., V_{\pi(k')})$ k'-uple does necessary belong to the list of possible k'-uples for the $(-c_1,...,-c_k)$ value of $(b_1,...,b_k)$.

For each tried $(V_{\pi(k'+1)}, ..., V_{\pi(n-m+k)})$ vector, we obtain in average $\frac{n!}{(n-k')!}$ p^{-k} $(V_{\pi(1)}, ..., V_{\pi(k')})$ values. Some of them have to be discarded because some values of $\{V_{\pi(1)}, ..., V_{\pi(k')}\}$ are already contained in $\{V_{\pi(k'+1)}, ..., V_{\pi(n-m+k)}\}$.

For each remaining $(V_{\pi(1)}, ..., V_{\pi(n-m+k)})$ vector, the still unsolved relations (k+1) to (m) provide successively one single possible value for the numbers $V_{\pi(n-m+k+1)}$ to $V_{\pi(n)}$.

At each stage of this process of extending a $(V_{\pi(1)}, ..., V_{\pi(n-m+k)})$ candidate to a $(V_{\pi(1)}, ..., V_{\pi(n)})$ solution, it has to be checked that the obtained values are non repeating and belong to the $\{V_1, ..., V_n\}$ set. This can be done in very few elementary operations.

The procedure described above finds all the existing solutions (i.e. the secret vector $(V_{\pi(1)}, ..., V_{\pi(n)})$ and the other solutions if there are some). The required memory for this step is negligible. The required time is about $\mathrm{Sup}\ (\ \dfrac{n!}{(m+k'-k)!}\ \dfrac{n!}{(n-k)!}\ p^{-k}\ ,\ \dfrac{n!}{(m+k'-k)!}\ )$ matrix vector products.

In Summary, we have shown that for each (k,k') pair $(0{\le}k{\le}m;\ 0{\le}k'{\le}n-m+k)$ an instance (p,A,V) of the PKP problem can be solved in time :

$$\frac{n!}{(n-k')!} + \mathrm{Sup}(\frac{n!}{(m+k'-k)!}\ \frac{n!}{(n-k)!}\ p^{-k},\ \frac{n!}{(m+k'-k)!}\ )\ \text{matrix-vector products} \quad (i)$$

(cf note below)

and space :

$$\frac{n!}{(n-k')!}\ \ k'\text{-vectors} \quad (ii).$$

Note : all the matrix vector products considered here involve a submatrix of A. Therefore, the cost of each such product can be reduced to very few elementary operations (mod p additions and accesses to arrays), at the expense of a marginal increase of the required space, by precomputing all the linear combinations modulo p of some subsets of the rows in A.

Discussion on the values of k and k'

The value $k=0$ corresponds to an exhaustive trial of all the $\dfrac{n!}{m!}$ possible $(V_{\pi(1)}, ..., V_{\pi(n-m)})$ values. The space cost is negligible and the time cost is about $\dfrac{n!}{m!}$. For larger values of k, precomputations on the k first equations may lead to an improved time cost, at the expense of increasing the required storage. Too large values of k are suboptimal, because the set of n-m+k variables involved in the equations (1) to (k) increases too much.

For a fixed value of k, the required amount of storage is an increasing function of k'; the time required is a decreasing function of k' as long as $k'\ \le (n-m+k)/2$ and $\dfrac{n!}{(n-k')!} \le p^k$ (the first condition says that the time spent on the precomputation should not exceed the time spent on step 2; the second condition says that the average number of k'-uples corresponding to a $(b_1, ...,b_k)$ value should be less than 1). Too large values of k' (such that the two above conditions are not realised) are suboptimal, because they lead to an increased memory cost without reducing the time cost.

4. Impact on the practical security of the PKP scheme

Table 1 gives the time and memory costs, calculated in using (i) and (ii), when n=32; m=16; p=251. Only the k values in the [0..9] interval and the k' values in the [0..15] interval have been considered. The value k=0 leads to a time cost of about 2^{73}, which is very close to the 2^{76} complexity of the best attack mentioned in [1]. The most interesting values are obtained for k=5; k'=8 (time :2^{60} **matrix vector products; memory 2^{38} k'-uples**) and k=6; k'=10 (time 2^{56}; memory 2^{47}).

The obtained complexity values are considerable, but the two above trade-offs cannot be regarded as strictly computationally infeasible, as it was the case for the attack mentioned in [1]. Therefore, the parameter values n=32; m=16; p=251 are not recommended for very secure applications.

Table 2 gives the time and memory costs corresponding to the parameter values n=64; m=37; p=251 for some of the k and k' parameter values. Some of the obtained time costs are substantially lower than the 2^{184} complexity of the best attack mentioned in [1]. For example, for k=8 and k'=11, the obtained time cost is 2^{137} matrix vector products, and the required storage is 2^{64} k'-uples. However, due to the very large values of the obtained time costs, the attacks summarised in Table 2 are computationally infeasible.

5. Concluding remarks

Independently of our work, the security of the PKP problem has also been investigated by J. Georgiades and the results are summarized in [2]. His method, which is based on the resolution of quadratic equations, is less efficient in time cost than the one described in this paper, but requires a negligible amount of storage to solve the PKP problem. We do not know whether both approaches can be combined efficiently.

References

[1] Adi Shamir, "An Efficient Identification Scheme Based on Permuted Kernels", Proceedings of Crypto'89, Springer Verlag.

[2] J. Georgiades, "Some Remarks on the Security of the Identification Scheme Based on the Permuted Kernel", Journal of Cryptology, Volume 5, Number 2, 1992.

k'= \ k=	0	1	2	3	4	5	6	7	8	9	space(k')
0	73	77	81	85	88	92	95	99	102	105	0
1	74	73	77	81	85	88	92	95	99	102	5
2	75	71	73	77	81	85	88	92	95	99	9
3	75	72	69	73	77	81	85	88	92	95	14
4	76	72	68	69	73	77	81	85	88	92	19
5	76	73	69	65	69	73	77	81	85	88	24
6	77	73	69	66	65	69	73	77	81	85	29
7	77	73	70	66	63	65	69	73	77	81	33
8	77	73	70	66	63	60	65	69	73	77	38
9	77	73	70	67	63	59	60	65	69	73	43
10	77	73	70	67	63	60	56	60	65	69	47
11	76	73	70	66	63	60	56	56	60	65	52
12	76	73	69	66	63	60	57	56	57	60	56
13	75	72	69	66	63	61	60	60	60	60	60
14	75	72	69	66	65	65	65	65	65	65	65
15	74	71	69	69	69	69	69	69	69	69	69

Table 1 : base 2 log of time cost t and space cost s when n=32; m=16; p≠251
(example : for k=6 and k'=10, t=2**56 and s=2**47)

k'= \ k=	0	1	2	3	4	5	6	7	8	9	10	11	12	13	14	15	space
0	152	157	163	168	173	178	183	188	193	198	202	207	212	216	221	226	0
1	153	152	157	163	168	173	178	183	188	193	198	202	207	212	216	221	6
2	154	151	152	157	163	168	173	178	183	188	193	198	202	207	212	216	11
3	154	152	149	152	157	163	168	173	178	183	188	193	198	202	207	212	17
4	155	152	150	147	152	157	163	168	173	178	183	188	193	198	202	207	23
5	155	153	150	148	147	152	157	163	168	173	178	183	188	193	198	202	29
6	156	153	151	148	145	147	152	157	163	168	173	178	183	188	193	198	35
7	156	154	151	149	146	143	147	152	157	163	168	173	178	183	188	193	41
8	157	154	152	149	146	144	142	147	152	157	163	168	173	178	183	188	47
9	157	154	152	149	147	144	142	142	147	152	157	163	168	173	178	183	53
10	157	155	152	150	147	145	142	139	142	147	152	157	163	168	173	178	58
11	157	155	152	150	148	145	142	140	137	142	147	152	157	163	168	173	64
12	157	155	153	150	148	145	143	140	138	136	142	147	152	157	163	168	70
13	157	155	153	150	148	145	143	140	138	135	136	142	147	152	157	163	76
14	157	155	153	150	148	146	143	141	138	136	133	136	142	147	152	157	81
15	157	155	153	150	148	146	143	141	138	136	133	131	136	142	147	152	87
16	157	155	153	150	148	146	143	141	138	136	133	131	131	136	142	147	93
17	157	155	153	150	148	146	143	141	139	136	134	131	129	131	136	142	98
18	157	155	152	150	148	146	143	141	139	136	134	131	129	126	131	136	104
19	157	154	152	150	148	145	143	141	138	136	134	131	129	126	126	131	109
20	156	154	152	150	148	145	143	141	138	136	134	131	129	126	124	126	115
21	156	154	152	149	147	145	143	140	138	136	133	131	129	126	124	122	120
22	155	153	151	149	147	145	142	140	138	136	133	131	129	127	126	126	126
23	155	153	151	149	146	144	142	140	138	135	133	132	131	131	131	131	131
24	154	152	150	148	146	144	142	140	138	137	136	136	136	136	136	136	136
25	154	152	150	148	146	144	142	142	142	142	142	142	142	142	142	142	142
26	153	151	149	148	147	147	147	147	147	147	147	147	147	147	147	147	147
27	153	153	152	152	152	152	152	152	152	152	152	152	152	152	152	152	152

Table 2 : base 2 log of time cost t and space cost s when n=64; m=37; p=251
(example : for k=8 and k'=11 , t=2**137 and s=2**64)

Massively Parallel Computation of Discrete Logarithms *

Daniel M. Gordon[†]

Kevin S. McCurley[‡]

Abstract

Numerous cryptosystems have been designed to be secure under the assumption that the computation of discrete logarithms is infeasible. This paper reports on an aggressive attempt to discover the size of fields of characteristic two for which the computation of discrete logarithms is feasible. We discover several things that were previously overlooked in the implementation of Coppersmith's algorithm, some positive, and some negative. As a result of this work we have shown that fields as large as $GF(2^{503})$ can definitely be attacked.

Keywords: Discrete Logarithms, Cryptography.

1 Introduction

The difficulty of computing discrete logarithms was first proposed as the basis of security for cryptographic algorithms in the seminal paper of Diffie and Hellman [4]. The discrete logarithm problem in a finite group is the following: given group elements g and a, find an integer x such that $g^x = a$. We shall write $x = \log_g a$, keeping in mind that $\log_g a$ is only determined modulo the multiplicative order of g. For general information on the discrete logarithm problem and its cryptographic applications, the reader may consult [9] and [11]. In this paper we shall report on some computations done for calculating discrete logarithms in the multiplicative group of a finite field $GF(2^n)$, and the lessons we learned from the computations. The computations that we carried out used a massively parallel implementation of Coppersmith's algorithm [2], combined with a new method of smoothness testing. Coppersmith's algorithm will be described in section 2, and our new method of smoothness testing will be described in section 2.2. The results of our calculations will be presented in section 3.

A great deal of effort (and CPU time!) has been expended on the cryptographically relevant problem of factoring integers, but comparatively little effort has gone into implementing discrete logarithm algorithms. The only published reports on computations of discrete logarithms in $GF(2^n)$ are in [1] and [2, 3]. Both papers report on the calculation of discrete logarithms in the field $GF(2^{127})$.

Odlyzko [11] has carried out an extensive analysis on Coppersmith's algorithm and projected the number of 32-bit operations required to deal with a field of a given size. A similar analysis was made by van Oorschot [13]. Many of their predictions are consistent with our experience, but there were some surprising discoveries that show their analysis to be quite

---

*This research was supported in part by the U.S. Department of Energy under contract number DE-AC04-76DP00789

[†]Department of Computer Science, University of Georgia, Athens, GA 30602. This work was begun while visiting Sandia National Laboratories

[‡]Sandia National Laboratories, Albuquerque, NM 87185

optimistic. We were able to complete most of the computation to compute discrete logarithms for fields of size up to $GF(2^{503})$, and can probably go at least a little bit further with our existing machines. The major limitation at this point seems to lie as much in the linear algebra as the equation generation, due to the large amount of computation time and storage needed to process equations for a large factor base.

Analyses of the type made by van Oorschot and Oldyzko can be extremely useful to chart the *increase* in difficulty of computing discrete logarithms as the field size increases. It is however almost impossible to get exact operation counts to within anything better than an order of magnitude using such an analysis. Among the reasons for this are:

- if a high-level language is used, then compilers vary widely in their ability to efficiently translate the code into machine instructions.

- even counting 32-bit operations is not enough, since the number of clock cycles may vary widely. On the nCUBE-2 that was used for most of our computation, 32-bit integer instructions take between 2 and 38 machine cycles.

- data cache misses can cost many operations (as many as 10 cycles on the Intel i860).

For these and other reasons, it is impossible to get very accurate estimates from analytic methods alone. The only reliable method is to actually implement the algorithms with careful attention to details, and measure the running time.

In the course of this work, we used a variety of machines for the computations. The parallel machines were all MIMD (multiple instruction, multiple data), and included

- a 1024 processor nCUBE-2, with four megabytes per processor,

- a 64 processor Intel iPSC/860, with 8-32 megabytes per processor,

- the 512-processor Intel Touchstone Delta, with 16 megabytes per processor.

We started out with the intention of using a Thinking Machines CM-2, but for technical reasons associated with the SIMD hardware and the system software, we found this to be uncompetitive. It also had the disadvantage that it required using a language specific to the machine, whereas the other machines could all accept standard C, with a few minor changes to accomodate differences in message passing syntax.

2 Coppersmith's algorithm

Coppersmith's algorithm belongs to a class of algorithms that are usually referred to as index calculus methods, and has three stages. In the first stage, we collect a system of linear equations (called relations) that are satisfied by the discrete logarithms of certain group elements belonging to a set called a factor base. In our case, the equations are really congruences modulo the order of the group, or modulo $2^n - 1$. In the second stage, we solve the set of equations to determine the discrete logarithms of the elements of our factor base. In the third stage, we compute any desired logarithm from our precomputed library of logarithms for the factor base.

For the Coppersmith algorithm, it is convenient that we construct our finite field $GF(2^n)$ as $GF(2)[x]/(f(x))$, where f is an irreducible polynomial of the form $x^n + f_1(x)$, with f_1 of small degree. Heuristic arguments suggest that this should be possible, and a search that we made confirms this, since it is possible to find an f_1 of degree at most 11 for all n up to 600, and it it is usually possible to find one of degree at most 7. For the construction of fields, it is also convenient to choose f so that the element x (mod $f(x)$) is primitive, i.e. of multiplicative order $2^n - 1$. As we shall explain later, there are other factors to be considered in the choice of f_1.

For a given polynomial f that describes the field, there is an obvious projection from elements of the field to the set of polynomials over GF(2) of degree at most n. In our case, we shall take as our factor base the set of field elements that correspond to the irreducible polynomials of degree at most B for some integer B to be determined later. Call a polynomial $B-smooth$ if all its irreducible factors have degrees not exceeding B. Let m be the cardinality of the factor base, and write g_i for an element of the factor base. We note that an equation of the form

$$\prod_{i=1}^{m} g_i^{e_i} \equiv x^t \pmod{f(x)}$$

implies a linear relationship of the form

$$\sum_{i=1}^{m} e_i \log_x g_i \equiv t \pmod{2^n - 1}.$$

In order to describe the first stage in the Coppersmith method, we shall require further notation. Let r be an integer, and define $h = \lfloor n2^{-r} \rfloor + 1$. To generate a relation, we first choose random relatively prime polynomials $u_1(x)$ and $u_2(x)$ of degrees at most d_1 and d_2, respectively. We then set $w_1(x) = u_1(x)x^h + u_2(x)$ and

$$w_2(x) = w_1(x)^{2^r} \pmod{f(x)}. \tag{1}$$

It follows from our special choice of $f(x)$ that we can take

$$w_2(x) = u_1(x^{2^r})x^{h2^r - n}f_1(x) + u_2(x^{2^r}), \tag{2}$$

so that $\deg(w_2) \leq \max(2^r d_1 + h2^r - n + \deg(f_1), 2^r d_2)$. If we choose d_1, d_2, and 2^r to be of order $n^{1/3}$, then the degrees of w_1 and w_2 will be of order $n^{2/3}$. If they behave as random polynomials of that degree (as we might expect), then there is a good chance that they will be B-smooth. If so, then from (1) we obtain a linear equation involving the logarithms of polynomials of degree $\leq B$.

An asymptotic analysis of the algorithm suggests that it is possible to choose the parameters so that the asymptotic running time of the first stage of the algorithm is of the form in such a way that the expected running time to complete stage one is of the form

$$\exp((c_2 + o(1))n^{1/3}\log^{2/3} n), \quad \text{where } c_2 < 1.405.$$

The system of equations generated by the first phase is relatively sparse, and there exist algorithms to solve the system that have an asymptotic running time of $O(m^{2+\epsilon})$ (see section 2.4). If such algorithms are used, then the asymptotic running time of the algorithm turns out to be the same as the first phase.

An analysis of the running time for the third stage (which we do not describe in detail here) suggest a running time of

$$\exp((c_3 + o(1))n^{1/3}\log^{2/3} n),$$

where $c_3 < 1.098$, so it takes less time than the first two stages.

The preceding statements pertain to the asymptotic running time, but give only a rough estimate of the time required in practice for actual cases.

2.1 Refinements of Stage 1.

Odlyzko has suggested several ways to speed up the performance of stage 1. None of these affect the asymptotic running time, but each of them may have some practical significance by speeding up the implementation by a factor of two or three. We shall not discuss these methods in great detail, but merely report on some of them.

Forcing a Factor Into w_1 **and** w_2 One method that was suggested by Odlyzko for improving the probability that w_1 and w_2 were smooth was by forcing them to contain at least one small degree factor. The method is described in complete detail in [11] and [13], but roughly speaking we fix polynomials v_1 and v_2 of degree at most B, and consider those (u_1, u_2) pairs for which w_1 and w_2 are divisible by v_1 and v_2 respectively. The (u_1, u_2) pairs with this property are described by a rather small set of linear equations modulo 2, and we can easily find such pairs by Gaussian elimination. For the size fields that we considered, the linear systems had fewer than 50 rows and equations, and a special purpose routine to solve these systems proved to be extremely efficient (rows could be added together by using two xor operations on 32-bit integers). One problem with this method is different v_1, v_2 pairs can lead to the same u_1, u_2 pairs, making it rather difficult to avoid duplication of effort. As far as we can tell, we were the first to implement this method, and our experience with it seemed to agree with the predictions made by Odlyzko.

Large Prime Variation   One well known method for speeding up the generation of equations is to also use equations that involve one irreducible polynomial of degree only slightly larger than B. The rationale for this is that these equations can be discovered essentially for free, and two such equations involving the same "large prime" can be combined to produce an equation involving only the irreducibles of degree at most B. Many such equations can be discovered by checking whether after removing the smooth part from a polynomial, the residual factor has small degree. After combining two such equations, the equations produced are on average twice as dense as the other equations, so they complicate the linear algebra in stage 2. Many of these equations can however be generated more or less for free, so we chose to use them in the calculations.

Double Large Prime Variation   Just as we can use equations involving only a single irreducible of degree slightly larger than B, we can also use equations having two "large prime" factors. This has been used to speed up the quadratic sieve integer factoring algorithm [8], and we might expect the same sort of benefit when it is applied to the Coppersmith algorithm. Many such equations can be produced from reporting those u_1, u_2 pairs that produced a w_1 and w_2 both of which contained a large prime factor.

Smoothness Testing   The most time-consuming part of the Coppersmith algorithm is the testing of polynomials for smoothness. At least two methods have been suggested for doing this, both of which are outlined in [11]. Of the two methods, we found the one used by Coppersmith to work faster for our implementation, and this was initially what we used. For this method, a polynomial $w(x)$ is tested for m-smoothness by computing

$$w'(x) \prod_{i=\lceil m/2 \rceil}^{m} (x^{2^i} + x) \pmod{w(x)}. \tag{3}$$

A faster method, using a polynomial sieve, will be outlined in Section 2.2.

Early Abort Strategy   One strategy that has been suggested for locating smooth integers is to search through random integers, initially dividing by small primes. At a certain point, we then check to see if the residual factor has moderate size, and abort the testing if it fails. It so happens that a random integer is more likely to be B-smooth from having many very small prime factors than it is from having just a few factors near B, and it follows that we should not spend a lot of time dividing by moderately large primes to test for smoothness. This strategy has come to be known as the "early abort" strategy, and the same heuristic reasoning carries over to the smoothness testing part of Coppersmith's algorithm. Odlyzko predicted that this may result in a speedup of a factor of two in the algorithm, but we never got around to implementing it. The major reason for this is that there seems to be no obvious way to combine this idea with sieving, and the latter gave a somewhat better speedup.

2.2 A Polynomial Sieve

Our first implementation of Coppersmith's algorithm used methods suggested previously by Odlyzko and Coppersmith to test polynomials for smoothness. After having carried out the computation for the case $n = 313$, we looked around for any variations that would speed up the smoothness testing. Drawing on the knowledge that sieving can be exploited to great advantage in integer factoring algorithms, we sought a way to use sieving to test many polynomials simultaneously for smoothness. Sieving over the integers is relatively efficient due to the fact that integers that belong to a fixed residue class modulo a prime lie a fixed distance apart, and it is very easy to increment a counter by this quantity and perform a calculation on some memory location corresponding to the set element.

For polynomials, the problem is slightly different, since we saw no obvious way of representing polynomials in such a way that representatives of a given residue class are a fixed distance apart. It turns out that this is not a great deterrent, since what is important is the ability to quickly move through the representatives, and for the data structures that we used, this can be done using the notion of a Gray code.

Polynomials over GF(2) of degree less than d can be thought of as the vertices of a d-dimensional hypercube, with the coefficient of x^i in a polynomial corresponding to the ith coordinate of a vertex. A Gray code gives a natural way to efficiently step through all such polynomials. The same applies to all polynomials that are divisible by a fixed polynomial g.

Let $G_1, G_2, \ldots, G_{2^d}$ be the standard binary reflected Gray code of dimension d. For any positive integer x, let $l(x)$ be the low-order bit of x, i.e. the integer i such that $2^i \| x$. Then we have (see, for example, [10]):

PROPOSITION 1. *The bit that differs in G_x and G_{x+1} is $l(x)$.*

This allows us to efficiently step through the Gray code. Let $s[0], \ldots, s[2^t - 1]$ be 8-bit memory locations corresponding to the u_2 of degree less than t in the obvious way (mapping $u_2(x)$ to $u_2(2)$). Figure 1 describes an algorithm which takes u_1, and finds all u_2 of degree less than t such that $w_1 = u_1 x^h + u_2$ is B-smooth.

Note that the inner loop consists of only two 32-bit operations, a shift to multiply g by x^i, and an exclusive-or to add gx^i to u_2, and one 8-bit add.

The actual implementation has a few additions. It checks for large primes, by reporting any pair for which $s[u_2] \geq (\text{degree}(w_1) + h - LP)$, where LP is the maximum degree of a large prime. A sieve by powers of irreducibles up to degree B is also done. Instead of calculating $u_1 x^h \bmod g$ each time to start sieving, $x^h \bmod g$ is saved for each g. Then to step from one u_1 to another, we only have to add a shift of $x^h \bmod g$ to the starting sieve location.

A sieve over polynomials w_2 would work similarly; the main difference is that initializing u_2 requires taking a fourth root, which slows things down. It turned out to be more efficient to test smoothness of each w_2 corresponding to a smooth w_1 individually, since only a small number of pairs u_1, u_2 survive the w_1 sieve (w_1 has much higher degree than w_2).

One reason that sieving works so well for the quadratic sieve algorithm is that it replaces multiple precision integer calculations with simple addition operations. We gain the same sort of advantage in Coppersmith's algorithm, by eliminating the need for many modular multiplications involving polynomials. The actual operation counts for sieving come out rather close to the operation counts given in [11] and [13], but in the case of sieving the operations are somewhat simpler, and the speedup is substantial.

The number of 32-bit operations to sieve a range of u_1, u_2 pairs is proportional to $\log B$ times the size of the range. This is because there are about $2^d/d$ irreducible polynomials of degree d, so the number of steps to sieve a range of l pairs is:

```
for i = 0 to 2^t - 1
    s[i] ← 0                                    /* initialize sieve locations */
for d = 1 to B
    dim ← max(t - d, 0)                         /* dimension of Gray code */
    for each irreducible g of degree d
        u_2 ← u_1 x^h mod g
        if degree(u_2) < t then
        for i = 1 to 2^dim
            s[u_2] ← s[u_2] + d
            u_2 ← u_2 + g x^{l(i)}              /* u_2 = u_1 x^h mod g + g G_i */
for i = 0 to 2^t - 1
    if s[i] ≥ (degree(u_1) + h - B) then print u_1, u_2
```

Figure 1: Pseudocode for sieve algorithm

$$\sum_{d=1}^{B} \sum_{\substack{g \text{ irreducible} \\ \deg(g)=d}} \left(c + \frac{l}{2^d} \right) \approx \sum_{d=1}^{B} \left(c + \frac{l}{2^d} \right) \frac{2^d}{d} \approx l \log B + \frac{c 2^{B+1}}{B},$$

where c represents the startup time for each irreducible. Each of these steps uses a fixed number of 32-bit operations (typically between 2 and 12, depending on the machine, compiler, and source code used). If l is sufficiently large, then the c operations performed for each irreducible become inconsequential. The time spent on finding the initial locations for sieving by each polynomial in the factor base can be made inconsequential by amortizing it over several sieving runs.

In comparison, the number of 32-bit operations needed to test a polynomial for smoothness using Coppersmith's method is at least $3Bh^2/32$ (see [13]), where $h = \lfloor n2^{-r} \rfloor + 1$ is the approximate degree of w_1. As n (and therefore B and h as well) become large, the advantage of using a polynomial sieve becomes overwhelming.

Note that the memory access patterns for the array $s[\cdot]$ in the sieving algorithm are somewhat chaotic, since the indices of consecutive values for u_2 are widely and irregularly dispersed. For processors such as the Intel i860 whose performance is heavily dependent on using memory caches, this severely limits the performance improvement gained from sieving. By contrast, the nCUBE processor is not so dependent on memory access patterns, and the improvement from sieving was more pronounced.

2.3 The choice of f_1

Once we were quite sure that our sieving code was giving completely reliable results, we were unpleasantly surprised that the number of relations discovered was not in agreement with the heuristic arguments given in [11] and [13], but was instead considerably smaller. This led us to reconsider the arguments there, in an attempt to produce more accurate predictions on the number of equations produced by examining a certain range of u_1 and u_2.

The assumption made in both [11] and [13] that w_1 and w_2 are smooth as often as a random polynomial of the same degree is not quite accurate. We shall provide several justifications for this statement, based on heuristic arguments showing ways that w_1 and w_2 (particularly w_2) deviate from behaviour of random polynomials. We have been unable to combine all of the effects we know of into an analytical method for accurately predicting these probabilities. Luckily, it is relatively simple to make random trials to estimate the actual probabilities.

For the cases that we shall be most interested in, w_2 has the form

$$x^T u_1(x)^4 f_1(x) + u_2(x)^4 \qquad (4)$$

where $T = 4h - n$ is 1 or 3, and $\gcd(u_1, u_2) = 1$. In the following discussion, g will be an irreducible polynomial of degree d.

First, note that if $g \mid u_1$, then $g \nmid u_2$, and therefore $g \nmid w_1$ and $g \nmid w_2$. Hence if $g \mid w_1$ or $g \mid w_2$, then $g \nmid u_1$. It follows that if if $g^e \mid w_2$ for some integer e, then

$$x^T f_1(x) \equiv (u_1^{-1} u_2)^4 \pmod{g^e}. \qquad (5)$$

Note that if $e \geq 2$ and $de > (T + \deg(f_1))$, then (5) is clearly impossible, since the right side reduces to a polynomial with only even exponents modulo g^2, whereas the left side will have odd powers since T is odd and $f_1(0) = 1$. Hence if $d \geq (T + deg(f_1))/2$, it follows that g^2 cannot divide w_2. This shows that w_2 is much more likely to be squarefree than a random polynomial, and therefore somewhat less likely to be smooth.

Another example of nonrandom behaviour from w_2 can be seen from examining the expected value of the degree of the power of an irreducible that divides w_2, compared to the expected power that divides a random polynomial. One can easily show that in some sense, a truly random polynomial will be divisible by an irreducible factor g to the e'th power with probability $1/2^{de}$, and will be exactly divisible by the e'th power with probability $(2^d - 1)/2^{d(e+1)}$. Hence the expected value of the degree of the power of g that divides a random polynomial is $d/(2^d - 1)$.

f_1	factorization	probability
$x^8 + x^5 + x^4 + x^2 + x + 1$	$(1 + x)^2(1 + x + x^3 + x^4 + x^6)$	0.002468
$x^8 + x^7 + x^5 + x^2 + x + 1$	$(1 + x)(1 + x^2 + x^3 + x^4 + x^7)$	0.002366
$x^9 + x^8 + x^5 + 1$	$(1 + x)^4(1 + x + x^2)(1 + x + x^3)$	0.002607
$x^{10} + x^7 + x^6 + x^3 + x^2 + 1$	$(1 + x)^2(1 + x^3 + x^5 + x^6 + x^8)$	0.001956
$x^{10} + x^9 + x^8 + x^2 + x + 1$	$(1 + x)^8(1 + x + x^2)$	0.002383

Table 1: Empirical probabilities that a (u_1, u_2) pair will produce a smooth w_2, for $n = 593$ and different choices of f_1. Tests based on examination of over five million random relatively prime pairs (u_1, u_2) of degrees 22 and 24, respectively.

The expected contribution to a polynomial w_2 is somewhat different. For the case where $g \nmid x^T f_1(x)$, an easy counting argument on residue classes modulo g shows that the probability that g divides w_2 is $(2^d - 1)/(2^{2d} - 1) = 1/(2^d + 1)$, so that the expected degree of the power of g dividing w_2 is $d/(2^d + 1)$, somewhat smaller than for a random polynomial. If $g^e \mid x^T f_1(x)$ for some integer $e \leq 4$, then g^e is automatically guaranteed to divide w_2 whenever $g \mid u_2$. If e is large for a small degree g, then this helps w_2 to be smooth, but if $e = 1$, then it makes w_2 less likely to be smooth.

A complete analysis of this situation is probably not worth the effort. In this paper, it suffices to illustrate the effects by considering the example of $n = 593$. The only f_1's of degree up to 10 for which $x^{593} + f_1$ is irreducible are in Table 1. Clearly the first two f_1's in the table have an advantage from having the smallest degrees, but the third and fifth have an advantage from the large power of $1 + x$ that divides them. The tradeoffs between these effects are not at all clear, but the results of the experiments show that the third f_1 gives a slight advantage, in spite of its larger degree. For the case of $n = 503$, it turned out that $f_1 = x^3 + 1$ was the best choice.

2.4 Linear Algebra

The solution of sparse linear systems over finite fields have received much less attention than the corresponding problem of solving sparse linear systems over the field of real numbers. The fundamental difference between these two problems is that issues involving numerical stability problems arising from finite precision arithmetic do not arise when working over a finite field. The only pivoting that is required is to avoid division by zero. Algorithms for the solution of sparse linear systems over finite fields include:

- standard Gaussian elimination.

- structured Gaussian elimination.

- Wiedemann's algorithm.

| | sparse matrix | | | dense matrix | | |
n	equations	unknowns	nonzeros	size	nonzeros	reduction
313	108736	58636	1615469	9195	633987	84%
401	117164	58636	2068707	16139	1203414	72%
503	434197	210871	10828595	78394	6394049	63%

Table 2: Results of structured Gaussian elimination for various n.

- Conjugate Gradient.

- Lanczos methods.

A description of these methods can be found in the paper by LaMacchia and Odlyzko [7], where they describe their experience in solving systems that arise from integer factoring algorithms and the computation of discrete logarithms over fields $GF(p)$ for a prime p. We chose to implement three of these algorithms: conjugate gradient, Wiedemann, and structured Gaussian elimination. For handling multiple precision integers we used the Lenstra-Manasse package. The original systems were reduced in size using the structured Gaussian elimination algorithm, after which the conjugate gradient or Wiedemann algorithm was applied to solve the smaller (and still fairly sparse) system.

This approach was used by LaMacchia and Odlyzko in [7] with great success. The structured Gaussian elimination reduced their systems by as much as 95%, leaving a small system that could easily be solved on a single processor. We were not as successful, due to a feature of the equations that Coppersmith's method produces. For the equations in [7], almost all the coefficients are ±1, and so during the Gaussian elimination most operations involve adding or subtracting one row from another. For our systems, half of the coefficients are multiples of 4, and so it is often necessary to multiply a row by ±4 before adding it to another. This caused the coefficients in the dense part of the matrix to grow rapidly.

This presented a dilemma. If the matrix coefficients are allowed to become large integers, then the arithmetic operations take considerably more time (and require considerable more complicated code). The alternative is to restrict which rows can be added to others, to keep the coefficients down to 32 bits. This results in a larger matrix, which also slows down stage 2. We elected to deal with the larger matrices. Table 2 gives results for partial gaussian elimination on several systems.

For the 127, 227, and 313 systems, we were able to solve the systems on a workstation (the last one took approximately ten days). The other systems were clearly too large to be solved on a single processor workstation, and the algorithm requires too much communication to effectively run on a network of workstations. We therefore wrote a parallel version (MIMD) of the conjugate gradient code. A single source program was written in C that would compile for Suns, the Intel iPSC/860, the Intel Delta Touchstone, and the nCUBE-2.

Parallelization of the algorithm was accomplished by distributing the matrix rows and columns across the processors. A matrix-vector multiply is then done by multiplying the rows held by the processor times the entire vector. After this operation, each processor communicates to every other processor (in a logarithmic manner) its contribution to the vector result. The distribution of the matrix rows was done by simply assigning the same number of rows to each processor. The structure of the matrix is such that each processor then gets essentially the same number of nonzero entries. For the distribution of the columns, this is certainly *not* the case, as the first few columns contain far more nonzeros than the last few columns. The columns of the matrix were then permuted in order to approximately balance the number of nonzeros assigned to each processor, and some processors ended up getting far more columns. This creates a slight imbalance in the communication phase, but is better than an imbalance in the computation phase.

Unfortunately, this approach suffered from a severe problem when scaled to a large number of processors, since the first column of the reduced 503 matrix contained 61166 nonzero entries, but a perfect load balance on 1024 processors would place $6394049/1024 \approx 6244$ nonzeros on each processor. Proper load balancing of the matrix multiplication would therefore have required that we divide columns between processors, and we were reluctant to modify the code for this due to the added complexity.

Instead, we chose to implement the Wiedemann method. This had the advantage that it required only multiplications of the coefficient matrix times a vector, not the multiplication of the transpose of the matrix. Once again, however, we discovered that there were scaling problems in moving to a large number of processors, since the amount of communication required for sharing results at the end of the distributed matrix-vector multiply increased at least with the logarithm of the number of processors, whereas the amount of computation decreases linearly with the number of processors. Hence when this code was run on 1024 processors of the nCUBE, it ran only slightly faster than it would run on 512 processors. For more dense matrices, the speedup would be larger, but so would the total runtime. This problem was even worse on the 512 processor Delta, where the bisection bandwidth of the machine is about 16% of that of the nCUBE, but the peak processor speed is about 10 times faster.

The communication that we used in each matrix-vector multiplication is often called an all-to-all broadcast, or global concatenation. For machines such as the nCUBE-2 and iPSC/860 that use a hypercube topology for their communications network, there is a fairly obvious algorithm for accomplishing the all-to-all broadcast in $\log(p)$ phases on p processors, passing a minimal amount of information, with no contention for communication channels. The Intel Delta Touchstone uses instead a 16×32 two-dimensional mesh topology. When we first ported the code from the iPSC/860 to the Delta, we were using an Intel-supplied library routine for the communication, but we found that the performance of the Intel routine was far from optimal on the Delta, and the result was that the Delta showed almost no speedup in moving to more processors. Subsequent to this, the second author worked with David Greenberg to develop code and algorithms that improved the performance of the all-to-all broadcast library routine (gcolx()) by a factor of 21. This work is reported in [5].

The Wiedemann algorithm requires the use of the Berlekamp-Massey algorithm for computing the minimal polynomial of the matrix. In contrast to the matrix-vector multiplications, this turned out to be quite easy to parallelize, since the core operations required are polynomial additions that are easily parallelized. The only difficulty arises from the fact that the degree steadily increases through the computation, requiring continual load balancing. Eventually the degree of the polynomials becomes large enough that this communication becomes insignificant, and all communication is between nearest-neighbor processors in the network topology, giving very good scalability to large parallel machines. In practice, the Berlekamp-Massey algorithm turned out to consume much less time than the matrix-vector multiplications.

To summarize, after we had invested a substantial amount of time in writing code for the various algorithms, we became aware that communication would be a severely limiting factor in the use of distributed memory parallel machines for solving the linear systems. Since then we have learned of other methods [6],[12] that might dramatically improve the performance. We believe that there remains substantial room for improvement in this area, using these and other ideas.

3 Results

We have completed the precomputation step required to compute discrete logarithms for the fields $GF(2^n)$ for $n = 227$, $n = 313$, and $n = 401$. Once this step has been completed, individual logarithms can be found comparatively easily. We have not bothered to implement the third phase yet, as we expect the running time for this to be substantially less than the first two phases.

The code for producing equations has gone through many revisions and removal of bugs. As a result, we ended up using much more computer time for producing the equations for 401 and 503 than would be required with our current version of the code. Moreover, most of our computations were carried out on the nCUBE-2, which has no queueing of jobs, and no priority system. We therefore wrote our own queueing system, and wrote some code for other users to kill our jobs. This extremely crude approach allowed us to aggressively consume computer time while at the same time allow other users to carry on their normal development activities. The unfortunate result is that many ranges of u_1, u_2 pairs were only partially completed before they were killed, so that very accurate statistics on the completed ranges are difficult to keep. After running the code for 503 for several months, we decided to go back and redo 401 with more care, to keep more accurate records and make an accurate measurement of the amount of calculation required.

For the case of $GF(2^{401})$, we chose to search through all u_1 of degree up to 20, and all u_2 of degree up to 22. The nCUBE-2 was able to process approximately 1.5×10^8 u_1, u_2 pairs per hour on a single processor. Using the full 1024 processors of our nCUBE-2, we could therefore carry out this calculation in approximately 111 hours, or just under 5 days. For comparison, a Sparcstation 2 is able to process approximately 6×10^8 u_1, u_2 pairs per hour, so a single Sun workstation would take approximately 19,000 days (or more realistically, 500 workstations would take just over a month).

Searching this range of u_1, u_2 pairs produced a total of 117,164 equations from a factor base of 58,636 polynomials (all irreducibles of degree up to 19). It also produced approximately 700,000 equations each of which involved only one "large prime" polynomial of degree 20 or 21, which we ended up ignoring due to previously mentioned difficulties with solving the linear system. Clearly there is a tradeoff to be made between producing more equations with a longer sieving phase, or spending more time on solving a harder system of equations. Since the sieving can be carried out in a trivially parallel manner, we opted to spend more time on this rather than claim the whole machine for a long dedicated period to solve a larger system of equations.

For the case of $n = 503$, we attempted to search all u_1 of degree up to 22 and all u_2 of degree up to 25 (again, some of this range was missed by killed jobs, but the percentage should be small). This range produced 165,260 equations over the factor base of 210,871 polynomials of degree up to 21. Combining pairs of equations involving a single irreducible of degree 22 or 23 brought the total up to 361,246 equations. We estimate that repeating this calculation would take approximately 44 days on the full 1024-processor nCUBE. In practice it took us several months due to the fact that we were trying to use idle time, and we never used the full machine. We later extended this calculation to produce a total of 434,197 equations, by running over some u_1 polynomials of degree 23.

The parallel conjugate-gradient code was able to solve the system of equations for $n = 313$ in 8.3 hours on 16 processors of a 64-processor Intel iPSC/860. The equations for $n = 401$ took approximately 33 hours on 32 processors.

Note that $2^{503} - 1$ factors as

$$\begin{aligned}
2^{503} - 1 &= 3213684984979279 \cdot 12158987054135300783 \\
&\quad \cdot 1873030665061080894263 \cdot p_4 \\
&= p_1 \cdot p_2 \cdot p_3 \cdot p_4,
\end{aligned}$$

where p_4 is a prime of 96 decimal digits. Solution of the system modulo $2^{503} - 1$ can thus be accomplished by solving four separate systems modulo these prime factors, and combined afterwards using the chinese remainder theorem. The only truly hard part is solving the system modulo p_4, since the individual operations are much slower and the amount of data to be communicated is also larger.

Our original projections for the solution of the 503 equations were too optimistic, since we underestimated the cost of communication. We have still not completed the solution of the 503 equations, but have now at least made timings of individual iterations to estimate the amount of time required. Timings that we have made on the Delta Touchstone and nCUBE-2 show that solution of the system modulo p_1 using the Wiedemann algorithm would take approximately 106 hours on 256 processors of the nCUBE for the matrix multiplications, and 38.4 hours on 512 processors of the Delta. The Berlekamp-Massey calculation would require less than two hours on each of these. For the prime p_4, we are unable to run the matrix-vector multiplications on the nCUBE with our current code due to memory limitations, but the time for matrix multiplications on the Delta is estimated at approximately 105 hours. Logistics have simply prevented us from reserving enough time on the machine to solve the equations in a single run (after all, the purpose of our project was to investigate the effectiveness of massively parallel computers and better algorithms, not to do real cryptanalysis).

4 Conclusion

We started out by repeating Coppersmith's calculation of discrete logarithms for $GF(2^{127})$. Our original goal was to determine whether it was possible to compute discrete logarithms for the field $GF(2^{593})$, which has been suggested for possible use in at least one existing cryptosystem. Odlyzko predicted that fields of size up to 521 should be tractable using the fastest computers available within a few years (exact predictions are difficult to make without actually carrying out an implementation). van Oorschot predicted that computing discrete logarithms in $GF(2^{401})$ should be about as difficult as factoring 100 digit numbers. Both predictions turned out to be reasonable.

We believe that 521 should now be possible to complete, albeit with the consumption of massive amounts of computing time. Discrete logarithms in $GF(2^{593})$ still seem to be out of reach. Sandia National Laboratories is scheduled to take delivery of an Intel Paragon machine in July 1993 whose peak speed is approximately 50 times the speed of the nCUBE-2 used for this work. Massively parallel machines are expected to be built in the next five years that will reach peak performance levels approximately 500 times faster than the 1024 processor nCUBE-2 that was our primary machine. Unfortunately, this peak speed will be harder to attain in future architectures, so the actual increase in speed for a given application is difficult to project. With a concerted effort on one of these faster machines, or further algorithmic improvements, computing discrete logarithms in $GF(2^{593})$ might be possible within the next 5-10 years. It would require a much larger factor base (we estimate at least the irreducibles up to degree 23, or 766150 polynomials). It would also be a computation of enormous proportions, and is not likely to be completed in the near future without further innovations.

Acknowledgment The authors wish to thank A.M. Odlyzko, Bruce Hendrickson, and Peter Montgomery for helpful comments in the course of this research.

References

[1] I. F. Blake, R. Fuji-Hara, R. C. Mullin, and S. A. Vanstone. Computing logarithms in fields of characteristic two. *SIAM Journal of Algebraic and Discrete Methods*, 5:276–285, 1984.

[2] D. Coppersmith. Fast evaluation of discrete logarithms in fields of characteristic two. *IEEE Transactions on Information Theory*, 30:587–594, 1984.

[3] D. Coppersmith and J. H. Davenport. An application of factoring. *Journal of Symbolic Computation*, 1:241–243, 1985.

[4] W. Diffie and M. Hellman. New directions in cryptography. *IEEE Transactions on Information Theory*, 22:472–492, 1976.

[5] David S. Greenberg and Kevin S. McCurley. Bringing theory to practice: The reality of interprocessor communication. unpublished manuscript, 1993.

[6] B. Hendrickson, Robert Leland, and Steve Plimpton. An efficient parallel algorithm for matrix-vector multiplication. Technical Report SAND92-2765, Sandia National Laboratories, 1992.

[7] B. A. LaMacchia and A. M. Odlyzko. Solving large sparse linear systems over finite fields. In *Advances in Cryptology - Proceedings of Crypto '90*, volume 537 of *Lecture Notes in Computer Science*, pages 109–133, New York, 1991. Springer-Verlag.

[8] A. K. Lenstra and Mark Manasse. Factoring with two large primes. In *Advances in Cryptology - Proceedings of Eurocrypt '90*, volume 473 of *Lecture Notes in Computer Science*, pages 72–82, New York, 1991. Springer-Verlag.

[9] Kevin S. McCurley. *The Discrete Logarithm Problem*, volume 42 of *Proceedings of Symposia in Applied Mathematics*, pages 49–74. American Mathematical Society, Providence, 1990.

[10] A. Nijenhuis and H.S. Wilf. *Combinatorial Algorithms*. Academic Press, New York, second edition, 1978.

[11] A. M. Odlyzko. Discrete logarithms in finite fields and their cryptographic significance. In *Advances in Cryptology (Proceedings of Eurocrypt 84)*, number 209 in Lecture Notes in Computer Science, pages 224–314, Berlin, 1985. Springer-Verlag.

[12] A.T. Ogielski and W. Aiello. Sparse matrix computations on parallel processor arrays. *SIAM Journal of Scientific and Statistical Computing*, 14:??–??, 1993.

[13] Paul C. van Oorschot. A comparison of practical public-key cryptosystems based on integer factorization and discrete logarithms. In Gustavus J. Simmons, editor, *Contemporary Cryptology: The Science of Information Integrity*, chapter 5, pages 289–322. IEEE Press, Piscataway, 1992.

A Quadratic Sieve on the n-Dimensional Cube

René Peralta *

Electrical Engineering and Computer Science Department
University of Wisconsin - Milwaukee

Abstract. Let N be a large odd integer. We show how to produce a long sequence $\{(X_i, Y_i)\}_{i=1}^{2^n}$ of integers modulo N which satisfy $X_i^2 \equiv Y_i$ modulo N, where $X_i > N^{1/2}$ and $|Y_i| < cN^{1/2}$. Our sequence corresponds to a Hamiltonian path on the n-dimensional hypercube C_n, where n is $\Theta(\log N / \log \log N)$. One application of these techniques is that, at each vertex of the hypercube, it is possible to search for equations of the form $U^2 \equiv V$ modulo N with V a product of small primes. The search is as in the quadratic sieve algorithm and therefore very fast. This yields a faster way of changing polynomials in the Multiple Polynomial Quadratic Sieve algorithm, since moving along the hypercube turns out to be very cheap.

1 Introduction

Given a large odd integer N, there is no known way of efficiently generating random congruences of the form $X^2 \equiv Y$ modulo N with Y substantially smaller than $N^{1/2}$. One reason for wanting to generate such congruences is that they can be used to factor N. The Continued Fraction Algorithm [2] factors N by generating many such congruences, choosing the ones for which Y factors over a small prime factor base FB, and then solving a linear system of equations in order to create one congruence $X^2 \equiv Z^2$ modulo N which, if $X \neq \pm Z$, yields a proper factor $p = GCD(X+Z, N)$ of N. An important bottleneck in the Continued Fraction Algorithm is the cost of testing whether Y factors over FB. This is done by trial division for each prime in the factor base. The Quadratic Sieve Algorithm [6] considers a sequence $\{(X_i, Y_i)\}_{i=1}^{M}$ of M pairs where $X_i = \lfloor \sqrt{N} \rfloor + i$ and $Y_i = X_i^2 - N$. Since Y_i is given by an integer quadratic polynomial, it is easy to predict which Y_i's will be divisible by a given prime p. The values of i which generate $Y_i \equiv 0 \bmod p$ lie on two arithmetic progressions $\alpha \pm kp$ and $\beta \pm kp$ ($k = 0, 1, \ldots$). The cost of avoiding trial division is that the Y_i's are of order $O(MN^{1/2})$ and therefore they are less likely to factor over FB than the Y's generated by the Continued Fraction Algorithm. However, avoiding trial division more than compensates for the increased size of the Y_i's. A variation on the Quadratic Sieve Algorithm is the Multiple Polynomial Quadratic Sieve [8] (MPQS), which uses several polynomials as a way to fight the increase in the size of the Y_i's. The latter is currently the algorithm of choice for factoring integers which are about one hundred digits long.

* Supported by National Science Foundation grant No. CCR-8909657

We show how to produce a long sequence $\{(X_i, Y_i)\}_{i=1}^{2^n}$ of integers modulo N which satisfy $X_i^2 \equiv Y_i$ modulo N, where $X_i > N^{1/2}$ and $| Y_i | < cN^{1/2}$. Our sequence corresponds to a Hamiltonian path on the n-dimensional hypercube C_n, where n is $\Theta(\log N / \log \log N)$. One application of these techniques is that, at each vertex of the hypercube, it is possible to search for equations of the form $U^2 \equiv V$ modulo N with V smooth. The search is as in the quadratic sieve algorithm and therefore very fast. This yields a factoring algorithm which is faster than the Multiple Polynomial Quadratic Sieve algorithm, since moving along the hypercube turns out to be very cheap. The asymptotics of the new algorithm are as in MPQS. Therefore it is not asymptotically as fast as the recently discovered Number Field Sieve algorithm [3, 1].

2   Generating "small" quadratic congruences

Let N be a large odd integer. We will use the symbol "$\equiv$" to denote modular congruence, and we will restrict the use of "$=$" to equality. Let s, t be such that

- $t = \prod_{j=1}^{n} p_j$, where the p_j's are distinct primes (n will be chosen later).
- The prime 2 may be among the p_j's if and only if $N \equiv 1$ modulo 4.
- N is a quadratic residue modulo each p_j.
- s satisfies $s^2 \equiv N$ modulo t^2 and $|s| < t^2$.

Lemma 1. *Let $c = t/N^{1/4}$. Let $x \equiv s/t$ modulo N and $y \equiv x^2$ modulo N where y is the member of the residue class of x^2 with smallest absolute value. Then $|y| < c_1 N^{1/2}$ where $c_1 = Max\{c^2 - \frac{1}{c^2}, \frac{1}{c^2}\}$.*

Proof. Since $s^2 \equiv N$ modulo t^2, we have $s^2 = kt^2 + N$ for some (possibly negative) integer k. Then $y \equiv \frac{s^2}{t^2} \equiv \frac{s^2 - N}{t^2} = \frac{kt^2 + N - N}{t^2} = k$, where congruence is modulo N. Thus we may choose $y = k$. Since $s^2 < t^4$ and $t = cN^{1/4}$, we have

$$|y| = |k| = \left| \frac{s^2 - N}{t^2} \right| \leq Max\{t^2 - \frac{N}{t^2}, \frac{N}{t^2}\} = N^{1/2} Max\{c^2 - \frac{1}{c^2}, \frac{1}{c^2}\}.\square$$

Thus, if $c \in (1, \sqrt{\frac{1+\sqrt{5}}{2}})$, then $|y| < N^{1/2}$ (the golden mean strikes again!). Also note that $Max\{c^2 - \frac{1}{c^2}, \frac{1}{c^2}\}$ is minimized at $c = 2^{1/4}$. For $c < 2^{1/4}$, the bound is $\frac{1}{c^2}$. For $c > 2^{1/4}$, the bound is $c^2 - \frac{1}{c^2}$.

By construction, there are 2^n square roots of N modulo t^2. Given the p_j's, it is a simple matter to compute one such root s_1. By the Chinese Remainder Theorem, we may think of s_1 as an $n-$tuple $(\alpha_1, \ldots, \alpha_n)$, where $\alpha_j^2 \equiv N$ modulo p_j^2. Then the complete set of square roots of N modulo t^2 is given by $(\pm\alpha_1, \ldots, \pm\alpha_n)$ for all choices of signs $\pm$. Any member of this set can be easily calculated as a sum $\sum_{j=1}^{n} \delta_j \alpha_j b_j$ modulo t^2 where

- δ_j is the sign at the $j-$th coordinate.
- b_j is the unique element of Z_{t^2} which is 1 modulo p_j^2 and 0 modulo p_i^2 for $j \neq i$.

Note that there are two possible values for each α_j. *We will choose α_j such that $b_j \alpha_j$ modulo t^2 is less than $\frac{t^2}{2}$.*

The maximum size of n can be estimated from the familiar relation $\sum_{p \le x} \ln p \sim x$, where the sum is over all primes p less than or equal to x. Since N is typically a quadratic residue modulo half of the first $2n$ primes we can estimate the maximum n from $2n \sim \pi(x)$ where x satisfies $\prod_{p < x} p = N^{1/2}$ (in this way, the product of the approximately n primes for which N is a quadratic residue is approximately $N^{1/4}$). This implies $\sum_{p \le x} \ln p \sim \frac{1}{2} \ln N$, and so $x \sim \frac{1}{2} \ln N$. Thus $n \sim \frac{1}{2} \pi(x) \sim \frac{1}{2} \pi(\frac{1}{2} \ln N) \sim \frac{1}{4} \frac{\ln N}{\ln \ln N - \ln 2} = \Theta(\log N / \log \log N)$.

Example : the RSA modulus

The 129-digit RSA modulus

$N_{RSA} = 114381625757888867669235779976146612010218296721242362562561$

$842935706935245733897830597123563958705058989075147599290026879543541$

is a quadratic residue modulo the 20 primes

$$\{2, 5, 17, 19, 29, 37, 41, 43, 47, 59, 79, 97, 101, 103, 107, 113, 131, 151, 157, 163\}.$$

Letting t be the product of all these primes except 79, we get $t \sim 1.01 N_{RSA}^{1/4}$. Thus, for this case we get $n = 19$.

3 Traversing the hypercube

The set of square roots of N modulo t^2 can be thought of as the n-dimensional hypercube, where we connect two roots if and only if they differ at exactly one sign. A Hamiltonian path on the hypercube is defined by a starting point s_1 and the sequence $\{k_i\}_{i=1}^{2^n - 1}$ of coordinate changes, e.g. $k_{130} = 8$ means the 130th move on the Hamiltonian path is a change of sign at coordinate 8. Let $\mu_i = +1$ if move i switches a $-$ sign for a $+$ sign and $\mu_i = -1$ if move i switches a $+$ sign for a $-$ sign. Let $\gamma_j \equiv \alpha_j b_j$ modulo t^2, where α_j, b_j are as defined in the previous section. Note that, by our choice of α_j, we have $0 < \gamma_j < \frac{t^2}{2}$ for all j. Then we may define the i-th square root of N modulo t^2 by

$$s_{i+1} = s_i + 2\mu_i \gamma_{k_i} - \omega_i t^2$$

where ω_i is a correction factor to make $s_i \in (0, t^2)$. Note that ω_i is always $-1, 0,$ or $+1$. The sequence of ω_i's can be easily computed from the few most significant bits of the γ_j's, and if simply allowed to be 0, the values of s_i will remain in a small interval (as shown by our next example).

An n-dimensional cube C_n is composed of two $(n-1)$-dimensional cubes $C_{n-1}^{(1)}, C_{n-1}^{(2)}$ whose vertices are connected in a 1-1 fashion. Thus, a simple way to traverse the n-cube is

 − traverse $C_{n-1}^{(1)}$;
 − move to $C_{n-1}^{(2)}$;
 − traverse $C_{n-1}^{(2)}$.

The recursive procedure works because on moving to $C_{n-1}^{(2)}$ the algorithm finds itself at a node which is, up to isomorphism, the same starting node as in $C_{n-1}^{(1)}$. *From now on we will assume the Hamiltonian path on the n-cube is generated by this procedure.*

Example : Traversing the 3-cube

A traversal of the 3-cube yields

i	1	2	3	4	5	6	7
k_i	1	2	1	3	1	2	1
μ_i	-1	-1	+1	-1	-1	+1	+1

Using $\omega_i = 0$ for all i, this table gives the following values for the s_i's.

$$s_2 = s_1 - 2\gamma_1$$
$$s_3 = s_2 - 2\gamma_2$$
$$= s_1 - 2(\gamma_1 + \gamma_2)$$
$$s_4 = s_3 + 2\gamma_1$$
$$= s_1 - 2\gamma_2$$
$$s_5 = s_4 - 2\gamma_3$$
$$= s_1 - 2(\gamma_2 + \gamma_3)$$
$$s_6 = s_5 - 2\gamma_1$$
$$= s_1 - 2(\gamma_1 + \gamma_2 + \gamma_3)$$
$$s_7 = s_6 + 2\gamma_2$$
$$= s_1 - 2(\gamma_1 + \gamma_3)$$
$$s_8 = s_7 + 2\gamma_1$$
$$= s_1 - 2\gamma_3$$

The value of s_6 could be larger in absolute value than $2t^2$, but no larger than $3t^2$. This illustrates the point that if the ω_i's are not used then s_i still remains in the interval $(-nt^2, nt^2)$. [2] Also note that the sequence of k_i's can be generated

---

[2] Different Hamiltonian paths on the hypercube might yield different bounds.

in linear time (to generate the sequence for C_n simply put n between two copies of the sequence for C_{n-1}).

Thus the integer recurrence

$$s_{i+1} = s_i + 2\mu_i\gamma_{k_i} - \omega_i t^2$$

together with

$$x_i \equiv (s_i/t) \text{ modulo } N \quad ; \quad y_i \equiv (x_i)^2 \text{ modulo } N$$

yield

$$|y_i| < Max\{c^2 - \frac{1}{c^2}, \frac{1}{c^2}\}N^{1/2}$$

if y_i is the member of the residue class of x_i^2 with smallest absolute value. We will now produce an integer recurrence for the y_i's.

Note that, modulo N,

$$\begin{aligned}
y_{i+1} &\equiv (s_{i+1}/t)^2 \\
&\equiv \left(\frac{s_i + 2\mu_i\gamma_{k_i} - \omega_i t^2}{t}\right)^2 \\
&\equiv \left[(\frac{s_i}{t})^2 \text{ modulo } N\right] + \frac{(2\mu_i\gamma_{k_i} - \omega_i t^2)^2}{t^2} + \frac{2s_i(2\mu_i\gamma_{k_i} - \omega_i t^2)}{t^2} \\
&\equiv y_i + \frac{(2\mu_i\gamma_{k_i} - \omega_i t^2)^2}{t^2} + \frac{2s_i(2\mu_i\gamma_{k_i} - \omega_i t^2)}{t^2}
\end{aligned}$$

Since $s_i^2 \equiv N$ modulo t^2 and $s_{i+1}^2 \equiv N$ modulo t^2 , we have $(2\mu_i\gamma_{k_i} - \omega_i t^2)^2 + 2s_i(2\mu_i\gamma_{k_i} - \omega_i t^2)$ is congruent to 0 modulo t^2. Thus

$$\frac{(2\mu_i\gamma_{k_i} - \omega_i t^2)^2}{t^2} + \frac{2s_i(2\mu_i\gamma_{k_i} - \omega_i t^2)}{t^2}$$

is an integer which is easily seen to be of order $N^{1/2}$. This means the integer recurrence

$$y_{i+1} = y_i + \frac{(2\mu_i\gamma_{k_i} - \omega_i t^2)^2}{t^2} + \frac{2s_i(2\mu_i\gamma_{k_i} - \omega_i t^2)}{t^2}$$

holds. By lemma 1, y_1 can be chosen so that $|y_1| < c_1 N^{1/2}$ where $c_1 = Max\{c^2 - \frac{1}{c^2}, \frac{1}{c^2}\}$. Again by lemma 1, the integer recurrence generates y_i's whose absolute value is less than $Max\{c^2 - \frac{1}{c^2}, \frac{1}{c^2}\}N^{1/2}$.

Now let us traverse the hypercube "modulo p", where p is a small prime. This simply means generating the sequence of y_i's modulo p. Assume p is not a factor of t. We may write

$$y_{i+1} = y_i + \Psi_i + s_i\Upsilon_i$$

$$s_{i+1} = s_i + \Delta_i$$

where

$$\Psi_i = \frac{(2\mu_i\gamma_{k_i} - \omega_i t^2)^2}{t^2} \quad ; \quad \Upsilon_i = \frac{2(2\mu_i\gamma_{k_i} - \omega_i t^2)}{t^2} \quad ; \quad \Delta_i = (2\mu_i\gamma_{k_i} - \omega_i t^2).$$

Note that Ψ_i, Υ_i and Δ_i can take on at most $6n$ values (μ_i can take on two values, ω_i can take on three values, and γ_{k_i} can take on n values). Thus, Ψ_i, Υ_i and Δ_i can be read from precomputed tables, of size $6n$ and indexed by μ_i, ω_i, k_i. Thus, computing (y_{i+1}, s_{i+1}) modulo p from (y_i, s_i) modulo p involves one multiplication and three additions modulo p. The cost of computing $s_{i+1} \bmod p$ from $s_i \bmod p$ is one addition modulo p. This fact will be used in section 4.

Note that precomputation is not possible if p divides t, since then Ψ_i and Υ_i may not be defined modulo p. More specifically, Ψ_i, Υ_i are not defined modulo p_{k_i}.

Also note that, if N is not a quadratic residue modulo p, then p does not divide y_i. This can be shown as follows: Suppose N is not a quadratic residue modulo p. Then p does not divide t^2 and therefore p divides $y_i = \frac{s_i^2 - N}{t^2}$ if and only if p divides $s_i^2 - N$. But if this was the case then $s_i^2 \equiv N$ modulo p, which would contradict the assumption that N is not a quadratic residue modulo p.

4 A factoring algorithm

The algorithm consists of visiting A vertices of the hypercube and, at each vertex s, finding the values of λ for which

$$z_\lambda = (s/t + \lambda t)^2 \bmod N \qquad (\lambda \in -M..M)$$

is B-smooth. The optimal values of A, B, and M will follow from the analysis of the algorithm. We actually do not compute the z_λ, but rather find the values of λ for which z_λ is B-smooth. So that the z_λ are "small", we will choose $t \sim \frac{N^{1/4}}{\sqrt{M}}$.

Notice that

$$z_\lambda \equiv (s/t)^2 + \lambda^2 t^2 + 2s\lambda \bmod N = \frac{s^2 - N}{t^2} + \lambda^2 t^2 + 2s\lambda.$$

Thus we set

$$\frac{s^2 - N}{t^2} + \lambda^2 t^2 + 2s\lambda \equiv 0 \bmod p.$$

This yields

$$\lambda \equiv (-s \pm \sqrt{N})t^{-2} \bmod p,$$

where $\sqrt{N}$ is a modular square root. Therefore z_λ is divisible by p for all $\lambda = kp + D_s$ and all $\lambda = kp + E_s$, where

- k is an integer;
- $D_s = (-s + \sqrt{N})t^{-2} \bmod p$;
- $E_s = (-s - \sqrt{N})t^{-2} \bmod p = D_s - 2\sqrt{N}t^{-2} \bmod p.$

Thus, the "good" λ modulo p are in arithmetic progressions. Therefore standard sieving techniques can be used to find those λ for which z_λ is B–smooth.

We may precompute $\sqrt{N} \bmod p$, $t^{-2} \bmod p$, and $-2\sqrt{N}t^{-2} \bmod p$. Therefore computing D_s and E_s involves

- one addition to compute $s \bmod p$ (see section 3);
- one addition and one multiplication to compute D_s;
- one addition to compute E_s.

Thus the total cost of moving from one vertex of the hypercube to another is, essentially, three additions and one multiplication modulo p for each prime in the factor base. This is much cheaper than the cost of changing polynomials in the Multiple Polynomial Quadratic Sieve.

We now show that the z_λ's are about $MN^{1/2}$ in absolute value. Recall that $t \sim \frac{N^{1/4}}{\sqrt{M}}$ and consider

$$z_\lambda \equiv (s/t + \lambda t)^2 \equiv (s/t)^2 + \lambda^2 t^2 + 2s\lambda \bmod N.$$

As in the proof of lemma 1 we have $(s/t)^2 \bmod N = k$ where $s^2 = kt^2 + N$. Since $s^2 < t^4 < N$, we have that k is negative. By lemma 1, $|(s/t)^2 \bmod N| = |k| \leq MN^{1/2}$. Assuming, for simplicity, that $t < \frac{N^{1/4}}{\sqrt{M}}$, we have $\lambda^2 t^2 \leq M^2\frac{N^{1/2}}{M} = MN^{1/2}$. Since $s < t^2$, we have $|2s\lambda| < 2\frac{N^{1/2}}{M}M = 2N^{1/2} << MN^{1/2}$. Thus z_λ is, essentially, the difference of two numbers in the range $0..MN^{1/2}$. We conclude that our algorithm is faster than the Multiple Polynomial Quadratic Sieve because

- vertices in the hypercube correspond to polynomials in MPQS.
- for each vertex, the cost of sieving $2M$ locations is the same in our algorithm as in MPQS.
- the size of the quadratic residues considered is, as in MPQS, about $MN^{1/2}$.
- changing polynomials is much more expensive than changing vertices of the hypercube. Therefore the optimal value for the size of the hypercube path is bigger than the optimal number of polynomials in MPQS. This means that M will be smaller in our algorithm and therefore $MN^{1/2}$ will be smaller. Thus, our algorithm will generate smaller quadratic residues than MPQS.

In practice there are many speedups to be included in an implementation of this factoring algorithm. All the enhancements described in [5] can be used with this algorithm. A rough estimate of how much faster this algorithm is than MPQS can be obtained as follows:

Let T be the running time of the algorithm, in terms of arithmetic operations on single-precision numbers. Let V be the cost of moving from one vertex to another. Let S be the cost of sieving at each vertex. Suppose we sieve modulo all primes less than B for which N is a quadratic residue. There are about $\frac{\pi(B)}{2}$ such primes. It takes four operations per prime to make a move on the hypercube. Therefore we can estimate V by $2\pi(B)$. We can estimate S by $1/2\sum_{p<B} 4M/p$ since for each prime p in the factor base about $4M/p$ locations of an accumulator

array need be updated. We can estimate this sum by $6M$. Thus our estimate for T is $A(S + V) = 2A\pi(B) + 6AM$.

Let $F(y, x)$ be the probability that a random number in Z_y factors over primes smaller than x. The number of quadratic residues considered by our algorithm is $2AM$, and each can be thought of as a random number in $Z_{\lfloor MN^{1/2} \rfloor}$. Thus about $2AMF(MN^{1/2}, B)$ of the quadratic residues will be B−smooth. Since we need about $\pi(B)/2$ smooth quadratic residues, we set

$$2AMF(MN^{1/2}, B) = \pi(B)/2.$$

Approximating $\pi(B)$ by $B/\ln B$ and $F(y, x)$ by $(\ln x/\ln y)^{\ln y/\ln x}$ (see [4]), our problem is to minimize

$$2AB/\ln B + 6AM$$

subject to

$$A = \frac{B}{4M\ln B} \left(\frac{\ln(MN^{1/2})}{\ln B} \right)^{\frac{\ln(MN^{1/2})}{\ln B}}$$

The solution to this optimization problem can be approximated numerically. For $N \sim 10^{100}$, optimal values are

$$A = 1.2 \cdot 10^5; B = 1.4 \cdot 10^8; M = 1.3 \cdot 10^7; T = 1.1 \cdot 10^{13}$$

Assuming the cost of changing polynomials in MPQS is $50\pi(B)$, [3] the numbers for MPQS are

$$A = 5.8 \cdot 10^3; B = 1.9 \cdot 10^8; M = 4.5 \cdot 10^8; T = 1.9 \cdot 10^{13}$$

Thus it appears that our techniques significantly improve on the running time of MPQS.

4.1 Remarks

1. The running-time predictions given above are very crude estimates. The true test of the running time of this algorithm will be its implementation.
2. In practice, the ω_i's defined in section 2 can be set to zero without a significant cost in the running time of the algorithm. Doing so has the advantage of diminishing the memory requirements of the algorithm.
3. In practice, the factors of t should not be small primes. This is because the numbers being sieved have a chance of $1/p$ of being divisible by p when p divides t (as opposed to $2/p$ when p is an odd prime in the factor base which does not divide t). The resulting loss of smoothness is significant for small p. Because of this, the t we use in practice may not have as many factors as the optimization of parameters requires.

---

[3] This is 25 times as expensive as in our algorithm. Changing polynomials in MPQS involves arithmetic with large numbers. Hence the cost will depend on the particular implementation of large number arithmetic. The number 25 was arrived at using "ln++", a c++ package developed at UWM.

4. Pomerance, Smith, and Tuler [7], and Montgomery (reported in [7]) propose ways of speeding up MPQS which are similar to the one proposed here. Their methods can be combined with the techniques being proposed here. It appears that doing so may further improve the running time of the algorithm.

5. The number of vertices of the hypercube to be visited by the factoring algorithm should be at most 2^{n-1}, where n is the number of factors of t. Otherwise duplication of polynomials occurs, since $(s/t+\lambda t)^2$ and $((t^2-s)/t+\lambda t)^2$ are essentially equivalent.

Acknowledgments

I am indebted to anonymous referees for numerous suggestions, including remarks 3 and 4 above.

References

1. L.M. Adleman. Factoring numbers using singular integers. In *Proceedings of the 23th Annual ACM Symposium on the Theory of Computing*, pages 64–71, 1991.
2. D.H. Lehmer and R.E. Powers. On factoring large numbers. *Bull. Amer. Math. Soc.*, 37:770–776, 1931.
3. A.K. Lenstra, H. W. Lenstra, M.S. Manasse, and J.M. Pollard. The number field sieve. In *Proceedings of the 22th Annual ACM Symposium on the Theory of Computing*, pages 564–572, 1990.
4. D. Knuth and L. Trabb Pardo. Analysis of a simple factorization algorithm. *Theoretical Computer Science*, 3:321–348, 1976.
5. A.K. Lenstra and M.S. Manasse. Factoring by electronic mail. In *Advances in Cryptology - proceedings of EUROCRYPT 89*, volume 434, pages 355–371. Springer-Verlag, 1990.
6. C. Pomerance. Analysis and comparison of some integer factoring algorithms. 154:89–139, 1982.
7. C. Pomerance, J.W. Smith, and R. Tuler. A pipeline architecture for factoring large integers with the quadratic sieve algorithm. *SIAM Journal on Computing*, 17(2):387–403, 1988.
8. R. Silverman. The multiple polynomial quadratic sieve. *Mathematics of Computation*, 48(177):329–339, 1987.

Efficient Multiplication on Certain Nonsupersingular Elliptic Curves

Willi Meier [1]   Othmar Staffelbach [2]

[1] HTL Brugg-Windisch
CH-5200 Windisch, Switzerland

[2] Gretag Data Systems AG
CH-8105 Regensdorf, Switzerland

Abstract

Elliptic curves defined over finite fields have been proposed for Diffie-Hellman type crypto systems. Koblitz has suggested to use "anomalous" elliptic curves in characteristic 2, as these are nonsupersingular and allow for efficient multiplication of points by an integer.

For anomalous curves E defined over $\mathbf{F}_2$ and regarded as curves over the extension field $\mathbf{F}_{2^n}$, a new algorithm for computing multiples of arbitrary points on E is developed. The algorithm is shown to be three times faster than double and add, is easy to implement and does not rely on precomputation or additional memory. The algorithm is used to generate efficient one-way permutations involving pairs of twisted elliptic curves by extending a construction of Kaliski to finite fields of characteristic 2.

1 Introduction

Elliptic curves defined over finite fields have been proposed for Diffie-Hellman type crypto systems [7,4] as well as for implementation of one-way permutations [2]. In particular, in [3] Koblitz has described the class of "anomalous" elliptic curves which in characteristic 2 have the following useful properties

1. They are nonsupersingular, so that one cannot use the Menezes-Okamoto-Vanstone reduction [6] of discrete logarithms from elliptic curves to finite fields.

2. Multiplication of points by an integer m can be carried out almost as efficiently as in the case of supersingular curves.

According to [3] an elliptic curve E defined over the field $\mathbf{F}_q$ is called anomalous if the trace of the Frobenius map $((x,y) \mapsto (x^q, y^q))$ is equal to 1. Equivalently, an elliptic curve over $\mathbf{F}_q$ is anomalous if and only if the number of $\mathbf{F}_q$-points is equal to q. As in [3] we will concentrate on curves in characteristic 2, and in particular on the anomalous curve

$$E : y^2 + xy = x^3 + x^2 + 1 \tag{1}$$

defined over $\mathbf{F}_2$. We will also consider its twist $\tilde{E}$ over $\mathbf{F}_2$, which is given by the equation $y^2 + xy = x^3 + 1$. Subsequently these curves will be considered over the extension fields $\mathbf{F}_{2^n}$. Hereby let E_n denote the $\mathbf{F}_{2^n}$-points of the curve E, and $\tilde{E}_n$ its twist over $\mathbf{F}_{2^n}$.

In applications, e.g., in a Diffie-Hellman key exchange, multiples mP of points P on the curve E_n have to be computed. In standard algorithms for multiplication, e.g, by double and add, this is reduced to a number of additions of points on E_n. Since these additions consume most of the computation time, it is desirable to have algorithms which need fewer additions on E_n. In [3] it is suggested to express multiplication by m as linear combinations of powers of the Frobenius map ϕ, as these can be computed by iterated squaring in $\mathbf{F}_{2^n}$ which, in a normal basis representation, is easily accomplished by shift operations. In [3] expansions of the form

$$m = \sum_j c_j \phi^j \tag{2}$$

are considered with $c_j \in \{0, \pm 1\}$. With this representation of m the computation of mP can be reduced to $l - 1$ additions where l is the number of nonzero terms in (2). Therefore it is desirable to have short expressions (2). The expansions given in [3] in the average have twice the length of the binary expansion of m.

In this paper we elaborate constructions of short expansions (2). In particular, in Section 2 we prove that there always exists an expansion $m = \sum_{j=0}^{n-1} c_j \phi^j$ of length n, where n is the degree of the extension field (Theorem 1). The proof of Theorem 1 leads to an efficient algorithm which produces expansions where half of the coefficients c_j are expected to be zero (Corollary 4).

Our construction exploits the fact that the endomorphism ring $\text{End}(E)$ of the curve E is related to the ring $\mathbf{Z}[\alpha] = \{a + b\alpha \mid a, b \in \mathbf{Z}\} \subset \mathbf{C}$, where $\alpha = (1 + \sqrt{-7})/2$. In particular we will reduce the problem of finding ϕ-expansions in $\text{End}(E)$ to finding α-expansions in $\mathbf{Z}[\alpha]$, where we make specific use of the rich algebraic structure of the ring $\mathbf{Z}[\alpha]$. The computational complexity of the reduction algorithm is of magnitude of a n-bit integer multiplication.

Since execution of ϕ^j is obtained almost for free, the ϕ-expansion of m allows to compute mP for an arbitrary point P on E_n with $n/2$ additions in the average. As the computation of the ϕ-expansion is negligible compared with a full multiplication by m on the curve, this results in an improvement by a factor 3 compared to double and add without using precomputation or additional memory. At this point we note that other methods have been proposed for accelerating this operation (see e.g., [1]). However these methods only apply if the point P is assumed to be fixed. Furthermore they need precomputation with this predefined point P (and additional memory). Observe for example that P cannot be assumed to be fixed in the second step of a Diffie-Hellman key exchange protocol.

Our results also apply to generate efficient one-way permutations based on elliptic curves. In [2] Kaliski has proposed a construction of one-way permutations involving pairs of twisted elliptic curves over $\mathbf{F}_p$ for large prime numbers p. It is easy to generalize the treatment in [2] to any extension field $\mathbf{F}_{p^n}$ of $\mathbf{F}_p$. In Section 3 we apply the construction to extension fields $\mathbf{F}_{2^n}$ in characteristic 2. The treatment in characteristic 2 differs from the treatment in odd characteristic. However the construction in characteristic 2 appears to be particularly attractive, as arithmetic can be carried out efficiently. On certain curves, arithmetic can be accelerated by using the ϕ-expansion of multiplication by m. Restriction to curves with short ϕ-expansion leaves enough freedom to find examples of curves with good cryptographic properties.

2   Frobenius Expansion of Multiplication by m

On an anomalous curve over $\mathbf{F}_q$, the Frobenius map ϕ satisfies the characteristic equation $T^2 - T + q = 0$. We will also consider the twist $\tilde{E}$ of E, whose Frobenius satisfies $T^2 + T + q = 0$. The number of $\mathbf{F}_q$-points on $\tilde{E}$ is $q + 2$. The "n-twist" $\tilde{E}_n$ is the twist of E regarded as curve over the extension field $\mathbf{F}_{q^n}$. Using the Weil conjecture (see [8, p. 136]), the number N_n of $\mathbf{F}_{q^n}$-points can be computed as

$$N_n = |\alpha^n - 1|^2 = |\beta^n - 1|^2 = 1 + q^n - \alpha^n - \beta^n, \tag{3}$$

where α and β in $\mathbf{C}$ are the roots of the characteristic equation $T^2 - T + q = 0$. The number $\tilde{N}_n$ of points on the twist $\tilde{E}_n$ is given by $\tilde{N}_n = |\alpha^n + 1|^2 = 1 + q^n + \alpha^n + \beta^n$. Equivalently, N_n and $\tilde{N}_n$ can be computed as $N_n = q^n + 1 - a_n$ and $\tilde{N}_n = q^n + 1 + a_n$, where $a_n = \alpha^n + \beta^n$ for $n \geq 2$ satisfies the recursion $a_n = a_{n-1} - q a_{n-2}$ with the initial values $a_0 = 2$ and $a_1 = 1$.

We now will concentrate on anomalous curves in characteristic 2, and in particular on the anomalous curve $E : y^2 + xy = x^3 + x^2 + 1$ defined over $\mathbf{F}_2$. Its twist over $\mathbf{F}_2$ is given by $\tilde{E} : y^2 + xy = x^3 + 1$. Let E_n denote the curve E regarded over the extension field $\mathbf{F}_{2^n}$, and $\tilde{E}_n$ its twist over $\mathbf{F}_{2^n}$.

Our aim in this section is to express multiplication by m as short linear combinations of powers of the Frobenius map ϕ, as this will lead to an efficient computation of multiples mP of arbitrary points on E_n. In [3] expansions of the form

$$m = \sum_j c_j \phi^j \tag{4}$$

are considered with $c_j \in \{0, \pm 1\}$. The expansions given in [3] in the average have twice the length of the binary expansion of m. On the other hand, from [5, p. 149] one concludes that there must be shorter expansions of the form

$$m = \sum_{j=0}^{n-1} a_j \phi^j, \tag{5}$$

possibly with larger coefficients, however. From [5] one can merely deduce that $|a_j| \leq 7$.

In the following theorem we show that one can construct expansions which simultaneously satisfy the conditions of (4) and (5).

Theorem 1 *For the anomalous curve $E : y^2 + xy = x^3 + x^2 + 1$ defined over $\mathbf{F}_2$, let E_n be the curve regarded over the extension field $\mathbf{F}_{2^n}$. Then on E_n multiplication by an integer m can be expressed as*

$$m = \sum_{j=0}^{n-1} c_j \phi^j, \tag{6}$$

with $c_j \in \{0, \pm 1\}$.

This theorem also holds for $\tilde{E}_n$. The proof proceeds in several steps. First observe that the Frobenius map satisfies the equation $\phi^2 - \phi + 2 = 0$, and that there is a natural homomorphism from the ring $\mathbf{Z}[\alpha] = \{a + b\alpha \,|\, a, b \in \mathbf{Z}\} \subset \mathbf{C}$ to the endomorphism ring $\text{End}(E)$ of E which maps $\alpha = (1 + \sqrt{-7})/2$ to ϕ. Thus, if we have an expansion $m = \sum_j c_j \alpha^j$ in $\mathbf{Z}[\alpha]$, we immediately get a corresponding expansion $m = \sum_j c_j \phi^j$ in $\text{End}(E)$. This means that $mP = \sum_j c_j \phi^j(P)$ for every point P on E_n. For finding such an expansion in $\mathbf{Z}[\alpha]$ we will make use of the algebraic structure of the ring $\mathbf{Z}[\alpha]$. Note that $\mathbf{Z}[\alpha]$ is an Euclidean domain with respect to the norm $N(a + b\alpha) = |a + b\alpha|^2 = (a + b\alpha)(a + b\overline{\alpha}) = a^2 + ab + 2b^2, a, b \in \mathbf{Z}$. For the proof of the theorem we will make use of the following stronger property.

Lemma 2 *For any $s, t \in \mathbf{Z}[\alpha]$, $t \neq 0$, there exist $q, r \in \mathbf{Z}[\alpha]$ such that $s = qt + r$ with*

$$N(r) \leq \frac{4}{7} N(t). \tag{7}$$

Proof. The elements of the ring $\mathbf{Z}[\alpha]$ form a lattice in $\mathbf{C}$, and the whole of $\mathbf{C}$ can be covered by triangles whose vertices are in $\mathbf{Z}[\alpha]$, as depicted in Figure 1. Consider the

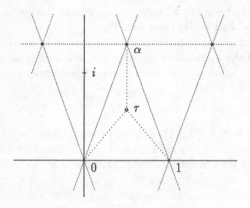

Figure 1: The lattice $\mathbf{Z}[\alpha]$.

triangle with vertices 0, 1 and α. The point $\tau = 1/2 + (3/(2\sqrt{7}))\, i$ is the center of the circumscribed circle of the triangle, as is easily verified by computing the distance of

τ to each vertex, that is $|\tau - 0| = |\tau - 1| = |\tau - \alpha| = 2/\sqrt{7}$. It follows that any other point in the triangle has distance less than $2/\sqrt{7}$ to some vertex. Since any point $z \in \mathbf{C}$ lies in some triangle, we conclude that for any complex number $z \in \mathbf{C}$ there is an element $u \in \mathbf{Z}[\alpha]$ with $N(z - u) \leq (2/\sqrt{7})^2 = 4/7$.

Now let $s, t \in \mathbf{Z}[\alpha]$ with $t \neq 0$. Consider the quotient $v = s/t$ computed in the quotient field of $\mathbf{Z}[\alpha]$, i.e., in the field $\mathbf{Q}(\alpha) = \{a + b\alpha \,|\, a, b \in \mathbf{Q}\} \subset \mathbf{C}$. Then, as discussed above, there is an element $q \in \mathbf{Z}[\alpha]$ with $N(v - q) \leq 4/7$, and $r = s - qt = t(v - q)$ has norm $N(r) = N(v - q)N(t) \leq (4/7)N(t)$, which implies that $q, r \in \mathbf{Z}[\alpha]$ have the properties as stated in the lemma. $\square$

Lemma 3 *For any $s \in \mathbf{Z}[\alpha]$ with norm $N(s) < 2^n$, $n \in \mathbf{N}$, there is an expansion*

$$s = \sum_{j=0}^{n-1} c_j \alpha^j \tag{8}$$

of length n with $c_j \in \{0, \pm 1\}$.

Proof. The proof is by induction on n. For $n = 1, 2$, consider the elements in $\mathbf{Z}[\alpha]$ with norm less than 4. These are the element 0 with norm 0, the elements ± 1 with norm 1 and the elements $\pm \alpha, \pm(1 - \alpha)$ with norm 2. For these elements the statement of the lemma holds as is seen by direct inspection.

Now consider $s \in \mathbf{Z}[\alpha]$ with $N(s) < 2^n, n > 2$. Since $\mathbf{Z}[\alpha]$ is an Euclidean domain, s can be expressed as

$$s = s'\alpha + c \tag{9}$$

with $N(c) < N(\alpha) = 2$, i.e., with $c \in \{0, \pm 1\}$. The idea is to reduce the problem of finding an expansion for s to the problem of finding an expansion for s'. If $c = 0$, i.e., if α divides s, the reduction (9) is unique. Otherwise, as α divides 2, there is always a reduction with $c = 1$ and another reduction with $c = -1$. If the reduction could be done such that $N(s') \leq N(s)/2 < 2^{n-1}$, the proof would easily be completed by induction. There are situations however, where there is no reduction with $N(s') \leq N(s)/2$, as we shall see below. We will distinguish between the following three cases:

1. *Non-critical case:* There is a reduction (9) with $N(s') < N(s)/2$.

2. *Semi-critical case:* There is a reduction (9) with $N(s') = N(s)/2$.

3. *Critical case:* There are only reductions (9) with $N(s') > N(s)/2$.

If α divides s, we have the reduction $s = s'\alpha$ with $c = 0$ and $N(s') = N(s)/2$, i.e., s is semi-critical. If α does not divide s, α is a divisor of both, $s - 1$ and $s + 1$. In this case the type of the reduction turns out to depend on the absolute value of the real part $\Re(s)$ of s:

1. **Non-critical case:** $|\Re(s)| \geq 1$. Assume for example that $\Re(s) \geq 1$, as illustrated for $s = s_1$ in Figure 2. Then $N(s - 1) < N(s)$, and we have the reduction $s = s'\alpha + 1$ with $N(s') = N(s - 1)/N(\alpha) < N(s)/2$. Similarly, if $\Re(s) \leq -1$, we have $s = s'\alpha - 1$ with $N(s') < N(s)/2$.

2. Semi-critical case: $|\Re(s)| = 1/2$. Assume for example that $\Re(s) = 1/2$, as illustrated for $s = s_2$ in Figure 2. Then $N(s-1) = N(s)$, and we have the reduction $s = s'\alpha + 1$ with $N(s') = N(s-1)/N(\alpha) = N(s)/2$. Similarly, if $\Re(s) = -1/2$, we have $s = s'\alpha - 1$ with $N(s') = N(s)/2$.

3. Critical case: $\Re(s) = 0$. This is illustrated for $s = s_3$ in Figure 2. Then, by Pythagoras' theorem, $N(s-1) = N(s+1) = N(s) + 1$, and we have the reductions $s = s'\alpha + 1$ and $s = s''\alpha - 1$ with

$$N(s') = N(s'') = \frac{N(s) + 1}{2} \tag{10}$$

Since $s'' - s' = 2/\alpha = 1 - \alpha$, either s' or s'' is not divisible by α. Assume that s' is not divisible by α. We claim that s' has a non-critical reduction. For this it suffices to show that $|\Re(s')| \geq 1$.

Since $\Re(s) = 0$, s must be of the form $s = a\sqrt{-7}$ for some odd integer $a \in \mathbf{Z}$. Then s' can be computed in $\mathbf{Q}(\alpha)$ as

$$s' = (s-1)\alpha^{-1} = (a\sqrt{-7} - 1)\frac{1}{4}(1 - \sqrt{-7}) = \frac{7a-1}{4} + \frac{a+1}{4}\sqrt{-7}.$$

It follows that $|\Re(s')| \geq 3/2$. Hence s' is non-critical. Similarly, s'' is non-critical if α does not divide s''.

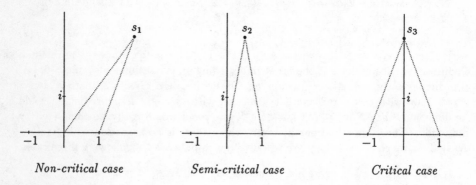

Figure 2:

Now the proof of the lemma is easily accomplished. In case that s has a non-critical or semi-critical reduction $s = s'\alpha + c$, we have $N(s') \leq N(s)/2 < 2^{n-1}$. By induction hypothesis, s' has an expansion in α of length $n - 1$, which yields an expansion of s in α of length n.

In case that s has a critical reduction $s = s'\alpha + c$, we have according to (10), $N(s') = (N(s) + 1)/2 \leq 2^{n-1}$. Since the inequality $N(s') \leq 2^{n-1}$ does not hold strictly, we cannot apply the induction hypothesis to s'. However, as discussed above, the reduction can be done such that s' has a non-critical reduction $s' = s''\alpha + c'$, i.e., $N(s'') < N(s')/2 \leq 2^{n-2}$. Thus $s = s''\alpha^2 + c'\alpha + c$, and by induction hypothesis, s''

has an expansion in α of length $n - 2$, which yields an expansion of s in α of length n. This completes the proof of the lemma. $\square$

Now we are in position to prove Theorem 1. As the curve E_n is regarded over the extension field $\mathbf{F}_{2^n}$, the Frobenius map satisfies the equation $\phi^n = 1$. It follows that for any two α-expansions which are congruent modulo $\alpha^n - 1$ the corresponding ϕ-expansions yield the same endomorphism on E_n. Therefore we compute the α-expansion of the remainder m' of the division of m by $\alpha^n - 1$,

$$m = q(\alpha^n - 1) + m', \tag{11}$$

where, according to Lemma 2, $N(m') \leq (4/7)N(\alpha^n - 1)$. To obtain a bound on $N(m')$ we compute (see formula (3))

$$N(\alpha^n - 1) = (\alpha^n - 1)(\beta^n - 1) = (\alpha\beta)^n - (\alpha^n + \beta^n) + 1 = 2^n + 1 - (\alpha^n + \beta^n) = N_n. \tag{12}$$

By Hasse's theorem (see [8, p.131]), $N_n \leq f(n) = 2^n + 1 + 2^{n/2+1}$, and for $n \geq 4$, $(4/7)f(n) < 2^n$, as $g(n) = 2^n - (4/7)f(n)$ is strictly increasing for $n \geq 1$ and strictly positive for $n = 4$. Hence for $n \geq 4$, $N(m') < 2^n$ and the theorem follows from Lemma 3. For $n \leq 3$ the statement of the theorem can be verified directly. $\square$

Note that an arbitrary element $s = a + b\alpha$ in $\mathbf{Z}[\alpha]$ is divisible by α if and only if a is even. Hence with probability $1/2$ this element has a reduction of the form $s = s'\alpha$, i.e., with $c = 0$. Continuing the reduction, it is to be expected that the intermediate results s' also have this property. This would imply that half of the coefficients c_j in (8) can be expected to be zero. This has been confirmed experimentally.

Corollary 4 (Experimental result) *In the expansion $m = \sum_{j=0}^{n-1} c_j \phi^j$ half of the coefficients c_j are expected to be zero.*

It is easy to compute the α-expansion of an arbitrary element $s = a + b\alpha \in \mathbf{Z}[\alpha]$. From the proof of Lemma 3 one can derive the following simple and efficient procedure which outputs c_j in ascending order for j.

```
While a ≠ 0 or b ≠ 0 do begin
    if a is even then
        c := 0;
    else begin
        if 2a + b ≠ 0 then c := sgn(2a + b);
        if 2a + b = 0 then begin
            if a ≡ 1  (mod 4) then c := -1;
            if a ≡ 3  (mod 4) then c := 1;
        end;
    end;
    x := (a - c)/2;   a := x + b;   b := -x;
    output(c);
end.
```

The problem of efficiently finding short ϕ-expansions of multplication by an arbitrary m was addressed by Koblitz in [3]. In the above procedure, the amount of work to perform the division (11) is roughly of the same magnitude as to perform the reduction. This is of magnitude of a n-bit integer multiplication, and is negligible in comparison with a full multiplication by m on the elliptic curve.

As execution of ϕ^j is obtained almost for free, according to Corollary 4, multiplication by m can be carried out with $n/2$ additions in the average. This results in an improvement by a factor 3 compared to double and add without using precomputation or additional memory.

The results of Theorem 1 and Corollary 4 may also be applied to the key exchange procedure suggested by H. Lenstra as mentioned in [3, p.285]. In this suggestion one chooses expansions $m = \sum_{j=0}^{n-1} c_j \phi^j$ where only a certain maximum number of coefficients c_j are allowed to be nonzero. However it is unclear which multiples are obtained when applying this restriction. Furthermore certain multiples could occur more than once which would result in a non uniform probability distribution of the chosen values of m, or in a non uniform distribution of the keys. Theorem 1 allows to obtain every multiple with the same probability by choosing m first and then making the reduction.

3 One-Way Permutations on Elliptic Curves in Characteristic 2

In [2] elliptic curves have been suggested as a tool for generating one-way permutations. Two constructions have been proposed in [2], one involving single elliptic curves and the other one involving pairs of twisted elliptic curves. Both constructions deal with curves over $\mathbf{F}_p$ for large prime numbers p. As already observed in [2], the elliptic curves used in the first construction are supersingular, so that the Menezes-Okamoto-Vanstone reduction [6] can be applied. The second construction applies to arbitrary elliptic curves over $\mathbf{F}_p$ for any odd prime number $p > 3$. It is easy to generalize the treatment in [2] to any extension field $\mathbf{F}_{p^n}$ of $\mathbf{F}_p$.

In this section we apply the second construction to extension fields $\mathbf{F}_{2^n}$ in characteristic 2. The treatment in characteristic 2 differs from the treatment in odd characteristic. However the construction in characteristic 2 appears to be particularly attractive for the following reasons.

1. Arithmetic in characteristic 2 can be carried out efficiently.

2. On certain curves, arithmetic can be accelerated by using the ϕ-expansion of multiplication by m.

3. Even restriction to anomalous curves leaves enough freedom to find curves with good cryptographic properties.

In the following all curves are considered to be defined over fields with characteristic 2. Recall that an elliptic curve in characteristic 2 is nonsupersingular if and only if the j-invariant is nonzero (see [8, p. 145]). The normal form of an elliptic curve E with $j(E) \neq 0$ is given by

$$y^2 + xy = x^3 + a_2 x^2 + a_6, \tag{13}$$

where $a_6 \neq 0$. If a_2, a_6 are in $\mathbf{F}_{2^n}$, the curve is defined over $\mathbf{F}_{2^n}$. The twist $\tilde{E}$ of E, up to isomorphism, is given by

$$y^2 + xy = x^3 + (a_2 + D)x^2 + a_6, \tag{14}$$

where $D \in \mathbf{F}_{2^n}$ is such that the polynomial $t^2 + t + D$ is irreducible over $\mathbf{F}_{2^n}$. Observe that E and $\tilde{E}$ are non-isomorphic over $\mathbf{F}_{2^n}$ but are isomorphic over $\mathbf{F}_{2^{n+1}}$. Now we prove the analogue of Lemma 4.1 in [2].

Lemma 5 *Every nonzero $x \in \mathbf{F}_{2^n}$ appears either as x-coordinate of exactly two points on E or as x-coordinate of exactly two points on $\tilde{E}$. The elliptic curve E together with its twist $\tilde{E}$ have order $2(2^n + 1)$, i.e., $\#E + \#\tilde{E} = 2(2^n + 1)$.*

Proof. For a fixed $x \neq 0$ the equation in y for (x, y) to be on E can be written as

$$t^2 + t + c = 0, \tag{15}$$

with $t = y/x$ and where $c = (x^3 + a_2 x^2 + a_6)/x^2$ is a constant. Similarly, with the same notation, the equation in y for (x, y) to be on $\tilde{E}$ is

$$t^2 + t + (c + D) = 0. \tag{16}$$

The equation $t^2 + t + c = 0$ has a solution if and only if c is in the image of the mapping $Q : \mathbf{F}_{2^n} \to \mathbf{F}_{2^n}, Q(t) = t^2 + t$. Since Q is a homomorphism of the additive group $\mathbf{F}_{2^n}$, with kernel $\mathbf{F}_2$, the image $\operatorname{im} Q$ is a subgroup of index 2 in $\mathbf{F}_{2^n}$. By assumption $t^2 + t + D$ is irreducible over $\mathbf{F}_{2^n}$, hence $D \notin \operatorname{im} Q$. As a consequence exactly one of the two elements c and $c + D$ is in $\operatorname{im} Q$. This implies that exactly one of the equations (15) and (16) has (two) solutions. Thus we conclude that every nonzero x appears either as x-coordinate of exactly two points on E or as x-coordinate of exactly two points on $\tilde{E}$, which implies the first part of the lemma.

For $x = 0$ we get the equation $y^2 = a_6$ for both curves. This equation always has exactly one solution, as squaring in F_{2^n} is a bijection. The latter holds as 2 is relatively prime to $|\mathbf{F}_{2^n}{}^*| = 2^n - 1$. Counting the points on E and $\tilde{E}$ we get $2(2^n - 1)$ points with $x \neq 0$, two points with $x = 0$ and the two points at infinity. This implies that $\#E + \#\tilde{E} = 2(2^n + 1)$ (which also follows from the Weil conjecture (3)). $\square$

Our aim is to identify the elements of E and $\tilde{E}$ with certain integers. For a given representation of the elements of $\mathbf{F}_{2^n}$ as residues modulo a fixed irreducible polynomial, we first identify the elements of $\mathbf{F}_{2^n}$ with the integers $0, 1, \ldots, 2^n - 1$ as follows: The polynomial $f(t) = c_{n-1} t^{n-1} + \ldots + c_1 t + c_0 \in \mathbf{F}_2[t]$ considered as element of $\mathbf{F}_{2^n}$ is identified with the integer $c_{n-1} 2^{n-1} + \ldots + c_1 2 + c_0$. This bijection defines an ordering of $\mathbf{F}_{2^n}$. This ordering is in no way compatible with the algebraic structure of the field, but we can use it to construct a map ℓ from $E \cup \tilde{E}$ to the integers.

First we define ℓ on E. For $x \neq 0$ suppose that $(x,y) \in E$. Then $(x, x+y)$ is the other point on E with the same x-coordinate. The idea in the definition of ℓ is to map the point with the smaller y-coordinate to the set $1, \ldots, 2^n - 1$ and the point with the larger y-coordinate to the set $2^n + 2, \ldots, 2^{n+1}$.

$$\ell(x,y) = 0 \qquad \text{if } x = 0 \text{ and } y = \sqrt{a_6} \tag{17}$$
$$\ell(x,y) = x \qquad \text{if } x \neq 0 \text{ and } y < x+y \tag{18}$$
$$\ell(x,y) = x + 2^n + 1 \quad \text{if } x \neq 0 \text{ and } y > x+y \tag{19}$$
$$\ell(\infty) = 2^n \tag{20}$$

The definition of ℓ on $\tilde{E}$ is similar.

$$\ell(x,y) = 2^n + 1 \qquad \text{if } x = 0 \text{ and } y = \sqrt{a_6} \tag{21}$$
$$\ell(x,y) = x \qquad \text{if } x \neq 0 \text{ and } y < x+y \tag{22}$$
$$\ell(x,y) = x + 2^n + 1 \quad \text{if } x \neq 0 \text{ and } y > x+y \tag{23}$$
$$\ell(\infty) = 2^{n+1} + 1 \tag{24}$$

Theorem 6 *Let $E : y^2 + xy = x^3 + a_2 x^2 + a_6$ be a nonsupersingular elliptic curve, and $\tilde{E} : y^2 + xy = x^3 + (a_2 + D)x^2 + a_6$ its twist over $\mathbf{F}_{2^n}$. Then the map ℓ as defined in (17) - (24) is a bijection from $E \cup \tilde{E}$ to the set of numbers $\{0, 1, \ldots, 2^{n+1} + 1\}$.*

Proof. According to Lemma 5 the set of possible nonzero x-coordinates of points on E and on $\tilde{E}$ are disjoint. Therefore ℓ, as defined in (17) - (24), is injective. Hence ℓ is bijective, as by Lemma 5 the two sets have the same cardinality. $\square$

We now assume that both curves E and $\tilde{E}$ are cyclic with generators $G \in E$ and $\tilde{G} \in \tilde{E}$. Let N denote the order of E. Then we define a map $f : \{0, \ldots, 2^{n+1} + 1\} \rightarrow \{0, \ldots, 2^{n+1} + 1\}$, as in [2] by

$$f(m) = \ell(mG) \quad \text{if } 0 \leq m < N \tag{25}$$
$$f(m) = \ell(m\tilde{G}) \quad \text{if } N \leq m < 2^{n+1} + 2 \tag{26}$$

As a consequence of Theorem 6 we obtain the following

Corollary 7 *Let $E : y^2 + xy = x^3 + a_2 x^2 + a_6$ be a nonsupersingular elliptic curve, and $\tilde{E} : y^2 + xy = x^3 + (a_2 + D)x^2 + a_6$ its twist over $\mathbf{F}_{2^n}$. If both curves E and $\tilde{E}$ are cyclic, then the function f as defined in (25) and (26) is a permutation of the set $\{0, \ldots, 2^{n+1} + 1\}$.*

As observed in [2], inverting the permutation f is equivalent to solving the discrete logarithm problem on the elliptic curves.

Our aim is to find practical examples where both curves E and $\tilde{E}$ are cyclic. At the same time the order of each curve should have at least one large prime divisor such that computation of discrete logarithms is supposed to be hard. A finite abelian group is cyclic if and only if the p-primary component of the group is cyclic for each prime p dividing the order of the group. For the p-primary component for $p = 2$ we have

Proposition 8 *For a nonsupersingular elliptic curve in characteristic 2 the 2-primary component is always cyclic.*

Proof. Let $P_0 = (x_0, y_0) \in E$ be a point of order 2, i.e. $2P_0 = 0$, or $P_0 = -P_0$. For $j(E) \neq 0$ the curve E has the normal form $y^2 + xy = x^3 + a_2 x^2 + a_6$, and the negative of a point $P = (x, y)$ is computed as $-P = (x, -y - x)$ (see [8, p. 58]). This implies that $y_0 = -y_0 - x_0$, hence $x_0 = -2y_0 = 0$ and $y_0 = \sqrt{a_6}$, i.e., there is only one point of order 2. $\square$

In order to guarantee that the p-primary component is cyclic for odd primes p we are looking for curves whose order is not divisible by p^2. For examples we concentrate on the anomalous curve $E : y^2 + xy = x^3 + x^2 + 1$ defined over $\mathbf{F}_2$ as discussed in Section 2. Thus denote by N_n the number of $\mathbf{F}_{2^n}$-points on E and by $\tilde{N}_n$ the number of $\mathbf{F}_{2^n}$-points on $\tilde{E}$. The degrees $n = 107$ and $n = 181$ of the extension fields turn out to be favourable in view of the desired criteria. The prime factorization of the corresponding orders N_n and $\tilde{N}_n$ are given as follows.

$$N_{107} = 2 \cdot 811296384146066692182851032212511$$
$$\tilde{N}_{107} = 4 \cdot 405648192073033356043634890037809$$

$$N_{181} = 2 \cdot 122719 \cdot 23531 \cdot 5306974831684643967309408891155993708352266943$$
$$\tilde{N}_{181} = 4 \cdot 1087 \cdot 12671 \cdot 115117 \cdot 307339 \cdot 1572431197704155598636826628289553813$$

The first example contains prime numbers with 32 decimal digits. This example is already mentioned in [3]. The second example contains prime numbers with 45 and 37 decimal digits, respectively.

References

[1] E. Brickell, D.M. Gordon, K.S. McCurley, D. Wilson, *Fast Exponentiation with Precomputation*, Eurocrypt'92, to appear.

[2] B.S. Kaliski, Jr., *One-way Permutations on Elliptic Curves*, Journal of Cryptology, Vol.3, No. 3, pp.187–199, 1991.

[3] N. Koblitz, *CM-Curves with Good Cryptographic Properties*, Advances in Cryptology—Crypto'91, Proceedings, pp. 279–287, Springer-Verlag, 1992.

[4] N. Koblitz, *Elliptic Curve Crypto Systems*, Math. of Computation, Vol. 48, pp. 203–209, 1987.

[5] N. Koblitz, *Hyperelliptic Cryptosystems*, Journal of Cryptology, Vol. 1, No. 3, pp. 139–150, 1989.

[6] A. Menezes, T. Okamoto, S.A. Vanstone, *Reducing Elliptic Curve Logarithms to Logarithms in a Finite Field*, Proceedings of the 23rd ACM Symp. Theory of Computing, 1991.

[7] V. Miller, *Use of Elliptic Curves in Cryptography*, Advances in Cryptology—Crypto'85, Proceedings, pp. 417–426, Springer-Verlag, 1986.

[8] J.H. Silverman, *The Arithmetic of Elliptic Curves*, Graduate Texts in Mathematics, Springer-Verlag, 1986.

Speeding up Elliptic Cryptosystems by Using a Signed Binary Window Method

Kenji Koyama,   Yukio Tsuruoka

NTT Communication Science Laboratories
Seikacho, Kyoto, 619-02 Japan

Abstract. The basic operation in elliptic cryptosystems is the computation of a multiple $d \cdot P$ of a point P on the elliptic curve modulo n. We propose a fast and systematic method of reducing the number of operations over elliptic curves. The proposed method is based on pre-computation to generate an adequate addition-subtraction chain for multiplier the d. By increasing the average length of zero runs in a signed binary representation of d, we can speed up the window method. Formulating the time complexity of the proposed method makes clear that the proposed method is faster than other methods. For example, for d with length 512 bits, the proposed method requires 602.6 multiplications on average. Finally, we point out that each addition/subtraction over the elliptic curve using homogeneous coordinates can be done in 3 multiplications if parallel processing is allowed.

1   Introduction

Elliptic curves over a finite field $\mathbf{F}_p$ or a ring $\mathbf{Z}_n$ can be applied to implement analogs [9] [11] [13] of the Diffie-Hellman scheme [4], ElGamal scheme [6] and RSA scheme [15], as well as primality testing [7] and integer factorization [12][13]. Cryptosystems based on elliptic curves, called *elliptic cryptosystems*, seem more secure than the original schemes. For example, it is conjectured that the low exponent attack on the RSA scheme cannot be analogously applied to the attack on the elliptic RSA scheme using a low multiplier [9]. The basic operation performed on an elliptic curve is the computation of a multiple $d \cdot P$ of a point P on the elliptic curve modulo n, which corresponds to the computation of x^d mod n. For a large n and d, the time complexity of elementary operations as well as the number of elementary operations are very high. Thus, reducing the number of such operations is important when implementing the above algorithms.

One solution is a so-called *binary method* [10] based on the *addition chain* [10] for multipliers d of $d \cdot P$ or exponents d of x^d. In general, an addition chain for a given d is a sequence of positive integers

$$a_0 \ (= 1) \rightarrow a_1 \rightarrow a_2 \rightarrow \cdots \rightarrow a_r \ (= d),$$

where r is the number of additions, and $a_i = a_j + a_k$, for some $k \leq j < i$, for all $i = 1, 2, \ldots, r$. The binary method is a systematic algorithm based on an addition chain with elements that are powers of 2, i.e. a two-valued binary representation of d. To evaluate $d \cdot P$ or x^d, the ordinary binary method without pre-computation requires $\frac{3}{2} \lfloor \log_2 d \rfloor$ multiplications on average. The ordinary binary method does not always guarantee the minimum number of multiplications (the shortest addition chain). Obtaining the shortest addition chain is a NP-complete problem [5]. There have been many studies on the computation of x^d [1] [2] [3] [8] [16] and a few studies on the computation of $d \cdot P$ [14] to achieve fast and efficient computation. Among the variants of the binary method attempted to speed up the computation of x^d, Bos and Coster [1] proposed a heuristic window method based on an *addition sequence*. The addition sequence is a generalized addition chain including the given set of values. In their algorithm, the two-valued binary representation of d is split into pieces (windows), and the value of each window is computed in shorter addition sequence.

An addition chain can be extended to an *addition-subtraction chain* [2] [10] [14], with a rule $a_i = a_j \pm a_k$ in place of $a_i = a_j + a_k$. This idea corresponds to the evaluation of x^d using multiplication and division. For integers, division (or the computation of a multiplicative inverse modulo n) is a costly operation, and implementing this idea does not seem feasible. The reason why elliptic curves are so attractive is that the division in $\mathbf{Z}_n$ is replaced by a subtraction, which has the same cost as an addition. An addition (subtraction) formula on elliptic curves does not contain a division in $\mathbf{Z}_n$ particularly when homogeneous coordinates are used. Thus, the addition-subtraction chain can be effectively applied to computations over elliptic curves.

This paper proposes a fast and systematic method of computing a multiple $d \cdot P$ of a point P on the elliptic curve modulo n. By increasing the average length of zero runs in a signed binary representation of d, the window method can be speeded up. The organization of this paper is as follows. Section 2 describes a new signed binary window method, clarifying the difference between previous methods and the new method, and analyzes the number of operations for the proposed method. Section 3 shows that the proposed method is faster than other methods. Elliptic curves over a finite field and a ring and the addition formula over elliptic curves are briefly reviewed in Section 4. Then, serial/parallel computations implemented in homogeneous coordinates and affine coordinates are compared.

2 New Method

The proposed method is a window method based on an adequately chosen signed binary representation of d. The new method is described, clarifying the differences between the previous window method [1] based on ordinary binary representation and the new one in this section. The window method is an extension of the 2^k-ary method. For a given number d, the window method consists of four phases: (1) representation of d, (2) splitting the representation into segments (windows), (3) computing all the segments, and (4) concatenating all the segments.

2.1 Representation

For a given number d, the original window method uses an ordinary (two-valued) binary representation $B : (b_\lambda, b_{\lambda-1}, \ldots, b_0)$, where $b_i \in \{0,1\}$, $\lambda = \lfloor \log_2 d \rfloor$. The proposed method uses the signed (three-valued) binary representation $T :$ $[t_{L-1}, \ldots, t_1, t_0]$, $t_{L-1} \neq 0$ for d satisfying $d = \sum_{i=0}^{L-1} t_i 2^i$, where $t_i \in \{\bar{1}, 0, 1\}$, and $\bar{1}$ denotes -1. Note that in ordinary binary representation B is uniquely determined for a given d, but T is not.

Morain-Olivos [14] and Jedwab-Mitchell [8] proposed algorithms to transform B into the equivalent T, for minimizing the weight (the number of non-zero digits) of T. Note that Morain-Olivos's method (MO method) is equivalent to Jedwab-Mitchell's method (JM method). We propose a new transform algorithm which increases the average length of zero runs in T, while minimizing the weight of T. The average length of the zero runs in specific T, denoted by $Z(T)$, is defined as follows.

$$Z(T) = \frac{1}{L} \sum_{i=0}^{L-1} z(i), \qquad z(i) = \begin{cases} 1 + z(i-1) & \text{if } t_i = 0 \\ 0 & \text{if } t_i \neq 0 \end{cases} \quad (0 \leq i \leq L-1),$$

where $z(-1) = 0$.

Let B' be a subsequence of B, and let T' be a subsequence of T. A rule for transforming B' to equivalent T' is as follows.

Transformation Rule

$B' : (1 \cdots b_i \cdots 1)$ can be transformed into $T' : [10 \cdots t_i \cdots \bar{1}]$, where $t_i = b_i - 1$.

Let $\#_0(B')$ be a number of zeroes in B', and let $\#_1(B')$ be a number of non-zero digits in B'. The weight of T' is estimated as $\#_1(T') = 2 + \sum |t_i| = 2 + \sum |b_i - 1| = 2 + \#_0(B')$. Thus, the weight decreases by the transformation if $\#_1(B') - \#_0(B') > 2$

The proposed transform algorithm inputs B in LSB first order and counts the difference $D(B') \equiv \#_1(B') - \#_0(B')$, and applies the transformation rule repeatedly to appropriate B' with $D(B') \geq 3$. The main difference between the proposed method and other methods is a threshold value (such as 3) to apply the transformation rule. MO method applies the rule to B' with $D(B') \geq 2$. Further, the output of both MO method and JM method are *sparse*, which means no two adjacent digits are nonzero. However, the output of the proposed method is not sparse. MO method and the proposed method generate T with same the same weight. Thus, the average length of zero runs of output of the proposed method is greater than that of MO method.

The proposed transform algorithm is shown below.

algorithm transform (**input** B: array, **output** T: array)
  begin
      $M := 0; J := 0; Y := 0; X := 0; U := 0; V := 0; W := 0; Z := 0;$
      while $X < \lfloor \log_2 d \rfloor$ **do begin**
          if $B[X] = 1$ **then** $Y := Y + 1$ **else** $Y := Y - 1;$
          $X := X + 1;$
          if $M = 0$ **then begin**
              if $Y - Z \geq 3$ **then begin**
                  while $J < W$ **do begin** $T[J] := B[J]; J := J + 1$ **end**;
                  $T[J] := -1;\ J := J + 1;\ V := Y;\ U := X;\ M := 1$
              end else if $Y < Z$ **then begin** $Z := Y; W := X$ **end**
          end else begin
              if $V - Y \geq 3$ **then begin**
                  while $J < U$ **do**
                      begin $T[J] := B[J] - 1; J := J + 1$ **end**;
                  $T[J] := 1;\ J := J + 1;\ Z := Y;\ W := X;\ M := 0$
              end else if $Y > V$ **then begin** $V := Y;\ U := X$ **end**
          end
      end;
      if $M = 0 \vee (M = 1 \wedge V \leq Y)$ **then begin**
          while $J < X$ **do begin** $T[J] := B[J] - M; J := J + 1$ **end**;
          $T[J] := 1 - M; T[J + 1] := M$
      end else begin
          while $J < U$ **do begin** $T[J] := B[J] - 1; J := J + 1$ **end**;
          $T[J] := 1;\ J := J + 1;$
          while $J < X$ **do begin** $T[J] := B[J]; J := J + 1$ **end**;
          $T[J] := 1; T[J + 1] := 0$
      end
      return T
  end

[Example] For a given $d = 25722562047811804942$, the binary representation for d is:

  B : (10110010011111000111001011110011000000100110001000101111100001110)

MO method transforms B into:

  T : $[10\bar{1}0\bar{1}0010100001\bar{1}001001\bar{1}010\bar{1}000\bar{1}010\bar{1}0000001010\bar{1}00010010\bar{1}0000\bar{1}00010010\bar{1}0]$

The proposed algorithm transforms B into:

  T : $[10110010100001\bar{1}001000\bar{1}\bar{1}0\bar{1}0000\bar{1}\bar{1}0\bar{1}00000010011000100011000\bar{1}0001001\bar{1}0]$

The average length of zero runs for MO method is 1.29 and that of the proposed algorithm is 1.42.

2.2 Splitting

The splitting phase is common to both ordinary binary representation B and signed binary representation T. Let w be the width of the window. B or T is split into segments with a length at most w. The following splitting procedure generates a list of all segments. For simplicity, the input array is represented by T.

> **procedure** split (**input** T: array, w: integer, **output** S: array)
>   Let segment list S be empty
>   **while** (length(T) $\geq w$)
>   **begin**
>   Let W be the left w digits of T.
>   Let R be T excluding W.
>   Let $\widetilde{W}$ be W excluding the right 0's.
>   Let $\widetilde{R}$ be R excluding the left 0's.
>   Add new segment $\widetilde{W}$ to segment list S
> $T := \widetilde{R}$.
>   **end**
>   Add last segment T to segment list S
>   **return** S

[Example] Assume T is the signed binary representation generated by the transform algorithm in the previous example. When $w = 4$, the splitting algorithm outputs the list of segments as

$$[\underline{1011}00\underline{101}0000\underline{1}00\underline{1}000\underline{1}\overline{1}0\overline{1}0000\underline{1}\overline{1}0\overline{1}0000000\underline{1001}\ 1000\underline{1}000\underline{11}0000\underline{1}000\underline{100\overline{1}}0]$$

where each block of underlined digits represents a segment. Note that the transform algorithm increases the run length of 0's in the segment gaps.

2.3 Computing the Segments

In B, the value of each segment is an odd positive integer up to $2^w - 1$. In T, if $w \geq 3$, the segment value never becomes $2^w - 1$ or $-(2^w - 1)$ because of the property of the transform algorithm. Thus, each segment value is an odd integer from $-(2^w - 3)$ to $2^w - 3$. The absolute values of all segments are obtained by the following simple addition sequence. ($i.e.$ $1, 2, 3, 5, 7, \ldots, 2^w - 3$)

$$a_0 = 1, \quad a_1 = 2, \quad a_2 = 3, \quad a_i = a_{i-1} + 2 \quad (3 \leq i \leq 2^{w-1} - 1)$$

Therefore, in T, all segment values can be computed by at most $2^{w-1} - 1$ additions. In B, all segment values can be computed by at most 2^{w-1} additions.

For the above example, segment values become $\{11, 5, 7, -13, -13, 9, 1, 1, 3, -1, 7\}$. Thus, all (absolute) segment values are computed by an addition sequence as $1, 2, 3, 5, \ldots, 13$ in 7 additions.

In reference [1], Bos and Coster computed all segments using a heuristic addition sequence. When the distribution of segment values is sparse, the heuristic method may be effective. However, if the distribution is dense a systematic method may be more effective. When $\lambda = 511$ and $w = 5$, the distribution becomes dense and consequently the proposed systematic method is more effective.

2.4 Concatenating and The Number of Operations

Concatenation requires doublings and non-doubling additions.

For example, for the split T in the above example, concatenation is achieved by:
$$dP = ((\cdots(((11P \cdot 2^{2+3} + 5P) \cdot 2^{4+4} - 7P) \cdot 2^{3+4} - 13P)\cdots) \cdot 2^{4+1} - P) \cdot 2^{3+4} + 7P) \cdot 2^1.$$

The inner most $11P$ corresponds to the most significant segment, and the exponent $2 + 3$ corresponds to the sum of the length 2 of the following window gap and the length 3 of the next segment.

Let L be the length of B or T. Note that L is $\lambda + 1$ for B and L is $\lambda + 1$ or $\lambda + 2$ for T. Let Z' be the average length of zero runs in the most significant windows for B or T. In other words, Z' is the average number of 0's deleted in W by the splitting algorithm in the beginning. Let Z'' be the average length of zero runs deleted in R by the splitting algorithm for B or T. The average length of the most significant segment is $w - Z'$. The number of doublings in concatenation is same as the length of T (or B) except for the most significant segment. Thus, the number of doublings in concatenation is $L - (w - Z')$ for B and T. The average number of segments becomes $L/(w + Z'')$, which corresponds to the number of non-doubling additions in concatenation.

Thus, on average, the window method requires R operations:

$$R = (L + Z' - w) + \frac{L}{w + Z''} + C,$$

where $C = 2^{w-1}$ for B, and $C = 2^{w-1} - 1$ for T.

2.5 Analysis of the number of operations

In this subsection, parameters L, Z', Z'' and w in the above expression R are analyzed.

The length of T is either $(\lambda + 1)$ or $(\lambda + 2)$. The transform algorithm outputs T of length $\lambda + 2$ with probability $1/4$. Thus, the average length of T, denoted by $\overline{L}$, is expressed by $\overline{L} = \frac{1}{4} \cdot (\lambda + 2) + \frac{3}{4} \cdot (\lambda + 1) = \lambda + 5/4$.

Let p be the probability that 0 occurs in B. If each digit in B is independent, a straight analysis results in $Z' = p(1 - p^w)/(1 - p)$ and $Z'' = p(1 - p^{(L-w)})/(1 - p)$ for B. If $p = 0.5$, then $Z' = 1 - 2^{-w}$ and $Z'' = 1 - 2^{(w-L)}$. If w is significant, then $Z'' \approx Z' \approx 1$ for B. For simplicity, let Z_B represent Z' and Z'' for B.

The expected value of $Z(T)$ for all possible T, denoted by Z_T, is analyzed as follows. The essence of the transform algorithm is represented by the automaton in Figure 1. The automaton inputs a sequence of bits$\{0, 1\}$ of B in LSB first order, and outputs $\{\overline{1}, 0, 1\}^*$.

In Figure 1, each arc is labeled by an input digit $b \in \{0, 1\}$. All output digits are determined by one of the following two functions.

$$f(b) = \begin{cases} b-1 & \text{if } s_3 < s_0, \\ b & \text{otherwise}, \end{cases} \qquad g(b) = \begin{cases} \overline{1} & \text{if } s_3 < s_0, \\ 1 & \text{otherwise}, \end{cases}$$

where the condition denoted by $s_3 < s_0$ means that s_3 is visited before s_0 by forthcoming transition. Solid arc corresponds to $f(b)$, and dotted arc corresponds to $g(b)$.

Assume input (*i.e.* B) comes from a memoryless binary information source with $p = 0.5$. Let z_i be an average length of zero runs at state s_i. Each value of z_i is obtained by solving the following equations.

$$\begin{cases} z_0 = (1/2)(1 + z_0), \\ z_1 = (1/2)(1 + z_0) + (1/2)Prob(s_3 < s_0|s_2)(1 + z_2), \\ z_2 = (1/2)(1 + z_3) + (1/2)Prob(s_0 < s_3|s_1)(1 + z_1), \\ z_3 = (1/2)(1 + z_3), \\ z_4 = (1/2)(1 + z_3) + (1/2)Prob(s_0 < s_3|s_5)(1 + z_5), \\ z_5 = (1/2)(1 + z_0) + (1/2)Prob(s_3 < s_0|s_4)(1 + z_4), \end{cases}$$

where $Prob(s_i < s_j|s_k)$ means the probability of the case of $(s_i < s_j)$ from state s_k. In the above equations, $Prob(s_3 < s_0|s_2) = Prob(s_0 < s_3|s_1) = Prob(s_0 < s_3|s_5) = Prob(s_3 < s_0|s_4) = (1/2) + (1/2)(1/4) + (1/2)(1/4^2) + \cdots \approx 2/3$. Thus, $z_0 = 1$, $z_1 = 2$, $z_2 = 2$, $z_3 = 1$, $z_4 = 2$, $z_5 = 2$.

Let p_i be a stationary probability of state s_i. All p_i are calculated by solving the equation $V = M \cdot V$ where V is the vector of all p_i and M is the given transition matrix. The result is $p_0 = 1/4$, $p_1 = 1/6$, $p_2 = 1/12$, $p_3 = 1/4$, $p_4 = 1/6$, $p_5 = 1/12$. Therefore, $Z_T = \sum_{i=0}^{5} p_i \cdot z_i = 3/2$ for the proposed method. Note that, $Z_T = 4/3$ for MO method and JM method.

In summary, using $\overline{L} = \lambda + 5/4$ and $Z_T = 3/2$, the average number of operations R for T, or R_T, is rewritten as:

$$R_T = (\lambda + \frac{11}{4} - w) + \frac{\lambda + \frac{5}{4}}{w + \frac{3}{2}} + 2^{w-1} - 1$$

The optimal value of window size w depends on the size of d. It is obtained by solving $\frac{\partial}{\partial w} R_T = 0$. For d with $\lambda = 511$, $w = 5$ is the optimal window size.

3 Comparison with other methods

Brickell [2] proposed a fast hardware implementation of computing $x^d \mod n$ using the precomputation of the multiplicative inverse $x^{-1} \mod n$.

Morain and Olivos [14] proposed an addition-subtraction chain algorithm based on a binary method. Their method obtains d_+, and d_- for $d(d = d_+ - d_-)$, and computes $d \cdot P$ as $(d_+ \cdot P) - (d_- \cdot P)$. In MO method, $d_+ \cdot P$ (and $d_- \cdot P$) are computed using the ordinary binary method. The average number of operations for MO method is $\frac{4}{3}\lambda + O(1)$.

Yacobi [16] applied the idea of data compression (Lempel-Ziv's incremental parsing algorithm) to splitting binary representation. The average number of operations in Yacobi's method is $\lambda + (\log(\lambda) - \log\log(\lambda))/2 + 1.5\lambda/\log(\lambda)$, where $\lambda = \lfloor \log_2 d \rfloor$. In his method, the segment size is initially small, and increases by parsing B. This method is inefficient for small d such as $\lambda = 511$

Bos and Coster [1] proposed a heuristics for an addition sequence and used a bigger window such as $w = 10$. This method requires an average of 605 operations for $\lambda = 511$. However, their method is based on heuristics.

A comparison of several addition(-subtraction) chain algorithms is shown in Table.1. From Table.1, the proposed method is seen to be faster than the other methods.

Table 1. The number of operations for d of 512 bit length

Method	Chain	Av.	Worst
Binary Method [10]	A	766.5	1022
Signed Binary Method [2] [8] [14]	A/S	681.7	768
Yacobi's Method [16]	A	635.1	—
Window Method ($w = 5$) [1]	A	609.3	630
Bos-Coster's Method [1]	A	605	—
Signed Binary Window Method ($w = 5$)	A/S	602.6	629

4 The Speed of Each Addition over Elliptic Curves

4.1 Elliptic Curves over a Finite Field and a Ring

Let K be a field of characteristic $\neq 2, 3$, and let $a, b \in K$ be two parameters satisfying $4a^3 + 27b^2 \neq 0$. An elliptic curve over K with parameters a and b is defined as the set of points (x, y) with $x, y \in K$ satisfying this equation on the affine plane

$$y^2 = x^3 + ax + b,$$

together with a special element denoted $\mathcal{O}$ and called the point at infinity [11]. Elliptic curves over the finite field $\mathbf{F}_p$ with p elements, for some prime p, are denoted by E_p. What makes elliptic curves interesting in cryptography and number-theoretic applications is the fact that an *addition operation* on the points of an elliptic curve E_p can be defined to make it an abelian group.

Elliptic curves over the ring $\mathbf{Z}_n$, where n is an odd composite square–free integer, can be defined in a similar way to E_p. For simplicity, let n be the product of two distinct large primes p and q as in the RSA scheme[15] and the KMOV scheme[9]. Addition on E_n, whenever it is defined, is equivalent to the group operation (defined by component) on $E_p \times E_q$. Thus, every point $P = (x, y)$ on E_n can be represented uniquely as a pair $[P_p, P_q] = [(x_p, y_p), (x_q, y_q)]$ where $P_p \in E_p$ and $P_q \in E_q$. Note that addition on E_n is undefined if and only if exactly one of the points P_p and P_q is the point at infinity. It is important to note that when all prime factors of n are large, it is extremely unlikely that the sum of two points on E_n is undefined.

4.2 Addition Formulae over Elliptic Curves

Let $P_1 = (x_1, y_1)$ and $P_2 = (x_2, y_2)$ be two points on the elliptic curve E_p. The point $P_3 = P_1 + P_2 = (x_3, y_3)$ is defined according to the following rules. If $P_1 = \mathcal{O}$, then $P_3 = P_1 + P_2 = P_2$. If $P_1 = -P_2$, that is, $x_1 = x_2$ and $y_1 = -y_2$, then $P_3 = P_1 + P_2 = \mathcal{O}$. When $P_1, P_2 \neq \mathcal{O}$, and $P_1 \neq -P_2$, an *addition formula* to find $P_3 = P_1 + P_2 = (x_3, y_3)$ is given below according to two cases: a non-doubling addition formula where $P_1 \neq P_2$ and a doubling formula $P_1 = P_2$ [11].

Non-doubling Addition Formula in Affine Coordinates

$$\begin{cases} x_3 = \lambda^2 - x_1 - x_2 \\ y_3 = \lambda(x_1 - x_3) - y_1, \end{cases} \tag{1}$$

where $\lambda = (y_2 - y_1)/(x_2 - x_1)$.

Doubling Formula in Affine Coordinates

$$\begin{cases} x_3 = \lambda^2 - 2x_1 \\ y_3 = \lambda(x_1 - x_3) - y_1, \end{cases} \tag{2}$$

where $\lambda = (3x_1^2 + a)/2y_1$.

Note that a subtraction to find $P_3 = P_1 - P_2$ is defined by changing the sign of y_2 in the addition formula $P_3 = P_1 + P_2$.

A point (x, y) on the affine plane is equivalent to a point (X, Y, Z) on the projective plane, where $x = X/Z$, $y = Y/Z$. That is, an elliptic curve is also defined as the set of points (X, Y, Z) in homogeneous coordinates satisfying the equation

$$ZY^2 = X^3 + aXZ^2 + bZ^3,$$

together with the point at infinity $(0,1,0)$. The non-doubling addition formula (1) and the doubling formula (2) in affine coordinates can be rewritten in homogeneous coordinates. Replace x_i with X_i/Z_i and y_i with Y_i/Z_i ($i = 1,2$) and reduce the fractions of x_3 and y_3 to a common denominator. Then, the resulting numerators of x_3 and y_3 become X_3 and Y_3, and the common denominator becomes Z_3. Let $P_1 = (X_1, Y_1, Z_1) \in E_p$, $P_2 = (X_2, Y_2, Z_2) \in E_p$, and $P_3 = (X_3, Y_3, Z_3) \in E_p$. The addition formulae to find $P_3 = P_1 + P_2$ in homogeneous coordinates are expressed as follows.

Non-doubling Addition Formula in Homogeneous Coordinates

$$
\begin{cases}
\begin{aligned}
X_3 =\ & X_1^4 Z_2^4 - 2X_1^3 X_2 Z_1 Z_2^3 + 2X_1 X_2^3 Z_1^3 Z_2 - X_1 Y_1^2 Z_1 Z_2^4 + 2X_1 Y_1 Y_2 Z_1^2 Z_2^3 \\
& - X_1 Y_2^2 Z_1^3 Z_2^2 - X_2^4 Z_1^4 + X_2 Y_1^2 Z_1^2 Z_2^3 - 2X_2 Y_1 Y_2 Z_1^3 Z_2^2 + X_2 Y_2^2 Z_1^4 Z_2,
\end{aligned} \\[4pt]
\begin{aligned}
Y_3 =\ & X_2^3 Y_2 Z_1^4 - 2X_2^3 Y_1 Z_1^3 Z_2 - Y_2^3 Z_1^4 Z_2 + 3X_1 X_2^2 Y_1 Z_1^2 Z_2^2 - 3X_1^2 X_2 Y_2 Z_1^2 Z_2^2 + \\
& 3Y_1 Y_2^2 Z_1^3 Z_2^2 + 2X_1^3 Y_2 Z_1 Z_2^3 - 3Y_1^2 Y_2 Z_1^2 Z_2^3 - X_1^3 Y_1 Z_2^4 + Y_1^3 Z_1 Z_2^4,
\end{aligned} \\[4pt]
Z_3 = -X_1^3 Z_1 Z_2^4 + 3X_1^2 X_2 Z_1^2 Z_2^3 - 3X_1 X_2^2 Z_1^3 Z_2^2 + X_2^3 Z_1^4 Z_2.
\end{cases} \tag{3}
$$

Doubling formula in Homogeneous Coordinates

$$
\begin{cases}
X_3 = 2Y_1 Z_1 (a^2 Z_1^4 + 6a X_1^2 Z_1^2 + 9X_1^4 - 8X_1 Y_1^2 Z_1), \\[4pt]
Y_3 = -a^3 Z_1^6 - 9a^2 X_1^2 Z_1^4 - 27a X_1^4 Z_1^2 + 12a X_1 Y_1^2 Z_1^3 - 27X_1^6 + 36X_1^3 Y_1^2 Z_1 - 8Y_1^4 Z_1^2, \\[4pt]
Z_3 = 8Y_1^3 Z_1^3.
\end{cases} \tag{4}
$$

By introducing moderate intermediate variables that are more moderate than ones in [9], addition formulae (3) and (4) can be revised to minimize the number of multiplications in serial processing. The revised addition formulae in homogeneous coordinates are:

Revised Non-doubling Addition Formula in Homogeneous Coordinates

$$
\begin{cases}
X_3 = VA \\[6pt]
Y_3 = U(V^2 X_1 Z_2 - A) - V^3 Y_1 Z_2, \\[6pt]
Z_3 = V^3 Z_1 Z_2,
\end{cases} \tag{5}
$$

where $U = Y_2 Z_1 - Y_1 Z_2$, $V = X_2 Z_1 - X_1 Z_2$, $A = U^2 Z_1 Z_2 - V^2 T$, $T = X_2 Z_1 + X_1 Z_2$.

Revised Doubling Formula in Homogeneous Coordinates

$$
\begin{cases}
X_3 = 2SH, \\[6pt]
Y_3 = W(4F - H) - 8E^2, \\[6pt]
Z_3 = 8S^3,
\end{cases} \tag{6}
$$

where $S = Y_1 Z_1$, $W = 3X_1^2 + aZ_1^2$, $E = Y_1 S$, $F = X_1 E$, $H = W^2 - 8F$.

Note that all of the above computations are modulo p or modulo n.

4.3   Performance Evaluation of Addition Formulae

Computations of the multiples of a point on the elliptic curves E_n can be performed in affine coordinates or homogeneous coordinates. The time complexity of the addition formulae implemented in these coordinates was compared. Each elementary addition over E_n was calculated using addition, subtraction, multiplication and division in $\mathbf{Z}_n$. For simplicity, addition, subtraction and special multiplication by a small constant such as $2y_1$ and $3(x_1^2)$ ware neglected because they are much faster than multiplication and division in $\mathbf{Z}_n$. In addition formulae (3)-(4) and (5)-(6) in homogeneous coordinates, contrary to addition formulae (1)-(2) in affine coordinates, the divisions in $\mathbf{Z}_n$ in each addition over E_n can be avoided. Computation in homogeneous coordinates requires 1 division $(1/Z_3)$ in $\mathbf{Z}_n$ to obtain both $x_3 = X_3/Z_3$ and $y_3 = Y_3/Z_3$ in the final stage of the chain. Note that division in $\mathbf{Z}_n$ can be implemented using the generalized Euclidean algorithm for computing the greatest common divisor.

A serial computation of non-doubling addition formula (1) requires two multiplications and one division in $\mathbf{Z}_n$. A serial computation of doubling formula (2) requires three multiplications and one division in $\mathbf{Z}_n$. A serial computation of non-doubling addition formula (5) required 15 multiplications in $\mathbf{Z}_n$. For the KMOV elliptic cryptosystem with $a = 0$, the computation of W in the doubling formula (6) can be simplified as $W = 3X_1^2$. Thus, a serial computation of doubling formula (6) requires 10 multiplications in $\mathbf{Z}_n$.

Assume that parallel processing of each addition over E_n is allowed in special hardware. For simplicity, the time for communication among processors is neglected. In affine coordinates, parallel processings of non-doubling addition and doubling require the same computational complexity as those in serial processing. Consider parallel processing of the addition formula in homogeneous coordinates. In general, parallel multiplication permits any polynomial of degree $2k$ to be computed in one step from the set of polynomials of degree k, where each step requires the time of one multiplication in $\mathbf{Z}_n$. The non-doubling addition formula (3) consists of polynomials of degree 8 with 6 variables, therefore, the values of X_3, Y_3, Z_3 can be obtained in 3 steps. The doubling formula (4) consists of polynomials of degree 6 with 4 variables (including a), therefore, the values of X_3, Y_3, Z_3 can be similarly obtained in 3 steps. That is, the related terms of degree 2 are computed in the first step, the related terms of degree 4 in the second step, and every term of degree 8 or 6 in the target polynomials in the third step.

Denote c be the ratio of the computation amount of division in $\mathbf{Z}_n$ to that of multiplication in $\mathbf{Z}_n$. Note that $c > 1$. Let R be the number of operations of addition formula in addition-subtraction chain. Assume that non-doubling additions occur with probability p_n and doublings occur with probability $(1 - p_n)$. For the proposed signed binary window method, we have $p_n \approx 1/6$ and $R \approx 602.6$ for $\lambda = 511$ as described in Section 2. Table 2 shows the numbers of multiplications in $\mathbf{Z}_n$ in serial/parallel processings in affine coordinates and homogeneous coordinates. From Table 2, we can observe that serial computation in homogeneous coordinates is faster than that in affine coordinates if $c > 8$

and $\lambda = 511$. Moreover, when $\lambda = 511$, parallel computation in homogeneous coordinates is always faster than that in affine coordinates.

Table 2.   The number of mult./div. in $\mathbf{Z}_n$ in the total chain

Processing	Coordinates	mult.	div.
Serial	Affine	$(2p + 3(1 - p))R$ $\approx 2.83 \cdot R$	R
	Homogeneous	$(15p + 10(1 - p))R$ $\approx 10.83 \cdot R$	1
Parallel	Homogeneous	$3 \cdot R$	1

When the multiplication chain is carried out based on alphabetically ordered factoring in formula (3), 17 processors are needed in the first step, 29 processors in the second step, and 24 processors in the third step. Since each processor can be used repeatedly, this multiplication system (or addition formula engine) requires 29 processors. Note that parallel computation of formula (4) requires less than 29 processors. As a result, each addition over the elliptic curve can be done in 3 multiplications if 29 parallel processors are used.

5   Conclusion

We have proposed a fast and systematic method of computing a multiple $d \cdot P$ over elliptic curves. This speeding up method is also applicable to computation in the group where the inverse operation is as fast as an ordinary operation. Furthermore, we pointed out that if parallel processing is allowed, each addition over the curve using homogeneous coordinates can be done in 3 multiplications.

References

1. Bos, J. and Coster, M: "Addition chain heuristics" *Proc. of CRYPTO'89* (1989).
2. Brickell, E.F.:"A fast modular multiplication algorithm with application to two key cryptography" *Proc. of CRYPTO'82* (1982).
3. Brickell, E.F., Gordon, D.M., McCurley, K.S., and Wilson, D.: "Fast exponentiation with precomputation " *Proc. of EUROCRYPT'92* (1992).
4. Diffie, W. and Hellman, M.E.: "New directions in cryptography", *IEEE Transactions on Information Theory*, Vol. 22, No. 6, (1976), pp. 644–654.
5. Downey, P. Leony, B. and Sethi, R:"Computing sequences with addition chains", *Siam J. Comput. 3* (1981) pp. 638–696.
6. ElGamal, T.:"A public key cryptosystem and a signature scheme based on the discrete logarithm", *IEEE Transactions on Information Theory*, Vol. 31, No. 4, (1985), pp. 469–472.

7. Goldwasser, S. and Killian, J.:"Almost all primes can be quickly certified", *Proc. 18th STOC*, Berkeley, (1986), pp. 316–329.

8. Jedwab, J. and Mitchell, C, J.:"Minimum weight modified signed-digit representations and fast exponentiation", *Electronics Letters* Vol. 25, No. 17, (1989), pp. 1171–1172.

9. Koyama, K. Maurer, U. Okamoto, T and Vanstone, S, A: "New public-key schemes based on elliptic curves over the ring $\mathbf{Z}_n$", *Proc. of CRYPTO'91* (1991).

10. Knuth, D.E.: "Seminumerical algorithm (arithmetic)" The Art of Computer Programming Vol.2, Addison Wesley, (1969).

11. Koblitz, N.:*A course in number theory and cryptography*, Berlin: Springer-Verlag, (1987).

12. Lenstra, H. W. Jr.: "Factoring integers with elliptic curves", *Ann. of Math. 126* (1987), pp. 649–673.

13. Montgomery, P.L.:"Speeding the Pollard and elliptic curve methods of factorization", Math. Comp. 48, (1987), pp. 243–264.

14. Morain, F. and Olivos, J.: "Speeding up the computations on an elliptic curve using addition-subtraction chains" Theoretical Informatics and Applications Vol. 24, No. 6 (1990) pp. 531–544.

15. Rivest, R.L. Shamir, A. and Adleman, L.:"A method for obtaining digital signatures and public-key cryptosystems", *Communications of the ACM*, Vol. 21, No. 2, (1978), pp. 120–126.

16. Yacobi, Y.: "Exponentiating faster with addition chains" *Proc. of EUROCRYPT'90* (1990).

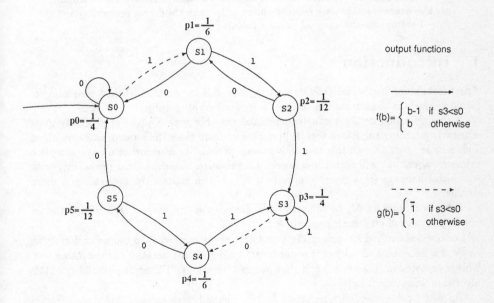

Figure.1   Automaton for the transform algorithm

On Generation of Probable Primes by Incremental Search

Jørgen Brandt and Ivan Damgård
Aarhus University

September 16, 1992

Abstract

This paper examines the following algorithm for generating a probable prime number: choose a random k bit odd number n, and test the numbers $n, n+2, \ldots$ for primality using t iterations of Rabin's test, until a probable prime has been found or some maximum number s of candidates have been tested.

We show an explicit upper bound as a function of k, t and s on the probability that this algorithm outputs a composite. From Hardy and Littlewoods prime r-tuple conjecture, an upper bound follows on the probability that the algorithm fails. We propose the entropy of the output distributrion as a natural measure of the quality of the output. Under the prime r-tuple conjecture, we show a lower bound on the entropy of the output distribution over the primes. This bound shows that as $k \to \infty$ the entropy becomes almost equal to the largest possible value.

Variants allowing repeated choice of starting points or arbitrary search length are also examined. They are guaranteed not to fail, and their error probability and output entropy can be analysed to some extent.

1   Introduction

Apart from being mathematically interesting, it is well-known that efficient generation of prime numbers is of extreme importance in modern cryptography.

Although prime numbers can be generated together with a proof of primality quite efficiently [8], using the Rabin test [10] remains in many cases the most practical method. This is true, even though this test allows some probability of error: it always accepts a prime number, but will sometimes accept a composite. The maximal probability with which this happens for a given composite is $1/4$, but in general the probability is much smaller.

Let $x = 2^k$, and let M_k be the set of odd numbers in the interval $[x/2..x[$, i.e. the set of odd numbers of bit length precisely k.

An obvious method for generating a probable prime number is to choose uniformly an element n from M_k and subject it to (at most) t independent iterations of the Rabin test. This is repeated until an n is found that passes t iterations. The error probability of this algorithm was studied in [4].

An often recommended alternative to this method uses incremental search from a randomly chosen starting point (see e.g. [6] or [9]). This alternative is more economical in its use of random bits, and can be optimized using test division etc. [2] to be significantly more efficient than the "uniform choice" method. Despite the practical advantages, an analysis of the incremental search method does not seem to have appeared before. Such an analysis will be the subject of this paper.

Before going into details, let us mention some related work: Bach [1] also uses the increment function in connection with primality testing. In his case, however, the increment function is used to generate the witnesses for the test, not the candidate numbers as in our case. Moreover, the results of [1] are concerned with the probability that the test accepts, *given that the input number is composite*, whereas we of course want the "reversed" probability, namely the probability that the input number is composite given that the test accepts. Finally, the general results of Cohen and Wigderson [3] on reducing the error probability of a BPP algorithm at the cost of a few more random bits would be applicable in our situation, more specifically to the Rabin test itself. This could be used to reduce the worst case probability of 1/4 dramatically. However, what the following shows is essentially that this is unecessary in our case: for most input numbers, the error probability is already exponentially small, and since we are interested in an algorithm for generating random prime numbers, not the decision problem as such, we only have to worry about the average case behavior.

We now give a more precise version of the algorithm we will look at. The description uses the notation above, and parameters s and t.

ALGORITHM PRIMEINC

1. Choose uniformly a number $n_0 \in M_k$. Put $n = n_0$.

2. Subject n to at most t iterations of the Rabin test. If n passes t iterations, output n and stop.

3. Otherwise (if n fails an iteration), put $n = n + 2$. If $n \geq n_0 + 2s$, output "fail" and stop, otherwise go to step 2.

It is well known that this algorithm can be optimized by test dividing by small prime numbers before applying the Rabin test to a candidate. [2] contains an analysis of the optimal number of primes to use. We have omitted this for simplicity because it is clear that, independently of the number of small primes used, the error probability of the optimized version would be at most that of PRIMEINC. This is because the test division can never reject a prime, and so can only give us a better chance of rejecting composites. It is clear that the error probability might even become much smaller by using test division, but this seems like a difficult problem to analyse.

2 The Error Probability

In this section we will look at the probability that PRIMEINC outputs a composite. Let E be the event that this happens, and let $q_{k,t,s} = Prob(E)$. Let $C_m \subset M_k$ be the set of composites for which the probability of passing the Rabin test is larger than 2^{-m}.

In [4], the following upper bound is proved on the size of C_m:

Lemma 1

If $m \leq 2\sqrt{k-1} - 1$, we have

$$|C_m|/|M_k| < a \sum_{j=2}^{m} 2^{m-j-(k-1)/j},$$

where $a = 10.32$.

Let $D_m = \{n \in M_k \mid [n..n+2s[\cap C_m \neq \emptyset\}$, for $m > 2$ and put (for convenience) $D_2 = \emptyset$. We clearly have:

Lemma 2

$D_m \subset D_{m+1}$ and $|D_m| \leq s \cdot |C_m|$.

This implies the following result:

Theorem 1

Let $s = c \cdot \log(x)$ for some constant c. Then

$$q_{k,t,s} \leq ck(0.5ck \sum_{m=3}^{M} P(C_m)2^{-t(m-1)} + 0.7\ 2^{-tM})$$

where $M \geq 3$, and $P(C_m)$ denotes the probability that a random number in M_k is in C_m.

Proof

Identify D_m with the event that the starting point n_0 is in D_m. Then by Lemma 2, $P(D_m) \leq s \cdot P(C_m)$. By the fact that $D_m \subset D_{m+1}$, we have

$$q_{k,t,s} = \sum_{m=3}^{M} P(E \cap (D_m \setminus D_{m-1})) + P(E \cap \neg D_M) \tag{1}$$

$$\leq \sum_{m=3}^{M} P(D_m)P(E \mid (D_m \setminus D_{m-1})) + P(E \mid \neg D_M) \tag{2}$$

Now consider the probability that E occurs given that some fixed $n_0 \notin D_m$ was chosen. Then no candidate we consider will be in C_m, and so for any composite candidate, the probability of accepting it will be at most 2^{-mt}. The probability of outputting a composite clearly is maximal when all candidates are composite, in which case we accept one of the cadidates with probability at most $s \cdot 2^{-mt}$. This means that

$$q_{k,t,s} \leq s^2 \sum_{m=3}^{M} P(C_m)2^{-t(m-1)} + s \cdot 2^{-tM} \tag{3}$$

$$\leq 0.5(ck)^2 \sum_{m=3}^{M} P(C_m)2^{-t(m-1)} + 0.7ck2^{-tM} \tag{4}$$

The theorem follows.

It is clear that we can get explicit bounds on $q_{k,t,s}$ by combining Lemma 1 and Theorem 1, where the value of M should be optimized to get the best possible result. Table 1 shows concrete results obtained for $c = 1, 5$ and 10. The following proposition gives a very rough idea of how the bound behaves for large k.

Proposition 1

For any constants c (where $c \log(x) = s$) and t, $q_{k,t,s}$ as a function of k satisfies

$$q_{k,t,s} \leq \delta k^3 2^{-\sqrt{k}}$$

for some constant δ.

Proof

It is sufficient to show the result for $t = 1$. From Lemma 1 we get that $P(C_m)$ is less than a constant times $2^m m 2^{-\sqrt{k}}$, by observing that $-j - (k-1)/j$ is always less than $-2\sqrt{k-1}$. The lemma now follows immediately by inserting this in the result of Theorem 1.

3 The Failure Probability

To get a good bound on the error probability, we should choose c to be as small as possible. The obvious disadvantage is that this is likely to increase the probability that the algorithm fails.

To say something conclusive about this, we should clearly know something about the gaps between consecutive primes. Unfortunately, no unconditional result is known about this that would be strong enough for our purposes. However, Gallagher [5] has shown, based on Hardy and Littlewoods prime r-tuple conjecture [7], the following result:

Lemma 3

Under the prime r-tuple conjecture, we have: For any constant λ, the number of n's in the interval $[1..x[$, such that the interval $[n..n + \lambda \log(x)]$ contains precisely k primes, is

$$x \frac{e^{-\lambda} \lambda^k}{k!} \quad as \quad x \to \infty$$

Let $d(n)$ denote the distance from n to the next, larger prime. Then Lemma 3 means in particular (by taking $k = 0$) that for n chosen uniformly between 1 and x, the probability that $d(n) > \lambda \log(x)$ is $exp(-\lambda)$, as $x \to \infty$. This is the only part of Gallaghers result we will need in the following. It implies that the expected distance to the next prime is $\log(x)$. We have confirmed this experimentally (see Figure 1).

The following lemma gives a corresponding result for the interval $[x/2..x[$.

c	k\t	1	2	3	4	5	6	7	8	9	10
1	100	0	7	13	18	22	26	29	32	34	36
	150	3	12	20	26	31	35	39	43	46	49
	200	5	17	25	33	39	44	49	53	57	61
	250	7	21	31	39	45	51	57	62	66	70
	300	9	24	35	44	51	58	64	70	75	79
	350	11	28	40	49	57	64	71	77	82	88
	400	13	31	44	54	63	70	77	84	90	95
	450	15	34	47	58	68	76	83	90	97	103
	500	17	37	51	63	72	81	89	96	103	110
	550	18	40	54	67	77	86	95	102	110	116
	600	20	42	58	70	81	91	100	108	115	123
5	100	0	3	9	14	18	21	24	27	29	32
	150	0	8	15	21	26	31	35	38	42	45
	200	1	12	21	28	34	39	44	48	52	56
	250	3	16	26	34	41	47	52	57	62	66
	300	5	20	31	39	47	53	59	65	70	75
	350	7	23	35	44	53	60	66	72	78	83
	400	9	26	39	49	58	66	73	79	85	91
	450	11	29	43	54	63	71	79	86	92	98
	500	13	32	46	58	68	77	85	92	99	105
	550	14	35	50	62	72	82	90	98	105	112
	600	16	38	53	66	77	86	95	103	111	118
10	100	0	2	7	12	16	19	22	25	27	30
	150	0	6	13	19	24	29	33	36	40	43
	200	0	10	19	26	32	37	42	46	50	54
	250	2	14	24	32	39	45	50	55	60	64
	300	4	18	29	37	45	51	57	63	68	73
	350	5	21	33	42	51	58	64	70	76	81
	400	7	24	37	47	56	64	71	77	83	89
	450	9	27	41	52	61	69	77	84	90	96
	500	11	30	44	56	66	75	83	90	97	103
	550	12	33	48	60	70	80	88	96	103	110
	600	14	36	51	64	75	84	93	101	109	116

Table 1: Shows $-\log_2$ of the upper bound on $q_{k,t,s}$ as a function of k and t and for $s = 1\log(x), 5\log(x)$ and $10\log(x)$.

Lemma 4

Assume the prime r-tuple conjecture holds, and that PRIMEINC is executed with $s = c \log(x)$. Let $p(c)$ be the probability that the algorithm fails. Then for any ϵ,

$$p(c) \leq 2exp(-2c) - exp(-2c - \epsilon) \quad as \quad x \to \infty$$

Proof

Put $\lambda = 2c$, and let $p_y(\lambda)$ be the probability that $d(n) > \lambda \log(x)$, when n is uniform between 1 and y. Then

$$p_x(\lambda) = 1/2(p(c) + p_{x/2}(\lambda))$$

The lemma now follows from Lemma 3, since for any $\epsilon > 0$ and all large enough x, $(\lambda + \epsilon) \log(x/2) > \lambda \log(x)$.

This means that for large x, the failure probability is essentially $exp(-2c)$, and certainly less than $2exp(-2c)$. By Theorem 1, the error probability increases at most quadratically with c; we can therefore choose values of c for which both error and failure probability are small.

As a realistic example, suppose we put $k = 300$ and $c = 10$. Then we fail with probability about 2^{-28}, or 1 in 200 million times, and with $t = 6$ we still get an error probability of at most 2^{-51}. It seems reasonable to make the error probability much smaller than the failure probability: a failure is detectable and can be recovered from, whereas as error is never detected, at least not by PRIMEINC itself.

4   The Output Distribution

This section is concerned with the quality of the output from PRIMEINC, in particular how the output is distributed over the possible primes. This is a critical point in cryptographic applications (e.g. RSA), where the output prime is to be kept secret. One should expect that an enemy knows which algorithm is being used to generate the primes, and it is natural to demand that this will not give him any significant advantage.

We suggest to use the entropy of the output distribution as a measure of its quality. This is natural, since it measures the enemy's uncertainty about the prime generated. From this point of view it is clear that the optimal output distribution is the uniform distribution over the primes in M_k, since it has maximal entropy. By the prime number theorem, this maximal value is about $H_u(x) = \log(x/2 \log(x))$ for large x. Below, we will show that the entropy of the distribution output by PRIMEINC is very close to $H_u(x)$ for large x.

Since we will be using the prime r-tuple conjecture directly in the following, we quote it here:

Prime r-tuple conjecture

For a fixed r-tuple of integers $d = (d_1, ..., d_r)$, let $\pi_d(x)$ be the number of n's less than or equal to x, such that $n + d_i$ are all primes. For a prime p, let $\nu_d(p)$ denote the number of distinct residue classes modulo p occupied by numbers in d. The conjecture now says that

$$\pi_d(x) \sim S_d \frac{x}{\log(x)^r} \quad as \quad x \to \infty$$

where

$$S_d = \prod_p (\frac{p}{p-1})^{r-1} \frac{p - \nu_d(p)}{p-1}$$

For $r = 1$, this is just the prime number theorem. For $d = (0, 2)$, it is a statement about prime twins, and here $S_{0,2} \approx 1.32$.

By an argument similar to that of Gallagher, one can show that under this conjecture, the following holds:

Lemma 5

Let $F_h(x)$ denote the number of primes p such that $p \leq x$ and $p - q \leq h$, where q is the largest prime less than p. Then for any constant λ,

$$F_{\lambda \log(x)}(x) = \frac{x}{\log(x)}(1 - e^{-\lambda})(1 + o(1))$$

as $x \to \infty$.

Proof

By definition of the π_d's, it is clear that $F_h(x)$ can be found using inclusion/exclusion:

$$F_h(x) = \sum_{1 \leq d_1 \leq h} \pi_{0,d_1}(x) - \sum_{1 \leq d_1 < d_2 \leq h} \pi_{0,d_1,d_2}(x) + \sum_{1 \leq d_1 < d_2 < d_3 \leq h} \pi_{0,d_1,d_2,d_3}(x) - \cdots$$

Using the prime r-tuple conjecture, we get that

$$F_h(x) \sim \sum_{1 \leq d_1 \leq h} S_{0,d_1} \frac{x}{\log(x)^2} - \sum_{1 \leq d_1 < d_2 \leq h} S_{0,d_1,d_2} \frac{x}{\log(x)^3} \cdots$$

as $x \to \infty$. Using a modification of Gallagher's method from [5], one can show that

$$\sum_{1 \leq d_1 < \cdots < d_r \leq h} S_{0,d_1,\cdots,d_r} \sim \frac{h^r}{r!}$$

as $h \to \infty$, with a error term of $O(h^{r-1/2+\epsilon})$ for any $\epsilon > 0$. Inserting this in the above and choosing $h = \lambda \log(x)$ for a constant λ gives us that

$$F_{\lambda \log(x)}(x) \sim \frac{x}{\log(x)}(\frac{\lambda}{1!} - \frac{\lambda^2}{2!} + \frac{\lambda^3}{3!} - \cdots)$$

which immediately implies the lemma.

This lemma shows that the gaps between consecutive primes loosely speaking follows an exponential distribution. It corresponds nicely with Gallaghers result since a

Poisson distribution in statistics results from counting random events with exponentially distributed gaps occurring in a certain time slot.

Let $F'_h(x)$ denote the number of primes p such that $x/2 < p \leq x$ and $p - q \leq h$, where q is the largest prime less than p. It is a trivial consequence of Lemma 5 that

$$F'_{\lambda \log(x)}(x) = \frac{x/2}{\log(x)}(1 - e^{-\lambda})(1 + o(1))$$

as $x \to \infty$.

We now return to PRIMEINC and consider the restriction of the output distribution to the cases where the algorithm does not fail. For simplicity, we will look at an "ideal PRIMEINC" that never accepts a composite; we let $H(x)$ denote the resulting distribution over the primes in M_k. By Proposition 1, the distribution of the real PRIMEINC only assigns a negligible amount of probability mass [1] to composite numbers, and hence the difference between the entropy of this distribution and $H(x)$ will be negligible.

Theorem 2

If ideal PRIMEINC is executed with $s = c \cdot \log(x)$ for any constant c, then under the prime r-tuple conjecture,

$$\frac{H(x)}{H_u(x)} \sim 1$$

as $x \to \infty$.

Proof

The number of starting points n_0 that lead to non-failure will be called N. For a prime p, let $d(p)$ be the distance to the largest prime smaller than p, and $P(p)$ the probability that p is produced as output. Then

$$P(p) = \begin{cases} d(p)/2N, & \text{if } d(p) \leq 2c\log(x) \\ 2c\log(x)/2N, & \text{if } d(p) > 2c\log(x) \end{cases}$$

Now choose $n + 1$ constants $0 = \lambda_0 < \lambda_1 < ... < \lambda_{n+1} = 2c$. By Lemma 5, the number of primes with $\lambda_i \log(x) \leq d(p) < \lambda_{i+1} \log(x)$ is

$$\frac{x}{2\log(x)}(e^{-\lambda_i} - e^{-\lambda_{i+1}})(1 + o(1))$$

This gives us the following:

$$\begin{aligned} H(x) &= \sum_{x/2 < p \leq x} P(p) \log(1/P(p)) \\ &= \sum_{i=0}^{n} \sum_{\lambda_i \log(x) \leq d(p) < \lambda_{i+1} \log(x)} P(p) \log(1/P(p)) + \sum_{d(p) \geq 2c\log(x)} P(p) \log(1/P(p)) \end{aligned}$$

---

[1] Here, a probability is called negligible if, as a function of k, it converges to 0 faster than any polynomial fraction

k	$H(x)$	$H_u(x)$	$H(x)/H_u(x)$
17	8.38	8.65	0.97
31	17.4	17.7	0.98
64	38.9	39.8	0.98
128	82.5	83.5	0.99
257	171	172	0.99

Table 2: Shows estimates for the entropy of the output from PRIMEINC with infinite search length compared with the maximal entropy value. The values are exact for $k = 17, 31$, and are based on a sample interval of length about 2^{27} for the rest of the values. Note that the entropy for any finite search length will be larger than the $H(x)$ value shown.

$$
\geq (1 + o(1)) \cdot \left(\sum_i \frac{\lambda_i \log(x)}{2N} \log\left(\frac{2N}{\lambda_{i+1} \log(x)}\right) \frac{x}{2\log(x)} (e^{-\lambda_i} - e^{-\lambda_{i+1}}) \right.
$$
$$
+ \left. \frac{2c\log(x)}{2N} \log\frac{2N}{2c\log(x)} \frac{x}{2\log(x)} e^{-2c} \right)
$$
$$
= (1 + o(1)) \frac{x}{4N} \left(\log\left(\frac{2N}{\log(x)}\right) \left(\sum_i \lambda_i (e^{-\lambda_i} - e^{-\lambda_{i+1}}) + 2ce^{-2c} \right) + A \right)
$$

where A is a constant. By Gallaghers result, $N \sim x/4(1 - e^{-2c})$ as $x \to \infty$. We therefore get that

$$
\frac{H(x)}{H_u(x)} \gtrsim \frac{1}{(1 - e^{-2c})} \left(\sum_i \lambda_i (e^{-\lambda_i} - e^{-\lambda_{i+1}}) + 2ce^{-2c} \right)
$$

By choosing a larger number of λ_i's with smaller intervals, we can make the sum over i arbitrarily close to $\int_0^{2c} y e^{-y} dy = 1 - e^{-2c} - 2ce^{-2c}$, which implies the result of the theorem since $H(x)/H_u(x)$ is always smaller than 1.

Thus, for large x, the entropy is very close to maximal independently of the value of c. In fact one can show by an argument similar to the one for Theorem 2 that if we take $c = \infty$, i.e. allow the algorithm to run indefinitely until a prime is found, the resulting entropy would still be close to maximal in the same sense as in Theorem 2. Taking any finite value of c means that we are limiting the probability for any single prime to a certain maximum. Intuitively, this should make the distribution closer to the uniform one, and so the entropy should be close to maximal for any finite value of c; Theorem 2 confirms this intuition.

We have done some numerical experiments to estimate how fast the convergence in Theorem 2 is. Table 2 shows the entropy for various values of $k = log_2(x)$ and an infinite search length. For $k \geq 64$ the values are estimates based on primes in an interval of length about 2^{27}. Already for small values of k, the entropy is close to maximal, and the value clearly tends to increase with increasing k, in accordance with Theorem 2.

To illustrate the exponential distribution of gaps between primes, we plotted the frequency of gaps between primes in various intervals against their length. A logarithmic scale was used, such that a straight line should result, according to the exponential distri-

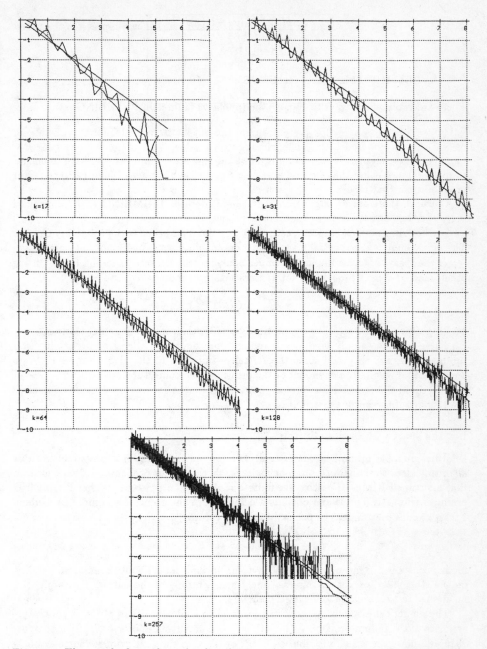

Figure 1: The graphs here show the distribution of gaps between primes in M_k and distances to the next prime from a random starting point. The "sawtooth" graphs represent the gaps. Results for $k = 17$ and 31 are exact, the rest are based on samples of length about 2^{27}. We plotted the distances divided by $\log(x)$ on the horizontal scale against the logarithm of the frequences of distances. This normalizes the graphs such that they can be directly compared. The distributions predicted by Lemma 4 and 5 are representred by the straight lines

bution. The result can be seen in Figure 1. The straight line shown represents the values predicted by Lemma 5.

5   Getting rid of Failures

There are two obvious ways to remove the (unlikely) failures in PRIMEINC:

Choose a new random starting point

- and try a new search. The question is, however, how many times we will have to do this to make (almost) sure that we find a probable prime, and what will happen to the error probability in that case?

Consider therefore an algorithm that simply iterates PRIMEINC with some fixed $k, t, s = c \cdot \log(x)$ until a probable prime is output.

By the prime number theorem, each starting point n_0 is prime with probability $O(k^{-1})$. Therefore, there is exponentially small (in k) probability that the number of iterations is larger than, say, k^2. This implies an upper bound on the expected running time. On the other hand, the error probability for k^2 iterations is at most

$$k^2 q_{k,t,s} \leq O(k^5 2^{-\sqrt{k}})$$

by Proposition 1, and so still asymptotically smaller than any polynomial fraction.

These arguments only give a very rough upper bound on the error probability and running time, but they add some theoretical justification to the algorithm, as they are unconditional results (independent of the prime r-tuple conjecture).

One disadvantage of this approach, however, is that is seems very difficult to analyse the entropy of the output distribution.

Let the search go on indefinitely

- until a probable prime has been found. Let $q_{k,t,\infty}$ denote the error probability of this algorithm. It is clear that $q_{k,t,\infty}$ can be no larger than the error probability of an algorithm that first runs PRIMEINC with some finite $s = c \log(x)$, outputs the number PRIMEINC produces (if one is found) and outputs a composite in the event of a failure. By Lemma 4, this implies that under the prime r-tuple conjecture,

$$q_{k,t,\infty} \leq p_{k,t,c\log(x)} + e^{-2c}(1 + o(1))$$

for any constant c. Thus we cannot say anything unconditional about the error probability, on the other hand we can in this case analyse the entropy of the output distribution, as mentioned in the remarks following Theorem 2.

The numerical evidence we collected leads us to conjecture that e^{-2c} is a good overestimate of the failure probability already for quite small values of k, say, larger than 64. Figure 1 shows the actual distribution of the distance to the next prime in some example intervals, compared with the values predicted by Lemma 4. Assuming e^{-2c} as an upper bound on the probability, we get that $q_{k,t,\infty} \leq p_{k,t,c\log(x)} + e^{-2c}$.

Using Theorem 1 and optimizing for the value of c, one can get concrete estimates for $q_{k,t,\infty}$. The results are shown in Table 3.

$k \backslash t$	1	2	3	4	5	6	7	8	9	10
100	0	4	9	13	16	20	22	24	27	29
150	1	8	14	20	24	28	32	35	38	41
200	3	12	19	26	31	36	40	44	48	52
250	4	15	24	31	37	43	48	53	57	61
300	6	18	28	36	43	49	55	60	65	70
350	8	21	32	41	49	55	61	67	73	78
400	9	24	36	45	53	61	67	73	79	85
450	11	27	39	49	58	66	73	80	87	92
500	12	29	43	53	63	72	79	86	93	99
550	14	32	46	58	67	76	84	92	99	105
600	15	35	49	61	72	81	90	98	105	111

Table 3: Shows $-\log_2$ of the upper bound on $q_{k,t,\infty}$ as a function of k and t *assuming* that the failure probability of PRIMEINC with $s = c\log(x)$ is at most e^{-2c}.

6 Conclusion

We have shown some explicit upper bounds on the error probability of the PRIMEINC algorithm. Together with the prime r-tuple conjecture, these bounds show that we can choose the maximal length of the search such that both the error probability and the failure probability are in practice negligible, even for quite small values of t.

Moreover, under the prime r-tuple conjecture, we have seen that the uncertainty about the prime produced by PRIMEINC is very close to maximal, for any value of c, including $c = \infty$. This strongly suggests that, compared to a uniform choice, there no significant loss of security when using PRIMEINC in cryptographic applications where secret primes are required.

For the case where one iterates PRIMEINC if no probable prime is found, we have seen unconditional results on the running time and error probability. In the case where in stead the search is allowed to go on indefinitely, the prime r-tuple conjecture implies results on both running time, error probability and entropy.

References

[1] E.Bach: *Realistic analysis of som randomized algorithms*, Proc. of STOC 87.

[2] J.Brandt, I.B.Damgård and P.Landrock: *Speeding up prime number generation*, Proc. of Asiacrypt 91, Springer Verlag Lecture Note Series.

[3] L.Cohen and A.Wigderson: *Dispersers, deterministic amplification, and weak random sources*, Proc. of FOCS 89.

[4] I.B.Damgård, P.Landrock and C.Pomerance: *Average case bounds for the strong probable prime test*, submitted to Math. Comp., available from authors.

[5] P.X.Gallagher: *On the distribution of primes in short intervals*, Mathematica 23 (1976) pp.4-9.

[6] J.Gordon: *Strong primes are easy to find*, Proc. of Crypto 84, Springer Verlag Lecture Note Series.

[7] G.H.Hardy and J.E.Littlewood: *Some problems of 'Partitio Numerorum': III. On the expression of a number as a sum of primes*, Acta Mathematica 44 (1922), pp.1-70.

[8] U.Maurer: *The generation of secure RSA products with almost maximal diversity*, Proc. of EuroCrypt 89, Springer Verlag Lecture Note Series.

[9] FIPS publications XX, Digital Signature Standard (DSS).

[10] M.O.Rabin: *Probabilistic algorithms for testing primality*, J. Number Theory 12 (1980) pp.128-138.

Kid Krypto

Michael Fellows[1] and Neal Koblitz[2]

[1] University of Victoria, Department of Computer Science, Victoria, B.C. V8W 3P6, Canada
[2] University of Washington, Department of Mathematics, Seattle, Washington 98195, U.S.A.

Abstract. Cryptographic ideas and protocols that are accessible to children are described, and the case is made that cryptography can provide an excellent context and motivation for fundamental ideas of mathematics and computer science in the school curriculum, and in other venues such as children's science museums. It is pointed out that we may all be doing "Kid Krypto" unawares. Crayon-technology cryptosystems can be a source of interesting research problems; a number of these are described.

1 Introduction

The purpose of this paper is to open a discussion of cryptography for children. The fruits of this discussion can serve several worthwhile purposes, such as:

(1) the popularization of cryptography with children and the general public through such forums as children's science museums,

(2) the enrichment and improvement of the school mathematics curriculum by providing a stimulating context for logical and mathematical modes of thinking, and

(3) the amusement and intellectual stimulation of researchers.

We hope to convince the reader that devising ways to present the fundamental ideas of cryptography to children not only makes it possible to expose children to some electrifying mathematics, but also can be stimulating for our own research and can give us a fresh perspective on what we do.

In the following sections of this paper we will describe some examples of cryptographic ideas and constructions that have been or could easily be presented to children. Some of these might be characterized as *pre-cryptography*, i.e., they involve certain elements of cryptography but do not yet constitute a sophisticated protocol. Others are fully developed cryptosystems. The following assertions summarize our outlook.

• By its very essence, cryptography is a most excellent vehicle for presenting fundamental mathematical concepts to children.

Cryptography can be broadly defined as "mathematics/computer science in the presence of an adversary." Implicit in any discussion of cryptography are elements of drama, of theater, of suspense. Few things motivate children as much as wanting to defeat the "bad guys" (or play the role of bad guys themselves). Children are in the business of decrypting the world of adults, and many of the video games that are now popular with children involve deciphering "clues" in order to achieve some goal.

Cryptography's ability to excite children has long been understood by advertisers of products like Rice Krispies and Crackerjacks. Many of us grew up quarreling with our siblings over who was going to get the decoder ring in the Crackerjacks box. Currently, boxes of Rice Krispies have on the back a "secret algorithm" age guessing game based on binary representation of integers.

• Kid Krypto is a source of interesting research problems.

We are essentially proposing a new criterion for deciding that a cryptosystem is worthy of attention: *accessibility*. As in the case of the more traditional criteria — efficiency and security — the search for cryptosystems that meet the accessibility standard naturally leads to interesting theoretical and practical questions. It is a new challenge to determine how much can really be done with minimal mathematical knowledge, and to find ways to present cryptographic ideas at a completely naive level. Moreover, experiences working with children have suggested some provocative problems in discrete mathematics and theoretical computer science, some examples of which will be described later.

A newly proposed cryptosystem might not be efficient enough or secure enough to compete in the realm of adult cryptography with those that already exist, but may nevertheless be of tremendous pedagogical value and merit our attention for that reason alone. In addition, it seems clear that interesting questions are likely to arise when we take a second look at some cryptosystems that have been too quickly forgotten.

Even a cryptosystem that can be broken in polynomial time may be of use in Kid Krypto, if the mathematics needed to break the system is less accessible than what is needed to implement the system. For example, some versions of the Perfect Code system explained below can be broken by linear algebra modulo m, but the system can be implemented using nothing more sophisticated than addition modulo m. In other words, we are proposing a new security hierarchy, with such notions as *accessible and secure for ages 5–10*, *accessible and secure for high school students*, etc.

• There is no sharp line between Kid Krypto and adult crypto, so it would be unwise for us to belittle the former.

After all, the security of all of our cryptosystems depends upon our assumed inability to perform certain mathematical tasks, such as discover a fast factoring algorithm. If a space alien from a very advanced civilization were to visit the earth, she might be surprised to find us using cryptosystems based on factoring and discrete log. Suppose that on her planet polynomial time algorithms to factor integers, find discrete logs in finite fields, and even find discrete logs on nonsupersingular elliptic curves have been known for centuries, and are routinely taught in high school. She would regard RSA, ElGamal, etc. as suitable only for pre-high school children in her culture.

In other words, in some sense *we are all doing Kid Krypto*, whether we know it or not.

• Kid Krypto is best done without computers.

This is *crayon-technology* cryptography. The tools needed are: pencils, a lot of paper, crayons of different colors, and perhaps some pieces of string or sticks. There is no material obstacle to introducing Kid Krypto in poor school districts as well as rich ones — in Watts and Soweto as well as in Santa Barbara and Scarsdale.

We see the absence of computers as a positive educational step. The public needs to understand that computer science is not about computers, in much the same way that cooking is not about stoves, and chemistry is not about glassware. What children need in order to become mathematically literate citizens is not early exposure to manipulating a keyboard, but rather wide-ranging experience working in a creative and exciting way with algorithms, problem-solving techniques and logical modes of thought.

Computers have been shamelessly oversold to teachers and school systems. In speaking to parents, teachers and school boards, many company representatives have taken the hard-sell approach: "If you don't buy our latest products you will be neglecting to prepare your children for the 21st century." Because of pressure from the companies and the media, computers have been fetishized to the extent that they threaten to become the Cargo Cult of the 21st century. Most of the time, computers serve as nothing more than an expensive distraction. The main beneficiaries of all the hype have been (1) computer hardware and software companies, and (2) educators who receive generous grants for the purpose of finding a way to use computers in the schools. Most schools would probably be better off if they threw their computers into the dumpster. It is our prediction that the Golly–Gee–Whiz–Look–What–Computers–Can–Do school of mathematical pedagogy will eventually come to be regarded as a disaster of the same magnitude as the "new math" rage of the 1960s.

2 Pre-Crypto

In order to present cryptography to young children, there are certain "building block" ideas which are useful to present first, and that are engaging in their own right. We point to three of these in particular, and describe how they can be simply presented.

(1) The notion of an algorithm, and of computational complexity.

(2) The notion of a one-way function.

(3) The notion of an information hiding protocol.

2.1 Algorithms and Complexity

There are now a tremendous number of delightful ways that the fundamental ideas of algorithmic procedure and computational complexity can be presented to young children. We mention here just a few of our favorites. The examples below have been tried out many times, with great success, with children sometimes as young as 5 or 6.

The first problem is Map Coloring. If you were to visit a first-grade classroom to share this lovely problem, you might very well arrive in a room full of children who are already coloring something anyway! You might tell the story of the poor Map-Colorer, trying to eke out a living with few crayons, and then pass out a map that needs to be colored. The definition of a proper coloring is visual, and can be illustrated with the maps at hand in the classroom. It is only a few minutes until most of the children understand the problem you have posed (finding out the minimum number of colors for the map you have passed out) and are puzzling away at it. As the children work to decrease the number of colors needed, you can display the "best known" solution so as to add to the excitement.

It is a good idea to come to the classroom with plenty of copies of 3 or 4 different maps. It is easy to generate a map that is two-colorable by overlaying closed curves. (Generating such a map is another topic the children may have fun thinking about). See Figure 1. In a typical first-grade classroom, children will figure out the algorithm for 2-coloring on their own, and they will see that it goes very quickly. It is easy enough to explain why it works: it has been called the "Have-to Algorithm" (if a country is red, then its neighbors have to be blue, and their neighbors have to be red, ...). Afterwards, you might distribute a map that requires 3 colors so that they can concretely contrast the 2-coloring experience with the apparent difficulty of finding a 3-coloring of a 3-colorable map.

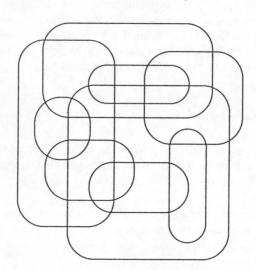

Fig. 1. Example of a 2-colorable map generated by overlaying closed curves

Another excellent topic for children is the problem of computing a Minimum

Weight Spanning Tree in a graph. Several efficient algorithms for solving this algorithmic problem are known and are routinely covered in college level courses on design and analysis of algorithms. The story we use to present the problem is meant to be entertaining, but it should be noted that there are many practical applications of this problem.

The children are given a map of Muddy City and told the story of its woes — cars disappearing into the mud after rainstorms, etc. The mayor insists that some of the streets must be paved, and poses the following problem. (1) Enough streets must be paved so that it is possible for everyone to travel from his or her house to anyone else's house by a route consisting only of paved roads, but (2) the paving should be accomplished at a minimum total cost, so that there will be funds remaining to build the town swimming pool. Thus, the children are asked to devise a paving scheme meeting requirement (1), connecting up the town by a network of paved roads, that involves a minimum total amount of paving. The cost of a paving scheme is calculated by summing the paving costs of the roads chosen for surfacing. For the map shown in Figure 2 a solution of total cost 23 can be found.

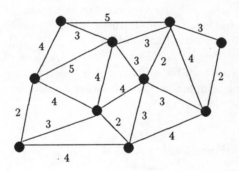

Fig. 2. Muddy City

The children work on the problem, usually in small groups, with the immediate objective of finding the best possible solution. This was typically recorded in a place that everyone could see. Students were asked to describe their strategies and ideas, both as they worked and in a concluding discussion. In classrooms where the students kept mathematics journals, they also wrote descriptions of the problem and of their ideas on how to solve it. These math journals were instituted with great success in a second-grade classroom and a fourth-grade classroom.

As part of the wrap-up discussion, we sometimes presented Kruskal's algorithm (one of several known algorithms for solving this problem efficiently). This method of finding an optimal solution consists simply of repeatedly paving a shortest street which does not form a cycle of paved streets, until no further paving is required. It is interesting that the children often discovered some of the essential elements of Kruskal's algorithm and could offer arguments supporting them. (Rediscovering Kruskal's algorithm is not the point, of course.)

This problem can be presented to children of ages 5–6 by using maps with distances marked by ticks rather than numerals, so that the total amount of paving can be figured by counting rather than by sums.

Minimum Dominating Set is another problem that can provide a nice illustration of the idea of computational complexity. Recall that a *dominating set* in a graph $G = (V, E)$ is a set of vertices $V' \subseteq V$ such that for every vertex x of G, either $x \in V'$ or x has a neighbor $y \in V'$.

The stories we have told for this problem generally run to the theme of *facilities location*. For example, in Tourist Town we want to place ice-cream stands at corners so that no matter which corner you might be standing on, you need only walk at most one block to get an ice-cream. See Figure 3 for an example of a small, somewhat difficult graph for which $\gamma = 6$.

We allow some time for the children to puzzle over the map of Tourist Town, gradually producing more efficient solutions. Often, none of them is able to find the optimal solution with only six ice-cream stands. The children usually get an intuitive sense that Tourist Town is harder than Muddy City; the former does not seem to lend itself to solution by a quick and simple algorithm. The contrast between these two problems — one solvable in polynomial time and the other apparently intractable — provides a concrete introduction to the notion of computational complexity. We will return to the subject of dominating sets (of a special kind) in Section 5.

2.2 One-Way Functions

After explaining that no one knows a good algorithm for Tourist Town, one can show that there is, however, a simple algorithm for "working backwards," i.e., starting with a set of vertices V' that is to become an efficient solution and constructing a Tourist Town $G = (V, E)$ around it. Namely, one uses a two-step process. First, one forms a number of "stars" made up of "rays" (edges) emanating from the vertices in V'. (Two rays from different vertices in V' are allowed to have a common endpoint.) This graph clearly has V' as a solution. Figure 4 below shows this step in the case of the Tourist Town example. The second step is to "disguise" this easy-to-solve graph by adding more edges. This clearly does not increase the number of vertices required in a dominating set, but it does make the original built-in solution harder to see.

In this way it seems to be relatively easy to generate graphs on a small number of vertices (e.g. 25-30), having a known dominating set of size $6 \leq \gamma \leq 10$, for which it is relatively difficult to work out a solution of size γ by hand. However, no mathematical results are presently known that quantify the computational difficulty of problems such as this for graphs of small size.

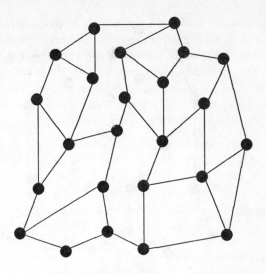

Fig. 3. Map of Tourist Town

This is a nice example of the idea of a one-way function. The children may look forward to trying out on their parents the process of creating a graph for which they secretly know a difficult-to-match solution. Open problem: can we sell this to Rice Krispies?

Remark 1. If the two-step "hidden solution" construction described above is modified by (1) in the first step, requiring that no two stars share a common vertex, and (2) in the second step, requiring that the additional disguising edges be added only between vertices not in V', then the hidden solution will be a *perfect code* in $G = (V, E)$. (A more precise definition of a perfect code will be given later.) This modified construction is useful for the Perfect Code public key cryptosystem described in Section 4.

Remark 2. In presenting the Dominating Set problem to children in El Salvador, the authors had to confront an example of the general problem of *cultural appropriateness* of the stories used to introduce these topics. We found that in El Salvador, as would be the case in many places in the world, the idea of minimizing the number of ice-cream stands makes no cultural sense whatsoever. So we changed the setting for the Dominating Set problem, presenting it by means of a story about minimizing the number of wells in order to achieve an efficient water supply for a village. Such a story is appropriate for a Third World context but would make no sense to children in the developed world.

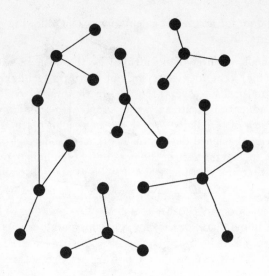

Fig. 4. The first step in the construction of Tourist Town: a configuration of stars

2.3 Information Hiding Protocols

A simple illustration of this is a method for computing the average allowance of children in the classroom, without revealing any individual's allowance. The protocol goes like this. The first person picks a large integer randomly, and adds to it her allowance. The sum is passed secretly to the second person, who adds to it her allowance, and so on. After all the allowances have been privately added in, the final sum is secretly passed by the last person to the first person, who subtracts her original secret large integer and computes the average for the group.

3 The Peruvian Coin Flip

One of the key issues we must face in designing crayon-technology cryptosystems is: what interesting functions can 7-year olds (for instance) compute reliably? That is, what sort of by-hand computing do we have available to work with?

With a little thought, we can see that interesting computations *can* be performed by young children to provide the computational engines for cryptosystems. For example, the outputs of Boolean circuits can be computed; finite-state automata and Mealy machines can be operated. (In principle, Turing machines can also be operated by paper and pencil, but our experience suggests that they are somewhat slow and unwieldy.) Cellular automata, if they are not too complicated, may offer another interesting possibility. Simple rewrite systems are another candidate for accessible calculations. The following protocol is based on Boolean circuits.

This protocol was first demonstrated by the authors with children in Peru (hence the name). The idea of trying out a crayon-technology cryptosystem in Peru seemed natural for several reasons. In the first place, the improvement of mathematics education is currently a hot topic of discussion among educators in Peru, as in much of the Third World. In the second place, developing countries (and international science development organizations such as the *Kovalevskaia Fund*) have a special interest in the possibility of enhancing math and computer science education in situations where machines are not available.

We first told a story to explain how the need for such a coin-flip protocol might arise. The women's soccer teams of Lima and Cuzco have to decide who gets to be the home team for the championship game. Alicia, representing Lima, and Berta, representing Cuzco, cannot spend the time and money to get together to flip a coin. So they agree to the following arrangement.

Working together, they construct a Boolean circuit made up of and-gates and or-gates (for simplicity, we allow only small and-gates and or-gates, and no not-gates). See Figure 5 for an example. In the construction process, each has an interest in ensuring enough complexity of the circuit so that the other will be unable to cheat (see below). The final circuit is public knowledge. Let n be the number of input bits, and let m be the number of output bits.

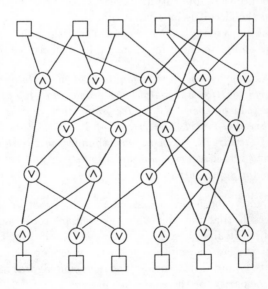

Fig. 5. A Boolean circuit for the Peruvian coin-flip

Alicia selects an arbitrary input string, which she keeps secret. She puts the string through the circuit, and sends Berta the output. Berta must then try to guess the *parity* of Alicia's input, i.e., the sum of its bits mod 2. If she guesses right, then the teams play in Cuzco. If her guess is wrong (which Alicia must demonstrate to her by revealing the input string), then they play in Lima.

Nothing in this description is hard to convey to a child of age 8 or above. Moreover, when we explain to the children the basic ingredient in the protocol (∧-gates and ∨-gates), we are talking about a really basic concept — perhaps *the* most basic concept — in formal logical thought. There is certainly as much justification for teaching about ∧-gates and ∨-gates as for long division and addition of fractions!

Remark. An alternative construction would be for Alicia and Berta each to construct a circuit with n input bits and m output bits. Both circuits would be public knowledge. Then Alicia would put her secret input through both circuits, and the final output would be the XOR of the outputs produced by the two circuits. This variant is "cleaner" in the sense that it avoids some interaction; but probably the first variant is easier to explain to kids. More importantly, the first variant is more fun, precisely because of the added interaction.

3.1 Cheating

Berta can cheat if she can invert the circuit, i.e., find the input (or inputs) that produce a given output. Alicia can cheat if she can find two inputs of opposite parity that produce the same output. It seems likely that both forms of cheating are infeasible if the circuit is large and complex.

If the circuit maps many-to-one, we claim that the ability to cheat in Berta's role implies the ability to cheat in Alicia's role. Namely, we have

Proposition 1. *Suppose we have a family C of many-to-one Boolean circuits, with the property that for any output the proportion of inputs in its preimage of given parity (odd or even) is bounded from below. Further suppose that one has an algorithm that inverts any circuit of C in time bounded by $f(n)$, where n is the size of the circuit. Then in time bounded by $kf(n) + p(n)$ (where p is a polynomial and k is a security parameter) one can find two inputs of opposite parity that give the same output.*

Proof. This result — both the statement and the proof — is completely analogous to the result that the ability to take square roots modulo a composite number n implies the ability to factor n. Namely, to find the two desired inputs, select one input at random, and then apply the inversion algorithm to its output. With probability bounded from below, the inversion algorithm will give a second input of different parity for the same output. □

On the other hand, we can entirely prevent Alicia from being able to cheat by choosing a circuit that maps inputs to outputs injectively, i.e., it effects an imbedding of $\{0,1\}^n$ into $\{0,1\}^m$. If we suppose that the circuit is complicated enough to behave like a random map, then the next proposition shows that it suffices to choose m somewhat larger than $2n$.

Proposition 2. *The probability that a random map from $\{0,1\}^n$ to $\{0,1\}^m$ is injective, is asymptotic to $1 - 2^{-(m-2n+1)}$ as $m - 2n \longrightarrow \infty$.*

Proof. This is a restatement of a well-known combinatorial result (the "birthday paradox"). $\qquad\qquad\square$

4 Perfect Code Cryptosystems and Molten Arithmetic

The public key system which we will describe in this section can be designed with different levels of accessibility and security. The simplest version, which will be described first, can be mastered by a child who understands only (1) the simplest properties of graphs, and (2) addition (say, modulo 2 or modulo 26). We shall then describe a more complicated version, appropriate for older children, using what we call *molten arithmetic*. The latter term refers to the fluidity in the definition of the cryptosystem. That is, the rules for building the system can be adjusted according to the level of accessibility and security desired.

We begin by considering a special kind of dominating set in a graph called a *perfect code*. In what follows, if u is a vertex of a graph $G = (V, E)$, then the notation $N[u]$ (the "neighborhood" of u) denotes the set of vertices which share an edge with u (including u itself).

Definition 3. A set of vertices $V' \subseteq V$ in a graph $G = (V, E)$ is said to be a *perfect code* if for every vertex $u \in V$ the neighborhood $N[u]$ contains exactly one vertex of V'.

Figure 6 below shows an example of a graph with a perfect code. The vertices of the perfect code are indicated by open circles.

Remark 1. Jan Kratochvíl has shown that the problem of determining whether a graph has a perfect code is NP-complete for d-regular graphs, for all $d \geq 3$ [3].

Remark 2. An interesting detour for kids along the way to our cryptosystem might be to investigate error-correcting codes. For example, let n be of the form $2^k - 1$, and let G be the hypercube graph, whose vertices are $\{0,1\}^n \subset \mathbf{R}^n$ and whose edges are the edges of the n-dimensional unit hypercube. Then a *binary Hamming code* of length $n = 2^k - 1$ and dimension $d = 2^k - k - 1$ corresponds to a perfect code of 2^d vertices in G. For example, when $k = 2$, the (unique) Hamming code is the pair of opposite vertices $(0,0,0)$ and $(1,1,1)$ on the ordinary cube.

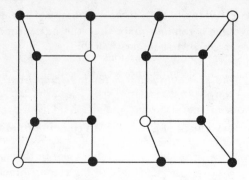

Fig. 6. Example of a perfect code in a cubic graph

4.1 Version A_1 of the Perfect Code Cryptosystem

This version is accessible to children of age 8. Suppose that the children have already mastered the Pre-Crypto topic *construction of a graph that has a well-disguised perfect code* (see the first remark of section 2.2). Now Alice wants to be able to receive an encrypted bit from Bobby. She constructs a graph $G = (V, E)$ with a perfect code V'. The graph G is her public key. Her private key is V'.

To send a bit b, Bobby makes a random assignment of 0's and 1's to all of the vertices of G except one. He then assigns either a 0 or 1 to the last vertex in such a way that the sum mod 2 over all of the vertices is equal to b. Next, he replaces the bit c_u assigned to each vertex u by a new bit c'_u determined by summing (mod 2) all of the bits that had been assigned to the neighboring vertices: $c'_u = \sum_{v \in N[u]} c_v$. He finally returns the graph to Alice with the bits c'_u annotating the vertices.

To decipher the message, Alice takes the sum of c'_u over the perfect code; that is, she has $b = \sum_{u \in V} c_u = \sum_{u \in V'} c'_u$, where the last equality follows from the definition of a perfect code.

4.2 Version A_2

The same as version A_1, but we make it more interesting by working modulo 26, so that Bobby can send Alice an enciphered letter $b \in \{A = 0, \ldots, Z = 25\}$.

Remark. Even if G is a complicated graph, both versions A_1 and A_2 of this cryptosystem can be broken in polynomial time using linear algebra (Gaussian elimination) modulo 2 (respectively, modulo 26). This will be shown later as a special case of a more general result. However, junior high school students have no more knowledge of how to do this than we have of how to factor integers in polynomial time. So with a judicious choice of G, versions A_1 and A_2 appear to be accessible and secure for junior high school.

4.3 Versions B and C

We now describe more elaborate versions which are harder to crack. We conjecture that version B is accessible and secure for high school students. Version C might be secure even for adults — at least we do not know how to break it.

First we need some notation and definitions. Given a graph $G = (V, E)$, we assign a variable denoted a_u to each vertex $u \in V$. Suppose that G has a perfect code V'. Let $x, y \in \mathbf{Z}$ and $m \in \mathbf{Z} \cup \{\infty\}$, $m \geq 2$. We let $\sigma(x, y, m)$ denote the substitution scheme which evaluates a polynomial in the a_u by setting $a_u = x$ if $u \in V'$ and $a_u = y$ otherwise, and performing the arithmetic modulo m (doing ordinary arithmetic in the case $m = \infty$).

By an *invariant expression* in a neighborhood $N[u]$ relative to the substitution scheme $\sigma(x, y, m)$ we mean a polynomial in the variables a_v, $v \in N[u]$, which evaluates to the same value irrespective of which vertex in $N[u]$ is in the perfect code. Here are some examples:

(1) For any substitution scheme, $\sum_{v \in N[u]} a_v$ is an invariant expression. More generally, any symmetric polynomial in the a_v, $v \in N[u]$, is an invariant expression relative to any substitution scheme.

(2) If the neighborhood $N[u]$ has 4 vertices, whose corresponding variables will be denoted a, b, c, d, and we have the substitution scheme $\sigma(2, 1, 3)$, then each of the following expressions is invariant: $ab + c + d$ (always evaluates to 1), $ab + ac + ad + a$ (always evaluates to 2), $ab + bc + cd + a + d$ (always evaluates to 1).

(3) If the neighborhood $N[u]$ has 4 vertices and we have the substitution scheme $\sigma(2, 1, \infty)$, then each of the following is invariant: $ab + c + d$, $ab + bc + cd + a + d$, $ab + cd$, $abc + d$.

We now describe versions B and C of the Perfect Code cryptosystem. In both cases the public key is the graph $G = (V, E)$ and the substitution scheme $\sigma(x, y, m)$ (i.e., a choice of x, y, m), the private key is the perfect code V', and the message Bobby wants to send is an integer b modulo m.

Version C is the most general. To send the message b, Bobby creates a large, complicated polynomial f from building blocks consisting of invariant expressions in neighborhoods of randomly selected vertices. This polynomial f must have two properties: (1) it evaluates to b under the substitution scheme $\sigma(x, y, m)$; and (2) someone who knows f but not how it was constructed from the building blocks would have great difficulty decomposing f into invariant expressions. Once Bobby constructs such an f, he sends it to Alice. Alice, who knows the perfect code, can correctly evaluate f without knowing how it decomposes into invariant expressions; she merely has to make the substitution $\sigma(x, y, m)$.

Version B is a special case of C. We use the substitution scheme $\sigma(1, 0, m)$ (m is arbitrary). To send the message b (a certain integer modulo m), Bobby chooses an arbitrary set I of subsets of vertices $S \subseteq V$ and a corresponding set of integers c_S such that $\sum_{S \in I} c_S \equiv b \pmod{m}$. He then forms the polynomial

$$f = \sum_{S \in I} c_S \prod_{u \in S} \sum_{v \in N[u]} a_v.$$

Since each inner sum evaluates to 1, the whole expression obviously evaluates to $\sum c_S = b$.

Remark. Versions A_1 and A_2 are special cases of version B where I consists of one-element sets $S = \{u\}$. Then $f = \sum_{u \in V} c_u \sum_{v \in N[u]} a_v = \sum_{u \in V} c'_u a_u$, where $c'_u = \sum_{v \in N[u]} c_v$.

4.4 Breaking Versions A_1, A_2 and B

Given a polynomial f in the variables a_u, we want to find an identity of the form

$$f = \sum_{S \in I} c_S \prod_{u \in S} \sum_{v \in N[u]} a_v.$$

By writing both sides as a sum of homogeneous terms, without loss of generality we may assume that on the left side of the equation f is homogeneous of total degree d, and on the right side of the equation I is the set of subsets $S \subset V$ of cardinality d. We regard the c_S as unknowns, and equate coefficients of each monomial on the left and right. There are $\binom{n}{d}$ unknowns c_S (here $n = \#V$ is the size of the graph), and there are $\binom{n+d-1}{d}$ monomials of total degree d, and hence $\binom{n+d-1}{d}$ equations. Although there are more equations than unknowns (except in the case $d = 1$), we know that there is a solution, because the f in version B was constructed as such a sum of products. The solution can be found by Gaussian elimination. (In practice, the system of equations will probably be very sparse, in which case special methods are available.)

Notice that if d is unbounded, then the time required to do the linear algebra is not polynomial in the size n of the graph. However, the time is polynomial in the size of the polynomial f that Bobby sends to Alice, unless he has some way of producing sparse polynomials f (polynomials f with mostly zero coefficients). But we know of no way systematically to produce sparse polynomials that are difficult to crack.

Remark 1. Any time the substitution scheme in use is $\sigma(1, 0, m)$ (as in version B), there is a simple way that Bobby can make the cryptosystem harder to break. Bobby knows that any monomial whose variables are not all in Alice's perfect code will evaluate to 0, and hence can be dropped from f before he sends f to Alice. Of course, Bobby does not know the perfect code. However, he knows that if a monomial contains two variables a_u and a_v corresponding to vertices which are at a distance ≤ 2 from one another, then those vertices cannot both be in her perfect code, and hence the monomial can be dropped.

Remark 2. Notice that the f in version B are actually invariant under any substitution scheme $\sigma(x, y, m)$. The f used in version C are much more general, since they are built up from expressions which need only be invariant under our one particular substitution scheme. Thus, the f in version C cannot, in general, be decomposed into building blocks made of symmetric polynomials in the variables in a neighborhood.

Remark 3. In implementing these cryptosystems, the youngsters have to search for invariant building blocks and then build up complicated f, using the distributive law and gathering similar terms so as to disguise the way f was formed. In this way Kid Krypto might add some excitement to the subject of polynomials, which is often presented in school in a dry, unmotivated manner. The decision as to what version of Perfect Code cryptography to use — how complicated to make the possible f — depends on the age of the children and their ability to keep track of a lot of data.

4.5 Cubic Graphs

One way to keep the level of difficulty under control is to use only regular graphs of degree 3. Then the invariant expressions in any neighborhood involve exactly 4 variables. The class of cubic graphs is still plenty complicated to support these cryptosystems — as mentioned before, determining whether a given cubic graph has a perfect code is NP-complete. We now describe a simple construction that gives a large class of cubic graphs having perfect codes. The construction is based on covering spaces of K_4, the complete graph on 4 vertices.

The construction is as follows. Let $n = 4n_0$ be the size of the cubic graph to be constructed. Select four sets of n_0 vertices each, which we denote A, B, C, D. Then randomly create six one-to-one correspondences between the sets: $A \approx B$, $A \approx C$, $A \approx D$, $B \approx C$, $B \approx D$, $C \approx D$. Draw edges between vertices that are associated under any of these six bijections. Let $G = (V, E)$ be the resulting graph. Notice that each neighborhood $N[u]$ contains exactly one vertex from each of the sets A, B, C, D; and each of these sets is a perfect code in G. The construction is completely general: every covering space of K_4 can be produced in this way. It is not presently known whether the problem of recovering such a vertex set partition for a graph that is known to be a cover of K_4 is difficult in the sense of average-case complexity. The problem of deciding whether an arbitrary graph is a cover of K_4, however, has been shown to be NP-complete [3].

5 Kid Krypto Research Problems

The project of sharing the subject of cryptography with children leads to a number of interesting research problems.

5.1 Accessible Combinatorial Cryptosystems

Kid Krypto gives us a reason to have another look at various proposals for cryptosystems based on simple combinatorics. For example, the public key system using reversible cellular automata proposed in [2] may have merit for Kid Krypto. Another combinatorially based cryptosystem was proposed by a group of researchers at Madras Christian College in India and the Hanoi Mathematical Institute in Vietnam. In [5], they show that a rewrite system — based on the word problem in a group — can be used to construct a public-key system. It would be interesting to try to adapt these ideas for Kid Krypto.

It is worthwhile to develop a variety of examples of Kid Kryptosystems. In that way one can convey some of the richness and interconnectedness of mathematics, and at the same time give oneself flexibility when using Kid Krypto in the classroom.

5.2 Other Fundamental Protocols

At this point, a number of fundamental cryptographic primitives are still unexplored from the Kid Krypto point of view. Can we find elegant and accessible implementations of *oblivious transfer, secure 2-party computation, secret sharing, zero-knowledge proof*, etc.?

5.3 The Complexity of Small Things

Crayon-technology cryptosystems work with mathematical objects that are essentially very small, mathematically speaking — such as graphs on fewer than 25 vertices, circuits of similar size, and two digit integers. From limited experience, it seems that it is relatively easy to generate small hard examples for the Minimum Dominating Set problem, but small hard examples of the 3-Coloring problem for planar graphs seem to be more difficult to generate. Is it possible to study this issue mathematically?

5.4 Breaking the Perfect Code Cryptosystem

Can the most general version of the Perfect Code system (version C) be broken in polynomial time?

5.5 Robustness Under Not Following Directions Properly

Classroom experiences seem inevitably to turn up intriguing questions in a playful vein. For example, the following question arose when the first author presented Map Coloring on one occasion. What is the minimum number of colors with which one can always color a planar map in a situation where one takes turns with an "incompetent helper" who is only assumed to color legally, but not necessarily judiciously? A bound of 33 was recently proved [4] for this lovely problem.

In presenting the Peruvian coin-flip to a junior high school audience, the authors encountered the situation where children attempted to evaluate the n-input/n-output circuit *upside down*. This leads to the following natural question, to which we do not know the answer. Let us suppose that all gates of our circuit have fan-out (as well as fan-in) of 2. (An alternative would be to allow large gates, i.e., gates with arbitrary fan-in and fan-out.) In addition, let us put ∨'s and ∧'s in the input gates in an arbitrary way, with the understanding that such a gate (with a fan-in of 1) leaves the input bit unchanged. Under these assumptions the circuit makes sense if the child turns it upside down, of course with each ∨-gate now becoming a ∧-gate and vice-versa. A natural question is whether it makes much difference (to a cheater) whether the circuit is right side up or upside down. More precisely, can one find a family of circuits which are easy to invert, but which when turned upside down are hard to invert? Can the problem of inverting the circuits in some presumably hard-to-invert family C be shown to be polynomial time equivalent to the problem of inverting the upside down circuits of C?

5.6 Physical Realizations of Cryptographic Protocols

Some cryptographic ideas can be effectively demonstrated by employing physical props. Such demonstrations can be useful in conveying the central concepts of cryptography to children and other mathematically unsophisticated audiences, such as in popular lectures. Although in this paper we have focused on the design of cryptosystems accessible to children that are fully mathematical and do not rely on physical props, cryptosystems based on physical primitives might also prove to be a source of interesting mathematics for children.

For example, a number of fundamental protocols, such as oblivious transfer and multi-party secure computation, can be nicely demonstrated by means of ordinary playing cards [1]. Note that these familiar physical objects have a number of cryptographically useful properties "built in": they have a convenient means of randomization (shuffling), are uniquely identifiable, and when face down are all indistinguishable. A number of entertaining research problems arise in constructing cryptosystems based on such physical primitives (see [1] for further discussion).

6 The Research Community and Mathematics Education

The past year has seen the inception of at least three major projects originating in the research communities of mathematics and theoretical computer science to develop engaging mathematical materials for children in the elementary grades: (1) The education projects associated with the Center for Discrete Mathematics and Theoretical Computer Science (DIMACS), located at Rutgers University. (2) The compendium project of the Association for Computing Machinery Special Interest Group on Algorithms and Computation Theory (SIGACT).

(3) The Megamath Project of the U.S. National Laboratories in Los Alamos, New Mexico.

All three of these projects are concerned with developing resource materials from the extensive treasury of accessible, active and applicable mathematics that has emerged in recent years in the intertwined subjects of discrete mathematics and computer science.

DIMACS now publishes the newsletter *In Discrete Mathematics* (the premiere issue is dated Nov. 1991) which contains articles on topics in discrete mathematics intended to be useful to teachers introducing discrete mathematics to their classes. The newsletter will also serve as a networking service and clearinghouse for ideas and materials related to discrete mathematics in education in the lower grades. Further information can be obtained from Joe Rosenstein (joer@math.rutgers.edu). Members of the cryptographic community who wish to contribute something would certainly be welcome.

The SIGACT compendium project was initiated at the business meeting at STOC in May, 1992. The newly formed SIGACT Committee on Education has as its first goal the production of a compendium of theoretical computer science topics and presentation strategies that may be useful in a variety of settings with children (for example, children's science museums). This project makes no commitment to any particular direction in school curriculum reform; rather, it is simply a collective effort at science popularization.

The Megamath Project of the U.S. Los Alamos National Labs intends to influence classroom practice, by making schoolwork more like the experience one has in a good science museum. That is, the goal is to bring to children in the classroom a live experience of mathematical science as something in which they can actively participate. Thus, the Megamath Project is looking into such things as (1) mathematics research problems accessible to children, (2) possible forums for children to present the results of their mathematical investigations, (3) extended projects for classroom investigation, (4) the classroom use of personal mathematics journals, and (5) opportunities for children to communicate with larger mathematical communities.

The three initiatives in discrete mathematics and computer science described above join other efforts involving research scientists in elementary education. These include the Mathematicians and Education Reform Network sponsored by the AMS and the NSF, and the Scientists in the Schools program of the Sandia U.S. National Research Laboratories. Many scientists are now looking for more direct ways to work with children and stimulate grade school educational reform. This seems to be an idea whose time has come.

Besides fitting in well with these initiatives, a program of classroom activity centered around Kid Krypto is an ideal way to implement the Curriculum Standards of the National Council of Teachers of Mathematics [NCTM 1989]. These standards, which stress the importance of mathematical thinking, problem-solving, communication, and connections between mathematics and the world, are a radical departure from earlier curriculum standards. (In the past, the mathematical curriculum was defined simply to consist of a list of topics.)

Moreover, the idea of presenting the mathematics of computers (without machines!) has proved to be attractive to organizations interested in promoting opportunities for women and minorities in science and technology, particularly in situations where funds for education are severely limited. One of the sponsoring organizations of the Los Alamos Megamath Project is the *American Association of Historically Black Colleges*. The *Kovalevskaia Fund* (a foundation for women in science in developing countries) has organized lectures and demonstrations on discrete mathematics in the classroom (including Kid Krypto) at universities in the Third World.

We believe that the cryptographic community has an important role to play in the ambitious curriculum reform projects articulated by the NCTM and other organizations. Kid Krypto includes a tremendous wealth of vivid, accessible, applicable, engaging and active mathematics in its treasury of ideas. The involvement of cryptologists and theoretical computer scientists in elementary education will have several effects — first and foremost in helping to clarify what computer science is about. Like any science, it is about *ideas*; it is not a Cargo Cult.

One of the purposes of this paper is to encourage the reader to become involved in developing Kid Krypto and related materials for children. This can be done, for example, through the SIGACT compendium project. Contributions to this effort can be communicated to the first author. Even rough ideas in rough form are solicited, and will be credited in the compendium publication.

References

1. Crépeau, C., Kilian, J.: Discreet solitary games. Manuscript, April 1992.
2. Kari, J.: Cryptosystems based on reversible cellular automata. Manuscript, August 1992.
3. Kratochvíl, J.: Perfect codes in general graphs. Monograph, Czechoslovakian National Academy of Sciences, Prague, 1991.
4. Kierstead, H., Trotter, T.: Planar graph coloring with an uncooperative partner. Manuscript, April 1992.
5. Siromoney, R., Jeyanthi, A., Do Long Van, Subramanian, K. G.: Public key cryptosystems based on word problem. ICOMIDC Symposium on the Mathematics of Computation, Ho Chi Minh City, April 1988.

On Defining Proofs of Knowledge

Mihir Bellare[1] and Oded Goldreich[2,*]

[1] High Performance Computing and Communications, IBM T.J. Watson Research Center, PO Box 704, Yorktown Heights, NY 10598, USA. e-mail: mihir@watson.ibm.com.

[2] Computer Science Department, Technion, Haifa, Israel. e-mail: oded@cs.technion.ac.il.

Abstract. The notion of a "proof of knowledge," suggested by Goldwasser, Micali and Rackoff, has been used in many works as a tool for the construction of cryptographic protocols and other schemes. Yet the commonly cited formalizations of this notion are unsatisfactory and in particular inadequate for some of the applications in which they are used. Consequently, new researchers keep getting misled by existing literature. The purpose of this paper is to indicate the source of these problems and suggest a definition which resolves them.

1 Introduction

The introduction of the concept of a "proof of knowledge" is one of the many conceptual contributions of the work of Goldwasser, Micali and Rackoff [14]. This fundamental work, though containing intuition and clues towards a definition of the notion of a "proof of knowledge," does not provide a formal definition of it. Furthermore, in our opinion, the commonly cited formal definitions, namely those of Feige, Fiat and Shamir [6] and Tompa and Woll [18], are not satisfactory, and, in particular, inadequate for some of the applications in which they have been used.

The purpose of this paper is two-fold. First, we would like to describe whence stem the flaws in the previous definitions and why these definitions do not suffice for some applications. We then propose a definition which we feel remedies these defects and also has other advantages.

We note that a definition which is much better than those of [6, 18] has appeared in the work of Feige and Shamir [7], but the community seems unaware of the fact that the definition in [7] is fundamentally different from, and preferable to, the one in [6] (in particular, this fact is not stated in [7]). The definition we present differs in many ways from that of [7] which we feel still has some conceptual problems. Yet both have in common the attempt to capture provers who convince with probabilities that are not non-negligible, thereby correctly addressing what we believe is one of the main flaws in the definitions of [6, 18].

* Research was partially supported by grant No. 89-00312 from the US-Israel Binational Science Foundation (BSF), Jerusalem, Israel.

Among the novel features of our new definition is that it allows us also to talk of the knowledge of machines which operate in super-polynomial-time. But this (and other novel features) we will discuss later; let us begin with the basics.

1.1 Basic approach in defining proofs of knowledge

Intuitively, a two-party protocol constitutes a "system for proofs of knowledge" if "whenever" one party (called the verifier) is "convinced"[3] then the other party (called the prover) indeed "knows" "something". The excessive use of quotation symbols in the condition of the above statement may provide some indication to the complexity of the notion. For simplicity, let us consider the special case in which the "object of knowledge" is a witness for membership of a common input in some predetermined language in NP. For example, let us consider the case in which the "object of knowledge" is a satisfying assignment for a CNF formula (given as input to both parties). Hence, a two-party protocol constitutes a "system for proofs of knowledge of satisfying assignments" if "whenever" the verifier is "convinced" then the prover indeed "knows" a satisfying assignment for the given formula. The clue to a formalization of "proofs of knowledge" is an appropriate interpretation of the phrases "whenever" and "knows" which appear in the condition. The phrase "convinced" has the straightforward and standard interpretation of accepting (i.e., entering a specified state in the computation).

Following [14] the interpretation of the phrases "whenever" and "knows" is as follows. Suppose for simplicity that the verifier is always convinced (i.e. after interaction with the prover the verifier always enters an accepting state). Saying that the prover "knows" a satisfying assignment means that it "can be modified" so that it outputs a satisfying assignment. The notion of "possible modifications of machine M" is captured by efficient algorithms that use M as an oracle. Hence, saying that the prover "knows" a satisfying assignment means that it is feasible to compute a satisfying assignment by using the prover as an oracle. Namely, there exists an efficient algorithm, called the *knowledge extractor*, that on input a formula ϕ and given oracle access to a good prover (i.e. a prover which always convince the verifier on common input ϕ) is able to output a satisfying assignment to ϕ. Indeed, this is exactly the interpretation given in works as [18, 6]. The problem is to deal with the general case in which the prover may convince the verifier with some probability $\epsilon < 1$. Again, for constant ϵ there is no problem and it can be required that even in this case the knowledge extractor succeeds in outputting a satisfying assignment in expected polynomial-time (or alternatively output such an assignment in polynomial time with probability exponentially close to 1). This interpretation is valid also if ϵ is any non-negligible function of the length of the input ϕ (a non-negligible function in n is a function which is asymptotically bounded from below by a function of the form n^{-c}, for some constant c). But *what should be required if the prover does not convince the verifier with non-negligible probability?* Most

---

[3] We have replaced the more intuitive but possibly misleading phrase "convinced that the prover knows something" by the neutral phrase "convinced".

previous formulations (e.g., [18, 6]) require nothing, and hence are unsatisfactory both from a conceptual point of view and from a practical point of view (i.e., in view of many known applications). In particular, this inadequacy often appears when "proofs of knowledge" are used as subprotocols inside larger protocols. In other words, the inadequate formulations of "proofs of knowledge" drastically limit their modular application in the construction of cryptographic protocols.

1.2 Provers which convince with probability that is not non-negligible

We start with an abstract justification of our claim that requiring nothing, in case the prover does not convince the verifier with non-negligible probability, is wrong. We first uncover the reason it has been believed that it is justified to require nothing. It has been believed that events which occur with probability which is not non-negligible can be ignored, just as events which occur with negligible probability can be ignored. However, a key observation, which has been overlooked by this argument, is that a sequence of probabilities can be neither negligible (i.e., smaller that n^{-c} for all $c > 0$ and all sufficiently large n's) nor non-negligible (i.e., bigger that n^{-c} for some $c > 0$ and all sufficiently large n). Hence, even if it were justified to require nothing in case the prover convinces the verifier with negligible probability, it is unjustified to require nothing in case the probability of being convinced is just not non-negligible!

To demonstrate what is wrong when we require nothing in case the prover does not convince the verifier with a non-negligible probability, we consider the following possibility. Suppose that there exist a prover and an infinite sequence of CNF formulae, $\{\phi_n : n \in \mathbb{N}\}$, such that the probability that the prover convinces the verifier on common input ϕ_n is n^{-k}, where n is the length of ϕ_n and k is the number of literals in the longest clause of ϕ_n. Furthermore, suppose that, for every $k > 0$, there exists infinitely many n's such that k is the number of literals in the longest clause of ϕ_n. An important observation is that the sequence of probabilities (defined by the above prover and formulae) is neither negligible (i.e., smaller that n^{-c} for all $c > 0$ and all sufficiently large n's) nor non-negligible (i.e., bigger that n^{-c} for some $c > 0$ and all sufficiently large n). Hence, previous definitions of "proof of knowledge" require nothing (or too little) with respect to the above prover. To appreciate the severity of the lack of requirement with respect to the above prover consider the following application. Suppose that each ϕ_n has a unique satisfying assignment, and that a "proof of knowledge of a satisfying assignment" is used as a subprotocol inside a protocol in which Alice will send Bob a satisfying assignment to ϕ_n if she is convinced by Bob that he already knows this assignment. We would like to argue that in this application Alice yields no knowledge to Bob (i.e., Alice is zero-knowledge). Using a reasonable definition of "proof of knowledge" one should be able to prove such a statement (and indeed using our definition such a proof can be presented). Yet, the zero-knowledge property of Alice can not be demonstrated using previous formulations of "proof of knowledge."[4]

A more concrete and practical setting can help to further clarify our point. It has been suggested to use a "proof of knowledge" as a subprotocol inside a multi-round encryption scheme secure against chosen ciphertext attack (cf. [8, Sec. 5] and [15, Sec. 5.4]). Namely, the decryption module returns a decryption of a chosen ciphertext only if "convinced" that the party asking for it already "knows it". (This is a special case of the application considered in the previous paragraph). Using previous formalizations of "proof of knowledge" it cannot be proved that the above "decryption module" is zero-knowledge (i.e., yields no knowledge) under a chosen ciphertext attack. Yet, the above decryption module is zero-knowledge and this zero-knowledge property (though not proven!) has been used to claim that the particular multi-round encryption scheme is secure against chosen message attack. We stress that the above mentioned encryption scheme is indeed secure under such attacks, it is just that its security has not been proven but rather "hand-waved", and that the essential flaw in the hand-waving is the fact that it is based on an inadequate formalization of proofs of knowledge.

The above example is very typical. In many (yet not all) applications of "proofs of knowledge" one relies on their meaningfulness with respect to arbitrary behavior of the prover. Yet as pointed out above, previous formalizations of "proof of knowledge" are meaningful only in case the prover convince the verifier with non-negligible probability. One should not make the mistake of saying that events which happen with probability that is not non-negligible can be ignored, since such probabilities are not negligible! Put in other words, negligible is not the negation of non-negligible!

To avoid confusion we stress that the definitions of [6] do suffice for the applications in their paper. Problems (as illustrated above) have arisen when these same definitions have (later) been used in other applications.

1.3 A few words about the definition presented in this paper

The most important aspect in which our definition (as well as the one of [7]) deviates from the previous ones is that there is no sharp distinction between provers based on whether they convince the verifier with non-negligible probability or not. In our case, the requirement is that the knowledge extractor always succeeds and that the average number of steps it performs is inversely proportional (via a polynomial factor) to the probability that the prover convinces the verifier.

Over and above this change, we have taken the opportunity to correct what we feel are other conceptual drawbacks of previous definitions (including [7]).

---

[4] Typically, the simulator for the zero-knowledge property uses the knowledge extractor (for the proof of knowledge) as a subroutine. However, previous formulations of "proof of knowledge" do not guarantee a knowledge extractor which handles the entire sequence of formulae. On the other hand, one cannot ignore the case in which something is sent by Alice since this case is not negligible.

Although these other changes are to some extent a matter of taste they are nonetheless important, and also enable us to obtain definitions that are more general than previous ones. As examples, a few such issues are discussed below; we refer the reader to §4 for more details as well as for a discussion of the many other points of difference.

All previous definitions refer only to provers which can be implemented by probabilistic, polynomial time programs (with auxiliary input). In some works it is even claimed that it makes no sense to talk of the knowledge of computationally unrestricted machines. We strongly disagree with such claims, and point out that previous definitions have considered only computationally restricted provers because of technical reasons. From a conceptual point of view it is desirable to have a "uniform" definition of proofs of knowledge which refers to all provers independently of their complexity, the probability they lead the verifier to accept, and so on. In fact, our definition has this property. A consequence of this property is that our definition enables one to talk of the "knowledge" of super-polynomial-time machines. For example, we are able to say in what sense the interactive proofs introduced by Shamir [17], in order to demonstrate that IP=PSPACE, constitute "proofs of knowledge."

Most proofs of knowledge (e.g., the proof of knowledge of an isomorphism used by [12] – see Appendix E) are constructed by iterating some "atomic" protocol. Typically, these atomic protocols have the property that one can easily lead the verifier to accept with some constant probability (say, 1/2) even when having no "knowledge" whatsoever. Yet, these atomic protocols do prove some "knowledge" of the prover, in case it is able to convince the verifier with higher probability. However, previous definitions of "proof of knowledge" were unable to capture this phenomenon; they were only able to say what it means for sufficiently (i.e. super-logarithmic) many iterations of these "atomic" protocols to be "proofs of knowledge." This belies the basic intuition and also precludes a modular approach to protocol design. We correct these weaknesses by showing how to measure the "knowledge error" of a proof, and then showing how composition reduces it.

A special case of our definition is when the knowledge error is zero. This special case is important is some applications. In particular, "proofs of knowledge with zero error" are important when using a proof of knowledge inside a zero-knowledge protocol so that one party sends some information only if he is convinced that the other party already knows it. A typical example is the zero-knowledge protocol for graph non-isomorphism of [12] (cf. §7.1). We stress that none of the previous definitions could handle "proofs of knowledge with zero error."

1.4 Organization

The main conventions used throughout the paper appear in §2. The new definition (of a proof of knowledge) appears in §3, and §4 contains a discussion of various aspects of this definition. This main part of the paper is augmented by Appendix A, in which previous definitions (of proofs of knowledge) are reviewed, and by §7 in which examples of the applications of the new definition are presented.

The rest of the paper addresses issues which are related to the definition of a proof of knowledge: §5 addresses the effect of repeating a proof of knowledge, and §6 presents an equivalent formulation of our definition of a proof of knowledge.

2 Preliminaries

Let $R \subseteq \{0,1\}^* \times \{0,1\}^*$ be a binary relation. We say that R is *polynomially bounded* if there exists a polynomial p such that $|y| \leq p(|x|)$ for all $(x,y) \in R$. We say that R is an NP relation if it is polynomially bounded and, in addition, there exists a polynomial-time algorithm for deciding membership in R.

If R is a binary relation we let $R(x) = \{\, y : (x,y) \in R \,\}$ and

$$L_R = \{\, x : \exists y \text{ such that } (x,y) \in R \,\}.$$

If $(x,y) \in R$ then we call y a *witness* for x.

The proof systems we define are two-party protocols. We model the players in these protocols not (as is common) as interactive machines, but rather as what we will call "interactive functions." The idea is to separate the computational aspect of the player from its input/output behaviour. We feel that this eases and clarifies the presentation of the (later) definitions.

Definition 1. An interactive function A associates to each $x \in \{0,1\}^*$ (common input) and $\eta \in \{0,1\}^*$ (prefix of a conversation) a probability distribution on $\{0,1\}^*$ which we denote by $A_x[\eta]$. We denote by $A_x(\eta)$ an element chosen at random from this distribution.

Intuitively, $A_x(\eta)$ is A's next message when the prefix of the conversation so far was η and the common input is x.

The two players in the protocols we will consider are called the *prover* and the *verifier*. Both are modeled as interactive functions. The interaction between prover P and verifier V on a common input x consists of a sequence of "moves" in each of which one player sends a message to the other. The players alternate moves, and for simplicity we will assume the prover moves first and the verifier last. We denote by α_i (resp. β_i) the random variable which is the message sent by the prover (resp. verifier) in his i-th move. We assume any prefix of a conversation can be uniquely parsed into its constituent messages. Then each message is specified by the prescribed interactive function as a function of the common input and previous messages. More precisely,

$$\alpha_i = P_x(\alpha_1\beta_1 \ldots \alpha_{i-1}\beta_{i-1}) \quad (i = 1, 2, \ldots)$$
$$\beta_i = V_x(\alpha_1\beta_1 \ldots \alpha_{i-1}\beta_{i-1}a_i) \, (i = 1, 2, \ldots).$$

These random variables are defined over the probabilistic choices of both interactive functions.

We will adopt the convention that there are special symbols which an interactive function may output to indicate things like acceptance or rejection. We assume there exists a function $t_V(\cdot)$ (the number of "rounds") such that the $t_V(x)$-th move of the verifier contains its verdict on acceptance or rejection. (For simplicity we restrict the number of rounds to be a function of the verifier and the common input, and do not allow it to depend on the prover. Yet this is without loss of generality). The *transcript of the interaction*, denoted $\mathrm{tr}_{P,V}(x)$, is the string valued random variable which records the conversation up to the verifier's verdict. That is, $\mathrm{tr}_{P,V}(x) = \alpha_1\beta_1\ldots\alpha_{t_V(x)}\beta_{t_V(x)}$. Note that the transcript of the interaction between a prover P and verifier V contains the sequence of message exchanged during the interaction, but not information which is available only to one party, such as its "auxiliary input" or its "internal coin tosses," unless these were sent to the other party.

Since we have assumed that the transcript contains the verifier's verdict on whether to accept or reject, we may, for each x, talk of the set of *accepting transcripts*, denoted $\mathrm{ACC}_V(x)$, and the set of *rejecting transcripts*, denoted $\mathrm{REJ}_V(x)$. Thus the "probability that the verifier accepts" is, by definition, $\Pr[\mathrm{tr}_{P,V}(x) \in \mathrm{ACC}_V(x)]$.

We stress that the definition of an interactive function makes no reference to its computational aspects. We may discuss the computational complexity of an interactive function in a natural way, namely by the complexity of a (probabilistic) Turing machine that computes it. In particular, we say that an interactive function A is computable in probabilistic polynomial time if there exists a probabilistic Turing machine which on input x, η outputs an element distributed uniformly in $A_x[\eta]$, and runs in time polynomial in the length of x.

For simplicity we will restrict the verifier's program to be computable in probabilistic, polynomial time. (We stress that we do not restrict the computational power of the party playing the role of the verifier.) We will also restrict the number of rounds (associated to this verifier program) to be a polynomially bounded, polynomial time computable function.

Sometimes we wish to discuss probabilistic, polynomial time players who receive an additional "auxiliary" input (such an input may be, for example, a witness for the membership of the common input in some predetermined NP language). We may capture such situations by thinking of the auxiliary input as being incorporated in the interactive function (i.e. the party's interaction on common input x and auxiliary y is captured by an oracle indexed by both x and y).

We will be interested in probabilistic machines which use interactive functions as oracles.

Definition 2. Let $K(\cdot)$ be a probabilistic oracle machine, and A an interactive function. Then $K^{A_x}(x)$ is a random variable describing the output of K with oracle A_x and input x, the probability being over the random choices of K and A.

The meaning of having A_x as an oracle is that K may specify a string η and, in one (special) step, obtain a random element from $A_x[\eta]$. We count the steps needed to specify η (and read the output), but the oracle invocation is just one step. It is understood that an invocation of the oracle on a string η returns a random element of $A_x[\eta]$, independently of any previous invocations of the oracle on other inputs.[5]

We call a function $f\colon \mathbb{N} \mapsto \mathbb{R}$ *negligible* if for all $c > 0$ and all sufficiently large n we have $f(n) < n^{-c}$. We call a function $f\colon \mathbb{N} \mapsto \mathbb{R}$ *non-negligible* if there exists $c > 0$ so that for all sufficiently large n we have $f(n) \geq n^{-c}$. We call $f\colon \{0,1\}^* \mapsto \mathbb{R}$ negligible if the function $n \mapsto \max_{x \in \{0,1\}^n} f(x)$ is negligible, and non-negligible if the function $n \mapsto \min_{x \in \{0,1\}^n} f(x)$ is non-negligible. As stressed above, non-negligible is not the negation of negligible but rather a very strong negation of it (and there exist functions which are neither negligible nor non-negligible).

3   A Definition of a Proof of Knowledge

Let $R \subseteq \{0,1\}^* \times \{0,1\}^*$ be a binary relation. Our aim is to define a "system of proofs of knowledge for R." For simplicity, we restrict our attention to polynomially bounded relations (and, unless otherwise stated, all relations in this paper are assumed to be such). Note that the most natural and important class of proofs of knowledge, namely those of "knowledge of a witness for an NP statement," correspond to the special case of NP relations.

The heart of the proof system is the verifier, which remains fixed for our entire discussion. This fixed verifier may interact with arbitrary provers, and we will relate the behavior of the verifier in these interactions with assertions concerning knowledge of the corresponding provers.

For the purpose of defining proofs of knowledge there is no need to restrict the verifier computationally, although in most applications one asks that it be probabilistic, polynomial time.

We make no assumptions concerning the possible provers (in contrast to previous formalizations). We don't even assume that they send messages that can be computed (say nothing about efficiently computed) from the information they receive (i.e., their initial input and in-coming messages). That is, provers are arbitrary interactive functions.

We wish to define the "knowledge of P about x which may be deduced from the interaction of P with V (on input x)". Clearly, this knowledge contains the transcript of the interaction. Yet, in case the interaction is accepting and this event is not incidental, one can say more on the knowledge of P. Namely, the ability of P to "often" lead the verifier to accept may say something about the knowledge of P. The crucial observation, originating in [14], is that the "knowledge of P about x (deduced by interaction)" can be captured by whatever can be *efficiently computed* on input x and access to the oracle P_x.

---

[5]   A stricter alternative is obtained by fixing the prover's sequence of coin tosses and treating it as auxiliary input to the prover. Note that all known "proofs of knowledge" satisfy also this more strict requirement. The fact that the strict requirement implies the main one can be shown by techniques similar to those used in Appendix C.

The phrase "efficiently computed on input x and access to an oracle P_x." is made precise in the definition of a "knowledge extractor." The straightforward approach is to require that the knowledge extractor is a probabilistic polynomial-time oracle machine. Indeed this is the approach taken in some previous works (if one translates their ideas to this slightly different setting). We will replace the strict requirement that the knowledge extractor works in polynomial-time by a more adaptive requirement which relates the running time of the knowledge extractor to the probability that the verifier is convinced. The advantages of this approach have already been discussed and will be further discussed below.

Let $p(x)$ be the probability that prover P convinces verifier V to accept on input x. In its simplest form, the requirement we impose is that the extractor succeed in outputting a witness in (expected) time proportional to $1/p(x)$. In actuality, we will introduce a "knowledge error function" $\kappa(\cdot)$ and ask that the extractor succeed in outputting a witness in (expected) time proportional to $1/(p(x) - \kappa(x))$. Intuitively, $\kappa(x)$ is the probability that the verifier might accept even if the prover did not in fact "know" a witness. We note that in applications $\kappa(x)$ is small, and often it is zero (cf. §4.4 and §5). The precise definition follows.

Definition 3. (System of proofs of knowledge) Let R be a binary relation, and $\kappa: \{0,1\}^* \to [0,1]$. Let V be an interactive function which is computable in probabilistic, polynomial time. We say that a V is a knowledge verifier for the relation R with knowledge error κ if the following two conditions hold.

- Non-triviality: There exists an interactive function P^* so that for all $x \in L_R$, all possible interactions of V with P^* on common input x are accepting (i.e. $\Pr[\text{tr}_{P^*,V}(x) \in \text{ACC}_V(x)] = 1$).
- Validity (with error κ): There exists a constant $c > 0$ and a probabilistic oracle machine K such that for every interactive function P and every $x \in L_R$, machine K satisfies the following condition:

  if $p(x) \stackrel{\text{def}}{=} \Pr[\text{tr}_{P,V}(x) \in \text{ACC}_V(x)] > \kappa(x)$ then, on input x and access to oracle P_x, machine K outputs a string from the set $R(x)$ within an expected number of steps bounded by

  $$\frac{|x|^c}{p(x) - \kappa(x)}.$$

The oracle machine K is called a universal knowledge extractor, and κ is called the knowledge error function.

The next section is devoted to remarks on various features of this definition.

4   Remarks

We discuss various features of our definition, with particular regard to how it differs from previous definitions.

4.1   Provers which convince with non-negligible probability

Suppose the knowledge error is negligible. Clearly, if the verifier accepts with non-negligible probability then the knowledge extractor runs in average polynomial in $|x|$ time. This conclusion yields essentially what [6, 18] have considered as sufficient. Yet, as we have argued, this conclusion by itself does not suffice.

4.2   The efficiency of the provers and verifier

For the purpose of *defining* proofs of knowledge, there is no need to restrict the prover to polynomial-time. This is a point on which we disagree with previous works which claimed that it makes no sense to talk of the knowledge of unrestricted machines. Our definition is presented without assuming anything about the power of the prover, and it is a corollary that machines with no time bounds may know facts which cannot be deduced in (say) double exponential time (and so on). In particular, as we will see (cf. §7.2), it is meaningful, under our definition, to say that the prover in Shamir's interactive proof system for a PSPACE-complete language "knows" an accepting computation of a polynomial-space machine. One the other hand, provers which succeed in convincing a verifier of their knowledge can be reasonably efficient. For example, they may be implemented by polynomial-time programs. Furthermore, all "reasonable" interactive proofs for languages in NP (and in particular the zero-knowledge ones [12]) can be convinced by probabilistic polynomial-time provers which get an NP-witness as auxiliary input. (However, membership in an NP language can be proven via Shamir's result that IP = PSPACE. The corresponding prover is unlikely to be implementable in polynomial-time).

Note that we do not ask that the verifier be a probabilistic polynomial time interactive Turing machine, but just that it be an interactive function computable by one. This distinction is conceptually useful when we consider applications such as the graph non-isomorphism protocol [12] in which the verifier (of the proof of knowledge) is the prover of the graph non-isomorphism protocol, and thus not a probabilistic polynomial time interactive Turing machine. However, the part of this prover's program which implements the verifier (of the proof of knowledge) is indeed computable in probabilistic polynomial time.

4.3   The knowledge extractor

What should not be given to the knowledge extractor. We deviate from some previous works in that we define the knowledge of the prover only with respect to what is publicly available (i.e., the common input x, access to an oracle for the prover, and possibly the transcript). Some other works define the knowledge of the prover with respect to the auxiliary information available to the prover as well as its sequence of coin tosses (which may[6] not be known to the verifier). To justify our choice we remind the reader that the definition of "proof of knowledge" is supposed to capture the knowledge of the prover *demonstrated by the*

---

[6] Using the term "may" is indeed an understatement!

interaction and not merely the knowledge of the prover. Hence, there seems to be little motivation and/or justification to talk about the knowledge of a machine with respect to something which is not known to the outside (i.e., verifier). In particular, only the common input (of the interaction) should be given as input to the knowledge extractor, and the auxiliary input or local coins of the prover should certainly not be given.

One thing that the knowledge extractor can do. In all examples we are aware of, the knowledge extractor proceeds by trying to find several (not more than polynomially many) related accepting transcripts. For example, the knowledge extractor presented in Appendix E tries to find a single accepting transcript in addition to the one given as input. Clearly such a knowledge extractor succeeds within an average number of steps which is inversely proportional to the density of the accepting transcripts (which is in other words the accepting probability). Note that if the proof of knowledge is zero-knowledge then a single accepting transcript (and in particular the one given as input) cannot suffice.

Universality of the knowledge extractor. In the above definition we require the existence of a universal knowledge extractor which works for all possible interactive functions P. Switching the quantifiers (i.e., requiring that for every interactive function P there exist a knowledge extractor K_P) would make little sense in practice since P in our conventions may depend on (non-uniform) auxiliary input of the "real" prover (cf. §2). However, the quantifiers may be switched if one considers only provers which are (uniform) interactive machines. For further discussion see the parenthetical subsection in [10, Sec. 4.1], which considers an analogous situation in the context of zero-knowledge. We stress that also in case the quantifiers are switched, the knowledge extractor (although it may depend on the prover) must be given *oracle access* to the prover. The reason being that the prover's program may be highly inefficient (and therefore cannot be "incorporated" into the extractor).

4.4   The knowledge error function

The knowledge error function is a novelty of our definition.[7] Let us see why it is important.

Typically, "proofs of knowledge" are constructed by repeating an "atomic" protocol sufficiently many times. An atomic protocol for graph isomorphism, for example, is the following (cf. [12]).

Example. The input is a pair of (isomorphic) graphs G_1 and G_2. The prover generates a *single* random isomorphic copy of G_1 which we call H, and sends H to the verifier. The latter responds with a random query $i \in \{1, 2\}$. The prover replies to i by presenting an isomorphism between G_i and H. The verifier accepts

---

[7]   Although the ideas in [5] may be interpreted as pointing to a similar notion.

if the permutation supplied by the prover is indeed an isomorphism between G_i and H.

Intuitively, this protocol does demonstrate some "knowledge" of an isomorphism between G_1 and G_2. Yet, previous definitions were unable to capture this fact; they were only able to show that sufficiently (i.e. super-logarithmic) many iterations of this protocol constituted a "proof of knowledge." This non-modular approach belies the basic intuition and is also not the natural approach to protocol design.

The introduction of the knowledge error function remedies these defects. In particular, we are able to capture "atomic" proofs of knowledge of the above type. Indeed, under our definition, the above is a proof of knowledge with knowledge error $1/2$. Furthermore, we are able to prove composition theorems which show how to reduce the knowledge error (cf. §5) and thus construct proofs of knowledge in a modular fashion.

Another motivation of the knowledge error function comes from cases where, for convenience, we have the verifier accept with some (usually small) probability even if the evidence supplied by the prover is not convincing. For example, we may do this to guarantee perfect completeness (i.e., the prover's ability to alway convince the verifier of valid statements). In such cases, the knowledge error can compensate for this small probability. The importance of this aspect of the knowledge error function, and the perfect completeness example, were pointed out to us by Feige (private communication, June 1992).

4.5 What about soundness?

We note that our definition makes no requirement for the case $x \notin L_R$. In particular, soundness (i.e., a bound on the prover's ability to lead the verifier to accept $x \notin L_R$) is not required. Consequently, a knowledge verifier for R does not necessarily define an interactive proof of membership in L_R. This is in contrast to previous definitions; they had the "validity" condition imply the soundness condition, so that the latter always held. We feel that our "decoupling" of soundness from validity is justified both conceptually and in the light of certain applications. Let us see why.

First, conceptually, it seems more natural to talk about extracting witnesses only when these witnesses exist. Furthermore, as long as one property is not known to imply the other it seems wrong to require the latter unless one really needs it.

Second, there are some natural applications (e.g., "zero-knowledge based" identification schemes) in which it is a-priori agreed that the protocol will be applied only to strings in some NP language (i.e., $x \in L_R \in \text{NP}$). Such applications are better modeled by our definition than by previous ones. To be concrete, consider the following identification scheme based on the hardness of quadratic residuosity.

Example. A user A (Alice), who wishes to be able to securely remote-login to a mainframe computer (which we denote by V because it plays the role of verifier) chooses at random a pair of large primes and multiplies them to get a modulus N_A. She also chooses $Y_A \in Z_{N_A}^*$ at random, sets $X_A = Y_A^2 \bmod N_A$, and gives the pair (N_A, X_A) to V. All this is performed once in a life-time, when Alice is identified by other means. Later, whenever Alice wishes to remote-login, she sends her name (A) to V, who responds by sending the pair (N_A, X_A). She now provides a (zero-knowledge) proof that she "knows" a square root of X_A mod N_A. Besides the fact that A can provide the proof (completeness) we require that if Bob $(B \neq A)$ were to attempt to remote-login as A then he (B) would fail. The point to note in (the formalization of) the latter requirement is that the interaction of B with V takes place on an input (namely (N_A, X_A)) which is in the underlying language L_R (the relation R here is $\{((N, X), Y) : Y^2 \equiv X \pmod N\}$ and the underlying language is $L_R = \{(N, X) : X$ is a square mod $N\}$). So it suffices to require that the interaction of B with V on inputs *in this language* "proves possession of a witness." *What happens on interactions on input not in the language is immaterial to the security of the identification scheme.* Thus the requirements for a secure (zero-knowledge based) identification scheme are more faithfully modeled by our Definition 3 than by previous definitions (which required that *any* proof of knowledge of a relation R be an interactive proof of membership in L_R).

We stress that we are not, of course, saying that soundness is *always* redundant. Rather, the above discussion justifies our choice not to make soundness a part of the definition of a proof of knowledge. In cases where soundness is necessary, it can be viewed as a separate, additional property that the knowledge verifier must satisfy. Furthermore, it is possible that some applications call for other kinds of conditions on $x \notin L_R$. One possibility, which we call *strong validity*, is discussed in Appendix B.

4.6 Relaxing the non-triviality requirement

The prover guaranteed by the non-triviality requirement must convince the verifier in all interactions of $x \in L_R$. This requirement, met in all known protocols, is not essential to the definition of a proof of knowledge. In general one may require that the existence of a prover that convinces the verifier, on input x, with probability $C(x)$. As far as polynomial-time (or even more powerful) verifiers are concerned any choice of a polynomial-time constructible bound, $C(\cdot)$, which is both non-negligibly greater than $\kappa(\cdot)$ and bounded above by $1 - 2^{-\text{poly}(\cdot)}$, is equivalent.[8] In fact, following the ideas in [9], one can eliminate the error probability in the completeness condition altogether and derive the definition as in the previous section. However, although the last transformation does preserve

---

[8] When saying that these choices are equivalent, as long as the above requirements are satisfied, we mean that existence of a verifier which satisfies one permissible bound yields the existence of another verifier which satisfies the second bound. Furthermore, the complexity both of the verifier and of the prover (meeting the completeness condition) is preserved (and so are zero-knowledge properties).

validity, it does not necessarily preserve the complexity of the prover and its zero-knowledge property.[9]

4.7   A word about computationally convincing proofs of knowledge

Some works (cf. [4, 5]) consider the situation in which the class of provers for which the protocol is supposed to be a "proof of knowledge" is restricted to the class of probabilistic, polynomial time interactive Turing machines with auxiliary input.[10] Typically, the protocols in question rely on the use of problems which are intractable for the prover(s). This is the case of *computationally convincing* (zero-knowledge) proofs, also known as *arguments* (cf. [3]).

Our definitions may be adapted to cover such settings as well. We would restrict the class of provers for which validity is required to hold to the class of interactive functions computable in probabilistic, polynomial time by interactive machines. We would, however, also relax slightly the validity requirement by asking that it only be true for sufficiently long inputs. More precisely, we would require that for each probabilistic, polynomial time computable interactive function P (prover) there exist a constant n_P such that for each $x \in L_R$ of length at least n_P, machine K satisfies the following condition:

if $p(x) \stackrel{\text{def}}{=} \Pr[\text{tr}_{P,V}(x) \in \text{ACC}_V(x)] > \kappa(x)$ then, on input x and access to oracle P_x, machine K outputs a string from the set $R(x)$ within an expected number of steps bounded by $|x|^c/(p(x) - \kappa(x))$.

In applications, $\kappa(x)$ could be set to $1/\text{poly}(x)$ for some specific $\text{poly}(\cdot)$. Alternatively, following [7], one can use $\kappa(\cdot)$ as a shorthand for "smaller than any function of the form $1/\text{poly}(\cdot)$". However, a much better alternative is to set $\kappa(\cdot)$ to be a specific negligible function (e.g., $\kappa(x) = 2^{-\sqrt[5]{|x|}}$) related to a specific intractability assumption concerning the computational problem on which the scheme is based (e.g., DLP is intractable with respect to algorithms which run in time $2^{\sqrt[5]{n}}$ on inputs of length n).

Some ideas on the subject of "computationally convincing proofs of knowledge" appear in the work of Brassard, Crépeau, Laplante and Léger [5]. Although they do not present definitions, it would appear these ideas bear many similarities to ours. We discuss their work in Appendix A.

The fact that some variations are needed to treat the case of "computationally convincing proofs of knowledge" has been pointed out to us by Feige (private communication, June 1992).

---

[9] In this context we note, however, that the zero-knowledge too may be preserved, as long as one is willing to make a complexity assumption, by further applying the transformation of [2].

[10] For simplicity we ignore the auxiliary inputs in this discussion. They can be treated as outlined in §2.

5   Reducing the knowledge error via repetitions

One of the reasons to introduce the knowledge error function is the theorems established here. We show that the knowledge error may be reduced by composition.

First we consider sequential composition. Here $m = m(x)$ independent copies of the original protocol are executed on input x, and the verifier accepts iff all copies are accepting (we stress that by "independent" we mean that the verifier acts in each of the copies independently of the others; of course we don't assume this about prospective provers). If κ was the knowledge error of the original protocol then the knowledge error the resulting protocol is essentially κ^m. The more precise statement follows.

Notational convention: by $\operatorname{poly}(\cdot)$ we mean any sufficiently large polynomial in the length of the input (string).

Required assumption: $y \in R(x)$ can be found (if such exists) in exponential-time (i.e., time $2^{\operatorname{poly}(|x|)}$). Finally, we assume of course that $m(x) \leq \operatorname{poly}(|x|)$.

Theorem 4. *Suppose that V is a knowledge verifier for the relation R with error $\kappa(\cdot)$. Let V_m denote the program that, on input x, sequentially executes the program V, on input x, for $m(x)$ times. Then V_m is a knowledge verifier for the relation R with error $\kappa_m(\cdot) \stackrel{\text{def}}{=} (1 + 1/\operatorname{poly}(\cdot)) \cdot \kappa(\cdot)^{m(\cdot)}$.*

The proof is in Appendix C.1.

With respect to error reduction via parallel repetitions we were only able to prove a statement concerning a special class of knowledge verifiers (which nonetheless contains all known verifiers). For further discussion see Appendix C.2.

Finally, we observe that tiny knowledge error can be eliminated.

Proposition 5. *Suppose that an element in $R(x)$, if such exists, can be found in time at most $t(x)$, given only x as input. Suppose V is a knowledge verifier for R with knowledge error smaller than $\frac{1}{2 \cdot t(x)}$. Then, V is a knowledge verifier for R with knowledge error 0.*

We omit the proof which uses methods similar to those used in Appendix B.

The resulting formulation (namely, knowledge error 0) is often the simplest way of thinking about proofs of knowledge: we are saying that the knowledge extractor succeeds in time $|x|^c/p(x)$, where $p(x)$ is as in Definition 3. Many proofs of knowledge (e.g., the one presented in Appendix E) are of this type.

6   An equivalent formulation of validity

Following is an equivalent formulation of the validity condition. The new formulation is inspired by (yet is quite different in many respects from) the definition in [7]. Let $p(x)$ be as in Definition 3. Instead of asking that the knowledge verifier always output $y \in R(x)$, we ask only that it output $y \in R(x)$ with a probability bounded below by $p(x) - \kappa(x)$, and otherwise output a special symbol, denoted $\perp$, indicating "failure to find $y \in R(x)$". However, whereas originally the extractor had expected time proportional to $1/(p(x) - \kappa(x))$, we now give it only expected polynomial time. More precisely, letting $\kappa\colon \{0,1\}^* \mapsto [0,1]$, we have the following.

- New validity (with error κ): We say that the verifier V satisfies *new validity with error κ* if there exists a probabilistic expected polynomial-time oracle machine K such that for every interactive function P and every $x \in L_R$ it is the case that $K^{P_x}(x) \in R(x) \cup \{\perp\}$ and

$$\Pr[K^{P_x}(x) \in R(x)] \geq \Pr[\mathrm{tr}_{P,V}(x) \in \mathrm{ACC}_V(x)] - \kappa(x) .$$

Proposition 6. *The new validity condition is equivalent to the one given in Definition 3.*

Here we give the proof for the case $\kappa(x) = 0$. The proof for the general case is more complex and is in Appendix D.

Suppose, first, that K is a knowledge extractor satisfying the new definition. We construct a knowledge extractor K' that, on input x repeatedly invokes K (on x) until $K(x) \neq \perp$. Clearly, K' always outputs a string in $R(x)$, halting in expected time $\mathrm{poly}(x)/\Pr[K(x) \in R(x)]$, which is bounded above by $\mathrm{poly}(x)/\Pr[\mathrm{tr}_{P,V}(x) \in \mathrm{ACC}_V(x)]$. Hence, K' satisfies the condition in Definition 3. Suppose, now, that K is a knowledge extractor satisfying Definition 3. We construct a knowledge extractor K' that, on input x first generates a random transcript (i.e., $\mathrm{tr}_{P,V}(x)$) and activates $K(x)$ if this transcript is accepting (i.e., in $\mathrm{ACC}_V(x)$). Otherwise, K' halts immediately outputting $\perp$. One can easily verify that K' runs in expected polynomial-time and outputs $y \in R(x)$ with probability exactly $\Pr[\mathrm{tr}_{P,V}(x) \in \mathrm{ACC}_V(x)]$.

7 Applications

Our formalization, as well as that of [7], do suffice to prove the security of those schemes for encryption secure against chosen-cyphertext attack which rely on zero-knowledge proofs of knowledge (cf. §1.2). However, we prefer to describe here two applications to which our definition of "proof of knowledge" can be applied, whereas all the previous formalizations fail. The first application is a modular description of the zero-knowledge proof for Graph Non-Isomorphism (of [12]) which uses a "proof of knowledge of an isomorphism" as a subprotocol. The second application is to Shamir's interactive proof for PSPACE.

7.1 Zero-Knowledge proof of Graph Non-Isomorphism

The second author first realized the inadequacy of previous formulations of "proofs of knowledge" when Leonid Levin insisted that the zero-knowledge interactive proof for Graph Non-Isomorphism (of [12]) should be presented in a modular manner.[11] As many people noticed, the intuition behind this zero-knowledge proof is that the verifier first proves to the prover that it "knows" an isomorphism between one of the input graphs and the query graph that it presents to the prover.[12] If the prover is convinced then it answers the query by indicating to which of the two input graphs the query graph is isomorphic. By doing so the prover yields no knowledge to the verifier, since the verifier "knows" to which of the two input graphs the query is isomorphic, yet the prover's answer supplies statistical evidence that the two input graphs are not isomorphic. This intuitive idea, taken from the Quadratic Non-Residousity zero-knowledge proof of [14], has indeed guided the development of the zero-knowledge proof system for GNI, but plays no part in the formal description and proof of correctness appearing in [12] (and [14]). Levin complained, rightfully, against this inelegant and non-modular approach. The second author's answer, at the time, was that an elegant proof which uses the subprotocol and its properties in a modular fashion is not possible due to lack of appropriate definitions.[13]

One definition that was lacking at the time was that of the information hiding property of the subprotocol used to prove "possession of knowledge". Specifically, that subprotocol, which consists of the parallel version of the zero-knowledge proof of Graph Isomorphism, is not known to be zero-knowledge (and in light of [11] it is unlikely that a proof that it is zero-knowledge can ever be given). Nevertheless, this subprotocol is "witness indistinguishable" (in the sense defined latter by Feige and Shamir [7]) and this property suffices to the soundness of the interactive proof of GNI. However this entire issue is irrelevant to the current paper.

The other definition that was lacking at that time was an *adequate* definition of a proof of knowledge. An adequate definition of a "proof of knowledge" is needed to ensure that if the GNI-prover is convinced that the GNI-verifier "knows" an isomorphism between the query graph and one of the input graphs then indicating to which input graph the query graph is isomorphic yields no knowledge to the GNI-verifier.[14] To this end, the simulator (constructed to meet the zero-knowledge clause) uses the knowledge extractor guaranteed by the definition of a "proof of knowledge". However, as pointed out above, previous definitions of "proof of knowledge" are useless in the case the GNI-prover is not convinced with non-negligible probability. It follows that the simulator will fail to construct the interactions in these cases which may occur with probability that is neither non-negligible nor negligible (see §1.2). In particular, consider the situation where for every $c > 0$ there exists an infinite sequence of inputs to the protocol such that on input of length n the GNI-prover is convinced with

---

[11] For sake of self-containment, this protocol is presented in Appendix E

[12] The prover in the zero-knowledge proof for GNI is the verifier in a "proof of knowledge of an isomorphism between two graphs"; whereas the verifier in the zero-knowledge proof for GNI is the party claiming and proving knowledge of an NP-witness for GI.

[13] It should be stressed that a proof of correctness of (the zero-knowledge property of) the protocol of does appear in [12]. The criticism points to the fact that the proof of correctness in [12] does not reflect the intuition just outlined.

[14] The reader may find it useful at this point to consult Appendix E.

probability n^{-c}.

On the other hand, one can show that the subprotocol "for proof of knowledge of isomorphism" (presented in [12] and Appendix E) constitutes a (sound) proof of knowledge, according to the definitions presented in §3. It follows that the running time of the knowledge extractor is inversely proportional to the probability that the GNI-prover is convinced. Hence, the simulator for the GNI-protocol will run in expected polynomial-time and produce a perfect simulation of the interaction. Furthermore, it can be easily shown that the GNI-prover *while playing the role of the GI-verifier in the proof of knowledge* yields no knowledge to the GNI-verifier (since its messages are generated in probabilistic polynomial-time from its inputs).

7.2   What does the prover of a PSPACE language know?

Using our definition, it is possible to say that the verifier in Shamir's interactive proof for a PSPACE-complete language L is a knowledge verifier for the relation R_L consisting of pairs (x, c) where c is the middle configuration in the computation of a fixed machine accepting $x \in L$. Hence, one can say that (in some meaningful sense) any prover which convinces this verifier (with, say, probability 1) on input x, does know an accepting computation on input x.

Let us show how a knowledge extractor may find the middle configuration. For the rest of this subsection, we assume that the reader is very familiar with the interactive proof for QBF as presented in [17, Section 5]. The standard reduction of a PSPACE language to QBF associates the middle configuration in an accepting poly-space computation with the first block of t existential quantifiers in the formula. So in the rest of this subsection we will consider only the problem of retrieving a sequence of truth-values so that assigning these values to the above mentioned variables yields value true for the resulting formula.

First, we consider a straightforward method for retrieving these t boolean values. This method does work in case the prover convinces the verifier with probability 1 (but will have to be modified to deal with arbitrary provers). First the knowledge extractor asks the oracle for the first message of the prover which is a pair (N, v_0), where N is a large prime and v_0 is a non-zero residue mod N (the value of the arithmetic expression mod N). Next, the knowledge extractor proceeds in t rounds. In the ith round, the extractor feeds the oracle the sequence $r_1, ..., r_{i-1} \in Z_N$ and gets the polynomial, p_i, which corresponds to the opening of the ith variable, when the previous $i-1$ variables are set to $r_1, ..., r_{i-1}$, respectively. The extractor then finds a $\mu_i \in \{0, 1\}$ so that $p_i(\mu_i)$ is not equal to zero modulo N (such μ_i must exist since $\sum_{\mu \in \{0,1\}} p_i(\mu) \equiv v_{i-1} \not\equiv 0$ (mod N)). Round i is completed by setting $r_i = \mu_i$ and $v_i = p_i(r_i)$.

In general the above method may fail as it relies too heavily on the answers of the prover on boolean r_i's. An alternative approach is to select the r_i's uniformly in Z_N. The problem is that the resulting residual arithmetic expression no longer reflects the truth value of the residual boolean formula. To solve the problem we need to find the polynomial resulting by setting the r_i's to μ_i's by examining the polynomials which result by random settings of the r_i's. To see how this can be done, we need to take a closer look at the formula used by Shamir and its arithmetization. It can be seen that the polynomial p_i received from the prover in round i has coefficients which are polynomials in r_1 through r_{i-1}. Denote by $c_{i,j}(r_1, ..., r_{i-1})$ the polynomial in r_1 through r_{i-1} representing the jth coefficient of p_i. The $c_{i,j}$'s are polynomials each of total degree at most $2(i-1) < 2t-1$, and we are interested in the values of $c_{i,j}(\sigma_1, ..., \sigma_{i-1})$. Using the ideas of [1] these values can be found via "interpolation" at $2t$ uniformly selected (yet dependent) points. Finally, we note that the knowledge extractor can tell whether it is given the correct polynomial at a point by carrying on the rest of the interactive proof using the oracle to the function P_x. Further details are omitted.

Acknowledgements

The second author thanks Leonid Levin for his interest in "proofs of knowledge" and his insistence that they have to be formalized in a sufficiently robust manner so that they can be used in applications such as the Graph Non-Isomorphism protocol.

We are grateful to Uri Feige for valuable criticisms of an earlier version of this paper. Specific credit to Feige's suggestions is given in the relevant places of the current manuscript.

References

1. D. Beaver, and J. Feigenbaum, "Hiding Instances in Multioracle Queries," *Proc. of the 7th STACS*, 1990, pp. 37-48.
2. M. Bellare, S. Micali and R. Ostrovsky, "The True Complexity of Statistical Zero-Knowledge," *Proceedings of the 22nd Annual ACM Symposium on the Theory of Computing*, ACM (1990), pp. 494-502.
3. G. Brassard, D. Chaum, and C. Crépeau, "Minimum Disclosure Proofs of knowledge," *JCSS*, Vol. 37, No. 2, 1988, pp. 156-189.
4. J. Boyar, C. Lund and R. Peralta, "On the Communication Complexity of Zero-Knowledge Proofs." 1989.
5. G. Brassard, C. Crépeau, S. Laplante and C. Léger, "Computationally Convincing Proofs of Knowledge," *Proc. of the 8th STACS*, 1991.
6. U. Feige, A. Fiat, and A. Shamir, "Zero-Knowledge Proofs of Identity", *Journal of Cryptology*, Vol. 1, 1988, pp. 77-94.
7. U. Feige, and A. Shamir, "Witness Indistinguishability and Witness Hiding Protocols," *Proceedings of the 22nd Annual ACM Symposium on the Theory of Computing*, ACM (1990), pp 416-426.
8. Z. Galil, S. Haber, and M. Yung, "Symmetric Public-Key Encryption", *Advances in Cryptology - Crypto85 proceedings*, Lecture Notes in Computer Science, Vol. 218, Springer-Verlag, 1986, pp. 128-137.
9. M. Furer, O. Goldreich, Y. Mansour, M. Sipser, and S. Zachos, "On Completeness and Soundness in Interactive Proof Systems", *Advances in Computing Research: a research annual*, Vol. 5 (S. Micali, ed.), pp. 429-442, 1989.
10. O. Goldreich, "A Uniform-Complexity Treatment of Encryption and Zero-Knowledge", *J. of Cryptology*, to appear.

11. O. Goldreich, and H. Krawczyk, "On Sequential and Parallel Composition of Zero-Knowledge Protocols", *17th ICALP*, Lecture Notes in Computer Science, Vol. 443, Springer-Verlag, 1990, pp. 268–282.

12. O. Goldreich, S. Micali, and A. Wigderson, "Proofs that Yields Nothing but Their Validity or All Languages in NP Have Zero-Knowledge Proof Systems", *JACM*, Vol. 38, No. 1, July 1991.

13. O. Goldreich, and Y. Oren, "Definitions and Properties of Zero-Knowledge Proof Systems", TR-610, Computer Science Dept., Technion, Haifa, Israel. Submitted to *Jour. of Cryptology*.

14. S. Goldwasser, S. Micali, and C. Rackoff, "The Knowledge Complexity of Interactive Proof Systems", *SIAM J. on Computing*, Vol. 18, No. 1, 1989, pp. 186-208.

15. S. Haber, "Multi-Party Cryptographic Computations: Techniques and Applications", PhD Dissertation, Computer Science Dept., Columbia University, Nov. 1987.

16. Y. Oren, "On the Cunning Power of Cheating Verifiers: Some Observations about Zero-Knowledge Proofs," *Proceedings of the 28th Annual IEEE Symposium on the Foundations of Computer Science*, IEEE (1987), pp. 462-471.

17. A. Shamir, "IP=PSPACE," *Proceedings of the 31st Annual IEEE Symposium on the Foundations of Computer Science*, IEEE (1990), pp. 11-15.

18. M. Tompa and H. Woll, "Random Self-Reducibility and Zero-Knowledge Interactive Proofs of Possession of Information," University of California (San Diego) Computer Science and Engineering Dept. Technical Report Number CS92-244 (June 1992). (Preliminary version in *Proceedings of the 28th Annual IEEE Symposium on the Foundations of Computer Science*, IEEE (1987), pp. 472-482.)

A   Previous Definitions of Proofs of Knowledge

For sake of self-containment we review below the definitions of "proof of knowledge" appearing in the literature. In general there are two generally cited formulations appearing in [6] and in [18]. In addition, there is the better (but lesser known) formulation of Feige and Shamir [7]. Finally, there is work on "computationally convincing proofs of knowledge" [4, 5].

"Proof of Knowledge" according to Feige, Fiat and Shamir [6]   The definition presented in [6] refers only to parties which work in probabilistic polynomial-time, yet may have auxiliary input (which is not necessarily generated efficiently). The knowledge extractor is given the prover's program and auxiliary input and may run the prover's program as a subroutine (yet being charged for the time).[15] The knowledge extractor is required to produce good output only for provers and inputs for which the prover has a non-negligible probability of convincing the verifier on that input. Specifically, it is required that

---

[15] The extractor may try to analyze the prover's program by other means but Feige, Fiat and Shamir claim that this does not make sense. In any case the knowledge extractors that they present only use the prover's program as a "black-box".

for every constant $a > 0$ there exists a probabilistic polynomial-time extractor M so that for all constants $b > 0$, all provers P, and all sufficiently large x, r, k, if $\Pr[(P, V)(x, r, k) = \text{ACC}] > |x|^{-a}$ then $\Pr[M(\text{desc}(P), x, r, k) \in R(x)] > 1 - |x|^{-b}$. ($\text{desc}(P)$ denotes the description of P).

The string k in the above definition denotes a-priori knowledge of P (given in the form of auxiliary input) where r denotes the prover's sequence of coin tosses. The fact that k is given to the knowledge extractor, though being indeed conceptually disturbing, can be justified in several applications (and in particular those in [6]). We stress that the definition of [6] does not guarantee one knowledge extractor which works regardless of the prover's success probability but rather a sequence of extractors each relevant for a different "measure" of non-negligence. As claimed in the our text this is conceptually unsatisfactory and inadequate for many applications in which a proof of knowledge is used as a subroutine. It should be said that "proofs of knowledge" are not used as subprotocols in [6], but rather as the "thing itself" (and hence our critic of their definition is only weakly relevant, if at all, to the results of that paper).

"Proof of Knowledge" according to Tompa and Woll [18] The definition presented in [18] differs slightly from the one of [6]. It allows the verifier to run for an arbitrary (not necessarily polynomial) amount of time. The running time of the knowledge extractor is polynomial in the length of the input and in the running time of the verifier. As explained in §4.3, we don't believe that this choice is justified. The knowledge extractor in the [18] definition is given as input the *prover's view* of the interaction with the verifier, which contains among other things the prover's auxiliary input (denoted k in the definition of [6] presented above). The requirement concerning the output of the verifier is that the event "on input x the verifier is convinced yet the knowledge extractor fails to find $y \in R(x)$" happens very rarely (i.e. with probability smaller than ϵ for some $\epsilon < 1$). The probability is taken over the random coin tosses of both parties (for any fixed input x and fixed auxiliary input k). Clearly, this definition suffers from all the disadvantages of the definition of [6] discussed above. Furthermore, if ϵ is indeed fixed, as suggested by the definition in [18], then protocols satisfying their definition are useless even in a stronger sense: the prover may convince the verifier with probability $\epsilon/2$ and yet the knowledge extractor is required nothing. Tompa and Woll were indeed aware of this point and seem to suggest to eliminate the problem by applying the protocol iteratively sufficiently many times. This is indeed a good suggestion. However, several problems remain. First a conceptual problem: their Lemma 3 (hereafter referred to as the *Composition Lemma*) indeed offers a useful tool, but it does not provide a general satisfactory definition of a "proof of knowledge". More annoying is the fact that the Composition Lemma constructs better protocols via *sequential* composition of worse ones. It is not clear (and furthermore it seems unlikely) that a parallel composition will have the same affect. Finally, the Composition Lemma is applicable only to relations R which are in BPP.

"Proof of Knowledge" according to Feige and Shamir [7] The definition presented in [7] looks similar to the one in [6], but in fact it is fundamentally different. The critical point is that the definition in [7] treats potential provers uniformly with respect to the probability they lead the verifier to accept. In this sense, the definition in [7] is similar to our definition. Specifically, the knowledge extractor, denoted M, runs in expected polynomial-time (rather than in strict polynomial-time as in [6]) and outputs an element of $R(x)$ with probability that is at most non-negligibly smaller than the probability that the verifier accepts on input x. Specifically, it is required that

there exists a probabilistic *expected* polynomial-time extractor M so that for all constants $b > 0$, all provers P, and all sufficiently large x, r, k,

$$\Pr[(P, V)(x, r, k) = \texttt{ACC}] > \Pr[M(\mathbf{desc}(P), x, r, k) \in R(x)] - |x|^{-b}$$

Consequently this definition does not suffer from the main criticism raised against the definition of [6]. However, it still suffers from the other problems such as the fact that k is given to M. Furthermore, it does not capture "knowledge" of super-polynomial-time provers.

Work on "computationally convincing proofs of knowledge". Brassard, Crépeau, Laplante and Léger [5] study "computationally convincing proofs of knowledge" (the "validity" condition refers only to probabilistic, polynomial-time provers). They do not present formal definitions so we found it difficult to compare their work to ours, but the ideas appear to have some relation. They too propose an "adaptive" requirement linking the running time of the extractor to the success of the prover. Specifically, they appear to consider a particular class of protocols, namely those consisting of k rounds, each of which contains a "challenge" (from verifier to prover) which the prover may correctly answer with probability $1/2$ if he correctly "guesses" a coin toss of the verifier. They require that the extractor succeed in time linear in $1/\varphi$, where $2^{-k} + \varphi$ is the "probability of undetected cheating." The quantity in quotes was not defined precisely, particularly for the case of the input being in the language, but if $2^{-k} + \varphi$ is interpreted as the probability that the verifier accepts, then it is like our definition with the knowledge error set to 2^{-k}.

Brassard et. al. [5] also raise some criticisms of the definitions of [6, 18], but their criticism is the opposite of ours: whereas we suggest that the previous definitions are too weak (and propose a stronger definition) they suggest that the previous definitions are already too strong.

B Soundness and Strong Validity

For completeness, we state here also the standard soundness condition (for interactive proof systems). We remind the reader that we view soundness as an additional property that a knowledge verifier may (or may not) satisfy.

Definition 7. (Additional possible properties of a system of proofs of knowledge) Let R be a binary relation, and suppose that V is a knowledge verifier for the relation R with knowledge error κ. We define two additional properties that V may satisfy:

- soundness: For every interactive function P, and for all $x \notin L_R$, most of the possible interactions of V with P on common input x are rejecting (i.e., $\Pr[\mathrm{tr}_{P,V}(x) \in \mathrm{ACC}_V(x)] < 1/2$).
- strong validity (with error κ): Let K be the universal knowledge extractor, and $c > 0$ be the constant guaranteed by the validity condition of Definition 3. Then, for every interactive function P and every $x \notin L_R$, machine K satisfies the following condition:

  if $p(x) \overset{\text{def}}{=} \Pr[\mathrm{tr}_{P,V}(x) \in \mathrm{ACC}_V(x)] > \kappa(x)$ then, on input x and access to oracle P_x, machine K outputs the special symbol $\perp$ within an expected number of steps bounded by

$$\frac{|x|^c}{p(x) - \kappa(x)}$$

As usual, the completeness (or non-triviality) and soundness conditions merely state that there is a gap between the probability that a prover may convince the verifier on $x \in L_R$ (which by the completeness condition is exactly 1) and the probability that a prover may convince the verifier on $x \notin L_R$ (which by the soundness condition is at most 1/2). Validity (resp., strong validity) is a more refined condition regarding the behavior of arbitrary provers on $x \in L_R$ (resp., arbitrary strings). Specifically, validity relates the probability that the prover convinces the verifier on $x \in L_R$ and the average time it takes the knowledge extractor to find a $y \in R(x)$ in the case $x \in L_R$. Strong validity is an analogous requirement regarding $x \notin L_R$. Validity, soundness, and strong validity are not always independent. Namely,

Proposition 8. *Validity and soundness imply strong validity for NP relations.*

The proof that follows is for the case $\kappa = 0$.

Recall that an NP relation is a polynomially bounded relation $R(\cdot, \cdot)$ which is decidable in polynomial time. Suppose an NP relation R possesses a knowledge verifier which (in addition) satisfies the soundness condition. Without loss of generality[16], we may assume the error probability in the soundness condition is at most $2^{-p(n)}$, where $p(\cdot)$ is a polynomial bounding the length of witnesses as a function of the length of the input. Let K be the universal knowledge extractor (satisfying the validity condition). Fix a deterministic procedure, with running-time $2^{p(n)} \cdot \mathrm{poly}(n)$, for deciding L_R (e.g., the one which scans through all possible witnesses for the given input).

---

[16] The error probability in the soundness condition may be reduced, as usual, by repetitions.

We construct a new knowledge extractor, denoted K', for the above proof of knowledge, satisfying also strong validity. On input x and oracle access to P_x, machine K' runs in parallel the extractor K (with input x and oracle P_x) and the decision procedure for L_R, fixed above. Suppose K halts before the decision procedure terminates, and yields an output y. Machine K' checks whether $R(x, y)$ is true (it can do this in polynomial time) and if so outputs y; otherwise it outputs $\bot$. On the other hand, suppose the decision procedure halts while K is still running. If the decision is negative ($x \notin L_R$) then K' outputs $\bot$; else it continues to run K to whatever outcome this might yield.

We note that the running time of K' is (within a polynomial factor of) that of K when $x \in L_R$, and at most (within a polynomial factor of) $2^{p(|x|)}$ otherwise. But in the latter case, the probability $p(x) = \Pr[\mathrm{tr}_{P,V}(x) \in \mathrm{ACC}_V(x)]$ is at most $2^{-p(|x|)}$, so that the running time of K' is expected $|x|^{O(1)}/p(x)$ in both cases. The fact that K' is a knowledge extractor for R which satisfies (validity and) strong validity follows.

Finally, we note that the above transformation preserves (upto polynomial factors) the running time of the knowledge verifier, and, as long as we do the error-reduction in a suitable way (for example, by serial composition), it also preserves zero-knowledge.

C Reducing the Knowledge Error via Repetitions

We prove the claims of §5. Let us first recall the notation and assumptions introduced there. By poly($\cdot$) we mean any sufficiently large polynomial in the length of the input (string). By assumption the messages of the verifier can be computed in polynomial-time, and $y \in R(x)$ can be found (if such exists) in exponential-time (i.e., time $2^{\mathrm{poly}(x)}$). Consequently, failure of the knowledge extractor occurring with exponentially small probability (i.e., probability $2^{-\mathrm{poly}(x)}$) can be ignored. Finally, we assume of course that $m(x) \leq \mathrm{poly}(x)$.

C.1 Reducing the Knowledge Error via Sequential Composition

Suppose that V is a knowledge verifier with error $\kappa(\cdot)$ for the relation R, and let K be a knowledge extractor witnessing this fact. Let V_m denote the program that, on input x, sequentially executes the program V, on input x, for $m(x)$ times. Theorem 4 asserts that V_m is a knowledge verifier with error $\kappa_m(\cdot) \stackrel{\text{def}}{=} (1+1/\mathrm{poly}(\cdot)) \cdot \kappa(\cdot)^{m(\cdot)}$ for the relation R. The theorem is proven by constructing a knowledge extractor, denoted K_m, as described below.

Suppose that P_m is a prover which, on input x, leads V_m to accept with probability $p_m(x) > \kappa_m(x)$. Loosely speaking, we observe that there exists an i, $0 \leq i \leq m(x) - 1$, and a partial transcript of i iterations so that, relative to this partial transcript, the $i + 1^{\text{st}}$ iteration is accepting with probability at least $\sqrt[m(x)]{p_m(x)}$. The idea is to use the guaranteed knowledge extractor, K, on the $i + 1^{\text{st}}$ iteration of V_m, relative to an appropriate partial i-iteration transcript. Details follow.

For simplicity, we assume here that all transcripts are equally likely. Let T_i denote the set of all possible partial transcripts of the first i iterations, and $A_i \subseteq T_i$ denote the set of partial (i-iteration) transcripts in which all the i iterations are accepting. Let $a_i \overset{\text{def}}{=} |A_i|/|T_i|$ ($a_0 \overset{\text{def}}{=} 1$). For every $\alpha \in A_i$, let $q(\alpha)$ denote the accepting probability of the $i+1^{\text{st}}$ iteration relative to a partial transcript α, and c_{i+1} denote the average of $q(\alpha)$ taken over all $\alpha \in A_i$.

The following sequence of claims lead to the construction of the knowledge extractor K_m.

Claim 1: for every i, $0 \leq i < m(x)$, it holds that $a_{i+1} = a_i \cdot c_{i+1}$.

Proof: Clearly,

$$|A_{i+1}| = \sum_{\alpha \in A_i} \frac{|T_{i+1}|}{|T_i|} \cdot q(\alpha)$$

$$= |T_{i+1}| \cdot \frac{|A_i|}{|T_i|} \cdot \frac{\sum_{\alpha \in A_i} q(\alpha)}{|A_i|}$$

and the claim follows. $\square$

Claim 2: there exists an i, $0 \leq i < m(x)$, such that

1. $c_{i+1} \geq \sqrt[m(x)]{p_m(x)}$.
2. $a_i \cdot (c_{i+1} - \kappa(x)) > \frac{p_m(x)}{\text{poly}(x)}$.

Proof: By Claim 1, $p_m(x) = \prod_{i=1}^{m(x)} c_i$, and Part (1) follows. Using $p_m(x) > \kappa_m(x)$, we get

$$c_{i+1} \geq \sqrt[m(x)]{1 + 1/\text{poly}(x)} \cdot \kappa(x)$$

$$= \left(1 + \frac{1}{\text{poly}(x)}\right) \cdot \kappa(x)$$

and hence $c_{i+1} - \kappa(x) \geq c_{i+1}/\text{poly}(x)$. Using $a_i \cdot c_{i+1} \geq p_m(x)$, Part (2) follows.
$\square$

Notation: Let i be as guaranteed by Claim 2, and denote $\delta_{i+1} \overset{\text{def}}{=} c_{i+1} - \kappa(x)$. Let $A_{i,t}$ denote the set of partial transcripts in A_i containing only partial transcripts relative to which the $i + 1^{\text{st}}$ iteration accepts with probability bounded below by $\kappa(x) + 2^t \delta_{i+1}/\text{poly}(x)$ and above by $\kappa(x) + 2^{t+1}\delta_{i+1}/\text{poly}(x)$, where poly($\cdot$) is a specific polynomial which depends on $m(\cdot)$ and the time required to find $y \in R(x)$. Namely,

$$A_{i,t} \overset{\text{def}}{=} \left\{ \alpha \in A_i : \kappa(x) + 2^t \cdot \frac{\delta_{i+1}}{\text{poly}(x)} \leq q(\alpha) < \kappa(x) + 2^{t+1} \cdot \frac{\delta_{i+1}}{\text{poly}(x)} \right\}$$

Claim 3: Let i and $A_{i,t}$ be as above. Then there exists an t, $1 \leq t < \text{poly}(x)$, such that $|A_{i,t}| \geq 2^{-t} \cdot |A_i|$.

Proof: Assume, on the contrary, that the current claim does not hold. Then

$$c_{i+1} < \kappa(x) + \frac{\delta_{i+1}}{\text{poly}(x)} + \sum_{t \geq 1} \frac{|A_{i,t}|}{|A_i|} \cdot \left(2^{t+1} \cdot \frac{\delta_{i+1}}{\text{poly}(x)} \right)$$

$$< \kappa(x) + \frac{\delta_{i+1}}{\text{poly}(x)} + \sum_{t=1}^{\text{poly}(x)} 2^{-t} \cdot \left(2^{t+1} \cdot \frac{\delta_{i+1}}{\text{poly}(x)} \right)$$

$$< \kappa(x) + \delta_{i+1}$$

$$= c_{i+1}$$

and contradiction follows. $\square$

Claim 4: There exists an i, $0 \leq i < m(x)$, and an j, $1 \leq j < \text{poly}(x)$, such that at least a 2^{-j} fraction of the $\alpha \in T_i$ satisfy

$$q(\alpha) > \kappa(x) + 2^j \cdot \frac{p_m(x)}{\text{poly}(x)}$$

Proof: Let i as guaranteed by Claim 2. Rephrasing Claim 3, we get that there exists an t, $1 \leq t < \text{poly}(x)$, such that at least a $2^{-t} \cdot a_i$ fraction of the $\alpha \in T_i$ satisfy $q(\alpha) > \kappa(x) + 2^t \cdot \delta_{i+1}/\text{poly}(x)$. Substituting $j = t + \log_2(1/a_i)$ and using Part (2) of Claim 2, the claim follows. $\square$

Using Claim 4, we are now ready to present the knowledge extractor K_m. Machine K_m runs in parallel $m(x) \cdot \text{poly}(x)$ copies of the following procedure, each with a different pair (i, j), $1 \leq i \leq m(x)$ and $1 \leq j \leq \text{poly}(x)$. By saying "run several copies in parallel" we mean execute these copies so that t steps are executed in each copy before step $t + 1$ is executed in any other copy[17].

The copy running with the pair (i, j), generates $M \stackrel{\text{def}}{=} 2^j \cdot \text{poly}(x)$ random partial transcripts of i-iterations, denoted $\gamma_1, ..., \gamma_M$, and runs M copies of the knowledge extractor K in parallel, each using a corresponding partial transcript (γ_k). The sub-procedure, indexed by the triple (i, j, k), uses the partial transcript γ_k to convert queries of the basic knowledge extractor (i.e., K) into queries concerning the $i + 1^{\text{st}}$ iteration. Namely, when K is invoked it asks queries to an oracle describing the messages of a prover interacting with V. However, K_m has access to an oracle describing prover P_m (which is supposedly interacting with V_m). Hence, K_m needs to simulate an oracle describing a basic prover (interacting with V), by using an oracle describing P_m. This is done by prefixing each query of K with the i-iteration partial transcript γ_k generated above.

To analyze the performance of K_m consider the copy of the procedure running with a pair (i, j) satisfying the conditions of Claim 4. If this is the case, then with very high probability (i.e., exponentially close to 1) at least one of the partial transcripts generated by this copy has the property that, relative to it,

---

[17] Actually, the condition can be related. For example, it suffices to require that at least t steps are executed in each copy before step $2 \cdot t$ is executed in any other copy.

the $i+1^{\text{st}}$ iteration accepts with probability at least $\kappa(x)+2^j p_m(x)/\text{poly}(x)$. It follows that the corresponding copy of the sub-procedure will halt, outputting $y \in R(x)$, within $\frac{\text{poly}(x)}{2^j \cdot p_m(x)}$ steps (on the average). Since the $(i,j)^{\text{th}}$ copy of the procedure consists of $2^j \cdot \text{poly}(x)$ copies of the sub-procedure running in parallel, this copy of the procedure will halt in expected time $\frac{\text{poly}(x)}{p_m(x)} < \frac{\text{poly}(x)}{p_m(x)-\kappa_m(x)}$. The entire knowledge extractor consists of polynomially many copies of the procedure, running in parallel, and hence it also runs in expected $\frac{\text{poly}(x)}{p_m(x)-\kappa_m(x)}$ time as required.

Remark: We believe that V_m is a knowledge verifier with error $\kappa(\cdot)^{m(\cdot)}$ for the relation R (rather than just being a knowledge verifier with error $(1+1/\text{poly}(\cdot)) \cdot \kappa(\cdot)^{m(\cdot)}$ for this relation). The difference is of little practical importance, yet we consider the question to be of theoretical interest.

C.2   Reducing the Knowledge Error via Parallel Composition

A fundamental problem with presenting a parallel analogue of the above argument is that we cannot fix a partial transcript for the other iterations while working with one selected iteration (which was possible and crucial to the proof used in the sequential case). Furthermore, even analyzing the profile of accepting transcripts is more complex.

As before, let $p_m(x)$ denote the accepting probability, here abbreviated by $p(x)$, and let $\delta(x) \overset{\text{def}}{=} p(x) - \kappa_m(x)$. Consider a $m(x)$-dimensional table in which the dimensions correspond to the $m \overset{\text{def}}{=} m(x)$ parallel executions, where the $(r_1, ..., r_m)$-entry in the table corresponds to the transcript when the verifier uses coin tosses r_1 in the first execution, r_2 in the second execution, and so on. Since a $p(x)$ fraction of the entries are accepting transcripts, it follows that there exists a dimension i so that at least a $\sqrt[m(x)]{p(x) - \delta(x)/2}$ fraction of the rows in the ith dimension contain at least $\delta(x)/2m(x)$ accepting entries. Furthermore, there exists a j, $0 \le j \le \log_2(\text{poly}(x)/\delta_m(x))$, so that at least a $2^j \cdot \sqrt[m(x)]{p(x) - \delta(x)/2}$ fraction of the rows in the ith dimension contain at least $\dfrac{p(x)-\delta(x)/2}{2^j \text{poly}(x) \cdot \sqrt[m(x)]{p(x)-\delta(x)/2}}$ accepting entries.

Getting back to the problem of using the knowledge extractor K (of the basic verifier V), we note that we need to simulate an oracle to K using an oracle describing P_m. The idea used in the sequential case is to augment all queries to the P-oracle by the same partial transcript. However, we can no longer guarantee high accepting probability for one execution relative to a fix transcript of the other (parallel) executions.

We can however treat the special case in which the basic knowledge extractor, K, operates by generating random transcripts and keeping a new transcript only if it satisfies some polynomial-time predicate with respect to the transcripts kept so far. Details omitted. We remark that the known knowledge extractors do operate in such a manner.

D   Equivalence of Two Formulations of Validity with Error

We now prove the equivalence of the definitions of validity with error given in Definition 3 and in §6, respectively. We assume that whenever $\Pr[\text{tr}_{P,V}(x) \in \text{ACC}_V(x)] > \kappa(x)$, we have $\Pr[\text{tr}_{P,V}(x) \in \text{ACC}_V(x)] > \kappa(x) + 2^{-\text{poly}(x)}$ as well. Alternatively, we may assume that there exist an exponential time algorithm for solving the relation R (i.e., finding $y \in R(x)$ if such exists within $2^{\text{poly}(x)}$ steps). The proof extends the argument presented in §6, for the special case $\kappa = 0$, yet in one direction an additional idea is required.

Let us start with the easy direction. Suppose that a verifier V satisfies validity with knowledge error $\kappa(\cdot)$ by the definition in §6. Let K be a knowledge extractor satisfying this definition. We construct a knowledge extractor K' that, on input x repeatedly invokes K (on x) until $K(x) \neq \perp$. Clearly, K' always outputs a string in $R(x)$, halting in expected time $\text{poly}(x)/\Pr[K(x) \in R(x)]$ which is bounded above by $\text{poly}(x)/(\Pr[\text{tr}_{P^*,V}(x) \in \text{ACC}_V(x)] - \kappa(x))$. Hence, K' satisfies the condition in Definition 3.

Suppose that a verifier V satisfies validity with knowledge error $\kappa(\cdot)$ by Definition 3, and let K be a knowledge extractor witnessing this fact. Let $c > 0$ be the constant satisfying the condition on the running-time of K. Namely, that its expected running-time is bounded above by $|x|^c/(\Pr[\text{tr}_{P,V}(x) \in \text{ACC}_V(x)] - \kappa(x))$. Assume, without loss of generality, that with very high probability (i.e., exponentially close to 1) K halts within at most $2^{\text{poly}(x)}$ steps[18]. We construct a knowledge extractor K' that, on input x runs $K(x)$ with the following modification. Machine K' proceeds in iterations, starting with $i = 1$, and terminating after at most $\text{poly}(x)$ iterations. In iteration i, machine K' executes $K(x)$ with time bound $2^i \cdot |x|^c$. If K halts with some output y then K' outputs y and halts. Otherwise (i.e., K does not halt within $2^i \cdot |x|^c$ steps), machine K' halts with probability $\frac{1}{2}$ with output $\perp$ and otherwise proceeds to iteration $i+1$. We stress that in all iterations, K uses the same internal coin tosses. In fact, we can record the configuration at the end of iteration i and consequently save half of the time spent in iteration $i+1$. Clearly, the expected running-time of $K'(x)$ is bounded above by

$$\sum_{i=1}^{\text{poly}(x)} \frac{1}{2^{i-1}} \cdot (2^i \cdot |x|^c) = \text{poly}(x)$$

We now evaluate the probability that, on input x, machine K' outputs $y \in R(x)$. It is guaranteed that, on input x, the extractor K outputs $y \in R(x)$ within $T(x) \leq |x|^c/(\Pr[\text{tr}_{P,V}(x) \in \text{ACC}_V(x)] - \kappa(x))$ steps on the average (and by hypothesis $T(x) < 2^{\text{poly}(x)}$). Hence, with probability at least $\frac{1}{2}$, on input x, machine K outputs $y \in R(x)$ within $2 \cdot T(x)$ steps. The probability that K' conducts $2 \cdot T(x)$ steps (i.e., K' reaches iteration $\log_2(T(x)/|x|^c)$) is $|x|^c/T(x) \geq \Pr[\text{tr}_{P,V}(x) \in \text{ACC}_V(x)] - \kappa(x)$. Hence, K' satisfies the condition in §6.

---

[18] This can be achieved by running the exponential time solver in parallel to K. Alternatively, assuming that if $\Pr[\text{tr}_{P,V}(x) \in \text{ACC}_V(x)] > \kappa(x)$ then $\Pr[\text{tr}_{P,V}(x) \in \text{ACC}_V(x)] > \kappa(x) + 2^{-\text{poly}(x)}$, we can implement a probabilistic exponential-time solver using K.

E The Zero-Knowledge proof of Graph Non-Isomorphism

Following is the basic ingredient of the zero-knowledge proof for Graph Non-Isomorphism (GNI) presented in [12].

Common input: Two graphs G_1 and G_2 of n vertices each.
Objective: In case the graphs are not isomorphic, supply (statistical) evidence to that affect.

Step V1: The *GNI-verifier* selects uniformly $i \in \{1, 2\}$, and a random isomorphic copy of G_i, hereafter denoted H and called the *query*, and sends H to the *GNI-prover*. Namely, H is obtained by selecting a random permutation π, over the vertex-set, and letting the edge-set of H consist of pairs $(\pi(u), \pi(v))$ for every pair (u, v) in the edge-set of G_i.

Step VP: The GNI-verifier "convinces" the GNI-prover that he (i.e., the GNI-verifier) "knows" an isomorphism between H and one of the input graphs. To this end the two parties execute a witness indistinguishable proof of knowledge (with zero error) for graph isomorphism. (Such a protocol is described below.) In that proof of knowledge the GNI-verifier acts as the prover while the GNI-prover acts as the verifier.

Step P1: If the GNI-prover is convinced by the proof given at step VP, then he finds j such that H is isomorphic to G_j, and sends j to the GNI-verifier. (If H is isomorphic to neither graphs or to both the GNI-prover sets $j = 1$; this choice is arbitrary.)

Step V2: If j (received in step P1) equals i (chosen in step V1) then the GNI-verifier accepts, else he rejects.

It is easy to see that if the input graphs are not isomorphic then there exists a GNI-prover which always convinces the GNI-verifier. This meets the completeness condition of interactive proofs. To show that some sort of soundness is achieved we use the witness indistinguishability of the subprotocol used in Step VP. Loosely speaking, it follows that no information about i is revealed to the GNI-prover and therefore if the input graphs are isomorphic then the GNI-verifier rejects with probability at least one half (no matter what the prover does).[19]

The demonstration that the GNI-prover is zero-knowledge is the place where the notion of proof of knowledge plays a central role. As required by the zero-knowledge condition we have to construct, for every efficient program playing the role of the GNI-verifier, an efficient simulator which outputs a distribution equal to that of the interaction of the verifier program with the GNI-prover. Following is a description of such a simulator. The simulator starts by invoking the verifier's program and obtaining a query graph, H, and a transcript of the execution of step VP (this is obtained when the simulator plays the role of the GNI-prover which is the knowledge-verifier in this subprotocol). If the transcript

---

[19] Reducing the cheating probability further can be done by iterating the above protocol either sequentially or in parallel. However, this is not our concern here.

is not accepting then the simulator halts and outputs it (thus perfectly simulating the real interaction). However, if the transcript is accepting the simulator must proceed (otherwise its output will not be correctly distributed). The simulator needs now to simulate step P1, but, unlike the real GNI-prover, the simulator does not "know" to which graph H is isomorphic. The key observation is that the simulator can obtain this information (i.e., the isomorphism) from the knowledge extractor guaranteed for the proof of knowledge (taking place in step VP), and once the isomorphism is found producing the rest of the interaction (i.e., the bit j) is obvious. Using our definition (of proof of knowledge with zero error), the simulator can find the isomorphism in expected $\text{poly}(n)/p(G_1, G_2, H)$ time, where $p(G_1, G_2, H)$ is the probability that the GNI-prover is convinced by the proof of knowledge in step VP. Since this module in the simulator is invoked only with probability $p(G_1, G_2, H)$, the simulator runs in expected polynomial-time, and the zero-knowledge property follows. *We stress that carrying out this plan is not possible when using any of the previous definitions of "proof of knowledge".*

To complete the description of the above protocol we present a (witness indistinguishable) proof of knowledge of Graph Isomorphism. This proof of knowledge can be easily adapted to a proof of knowledge of an isomorphism between the first input graph and one of the other two input graphs.

Common input: Two graphs H and G of n vertices each.

Objective: In case the graphs are isomorphic, the GI-prover has to "prove knowledge of ψ", where ψ is an isomorphism between H and G.

Note: In our application the GNI-verifier plays the role of the GI-prover, while the GNI-prover plays the role of the GI-verifier.

Notation: Let $t \stackrel{\text{def}}{=} t(n) \stackrel{\text{def}}{=} n^2$.

Step p1: The *GI-prover* selects uniformly t random isomorphic copies of H, denoted $K_1, ..., K_t$ and called the *mediators*, and sends these graphs to the *GI-verifier*. Namely, K_i is obtained by selecting a random permutation π_i over the vertex-set, and letting the edge-set of K_i consist of pairs $(\pi_i(u), \pi_i(v))$ for every pair (u, v) in the edge-set of H.

Step v1: The GI-verifier selects uniformly a subset S of $\{1, 2, ..., t\}$ and sends S to the GI-prover.

Step p2: For every $i \in S$, the *GI-prover* sets $a_i = \pi_i$, where π_i is the permutation selected in step p1 to form K_i. For every $i \in \{1, ..., t\} - S$, the *GI-prover* sets $a_i = \pi_i \psi$, where π_i is as before, ψ is the isomorphism between G and H (known to the GI-prover), and $\pi\psi$ denotes composition of permutations (or isomorphisms). The GI-prover sends $a_1, a_2, ..., a_t$ to the GI-verifier.

Step v2: The GI-verifier checks if, for every $i \in S$, the permutation a_i (supplied in step p2) is indeed an isomorphism between the graphs H and K_i. In addition, the GI-verifier checks if, for every $i \in \{1, 2, ..., t\} - S$, the permutation a_i (supplied in step p2) is indeed an isomorphism between the graphs G and K_i. If both conditions are satisfied (i.e., all t permutations are indeed what they are supposed to be) then the GI-verifier accepts (i.e., is convinced that the GI-prover knows an isomorphism between G and H).

One can show that the above GI-verifier constitutes a knowledge-verifier (with zero error) for Graph Isomorphism. This is done by considering all possible choices of $S \subseteq \{1, 2, ..., t\}$ for a fixed set of mediators $K_1, ..., K_t$. Denote by s the number of subsets S for which the GI-verifier accepts. A knowledge extractor, given one accepting interaction (i.e., containing a good S) tries to find another one (i.e. a good subset different from S). Having two good subsets clearly yields an isomorphism between G and H (i.e., using any index in the symmetric difference between the good subsets). Clearly, if $s = 1$ then there exists no good subset other than S. In this case the extractor finds an isomorphism by exhaustive search (which is always performed in parallel to the attempts of the extractor to find a different good subset). The exhaustive search requires less than 2^t steps, but dominates the total running time only in case $s = 1$ (in which case the accepting probability is $1/2^t$). Yet, for any $s > 1$, the expected number of tries required to find a different good subset is

$$\frac{1}{(s-1)/(2^t - 1)} < \frac{2^t}{s-1} \leq \frac{2 \cdot 2^t}{s}$$

(the last inequality follows from $s \geq 2$). Since $s/2^t$ is the probability that the GI-verifier accepts, the extractor described above indeed runs in expected time inversely proportional to the accepting probability of the GI-verifier. Our claim follows.

Public Randomness
in
Cryptography[*]

Amir Herzberg[1] and Michael Luby[2]

[1] I.B.M. T.J. Watson, Yorktown Heights, NY 10598
[2] International Computer Science Institute, U.C. Berkeley, Berkeley, California 94704

Abstract. The main contribution of this paper is the introduction of a formal notion of public randomness in the context of cryptography. We show how this notion affects the definition of the security of a cryptographic primitive and the definition of how much security is preserved when one cryptographic primitive is reduced to another. Previous works considered the public random bits as a part of the input, and security was parameterized in terms of the total length of the input. We parameterize security solely in terms of the length of the private input, and treat the public random bits as a separate resource. This separation allows us to independently address the important issues of how much security is preserved by a reduction and how many public random bits are used in the reduction.

To exemplify these new definitions, we present reductions from weak one-way permutations to one-way permutations with strong security preserving properties that are simpler than previously known reductions.

1 Introduction

Over the years, randomness has proved to be a powerful algorithmic resource, i.e. randomized algorithms that are simpler, or more efficient, or both, than any known deterministic algorithm have been developed for a variety of problems. Randomness has also proved to be a powerful resource in the construction of cryptographic primitives based on other primitives, e.g., the randomized reductions from weak one-way functions to one-way functions and the reductions from one-way functions to pseudo-random generators. The source of randomness used in these reductions is typically *public*, in the sense that the random bits are accessible to all parties enacting the primitive and to any adversary trying to break the primitive. However, up till now, the distinction between the private part of the input and the public random bits has been blurred.

The main contributions of this paper are to formally introduce the notion of public randomness, to introduce appropriate generalizations of the definitions of

[*] Research supported in part by National Science Foundation operating grant CCR-9016468 and grant No. 89-00312 from the United States-Israel Binational Science Foundation (BSF)

cryptographic primitives that use public randomness and, perhaps most importantly, to modify the definition of what it means to reduce one cryptographic primitive to another by allowing public randomness to be used in the reduction. In terms of generalizing the definition of cryptographic primitives to include public randomness, the main advantage is that the security of a primitive can now be parameterized, as it should be, solely in terms of the length of the private part of the input, and not at all in terms of the public random bits. In terms of reductions, the main advantage is that we can now separately consider the two issues of how much security is preserved by the reduction and how much public randomness is used in the reduction.

As particular examples of how a primitive that uses public randomness can be defined, we extend the definitions of one-way functions and pseudo-random generators to include public random bits. Generalizations along the same lines for many other cryptographic primitives can be made, including those related to public key cryptography.

As particular examples of how the new definitions of reductions using public randomness work, we provide reductions that use public randomness from weak one-way permutations to one-way permutations. Following [1], our prime concern is the security preserving properties of the reduction, i.e., how much of the security of the weak one-way permutation is transferred to the one-way permutation. However, unlike [1], we consider the security as a function solely of the length of the private input, which does not include the public random bits. We show reductions that preserve security in a very strong sense, which is stronger than that of the reduction due to [1] (under the new definitions). We begin with a very simple reduction (much simpler than that found in [1]), which uses a large number of public random bits. Through a sequence of increasingly intricate reductions, we converge on a reduction that is a slight modification of the reduction due to [1]. Both the reduction of [1] and our improvement use only a linear number of public random bits.

Another simple reduction from a weak one-way permutation to a one-way permutation was developed recently and independently by Phillips [2]. Phillips showed that his reduction preserves security somewhat better than the reduction of [1], when considering the randomness as a part of the input. However, our new definitions of security preserving reductions with public randomness reveal that Phillips' reduction actually preserves security as well as our reductions, i.e. much better than [1]. Phillips' reduction uses more public random bits ($O(n \log(n))$) than our best reduction.

A full development and details of this work can be found in [3].

2 Definitions

2.1 Basic Notation

If S is a set then $\sharp S$ is the number of elements in S. Let x and y be bit strings. We let $\|x\|$ denote the length of x. We let $\langle x, y \rangle$ denote the sequence of two strings

x followed by y, and when appropriate we also view this as the concatenation of x and y. When $\langle x, y \rangle$ is the input to a function f, we write this as $f(x, y)$. We let x_i denote the ith bit of x. Let $x \in \{0,1\}^n$ and let $S \subset \{1, \ldots, n\}$. We let x_S denote the subsequence of bits in x indexed by S, e.g. $x_{\{1,\ldots,i\}}$ denotes the first i bits of x, $x_{\{i+1,\ldots,n\}}$ denotes all but the first i bits of x, and thus $x = \langle x_{\{1,\ldots,i\}}, x_{\{i+1,\ldots,n\}} \rangle$.

If x and y are bit strings, each of length l, then $x \oplus y$ is the vector sum mod 2 (i.e. bit wise parity) of x and y, i.e. $(x \oplus y)_i = (x_i + y_i) \bmod 2$.

An $m \times n$ bit matrix x is indicated by $x \in \{0,1\}^{m \times n}$. We write $x_{i,j}$ to refer to the (i, j) in x. We can also view x as a sequence $x = \langle x_1, \ldots, x_m \rangle$ of m strings, each of length n, where x_i is the ith row of the matrix, or as a string $x \in \{0,1\}^{mn}$ of length mn, which is the concatenation of the rows of the matrix.

If a is a number, then $|a|$ is the absolute value of a, $\lceil a \rceil$ is the smallest integer greater than or equal to a, $\log(a)$ is the logarithm base two of a. If a number is an input to or an output of an algorithm, the assumption is that it is presented in binary notation.

In general, we use capital letters to denote random variables and random events. When S is a set we use the notation $X \in_{\mathcal{U}} S$ to mean that X is a random variable uniformly distributed in S, and $x \in_{\mathcal{U}} S$ indicates that x is a fixed element of S chosen uniformly.

2.2 Public Randomness

A source of random bits is *public* for a primitive if it can be read by all parties enacting the primitive and by any adversary trying to break the primitive. The public random string is always chosen uniformly. We use ";" to keep the public random string separated from other strings in a list of strings, e.g. if y is the value of the public random string and x is the input to some function f, then we write $f(x; y)$ to indicate the evaluation of f on input x with respect to y (note that the value of f depends both on the input x and on the public random string y), and we write $\langle f(x; y); y \rangle$ to indicate the pair of strings $f(x; y)$ and y.

Although the public random bits are known to an adversary, it turns out that these bits often plays a crucial role in ensuring that the primitive is secure.

2.3 Security

The security of a primitive quantifies how secure the primitive is against attacks by an adversary trying to break the primitive. The important question to consider is "What does the security measure?" Intuitively, the security of a primitive is a measure of the minimal computational resources needed by any adversary to break the primitive. There are two natural computational resources we consider; the maximal total time T that the adversary runs and the success probability δ of the adversary. Both T and δ are stated with respect to a given input instance to the adversary, and their definitions are primitive dependent.

A trivial strategy to increase the success probability δ is to run the adversary again. This doubles the running time, but also almost doubles the success probability (especially if it is low). This motivates us to simplify the analysis by comparing only the ratios between the success probabilities and the running times of different adversaries. An additional simplification is to consider the ratio between the success probability $\delta(n)$ and the maximal running time $T(n)$, both over all private inputs of length n. Without much loss in generality, we hereafter assume that an adversary A always runs for the same amount of time $T(n)$ on all inputs parameterized by n.

Definition (achievement ratio): The achievement ratio of an adversary A for a primitive f is defined as $\frac{\delta(n)}{T(n)}$, where $T(n)$ is the running time of A and $\delta(n)$ is the success probability of A for f on private inputs of length n.

Definition (breaking adversary and security): An adversary A is $R_f(n)$-breaking for a primitive f if the achievement ratio $\frac{\delta(n)}{T(n)}$ of A for f satisfies $\frac{\delta(n)}{T(n)} \geq R_f(n)$ for infinitely many $n \in \mathcal{N}$. The primitive f is $(1 - R_f(n))$-secure if there is no $R_f(n)$-breaking adversary for f.

Intuitively, 0-secure means totally insecure, whereas 1-secure means totally secure. We would like the primitive to be harder to break than it is to use. For example, suppose f is a $(1 - R_f(n))$-secure one-way function, where for all constants c, $R_f(n) < \frac{1}{n^c}$ for sufficiently large n. Then f can be computed in polynomial time, whereas a polynomial time algorithm can only invert f with inverse polynomial probability for finitely many values of $n \in \mathcal{N}$.

Allowing the security of a primitive to be parameterized is important because different implementations of primitives may achieve different levels of security, which may offer different tradeoffs between efficiency and security. We note that inverse polynomial security (e.g. $R_f(n) = \frac{1}{n^{100}}$) means that the primitive may be broken by a polynomial adversary, so we expect that many applications would require higher security. For example, it may be that a particular function f is a $(1 - R_f(n))$-secure one-way function, where $R_f(n) = \frac{1}{2^{\log^5(n)}}$, or even better with $R_f(n) = \frac{1}{2^{\sqrt{n}}}$. The statement that f is secure with respect to either of these bounds is quantifiably stronger than the statement that it is secure with respect to an inverse polynomial bound. On the other hand, for any function computable in time $T(n)$ there is an inverting adversary that runs in $T(n) \cdot 2^n$ time, and thus there is no $(1 - \frac{1}{T(n) \cdot 2^n})$-secure one-way function.

2.4 Primitives with Public Randomness

Definition (standard function): A function $f(x; y)$ is called a *standard function* with *length relationship* $\|x\| = n$, $\|y\| = l(n)$, $\|f(x; y)\| = m(n)$ if

- $f(x; y)$ is computable in polynomial time.
- If $\|x\| = n$ then $\|y\| = l(n)$ and $\|f(x; y)\| = m(n)$, where both $l(n)$ and $m(n)$ are polynomial in n.

We now give the definitions of primitives using public randomness.

Definition (one-way function with public random bits): Let $f(x; y)$ be a standard function with length relationship $\|x\| = n$, $\|y\| = l(n)$, $\|f(x; y)\| =$

$m(n)$. Let $X \in_{\mathcal{U}} \{0,1\}^n$ and $Y \in_{\mathcal{U}} \{0,1\}^{l(n)}$. The *success probability* of adversary A for f is

$$\delta(n) = \Pr_{X,Y}[f(A(f(X;Y);Y);Y) = f(X;Y)].$$

The *running time* $T(n)$ of adversary A for f is the maximum over all $x \in \{0,1\}^n$ and $y \in \{0,1\}^{l(n)}$ of the running time of A on input $\langle f(x;y);y\rangle$. Then, f is a $(1 - R_f(n))$-secure one-way function if there is no $R_f(n)$-breaking adversary for f.

Definition (one-way permutation with public random bits): Let $f(x;y)$ be a standard function with length relationship $\|x\| = n$, $\|y\| = l(n)$, $\|f(x;y)\| = m(n)$. Then, f is a $(1 - R_f(n))$-secure one-way permutation if f is a $(1 - R_f(n))$-secure one-way function and $m(n) = n$ and x is uniquely determined by $f(x;y)$ and y.

Definition (pseudo-random generator with public random bits): Let $g(x;y)$ be a standard function with length relationship $\|x\| = n$, $\|y\| = l(n)$, $g(x) = m(n)$, where $m(n) > n$. The *stretching parameter* of $g(x;y)$ is $m(n) - n$. Let $X \in_{\mathcal{U}} \{0,1\}^n$, $Y \in_{\mathcal{U}} \{0,1\}^{l(n)}$ and $Z \in_{\mathcal{U}} \{0,1\}^{m(n)}$. The *success probability* (distinguishing probability) of adversary A for g is

$$\delta(n) = \Pr_{X,Y}[A(g(X;Y);Y) = 1] - \Pr_{Z,Y}[A(Z;Y) = 1].$$

The *running time* $T(n)$ of adversary A for g is the maximum over all $z \in \{0,1\}^{m(n)}$ and $y \in \{0,1\}^{l(n)}$ of the running time of A on input $\langle z;y\rangle$. Then, g is a $(1 - R_g(n))$-secure pseudo-random generator if there is no $R_g(n)$-breaking adversary for g.

Example : To exemplify the difference between the traditional definition of a one-way function and the definition introduced here with public randomness, consider the subset sum problem. A one-way function based on the difficulty of this problem can be defined in two ways; without public random bits and with public random bits. Let $b \in \{0,1\}^n$ and let $a \in \{0,1\}^{n \times n}$. In the first definition without public random bits the function is

$$f(a,b) = \langle a, \sum_{i=1}^{n} b_i \cdot a_i \rangle.$$

The security is parameterized by the input length $N = n^2 + n$. In the second definition, a is the public random string and the function is defined as

$$f(b;a) = \sum_{i=1}^{n} b_i \cdot a_i.$$

In this case, the security is parameterized by the length of the private input b, which is simply n. Note that in both cases, the actual security of f is based on exactly the same thing, i.e. when a and b are chosen uniformly then given a and $\sum_{i=1}^{n} b_i \cdot a_i$ there is no fast adversary that can find on average a $b' \in \{0,1\}^n$ such that $\sum_{i=1}^{n} b'_i \cdot a_i = \sum_{i=1}^{n} b_i \cdot a_i$. The only difference is how the security is parameterized. Intuitively, security should be parameterized in terms of what is hidden from the adversary, and not in terms of the overall amount of randomness available to the function. The first definition parameterizes the security in terms of the overall amount of randomness available to the function, i.e. security is parameterized in terms of the length of b *plus* the length of a. The parameter of security in the second definition is the length of b, where b is what is really secret.

Intuitively, a weak one-way function f is a function such that it is hard to find an inverse of $f(x)$ for some significant but perhaps not very large fraction of $x \in \{0,1\}^n$ (the 'hard set'). (In contrast, for a one-way function it is hard to find an inverse of $f(x)$ for all but an insignificant fraction of the $x \in \{0,1\}^n$.) We only give the traditional definition (not using public randomness); the definition using public randomness is straightforward.

Definition (weak one-way function): Let $f(x)$ be a standard function with length relationship $\|x\| = n$, $\|f(x)\| = l(n)$. The *weakness parameter* of f is a function $s(n)$ such that $s(n) \geq \frac{1}{n^c}$ for some constant c. The time bound and success probability of an adversary A for f are defined exactly the same way as for a one-way function. An adversary A is $R_f(n)$-breaking for $s(n)$-weak f if there is a subset H_n of $\{0,1\}^n$ of measure at least $s(n)$ such that $R_f(n) \leq \frac{\delta_H(n)}{T_H(n)}$, where $\delta_H(n)$ is the average success probability over $H(n)$ and $T_H(n)$ is the maximal running time over $H(n)$. A function f is a $(1 - R_f(n))$-secure $s(n)$-weak one-way function if there is no $R_f(n)$-breaking adversary for $s(n)$-weak f.

Example : Define $f(x,y) = xy$, where $x,y \in \{2, \ldots, 2^n - 1\}$. The problem of inverting $f(x,y)$ consists of finding $x', y' \in \{2, \ldots, 2^n - 1\}$ such that $x'y' = xy$. Let $X, Y \in_{\mathcal{U}} \{2, \ldots, 2^n - 1\}$ be independent random variables. On average, $f(X,Y)$ is rather easy to invert. For instance, with probability $\frac{3}{4}$, XY is an even number, in which case setting $x' = 2$ and $y' = \frac{XY}{2}$ inverts $f(X,Y)$. However, with probability approximately $1/n^2$ both X and Y are prime n-bit numbers. If there is no adversary that can factor the product of a pair of random n-bit prime numbers in time $\frac{1}{R_f(2n)}$ on average then f is a $(1 - R_f(2n))$-secure $\frac{1}{n^2}$-weak one-way function.

3   Reductions

All of the results presented in this paper involve a reduction ¿from one type of cryptographic primitive to another. In this section, we give a formal definition of reduction. We only define a reduction in the case when both cryptographic primitives are standard functions.

Central to the definition of a reduction is the notion of an oracle Turing machine.

Definition (oracle Turing machine): An *oracle Turing machine* is a randomized Turing machine S whose behavior is not fully specified. The behavior is not fully specified in the sense that S, in the course of its execution, interactively makes calls (hereafter described as oracle calls) to and receives corresponding outputs from an algorithm that is not part of the description of S. We let S^A

denote the fully specified Turing machine described by S using algorithm A to compute the oracle calls.

Note that although the running time of S is not defined, the running time of S^A is defined. Also, if A is a Turing machine then so is S^A.

Let f be a generic instance of the first primitive, where $f(x)$ is a standard function with length relationship $\|x\| = n$ and $\|f(x)\| = l(n)$. Let $X \in_{\mathcal{U}} \{0,1\}^n$. There are two parts to a reduction: (1) an oracle Turing machine P that efficiently converts $f(X)$ into an instance $g(Y)$ of the second primitive, where g is a standard function and Y is the polynomially samplable probability distribution on inputs to g; (2) an oracle Turing machine S that is the guarantee that the security of $f(X)$ is passed on to $g(Y)$. The security guarantee is of the form that if A is a breaking adversary for $g(Y)$ then S^A is a breaking adversary for $f(X)$. More formally,

Definition (reduction): We say that there is a reduction from *primitive_1* to *primitive_2* if there are two oracle Turing machines P and S with the following properties. Given any instance f of *primitive_1*, P^f is an instance g of *primitive_2*. Given any $R_g(n)$-breaking adversary A for g, S^A is a $R_f(n)$-breaking adversary for f.

The reduction guarantees that there is no $R_g(n)$-breaking adversary for g as long as there is no $R_f(n)$-breaking adversary for f. To have the reduction inject as much of the security of f as possible into g, we would like $R_f(n)$ to be as large as possible with respect to $R_g(n)$, e.g., $R_f(n) = R_g(n)$.

To give a rough measure of the amount of security a reduction preserves, we make the following definitions. Note that in all definitions the reduction has an overhead of $\frac{1}{n^c}$. However, typically $R_g(n) \ll \frac{1}{n^c}$ and it is therefore the dominant factor.

Definition (preserving reductions): The reduction from *primitive_1* to *primitive_2* is said to be

- *slightly preserving* if there are constants $\alpha \geq 1$, $\beta \geq 1$ and $c \geq 0$ such that

$$R_f(n) \geq \frac{R_g(n^\alpha)^\beta}{n^c}.$$

- *polynomially preserving* if there are constants $\beta \geq 1$ and $c \geq 0$ such that

$$R_f(n) \geq \frac{R_g(n)^\beta}{n^c}.$$

- *linearly preserving* if there is a constant $c \geq 0$ such that

$$R_f(n) \geq \frac{R_g(n)}{n^c}.$$

For a linearly preserving reduction, $R_f(n)$ is linearly lower bounded by $R_g(n)$, and for a polynomially preserving reduction, $R_f(n)$ is polynomially lower bounded by $R_g(n)$ (in both cases there is also a polynomial in n factor). On the other hand, for a slightly preserving reduction the lower bound on $R_f(n)$ can be much

428

weaker than any polynomial factor in $R_g(n)$. For this reason, a linearly preserving reduction is more desirable than a polynomially preserving reduction which in turn is more desirable than a slightly preserving reduction.

Consider a reduction from a one-way function f to a pseudo-random generator g and suppose we want the reduction to guarantee that g is $(1-R_g(n))$-secure. The difference between these types of guarantees isn't so important when $R_g(n)$ is not too small, e.g., if $R_g(n)$ is inverse polynomial in n then all types guarantee that $R_f(n)$ is inverse polynomial in n, and thus g is $(1 - R_g(n))$-secure if there is no polynomial time adversary that can invert f with inverse polynomial probability. However, the difference between these types of guarantees increases dramatically as $R_g(n)$ goes to zero at a faster rate, which is expected in most applications. To see the dramatic differences between the strengths of the reductions, consider the case when $R_g(n) = 2^{-n^{1/2}}$ and $\alpha = \beta = 2$ and $c = 0$. For a linearly preserving reduction, g is $(1 - R_g(n))$-secure if there is no $R_f(n) = 2^{-n^{1/2}}$-breaking adversary for f. For a polynomially preserving reduction, g is $(1 - R_g(n))$-secure if there is no $R_f(n) = 2^{-2n^{1/2}}$-breaking adversary for f. For a slightly preserving reduction, g is $(1 - R_g(n))$-secure if there is no $R_f(n) = 2^{-2n}$-breaking adversary for f. Note that in this case $R_f(n)$ is the $2n^{1/2}$ power of $R_g(n)$, which is not at all polynomial in $R_g(n)$. In fact, for trivial reasons there is a 2^{-2n}-breaking adversary for f, and thus the slightly preserving reduction does not guarantee that g is $(1 - 2^{-n^{1/2}})$-secure no matter how secure f is.

Because of the tremendous superiority of a linearly preserving over a polynomially preserving reduction over a slightly preserving reduction, it is important to design the strongest reduction possible. Some of the most important work (both theoretically and practically) left to be done is to find stronger preserving reductions between cryptographic primitives than are currently known, e.g. the strongest reductions known from a one-way function to a pseudo-random generator and from a weak one-way function to a one-way function (in the general case) are only slightly preserving.

It turns out that the primary quantity that determines the strength of the reduction is the ratio $\frac{n}{n'}$, where n is the length of the private part of the input for g and n' is the length of the private part of the input for calls to f when computing g. The bigger this ratio the more the loss in security. The best case is when these two lengths are equal or nearly equal. The reason for this is that typically the achievement ratio for S^A is either linear or polynomial in the achievement ratio $R_g(n)$ for A, and S^A breaks one of the calls to f on inputs of length n', and thus $R_f(n')$ is either linear or polynomial in $R_g(n)$. For example, if $n' = n$ and $R_f(n') = R_g(n)$ then the reduction is linearly preserving. Slightly weaker, if $n' = \epsilon n$ for some constant $\epsilon > 0$ and $R_f(n') = R_g(n)^\beta$ for some constant $\beta > 1$ then the reduction is polynomially preserving. This can be seen as follows. Even in the worst case, when $R_g(n) = \frac{1}{2^n}$, it is easy to verify that $R_g(n) = R_g(n'/\epsilon) \leq R_g(n')^{1/\epsilon}$. Thus, $R_f(n') \leq R_g(n')^{\beta/\epsilon}$. If n' is substantially smaller than n (but still polynomial in n), then the reduction is typically only slightly preserving.

4 The Reductions

We describe several linearly preserving reductions from a weak one-way permutation to a one-way permutation. All of the reductions work only for functions that are permutations.[3] In [4], Yao describes a reduction from a general weak owf to a one-way function, but the reduction is only slightly preserving. A good research problem is to design a linearly preserving (or even polynomially preserving) reduction without any restriction on the weak one-way function.

In all the reductions, we assume that the weak one-way function doesn't use a public random string. Only minor modifications need be made to handle the case when the weak one-way function uses public randomness.

All of the reductions share a common approach, and each reduction builds on the ideas developed in previous reductions. For completeness, we first describe a general reduction from a weak one-way function to a one-way function.

Reduction 1 [Yao] : Let $f(x)$ be a $s(n)$-weak one-way function, where $x \in \{0,1\}^n$. Let $N = \frac{n}{s(n)}$, let $y \in \{0,1\}^{N \times n}$ and define the one-way function

$$g(y) = \langle f(y_1), \ldots, f(y_N) \rangle.$$

Theorem 1 [Yao] : Reduction 1 is a slightly preserving reduction from a $s(n)$-weak one-way function f to one-way function g. More precisely, there is an oracle algorithm S such that if A is an $R_g(nN)$-breaking adversary for $g(y)$ then S^A is a $R_f(n)$-breaking adversary for $s(n)$-weak $f(x)$, where $R_f(n) = \frac{R_g(nN)}{nN}$.

Note that $s(n)$ must be at least inverse polynomial in n for the reduction to be even slightly preserving. This is because it is necessary for n to be a polynomial fraction of N, and $N = \frac{n}{s(n)}$.

4.1 A simple linearly preserving reduction

An important observation about Reduction 1 is that g doesn't use any public random bits beyond what is used by f. The reason the reduction is only slightly preserving is that g partitions its private input into many small strings and uses each of these strings as a private input to f. This can be thought of as a parallel construction, in the sense that the calls to f are on independent inputs and thus all calls to f can be computed simultaneously. The linearly preserving reduction given here is similar in its basic structure to Reduction 1. The main difference is that instead of partitioning the private input of g into N private inputs of length n for f, the private input to g is a single string $x \in \{0,1\}^n$, and the public random string is used to generate N inputs of length n to f sequentially.

Reduction 2 : Let $f(x)$ be a $s(n)$-weak one-way permutation, where $x \in \{0,1\}^n$. Let $N = \frac{n}{s(n)}$, let $\pi \in \{0,1\}^{N \times n}$ and define the one-way permutation

$$g(x; \pi) = y_{N+1}$$

---

[3] These reductions can be extended to the important case of regular functions, which is more general than permutations but still not the general case. A function is regular if each point in the range of the function has the same number of preimages.

where $y_1 = x$ and, for all $i = 2, \ldots, N + 1$, $y_i = \pi_{i-1} \oplus f(y_{i-1})$.

Theorem 2 : Reduction 2 is a linearly preserving reduction from a $s(n)$-weak one-way permutation f to one-way permutation g. More precisely, there is an oracle algorithm S such that if A is an $R_g(n)$-breaking adversary for $g(x; \pi)$ then S^A is a $R_f(n)$-breaking adversary for $s(n)$-weak $f(x)$, where $R_f(n) = \frac{R_g(n)}{nN}$.

The proof of Theorem 2 is similar in spirit to the proof of the Theorem 1. We only describe the oracle algorithm S. Suppose that A is an adversary with time bound $T(n)$ and success probability $\delta(n)$ for g, and thus the achievement ratio is $\frac{\delta(n)}{T(n)}$. A on input $g(x; \pi)$ and π finds x with probability $\delta(n)$ when $x \in_{\mathcal{U}} \{0,1\}^n$ and $\pi \in_{\mathcal{U}} \{0,1\}^{N \times n}$. The oracle machine described below has the property that S^A inverts f on inputs of length n with probability at least $1 - s(n)$, where the time bound for S^A is $\frac{nNT(n)}{\delta(n)}$. The input to S^A is $f(x)$ where $x \in_{\mathcal{U}} \{0,1\}^n$.

Adversary S^A on input $f(x)$: .

Repeat $\frac{nN}{\delta(n)}$ times

Randomly choose $i \in_{\mathcal{U}} \{2, \ldots, N+1\}$.

Randomly choose $\pi \in_{\mathcal{U}} \{0,1\}^{N \times n}$.

Let $y_i = f(x) \oplus \pi_{i-1}$.

Compute $y_{i+1} = \pi_i \oplus f(y_i), \ldots, y_{N+1} = \pi_N \oplus f(y_N)$.

Compute $v_0 = A(y_{N+1}; \pi)$.

Compute $v_1 = \pi_0 \oplus f(v_0), \ldots, v_{i-1} = \pi_{i-2} \oplus f(v_{i-2})$.

if $f(v_{i-1}) = f(x)$ then output v_{i-1}.

4.2 A linearly preserving reduction using less randomness

Although Reduction 2 is linearly preserving, it does have the drawback that the length of the public random string is rather large, and even worse this length depends linearly on the weakness parameter $s(n)$ of the weak one-way function. In this subsection, we describe a linearly preserving reduction that uses a much shorter public random string.

The overall structure of the reduction is the same as Reduction 2. The difference is that we use many fewer public random strings of length n in a recursive way to produce the almost random inputs to f. The reduction is in two steps. In the first step we describe a linearly preserving reduction from a $s(n)$-weak one-way permutation f to a $\frac{1}{2}$-weak one-way permutation g. The second step reduces g to a one-way permutation h using the construction given in Reduction 2.

Reduction 3 : Let $f(x)$ be a $s(n)$-weak one-way permutation, where $x \in \{0,1\}^n$. Let $l = \lceil \log_{3/2}(2/s(n)) \rceil$ and let $N = 2^l$. Let $\pi \in \{0,1\}^{l \times n}$. Define

$$g(x; \pi_1) = f(\pi_1 \oplus f(x)).$$

For all $i = 2, \ldots, l$, recursively define

$$g(x; \pi_{\{1,\ldots,i\}}) = g(\pi_i \oplus g(x; \pi_{\{1,\ldots,i-1\}}); \pi_{\{1,\ldots,i-1\}}).$$

Theorem 3 : Reduction 3 is a linearly preserving reduction from a $s(n)$-weak one-way permutation $f(x)$ to $\frac{1}{2}$-weak one-way permutation $g(x;\pi)$. More precisely, there is an oracle algorithm S such that if A is an $R_g(n)$-breaking adversary for $\frac{1}{2}$-weak $g(x;\pi)$ then S^A is a $R_f(n)$-breaking adversary for $s(n)$-weak f, where $R_f(n) = \frac{R_g(n)}{nN}$.

The final step in the reduction is to go from weak one-way permutation g with weakness parameter $\frac{1}{2}$ to a one-way permutation h using Reduction 2, except now g has weakness parameter $\frac{1}{2}$ and uses a public random string of length $m = O(n \log(1/s(n)))$. Thus, when using Reduction 2 to go from g to h, we set $N = \log(1/R_g(n)) \leq n$ and partition the public random string into N blocks of length $n + m$. Thus, the overall reduction uses $O(n^2 \log(1/s(n)))$ public random bits, as opposed to $O\left(\frac{n^2}{s(n)}\right)$ for Reduction 2. It is not hard to verify that the overall reduction from f to h is linearly preserving.

4.3 A linearly preserving reduction using expander graphs

The work described in [1] gives a polynomially preserving reduction from a weak one-way permutation to a one-way permutation that uses only a linear amount of public randomness. As briefly described below, their reduction can be modified in minor ways to yield a linearly preserving reduction ¿from a weak one-way permutation f to a one-way permutation h that uses only a linear number of public random bits overall.

As in Reduction 3, the reduction is in two steps: The first step is a linearly preserving reduction from a $s(n)$-weak one-way permutation f to a $\frac{1}{2}$-weak one-way permutation g and the second step reduces g to a one-way permutation h. As in Reduction 3, the first step is recursive and uses $O(\log(s(n))$ independent public random strings, but they are each of constant length instead of length n. The idea is to define a constant degree expander graph with vertex set $\{0,1\}^n$, and then each string is used to select a random edge out of a vertex in the expander graph. The second step is iterative, but uses only an additional $O(n)$ public random bits. These $O(n)$ public random bits are used to define a random walk of length $O(n)$ on a related expander graph.

The overall number of public random bits used in the entire reduction ¿from f to h is only linear. The way [1] describes the reduction, the one-way permutation f is applied to inputs of different lengths (all within a constant multiplicative factor of each other) to yield h. For this reason, as they describe their reduction it is only polynomially preserving, even with respect to the new definitions. Minor modifications to their reduction yields an alternative reduction where all inputs to f are of the same length as the private input to h. It can be shown that the alternative reduction with respect to the new definitions is linearly preserving.

Acknowledgements

We wish to thank Oded Goldreich, Hugo M. Krawczyk and Rafail Ostrovsky for their comments.

432

References

1. Goldreich, O., Impagliazzo, R., Levin, L., Venketesan, R., Zuckerman, D., "Security Preserving Amplification of Hardness", Proceedings of the 31st IEEE Symposium on Foundations of Computer Science, pp. 318-326, 1990.
2. Phillips, S. J., "Security Preserving Hardness Amplification Using PRGs for Bounded Space - Preliminary Report", unpublished manuscript, July 1992.
3. Luby, M., "Pseudorandomness and Applications", monograph in progress.
4. Yao, A., "Theory and applications of trapdoor functions", Proceedings of the 23rd IEEE Symposium on Foundations of Computer Science, pp. 80-91, 1982.

Necessary and Sufficient Conditions for Collision-Free Hashing

Alexander Russell*
acr@theory.lcs.mit.edu

Laboratory for Computer Science
545 Technology Square
Massachusetts Institute of Technology
Cambridge, MA 02139 USA

Abstract. This paper determines an exact relationship between collision-free hash functions and other cryptographic primitives. Namely, it introduces a new concept, the pseudo-permutation, and shows that the existence of collision-free hash functions is equivalent to the existence of claw-free pairs of pseudo-permutations. When considered as one bit contractors (functions from $k + 1$ bits to k bits), the collision-free hash functions constructed are more efficient than those proposed originally, requiring a single (claw-free) function evaluation rather than k.

1 Introduction

Hash functions with various cryptographic properties have been studied extensively, especially with respect to signing algorithms (see [2, 3, 4, 10, 12, 14, 15]). We focus on the most natural of these functions, the *collision-free* hash functions. A hash function h is collision-free if it is hard for any efficient algorithm, given h and 1^k, to find a pair (x, y) so that $|x| = |y| = k$ and $h(x) = h(y)$. These functions were first carefully studied by Damgård in [2]. Given the interest in these functions, we would like to determine necessary and sufficient conditions for their existence in terms of other, simpler, cryptographic machinery.

There has been recent attention to the minimal logical requirements for other cryptographic primitives. Rompel (in [12]), improving a construction of Naor and Yung (in [10]), shows that the existence of secure digital signing systems (in the sense of [5]) is equivalent to the existence of one-way functions. Impagliazzo, Levin, and Luby (in [7]) and Håstad (in [6]) demonstrate the equivalence of the existence of pseudo-random number generators (see [1, 13]) and the existence of one-way functions.

Damgård (in [2]), distilling arguments of Goldwasser, Micali, and Rivest (in [5]), shows that the existence of another cryptographic primitive, a *claw-free* pair of permutations, is sufficient to construct collision-free hash functions. A pair of permutations (f, g) is *claw-free* if it is hard for any efficient algorithm, given (f, g) and 1^k, to find a pair (x, y) so that $|x| = |y| = k$ and $f(x) =$

---

* Supported by a NSF Graduate Fellowship

$g(y)$. Comparing the definitions of collision-free hash functions and claw-free pairs of permutations, it seems unlikely that the existence of claw-free pairs of permutations is necessary for the existence of collision-free hash functions because the hash functions have no explicit structural properties that reflect the condition of permutativity in the claw-free pairs of permutations. Our paper relaxes this condition of permutativity and defines a natural object, the existence of which is necessary and sufficient for the existence of a family of collision-free hash functions.

We define a new concept, the *pseudo-permutation*. A function $f : S \to S$ is a pseudo-permutation if it is computationally indistinguishable from a permutation. For this "indistinguishability" we require that it be hard for any efficient algorithm, given the function f and 1^k, to compute a quickly verifiable proof of non-injectivity, i.e. a pair (x, y) where $|x| = |y| = k, x \neq y$, and $f(x) = f(y)$. The main contribution of our paper is that the existence of a collection of claw-free pairs of pseudo-permutations is equivalent to the existence of a collection of collision-free hash functions. This fact shows that nontrivial "claw-freeness" is essential to collision-free hashing and also weakens the assumptions necessary for their existence.

In §2 we describe our notation and define some cryptographic machinery. In §3 we present our main theorem. In §4 we consider the efficiency of our construction. Finally, in §5, we discuss an open problem and the motivation for this research.

2 Notation and Definitions

We adopt the following class of *expected polynomial time Turing machines* as our standard class of "efficient algorithms" (see [9] for a precise definition and discussion of this class).

Definition 1. Let $\mathcal{EA}$, our class of efficient algorithms, be the class of probabilistic Turing machines (with output) running in expected polynomial time. We consider these machines to compute probability distributions over Σ^*. For $M \in \mathcal{EA}$ we use the notation $M[w]$ to denote both the probability space defined by M on w over Σ^* and an element selected according to this space.

For simplicity, let us fix a two letter alphabet $\Sigma = \{0, 1\}$. The consequences of a larger alphabet will be discussed in §4. 1^k denotes the concatenation of k 1's. $\mathcal{Q}[x]$ denotes the class of polynomials over the rationals. Borrowing notation from [4], if S is a probability space, $x \leftarrow S$ denotes the assignment of x according to S. If $p(x_1, \ldots, x_k)$ is a predicate, then $\Pr[x_1 \leftarrow S_1, \ldots, x_k \leftarrow S_k : p(x_1, \ldots, x_k)]$ denotes the probability that p will be true after the ordered assignment of x_1 through x_k.

Definition 2. A **collection of claw-free functions** is a collection of function tuples $\{(f_i^0, f_i^1) | i \in I\}$ for some index set $I \subseteq \Sigma^*$ where $f_i^j : \Sigma^{|i|} \to \Sigma^{|i|}$ and:

CF1. *[accessibility]* there exists a generating algorithm $G \in \mathcal{EA}$ so that $G[1^n] \in \{0,1\}^n \cap I$.

CF2. *[efficient computability]* there exists an computing algorithm $C \in \mathcal{EA}$ so that for $i \in I, j \in \{0,1\}$, and $x \in \Sigma^{|i|}, C[i,j,x] = f_i^j(x)$.

CF3. *[claw-freedom]* for all claw finding algorithms $A \in \mathcal{EA}$, $\forall P \in \mathcal{Q}[x]$, $\exists k_0$, $\forall k > k_0$,

$$\Pr[i \leftarrow G[1^k], (x,y) \leftarrow A[i] : f_i^0(x) = f_i^1(y)] < \frac{1}{P(k)}$$

If (f^0, f^1) is a member of a collection of claw-free pairs, then (f^0, f^1) is called a *claw-free pair* and a pair (x,y) so that $f^0(x) = f^1(y)$ is called a *claw* of (f^0, f^1).

This definition, from a cryptographic perspective, requires nothing of the function pairs involved unless they have overlapping images. One way to require that the functions have overlapping images is to require that the functions be permutations. This yields the following object, originally defined in [5] and then in this form by [2].

Definition 3. **A collection of claw-free permutations** is a collection of claw free functions $\{(f_i^0, f_i^1) | i \in I\}$ where each f_i^j is a permutation.

Although the intractability of certain number theoretic problems implies the existence of a collection of claw-free pairs of permutations[2], the existence of one-way permutations is not known to be enough.[3]

Definition 4. **A collection of pseudo-permutations** is a collection of functions $\{f_i | i \in I\}$ for some index set $I \subseteq \Sigma^*$ where $f_i : \Sigma^{|i|} \to \Sigma^{|i|}$ and:

ψP1. *[accessibility]* there exists a generating algorithm $G \in \mathcal{EA}$ so that $G[1^n] \in \{0,1\}^n \cap I$.

ψP2. *[efficient computability]* there exists a computing algorithm $C \in \mathcal{EA}$ so that for $i \in I$ and $x \in \Sigma^{|i|}, C[i,x] = f_i(x)$.

ψP3. *[collapse freedom]* for all collapse finding algorithms $A \in \mathcal{EA}$, $\forall P \in \mathcal{Q}[x]$, $\exists k_0, \forall k > k_0$

$$\Pr[i \leftarrow G[1^k], (x,y) \leftarrow A[i] : f_i(x) = f_i(y) \wedge x \neq y] < \frac{1}{P(k)}$$

If a function f is a member of a collection of pseudo-permutations it is called a *pseudo-permutation* and a pair (x,y) where $f(x) = f(y)$ and $x \neq y$ is called a *collapse* of f. Property ψP3 means that it is hard for an efficient algorithm to produce a quickly verifiable proof that f is not a permutation. In the definition above, this proof is specifically required to be a proof of non-injectivity: a collapse. One might also prove that a function $f : S \to S$ is

---

[2] In [5] the intractability of factoring is shown to be sufficient. In [2], the construction of [5] is extended and the intractability of the discrete log is also shown to be sufficient.

[3] [11] discusses algebraic forms of one way permutations sufficient for claw-free permutations.

not a permutation by producing a proof of non-surjectivity: an element in $S -$ $\mathrm{Im} f$. We require the former because of the difference in computational resources necessary to verify these proofs: a proof of non-injectivity may be verified with two function applications whereas a proof of non-surjectivity requires evaluation of f at every point in the domain. Like the definition for claw-free functions, the above definition requires nothing cryptographically of the functions involved unless $|\mathrm{Im}\, f_i| < |\mathrm{Dom}\, f_i|$. If the functions in the collection are injective, then ψP3 is vacuously true.

Pseudo-permutations are a reasonable replacement for permutations in a cryptographic setting; for example, the entire signing algorithm of Naor and Yung (in [10]) may be implemented with one-way[4] pseudo-permutations rather than one-way permutations.

Definition 5. **A collection of claw-free pseudo-permutations** is a collection of claw-free functions $\{(f_i^0, f_i^1)|i \in I\}$ so that both $\{f_i^0|i \in I\}$ and $\{f_i^1|i \in I\}$ are collections of pseudo-permutations.

Collections of claw-free pseudo-permutations gather their cryptographic structure from the tension between two otherwise weak definitions. If the pseudo-permutations lack cryptographic richness (so that they are very close to permutations) then the intersection of their images must be large and there must be many claws, imparting richness by virtue of claw-freedom. If, instead, the pair has few claws, then the images of the two functions must be nearly disjoint (and so, small) so that the functions themselves are cryptographically rich by virtue of their many collapses.

3 The Structure of Collision-Free Hash Functions

Definition 6. **A collection of collision-free hash functions** is a collection of functions $\{h_i|i \in I\}$ for some index set $I \subseteq \Sigma^*$ where $h_i : \Sigma^{|i|+1} \to \Sigma^{|i|}$ and:

H1. *[accessibility]* there exists a generating algorithm $G \in \mathcal{EA}$ so that $G[1^n] \in \{0,1\}^n \cap I$.

H2. *[efficient computability]* there exists a computing algorithm $C \in \mathcal{EA}$ so that for $i \in I$, and $w \in \Sigma^{|i|+1}$, $C[i, w] = h_i(w)$.

H3. *[collision-freedom]* for all collision generating algorithms $A \in \mathcal{EA}, \forall P \in \mathcal{Q}[x], \exists k_0, \forall k > k_0$

$$\Pr[i \leftarrow G[1^k], (x, y) \leftarrow A[i] : h_i(x) = h_i(y) \wedge x \neq y] < \frac{1}{P(k)}$$

If h is a member of a collection of collision-free hash functions then h is called a *collision-free hash function* and a pair (x, y) where $h(x) = h(y)$ and $x \neq y$ is called a *collision* of h.

---

[4] This is a collection of pseudo-permutations which are hard to invert in the sense of one-way functions.

The notion of a polynomial separator will be used in the following proof. For the purposes of this paper, a separator is a pair of bijections from Σ^k into Σ^{k+1} so that their images have no intersection. (Because $|\Sigma| = 2$, their images cover Σ^{k+1}.)

Definition 7. **A collection of polynomial separators** is a collection of function pairs $\{(\sigma_i^0, \sigma_i^1)|i \in I\}$ for some index set $I \subseteq \Sigma^*$ where $\sigma_i^j : \Sigma^{|i|} \to \Sigma^{|i|+1}$ for $j \in \{0, 1\}$ and:

PS1. *[accessibility]* there exists a generating algorithm $G \in \mathcal{EA}$ so that $G[1^n] \in \{0, 1\}^n \cap I$.

PS2. *[injectivity]* σ_i^0 and σ_i^1 are injective.

PS3. *[disjointness]* $\operatorname{im} \sigma_i^0 \cap \operatorname{im} \sigma_i^1 = \emptyset$

PS4. *[efficient computability]* there exists a computing algorithm $C \in \mathcal{EA}$ so that for $i \in I, w \in \Sigma^{|i|}$, and $j \in \{0, 1\}, C[i, j, w] = \sigma_i^j(w)$.

With each such collection, we associate a collection of inverses $\{\iota_i\}$ where $\iota_i : \Sigma^{|i|+1} \to \Sigma^{|i|}$ and $\iota_i \circ \sigma_i^0 = \iota_i \circ \sigma_i^1 = \operatorname{id}_{\Sigma^{|i|}}$ and a collection of image deciders $\{\delta_i\}$ where $\delta_i : \Sigma^{|i|+1} \to \{0, 1\}$ and $\forall w \in \Sigma^{|i|+1}, w \in \operatorname{im} \sigma_i^{\delta_i(w)}$.

The collection is said to have a **polynomial inverse** if the collection of inverses is so that $\exists C_\iota \in \mathcal{EA}, \forall w \in \Sigma^{|i|+1}, \forall i \in I, C_\iota[w, i] = \iota_i(w)$. If a collection is so endowed, then it is clear that the image deciders are also efficiently computable.

Construction of a family of polynomial separators with a polynomial inverse is easy: the $append_0 : x \mapsto x0$ and $append_1 : x \mapsto x1$ functions, for example.

Theorem 8. *There exists a collection of collision-free hash functions iff there exists a collection of claw-free pairs of pseudo-permutations.*

Proof. ($\Rightarrow$) Let $\{h_i | i \in I\}$ be a collection of collision-free hash functions and let $\{(\sigma_i^0, \sigma_i^1) | i \in I\}$ be a collection of polynomial separators (unrelated to the hash functions, but over the same index set). Define the collection $\{(f_i^0, f_i^1) | i \in I\}$ so that

$$f_i^j = h_i \circ \sigma_i^j \text{ for } j \in \{0, 1\}$$

We show that the collection of functions so defined is a collection of claw-free pseudo-permutations. Properties *CF1* and *CF2* are immediate. Assume that property *CF3* does not hold, that is $\exists A \in \mathcal{EA}, \exists P \in \mathcal{Q}[x], \forall k_0, \exists k > k_0$,

$$\Pr[i \leftarrow G[1^k], (x, y) \leftarrow A[i] : f_i^0(x) = f_i^1(y)] \geq \frac{1}{P(k)}$$

Let (x, y) be a claw for (f_i^0, f_i^1), then $f_i^0(x) = f_i^1(y) \Rightarrow h_i(\sigma_i^0(x)) = h_i(\sigma_i^1(y))$, but $\operatorname{im} \sigma_i^0 \cap \operatorname{im} \sigma_i^1 = \emptyset$ so that $\sigma_i^0(x) \neq \sigma_i^1(y)$ and a collision has been found for h_i. Then, given this claw generating algorithm A we can construct a collision generating algorithm A' succeeding with identical probability as A, violating *H3*. Therefore, *CF3* holds.

To show that $\{f_i^j | i \in I\}$ for each $j \in \{0,1\}$ are collections of pseudo-permutations, we verify properties $\psi P1 - \psi P3$ for each. $\psi P1$ and $\psi P2$ are immediate. Suppose, for contradiction, that property $\psi P3$ is not satisfied, so that $(\exists j \in \{0,1\},) \exists A \in \mathcal{EA}, \exists P \in Q[x], \forall k_0, \exists k > k_0$

$$\Pr[i \leftarrow G[1^k], (x,y) \leftarrow A[i] : f_i^j(x) = f_i^j(y)] \geq \frac{1}{P(k)}$$

Let (x,y) be a collapse of f_i^j, so that $f_i^j(x) = f_i^j(y)$ and $x \neq y$. Then $\sigma_i^j(x) \neq \sigma_i^j(y)$ because σ_i^j is injective, so that $(\sigma_i^j(x), \sigma_i^j(y))$ is a nontrivial collision of h_i (because $f_i^j = h_i \circ \sigma_i^j$). Then, given this collapse generating algorithm A we can construct a collision generating algorithm A' succeeding with identical probability as A, violating $H3$. Therefore, $\psi P3$ holds.

($\Leftarrow$) Let $\{(f_i^0, f_i^1) | i \in I\}$ be a collection of claw-free pairs of pseudo-permutations and let $\{(\sigma_i^0, \sigma_i^1) | i \in I\}$ be a collection of polynomial separators with inverses $\{\iota_i | i \in I\}$ and image deciders $\{\delta_i | i \in I\}$. Then define $\{h_i | i \in I\}$ so that

$$h_i(x) = f_i^{\delta_i(x)}(\iota_i(x))$$

We show that $\{h_i | i \in I\}$ is a collection of collision-free hash functions. Properties $H1$ and $H2$ are immediate. Assume, for contradiction, that property $H3$ is not satisfied, that is $\exists A \in \mathcal{EA}, \exists P \in Q[x], \forall k_0, \exists k > k_0$

$$\Pr[i \leftarrow G[1^k], (x,y) \leftarrow A[i] : h_i(x) = h_i(y) \wedge x \neq y] \geq \frac{1}{P(k)}$$

so that $\forall k_0, \exists k > k_0$

$$\Pr[i \leftarrow G[1^k], (x,y) \leftarrow A[i] : h_i(x) = h_i(y) \wedge x \neq y \wedge \delta_i(x) = \delta_i(y)] \geq \frac{1}{2P(k)} \bigvee$$

$$\Pr[i \leftarrow G[1^k], (x,y) \leftarrow A[i] : h_i(x) = h_i(y) \wedge x \neq y \wedge \delta_i(x) \neq \delta_i(y)] \geq \frac{1}{2P(k)}$$

and we encounter at least one of two possibilities:

1. $\forall k_0, \exists k > k_0$

$$\Pr[i \leftarrow G[1^k], (x,y) \leftarrow A[i] : h_i(x) = h_i(y) \wedge x \neq y \wedge \delta_i(x) = \delta_i(y)] \geq \frac{1}{2P(k)}$$

2. $\forall k_0, \exists k > k_0$

$$\Pr[i \leftarrow G[1^k], (x,y) \leftarrow A[i] : h_i(x) = h_i(y) \wedge x \neq y \wedge \delta_i(x) \neq \delta_i(y)] \geq \frac{1}{2P(k)}$$

In the event of 1 above, the algorithm A generates collisions (x, y) where $\delta_i(x) = \delta_i(x)$. In this case, for at least one $j \in \{0, 1\}, \forall k_0, \exists k > k_0$

$$\Pr[i \leftarrow G[1^k], (x, y) \leftarrow A[i] : h_i(x) = h_i(y) \wedge x \neq y \wedge j = \delta_i(x) = \delta_i(y)] \geq \frac{1}{4P(k)}$$

Given a collision of this sort, $h_i(x) = h_i(y) \Rightarrow f_i^j(x) = f_i^j(y)$ which is a collapse of f_i^j. Then, given algorithm A, we may produce another algorithm A' which produces a collapse of f_i^j with success related to the success of A by a constant, violating $\psi P3$.

In the event of 2 above, the algorithm A generates collisions (x, y) where $\delta_i(y) \neq \delta_i(x)$. A collision of this sort produces a claw because $h_i(x) = h_i(y) \Rightarrow f_i^{\delta_i(x)}(\iota_i(x)) = f_i^{\delta_i(y)}(\iota_i(y))$. Then, with algorithm A, we may construct a claw generating algorithm A' which produces claws with success related to the success of A by a constant, violating $CF3$.

A pair of separators partitions Σ^{k+1} into two equal sized subsets (the images of the separators). We couple the definition of collision-free hash functions with the definition of polynomial separators to define a class of hash functions where every collision occurs across the partition boundary — then $h(x) = h(y)$ implies that x and y are in the images of different separators.

Definition 9. A collection of separated collision-free hash functions is a collection of function tuples $\{(h_i, \sigma_i^0, \sigma_i^1) | i \in I\}$ so that $\{h_i | i \in I\}$ forms a collection of collision-free hash functions, $\{(\sigma_i^0, \sigma_i^1) | i \in I\}$ forms a collection of polynomial separators, and

SH. *[separation]* $\forall j \in \{0, 1\}$, $h_i|_{\mathbf{im}\,\sigma_i^j}$, the restriction of h_i to $\mathbf{im}\,\sigma_i^j$, is bijective. Equivalently, $h_i(x) = h_i(y) \Rightarrow \delta_i(x) \neq \delta_i(y)$, where $\{\delta_i | i \in I\}$ is the collection of image deciders for the separators.

The existence of a collection of separated collision-free hash functions is equivalent to the existence of a collection of claw-free pairs of permutations.

Theorem 10. *There exists a collection of claw-free permutations iff there exists a collection of separated collision-free hash functions.*

Proof. This proof is omitted due its similarity with the previous proof.

The collision-free hash functions constructed in the two theorems above naturally inherit properties from the primitives with which they are constructed. If, for example, the claw-free pairs of (pseudo-) permutations are *trapdoor* functions, then the hash functions constructed share this property. It is not clear that the original hash functions constructed in [2] offer inheritance of this sort.

Toshiya Itoh [8] has pointed out that in the above constructions, the demand of claw-freedom can be replaced in an appropriate way with the demand of "distinction intractibility" as discussed in [14].

It is not hard to show that by composition the above collections of hash functions can be used to construct families of collision-free hash functions $\{h_i : i \in I\}$ where $h_i : \Sigma^{P(|i|)} \rightarrow \Sigma^{|i|}$ for any polynomial $P \in \mathcal{Q}[x]$ where $\forall x \in \mathcal{N}, P(x) > x$.

4 Comments on Efficiency

The ($\Leftarrow$) part of theorem 8 constructs a family of collision-free hash functions which are one bit contractors (functions from Σ^{k+1} to Σ^k) and require 1 claw-free function evaluation to compute. Building a family of contractors by applying the construction in [2] yields hash functions which require k evaluations of the underlying claw-free functions. For the case of one-bit contractors, then, the above construction is substantially more efficient.

In general, to construct hash functions from $\Sigma^{P(k)}$ to Σ^k (for a polynomial P) one can do better than naive composition. Using arguments similar to those of Damgård in [2], the construction above can be altered to yield hash functions from $\Sigma^{P(k)}$ to Σ^k which require $P(k) - k$ evaluations of the underlying claw-free functions on k bit arguments. The collection constructed in [2] of the same sort requires $P(k)$ evaluations, so the efficiency improvement in this case is only an additive factor of k.

In [2], Damgård shows that expanding the size of the alphabet (and using claw-free *tuples* of functions) can reduce the number of required claw-free function evaluations by a multiplicative constant factor. This same procedure is applicable to our above construction.

5 An Open Problem

The motivation for this research is the following open problem: Is the existence of one-way functions sufficient for the existence of collision-free hash functions? We believe this to be the case, and that this paper represents a step towards proving this goal by demonstrating the equivalence between collision-free hashing and a primitive not requiring pure cryptographic permutativity.

6 Acknowledgements

We gratefully acknowledge the keen guidance of Silvio Micali, who originally suggested this problem. We also acknowledge Ravi Sundaram for several helpful discussions.

References

1. Manual Blum and Silvio Micali. How to generate cryptographically strong sequences of pseudo-random bits. *SIAM Journal of Computing*, 13(4):850–864, November 1984.

2. Ivan Damgård. Collision free hash functions and public key signature schemes. In *Proceedings of EUROCRYPT '87*, volume 304 of *Lecture Notes in Computer Science*, pages 203–216, Berlin, 1988. Springer-Verlag.

3. Alfredo De Santis and Moti Yung. On the design of provably-secure cryptographic hash functions. In *Proceedings of EUROCRYPT '90*, volume 473 of *Lecture Notes in Computer Science*, pages 412 – 431, Berlin, 1990. Springer-Verlag.

4. Oded Goldreich, Shafi Goldwasser, and Silvio Micali. How to construct random functions. *Journal of the Association for Computing Machinery*, 33(4):792–807, October 1986.

5. Shafi Goldwasser, Silvio Micali, and Ronald L. Rivest. A digital signature scheme secure against adaptive chosen-message attack. *SIAM Journal of Computing*, 17(2):281–308, April 1988.

6. J. Håstad. Pseudo-random generators under uniform assumptions. In *Proceedings of the Twenty Second Annual ACM Symposium on Theory of Computing*, pages 395–404. ACM, 1990.

7. Russell Impagliazzo, Leonid A. Levin, and Michael Luby. Pseudo-random generation from one-way functions. In *Proceedings of the Twenty First Annual ACM Symposium on Theory of Computing*, pages 12–24. ACM, 1989.

8. Toshiya Itoh. Personal comminucation, August 1992.

9. Leonid A. Levin. Average case complete problems. *SIAM Journal on Computing*, 15:285–286, 1986.

10. M. Naor and M. Yung. Universal one-way hash functions and their cryptographic applications. In *Proceedings of the Twenty First Annual ACM Symposium on Theory of Computing*, pages 33–43. ACM, 1989.

11. Wakaha Ogata and Kaoru Kurosawa. On claw free families. In *Proceedings of ASIACRYPT '91*, 1991.

12. John Rompel. One-way functions are necessary and sufficient for secure signatures. In *Proceedings of the Twenty Second Annual ACM Symposium on Theory of Computing*, pages 387–394. ACM, 1990.

13. A. Yao. Theory and applications of trapdoor functions. In *Proceedings of the Twenty Third IEEE Symposium on Foundations of Computer Science*, pages 80–91. IEEE, 1982.

14. Yuliang Zheng, Tsutomu Matsumoto, and Hideki Imai. Duality between two cryptographic primitives. In *Proceedings of the Eighth International Conference on Applied Algebra, Algebraic Algorithms and Error-Correcting Codes*, volume 508 of *Lecture Notes in Computer Science*, pages 379–390, Berlin, 1990. Springer-Verlag.

15. Yuliang Zheng, Tsutomu Matsumoto, and Hideki Imai. Structural properties of one-way hash functions. In *Proceedings of CRYPTO '90*, volume 537 of *Lecture Notes in Computer Science*, pages 285–302, Berlin, 1990. Springer-Verlag.

Certifying Cryptographic Tools: The Case of Trapdoor Permutations

Mihir Bellare[1] and Moti Yung[2]

[1] High Performance Computing and Communications, IBM T.J. Watson Research Center, PO Box 704, Yorktown Heights, NY 10598. e-mail: mihir@watson.ibm.com.

[2] IBM Research, IBM T.J. Watson Research Center, PO Box 704, Yorktown Heights, NY 10598. e-mail: moti@watson.ibm.com.

Abstract. In cryptographic protocols it is often necessary to verify/certify the "tools" in use. This work demonstrates certain subtleties in treating a family of trapdoor permutations in this context, noting the necessity to "check" certain properties of these functions. The particular case we illustrate is that of non-interactive zero-knowledge. We point out that the elegant recent protocol of Feige, Lapidot and Shamir for proving NP statements in non-interactive zero-knowledge requires an additional certification of the underlying trapdoor permutation, and suggest a certification method to fill this gap.

1 Introduction

Primitives such as the RSA function, the discrete log function, or, more generally, any trapdoor or one-way function, have applications over and above the "direct" ones to public-key cryptography. Namely, they are also (widely) used as "tools" in the construction of (often complex) cryptographic protocols.

This paper points to the fact that in this second kind of application, some care must be exercised in the manner in which the "tool" is used. Checks might be necessary that are not necessary in public-key applications.

The need for such checks arises from the need to consider adverserial behavior of parties in a cryptographic protocol. Typically, the problem is that one cannot trust a party to "correctly" create the tool in question. For example, suppose a party A is supposed to give another party B a modulus N product of two primes, and an RSA exponent e, to specify an RSA function. On receipt of a number N and an exponent e, it might be important that the receiver know that e is indeed an RSA exponent (i.e. relatively prime to the Euler Phi Function of N). This is because the use of RSA in the protocol might be such that making e not an RSA exponent could give A an advantage (such applications do exist). Such a problem is not present in public-key applications, where, if I wish, for example, to construct a digital signature scheme based on RSA, I put in my public file a modulus N (which I have chosen to be the product of two primes) and an RSA exponent e (and I keep secret the primes). The question of my choosing e to not be an RSA exponent does not arise because it is not to my advantage to do so.

Protocols address this issue in several ways. Often, they incorporate additional sub-protocols which "certify" that the "tool" used is indeed "correct."

In applications, these sub-protocols usually need to be zero-knowledge ones. In most applications, such sub-protocols may be simply realized, by using, say, the result of [GMW]. But we note that this is not always the case. For example (cf. [BMO]), if we are trying to construct statistical ZK proofs, then we cannot use [GMW] to certify the tools because the latter yields only computational ZK. The issue must then be settled by other means. .

Sometimes, we note, the issue does not arise; this is the case, for example, if the tool is a one-way function, because a one-way function is a single object, a map from $\{0,1\}^*$ to $\{0,1\}^*$, specified by a string known to everyone. At other times, stronger assumptions about the primitive might be made. An example of this is the use, in protocols, of the "certified discrete log assumption" (as opposed to the usual "discrete log assumption").

The particular instance of this issue that we focus on in this paper is the use of trapdoor permutations in non-interactive zero-knowledge (NIZK) proofs. We point out that the elegant recent NIZK protocol of Feige, Lapidot and Shamir [FLS] makes the (implicit) assumption that the trapdoor permutation is "certified." We note that this assumption is not valid for standard (conjectured) trapdoor permutations like RSA or those of [BBS] (and so their protocol cannot be instantiated with any known (conjectured) trapdoor permutation). We suggest a certification method to fill this gap (so that *any* trapdoor permutation truly suffices, and RSA or the construction of [BBS] may be used). Our certification method involves a NIZK proof that a function is "almost" a permutation, and might be of independent interest.

Below we begin by recalling the notions of trapdoor permutations and NIZK proofs. We then discuss the FLS protocol and indicate the source of the problem. We then, briefly, discuss our solution. Later sections specify the definitions and our solution in more detail.

1.1 Trapdoor Permutations

Let us begin by recalling, in some detail, the definition of a trapdoor permutation generator (cf. [BeMi]), and seeing what it means for such a generator to be certified.

A *trapdoor permutation generator* is a triplet of polynomial time algorithms (G, E, I) called the *generating, evaluating,* and *inverting* algorithms, respectively. The generating algorithm is probabilistic, and on input 1^n outputs a pair of n-bit strings $(f^*, \bar{f}^*)$, describing, respectively, a trapdoor permutation and its inverse. If x, y are n-bit strings, then so are $E(f^*, x)$ and $I(\bar{f}^*, y)$. Moreover, the maps $f, \bar{f}: \{0,1\}^n \rightarrow \{0,1\}^n$ specified by $f(x) = E(f^*, x)$ and $\bar{f}(y) = I(\bar{f}^*, y)$ are permutations of $\{0,1\}^n$, and $\bar{f} = f^{-1}$. Finally, f is "hard to invert" without knowledge of $\bar{f}$. (We refer the reader to §2.2 for more precise definitions).

Fix a trapdoor permutation generator (G, E, I). We call an n-bit string f^* a *trapdoor permutation* if there exists some n-bit string $\bar{f}^*$ such that the pair $(f^*, \bar{f}^*)$ has a non-zero probability of being obtained when we run G on input 1^n. It is important to note that not every n-bit string f^* is a trapdoor permutation. In fact, the set of n-bit strings which are trapdoor permutations may be a very sparse subset of $\{0,1\}^n$, and perhaps not even recognizable in polynomial time. If it *is* recognizable in polynomial time, we say the generator is *certified* (that is, the trapdoor permutation generator (G, E, I) is said to be certified if there exists a polynomial time algorithm which, on input a string f^*, outputs 1 iff f^* is a trapdoor permutation).

We note that certification is a lot to ask for. Consider our two main (conjectured) examples of trapdoor permutation generators: RSA [RSA], and the factoring based generator of Blum, Blum and Shub [BBS]. Neither is likely to be certified. This is because, in both cases, certification would need the ability to recognize in polynomial time the class of integers which are a product of (exactly) two (distinct) primes.

The importance of certification arises, as will see, from the use of trapdoor permutations as "tools" in protocols. Typically, one party (for example, the prover) gives the other party (for example, the verifier) a string f^* which is supposed to be a trapdoor permutation. For security reasons he may not wish to reveal (as proof that it is indeed one) the string $\bar{f}^*$, but may nonetheless need to convince the verifier that f^* is indeed a trapdoor permutation. This is clearly easy if the underlying generator is certified. If the generator is not certified, the protocol itself must address the task of giving suitable conviction that f^* is really a trapdoor permutation. In interactive protocols this is usually (but not necessarily always!) easy. As we will see, the issue is more complex in the non-interactive case.

1.2 Non-Interactive Zero-Knowledge Proofs

The setting we focus on in this paper is that of non-interactive zero-knowledge (NIZK) proof systems. NIZK is an important notion for cryptographic systems and protocols which was introduced by Blum, Feldman, and Micali [BFM] and Blum, De Santis, Micali and Persiano [BDMP]. There are numerous applications. In particular, Naor and Yung show how to use NIZK proofs to implement public-key cryptosystems secure against chosen-ciphertext attack [NaYu], and Bellare and Goldwasser present a novel paradigm for digital signatures based on NIZK proofs [BeGo].

The model is as follows. The prover and verifier have a common input w and also share a random string (of length polynomial in the length of w). We call this string the *reference* string, and usually denote it by σ. The prover must convince the verifier of the membership of w in some fixed underlying NP language L. To this end, the prover is allowed to send the verifier a single message, computed as a function of w and σ (in the case where $w \in L$, we also give the prover, as an auxiliary input, a witness to the membership of w in L). We usually denote this message by p. The verifier (who is polynomial time) decides whether or not to accept as a function of w, σ and p. We ask that there exist a prover who can convince the verifier to accept $w \in L$, for all random strings σ (this is the *completeness* condition). We ask that for any prover, the probability (over the choice of σ) that the verifier may be convinced to accept when $w \notin L$ is small (this is the soundness condition). Finally, we ask the the proof provided by the prover of the completeness condition (in the case $w \in L$) be zero-knowledge, by requiring the existence of an appropriate "simulator." For a more complete specification of what it means to be a NIZK proof system, we refer the reader to §2.3.

We will focus here on protocols with efficient provers. That is, we want the prover of the completeness condition (we call it the "honest" prover) to run in polynomial (in $n = |w|$) time.

We note that we are considering what are called "single-theorem" or "bounded" NIZK proof systems. The primitive of importance in applications is the "many-theorem" proof system (cf. [BFM, BDMP]). However, the former is known to imply the latter, given the existence of one-way functions [DeYu, FLS]. So we may, wlog, stick to the former.

1.3 The Need for Certification in the FLS Protocol

Feige, Lapidot and Shamir [FLS] recently presented an elegant NIZK proof system based on the existence of trapdoor permutations. The assumption, implicit in their analysis, is that the underlying trapdoor permutation generator is certified. Here we indicate whence arises the need for this certification. Once we have identified the source of the problem, we will discuss how we propose to solve it.

Let L be a language in NP, and let (G, E, I) be a trapdoor permutation generator. Fix a common input $w \in \{0,1\}^n$, and let σ denote the reference string. We will describe how the prover and verifier are instructed to operate under the FLS protocol. First, however, we need some background and some notation.

First, note that even if f^* is not a trapdoor permutation, we may assume, wlog, that $E(f^*, x)$ is n-bits long. Thus, f^* does specify (via E) a map from $\{0,1\}^n$ to $\{0,1\}^n$; specifically, the map given by $x \mapsto E(f^*, x)$. We call this map the function specified by f^* under E, and will denote it by f. Of course, if f^* is a trapdoor permutation then f is a permutation.

If x and r are n-bit strings then $H(x, r)$ denotes the dot product, over $GF(2)$, of x and r (more precisely, $H(x, r) = \bigoplus_{i=1}^{n} x_i r_i$). The theorem of Goldreich and Levin [GoLe] says that H is a "hard-core" predicate for (G, E, I). Very informally, this means the following. Suppose we run G (on input 1^n) to get $(f^*, \bar{f}^*)$, select x and r at random from $\{0,1\}^n$, and let $y = f(x)$. Then, given y and r, the task of predicting $H(x, r)$, and the task of finding x, are equally hard.

We are now ready to describe the protocol.

The protocol first asks that the prover P run G on input 1^n to obtain a pair $(f^*, \bar{f}^*)$. P is then instructed to send f^* to V (while keeping $\bar{f}^*$ to himself).

And the problem is right here, in this first step. The analysis of [FLS] assumes that the prover performs this step correctly. This may be justified under the assumption that the trapdoor permutation generator is certified. If the generator is not certified, a cheating prover could, when $w \notin L$, select, and send to the verifier, an n-bit string which is *not* a trapdoor permutation. As we will see, this could compromise the soundness of the protocol. Let us proceed.

Once the prover has supplied f^*, the reference string is regarded as a sequence $\sigma = y_1 r_1 \ldots y_l r_l$ of l blocks of size $2n$, where $l = l(n)$ is a (suitable) polynomial (block i consists of the pair of n bit strings $y_i r_i$). We say that the prover "opens block i with value b_i" if he provides the verifier with an n-bit string x_i such that $f(x_i) = y_i$ and $H(x_i, r_i) = b_i$. The prover now opens certain blocks of the random string (and the protocol specifies how an honest prover should choose which blocks to open). Based on the values of the opened blocks, their relative locations in the reference string, and the common input, the verifier decides whether or not to accept. Exactly how he does this is not relevant to our discussion. Exactly how the honest prover is supposed to decide which blocks to open (which he does as a function of the block, the common input, and his witness to the membership of the common input in L) is also not relevant to our discussion. What is important to note is that the soundness condition relies on the assumption that, with f^* fixed, there exists a *unique* way to open any given block. If it is possible for the prover to open a block with value either 0 or 1, then the soundness of the FLS protocol is compromised.

The assumption that there is (one and) only one way to open a block is justified if f^* is a trapdoor permutation, because, in this case, f is a permutation. However, if f^* is not a trapdoor permutation, then f may not be a permutation, and in such a case, the possibility exists that blocks may be opened with values of the prover's choice.

We note that the gap is not an academic one. Considering concrete cases, such as the use of RSA or the trapdoor permutations based on quadratic residuosity that are suggested by [BBS], we see that the prover may indeed cheat.

The solution that first suggests itself is that the prover prove (in NIZK) that he really got f by running the generator G (this is an NP statement). The problem is, however, that to prove this new statement itself requires the use of a trapdoor permutation, and we are only chasing our tail.

We note that the whole problem would not arise if we were using a one-way permutation (rather than a trapdoor one) because, as we said above, a one-way permutation is a single object which both parties know à priori. Yet for the sake of maintaining the efficiency of the prover, we cannot use one-way permutations.

Remark. Note that in the above NIZK proof, a (cheating) prover may choose f^* as a function of the random string. But, as pointed out in [FLS], this causes no difficulties. We may assume, in the analysis, that the reference string is chosen after f^* is fixed; later we apply a simple transformation which results in the proof system being secure even if f^* was chosen as a function of σ. We will deal with this issue explicitly when it arises.

1.4 Our Solution

Let f^* denote the n-bit string provided by the prover in the first step of the FLS protocol, as described above. As that discussion indicates, soundness does not really require that f be a trapdoor permutation. All that it requires is that f be a permutation. So it would suffice to certify this fact.

To certify that a map from $\{0,1\}^n$ to $\{0,1\}^n$ is a permutation seems like a hard task (it is a coNP statement). What we will do is certify it is "almost" a permutation, and then show that this suffices.

More precisely, let us call f an ϵ-permutation if at most an ϵ fraction of the points in $\{0,1\}^n$ have more than one pre-image under f. We show that on common input f^*, and access to a common (random) reference string of length $\epsilon^{-1} \cdot n$, the prover can provide the verifier with a non-interactive, zero-knowledge proof that that f is an ϵ-permutation. For a more precise statement of the theorem and its proof we refer the reader to §3.

We then show that adding this step to augment a multitude of independent FLS protocol instances yields a NIZK proof system (for any NP language) given the existence of any (not necessarily certified) trapdoor permutation generator. A complete proof of this fact is in §4. We note that this proof is in fact quite independent of the details of the FLS protocol and can be understood without a deep knowledge of the techniques of that paper.

2 Preliminaries

We begin by summarizing some basic notation and conventions which are used throughout the paper. We then discuss trapdoor permutations and say what it means for them to be "certified." Finally, we recall the definition, and some basic properties, of non-interactive zero-knowledge proof systems.

2.1 Notation and Conventions

We use the notation and conventions for probabilistic algorithms that originated in [GMR].

We emphasize the number of inputs received by an algorithm as follows. If algorithm A receives only one input we write "$A(\cdot)$"; if it receives two we write "$A(\cdot, \cdot)$", and so on. If A is a probabilistic algorithm then, for any input i the notation $A(i)$ refers to the probability space which to the string σ assigns the probability that A, on input i, outputs σ.

If S is a probability space we denote its support (the set of elements of positive probability) by $[S]$.

If $f(\cdot)$ and $g(\cdot, \cdots)$ are probabilistic algorithms then $f(g(\cdot, \cdots))$ is the probabilistic algorithm obtained by composing f and g (i.e. running f on g's output). For any inputs $x, y, \ldots$ the associated probability space is denoted $f(g(x, y, \cdots))$.

If S is a probability space then $x \xleftarrow{R} S$ denotes the algorithm which assigns to x an element randomly selected according to S. In the case that $[S]$ consists of only one element e we might also write $x \leftarrow e$.

For probability spaces $S, T, \ldots$, the notation

$$\Pr\left[p(x, y, \cdots) \; : \; x \xleftarrow{R} S \; ; \; y \xleftarrow{R} T \; ; \; \cdots \right]$$

denotes the probability that the predicate $p(x, y, \cdots)$ is true after the (ordered) execution of the algorithms $x \xleftarrow{R} S$, $y \xleftarrow{R} T$, etc.

Let f be a function. The notation

$$\{ f(x, y, \cdots) : x \overset{R}{\leftarrow} S ; \ y \overset{R}{\leftarrow} T ; \ \cdots \}$$

denotes the probability space which to the string σ assigns the probability

$$\Pr \left[\sigma = f(x, y, \cdots) \ : \ x \overset{R}{\leftarrow} S ; \ y \overset{R}{\leftarrow} T ; \ \cdots \right] .$$

When we say that a function is computable in polynomial time, we mean computable in time polynomial in the length of its first argument.

We will be interested in families of efficiently computable functions of polynomial description. The following definition will be a convenient way of capturing them.

Definition 1. Let $E(\cdot, \cdot)$ be a polynomial time computable function. We say that E specifies an efficiently computable family of functions if for each $n > 0$ and each $f^*, x \in \{0, 1\}^n$ it is the case that $|E(f^*, x)| = n$. Let $n > 0$ and $f^* \in \{0, 1\}^n$. The function specified by f^* under E is the map from $\{0, 1\}^n$ to $\{0, 1\}^n$ given by $x \mapsto E(f^*, x)$.

2.2 Trapdoor Permutations and Certified Ones

Let us present a precise definition of trapdoor permutations and see what it means for them to be "certified." The definition that follows is from Bellare and Micali [BeMi].

Definition 2. (Trapdoor Permutation Generator) Let G be a probabilistic, polynomial time algorithm, and let E, I be polynomial time algorithms. We say that (G, E, I) is a trapdoor permutation generator if the following conditions hold:

- Generation: For every $n > 0$, the output of G on input 1^n is a pair of n bit strings.

- Permutation: For every $n > 0$ and $(f^*, \bar{f}^*) \in [G(1^n)]$, the maps $E(f^*, \cdot)$ and $I(\bar{f}^*, \cdot)$ are permutations of $\{0, 1\}^n$ which are inverses of each other (that is, $I(\bar{f}^*, E(f^*, x)) = x$ and $E(f^*, I(\bar{f}^*, y)) = y$ for all $x, y \in \{0, 1\}^n$).

- Security: For all probabilistic polynomial time (adversary) algorithms $A(\cdot, \cdot, \cdot)$, for all constants c and sufficiently large n, it is the case that

$$\Pr \left[E(f^*, x) = y \ : \ (f^*, \bar{f}^*) \overset{R}{\leftarrow} G(1^n) ; \ y \overset{R}{\leftarrow} \{0, 1\}^n ; \ x \overset{R}{\leftarrow} A(1^n, f^*, y) \right]$$

is at most n^{-c}.

We call G, E, I the generating, evaluating and inverting algorithms, respectively.

The standard (conjectured) "trapdoor permutations," such as RSA [RSA] and the factoring based ones of Blum, Blum and Shub [BBS], do fit this definition, after some minor transformations (the need for these transformations arises from the fact that these number theoretic functions have domain Z_N^* rather than $\{0, 1\}^n$; we refer the reader to [BeMi] for details).

If a trapdoor permutation generator (G, E, I) is fixed and $(f^*, \bar{f}^*) \in [G(1^n)]$ for some $n > 0$, then, in informal discussion, we call f^* a trapdoor permutation. It is important to note that not every n bit string f^* is a trapdoor permutation: it is only one if there exists some $\bar{f}^*$ such that $(f^*, \bar{f}^*) \in [G(1^n)]$. In fact, the set of ($n$ bit strings which are) trapdoor permutations may be a fairly sparse subset of $\{0, 1\}^n$, and, in general, may not be recognizable in polynomial (in n) time. If a trapdoor permutation generator *does* have the special property that it is possible to recognize a trapdoor permutation in polynomial time then we say that this generator is *certified*. The more formal definition follows.

Definition 3. Let (G, E, I) be a trapdoor permutation generator. We say that (G, E, I) is certified if the language

$$L_{G,E,I} = \bigcup_{n \geq 1} \{ f^* \in \{0, 1\}^n : \exists \bar{f}^* \in \{0, 1\}^n \text{ such that } (f^*, \bar{f}^*) \in [G(1^n)] \}$$

is in BPP.

We note that standard (conjectured) trapdoor permutation generators are (probably) *not* certified. In particular, RSA is (probably) not certified, and nor is the trapdoor permutation generator of Blum, Blum and Shub [BBS]. This is because, in both these cases, the (description of) the trapdoor permutation f^* includes a number which is a product of two primes, and there is (probably) no polynomial time procedure to test whether or not a number is a product of two primes.

The importance of certification stems, as we have seen, from applications in which one party (for example, the prover) gives the other party (for example, the verifier) a string f^* which is supposed to be a trapdoor permutation. For security reasons he may not wish to reveal (as proof that it is indeed one) the string $\bar{f}^*$, but may nonetheless need to convince the verifier that f^* is indeed a trapdoor permutation. In particular, the (implicit) assumption in [FLS] is that the trapdoor permutation generator being used is certified. As the above indicates, this means that their scheme cannot be instantiated with RSA or the trapdoor permutations of [BBS]. In later sections we will show how to extend their scheme so that any (not necessarily certified) trapdoor permutation generator suffices (so that RSA or the generator of [BBS] may in fact be used).

We note that if (G, E, I) is a trapdoor permutation generator, $f^* \in \{0, 1\}^n$, and $x \in \{0, 1\}^n$ then we may assume, without loss of generality, that $E(f^*, x)$ is an n-bit string. Hence $E(f^*, \cdot)$ does specify some map from $\{0, 1\}^n$ to $\{0, 1\}^n$, even if f^* is not a trapdoor permutation. That is, in the terminology of Definition 1, we may assume, without loss of generality, that the algorithm E specifies an efficiently computable family of functions. Of course, the map $E(f^*, \cdot)$ need not be a permutation on $\{0, 1\}^n$.

2.3 Non-Interactive Zero-knowledge Proof Systems

We will consider non-interactive zero-knowledge proof systems for NP. It is helpful to begin with the following terminology.

Definition 4. Let $\rho(\cdot, \cdot)$ be a binary relation. We say that ρ is an NP-relation if it is polynomial time computable and, moreover, there exists a polynomial p such that $\rho(w, \bar{w}) = 1$ implies $|\bar{w}| \leq p(|w|)$. For any $w \in \{0,1\}^*$ we let $\rho(w) = \{ \bar{w} \in \{0,1\}^* : \rho(w, \bar{w}) = 1 \}$ denote the witness set of w. We let $L_\rho = \{ w \in \{0,1\}^* : \rho(w) \neq \emptyset \}$ denote the language defined by ρ. A witness selector for ρ is a map $W : L_\rho \rightarrow \{0,1\}^*$ with the property that $W(w) \in \rho(w)$ for each $w \in L_\rho$.

Note that a language L is in NP iff there exists an NP-relation ρ such that $L = L_\rho$.

We recall the definition of computational indistinguishability of ensembles. First, recall that a function $\delta : \{0,1\}^* \rightarrow \mathbb{R}$ is *negligible* if for every constant d there exists an integer n_d such that $\delta(w) \leq |w|^{-d}$ for all w of length at least n_d.

Definition 5. An ensemble indexed by $L \subseteq \{0,1\}^*$ is a collection $\{E(w)\}_{w \in L}$ of probability spaces (of finite support), one for each $w \in L$. Let $\mathcal{E}_1 = \{E_1(w)\}_{w \in L}$ and $\mathcal{E}_2 = \{E_2(w)\}_{w \in L}$ be ensembles over a common index set L. We say that they are (computationally) indistinguishable if for every family $\{D_w\}_{w \in L}$ of non-uniform, polynomial time algorithms, the function

$$\delta(w) \stackrel{\text{def}}{=} \left| \Pr\left[D_w(v) = 1 \ : \ v \stackrel{R}{\leftarrow} E_1(w) \right] - \Pr\left[D_w(v) = 1 \ : \ v \stackrel{R}{\leftarrow} E_2(w) \right] \right|$$

is negligible.

The definition that follows is based on that of Blum, De Santis, Micali and Persiano [BDMP]. However, we state the zero-knowledge condition differently; specifically, we use the notion of a witness selector to state the zero-knowledge condition in terms of the standard notion of computational indistinguishability, whereas in [BDMP] the zero-knowledge condition makes explicit reference to "distinguishing" algorithms. The two formulations are, of course, equivalent (but we feel this one is a little simpler because of its "modularity.")

Definition 6. Let ρ be an NP-relation and let $L = L_\rho$. Let P be a machine, V a polynomial time machine, and S a probabilistic, polynomial time machine. We say that (P, V, S) defines a non-interactive zero-knowledge proof system (NIZK proof system) for ρ if there exists a polynomial $l(\cdot)$ such that the following three conditions hold.

- Completeness: For every $w \in L$ and $\bar{w} \in \rho(w)$,

$$\Pr\left[V(w, \sigma, p) = 1 \ : \ \sigma \stackrel{R}{\leftarrow} \{0,1\}^{l(n)} \, ; \, p \leftarrow P(w, \bar{w}, \sigma) \right] = 1 \, ,$$

  where $n = |w|$.

- Soundness: For every machine $\widehat{P}$ and every $w \notin L$,

$$\Pr\left[V(w, \sigma, p) = 1 \ : \ \sigma \stackrel{R}{\leftarrow} \{0,1\}^{l(n)} \, ; \, p \leftarrow \widehat{P}(w, \sigma) \right] \leq \tfrac{1}{2} \, ,$$

  where $n = |w|$.

- Zero-knowledge: Let W be any witness selector for ρ. Then the following two ensembles are (computationally) indistinguishable:

  (1) $\{S(w)\}_{w \in L}$

  (2) $\{ (\sigma, p) : \sigma \xleftarrow{R} \{0,1\}^{l(|w|)} ; \; p \leftarrow P(w, W(w), \sigma) \}_{w \in L}.$

We call P the prover, V the verifier and S the simulator. The polynomial l is the length of the reference string. We say that P is efficient if it is polynomial time computable.

We call σ the "common random string" or the "reference string."

The choice of $1/2$ as the error-probability in the soundness condition is not essential. Given any polynomial $k(\cdot)$, the error-probability can be reduced to $2^{-k(n)}$ by running $k(n)$ independent copies of the original proof system in parallel and accepting iff all sub-proofs are accepting.

A stronger definition (cf. [BDMP]) asks that in the soundness condition the adversary $\widehat{P}$ be allowed to select a $w \notin L$ as a function of the reference string. This definition is, however, implied by the one above. More precisely, given (P, V, S) satisfying the above definition, one can construct (P', V', S') satisfying the more stringent definition, by a standard trick. Hence, we will stick to the simple definition.

We note we are considering what have been called "single-theorem" or "bounded" NIZK proof systems. That is, the given reference string can be used to prove only a single theorem. The primitive of importance in applications (cf. [BeGo, NaYu]) is the "many-theorem" proof system. However, De Santis and Yung [DeYu], and Feige, Lapidot and Shamir [FLS], have shown that the existence (for some NP-complete relation) of a bounded NIZK proof system with an efficient prover implies the existence (for any NP-relation) of a many-theorem NIZK proof system (with an efficient prover), as long as one-way functions exist. Hence, given that the (bounded) NIZK proof systems we construct do have efficient provers, we may, without loss of generality, stick to the bounded case.

3 A NIZK Proof that a Map is Almost a Permutation

Suppose E specifies an efficiently computable family of functions (cf. Definition 1), and suppose $f^* \in \{0,1\}^n$ for some $n > 0$. We address in this section the problem of providing a NIZK proof that the function specified by f^* under E is "almost" a permutation.

We note that although this problem is motivated by the need to fill the gap in the FLS protocol (cf. §1.3), the results of this section might be of interest in their own right. Thus, we prefer to view them independently, and will make the link to [FLS] in the next section.

In addressing the task of providing a NIZK proof that the function specified by f^* under E is "almost" a permutation, we must begin by clarifying two things. First, we need to say what it means for a function $f: \{0,1\}^n \rightarrow \{0,1\}^n$ to be "almost" a permutation. Our definition, of an ϵ-permutation, follows. Second, we must also say what we mean, in this context, by an NIZK proof (because the problem is not one of language membership). This is clarified below and in the statement of Theorem 8.

Let us begin with the definition. It says that f is an ϵ permutation if at most an ϵ fraction of the points in $\{0,1\}^n$ have more than one pre-image (under f). More formally, we have the following.

Definition 7. Let $n > 0$ and $f\colon \{0,1\}^n \to \{0,1\}^n$. The collision set of f, denoted $C(f)$, is $\{ y \in \{0,1\}^n : |f^{-1}(y)| > 1 \}$. Let $\epsilon \in [0,1]$. We call f an ϵ-permutation if $|C(f)| \leq \epsilon 2^n$.

We will now turn to the NIZK proof. The formal statement and proof of the theorem follow. Let us begin, however, by saying, informally, what we achieve, and giving the idea.

We fix E specifying an efficiently computable family of functions, and we fix a map $\epsilon\colon \{0,1\}^* \to (0,1]$. We consider a prover and verifier who share a (random) reference string and have as common input a string $f^* \in \{0,1\}^n$. If f (the function specified by f^* under E) is a permutation then the prover can convince the verifier to accept (this is the completeness condition). If f is not an $\epsilon(n)$-permutation, then the verifier will usually reject (this is the soundness condition).

We note the gap between these two conditions: we are guaranteed nothing if f *is* an $\epsilon(n)$-permutation (but not a permutation). This is one way in which this "proof system" differs from proofs of language membership, where there are only two possibilities: either the input is in the language (and completeness applies) or it is not (and soundness applies).

In addition, when f is a permutation, the interaction yields no (extra) knowledge to the verifier. This is formalized, as usual, by requiring the existence of an appropriate "simulator."

The idea is very simply stated. Let σ be the reference string, which we think of as divided into blocks of size n. If f is not an $\epsilon(n)$-permutation, then each block has probability at most $1 - \epsilon(n)$ of being in the range of f. So if we ask the prover to provide the inverse of f on $\epsilon^{-1}(n)$ different blocks, then he can succeed with probability at most $(1 - \epsilon(n))^{\epsilon^{-1}(n)} \leq 1/2$. Moreover, a collection of pre-images of f on random points provide no information about (the easily computed) f, so the proof is zero-knowledge.

Theorem 8. *Let E specify an efficiently computable family of functions. Let $\epsilon\colon N \to (0,1]$, and assume ϵ^{-1} is polynomially bounded and polynomial time computable. Then there is a polynomial time oracle machine A, a polynomial time machine B, and a probabilistic, polynomial time machine M such that the following three conditions hold.*

- Completeness: *Let $n > 0$ and $f^* \in \{0,1\}^n$. Let f denote the function specified by f^* under E. Suppose f is a permutation. Then*

$$\Pr\left[B(f^*,\sigma,p) = 1 \,:\, \sigma \stackrel{R}{\leftarrow} \{0,1\}^{\epsilon^{-1}(n)\cdot n} \,;\, p \leftarrow A^{f^{-1}}(f^*,\sigma) \right] = 1 .$$

Here $A^{f^{-1}}$ denotes A with oracle f^{-1}.

- Soundness: *Let $n > 0$ and $f^* \in \{0,1\}^n$. Let f denote the function specified by f^* under E. Suppose f is not a $\epsilon(n)$-permutation. Then for any function $\widehat{P}$,*

$$\Pr\left[B(f^*,\sigma,p) = 1 : \sigma \xleftarrow{R} \{0,1\}^{\epsilon^{-1}(n)\cdot n} ; p \leftarrow \widehat{P}(f^*,\sigma)\right] \leq \tfrac{1}{2}.$$

- Zero-knowledge: *Let $n > 0$ and $f^* \in \{0,1\}^n$. Let f denote the function specified by f^* under E, and suppose f is a permutation. Then the distributions $M(f^*)$ and $\{(\sigma,p) : \sigma \xleftarrow{R} \{0,1\}^{\epsilon^{-1}(n)\cdot n} ; p \leftarrow A^{f^{-1}}(f^*,\sigma)\}$ are equal.*

Proof: We specify the algorithm for verifier. Let $f^* \in \{0,1\}^n$ and let $\sigma = \sigma_1 \ldots \sigma_{\epsilon^{-1}(n)}$ where each σ_i has length n. Let f denote the function specified by f^* under E. On input f^*, σ, and a string p, the verifier B rejects if the length of p is not $\epsilon^{-1}(n) \cdot n$. Otherwise, it partitions p into consecutive blocks of size n. We denote the i-th block by p_i, so that $p = p_1 \ldots p_{\epsilon^{-1}(n)}$. Then B accepts iff for each $i = 1, \ldots, \epsilon^{-1}(n)$ it is the case that $f(p_i) = \sigma_i$.

Next we specify the prover A. Let $f^* \in \{0,1\}^n$ and let $\sigma = \sigma_1 \ldots \sigma_{\epsilon^{-1}(n)}$ where each σ_i has length n. Let f denote the function specified by f^* under E, and suppose f is a permutation. On input f^* and σ, and given f^{-1} as oracle, A sets $p_i = f^{-1}(\sigma_i)$ for each $i = 1, \ldots, \epsilon^{-1}(n)$. It then sets $p = p_1 \ldots p_{\epsilon^{-1}(n)}$ and outputs p. It is easy to see that the completeness condition is true.

We now check the soundness condition. Let $f^* \in \{0,1\}^n$ and let f denote the function specified by f^* under E. We recall that $C(f) = \{y \in \{0,1\}^n : |f^{-1}(y)| > 1\}$ is the collision set of f. Let $D(f) = \{y \in \{0,1\}^n : |f^{-1}(y)| = 0\}$ be the set of n bit strings not in the range of f. Note that $|D(f)| \geq |C(f)|$. We let $\delta(n) \stackrel{\text{def}}{=} |D(f)|/2^n$ denote the density of $D(f)$. Now assume f is not a $\epsilon(n)$-permutation. Then $|C(f)| \geq \epsilon(n)2^n$, and thus $\delta(n) \geq \epsilon(n)$. For any fixed string $\sigma = \sigma_1 \ldots \sigma_{\epsilon^{-1}(n)}$, the following are clearly equivalent:

- There exists a string p such that $B(f^*,\sigma,p) = 1$
- For each $i = 1, \ldots, \epsilon^{-1}(n)$ it is the case that σ_i is in the range of f.

However, if σ is chosen at random, then for each $i = 1, \ldots, \epsilon^{-1}(n)$, the probability that σ_i is in the range of f is at most $1 - \delta(n)$, independently for each i. So for any $\widehat{P}$,

$$\Pr\left[B(f^*,\sigma,p) = 1 : \sigma \xleftarrow{R} \{0,1\}^{\epsilon^{-1}(n)\cdot n} ; p \leftarrow \widehat{P}(f^*,\sigma)\right] \leq [1-\delta(n)]^{\epsilon^{-1}(n)}$$

$$\leq [1-\epsilon(n)]^{\epsilon^{-1}(n)}$$

$$\leq \tfrac{1}{2}.$$

We now specify M. Let $f^* \in \{0,1\}^n$ and let f denote the function specified by f^* under E. Suppose f is a permutation. On input f^*, the machine M

picks $\tau_1, \ldots, \tau_{\epsilon^{-1}(n)} \in \{0,1\}^n$ at random and sets $\sigma_i = f(\tau_i)$, for each $i = 1, \ldots, \epsilon^{-1}(n)$. It sets $p = \tau_1 \ldots \tau_{\epsilon^{-1}(n)}$ and outputs (σ, p). The zero-knowledge is easy to check. ∎

We note that, in the above, we are thinking of f^* as being the common input, and the reference string is chosen at random independently of f^*. Of course, in our application, the prover may choose f^* as a function of the reference string. This, however, is easily dealt with by a standard trick, and so, for the moment, we focus on the case presented here. When we put everything together (cf. Theorem 12) we will return to this issue and show explicitly how to deal with it, given what we establish here.

We note also that no cryptographic assumptions were needed for the above proof, and the zero-knowledge is "perfect."

4 Using the Certification Procedure

In this section we show how the certification procedure of Theorem 8 can be combined with the results of [FLS] to yield a NIZK proof system for any NP-relation. We stress that the argument we present here depends little on the specifics of the protocol of [FLS], and our proof does not presume familiarity with that paper. We begin by extending Definition 7 with the following terminology.

Definition 9. Let $n > 0$ and $f \colon \{0,1\}^n \to \{0,1\}^n$. Let $\sigma = \sigma_1 \ldots \sigma_l$ for some $l \in \mathbb{N}$, where each σ_i has length n. We say that σ is f-bad if there is an $i \in \{1, \ldots, l\}$ such that $\sigma_i \in C(f)$. We denote by $C_l(f)$ the set of all ln-bit strings which are f-bad.

We now state, without proof, a lemma which can be derived from [FLS]. The formal statement follows, but, since it is rather long, let us first try to give an informal explanation of what it says.

Briefly, we show how to "measure" the "additional" error incurred by the [FLS] protocol in the case that the function being used is not a permutation. More precisely, we fix a trapdoor permutation generator (G, E, I) and an NP-relation ρ. In order to make explicit the role played by the function used in the proof, we consider an interaction in which the common input is a pair (w, f^*) of n-bit strings. The prover wishes to convince the verifier that $w \in L \overset{\text{def}}{=} L_\rho$, using f^* as a "tool." We do not, à priori, know whether or not f^* is a trapdoor permutation.

The completeness condition (below) says that if $w \in L$, then, assuming f^* really is a trapdoor permutation, the prover can convince the verifier that $w \in L$. Moreover, the zero-knowledge condition says this proof is zero-knowledge. The part we are really concerned with, however, is the soundness condition.

The soundness condition says that if $w \notin L$ then the probability that a prover can convince the verifier to accept is bounded by a small error $(1/4)$ plus a quantity that depends on f^*. Specifically, this quantity is the probability that the reference string is f-bad (cf. Definition 9), where f is the function specified by f^* under E.

A priori, this quantity may be large. Once we have stated the lemma, we will show how to use the results of the previous section to decrease it.

Lemma 10. *Let (G, E, I) be a trapdoor permutation generator. Let ρ be an NP-relation, and let $L = L_\rho$. Then there exists a polynomial time machine $\bar{A}$, a polynomial time machine $\bar{B}$, a probabilistic, polynomial time machine $\bar{M}$, and a polynomial $l(\cdot)$ such that the following three conditions hold.*

- Completeness: *For every $w \in L$, every $\bar{w} \in \rho(w)$, and every $(f^*, \bar{f}^*) \in [G(1^n)]$,*

$$\Pr\left[\, \bar{B}(w, \sigma, f^*, p) = 1 \,:\, \sigma \stackrel{R}{\leftarrow} \{0,1\}^{l(n)\cdot n} \,;\, p \leftarrow A(w, \bar{w}, \sigma, f^*, \bar{f}^*)\right] = 1\,,$$

where $n = |w|$.

- Soundness: *For every machine $\widehat{P}$, every $w \notin L$, and every $f^* \in \{0,1\}^n$,*

$$\Pr\left[\, \bar{B}(w, \sigma, f^*, p) = 1 \,:\, \sigma \stackrel{R}{\leftarrow} \{0,1\}^{l(n)\cdot n} \,;\, p \leftarrow \widehat{P}(w, \sigma, f^*)\right]$$

$$\leq \tfrac{1}{4} + \Pr\left[\, \sigma \in C_{l(n)}(f) \,:\, \sigma \stackrel{R}{\leftarrow} \{0,1\}^{l(n)\cdot n}\right]\,,$$

where $n = |w|$ and f denotes the function specified by f^* under E.

- Zero-knowledge: *Let W be any witness selector for ρ. Then the following two ensembles are (computationally) indistinguishable:*

(1) $\{\, (\sigma, f^*, p) : (f^*, \bar{f}^*) \stackrel{R}{\leftarrow} G(1^{|w|}) \,;\, (\sigma, p) \stackrel{R}{\leftarrow} \bar{M}(w, f^*, \bar{f}^*) \,\}_{w \in L}$

(2) $\{\, (\sigma, f^*, p) : \sigma \stackrel{R}{\leftarrow} \{0,1\}^{l(|w|)\cdot|w|} \,;\, (f^*, \bar{f}^*) \stackrel{R}{\leftarrow} G(1^{|w|}) \,;$
$\qquad p \leftarrow \bar{A}(w, W(w), \sigma, f^*, \bar{f}^*) \,\}_{w \in L}.$

We note that the statement of the above lemma makes no explicit reference to the methods underlying the proof of [FLS]. Our previous discussions should indicate whence, in the light of the [FLS] protocol, arises the "extra" term in the soundness condition, but this is not relevant to the present discussion: everything we need is captured by the statement of the lemma (and we refer the reader to [FLS] for its proof).

We now show how to remove this extra f^* dependent term in the soundness condition by having the prover certify (using the proof system of Theorem 8) that f is almost a permutation. The lemma that follows provides the formal statement and proof, but let us first say, informally, what is happening.

On common input (w, f^*), we have the prover give the proof of Lemma 10, and also, using a separate part of the reference string, run the procedure of Theorem 8. The verifier accepts iff both of these proofs are accepted (by their respective verifiers). The completeness and zero-knowledge conditions stay the same as in Lemma 10 (except that the reference string is longer, indicated by using a different symbol for its length); clearly, this is because the additional proof cannot hurt them. The soundness condition, however, now becomes more like a "real" soundness condition in that the "extra" term of Lemma 10 has disappeared.

In the proof of the new soundness condition, we will have to consider two cases. First, we assume that f is "almost" a permutation, and show that in this case the "extra" term from the soundness condition of Lemma 10 is small. Second, we assume that f is not "almost" a permutation, and use the fact that we are guaranteed rejection (with high probability) by the soundness condition of Theorem 8.

Lemma 11. *Let (G, E, I) be a trapdoor permutation generator. Let ρ be an NP-relation and let $L = L_\rho$. Then there exists a polynomial time machine A', a polynomial time machine B', a probabilistic, polynomial time machine M', and a polynomial $m(\cdot)$ such that the following three conditions hold.*

- Completeness: *For every $w \in L$, every $\bar{w} \in \rho(w)$, and every $(f^*, \bar{f}^*) \in [G(1^n)]$,*

$$\Pr\left[B'(w, \sigma, f^*, p) = 1 : \sigma \stackrel{R}{\leftarrow} \{0,1\}^{m(n) \cdot n} ; p \leftarrow A(w, \bar{w}, \sigma, f^*, \bar{f}^*) \right] = 1 ,$$

where $n = |w|$.

- Soundness: *For every machine $\widehat{P}$, every $w \notin L$, and every $f^* \in \{0,1\}^n$,*

$$\Pr\left[B'(w, \sigma, f^*, p) = 1 : \sigma \stackrel{R}{\leftarrow} \{0,1\}^{m(n) \cdot n} ; p \leftarrow \widehat{P}(w, \sigma, f^*) \right] \leq \tfrac{1}{2} ,$$

where $n = |w|$.

- Zero-knowledge: *Let W be any witness selector for ρ. Then the following two ensembles are (computationally) indistinguishable:*

   (1) $\{ (\sigma, f^*, p) : (f^*, \bar{f}^*) \stackrel{R}{\leftarrow} G(1^{|w|}) ; (\sigma, p) \stackrel{R}{\leftarrow} M(w, f^*, \bar{f}^*) \}_{w \in L}$

   (2) $\{ (\sigma, f^*, p) : \sigma \stackrel{R}{\leftarrow} \{0,1\}^{m(|w|) \cdot |w|} ; (f^*, \bar{f}^*) \stackrel{R}{\leftarrow} G(1^{|w|}) ;$
   $p \leftarrow A(w, W(w), \sigma, f^*, \bar{f}^*) \}_{w \in L}.$

Proof: Let $\bar{A}, \bar{B}, \bar{M}$ be the machines, and l the polynomial, specified by Lemma 10. Let $\epsilon(\cdot) = 1/(4l(\cdot))$. We apply Theorem 8 (with the algorithm E being the evaluating algorithm of our trapdoor family) to get a triplet of machines A, B, M satisfying the conditions of that theorem. We let $m(\cdot) = \epsilon^{-1}(\cdot) + l(\cdot) = 5l(\cdot)$.

Notation: If σ is a string of length $m(n) \cdot n$, then $\sigma[1]$ denotes the first $\epsilon^{-1}(n) \cdot n = 4l(n) \cdot n$ bits and $\sigma[2]$ denotes the last $l(n) \cdot n$ bits.

We now specify the algorithm for the verifier B'. Let $f^* \in \{0,1\}^n$ and let σ be a string of length $m(n) \cdot n$. On input f^*, σ, and a string p, the verifier B rejects if $|p| < \epsilon^{-1}(n) \cdot n$. Otherwise, it accepts if and only if

$$B(f^*, \sigma[1], p[1]) = 1 \quad \text{and} \quad \bar{B}(w, \sigma[2], f^*, p[2]) = 1 ,$$

where $p[1]$ denotes the first $\epsilon^{-1}(n) \cdot n$ bits of p and $p[2]$ denotes the rest.

Next we specify A'. Let $w \in L$ and $\bar{w} \in \rho(w)$. Let $n = |w|$. Let $(f^*, \bar{f}^*) \in [G(1^n)]$. Let σ be a string of length $m(n) \cdot n$. On input $w, \bar{w}, \sigma, f^*, \bar{f}^*$, the machine A' sets $p[1] = A^{f^{-1}}(f^*, \sigma[1])$ (note that A' can obtain this output in polynomial time because, using $\bar{f}^*$, it can compute f^{-1} in polynomial time). It then sets $p[2] = \bar{A}(w, \bar{w}, \sigma[2], f^*, \bar{f}^*)$. Finally it sets $p = p[1]p[2]$ and outputs p. The fact that the completeness condition holds follows from the respective completeness conditions of Lemma 10 and Theorem 8.

Now for the interesting part, namely the soundness condition. Suppose $w \notin L$. Let $n = |w|$ and let $f^* \in \{0,1\}^n$. Let f denote the function specified by f^* under E. We split the proof into two cases.

Case 1: f is a $\epsilon(n)$-permutation.

By assumption, $|C(f)| \le \epsilon(n)2^n$. So

$$\Pr\left[\sigma[2] \in C_{l(n)}(f) \ : \ \sigma[2] \xleftarrow{R} \{0,1\}^{l(n) \cdot n}\right] \le \epsilon(n)l(n) = \tfrac{1}{4} .$$

By the soundness condition of Lemma 10 it follows that for every machine $\hat{P}$,

$$\Pr\left[\bar{B}(w, \sigma, f^*, p[2]) = 1 \ : \ \sigma[2] \xleftarrow{R} \{0,1\}^{l(n) \cdot n} \ ; \ p[2] \leftarrow \hat{P}(w, \sigma, f^*)\right]$$
$$\le \tfrac{1}{4} + \tfrac{1}{4}$$
$$= \tfrac{1}{2} .$$

The soundness condition follows from the definition of B'. Let us proceed to the next case.

Case 2: f is not a $\epsilon(n)$-permutation.

The soundness condition of Theorem 8 implies that for any function $\hat{P}$,

$$\Pr\left[B(f^*, \sigma[1], p[1]) = 1 \ : \ \sigma[1] \xleftarrow{R} \{0,1\}^{\epsilon^{-1}(n) \cdot n} \ ; \ p \leftarrow \hat{P}(f^*, \sigma[1])\right] \le \tfrac{1}{2} .$$

The soundness condition then follows directly from the definition of B'. This completes the proof of the soundness condition.

The zero-knowledge, again, follows immediately from Lemma 10 and Theorem 8. Let $w \in L$ and let $n = |w|$. Let $(f^*, \bar{f}^*) \in [G(1^n)]$. On input $w, f^*, \bar{f}^*$, machine M' runs M on input f^* to get an output $(\sigma[1], p[1])$. It then runs $\bar{M}$ on input $w, f^*, \bar{f}$ to get an output $(\sigma[2], p[2])$. It sets $\sigma = \sigma[1]\sigma[2]$ and $p = p[1]p[2]$ and outputs (σ, p). $\blacksquare$

One more step is needed to derive from Lemma 11 the existence of NIZK proof systems for any NP-relation (given the existence of a trapdoor permutation generator). Namely, the interaction must be on input w (alone); the prover must be allowed to select f^* (which in Lemma 11 is part of the common input) not only as a function of w but also as a function of the reference string. Clearly, in the completeness condition, we may simply ask the prover to select f^* by running the generation algorithm G. Any problems that arise will be in the soundness condition, where a cheating prover will take full advantage of the freedom to choose f^* as a function of the reference string.

For $w \notin L$, we may use the following "trick" (a standard probabilistic one, used, for the same purpose, in [BDMP] and [FLS]). For each fixed $f^* \in \{0,1\}^n$, we reduce the probability that the verifier accepts the interaction on inputs (w, f^*) to $2^{-(n+1)}$, by parallel repetition. It follows that the probability that there exists a string $f^* \in \{0,1\}^n$ such that the verifier accepts on input (w, f^*) is at most $2^n \cdot 2^{-(n+1)} = 1/2$. Details are below.

Theorem 12. *Let ρ be an NP-relation. Suppose there exists a trapdoor permutation generator. Then ρ possesses a non-interactive zero-knowledge proof system with an efficient prover.*

Proof: Let (G, E, I) be a trapdoor permutation generator. Let A', B', M' be the machines, and m the polynomial, specified by Lemma 11. Let $l(n) = m(n) \cdot n(n+1)$. We construct P, V, S satisfying the conditions of Definition 6.

Notation: If σ is a string of length $l(n)$ then we think of it as partitioned into $n+1$ blocks, each of length $m(n) \cdot n$, and denote the i-th block by $\sigma[i]$ $(i = 1, \ldots, n+1)$.

We may assume, without loss of generality, that there is a polynomial t such that $B'(w, \cdot, \cdot, p) = 1$ only if $|p| = t(|w|)$. Let $L = L_\rho$. We specify V. Let $w \in L$ and $\sigma \in \{0,1\}^{l(n)}$. On input w, σ, p, machine V rejects if $|p| \neq n + (n+1)t(n)$. Otherwise, it sets f^* to the first n bits of p and p' to the rest. It further sets $p'[i]$ to the i-th $t(n)$-bit block of p' $(i = 1, \ldots, n+1)$. Now V accepts iff for each $i = 1, \ldots, n+1$ it is the case that $B'(w, \sigma[i], f^*, p'[i]) = 1$.

We now specify P. Let $w \in L$ and $\bar{w} \in \rho(w)$. Let $n = |w|$, and let $\sigma \in \{0,1\}^{l(n)}$. P runs G to obtain a (random) pair $(f^*, \bar{f}^*) \in [G(1^n)]$. It sets $p'[i] = A'(w, \bar{w}, \sigma[i], f^*, \bar{f}^*)$ for $i = 1, \ldots, n+1$, and sets $p' = p'[1] \ldots p'[n+1]$. Finally it sets $p = f^* \cdot p'$ ("." denotes concatenation) and outputs p. The completeness condition (as required by Definition 6) follows from the completeness condition of Lemma 11.

Next we check the soundness condition. Suppose $w \notin L$. Let $n = |w|$ and let $f^* \in \{0,1\}^n$. Let $\sigma \in \{0,1\}^{l(n)}$. We say that σ is f^*-bad if there exists an $i \in \{1, \ldots, n+1\}$ and a string $q \in \{0,1\}^{t(n)}$ such that $B'(w, \sigma[i], f^*, q) = 1$. The soundness condition of Lemma 11 implies that

$$\Pr\left[\sigma \text{ is } f^*\text{-bad} : \sigma \xleftarrow{R} \{0,1\}^{l(n)} \right] \leq 2^{-(n+1)}.$$

Now let us say that a string $\sigma \in \{0,1\}^{l(n)}$ is bad if there exists an n-bit string f^* such that σ is f^*-bad. It follows that

$$\Pr\left[\sigma \text{ is bad} : \sigma \xleftarrow{R} \{0,1\}^{l(n)}\right] \leq 2^n \cdot 2^{-(n+1)} = \tfrac{1}{2} .$$

This implies the soundness condition (as required by Definition 6).

Finally, we specify the simulator. Let $w \in L$ and let $n = |w|$. On input w, the simulator S runs G on input 1^n to obtain a (random) pair $(f^*, \bar{f}^*) \in [G(1^n)]$. For $i = 1, \ldots, n+1$ it runs M' on input $w, f^*, \bar{f}^*$ to get an output $(\sigma[i], p'[i])$. It sets $\sigma = \sigma[1] \ldots \sigma[n+1]$ and $p' = p'[1] \ldots p'[n+1]$. It then sets $p = f^* . p'$ and outputs (σ, p). The zero-knowledge (as required by Definition 6) can be argued based on the zero-knowledge condition of Lemma 11. We omit the details. ∎

In particular, NIZK proof systems are constructible based on RSA.

Combining Theorem 12 with the result of [NaYu] yields the following.

Corollary 13. *Suppose there exists a trapdoor permutation generator. Then there exists an encryption scheme secure against chosen-ciphertext attack.*

Similarly, combining Theorem 12 with the result of [BeGo] yields the following.

Corollary 14. *Suppose there exists a trapdoor permutation generator. Then there exists an implementation of the signature scheme of [BeGo].*

References

[BeGo]   M. Bellare and S. Goldwasser. *New Paradigms for Digital Signatures and Message Authentication Based on Non-Interactive Zero-Knowledge Proofs.* Advances in Cryptology – CRYPTO 89. Lecture Notes in Computer Science, Vol. 435, Springer Verlag.

[BeMi]   M. Bellare and S. Micali. *How to Sign Given any Trapdoor Permutation.* JACM, Vol. 39, No. 1, January 1992, pp. 214-233. (Preliminary version in Proceedings of the 20th STOC, 1988).

[BMO]   M. Bellare, S. Micali and R. Ostrovsky. *The True Complexity of Statistical Zero-Knowledge.* Proceedings of the 22nd Annual ACM Symposium on the Theory of Computing, 1990.

[BBS]   L. Blum, M. Blum, and M. Shub. *A Simple Unpredictable Pseudo-Random Number Generator.* SIAM Journal on Computing, Vol. 15, No. 2, May 1986, pp. 364-383.

[BDMP]   M. Blum, A. De Santis, S. Micali, and G. Persiano, *Non-Interactive Zero-Knowledge Proof Systems*, SIAM Journal on Computing, Vol. 20, No. 6, December 1991,pp. 1084-1118.

[BFM]   M. Blum, P. Feldman, and S. Micali, *Non-Interactive Zero-Knowledge Proof Systems and Applications,* Proceedings of the 20th Annual ACM Symposium on Theory of Computing, 1988.

[DeYu]   A. De Santis and M. Yung. *Cryptographic Applications of the Metaproof and Many-prover Systems.* Advances in Cryptology – CRYPTO 90. Lecture Notes in Computer Science, Vol. 537, Springer-Verlag.

[FLS]   U. Feige, D. Lapidot, and A. Shamir. *Multiple Non-Interactive Zero-Knowledge based on a Single Random String.* Proceedings of the 31st Annual IEEE Symposium on Foundations of Computer Science, 1990.

[GMW]   O. Goldreich, S. Micali, and A. Wigderson. *Proofs that Yield Nothing but their Validity and a Methodology of Cryptographic Design.* JACM, July 1991. (Preliminary version in the 27th FOCS, 1986).

[GoLe]   O. Goldreich and L. Levin. *A Hard-Core Predicate for all One-Way Functions.* Proceedings of the 21st Annual ACM Symposium on the Theory of Computing, 1989.

[GMR]   S. Goldwasser, S. Micali, and R. Rivest. *A Digital Signature Scheme Secure Against Adaptive Chosen-Message Attacks.* SIAM Journal on Computing, Vol. 17, No. 2, April 1988, pp. 281-308.

[NaYu]   M. Naor and M. Yung. *Public Key Cryptosystems secure against chosen-ciphertext attacks.* Proceedings of the 22nd Annual ACM Symposium on the Theory of Computing, 1990.

[RSA]   R. Rivest, A. Shamir, and L. Adleman. *A Method for Obtaining Digital Signatures and Public-Key Cryptosystems.* Communications of the ACM, Vol. 21, No. 2, February 1978, pp. 120-26.

Protocols for Secret Key Agreement
by Public Discussion
Based on Common Information

Ueli M. Maurer

Institute for Theoretical Computer Science
ETH Zürich
CH-8092 Zürich, Switzerland
Email address: maurer@inf.ethz.ch

Abstract. Consider the following scenario: Alice and Bob, two parties who share no secret key initially but whose goal it is to generate a (large amount of) information-theoretically secure (or unconditionally secure) shared secret key, are connected only by an insecure public channel to which an eavesdropper Eve has perfect (read) access. Moreover, there exists a satelite broadcasting random bits at a very low signal power. Alice and Bob can receive these bits with certain bit error probabilities ϵ_A and ϵ_B, respectively (e.g. $\epsilon_A = \epsilon_B = 30\%$) while Eve is assumed to receive the same bits much more reliably with bit error probability $\epsilon_E \ll \epsilon_A, \epsilon_B$ (e.g. $\epsilon_E = 1\%$). The errors on the three channels are assumed to occur at least partially independently. Practical protocols are discussed by which Alice and Bob can generate a secret key despite the facts that Eve possesses more information than both of them and is assumed to have unlimited computational resources as well as complete knowledge of the protocols.

The described scenario is a special case of a much more general setup in which Alice, Bob and Eve are assumed to know random variables X, Y and Z jointly distributed according to some probability distribution P_{XYZ}, respectively. The results of this paper suggest to build cryptographic systems that are provably secure against enemies with unlimited computing power under realistic assumptions about the partial independence of the noise on the involved communication channels.

1. Introduction

One of the fundamental problems in cryptography is the transmission of a message M from a sender (referred to as Alice) to a receiver (Bob) over an insecure communication channel such that an enemy (Eve) with access to this channel is unable to obtain useful information about M.

In the classical model of a cryptosystem introduced by Shannon [9], Eve has perfect access to the insecure channel; thus she is assumed to receive an identical copy of the ciphertext C received by the legitimate receiver Bob, where C is obtained as a function of the plaintext message M and a secret key K shared by Alice and Bob. Shannon defined a cipher system to be perfect if the ciphertext is statistically independent of the plaintext or, in information-theoretic terms, if the ciphertext gives no information about the plaintext:

$$I(M;C) \ = \ 0.$$

When a perfect cipher is used to encrypt a message M, an enemy can do no better than guess M without even looking at the ciphertext C.

It is assumed that the reader is familiar with the fundamentals of information theory, in particular with the entropy $H(X)$ of a random variable X, the conditional entropy of X given Y, $H(X|Y)$, and the mutual information between X and Y defined as $I(X;Y) = H(X) - H(X|Y)$. We refer to [4] for an introduction to information theory.

Shannon gave as a simple example of a perfect cipher the well-known one-time pad which is completely impractical for most applications where only a short secret key is available. Shannon proved the pessimistic result that perfect secrecy can be achieved only when the secret key is at least as long as the plaintext message or, more precisely, when

$$H(K) \ \geq \ H(M). \tag{1}$$

Almost all presently-used ciphers are based on Shannon's model but have only a short secret key; they can therefore theoretically be broken, for instance by an exhaustive key search. The goal of designing such a practical cipher is to guarantee that there exists no efficient algorithm for breaking it, for a reasonable definition of breaking. However, for no existing cipher can the computational security be proved without invoking an unproven intractability hypothesis.

Perfect secrecy on the other hand is often prejudged as being impractical because of Shannon's pessimistic inequality (1). It is one of the goals of this paper to relativize this pessimism by pointing out that Shannon's apparently innocent assumption that, except for the secret key, the enemy has access to precisely the same information as the legitimate receiver, is much more restrictive than has generally been realized.

The key to perfect secrecy without a shared secret key K satisfying (1) is to modify Shannon's model such that the enemy cannot receive precisely (albeit almost) the same information as the legitimate receiver. Two previous approaches based on this idea are quantum cryptography introduced by Wiesner and put forward by Bennett, Brassard *et al.* [1], and Maurer's randomized cipher [7] which makes use of a public random string that is too long to be read entirely in feasible time. Both these approaches are impractical at present.

Another approach is due to Wyner [11] and subsequently Csiszár and Körner [5] who considered a scenario in which the enemy Eve is assumed to receive messages transmitted by the sender Alice over a channel that is noisier than the legitimate receiver Bob's channel. The assumption that Eve's channel is worse than the main channel is unrealistic in general. The results of this paper demonstrate that this unrealistic assumption is unnecessary if Alice and Bob can also communicate over a completely insecure public channel.

In this paper, the broadcast channel scenario is generalized to a scenario where Alice, Bob and Eve know random variables X, Y and Z, respectively, jointly distributed according to some probability distribution P_{XYZ}, and where Alice and Bob can also communicate over a public channel.

Note that the need for a public channel entails no significant loss of practicality in a cryptographic context because the channel need not provide secrecy. It is assumed, however, that all messages sent over the public channel can be received by Eve without error, but that she cannot modify messages or introduce fraudulent messages without being detected. If this last assumption cannot realistically be made, authenticity and data integrity can be ensured by using an unconditionally secure authentication scheme, for instance that of [10] based on universal hashing, which requires that Alice and Bob share a short secret key initially. In this case, the purpose of our protocols is to stretch (rather than to generate) a secret key unconditionally securely. Part of the generated key can be used for authentication in a subsequent instance of the protocol.

The use of a public channel by two parties for extracting a secret key from an initially shared partially secret string was previously considered by Leung-Yan-Cheong [6] and independently by Bennett, Brassard and Robert [3].

This paper is concerned with key distribution as well as encryption. An unconditionally secure shared secret key generated by one of our protocols can be used as the key sequence in the one-time pad, thus achieving (virtually) perfect secrecy of the transmitted messages.

2. Secret Key Agreement by Public Discussion

Consider the following general key agreement problem. Assume that Alice, Bob and Eve know random variables X, Y and Z, respectively, with joint probability distribution P_{XYZ}, and that Eve has no information about X and Y other than through her knowledge of Z. More precisely, $I(XY;T|Z) = 0$ where T summarizes Eve's complete information about the universe. X, Y and Z take on values in some finite alphabets $\mathcal{X}, \mathcal{Y}$ and $\mathcal{Z}$, respectively. Alice and Bob share no secret key initially (other than possibly a short key required for guaranteeing authenticity and integrity of messages sent over the public channel), but are assumed to know P_{XYZ}. In particular, the protocol and the codes used by Alice and Bob are known to Eve. Every message communicated between Alice and Bob can be intercepted by Eve, but it is assumed that Eve cannot insert fraudulent messages nor modify messages on this public channel without being detected.

Alice and Bob use a protocol in which at each step either Alice sends a message to Bob depending on X and all the messages previously received from Bob, or vice versa (with X replaced by Y). Without loss of generality, we consider only protocols in which Alice sends messages at odd steps $(C_1, C_3, \ldots)$ and Bob sends messages at even steps $(C_2, C_4, \ldots)$. Moreover, we can restrict the analysis to deterministic protocols since a possible randomizer which Alice's and/or Bob's strategy and messages might depend on can be considered as part of X and Y, respectively. In other words, Alice and Bob can without loss of generality extend their known random variables X and Y, respectively, by random bits that are statistically independent of X, Y and Z. At the end of the t-step protocol, Alice computes a key S as a function of X and $C^t \triangleq [C_1, \ldots, C_t]$ and Bob computes a key S' as a function of Y and C^t. Their goal is to maximize $H(S)$ under the conditions that S and S' agree with very high probability and that Eve has very little information about S. More formally,

$$H(C_i|C^{i-1}X) = 0 \tag{2}$$

for odd i,

$$H(C_i|C^{i-1}Y) = 0 \tag{3}$$

for even i,

$$H(S|C^t X) = 0 \tag{4}$$

and

$$H(S'|C^t Y) = 0, \tag{5}$$

and it is required that

$$P[S \neq S'] \leq \epsilon \tag{6}$$

and

$$I(S; C^t Z) \leq \delta \tag{7}$$

for some specified (small) δ and ϵ.

By Fano's Lemma (cf. [4], p. 156) condition (6) implies that

$$H(S|S') \leq h(\epsilon) + \epsilon \log_2(|\mathcal{S}| - 1) \tag{8}$$

where $|\mathcal{S}|$ denotes the number of distinct values that S takes on with non-zero probability. Note that $H(S|S') \to 0$ as $\epsilon \to 0$.

If one requires that $P[S \neq S'] = 0$ and $I(S; C^t) = 0$ (i.e., that $\epsilon = 0$ in (6) and $\delta = 0$ in (7)) it appears obvious that $I(X; Y)$ is an upper bound on $H(S)$. It appears to be similarly obvious that $H(S) \leq I(X; Y|Z) = I(XZ; YZ) - H(Z)$ because even under the assumption that Alice and Bob could learn Z, the remaining information shared by Alice and Bob is an upper bound on the information they can share in secrecy. The following theorem, which is proved in [8], summarizes these results.

Theorem 1. *For every key agreement protocol satisfying (2)-(5),*

$$H(S) \leq I(X; Y|Z) + H(S|S') + I(S; C^t Z).$$

In particular,

$$H(S) \leq I(X; Y) + H(S|S') + I(S; C^t).$$

The following corollary follows from Theorem 1, inequality (8) and from $I(S; C^t) \leq I(S; C^t Z)$. It should be pointed out that $I(X; Y) < I(X; Y|Z)$ is possible.

Corollary 2. *For every key agreement protocol satisfying (2)-(7),*

$$H(S) \leq \min[I(X; Y), I(X; Y|Z)] + \delta + h(\epsilon) + \epsilon \log_2(|\mathcal{S}| - 1).$$

3. The Secret Key Rate

In order to be able to prove lower bounds on the achievable size of a key shared by Alice and Bob in secrecy we need to make more specific assumptions about the distribution P_{XYZ}. One natural assumption is that the random experiment generating XYZ is repeated many times independently: Alice, Bob and Eve receive $X^N = [X_1, \ldots, X_N]$, $Y^N = [Y_1, \ldots, Y_N]$ and $Z^N = [Z_1, \ldots, Z_N]$, respectively, where

$$P_{X^N Y^N Z^N} = \prod_{i=1}^{N} P_{X_i Y_i Z_i}$$

and where $P_{X_i Y_i Z_i} = P_{XYZ}$ for $1 \leq i \leq N$.

For such a scenario of independent repetitions of a random experiment, which is well motivated by models such as discrete memoryless sources and channels previously considered in information theory, the quantity that appears to be of most interest from an information-theoretic point of view is defined below.

Definition. The *secret key rate of X and Y with respect to Z*, denoted $S(X;Y\|Z)$, is the maximum rate at which Alice and Bob can agree on a secret key S while keeping the rate at which Eve obtains information arbitrarily small, i.e., it is the maximal R such that for every $\epsilon > 0$ there exists a protocol for sufficiently large N satisfying (2)-(6) with X and Y replaced by X^N and Y^N, respectively, satisfying

$$\frac{1}{N} I(S; C^t Z^N) \leq \epsilon,$$

and achieving

$$\frac{1}{N} H(S) \geq R - \epsilon.$$

Before deriving lower bounds on $S(X;Y\|Z)$ we state the following theorem, which is an immediate consequence of Corollary 2.

Theorem 3. *The secret key rate of X and Y with respect to Z is upper bounded by*

$$S(X;Y\|Z) \leq \min[I(X;Y), \ I(X;Y|Z)].$$

The following theorem (cf. [8] for a proof) states a nontrivial lower bound on the secret key rate. If it is either the case that Eve has less information about Y than Alice or, by symmetry, less information about X than Bob, then such a difference of information can be exploited.

Theorem 4. *The secret key rate of X and Y with respect to Z is lower bounded by*

$$S(X;Y\|Z) \geq \max[I(Y;X) - I(Z;X), \ I(X;Y) - I(Z;Y)].$$

Theorem 4 demonstrates that the upper bound in Theorem 3 is tight if either $P_{YZ|X} = P_{Y|X} \cdot P_{Z|X}$ or $P_{XZ|Y} = P_{X|Y} \cdot P_{Z|Y}$. The lower bound of Theorem 4 is not tight in general as will be demonstrated in the next section. In particular, the lower bound of Theorem 4 is 0 for the situation described in the abstract of the paper. There exist protocols with several rounds of interaction between Alice and Bob which are superior to single-round protocols like the one used in the proof of Theorem 4 (cf. [8]).

4. Binary Symmetric Random Variables

In this section the case of symmetrically distributed binary random variables is considered. One way of generating such a set X, Y, Z is by generating a random bit R according to

$$P_R(0) = P_R(1) = 1/2 \tag{9}$$

and "sending" R over three *independent* binary symmetric channels C_A, C_B and C_E with error probabilities ϵ_A, ϵ_B and ϵ_E, respectively, i.e., P_{XYZ} is defined by

$$P_{XYZ|R} = P_{X|R} \cdot P_{Y|R} \cdot P_{Z|R} \tag{10}$$

where $P_{X|R}(x, r) = 1 - \epsilon_A$ if $x = r$ and ϵ_A else, $P_{Y|R}(y, r) = 1 - \epsilon_B$ if $y = r$ and ϵ_B else and $P_{Z|R}(z, r) = 1 - \epsilon_E$ if $z = r$ and ϵ_E else.

Consider now an arbitrary probability distribution P_{XYZ} over $\{0, 1\}^3$ satisfying the symmetry condition

$$P_{XYZ}(x, y, z) = P_{XYZ}(\overline{x}, \overline{y}, \overline{z}) \tag{11}$$

for $x, y, z \in \{0, 1\}$, where $\overline{c}$ denotes the complement of a binary variable c. Note that condition (11) implies that X, Y and Z are symmetrically distributed. One can prove (see [8]) that every set X, Y and Z of random variables satisfying (11) and for which not exactly for one of the pairs $[X, Y], [X, Z]$ and $[Y, Z]$ the two random variables are statistically independent, can be generated according to (9) and (10) for some ϵ_A, ϵ_B and ϵ_E.

As one realistic scenario where X, Y and Z with probability distribution P_{XYZ} satisfying (11) are available for two parties and an enemy, consider a satellite broadcasting random bits at a very low signal-to-noise ratio such that even an enemy Eve with a receiving antenna that is much larger and more sophisticated than Alice's and Bob's antenna cannot receive the bits without error. Note that P_{XYZ} satisfies the given condition also when the channels C_A, C_B and C_E are dependent, as one would realistically have to assume. The following theorem has been proved in [8].

Theorem 5. *Let X, Y and Z be binary random variables generated according to (9) and (10). Then*

$$S(X; Y \| Z) \geq \max[h(\epsilon_A + \epsilon_E - 2\epsilon_A \epsilon_E), h(\epsilon_B + \epsilon_E - 2\epsilon_B \epsilon_E)] - h(\epsilon_A + \epsilon_B - 2\epsilon_A \epsilon_B).$$

The lower bound of Theorem 5 vanishes unless either $\epsilon_A < \epsilon_E$ or $\epsilon_B < \epsilon_E$, i.e., unless either Alice's or Bob's channel is superior to Eve's channel. It is somewhat surprising that even when Eve's channel is much more reliable that both Alice's and Bob's channel, secret key agreement is possible.

The proof of Theorem 4 in [8] illustrates that by sending $X_i + V_i$ over the public channel, where X_i is the ith random bit received by Alice and where addition is modulo 2, Alice can send the bit V_i over a conceptual broadcast channel to Bob and Eve such that Bob receives V_i as if it were sent over a cascade of Alice's and Bob's channel (bit error probability $\epsilon_A + \epsilon_B - 2\epsilon_A\epsilon_B$) and Eve receives V_i as if it were sent over a cascade of Alice's and Eve's channel (bit error probability $\epsilon_A + \epsilon_E - 2\epsilon_A\epsilon_E$).

In order to share a secret key with Bob, Alice randomly selects a codeword V^N from the set of codewords of an appropriate error-correcting code $\mathcal{C}$ with codewords of length N and sends it to Bob (and also to Eve) over the described conceptual broadcast channel. The key to achieving a positive secret key rate even if both $\epsilon_A > \epsilon_E$ and $\epsilon_B > \epsilon_E$ is for Bob to accept a received word only if he can make a very reliable decision about the codeword sent by Alice, i.e., if it is very close to some codeword of the code $\mathcal{C}$, i.e., if the Hamming distance to a codeword is much smaller than the number of errors correctable by an optimal decoder for the code. For each received block Bob announces over the public channel whether he accepts or rejects it.

The key observation in the above protocol is that although Eve receives codewords V^N more reliably than Bob on the average, her conceptual channel may nevertheless be worse (for appropriate choices of a code $\mathcal{C}$ and for an appropriate reliability decision) than Bob's channel, if one averages only over those instances accepted by Bob. Because consecutive uses of the channel are independent, the words discarded by Bob are also useless for Eve.

The special case of a repeat code was considered in [8]. Alice sends each bit N times over the conceptual channel, and Bob accepts a received word if and only if all the bits are equal. Although this scheme demonstrates that secret key agreement is possible even if $\epsilon_A > \epsilon_E$ and $\epsilon_B > \epsilon_E$, it is extremely inefficient when ϵ_E is considerably smaller than both ϵ_A and ϵ_B. The reason is that in order to arrive at a situation where Bob's channel is better than Eve's channel if averaged over those instances accepted by Bob, a large block length N must be used in which case the probability that no error occurs within a block and thus the block is accepted by Bob can be extremely small. It is one of the purposes of this paper to describe protocols that are much more efficient than the protocol discussed in [8].

An important observation towards improving the key agreement rate is that several rounds of a protocol as described above can be used by Alice and Bob to continuously increase the reliability of the shared string at the expense of shrinking it. In a first step, and even in some subsequent steps, it is not required that Bob knows Alice's bits more reliably than Eve; it is sufficient that Eve's advantage is reduced in every step. Hence using several protocol steps with short blocks allows to achieve comparable bit error probabilities for the finally shared string as if a long repeat code were used, but with a much larger rate.

Consider as an example a simple $N = 3$ repeat code. Bob accepts a received block of length 3 if and only if all three bits agree, and announces which blocks he accepts. The probability of accepting a block is $\geq 1/4$; hence the strings held by Alice and Bob are shrunk by this step by at most a factor 12. Alice and Bob can use the same step on the resulting string repeatedly, each time decreasing its length by at most a factor 12 while increasing the bit agreement probability. It is straight-forward to verify that when k steps are used, Bob's and Eve's bit error probabilities when guessing the bits of Alice's final string are precisely the same as if a repeat code of length 3^k had been used in the above described basic protocol, but that the expected rate at which random secret key bits are extracted is exponentially larger in the new protocol.

Example. Let $\epsilon_A = \epsilon_B = 0.47$ and let Eve's channel be 100 times less noisy, i.e., have 100 times greater capacity. From $1 - h(\epsilon_E) = 100 \cdot (1 - h(\epsilon_A))$ we obtain $\epsilon_E = 0.2093$. A repeat code of length 243 yields bit error probabilities 0.148 and 0.193 for Bob and Eve, but the probability that a block is accepted by Bob is not significantly larger than 2^{-242}. On the other hand, 5 consecutive applications of the described step with a code of length 3 allow to achieve the same bit error probabilities, but only an expected number of at most $12^5 < 250.000$ (actually much less) bits are required for generating one bit shared with the mentioned bit error probabilities.

Of course, additional protocol steps are required for exploiting the advantage over Eve achieved by this protocol and reducing the bit error probability of the final shared string. For example, error correcting codes can be used to remove the errors between Alice's and Bob's string, and universal hashing as described in [3] can be used to reduce Eve's information.

It should be pointed out that for given assumed ratios of the noise power on the three channels, the signal power is a free parameter; thus ϵ_A can be chosen arbitrarily. The larger ϵ_A, the smaller is the signal power and hence the larger can the satelite's bit transmission rate be chosen.

The use of repeat codes as described above, and more generally of linear error-correcting codes, is equivalent to the exchange of parity checks of the stored string over the public channel, without generating and encoding random bits, and using as a new string some orthogonal parity checks. Reconciliation protocols based on the exchange of parity checks were also discussed in [2].

A further improvement over the basic use of repeat codes described above is for Bob to also accept instances for which a decision about the bit sent by Alice is less reliable than if N identical bits were received. In such a scenario, Bob informs Alice (and Eve) about the number of errors he has received in a block, assuming that his majority decision is correct.

References

[1] C.H. Bennett, F. Bessette, G. Brassard, L. Salvail and J. Smolin, Experimental quantum cryptography, *Journal of Cryptology*, Vol. 5, No. 1, 1992, pp. 3-28.

[2] C.H. Bennett, G. Brassard and J.-M. Robert, How to reduce your enemy's information, *Advances in Cryptology – Crypto '85*, Springer Verlag, New York, pp. 468-476.

[3] C.H. Bennett, G. Brassard and J.-M. Robert, Privacy amplification by public discussion, *SIAM Journal on Computing*, Vol. 17, No. 2, 1988, pp. 210-229.

[4] R.E. Blahut, *Principles and Practice of Information Theory*, Reading, MA: Addison-Wesley, 1987.

[5] I. Csiszár and J. Körner, Broadcast channels with confidential messages, *IEEE Transactions on Information Theory*, Vol. 24, No. 3, 1978, pp. 339-348.

[6] S.K. Leung-Yan-Cheong, Multi-user and wiretap channels including feedback, Tech. Rep. No. 6603-2, Stanford University, Information Systems Lab., July 1976.

[7] U.M. Maurer, Conditionally-perfect secrecy and a provably-secure randomized cipher, *Journal of Cryptology*, Vol. 5, No. 1, 1992, pp. 53-66.

[8] U.M. Maurer, Secret key agreement by public discussion from common information, to appear in *IEEE Transactions on Information Theory*.

[9] C.E. Shannon, Communication theory of secrecy systems, *Bell System Technical Journal*, Vol. 28, Oct. 1949, pp. 656-715.

[10] M.N. Wegman and J.L. Carter, New hash functions and their use in authentication and set equality, *Journal of Computer and System Sciences*, Vol. 22, 1981, pp. 265-279.

[11] A.D. Wyner, The wire-tap channel, *Bell System Technical Journal*, Vol. 54, No. 8, 1975, pp. 1355-1387.

Perfectly-Secure Key Distribution for Dynamic Conferences

Carlo Blundo[1,*], Alfredo De Santis[1,*], Amir Herzberg[2],
Shay Kutten[2], Ugo Vaccaro[1,*], Moti Yung[2]

[1] Dipartimento di Informatica ed Applicazioni, Università di Salerno, 84081 Baronissi
(SA), Italy.
[2] IBM T.J. Watson Research Center, Yorktown Heights, NY 10598, USA.

Abstract. A key distribution scheme for dynamic conferences is a method by which initially an (off-line) trusted server distributes private individual pieces of information to a set of users. Later any group of users of a given size (a dynamic conference) is able to compute a common secure key. In this paper we study the theory and applications of such perfectly secure systems. In this setting, *any* group of t users can compute a common key by each user computing using only his private piece of information and the *identities* of the other $t-1$ group users. Keys are secure against coalitions of up to k users, that is, even if k users pool together their pieces they cannot compute anything about a key of any t-size conference comprised of other users.

First we consider a non-interactive model where users compute the common key without any interaction. We prove a lower bound on the size of the user's piece of information of $\binom{k+t-1}{t-1}$ times the size of the common key. We then establish the optimality of this bound, by describing and analyzing a scheme which *exactly* meets this limitation (the construction extends the one in [2]). Then, we consider the model where interaction is allowed in the common key computation phase, and show a *gap* between the models by exhibiting an interactive scheme in which the user's information is only $k+t-1$ times the size of the common key. We further show various applications and useful modifications of our basic scheme. Finally, we present its adaptation to network topologies with neighborhood constraints.

1 Introduction

Key distribution is a central problem in cryptographic systems, and is a major component of the security subsystem of distributed systems, communication systems, and data networks. The increase in bandwidth, size, usage, and applications of such systems is likely to pose new challenges and to require novel

* Partially supported by Italian Ministry of University and Research (M.U.R.S.T.) and by National Council for Research (C.N.R.) under grant 91.02326.CT12.

ideas. A growing application area in networking is "conferencing" a group of entities (or network locations) collaborate privately in an interactive procedure (such as: board meeting, scientific discussion, a task-force, a classroom, or an bulletin-board). In this work we consider perfectly-secure key distribution for conferences. (Note that key distribution for two-party communication (session-keys) is a special case of conferences of size two).

If users of a group (a conference) wish to communicate in a network using symmetric encryption, they must share a common key. A key distribution scheme (denoted KDS for short) is a method to distribute initial private pieces of information among a set of users, such that each group of a given size (or up to a given size) can compute a common key for secure conference. This information is generated and distributed by a trusted server which is active only at the distribution phase.

Various key distribution schemes have been proposed so far, mainly to pairs of users (session keys). A basic and straightforward perfectly-secure scheme (which is useful in small systems) consists of distributing initial keys to users in such a way that each potential group of users shares a common key. In the case of session keys, if n is the number of users, the server has to generate $n(n-1)/2$ keys and each user holds $n-1$ keys, one for each possible communication. When n gets large it becomes problematic or even impossible to manage all keys. This is known as the n^2 problem. For conferences, when we allow all possible subsets of a given size to join together (what we call the dynamic conference setting), the number of keys becomes prohibitively large.

Given the high complexity of such a distribution mechanism, a natural step is to trade complexity for security. We may still require that keys are perfectly secure, but only with respect to an adversary controlling coalitions of a limited size. This novel approach was initiated by Blom [2] for the case of session keys (other related schemes are given in [10, 14]). We are motivated by Blom's (somewhat forgotten) pioneering work. We consider key-distribution for dynamic conferences and study the theory and applications of such systems. Our scheme has two parameters: t, the size of the conference (group), and k, the size of adversary coalitions. Another characteristic of such schemes is whether they are interactive (users discuss during common-key establishment phase) or non-interactive.

1.1 The results

We give a precise model of our setting and then we analyze and design perfectly-secure key distribution schemes for dynamic conferences. We show the following:

1. **Lower bound**: We consider the non-interactive model and prove that the size of the piece of a user's information is at least $\binom{k+t-1}{t-1}$ times the size of the common key.
2. **Matching upper bound**: We propose a concrete scheme and show that it indeed gives pieces of this size, thus establishing the optimality of the bound.
3. **Gap**: We compare the interactive to the non-interactive settings. We show an interactive scheme where the user's information is only $k+t-1$ times the

size of the common key, proving a separation between the interactive and the non-interactive cases.

4. **Constrained Conferencing**: In Section 7 we present modifications of the schemes to systems in which conferences are generated according to neighborhood constraints (of the network communication graph).

5. **Applications**: We then extend the ideas to show numerous applications and uses of the scheme, such as: hierarchical key distribution schemes, asymmetric user-population, access-control validation, partial key revocation, etc.

Our analysis applies information-theory and its basic notions of entropy and mutual information, as well as their conditional versions. In Section 2 we review these notions and present basic equations to be used in the analysis.

1.2 Related work

The two common approaches to key distribution, taken in order to reduce the inherent complexity of the basic straightforward scheme are schemes based on public-key cryptography [5] or on an authentication server [19]. Numerous suggestions for key distribution schemes based on computational assumptions are known, as well as a number of suggestions for conference keys. We note that "Merkle's puzzles" [17] is also a pioneering key generation scheme which is computational, for a *seemingly negative* result concerning such methods see [11]. The interactive model is related to (but different from) the recent models basing perfectly-secure common key generation on an initial card deal [6, 7]. Blom's innovative method (and thus our setting) is a key distribution which is ID-based that predated the formal definition of this notion by Shamir [21]; his technical tool was MDS linear codes. Later, Matsumoto and Imai [16] extended the work of [2] to general symmetric functions, and systematically defined key distribution schemes based on such general function; our scheme can actually be viewed as a special case of their general system. (Another related recent work is in [23]). Fiat and Naor have suggested recently a key distribution scheme which is not algebraic, and Alon has given a lower bound for their scheme [18]. Remark: finally we note that various suggestions for computational key distribution in different settings (e.g., [15, 20, 25, 24, 8]) and conferencing (e.g., [12, 3, 22]) have appeared in the last years, (mainly in the Crypto and Eurocrypt conferences proceedings series).

Organization: In Section 2 we recall the definition of the entropy and some of its property. In Section 3 we formally describe the model of a KDS in terms of the entropy. In Section 4 we prove the lower bound on the entropy of each user in a k-secure t-conference KDS. In Section 5 we then describe and analyze the actual schemes for k-secure t-conference KDS. In Section 6 we show how interaction can be used to dramatically decrease the amount of information held by each user. In Section 7 we present another result: a protocol to realize a conference KDS when not all of pairs of users are able to communicate. In Section 8 we present applications, in particular the scheme can be combined with authentication procedures, as the ID of the owner and other meaning attached to a key owner can be naturally supported by such a system.

2   Background

In this part we review the information theoretic concepts we are going to use. For a complete treatment of the subject the reader is advised to consult [4] and [9].

Given a probability distribution $\{p(x)\}_{x \in X}$ on a set X, we define the *entropy* of X, $H(X)$, as

$$H(X) = -\sum_{x \in X} p(x) \log p(x)^2.$$

The entropy $H(X)$ is a measure of the average information content of the elements in X or, equivalently, a measure of the average uncertainty one has about which element of the set X has been chosen when the choices of the elements from X are made according to the probability distribution $\{p(x)\}_{x \in X}$. It is well known that $H(X)$ is a good approximation to the average number of bits needed to faithfully represent the elements of X. The following property of $H(X)$ can somehow illustrate the soundness of our first claim:

$$0 \leq H(X) \leq \log |X|, \tag{1}$$

where $H(X) = 0$ if and only if there exists $x_0 \in X$ such that $p(x_0) = 1$; $H(X) = \log |X|$ if and only if $p(x) = 1/|X|, \forall x \in X$.

Given two sets X and Y and a joint probability distribution $\{p(x, y)\}_{x \in X, y \in Y}$ on their cartesian product, the *conditional entropy* $H(X|Y)$, also called the equivocation of X given Y, is defined as

$$H(X|Y) = -\sum_{y \in Y} \sum_{x \in X} p(y)p(x|y) \log p(x|y).$$

The conditional entropy can be written as $H(X|Y) = \sum_{y \in Y} p(y) H(X|Y = y)$ where $H(X|Y = y) = -\sum_{x \in X} p(x|y) \log p(x|y)$ can be interpreted as the average uncertainty one has about which element of X has been chosen when the choices are made according to the probability distribution $p(x|y)_{x \in X}$, that is, when it is known that the value chosen ¿from the set Y is y. From the definition of conditional entropy it is easy to see that

$$H(X|Y) \geq 0. \tag{2}$$

If we have $n + 1$ sets $X_1, \ldots, X_n, Y$ the entropy of $X_1 \ldots X_n$ given Y can be written as

$$H(X_1 \ldots X_n|Y) = H(X_1|Y) + H(X_2|X_1Y) + \cdots + H(X_n|X_1 \ldots X_{n-1}Y) \tag{3}$$

---

[2] All logarithms in this paper are of base 2

The *mutual information* between X and Y is defined by

$$I(X;Y) = H(X) - H(X|Y) \tag{4}$$

and enjoys the following properties:

$$I(X;Y) = I(Y;X), \tag{5}$$

and $I(X;Y) \geq 0$, from which one gets

$$H(X) \geq H(X|Y) \tag{6}$$

with equality if and only if X and Y are independent. Given sets X, Y, Z and a joint probability distribution on their cartesian product, the *conditional mutual information* between X and Y given Z can be written as

$$I(X;Y|Z) = H(X|Z) - H(X|Z\,Y). \tag{7}$$

Since a property of the conditional mutual information is $I(X;Y|Z) \geq 0$ we get

$$H(X|Z) \geq H(X|Z\,Y). \tag{8}$$

3   The Model

In this section we present the key distribution problem and model. A key distribution scheme (indicated by KDS for short) distributes some information among a set of users, so that any t of them can join and generate a secure key. We assume a trusted off-line server active only at initiation (unlike an on-line server approach put forth in [19] which we call server-based KDS). We say the system is k-secure if any k users, pooling together their pieces, have no information on keys they should not know. These schemes can be further classified into two categories: interactive (where users are engaged in a protocol, prior to usage of the common key), and non-interactive where keys are generated privately by the individuals. Next, we formally define non-interactive key distribution schemes. Our definition of security is based on the notion of entropy and is thus unconditional.

Let $\mathcal{U} = \{U_1, \ldots, U_n\}$ be a set of users. The algorithm used by the server to generate the pieces of information, that will be distributed to the users, is randomized. The server generates the vector $(u_1, u_2, \ldots, u_n)$ according to some probability distribution on the cartesian product $U_1 \times \cdots \times U_n$. The piece u_i denotes the information given by the server to user U_i. In order to simplify notation we denote by U_i both the user U_i and the random variable induced by the value u_i, and by $S_{i_1,\ldots,i_t}$ we denote both the set of common keys among users $U_{i_1}, \ldots, U_{i_t}$ and the random variable induced by these common keys. Each user U_{i_j} can deterministically compute, on input only u_{i_j} and $i_1, \ldots, i_{j-1}, i_{j+1}, \ldots, i_t$, his common keys $s_{\sigma(i_1),\ldots,\sigma(i_t)}$, for all permutations $\sigma : \{i_1, i_2, \ldots, i_t\} \rightarrow \{i_1, i_2, \ldots, i_t\}$, to be used with users $U_{i_1}, \ldots, U_{i_{j-1}}, U_{i_{j+1}}, \ldots, U_{i_t}$. Each common key $s_{i_1,\ldots,i_t}$ is generated according to a probability distribution $\{p(s_{i_1,\ldots,i_t})\}_{i_1,\ldots,i_t}$, induced by the fact that each user calculates deterministically the common key by using the initial information received ¿from the server, which has been generated by a randomized algorithm. The probability $p(s_{i_1,\dots,i_t})$ denotes the *a priori* probability that the common key among users $U_{i_1},\dots,U_{i_t}$ is $s_{i_1,\dots,i_t}$.

The maximum value that the security parameter k can take in any t-conference KDS for n users is $n-t$ since any adversary coalition can contain at most $n-t$ users. Formally we define a k-secure t-conference key distribution scheme for n users as follows.

Definition 3.1 *Let* $\mathcal{U}$ *be a set of users and let* k, $k \le |\mathcal{U}| - t$, *be an integer. A non-interactive key distribution scheme for* $\mathcal{U}$ *is* k-*secure if*

1. *Each* t-*uple of users can non-interactively compute the common key.*
   For all $U_{i_1},\dots,U_{i_t} \in \mathcal{U}$, it holds $p(s_{i_1,\dots,i_t}|u_{i_1}) = \cdots = p(s_{i_1,\dots,i_t}|u_{i_t}) = 1$.
2. *Any group of* k *users have no information on a key they should not know.*
   For all $U_{i_1},\dots,U_{i_t}, U_{j_1},\dots,U_{j_k} \in \mathcal{U}$ such that $j_1,\dots,j_k \notin \{i_1,\dots,i_t\}$, it holds
   $$p(s_{i_1,\dots,i_t}|u_{j_1},\dots,u_{j_k}) = p(s_{i_1,\dots,i_t}).$$

Property 1. means that given the value held by the user U_{i_l}, $l = 1,2,\dots,t$, a unique value of the common key exists. Property 2. states that the probability that the common key among users $U_{i_1},\dots,U_{i_t}$ is $s_{i_1,\dots,i_t}$ given the information held by users $U_{j_1},\dots,U_{j_k}$ is equal to the *a priori* probability that the common key is $s_{i_1,\dots,i_t}$. This means that random variables $S_{i_1,\dots,i_t}$ and $U_{j_1} \times \cdots \times U_{j_k}$ are statistically independent, so the values $u_{j_1},\dots,u_{j_k}$ reveal no information on the common key $s_{i_1,\dots,i_t}$. By using the entropy function it is possible to give an equivalent definition of a k-secure non-interactive t-conference KDS.

Definition 3.2 *Let* $\mathcal{U} = \{U_1,\dots,U_n\}$ *be a set of users and let* k, $k \le n - t$, *be an integer. A non-interactive* t-*conference key distribution scheme for* $\mathcal{U}$ *is* k-*secure if*

1'. *Each* t *users can non-interactively compute the common key.*
   For all different $i_1,\dots,i_t \in \{1,2,\dots,n\}$, $H(S_{i_1,\dots,i_t}|U_{i_1}) = \cdots = H(S_{i_1,\dots,i_t}|U_{i_t}) = 0$.
2'. *Any group of* k *users have no information on a key they should not know.*
   For all users $U_{j_1},\dots,U_{j_k}$ such that $j_1,\dots,j_k \notin \{i_1,\dots,i_t\}$, $H(S_{i_1,\dots,i_t}|U_{j_1}\dots U_{j_k}) = H(S_{i_1,\dots,i_t})$.

Notice that $H(S_{i_1,\dots,i_t}|U_{i_1}) = \cdots = H(S_{i_1,\dots,i_t}|U_{i_t}) = 0$, for all different $i_1,\dots,i_t \in \{1,2,\dots,n\}$, means that each set of values held by the user U_{i_l}, $l = 1,2,\dots,t$, corresponds to a unique value of the common key. In fact, by definition, $H(S_{i_1,\dots,i_t}|U_{i_l}) = 0$ is equivalent to the fact that for all $u_{i_l} \in U_{i_l}$ with $p(u_{i_l}) > 0$, a unique value $s_{i_1,\dots,i_t} \in S_{i_1,\dots,i_t}$ such that $p(s_{i_1,\dots,i_t}|u_{i_l}) = 1$ exists. Moreover, $H(S_{i_1,\dots,i_t}|U_{j_1}\dots U_{j_k}) = H(S_{i_1,\dots,i_t})$ is equivalent to saying that $S_{i_1,\dots,i_t}$ and $U_{j_1} \times \cdots \times U_{j_k}$ are statistically independent, i.e., for all $(u_{j_1},\dots,u_{j_k}) \in U_{j_1} \times \cdots \times U_{j_k}$, we have $p(s_{i_1,\dots,i_t}|u_{j_1},\dots,u_{j_k}) = p(s_{i_1,\dots,i_t})$.

Property 1'. in Definition 3.2 states that any t users can compute the same common key. Actually, each user U_i can calculate $t!$ keys for the same conference. Property 1'. does not say anything on the relationship among these $t!$ keys: all $t!$ keys could be equal so one key uniquely determines the other keys, that is $H(S_{\sigma(i_1),\ldots,\sigma(i_t)}|S_{i_1,\ldots,i_t}) = 0$, for all permutation $\sigma : \{i_1, i_2, \ldots, i_t\} \rightarrow \{i_1, i_2, \ldots, i_t\}$; or the keys could be all different and given one key we do not know anything on the other keys, that is $H(S_{\sigma(i_1),\ldots,\sigma(i_t)}|S_{i_1,\ldots,i_t}) = H(S_{\sigma(i_1),\ldots,\sigma(i_t)})$. Our lower bounds are valid in both cases, since they are based only on Property 1'. and 2'.. On the other hand, in this paper all schemes that realize k-secure t-conference KDS are symmetric, that is schemes in which the common key is symmetric: $s_{i_1,\ldots,i_t} = s_{\sigma(i_1),\ldots,\sigma(i_t)}$ for all permutations $\sigma : \{i_1, i_2, \ldots, i_t\} \rightarrow \{i_1, i_2, \ldots, i_t\}$.

Definition 3.2 does not say anything on the entropies of random variables $S_{i_1,\ldots,i_t}$ and $S_{i'_1,\ldots,i'_t}$. For example, we could have either $H(S_{i_1,\ldots,i_t}) > H(S_{i'_1,\ldots,i'_t})$ or $H(S_{i_1,\ldots,i_t}) \leq H(S_{i'_1,\ldots,i'_t})$. Our results apply for the general case of arbitrary entropies on keys, but for clarity we often state our results for the simpler case that all entropies on keys are equal, i.e. $H(S_{i_1,\ldots,i_t}) = H(S_{i'_1,\ldots,i'_t})$ for all t-uples of users $(U_{i_1}, \ldots, U_{i_t})$ and $(U_{i'_1}, \ldots U_{i'_t})$, and we denote this entropy by $H(S)$.

The next simple lemma proves that if a t-conference KDS is k-secure then it is k'-secure for all integers $k' < k$.

Lemma 3.1 *Let $\mathcal{U} = \{U_1, \ldots, U_n\}$ be a set of users and let k, $k \leq n - t$, be an integer. In any k-secure key distribution scheme for $\mathcal{U}$, for any integer $k' < k$ it holds*

$$H(S_{i_1,\ldots,i_t}|U_{j_1}\ldots U_{j_{k'}}) = H(S_{i_1,\ldots,i_t}).$$

For all users $U_{i_1}, \ldots, U_{i_t}, U_{j_1}, \ldots, U_{j_{k'}}$ such that $j_1, \ldots, j_{k'} \notin \{i_1, \ldots, i_t\}$.

Proof : From 2'. of Definition 3.2 we have $H(S_{i_1,\ldots,i_t}) = H(S_{i_1,\ldots,i_t}|U_{j_1}\ldots U_{j_k})$. From (8), one gets

$$H(S_{i_1,\ldots,i_t}|U_{j_1}\ldots U_{j_k}) \leq H(S_{i_1,\ldots,i_t}|U_{j_1}\ldots U_{j_{k'}}) \leq H(S_{i_1,\ldots,i_t}).$$

Thus, $H(S_{i_1,\ldots,i_t}|U_{j_1}\ldots U_{j_{k'}}) = H(S_{i_1,\ldots,i_t})$. $\qquad\qquad\Box$

From Lemma 3.1 one has that Property 2'. can be equivalently written as

2''. *Any group of $k' \leq k$ users have no information on a key they should not know.*
   For all users $U_{i_1}, \ldots, U_{i_t}, U_{j_1}, \ldots, U_{j_{k'}}$ such that $j_1, \ldots, j_{k'} \notin \{i_1, \ldots, i_t\}$, it holds
   $H(S_{i_1,\ldots,i_t}|U_{j_1}\ldots U_{j_{k'}}) = H(S_{i_1,\ldots,i_t})$.

4 Lower Bound: Conference Key Distribution

In this section we prove a lower bound on the size of user's information for a k-secure t-conference KDS. Let $U_{i_1} \ldots U_{i_t}$ be t users and let $A = \{j_1, \ldots, j_t\}$ be a set of t indices. With S_A we denote both the set of common keys among the users $U_{i_1} \ldots U_{i_t}$ and the random variable induced by these common keys, and with U_A we denote both the set of users $\{U_{i_1}, \ldots, U_{i_t}\}$ and the random variable induced by the value $u_{i_1}, \ldots, u_{i_t}$.

In a k-secure t-conference KDS the knowledge of k keys does not convey any information on another key. This is formalized by next lemma.

Lemma 4.1 *Let* $\mathcal{U} = \{U_1, \ldots, U_n\}$ *be a set of n users and let r and k, $k \leq n-t$, be integers. Let* $X, Y_1, \ldots, Y_r, Z$ *be subsets of* $\{1, 2, \ldots, n\}$ *such that* $|Z| = k$, $Z \cap X = \emptyset$, $Z \cap Y_i \neq \emptyset$ *and* $|X| = |Y_i| = t$, *for* $i = 1, \ldots, r$. *Then, in any k-secure t-conference key distribution scheme for* $\mathcal{U}$

$$H(S_X | S_{Y_1} \ldots S_{Y_r}) = H(S_X).$$

Proof : From (6) we have $H(S_X) \geq H(S_X | S_{Y_1} \ldots S_{Y_r})$. To prove the lemma it is enough to prove that $H(S_X | S_{Y_1} \ldots S_{Y_r}) \geq H(S_X)$. Note that $Z \cap X = \emptyset$.

First note that the conditional mutual information between S_X and $S_{Y_1} \ldots S_{Y_r}$ given U_z is

$$
\begin{aligned}
I(S_{Y_1} \ldots S_{Y_r} ; S_X | U_z) &= H(S_{Y_1} \ldots S_{Y_r} | U_z) - H(S_{Y_1} \ldots S_{Y_r} | U_z S_X) \text{ (from (7))} \\
&\leq H(S_{Y_1} \ldots S_{Y_r} | U_z) \text{ (from (2))} \\
&\leq \sum_{l=1}^{r} H(S_{Y_l} | U_z) \text{ (from (3) and (8))} \\
&\leq 0 \text{ (from (8) and 1'. of Definition 3.2)}
\end{aligned}
$$

Since the mutual information is non-negative we have

$$I(S_{Y_1} \ldots S_{Y_r} ; S_X | U_z) = 0$$

From (5) it follows $I(S_X ; S_{Y_1} \ldots S_{Y_r} | U_z) = I(S_{Y_1} \ldots S_{Y_r} ; S_X | U_z)$ and thus

$$H(S_X | U_z) = H(S_X | U_z S_{Y_1} \ldots S_{Y_r}). \tag{9}$$

Finally, one gets

$$
\begin{aligned}
H(S_X | S_{Y_1} \ldots S_{Y_r}) &\geq H(S_X | U_z S_{Y_1} \ldots S_{Y_r}) \text{ (from (6))} \\
&= H(S_X | U_z) \text{ (from (9))} \\
&= H(S_X) \text{ (from 2'. of Definition 3.2)}
\end{aligned}
$$

which proves the lemma. $\qquad\qquad\square$

We assume that all keys have the same entropy, i.e. $H(S_{j_1,\dots,j_t}) = H(S)$ for all different $j_1,\dots,j_t$. Next theorem states a lower bound on the size of information held by each user.

Theorem 4.1 *Let $\mathcal{U}$ be a set of n users and let k, $k \le n - t$, be an integer. In any k-secure t-conference key distribution scheme, the entropy $H(U_i)$ of each user U_i satisfies*

$$H(U_i) \ge \binom{k+t-1}{t-1} H(S).$$

Proof : Consider the set of indices $I = \{j_1,\dots,j_{k+t-1}\}$ and an index i such that $i \notin I$. Let $m = \binom{k+t-1}{t-1} - 1$. Construct $A, B_1,\dots,B_m, C$ as follows. Set C is equal to $C = \{j_1,\dots,j_k\}$, set A is equal to $A = \{i, j_{k+1},\dots,j_{k+t-1}\}$, and, finally, set B_l, for $l = 1,\dots,m$ is constructed taking the element i along with any $(t-1)$ elements from the set I, with the exception of $\{j_{k+1},\dots,j_{k+t-1}\}$, that is,

$$B_l \in \Big\{ \{i, x_1,\dots,x_{t-1}\} | x_1,\dots,x_{t-1} \in I, \ \{x_1,\dots,x_{t-1}\} \ne \{j_{k+1},\dots,j_{k+t-1}\} \Big\}.$$

We have

$$H(U_i) = H(S_{B_1}\dots S_{B_m} S_A) - H(S_{B_1}\dots S_{B_m} S_A | U_i) + H(U_i | S_{B_1}\dots S_{B_m} S_A)$$
$$\text{(from (4) and (5))}$$

$$\ge H(S_{B_1}\dots S_{B_m} S_A) - \sum_{l=1}^{m} H(S_{B_l}|U_i) - H(S_A|U_i) + H(U_i|S_{B_1}\dots S_{B_m} S_A)$$
$$\text{(from (3) and (8))}$$

$$= H(S_{B_1}\dots S_{B_m} S_A) + H(U_i|S_{B_1}\dots S_{B_m} S_A) \quad \text{(from 1'. of Definition 3.2)}$$
$$\ge H(S_{B_1}\dots S_{B_m} S_A) \quad \text{(from (2))}$$
$$= H(S_{B_1}) + H(S_{B_2}|S_{B_1}) + \dots + H(S_{B_m}|S_{B_1}\dots S_{B_{m-1}}) + H(S_A|S_{B_1}\dots S_{B_m})$$
$$\text{(from (3))}$$

Sets $Z = A$, $X = C$, $Y_l = B_l$ for $l = 1,\dots,m$ satisfy the hypothesis of Lemma 4.1. Thus we have $H(S_A|S_{B_1}\dots S_{B_m}) = H(S_A)$. Moreover, for each h, $1 \le h \le m$, sets $X = B_h$, $Z = I \setminus B_h$ and $Y_l = B_l$, for $l = 1,\dots,h-1$, satisfy the hypothesis of Lemma 4.1. Thus, $H(S_{B_h}|S_{B_1}\dots S_{B_{h-1}}) = H(S_{B_h})$ and,

$$H(U_i) \ge H(S_{B_1}) + H(S_{B_2}) + \dots + H(S_{B_m}) + H(S_A)$$
$$= (m+1)H(S)$$
$$= \binom{k+t-1}{t-1} H(S)$$

Hence the theorem follows. $\quad\square$

A particular case of Theorem 4.1 is when $t = 2$ and $k = n - 2$. In this case the key of a pair of users cannot be computed (even one of its bits cannot be computed) by an adversary coalition of the other $n - 2$ users. Each user holds at least $n - 1$ pieces of information of size equal to the size of the common key. The total number of pieces of information held by all users is at least $n(n - 1)$. This is the well know problem of n^2 keys. The bound $H(U_i) \geq \binom{k+t-1}{t-1} H(S)$ is achieved by the protocol we next propose.

5 Protocols for Key Distribution

In this section we design and analyze protocols for k-secure t-conference key distribution which are applicable to hierarchical KDS as well (as will be later explained). The scheme we propose when applied to 2-party KDS is a particular case of the Blom's scheme [2] based on MDS linear codes, and, in particular based on polynomials.

Blom's protocol for a k-secure (2-conference) KDS for n users is as following. Let G be a (publicly known) generator matrix of a $(n, k + 1)$ MDS linear code over $GF(q)$ (see [13] for definitions and analysis of such codes) and let D be a secret random matrix with elements in $GF(q)$. From the matrices G and D, construct a $n \times n$ symmetric matrix K whose entries will be the users' keys. The matrix K is equal to $K = (DG)^T G$. The information given to user U_i consists of the row i of $(DG)^T$. If user U_i wants to communicate with user U_j then he computes the inner product of the held vector with the column j of G and he obtains the common key $s_{i,j} = K(i, j)$.

We propose the following protocol (to be extendible to various other applications in the sequel) for a k-secure t-conference KDS. Let $P(x_1, \ldots, x_t)$ be a symmetric polynomial in t variables of degree k with coefficients over $GF(q)$, $q > n$, that is, $P(x_1, \ldots, x_t) = P(x_{\sigma(1)}, \ldots, x_{\sigma(t)})$ for all permutations $\sigma : \{1, 2, \ldots, t\} \to \{1, 2, \ldots, t\}$. To each user U_i the server gives the polynomial $f_i(x_2, \ldots, x_t) = P(i, x_2 \ldots, x_t)$, that is the polynomial obtained by evaluating $P(x_1, \ldots, x_t)$ at $x_1 = i$. If users $U_{j_1}, \ldots, U_{j_t}$ want to set up a conference key then each user U_{j_i} evaluates $f_{j_i}(x_2, \ldots, x_t)$ at $(x_2, \ldots, x_t) = (j_1, \ldots, j_{i-1}, j_{i+1}, \ldots, j_t)$. The conference key is equal to $s_{j_1, \ldots, j_t} = P(j_1, \ldots, j_t)$.

As we mentioned above, when $t = 2$ our scheme is a particular case of Blom's scheme. Indeed, the generator matrix G of the MDS code is constructed by setting the entry $G(i, j)$ to j^{i-1}.

Theorem 5.1 *In the scheme based on symmetric polynomial, if all coefficients of the symmetric polynomial in t variables of degree k are uniformly chosen in $GF(q)$, then the t-conference key distribution scheme is k-secure, and optimal.*

The scheme proposed meets the bound provided by Theorem 4.1, when all coefficients are uniformly chosen. Indeed, in a symmetric polynomial $P(x_1, \ldots, x_r)$ the coefficient $a_{i_1, \ldots, i_r}$ is equal to $a_{\sigma(i_1), \ldots, \sigma(i_r)}$, for all permutations $\sigma : \{i_1, i_2, \ldots, i_r\} \to \{i_1, i_2, \ldots, i_r\}$. Thus, the number of coefficients of a symmetric polynomial in r variables of degree k is equal to the number of possible ways of choosing with repetitions r elements (corresponding to indices $i_1, \ldots, i_r$) from a set of $k+1$ elements (each i_j can assume $k+1$ values). This is equal to $\binom{k+r}{r}$.

6 Non-Interactive versus Interactive Schemes

In Section 4 we proved that in a non-interactive k-secure t-conference KDS, for each user U_i it holds $H(U_i) \geq \binom{k+t-1}{t-1} H(S)$. In this section we prove that if we allow interaction among users (not with the server!) to set up a common key, then the lower bound can be beaten!

The idea of the protocol is the following. We construct a non-interactive $(k+t-2)$-secure 2-conference KDS using the protocol in [2]. Given a group of t users that want to compute a conference key, the user with the largest identity in the group chooses as conference key a random value in $GF(q)$. Then he sends this value to the other $t-1$ users by using the $(k+t-2)$-secure 2-conference KDS. More formally the protocol for users $U_1, \ldots U_n$, is the following (based on the scheme presented above).

1. The server chooses a symmetric polynomial $P(x, y)$ of degree $k+t-2$, with coefficients over $GF(q)$, $q > n$, by randomly choosing its coefficients.
2. To each user U_i the server gives the polynomial $f_i(y) = P(i, y)$, that is the polynomial obtained by evaluating $P(x, y)$ at $x = i$.
3. If users $U_{i_1}, \ldots, U_{i_t}$, where $i_1 < i_2 < \cdots < i_t$, want to set up a conference key, then:
   3.1 User U_{i_t} randomly chooses a secret key s in $GF(q)$.
   3.2 User U_{i_t} evaluates the polynomial $f_{i_t}(y)$ at $y = i_l$, for $l = 1, \ldots, t-1$, and, then, he computes temporary keys $s_{i_t, i_l} = f_{i_t}(i_l)$ (which is equal to $P(i_t, i_l)$).
   3.3 User U_{i_t} sends to user U_{i_l} the value $\alpha_l = s_{i_t, i_l} \otimes s$, for $l = 1, \ldots, t-1$, where $\otimes$ is the bitwise xor.
   3.4 For $l = 1, \ldots, t-1$:
   User U_{i_l}, first computes $s_{i_t, i_l} = s_{i_l, i_t} = f_{i_l}(i_t)$ (which is equal to $P(i_l, i_t) = P(i_t, i_l)$). Then, U_{i_l} computes s by taking the bitwise xor between s_{i_t, i_l} and the value α_l received by U_{i_t}.

The above protocol is k-secure, since the KDS that is established at steps 1 and 2 is $(k+t-2)$-secure.

In the above protocol only $k+t-1$ elements of $GF(q)$ are distributed by the server and kept by each user.

This, proves a separation between the interactive and the non-interactive case for information-theoretically key distribution schemes for dynamic conferences.

7 Conference Key Distribution and Communication Graph

In a non-interactive 2-conference KDS for n users each pair of users is able to compute a common key. It can be the case that some pairs of users will never need to compute a common key. This situation can arise when a computer network has a topology which is not the complete graph; here each computer takes the place of a user in a KDS, and two computers can communicate if and only if there is a link between them. As an example, consider a ring of n computers $\mathcal{R} = \{C_0, C_1, \ldots, C_{n-1}\}$: computer C_i can communicate with only two computers, C_{i-1} and C_{i+1} (arithmetic on indices is modulo n) so it will never need to compute a common key with C_{i+2}.

In this section we analyze this situation.

Let $\mathcal{U} = \{U_1, \ldots, U_n\}$ be a set of users. A *communication structure* $\mathcal{C}$ is a subset of $\mathcal{U} \times \mathcal{U}$. The communication structure contains all pairs of users for which the server has to provide a common key. A convenient way to represent a communication structure is by a graph G, in which each vertex U_i corresponds to user U_i, and there is an edge (U_i, U_j) if and only if $(U_i, U_j) \in \mathcal{C}$. We call the graph associated to a communication structure the *communication graph*.

Definition 3.2 can be extended to a key distribution scheme for any communication structure $\mathcal{C}$, as follows.

Definition 7.1 *Let* $\mathcal{U} = \{U_1, \ldots, U_n\}$ *be a set of users, let* $k \leq n - 2$, *be an integer, and let* $\mathcal{C} \subseteq \mathcal{U} \times \mathcal{U}$ *be a communication structure. A non-interactive key distribution scheme for* $\mathcal{C}$ *is* k-*secure if*

1. *Each pair of users in* $\mathcal{C}$ *can non-interactively compute the common key. For all* $(U_i, U_j) \in \mathcal{C}$, $H(S_{i,j}|U_i) = H(S_{i,j}|U_j) = 0$.
2. *Any group of* k *users have no information on a key they should not know. For all users* $U_i, U_j, U_{i_1}, \ldots, U_{i_k}$ *such that* $i, j \notin \{i_1, \ldots, i_k\}$, $H(S_{i,j}|U_{i_1} \ldots U_{i_k}) = H(S_{i,j})$.

Now, we describe a k-secure (2-conference) KDS for a communication structure $\mathcal{C}$. First, we do not take into account the communication structure and construct a k-secure KDS for all users as if each pair has to compute a common key. User U_i could receive more information than needed. If the degree of vertex U_i in the communication graph is less than k, then the piece of information given to U_i could consist of only the actual keys he needs for communicating.

Below we describe a non-interactive k-secure key distribution scheme for a communication structure $\mathcal{C}$. In the following, $deg(U_i)$ denotes the cardinality of the set $\{U_j|(U_i, U_j) \in \mathcal{C}\}$.

1. The server chooses a symmetric polynomial $P(x, y)$ of degree k with coefficients over $GF(q)$, $q > n$, by randomly choosing its coefficients.
2. To each user U_i, the server gives the following pieces of information:
   2.1 If $deg(U_i) > k$ then the server gives to user U_i the polynomial $f_i(y) = P(i, y)$, that is the polynomial obtained by evaluating $P(x, y)$ at $x = i$.
   2.2 If $deg(U_i) \leq k$ and $U_{i_1}, \ldots, U_{i_m}$, where $m = deg(U_i)$, are the adjacent vertices of U_i in the communication graph G, then the server gives to user U_i the pieces $\alpha_j = P(i, i_j)$, where $j = 1, \ldots, m$.

This protocol is k-secure. The proof is analogous to the proof of Theorem 5.1.

Theorem 7.1 *The above described non-interactive key distribution scheme for a communication structure $\mathcal{C}$ is k-secure.*

It is easy to see that in previous protocol each user U_i receives $\min\{k + 1, deg(u_i)\}$ pieces of information, that is the size of the information he has is $\min\{k + 1, deg(u_i)\}$ the size of the common key. The following theorem proves that the protocol is optimal with respect to the size of the information held by each user. In the following theorem we suppose that all keys have the same entropy, i.e. $H(S_{i,j}) = H(S)$ for all i and j.

Theorem 7.2 *Let $\mathcal{U} = \{U_1, \ldots, U_n\}$ be a set of users, let k, $k \leq n - 2$, be an integer, and let G be a communication graph on $\mathcal{U}$. In any k-secure key distribution scheme for G, the entropy $H(U_i)$ of each user U_i satisfies*

$$H(U_i) \geq \mu \cdot H(S),$$

where $\mu = \min\{k + 1, deg(u_i)\}$.

Proof : Let $(U_i, U_{j_1}), \ldots, (U_i, U_{j_\mu})$ be elements of the communication structure described by graph G. That is, the server has to provide a common key for such pairs of users. Then, one has

$$H(U_i) = H(S_{i,j_1} \ldots S_{i,j_\mu}) - H(S_{i,j_1} \ldots S_{i,j_\mu}|U_i) + H(U_i|S_{i,j_1} \ldots S_{i,j_\mu})$$

$$\text{(from (4) and (5))}$$

$$\geq H(S_{i,j_1} \ldots S_{i,j_\mu}) - \sum_{l=1}^{\mu} H(S_{i,j_l}|U_i) + H(U_i|S_{i,j_1} \ldots S_{i,j_\mu}) \quad \text{(from (3) and (8))}$$

$$= H(S_{i,j_1} \ldots S_{i,j_\mu}) + H(U_i|S_{i,j_1} \ldots S_{i,j_\mu}) \quad \text{(from 1. of Definition 7.1)}$$

$$\geq H(S_{i,j_1} \ldots S_{i,j_\mu}) \quad \text{(from (2))}$$

$$= H(S_{i,j_1}) + H(S_{i,j_2}|S_{i,j_1}) + \cdots + H(S_{i,j_\mu}|S_{i,j_1} \ldots S_{i,j_{\mu-1}}) \quad \text{(from (3))}$$

$$= H(S_{i,j_1}) + H(S_{i,j_2}) + \cdots + H(S_{i,j_\mu}) \quad \text{(from Lemma 4.1)}$$

$$= \mu H(S) \qquad \qquad \qquad \square$$

Analogously to KDSs, in t-conference KDS we can consider the case when not all the t-tuples of users need to set up a common key. Let $\mathcal{U} = \{U_1, \ldots, U_n\}$ be a set of users. A t-*communication structure* $\mathcal{C}_t$ is a subset of $\mathcal{U}^t$. The communication structure contains all t-tuple of users for which the protocol has to provide a conference key. A convenient way to represent a t-communication structure is by an hypergraph H in which each vertex U_i corresponds to user U_i, and there is a hyperedge $(U_{i_1}, \ldots, U_{i_t})$ if and only if $(U_{i_1}, \ldots, U_{i_t}) \in \mathcal{C}_t$. We will call the hypergraph associated with a t-communication structure the *communication hypergraph*. Definition 7.1, the previously described protocol, and Theorem 7.2 can be extended to a key distribution scheme for any t-communication structure $\mathcal{C}_t$.

8 Applications: Authentication and Master Keys

The polynomial-based scheme proposed applies to settings where a limited coalition of up to a certain security parameter k of adversaries are expected. A basic application is a secure conference key generation. The setting is ideal for the case of a master key generation (to derive further temporal keys), or authentication of conference members based on conventional cryptosystems using the key in authentication protocols (such as the ones described in [1]) and without the need of going to an on-line server (as in [19]). For authentication applications it has a necessary and elegant feature as it connects the IDs of parties to the authentication master key (an ID-based authentication method). Further, additional authenticating information can be attached (as explained in the following sub-sections). The advantage of the system ¿from operational point of view is the disposal of the necessity to contact an on-line remote server, the alternative cost is, naturally, the on-line key computation (evaluation) cost, (this can be somewhat reduced if keys are cached).

8.1 Mixed User Groups

It may be desired to have an asymmetric protocol where the two parties should not be considered equal. For example, one party is a server, and the other a client (e.g., a server-user model). The protocol, in this case, will not only authenticate the name of the user (say), but also the fact that it is an entity with a status of user (rather than a server); users will not be able to claim to be servers. In this case we can modify the scheme to use asymmetric polynomials. This asymmetric scheme can be used to define status (type) of users in various security domains.

8.2 Two-level hierarchical polynomial

Another use for the scheme is for a hierarchical transfer of trust. This can be done either in the symmetric or in the asymmetric polynomial methods. Let us demonstrate here a two level hierarchy of authority servers (domains) and users. The system's polynomial has four (sets of) variables $Q(x, y, z, w)$. $Q(x, y, \cdot, \cdot) = Q(y, x, \cdot, \cdot)$ and $Q(\cdot, \cdot, z, w) = Q(\cdot, \cdot, w, z)$. The first half of variables are to be evaluated under the servers' names and the later half to be evaluated under the users' names. This gives an identification of both the user and its domain (server) in an authentication process. This can be extended to a few levels.

   the symmetric polynomial

8.3 Uses for internetworking

In an inter-enterprise environment, using the above method — an organization (company) can issue permits (authentication polynomials) to its own employees, without knowing the main polynomial. Whenever an employee of this company uses the network, it is clear that he indeed has received its authorization ¿from that company (since he must send C, otherwise he will not be able to authenticate itself). Moreover, if it is desired to revoke the permit of this company, it is not necessary to revoke the permit of each of its employees separately, rather revoke the server's authorization and eliminate the right to its users.

8.4 Additional control variables

A multi-variate polynomial may have additional uses. Additional meanings can be assigned to a few additional variables, for example:

- Time-stamp: The polynomial can be evaluated at a specific date by the distributor. The entity using it will have to specify the date it received it (otherwise it will not be able to generate to authentication key). Thus , validity and expiration can easily be decided.
- Group membership: Members of a specific group will be given private polynomials evaluated also under the name of the group (while others will be given the polynomial evaluated under the names of other groups).
- Permission to access a certain resource for access-control mechanism can be embedded in the private polynomial computation.

To conclude, we have modeled, analyzed, and designed dynamic optimal conference key distribution schemes, presented the advantage of interaction in this setting, and presented modifications and essential applications.

References

1. R. Bird, I. Gopal, A. Herzberg, P. Jansen, S. Kutten, R. Molva and M. Yung *Systematic Design of Two-Party Authentication*, Advances in Cryptology: Proceedings of Crypto 91, Lecture Notes in Computer Science, vol. 576, Springer-Verlag, Berlin, 1991.
2. R. Blom, *An Optimal Class of Symmetric Key Generation Systems*, Advances in Cryptology: Proceedings of Eurocrypt 84, Lecture Notes in Computer Science, vol. 209, Springer-Verlag, Berlin, 1984, pp. 335–338.
3. E. Brickell, P.J. Lee and Y. Yacobi, *Secure Audio Conferencing*, Advances in Cryptology: Proceedings of Crypto 87, Lecture Notes in Computer Science, vol. 239, Springer-Verlag, Berlin, 1987, pp. 418–426.
4. I. Csiszár and J. Körner, *Information Theory. Coding theorems for discrete memoryless systems,* Academic Press, 1981.
5. W. Diffie and M.E. Hellman, *New Direction in Cryptography*, IEEE Transaction on Information Theory, vol. 22, no. 6, December 1976, pp. 644–654.
6. M.J. Fischer, M.S. Paterson and C. Rackoff, *Secure Bit Transmission Using a Random Deal of Cards*, in *Distributed Computing and Cryptography*, AMS, 1991, pp. 173–181.
7. M.J. Fischer and R.N. Wright, *Multiparty Secret Key Exchange Using a Random Deal of Cards*, Advances in Cryptology: Proceedings of Crypto 91, Lecture Notes in Computer Science, vol. 576, Springer-Verlag, Berlin, 1991, pp. 141–155.

8. W. Fumy and M. Munzert, *A Modular Approach to Key Distribution*, Advances in Cryptology: Proceedings of Crypto 90, Lecture Notes in Computer Science, vol. 537, Springer-Verlag, Berlin, 1990, pp. 274–283.

9. R. G. Gallager, *Information Theory and Reliable Communications*, John Wiley & Sons, New York, NY, 1968.

10. L. Gong and D.J. Wheeler, *A Matrix Key-Distribution Scheme*, Journal of Cryptology, vol. 2, 1990, pp. 51–59.

11. R. Impagliazzo and S. Rudich, *Limits on the Provable Consequences of One-Way Permutations*, 21-st STOC proceedings, May 1989, pp. 44–61.

12. K. Koyama and K. Ohta, *Identity-based Conference Key Distribution*, Advances in Cryptology: Proceedings of Crypto 87, Lecture Notes in Computer Science, vol. 239, Springer-Verlag, Berlin, 1987, pp. 175–184.

13. F.J. MacWilliams and N.J.A. Sloane, *The Theory of Error Correcting Codes*, North-Holland, New York, 1988.

14. T. Matsumoto and H. Imai, *On the Key Predistribution System: A Practical Solution to the Key Distribution Problem*, Advances in Cryptology: Proceedings of Crypto 87, Lecture Notes in Computer Science, vol. 239, Springer-Verlag, Berlin, 1987, pp. 185–193.

15. K.S. McCurley, *A Key Distribution System Equivalent to Factoring*, Journal of Cryptology, vol. 1, 1988, pp. 95–105.

16. U. Maurer and Y. Yacobi, *Non-interactive Public-Key Cryptography*, Advances in Cryptology: Proceedings of Eurocrypt 91, Lecture Notes in Computer Science, vol. 547, Springer-Verlag, Berlin, 1991, pp. 498–507.

17. R. C. Merkle, *Secure Communication over Insecure Channels*, Communications of the ACM, vol. 21, Apr. 1978, pp. 294–299.

18. Fiat, Naor; and Alon (personal communication).

19. R. M. Needham and M. D. Schroeder, *Using Encryption for Authentication in Large Networks of Computers*, Communications of the ACM, vol. 21, Dec. 1978, pp. 993–999.

20. E. Okamoto and K. Tanaka, *Key Distribution System Based on Identification Information*, IEEE Journal on Selected Areas in Communications, vol. 7, no. 4, May 1989, pp. 481–485.

21. A. Shamir, *Identity-based Cryptosystems and Signature Scheme*, Proceedings of Crypto 84, pp. 47–53.

22. D.G. Steer, L. Strawczynsji, W. Diffie and M Wiener, *A Secure Audio Teleconferencing System*, Advances in Cryptology: Proceedings of Crypto 89, Lecture Notes in Computer Science, vol. 403, Springer-Verlag, Berlin, 1990, pp. 518–528.

23. S.. Tsujii and J. Chao, *A New ID-based Key Sharing Scheme*, Advances in Cryptology: Proceedings of Crypto 91, Lecture Notes in Computer Science, vol. 576, Springer-Verlag, Berlin, 1991, pp. 288–299.

24. Y. Yacobi, *A Key Distribution Paradox*, Advances in Cryptology: Proceedings of Crypto 90, Lecture Notes in Computer Science, vol. 537, Springer-Verlag, Berlin, 1990, pp. 268–273.

25. Y. Yacobi and Z. Shmueley, *On Key Distribution Systems*, Advances in Cryptology: Proceedings of Crypto 89, Lecture Notes in Computer Science, vol. 435, Springer-Verlag, Berlin, 1990, pp. 344–355.

Differential Cryptanalysis of
the Full 16-round DES

Eli Biham

Computer Science Department
Technion - Israel Institute of Technology
Haifa 32000, Israel

Adi Shamir

Department of Applied Mathematics and Computer Science
The Weizmann Institute of Science
Rehovot 76100, Israel

Abstract

In this paper we develop the first known attack which is capable of breaking the full 16 round DES in less than the 2^{55} complexity of exhaustive search. The data analysis phase computes the key by analyzing about 2^{36} ciphertexts in 2^{37} time. The 2^{36} usable ciphertexts are obtained during the data collection phase from a larger pool of 2^{47} chosen plaintexts by a simple bit repetition criteria which discards more than 99.9% of the ciphertexts as soon as they are generated. While earlier versions of differential attacks were based on huge counter arrays, the new attack requires negligible memory and can be carried out in parallel on up to 2^{33} disconnected processors with linear speedup. In addition, the new attack can be carried out even if the analyzed ciphertexts are derived from up to 2^{33} different keys due to frequent key changes during the data collection phase. The attack can be carried out incrementally with any number of available ciphertexts, and its probability of success grows linearly with this number (e.g., when 2^{29} usable ciphertexts are generated from a smaller pool of 2^{40} plaintexts, the analysis time decreases to 2^{30} and the probability of success is about 1%).

1   Introduction

The Data Encryption Standard (DES) is the best known and most widely used cryptosystem for civilian applications. It consists of 16 rounds of substitution and permutation operations, carried out under the control of a 56 bit key (see [6] for further details). It was adopted as a US national standard in the mid 70's, and had been extensively analyzed for over 15 years. However, no attack which is faster than exhaustive search (whose complexity is 2^{55} due to a simple complementation property that halves the number of searched keys) has ever been reported in the open literature.

The lack of progress in the cryptanalysis of the full DES led many researchers to analyse simplified variants of DES, and in particular variants of DES with fewer than 16 rounds. Chaum and Evertse[4] described an attack on reduced variants of DES, whose complexity is 2^{54} for the six-round variant. They showed that their attack is not applicable to variants with eight or more rounds. Davies[5] devised a known plaintext attack whose application to DES reduced to eight rounds analyzes 2^{40} known plaintexts and has time complexity 2^{40}. This attack is not applicable to the full 16-round DES since it has to analyze more than the 2^{64} possible plaintexts. The most successful attack on reduced variants of DES was the method we called differential cryptanalysis [1], which could break variants of DES with up to 15 rounds faster than via exhaustive search. However, for the full 16-round DES the complexity of the attack was 2^{58}, which was slower than exhaustive search. Similar attacks were used to cryptanalyze a large number of DES-like cryptosystems and hash functions [2,3].

In this paper we finally break through the 16-round barrier. We develop an improved version of differential cryptanalysis which can break the full 16-round DES in 2^{37} time and negligible space by analyzing 2^{36} ciphertexts obtained from a larger pool of 2^{47} chosen plaintexts. An interesting feature of the new attack is that it can be applied with the same complexity and success probability even if the key is frequently changed and thus the collected ciphertexts are derived from many different keys. The attack can be carried out incrementally, and one of the keys can be computed in real time while it is still valid. This is particularly important in attacks on bank authentication schemes, in which the opponent needs only one opportunity to forge a multi-million dollar wire transfer, but has to act quickly before the next key changeover invalidates his message.

2   The New Attack

The reader is assumed to be familiar with the general concept of differential cryptanalysis, and in particular with the definitions and notations introduced in [1]. As usual, we ignore the initial permutation IP and final permutation IP^{-1} of DES, since they have no effect on our analysis.

The old attack on the 15-round variant of DES was based on the following two-round iterative characteristic:

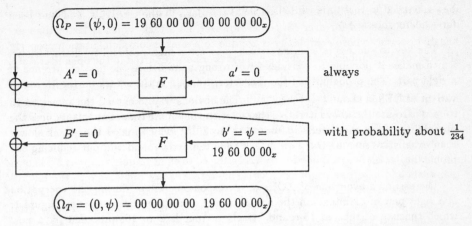

The 13-round characteristic results from iterating this characteristic six and a half times and it's probability is about $2^{-47.2}$. The attack used this characteristic in rounds 1 to 13, followed by a 2R-attack on rounds 14 to 15. Any pair of plaintexts which gives rise to the intermediate XORs specified by this characteristic is called a right pair. The attack tries many pairs of plaintexts, and eliminates any pair which is obviously wrong due to its known input and output values. However, since the cryptanalyst cannot actually determine the intermediate values, the elimination process is imperfect and leaves behind a mixture of right and wrong pairs.

In earlier versions of differential cryptanalysis, each surviving pair suggested several possible values for certain key bits. Right pairs always suggest the correct value for these key bits (along with several wrong values), while wrong pairs suggest random values. When sufficiently many right pairs are analyzed, the correct value (signal) overcomes the random values (noise) by becoming the most frequently suggested value. The actual algorithm is to keep a separate counter for the number of times each value is suggested, and to output the index of the counter with the maximal final value. This approach requires a huge memory (with up to 2^{42} counters in the attack on the 15-round variant of DES), and has a negligible probability of success when the number of analyzed pairs is reduced below the threshold implied by the signal to noise ratio.

In the new version of differential cryptanalysis, we work somewhat harder on each pair, and suggest a list of complete 56-bit keys rather than possible values for a subset of key bits. As a result, we can immediately test each suggested key via trial encryption, without using any counters. These tests can be carried out in parallel on disconnected processors with very small local memories, and the algorithm is guaranteed to discover the correct key as soon as the first right pair is encountered. Since the processing of different pairs are unrelated, they can be generated by different keys at different times due to frequent key changes, and the discovery of a key can be announced in real time while it is still valid (e.g., in order to forge authenticators for banking messages).

The key to success in such an attack is to use a high probability characteristic, which makes it possible to consider fewer wrong pairs before the first occurrence of a right pair. The probability of the characteristic used in the attack on the 15-round variant of DES is about $\left(\frac{1}{234}\right)^6 = 2^{-47.2}$. The obvious way to extend the attack to 16 rounds is to use the above iterative characteristic one more time, but this reduces the probability of the characteristic from $2^{-47.2}$ to $2^{-55.1}$, which makes the attack slower than exhaustive search. Our new attack adds the extra round without reducing the probability at all.

The assumed evolution of XORs of corresponding values during the encryption of a right pair of plaintexts in the new 16-round attack are summarized in Figure 1, which consists of the old 15-round attack on rounds 2 to 16, preceded by a new round 1.

Our goal is to generate without loss of probability pairs of plaintexts whose XORed outputs after the first round are the required XORed inputs $(\psi, 0)$ into the 13-round characteristic of rounds 2 to 14. Let P be an arbitrary 64-bit plaintext, and let $v_0, \ldots, v_{4095}$ be the 2^{12} 32-bit constants which consist of all the possible values at the 12 bit positions which are XORed with the 12 output bits of S1, S2 and S3 after the first round, and 0 elsewhere. We now define a structure which consists of 2^{13} plaintexts:

$$P_i = P \oplus (v_i, 0) \qquad \bar{P}_i = (P \oplus (v_i, 0)) \oplus (0, \psi) \qquad \text{for } 0 \leq i < 2^{12}$$
$$T_i = \text{DES}(P_i, K) \qquad \bar{T}_i = \text{DES}(\bar{P}_i, K)$$

The plaintext pairs we are interested in are all the pairs P_i, $\bar{P}_j$ with $0 \leq i, j < 2^{12}$. There are 2^{24} such plaintext pairs, and their XOR is always of the form (v_k, ψ), where each v_k occurs exactly 2^{12} times. Since the actual processing of the left half of P and of the left half of P XORed with ψ in the first round under the actual key creates a XORed value after the first round which can be non-zero only at the outputs of S1, S2 and S3, this XORed value is one of the v_k. As a result, for exactly 2^{12} of the plaintext pairs, the output XOR of the first F-function is exactly cancelled by XORing it with the left half of the plaintext XOR, and thus the output XOR of the first round (after swapping the left and right halves) is the desired input XOR $(\psi, 0)$ into the iterative characteristic. Therefore, each structure has a probability of about $2^{12} \cdot 2^{-47.2} = 2^{-35.2}$ to contain a right pair.

The problem in this approach is that we do not know the actual value of v_k, which cancels the output XOR of the first F-function, and thus we do not know on which 2^{12} plaintext pairs to concentrate. Trying all the 2^{24} possible pairs takes too long, but we can use their cross-product structure to isolate the right pairs among them in just 2^{12} time. In any right pair, the output XOR after 16 rounds should be zero at the outputs of the five S-boxes S4, ..., S8 (i.e., , at 20 bit positions). We can thus sort

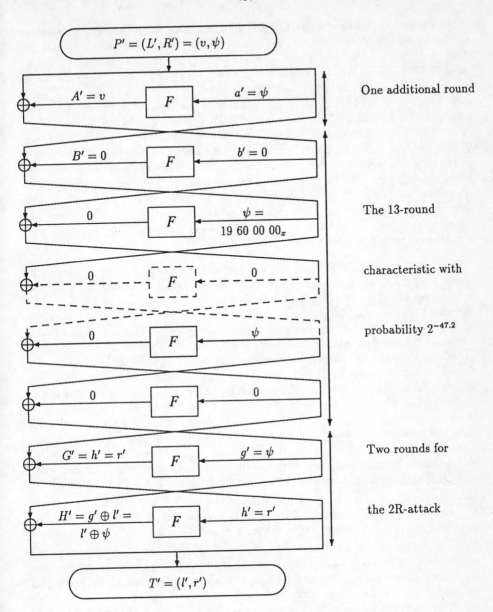

Figure 1. The extension of the attack to 16 rounds

(or hash) the two groups of 2^{12} ciphertexts T_i, $\bar{T}_j$ by these 20 bit positions, and detect all the repeated occurrences of values among the 2^{24} ciphertext pairs in about 2^{12} time. Any pair of plaintexts which fails this test has a non-zero ciphertext XOR at those 20 bit positions, and thus cannot be a right pair by definition. Since each one of the 2^{24} possible pairs passes this test with probability 2^{-20}, we expect about $2^4 = 16$ pairs to survive. By testing additional S boxes in the first, fifteenth, and sixteenth rounds and eliminating all the pairs whose XOR values are indicated as impossible in the pairs XOR distribution tables of the various S boxes, we can discard about 92.55% of these surviving pairs[1] leaving only $16 \cdot 0.0745 = 1.19$ pairs per structure as the expected output of the data collection phase. All these additional tests can be implemented by a few table lookup operations into small precomputed tables, and their time complexity is much smaller than the time required to perform one trial encryption during an exhaustive search. Note that this filtering process removes only wrong pairs but not all of them and thus the input of the data analysis phase is still a mixture of right and wrong pairs.

The data analysis phase of previous differential cryptanalytic attacks used huge arrays of up to 2^{42} counters to find the most popular values of certain key bits. The new variant of differential attack described in this paper uses only negligible space. We want to count on all the key bits simultaneously but cannot afford the huge array of 2^{56} counters. Instead, we immediately try each suggested value of the key. A key value is suggested when it can create the output XOR values of the last round as well as the expected output XOR of the first round and the fifteenth round for the particular plaintext pairs and ciphertext pairs. In the first round and in the fifteenth round the input XORs of S4 and S5, ..., S8 are always zero. Due to the key scheduling algorithm, all the 28 bits of the left key register are used as inputs to the S boxes S1, S2 and S3 in the first and the fifteenth rounds and S1, ..., S4 in the sixteenth round. Only 24 bits of the right key register are used in the sixteenth round. Thus, $28 + 24 = 52$ key bits enter these S boxes. $\frac{2^{-32}}{0.8^8}$ of the choices of the 52-bit values remain by comparing the output XOR of the last round to its expected value and discarding the ones whose values are not possible and $\frac{2^{-12}}{\frac{14}{16} \cdot \frac{13}{16} \cdot \frac{15}{16}}$ of the remaining ones remain by comparing the output XOR of the three S boxes in the first round to its expected value. A similar fraction of the remaining 52-bit values remain by analyzing the three S boxes in the fifteenth round. Each analyzed pair suggests about $2^{52} \cdot \frac{2^{-32}}{0.8^8} \cdot \frac{2^{-12}}{\frac{14}{16} \cdot \frac{13}{16} \cdot \frac{15}{16}} \cdot \frac{2^{-12}}{\frac{14}{16} \cdot \frac{13}{16} \cdot \frac{15}{16}} = 0.84$ values for these 52 bits of the key, and each one of them corresponds to 16 possible values of the full 56-bit key. Therefore, each structure suggests about $1.19 \cdot 0.84 \cdot 16 = 16$ choices for the whole key. By peeling up two additional rounds we can verify each such key by performing about one quarter of a DES encryption (i.e., executing two rounds for each one of the two members of the pair), leaving only about 2^{-12} of the choices of the key. This filtering costs about

---

[1] A fraction of about $\left(\frac{14}{16} \cdot \frac{13}{16} \cdot \frac{15}{16}\right)^2 \cdot 0.8^8 = 0.0745$ of these pairs remain and thus a fraction of about 0.9255 of them are discarded. The input XOR values of the S boxes in the first and the fifteenth rounds of right pairs are known and fixed, and thus we use the fraction of non-zero entries of the corresponding lines in the pairs XOR distribution tables whose values are $\frac{14}{16}$, $\frac{13}{16}$ and $\frac{15}{16}$, rather than the fraction of the non-zero entries in the whole tables, which is approximated by 0.8.

		K16								
		Left Key Register				Right Key Register				
		S1	S2	S3	S4	X	S5	S6	S7	S8	X
K1	S1		2	1	1	2					
	S2	2		1	2	1					
	S3	2			3	1					
	S4	2	3	1							
	X		1	3							
	S5							1	2	2	1
	S6						3		2	1	
	S7							2		2	2
	S8						2	3			1
	X						1		2	1	

Table 1. The number of common bits entering the S boxes in the first round (K1) and in the sixteenth round (K16)

$16 \cdot \frac{1}{4} = 4$ equivalent DES operations. Each remaining choice of the 56-bit key is verified via trial encryption of one of the plaintexts and comparing the result to the corresponding ciphertext. If the test succeeds, there is a very high probability that this key is the right key. Note that the signal to noise ratio of this counting scheme is $S/N = \frac{2^{52} \cdot 2^{-47.2}}{1.19/2^{12} \cdot 0.84} = 2^{16.8}$.

This data analysis can be carried out efficiently by carefully choosing the order in which we test the various key bits. We first enumerate all the possible values of the six key bits of $S4_{Kh}$, and eliminate any value which does not give rise to the expected XOR of the four output bits from this S box. This leaves four out of the 64 possibilities in average. Table 1 shows the number of common bits entering the S boxes in the first round and in the sixteenth round. The notation X denotes the bits which are not used in the specific subkey. We see that three of the bits of $S4_{Kh}$ are shared with $S3_{Ka}$. We complete the three missing bits of $S3_{Ka}$ in all possible ways, and reduce the average number of possibilities to two. Two bits of $S1_{Kh}$ are shared with $S3_{Ka}$. By completing the four missing bits of $S1_{Kh}$ and then the two missing bits of $S2_{Ka}$ we can reduce the average number of possibilities to about half. After completing the 13 remaining bits of the left key register in a similar way, the average number of values suggested for this half of the key is one.

To compute bits from the right key register, we first extract actual S box bits from their assumed XORed values. In the fifteenth round we know the input XORs and the output XORs of S1, S2 and S3. We can thus generate about 4–5 candidate inputs for each one of these S boxes, and deduce the corresponding bits in g by XORing with the known bits of the left key register. In a similar way, we can calculate the outputs of the S boxes S1, S2, S3 and S4 in the sixteenth round, XOR these bits of H with the known bits of the left half of the ciphertext l and get 16 bits of g, from which two bits enter S1, two bits enter S2 and three bits enter S3 in the fifteenth round.

By comparing these bit values to the candidate inputs of the S boxes we end up with about one candidate input for S1, one for S2, and only about half of the trials would result with a candidate input for S3. We can now deduce all the bits of g which enter these three S boxes and deduce the corresponding bits of H by $H = g \oplus l$. Two of these bits are outputs of S5, two bits are outputs of S6, three are outputs of S7 and one is output of S8. For each of these four S boxes we know the input XOR and the output XOR, and can deduce about 4–5 possible inputs. Since we also know actual output bits, the number of possible inputs is reduced to about one for S5 and S6, two for S8, but only half of the trials would result with a candidate for S7. We can deduce 24 out of the 28 bits of the right key register by XORing the 24 computed bits at the inputs of these four S boxes with the expanded value of the known right half of the ciphertext.

We can now summarize the performance of the new attack in the following way. Each structure contains a right pair with probability $2^{-35.2}$. The data collection phase encrypts a pool of about 2^{35} structures, which contain about $2^{35} \cdot 2^{13} = 2^{48}$ chosen plaintexts, from which about $2^{35} \cdot 1.19 = 2^{35.25}$ pairs ($2^{36.25}$ ciphertexts) remain as candidate inputs to the data analysis phase. The probability that at least one of them is a right pair is about 58%, and the analysis of any right pair is guaranteed to lead to the correct key. The time complexity of this data analysis phase is about $2^{35} \cdot 4 = 2^{37}$ equivalent DES operations.

In order to further reduce the number of chosen plaintexts, we can use the quartet method of [1]. Since the basic collection of plaintexts in the new attack is a structure rather than a pair, we create metastructures which contain 2^{14} chosen plaintexts, built from two structures which correspond to the standard iterative characteristic and from two structures which correspond to the following iterative characteristic:

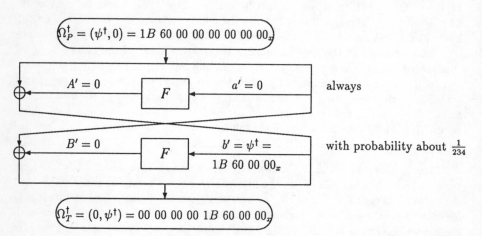

This characteristic has the same probability as the previous one. With these metastructures, we can obtain four times as many pairs from twice as many plaintexts, and thus reduce the number of chosen plaintexts encrypted in the data collection phase from 2^{48} to 2^{47}.

Rounds	Chosen Plaintexts	Analyzed Plaintexts	Complexity of Analysis	Best Previous Time	Space
8	2^{14}	4	2^9	2^{16}	2^{24}
9	2^{24}	2	2^{32}	2^{26}	2^{30}
10	2^{24}	2^{14}	2^{15}	2^{35}	—
11	2^{31}	2	2^{32}	2^{36}	—
12	2^{31}	2^{21}	2^{21}	2^{43}	—
13	2^{39}	2	2^{32}	2^{44}	2^{30}
14	2^{39}	2^{29}	2^{29}	2^{51}	—
15	2^{47}	2^7	2^{37}	2^{52}	2^{42}
16	2^{47}	2^{36}	2^{37}	2^{58}	—

Table 2. Summary of the new memoryless results on DES

The general form of the new attack can be summarized in the following way: Given a characteristic with probability p and signal to noise ratio S/N for a cryptosystem with k key bits, we can apply a memoryless attack which encrypts $\frac{2}{p}$ chosen plaintexts in the data collection phase and has complexity of $\frac{2^k}{S/N}$ trial encryptions during the data analysis phase. The number of chosen plaintexts can be reduced to $\frac{1}{p}$ by using appropriate metastructures, and the effective time complexity can be reduced by a factor of $f \leq 1$ if a tested key can be discarded by carrying out only a fraction f of the rounds. Therefore, memoryless attacks can be mounted whenever $p > 2^{1-k}$ and $S/N > 1$. The memoryless attacks require fewer chosen plaintexts compared to the corresponding counting schemes, but if the signal to noise ratio is too low or if the number of the key bits on which we count is small, the time complexity of the data analysis phase may be higher than the corresponding complexity of the counting scheme.

In the attack described in this paper, $p = 2^{-47.2}$, $k = 56$, $f = \frac{1}{4}$ and $S/N = 2^{16.8}$. Therefore, the number of chosen plaintexts is $\frac{2}{p} = 2^{48.2}$ which can be reduced to $\frac{1}{p} = 2^{47.2}$ by using metastructures, and the complexity of the data analysis phase is $2^{37.2}$ equivalent DES operations.

The performance of the new attack for various numbers of rounds is summarized in Table 2. Variants with an even number of rounds n have a characteristic with probability $p = \left(\frac{1}{234}\right)^{(n-4)/2}$, require p^{-1} chosen plaintexts, and analyze $p^{-1} \cdot 2^{-10.75}$ plaintexts in time complexity $p^{-1} \cdot 2^{-10}$. The known plaintext variant of the new attack needs about $2^{31.5} \cdot p^{-0.5}$ known plaintexts (using the symmetry of the cryptosystem which makes it possible to double the number of known encryptions by reversing the roles of the plaintexts and the ciphertexts). Variants with an odd number of rounds n have a characteristic with probability $p = \left(\frac{1}{234}\right)^{(n-3)/2}$, require p^{-1} chosen plaintexts, and analyze $p^{-1} \cdot 2^{-40.2}$ plaintexts in time complexity $p^{-1} \cdot 2^{-10}$. For such odd values of n, if $p > 2^{-40.2}$ then the number of analyzed plaintexts is two and the complexity of the data analysis phase is 2^{32}. However, using about four times as many chosen plaintexts, we can use the clique algorithm (described in [1]) and reduce the time complexity of the data analysis phase to less than a second on a personal computer. The known plaintext attacks need about $2^{32} \cdot p^{-0.5}$ known plaintexts (in this case the symmetry does not help). The application of the known plaintext attack to eight rounds needs a pool of $2^{38.5}$ known plaintexts. The application to 12 rounds needs a pool of $2^{47.2}$ known plaintexts. The application to 15 rounds needs a pool of $2^{55.6}$ known plaintexts and the application to the full 16-round DES needs a pool of $2^{55.1}$ known plaintexts. This is slightly worse than the 2^{55} complexity of exhaustive search (which in the case of a known plaintext attack requires about 2^{33} plaintexts in order to generate a complementary pair via the birthday paradox).

This specific attack is not directly applicable to plaintexts consisting solely of ASCII characters since such plaintexts cannot give rise to the desired XOR differences. By using several other iterative characteristics we can attack the full 16-round DES with a pool of about 2^{49} chosen ASCII plaintexts (out of the 2^{56} possible ASCII plaintexts).

References

[1] Eli Biham, Adi Shamir, *Differential Cryptanalysis of DES-like Cryptosystems*, Journal of Cryptology, Vol. 4, No. 1, pp. 3–72, 1991. The extended abstract appears in Advances in cryptology, proceedings of CRYPTO'90, pp. 2–21, 1990.

[2] Eli Biham, Adi Shamir, *Differential Cryptanalysis of Feal and N-Hash*, technical report CS91-17, Department of Applied Mathematics and Computer Science, The Weizmann Institute of Science, 1991. The extended abstract appears in Advances in cryptology, proceedings of EUROCRYPT'91, pp. 1–16, 1991.

[3] Eli Biham, Adi Shamir, *Differential Cryptanalysis of Snefru, Khafre, REDOC-II, LOKI and Lucifer*, technical report CS91-18, Department of Applied Mathematics and Computer Science, The Weizmann Institute of Science, 1991. The extended abstract appears in Advances in cryptology, proceedings of CRYPTO'91, 1991.

[4] David Chaum, Jan-Hendrik Evertse, *Cryptanalysis of DES with a reduced number of rounds, Sequences of linear factors in block ciphers*, Advances in cryptology, proceedings of CRYPTO'85, pp. 192–211, 1985.

[5] D. W. Davies, private communication.

[6] National Bureau of Standards, *Data Encryption Standard*, U.S. Department of Commerce, FIPS pub. 46, January 1977.

Iterative Characteristics of DES and s²-DES

Lars Ramkilde Knudsen

Aarhus University
Computer Science Department
Ny Munkegade
DK-8000 Aarhus C.

Abstract. In this paper we show that we are close at the proof that the type of characteristics used by Biham and Shamir in their differential attack on DES [3] are in fact the best characteristics we can find for DES. Furthermore we show that the criteria for the construction of DES-like S-boxes proposed by Kim [6] are insufficient to assure resistance against differential attacks. We show several good iterative characteristics for these S-boxes to be used in differential attacks. Finally we examine the probabilities of the two characteristics used by Biham and Shamir in [3]. We found that for some keys we do not get the probabilities used in the attack. We suggest the use of 5 characteristics instead of two in the attack on DES.

1 Introduction

In 1990 Eli Biham and Adi Shamir introduced *differential cryptanalysis*, a chosen plaintext attack on block ciphers that are based on iterating a cryptographically weak function r times (e.g. the 16-round Data Encryption Standard (DES)). The method proved strong enough to break several cryptosystems, Lucifer, GDES, Feal-4, Feal-8, Snefru a.o. and DES with a reduced number of rounds, i.e. less than 16 rounds [1, 2, 4].

In december 1991 Biham and Shamir published an improved differential attack that is capable of breaking the full 16-round DES [3]. The attack needs 2^{47} chosen plaintexts. The heart in differential attacks is the finding and the use of characteristics. In their attack Biham and Shamir use 2-round iterative characteristics. These characteristics are believed to be the best characteristics for an attack on 16-round DES, but so far no proof of this has been published in the open literature. We are close to the conclusion that this is in fact the case.

After the breaking of the full 16-round DES the question is if we can redesign DES to withstand this kind of attack. There has been a huge research on DES, since its publication in the mid 70's. Some of this work has been concentrating on the design of secure S-boxes. In [6] Kwangjo Kim provides a way of constructing DES-like S-boxes based on boolean functions satisfying the SAC (Strict Avalanche Criterion). Kim lists 5 criteria for the constructions, including "Resistance against differential attacks". Furthermore 8 concrete examples of these S-boxes, the s²-DES S-boxes, are listed. The cryptosystem s²-DES is obtained by replacing all the 8 DES S-boxes by the 8 s^2-DES S-boxes, keeping everything else as in DES. It is suggested that s^2-DES withstands differential attacks better than DES. We show that this is indeed not the case. The conlusion is that Kims 5 criteria for the construction of DES-like S-boxes are insufficient to assure resistance against differential attacks.

In [1] Biham and Shamir observed that the probability of the two characteristics used in [3] will split into two depending on the values of certain keybits. In [3] this phenomena is not considered, and the estimates of complexity are calculated using average probabilities. This means that for some keys we will need more chosen plaintexts as stated in [3]. We think that exact probabilities should be used in the estimates of complexity and suggest the use of 3 additional characteristics to lower the need for chosen plaintexts for a successful attack.

In section 2 we show different models of iterative characteristics for DES and s^2-DES to be used in differential attacks. In section 3 and 4 we show concrete examples of these characteristics for DES and s^2-DES, the probabilities all being average values. In section 5 we consider the exact probabilities of iterative characteristics for DES.

2 Iterative characteristics for DES and s^2-DES

We expect the reader to be familiar with the general concepts of differential cryptanalysis and refer to [1, 8] for further details. In DES and s^2-DES equal inputs (to the F-function) always lead to equal outputs. This means that an inputxor equal to zero leads to an outputxor equal to zero with probability 1. This is the best combination of input/outputxors. In finding the best characteristics we therefore try to maximize the number of these *zero-rounds*. In the following we will show different models of iterative characteristics for DES and s^2-DES. In section 3 and 4 we will justify the usability of the models by showing concrete examples of these in DES and s^2-DES.

2.1 2-round iterative characteristics

Two consecutive zero-rounds in a characteristic of DES-like cryptosystems lead to equal inputs and outputs of all rounds. We get equal plaintexts resulting in equal ciphertexts, a trivial fact. The maximum occurrences of zero-rounds therefore is every second round. This situation evolves by using the 2-round characteristic as in [1]. In the following we will use this notation:

$$(\Phi, 0)$$
$$0 \leftarrow 0 \quad \text{prob. 1}$$
$$0 \leftarrow \Phi \quad \text{prob. something}$$
$$(0, \Phi)$$

for the 2-round iterative characteristic.

2.2 3-round characteristics

In [7] Knudsen found that the best differential attack on LOKI89 [9] was based on a 3-round *fixpoint* characteristic. A fixpoint is an inputxor that can result in itself as an outputxor. Instead of looking for fixpoints we should in general look for, what we call, **twinxors**.

Definition 1 *Twinxors, Γ and Φ, are xors for which $\Phi \leftarrow \Gamma$ and $\Gamma \leftarrow \Phi$, both combinations with a positive probability.*

[1] With twinxors we can build the following 3-round characteristic :

$$
\begin{array}{ccccl}
& (\Gamma, 0) & & & \\
0 & \leftarrow & 0 & & \text{prob. 1} \\
\Phi & \leftarrow & \Gamma & & \text{some prob.} \\
\Gamma & \leftarrow & \Phi & & \text{some prob.} \\
& (0, \Phi) & & &
\end{array}
$$

The characteristic is in fact only "half" an iterative characteristic. Concatenated with the characteristic with rounds no. 2 and 3 interchanged we obtain:

$$
\begin{array}{ccccl}
& (\Gamma, 0) & & & \\
0 & \leftarrow & 0 & & \text{prob. 1} \\
\Phi & \leftarrow & \Gamma & & \text{some prob.} \\
\Gamma & \leftarrow & \Phi & & \text{some prob.} \\
0 & \leftarrow & 0 & & \text{prob. 1} \\
\Gamma & \leftarrow & \Phi & & \text{some prob.} \\
\Phi & \leftarrow & \Gamma & & \text{some prob.} \\
& (0, \Gamma) & & &
\end{array}
$$

In that way we get a 6-round iterative characteristic. Still we choose to call the 3-round characteristic an iterative characteristic.

2.3 4-round characteristic

As for the 3-round characteristic we look for a 4-round characteristic, which extended to 8 rounds becomes an iterative characteristic. It must have the following form:

$$
\begin{array}{ccccl}
& (\Gamma, 0) & & & \\
0 & \leftarrow & 0 & & \text{prob. 1} \\
\Phi & \leftarrow & \Gamma & & \text{some prob.} \\
\Gamma \oplus \Psi & \leftarrow & \Phi & & \text{some prob.} \\
\Phi & \leftarrow & \Psi & & \text{some prob.} \\
& (0, \Psi) & & &
\end{array}
$$

It means that we have to find two inputxors Ψ and Γ both resulting in Φ and Φ resulting in the (xor-)difference between Ψ and Γ.

---

[1] The best twinxors for LOKI89 is obtained with $\Phi = \Gamma = 00400000_x$, i.e. fixpoints.

2.4 Longer characteristics

We can of course continue the search for n-round characteristics, $n > 4$. For every time we go one round further, we compare the characteristic we are now looking for with the best characteristic, we have found so far. We can easily find the best non-trivial input/outputxor combination in the *pairs xor distribution table*. From this probability we calculate the maximum number of different inputs to S-boxes we can have for the characteristic to be better than the one we have found.

By looking closer at the possible xor-combinations and the overall architecture of the cryptosystem we can calculate the minimum number of different inputs to S-boxes we must have for the particular characteristic. Using this minimum and the above maximum we find the possible combinations of input- and outputxors in the characteristic and compare the probability with the other characteristics we have found.

Of course characteristics do not have to contain a zero-round. Before making any conclusions about the best possible characteristic, we must check whether good characteristics of this kind exist.

3 DES

3.1 Properties

The following 5 properties of the DES S-boxes are well known.

1. No S-box is a linear of affine function.
2. Changing one bit in the input to an S-box results in changing at least two output bits.
3. The S-boxes were chosen to minimize the difference between the number of 1's and 0's when any single bit is held constant.
4. $S(\mathbf{x})$ and $S(\mathbf{x} \oplus (001100))$ differ in at least two bits.
5. $S(\mathbf{x}) \neq S(\mathbf{x} \oplus (11ef00))$ for any e and f.

A DES S-box consists of 4 rows of 4-bit bijective functions. The input to an S-box is 6 bits. The left outermost bit and the right outermost bit (the row bits) determine through which function the four remaining bits (the column bits) are to be evaluated. This fact gives us a 6'th property of the DES S-boxes important for differential cryptanalysis.

6. $S(\mathbf{x}) \neq S(\mathbf{x} \oplus (0abcd0))$ for any a, b, c and d, $abcd \neq 0000$.

The inner input bits for an S-box are input bits that do not affect the inputs of other S-boxes. We have two inner input bits for every S-box. Because of the P-permutation we have the following property also important for differential cryptanalysis.

- The inner input bits for an S-box, Si, come from S-boxes, whose inner input bits cannot come from Si.

Example: The inner input bits for $S1$ come from $S2$ and $S5$, whose inner input bits come from $S3$ and $S7$ respectively $S2$ and $S6$.

3.2 2-round iterative characteristics

As stated in [1, 3] the best characteristics for a differential attack on 16-round DES is based on a 2-round iterative characteristic. The following theorem was already proven in [5]. We give the proof in a different manner.

Theorem 1 *If two inputs to the F-function result in equal outputs, the inputs must differ in at least 3 neighbouring S-boxes.*

Proof: If the inputs differ only in the input to one S-box the expanded inputxor must have the following form: $00ab00$ (binary), where $ab \neq 00$. Because of properties 2 and 4 above, these inputs cannot give equal outputs. This also tells us that the inputs must differ in neighbouring S-boxes. If the inputs differ in only two neighbouring S-boxes, Si and $S(i + 1)$, the two inputxors must have the following forms: $Si : 00abcd$ and $S(i + 1) : cdef00$. Now

$cd \neq 00$, because of properties 2 and 4.
$cd \neq 01$, because of property 6 for $S(i + 1)$.
$cd \neq 10$, because of property 6 for $S(i)$.
$cd \neq 11$, because of property 5 for $S(i + 1)$.

□

We have several 2-round iterative characteristics for DES, where the inputs differ in three neighbouring S-boxes. By consulting the *pairs XOR distribution table* for the 8 S-boxes we easily find the best possibilities. The two best of these are used in [3] to break the full 16-round DES using 2^{47} chosen plaintexts. The probability of the two characteristics is $\frac{1}{234}$ for the two rounds.

3.3 3-round iterative characteristics

The highest probability for a non trivial input/outputxor combination in DES is $\frac{1}{4}$. Because $(\frac{1}{4})^x \geq (\frac{1}{234})^{1.5} \Rightarrow x < 6$, there can be different inputs to at most 5 S-boxes for the two nonzero round together. Because of the P-permutation in DES, see Section 3.1., Φ and Γ must differ in the inputs to at least two S-boxes each. Property 2 of the S-boxes implies that at least one additional S-box have different inputs, making Φ and Γ together differ in the inputs to at least 5 S-boxes. The proof is given in the Appendix. For DES the best twinxors, which differ in the inputs to 5 S-boxes are: $\Phi = 31200000_x$ and $\Gamma = 00004200_x$. The probability for the 3-round iterative characteristic is $2^{-18.42}$. This probability is very low and there is in fact twinxors, which together differ in the inputs to 6 S-boxes with a higher probability, $\Phi = 03140000_x$ and $\Gamma = 00004014_x$. The probability for the 3-round iterative characteristic is $2^{-18.1}$. Both characteristics have a probability too low to be used in a successful differential attack.

3.4 4-round iterative characteristics

There can be different inputs to at most 7 S-boxes, because $(\frac{1}{4})^x \geq (\frac{1}{234})^2 \Rightarrow x < 8$, however there is no 4-round iterative characteristics for DES with a probability higher than for best 2-round iterative characteristic concatenated with itself. The proof is tedious and is given in the Appendix.

3.5 Longer characteristics

We believe that it can be proven that we cannot find n-round iterative characteristics, $n > 4$, with probabilities higher than for the best 2-round iterative characteristic concatenated with itself $\frac{n}{2}$ times. To obtain this for a 5-round iterative characteristic there can be different inputs to at most 9 S-boxes, as $(\frac{1}{4})^x \geq (\frac{1}{234})^{2.5} \Rightarrow x < 10$. It seems impossible that we can find such a characteristic different in the inputs to 9 S-boxes and all combinations with a probability close to the highest possible of $\frac{1}{4}$. If we go one round further to a 6-round iterative characteristic the doubt will be even bigger. Before making any conclusions for the best differential attack on DES using characteristics, we must also check that no non iterative characteristics exist, as stated in Section 2.4. These proofs are a topic for further research.

4 s²-DES

4.1 Properties

Kims s²-DES S-boxes do not have the DES properties 2, 4 and 5. They do have a property though that is part of property 2 for the DES S-boxes.

4a. $S(\mathbf{x}) \neq S(\mathbf{x} \oplus (a0000b))$ for $ab \neq 00$.

As the s²-DES S-boxes are build as 4 rows of 4-bit bijective functions, they have property 6 like the DES S-boxes.

4.2 2-round characteristics

Because of property 6 there is no 2-round iterative characteristic for Kims s²-DES S-boxes where the inputs differ only in one S-box, however the lack of property 5 enables us to build a 2-round iterative characteristic where the inputs differ in two neighbouring S-boxes. We have

$$0_x \leftarrow 00000580_x \text{ with prob. } \frac{8*10}{64*64} \simeq \frac{1}{51}$$

Extending this characteristic to 15-rounds yields a probability of $2^{-39.7}$. Using the original attack by Biham and Shamir [1] we will need about 2^{42} chosen plaintexts for a successful differential attack. To do a similar attack as by Biham and Shamir in [3] we construct a 13-round characteristic with probability 2^{-34}. The megastructures used in the attack will consist of 2^9 plaintexts and we will need a total of about 2^{35} chosen plaintexts for the attack. This being said without having studied the attack in details. The above characteristic is not the only 2-round iterative characteristic for s²-DES that is better than the best 2-round iterative characteristics for DES. We have several others, the two secondbest characteristics both with probability $\frac{6*10}{64*64} \simeq \frac{1}{68}$ are based on the combinations: $0_x \leftarrow 07e00000_x$ and $0_x \leftarrow 5c000000_x$.

4.3   3-round characteristics

The best non-trivial input/outputxor combination in s^2-DES has probability $\frac{1}{4}$. Therefore there can be at most 4 S-boxes with different inputs in the 3 rounds all together, as $(\frac{1}{4})^x \geq (\frac{1}{51})^{1.5} \Rightarrow x < 5$. As with DES, because of the P-permutation, Φ and Γ must differ in the inputs to at least two S-boxes each. Unlike for DES it is possible for two inputs different in only 1 bit to result in two outputs different in 1 bit. Therefore we can build a 3-round characteristic with $\Phi = 04040000_x$ and $\Gamma = 00404000_x$. The probability for the characteristic is $\frac{8*6*4*10}{64^4} \simeq 2^{-13.5}$. This is the best 3-round characteristic we have found for s^2-DES. We can build a 13-round characteristic to be used as in the attack in [3]. The probability for the characteristic is $2^{-52.5}$. However we can use the combinations from the 3-round characteristic to build 6-round "half"-iterative characteristics, which are better, as we will show later.

4.4   4-round characteristics

There can be at most 5 S-boxes with different inputs, because $(\frac{1}{4})^x \geq (\frac{1}{51})^2 \Rightarrow x < 6$, and again we exploit the fact that s^2-DES S-boxes do not have property 2. We construct a 4-round characteristic based on the following combinations:

$$00000002_x \leftarrow 0000006e_x \text{ with prob. } \frac{8*10}{64*64}$$
$$00080000_x \leftarrow 00020000_x \text{ with prob. } \frac{8}{64}$$
$$00000002_x \leftarrow 0000002e_x \text{ with prob. } \frac{6*10}{64*64}$$

We have $P(00000002_x) = 00020000_x$ and $P(00080000)_x = 00000040_x = 0000006e_x \oplus 0000004e_x$. The total probability for the 4-round characteristic is $2^{-14.77}$. Extended to 13 rounds we obtain a probability of $2^{-44.3}$.

4.5   Longer characteristics

A 5-round iterative characteristic will have to differ in the inputs to at least 6 S-boxes. However we can find 6-round iterative characteristics also different in the inputs to only 6 S-boxes as indicated above. The P-permutation makes it impossible to have $\Phi \rightarrow \Gamma$ and $\Gamma \rightarrow \Phi$, where both Φ and Γ differ only in the inputs to one S-box. However it is possible to have Φ, Γ, Ψ and Ω, all four different only in the input to one S-box and such that $\Phi \rightarrow \Gamma$, $\Gamma \rightarrow \Psi$, $\Psi \rightarrow \Omega$ and $\Omega \rightarrow \Phi$. We use this observation to construct a 6-round characteristic:

$$
\begin{array}{rcll}
\multicolumn{4}{c}{(\Phi\,,\,0)} \\
0 & \leftarrow & 0 & \text{prob. 1} \\
\Gamma & \leftarrow & \Phi & \text{some prob..} \\
\Psi & \leftarrow & \Gamma & \text{some prob.} \\
\Gamma \oplus \Omega & \leftarrow & \Phi \oplus \Psi & \text{some prob.} \\
\Phi & \leftarrow & \Omega & \text{some prob.} \\
\Omega & \leftarrow & \Psi & \text{some prob.} \\
\multicolumn{4}{c}{(0\,,\,\Psi)} \\
\end{array}
$$

With $\Phi = 04000000_x$, $\Gamma = 00004000_x$, $\Psi = 00040000_x$ and $\Omega = 00400000_x$ we get a total probability for the 6-round characteristic of $\frac{8*10*8*6*4*6}{64^6} \simeq 2^{-19.5}$. Extended to 13 rounds the probability becomes 2^{-39}. Starting with $(\Gamma, 0)$ we get a similar 6-round characteristic with probability $2^{-19.5}$. Starting with $(\Psi, 0)$ or $(\Omega, 0)$ yields a 6-round characteristic with probability $2^{-19.8}$. These 6-round characteristics differ in the inputs to 6 S-boxes, that is, different inputs to one S-box per round in average.

If we try to construct n-round iterative characteristics, $n > 6$, we find that we will get more than one S-box difference per round in average.

4.6 Conclusion on Kims s²-DES S-boxes.

The above illustrates that we have to ensure that DES-like S-boxes have the six properties listed in section 3.1. The fact that for s²-DES two inputs different only in the inputs to 2 neighbouring S-boxes can result in equal outputs enables us to build 2-round iterative characteristic more than 4 times as good as the best 2-round characteristic for DES. The fact that two S-box inputs different in only one bit can result in outputs different in one bit enables us to construct a 4-round and a 6-round iterative characteristic both better for differential attacks on s²-DES than the 2-round characteristic for DES. Furthermore we must check that there is no 2-round iterative characteristic where only 3 neighbouring S-boxes differ in the inputs with a too high probability. For the s²-DES S-boxes the best such characteristic is based on the combination $dc000002_x \leftarrow 0_x$. It has probability $\frac{10*10*14}{64^3} \simeq \frac{1}{187}$. This is higher than the best 2-round characteristic for DES and illustrates that we should also consider this in the construction of DES-like S-boxes.

5 Probabilities of iterative characteristics

5.1 DES

As stated earlier the best characteristics for a differential attack on DES are based on 2-round iterative characteristics. The two best of these have the following inputxors in the second round: $\Phi = 19600000_x$ and $\Gamma = 1b600000_x$. Both xors lead to equal outputs with probability $\frac{1}{234}$. However this probability is only an "average" probability. As stated in [1, section 6.5], if the sixth keybit used in S2 is different from the second keybit used in S3 the probability for Φ increases to $\frac{1}{146}$ and the probability for Γ decreases to $\frac{1}{585}$. If the two keybits are equal the probabilities will be interchanged. We call these keybits, **critical** keybits for Φ and Γ. In their attack on DES [3] Biham and Shamir use these two characteristics to build 13-round characteristics, where six rounds have inputxor Φ or Γ. The probability is claimed to be $(\frac{1}{234})^6 \simeq 2^{-47.22}$. But depending on the values of the six pairs of critical keybits the probability for Φ will vary from $(\frac{1}{146})^6 \simeq 2^{-43.16}$ to $(\frac{1}{585})^6 \simeq 2^{-55.16}$ and the other way around for Γ. Using both characteristics as in [3] we are ensured to get one characteristic with a probability of

Table 1. The probabilities for the best 13-round characteristic obtained by using the 2 characteristics Φ and Γ.

#Keys ($\log_2$)	Probability ($\log_2$)
51.00	-43.16
53.58	-45.16
54.88	-47.16
54.30	-49.16

at least $(\frac{1}{146*585})^3 \simeq 2^{-49.16}$. Table 1 shows the probabilities and for how many keys they will occur.

It means that for one out of 32 keys, we will get a 13-round characteristic with the highest probability and for about one out of three keys we will get the lowest probability. We found that for other 2-round iterative characteristics the probability splits into more than one depending on equality/inequality of certain critical keybits. It turns out that we can find 2-round iterative characteristics for which the best of these probabilities is better than for the lowest for Φ and Γ. For the 2-round characteristic (with inputxor) 00196000_x we have only one probability. It means that regardless of the key values this characteristic will have a probability of $\frac{1}{256}$. Table 2 shows the probabilities for Φ and Γ and for the 2-round iterative characteristics, whose best probability is higher than $\frac{1}{256}$.

Table 2. Exact probabilities for 11 characteristics.

Characteristic	Probabilities (1/n)	Average Prob.(1/n)
19600000_x	146, 585	234
$1b600000_x$	585, 146	234
00196000_x	256	256
$000003d4_x$	210, 390	273
$4000001d_x$	205, 1024	341
19400000_x (+)	0, 195	390
$1b400000_x$ (+)	195, 0	390
40000019_x ($)	248, 390, 744, 1170	455
$4000001f_x$ ($)	248, 390, 744, 1170	455
$1d600000_x$ (+)	205, 512, 819, 2048	468
$1f600000_x$ (+)	205, 512, 819, 2048	468

It seems unlikely that we can find n-round characteristic, $n > 2$, for which the exact probabilities will be higher than for the above mentioned 2-round iterative characteristics. The subkeys in DES are dependent, therefore some keybits might be critical for one characteristic in one round and for another characteristic in another round. For example by using characteristic 19400000_x we have the two probabilities $\frac{1}{195}$ and 0. But this division of the probability depends on the values of the same critical keybits as for Φ and Γ and we would get a probability of

$\frac{1}{146}$ for either Φ or Γ. The characteristics marked with (+) in Table 2 depends on the values of the same critical keybits as for Φ and Γ. Doing an attack on DES similar to the one given in [3], this time using the first 5 of the above characteristics will give us better probabilities for a 13-round characteristic. Table 3 shows the best probabilities and for how many keys these will occur. The above

Table 3. The probabilities for the best 13-round characteristic obtained by using 5 characteristics.

#Keys ($\log_2$)	Probability ($\log_2$)
51.00	-43.16
53.58	-45.16
49.64	-46.07
49.64	-46.29
54.88	-47.16
50.90	-47.18
54.10	-48.00

probabilities are calculated by carefully examining the critical keybits for the 5 characteristics in the rounds no. 3, 5, 7, 9, 11 and 13, i.e. the rounds where we will expect the above inputxors to be. By using the two characteristics in Table 2 marked with ($) in addition would yield slightly better probabilities. However the best probability we would get by using these characteristics is $(\frac{1}{248})^6 \simeq 2^{-47.7}$ and it would occur only for a small number of keys.

As indicated in Table 3 we are ensured to get a characteristic with a probability of at least 2^{-48}. However the megastructures of plaintexts and analysis will become more complex. Whether using 5 characteristics instead of two will dramatically increase the complexity of the analysis remains an open question.

5.2   s²-DES

The best characteristic for an attack on s²-DES is, as we saw earlier, a 2-round iterative characteristic with (average) probability of $\frac{1}{51}$. The exact probabilities of this characteristic is $\frac{1}{57}$ and $\frac{1}{46}$ making the probability for a 13-round characteristic vary from 2^{-35} to 2^{-33}. It means that even in the worst case the characteristic is far better than the best characteristics for DES.

A   Appendix

In this section we give the proofs of the claims given in Sect. 3.3 and 3.4. Notation: Let Γ be an xor-sum of two inputs Y, Y^* to the F-function. Then $\Delta S(\Gamma)$ is the set of S-boxes, whose inputs are different after the E-expansion of Y and Y^*. Furthermore $\#\Delta S(\Gamma)$ denotes the number of S-boxes in $\Delta S(\Gamma)$. Example: Let $\Gamma = 0f000000_x$ (hex), then $\Delta S(\Gamma) = \{S1, S2, S3\}$ and $\#\Delta S(\Gamma) = 3$.

Note that xor-addition is linear in both the E-expansion and the P-permutation of DES. In the proofs below the following Tables and lemmata are used. Table 4 shows for each of the 8 S-boxes, which S-boxes are affected by the output of the particular S-box. Numbers with a subscript indicate that the particular bit affects one S-box directly and another S-box via the E-expansion. Example: If the output of $S1$ is 6_x (hex), then S-boxes 5 and 6 are directly affected and S-box 4 is affected after the E-expansion in the following round. Table 5 shows the *reverse* of Table 4, i.e. for every S-box it is shown which S-boxes from the preceding round affect the input.

Table 4. Where the bits from an S-box goes to

$S1 \rightarrow 3\,_2$	$5\,_4$	6	8		
$S2 \rightarrow 4\,_3$	$7\,_8$	1	5		
$S3 \rightarrow 6\,_7$	$4\,_5$	8	2		
$S4 \rightarrow 7$	$5\,_6$	3	$1\,_8$		
$S5 \rightarrow 2\,_3$	4	$7\,_6$	1		
$S6 \rightarrow 1\,_2$	$8\,_7$	3	5		
$S7 \rightarrow 8\,_1$	$3\,_4$	6	2		
$S8 \rightarrow 2\,_1$	7	4	$6\,_5$		

Table 5. Where the bits for an S-box come from

S1	S2	S3	S4	S5	S6	S7	S8
4 2 5 6	8 3 7 5	1 4 6 7	2 5 8 3	1 2 6 4	8 7 1 3	5 4 8 2	6 3 1 7

The next five lemmata follow from Table 4 and 5.

Lemma 1 *The six bits that make the input for an S-box, Si, come from six distinct S-boxes and not from Si itself.*

Lemma 2 *The middle six input bits for two neighbouring S-boxes come from six distinct S-boxes.*

Lemma 3 *The middle ten input bits for three neighbouring S-boxes come from all 8 S-boxes. Six of the ten bits come from six distinct S-boxes and four bits come from the remaining two S-boxes.*

Lemma 4 *The middle two bits in the input of an S-box Si, the inner input bits, come from two S-boxes, whose inner input bits cannot come from Si.*

Lemma 5 *Let Φ and Γ be two input sums, where $\Phi \rightarrow \Gamma$. If $\#\Delta S(\Phi) = \#\Delta S(\Gamma) = 2$ then for at least one S-box of $\Delta S(\Gamma)$ the inputs differ in only one bit.*

Theorem 2 *For twinxors, Γ and Φ, i.e. $\Gamma \rightarrow \Phi$ and $\Phi \rightarrow \Gamma$, the inputs to at least 5 S-boxes are different. That is, $\#\Delta S(\Gamma) + \#\Delta S(\Phi) \geq 5$.*

Proof: 1. $\#\Delta S(\Gamma) = 1$. The inputs to $\Delta S(\Gamma)$ differ in the inner input bits, i.e. at most two bits. Because of properties 2 and 4 of the DES S-boxes $\#\Delta S(\Phi) \geq 2$. The inputs of $\Delta S(\Phi)$ differ in at most one bit each. Because of property 2 the outputs of Φ differ in at least four bits. Therefore $\Phi \not\rightarrow \Gamma$.

2. $\#\Delta S(\Gamma) = 2$. Because of the symmetry of the characteristic we have immediately $\#\Delta S(\Phi) \geq 2$. There are two cases to consider:

a. $\Delta S(\Gamma)$ are not neighbours. Because of properties 2 and 4 the outputs of both S-boxes in $\Delta S(\Gamma)$ will differ in at least two bits, making $\#\Delta S(\Phi) \geq 3$ according to Table 4.

b. $\Delta S(\Gamma)$ are neighbours. From Lemma 2 it follows that the outputs of $\Delta S(\Phi)$ differ in at most one bit each. Property 2 requires the inputs of $\Delta S(\Phi)$ to differ in at least two bits each. From Table 4 it follows that the only way two neighbouring S-boxes in Γ can make the inputs of $\Delta S(\Phi)$ differ in at least two bits each, is when $\#\Delta S(\Phi) = 3$. This is however not possible for all two neighbouring S-boxes. For example let $\Delta S(\Gamma) = \{S5, S6\}$, then it is possible to get $\Delta S(\Phi) = \{S1, S2, S3\}$ where for each S-box the inputs differ in two bits. But for $\Delta S(\Gamma) = \{S1, S2\}$ there will always be at least one S-box in $\Delta S(\Phi)$, whose inputs differ in only one bit.

3. $\#\Delta S(\Gamma) \geq 3$. Because of the symmetry of twinxors $\#\Delta S(\Phi) \geq 2$. $\qquad\square$

We want to show that there is no 4-round iterative characteristic with a probability higher than the best 2-round iterative characteristic concatenated with itself. First we prove

Theorem 3 *For a 4-round iterative characteristic with input sums Γ, Φ and Ψ, see Section 2.2,*
$$\#\Delta S(\Gamma) + \#\Delta S(\Phi) + \#\Delta S(\Psi) \geq 7.$$
Furthermore, for at least one of the input sums, the inputs to three neighbouring S-boxes differ.

Proof: We are looking for input sums Γ, Φ and Ψ, such that $\Gamma \rightarrow \Phi$, $\Psi \rightarrow \Phi$ and $\Phi \rightarrow \Gamma \oplus \Psi$. Note that $\Delta S(\Gamma) \cap \Delta S(\Psi) \neq \emptyset$ and that if $\Delta S(\Gamma)$ are neighbours then so are $\Delta S(\Psi)$.

1. $\#\Delta S(\Gamma) = 1$. From the proof of Theorem 2 we have $\#\Delta S(\Phi) \geq 2$, and each of the inputs to those S-boxes differ in exactly one bit.

a. $\#\Delta S(\Phi) = 2$. The S-boxes in $\Delta S(\Phi)$ are not neighbours and the inputs differ in one inner input bit, therefore each of the outputs differ in at least two bits. From a close look at Table 4 it follows that if $\Delta S(\Gamma) = S7$ then it is possible to get $\#\Delta S(\Psi) = 3$, but then for one S-box $\in \Delta S(\Psi)$, not $S7$, the inputs differ in only one bit, an inner input bit. If $\Delta S(\Gamma) \neq S7$ then $\#\Delta S(\Psi) \geq 4$ and for at least one S-box, not $\Delta S(\Gamma)$, the inputs differ in only one bit. Therefore $\Psi \not\rightarrow \Phi$.

b. $\#\Delta S(\Phi) \geq 3$. The outputs for every S-box of $\Delta S(\Phi)$ differ in at least two bits. It follows easily from Table 4 that $\#\Delta S(\Gamma \oplus \Psi) \geq 4$. Since $\Delta S(\Gamma) \subseteq \Delta S(\Psi)$, $\#\Delta S(\Psi) \geq 4$.

2. $\#\Delta S(\Gamma) = 2$. By the symmetry of the characteristic $\#\Delta S(\Psi) \geq 2$ and therefore $\#\Delta S(\Phi) \leq 3$. There are two cases to consider:

a. $\Delta S(\Gamma)$ are not neighbours. Because of properties 2 and 4 $\#\Delta S(\Phi) \geq 3$ leaving only the possibility that $\#\Delta S(\Psi) = 2$ and $\#\Delta S(\Phi) = 3$. The S-boxes in $\Delta S(\Phi)$ must be neighbours. If not, let Si be an isolated S-box, different in the inputs in only inner bits. The outputs of Si differ in at least two bits, that must go to the inner bits of the two S-boxes in $\Delta S(\Gamma)$, since $\Delta S(\Gamma) = \Delta S(\Psi)$. But that is not possible according to Lemma 4.

b. $\Delta S(\Gamma)$ are neighbours.

    i) $\#\Delta S(\Phi) = 1$. The outputs of $\Delta S(\Phi)$ differ in at least two bits. From Table 4 it follows easily that for at least one S-box $\in \Delta S(\Psi)/\Delta S(\Gamma)$ the inputs differ in only one bit and $\Psi \not\to \Phi$.

    ii) $\#\Delta S(\Phi) = 2$. Assume that $\#\Delta S(\Psi) = 2$. If $\Delta S(\Gamma) = \Delta S(\Psi)$ then the outputs of $\Delta S(\Phi)$ can differ in at most one bit each, according to Lemma 2. But by Lemma 5, the inputs of at least one S-box in $\Delta S(\Phi)$ differ in only one bit, a contradiction by property 2. Therefore $\Delta S(\Gamma) \neq \Delta S(\Psi)$. Since $\Delta S(\Gamma) \cap \Delta S(\Psi) \neq \emptyset$ and $\Delta S(\Gamma)$ are neighbours we must have $\Delta S(\Gamma) = \{S(i-1), Si\}$ and $\Delta S(\Psi) = \{S(i), S(i+1)\}$ or vice versa. The outputs from $S(i-1)$ in Γ must be equal as must the outputs from $S(i+1)$ in Ψ. Therefore $\Gamma \oplus \Psi$ must have the following form (before the expansion):

$$S(i-1) \,\|\, S(i) \,\|\, S(i+1) = 0xyz \,\|\, 1**1 \,\|\, 0vw0 \,,$$

where '*' is any bit, $xyz \neq 000$ and $vw \neq 00$. From Table 5 it follows that $\Phi \not\to \Gamma \oplus \Psi$ for $\#\Delta S(\Phi) = 2$ and therefore $\#\Delta S(\Psi) \geq 3$.

    iii) $\#\Delta S(\Phi) = 3$. Then $\#\Delta S(\Psi) = 2$. If $\Delta S(\Gamma) \neq \Delta S(\Psi)$ then the differences in the inputs to Φ is the effect of one S-box. For every S-box in $\Delta S(\Phi)$ the inputs differ in only one bit, therefore $\Phi \not\to \Gamma \oplus \Psi$. By similar reasoning we find that for both S-boxes in $\Delta S(\Gamma)$ the outputs have to differ. Furthermore $\Delta S(\Phi)$ are neighbours. Assume that they are not. Then the outputs of the isolated S-box differ in at least two bits and from Table 4 it follows that they affect at least 2 not neighbouring or 3 neighbouring S-boxes, a contradiction with $\Delta S(\Gamma) = \Delta S(\Psi)$.

3. $\#\Delta S(\Gamma) = 3$. Because of the symmetry in the characteristic we already covered the cases where $\Delta S(\Psi) < 3$. Therefore $\#\Delta S(\Gamma) = \#\Delta S(\Psi) = 3$ and $\#\Delta S(\Phi) = 1$. $\Delta S(\Gamma)$ must be neighbours. Furthermore $\Delta S(\Gamma) = \Delta S(\Psi)$ otherwise $\Phi \not\to \Gamma \oplus \Psi$. $\qquad\qquad\Box$

Theorem 4 *There are no 4-round iterative characteristics with a probability higher than $(\frac{1}{234})^2$.*

Proof: From the proof of Theorem 3 we find that to have a 4-round iterative characteristic, the inputs to seven S-boxes must be different in the three nonzero rounds. Furthermore for at least one round the inputs to three neighbouring S-boxes must be different. There are three cases to consider. Case A: By Lemma

	$\Delta S(\Gamma)$	$\Delta S(\Phi)$	$\Delta S(\Psi)$
Case A	2	2	3
Case B	2	3	2
Case C	3	1	3

5 we know that for at least one S-box in $\Delta S(\Phi)$ the inputs differ in only one bit. Furthermore for at least one of the three neighbouring S-boxes in $\Delta S(\Psi)$ the outputs must be equal, otherwise $\Gamma \not\to \Phi$. There are two cases to consider:

1. For both S-boxes in $\Delta S(\Phi)$ the inputs differ in only one bit. By property 2 the outputs differ in at least two bits each. For every three neighbouring S-boxes in Ψ we know the only two possible S-boxes of $\Delta S(\Phi)$ by Lemma 3 and Table 5. Example: If $\Delta S(\Psi) = \{S1, S2, S3\}$ then $\Delta S(\Phi) = \{S5, S6\}$. Furthermore the outputs of either $S1$ or $S3$ must be equal.

   We have eight triples of three neighbouring S-boxes in Ψ to examine and from Table 4 and 5 it follows that there are only three possible values for $\Delta S(\Psi)$ and $\Delta S(\Phi)$. From the *pairs xor distribution table* we find that the best combination for $\Psi \to \Phi$ has probability $\frac{8 \times 12 \times 10}{64^3}$. But then the probability for a 4-round iterative characteristic $P(4R) \leq \frac{1}{4^4} \times \frac{8 \times 12 \times 10}{64^3} < (\frac{1}{234})^2$.

2. For one of the S-boxes in $\Delta S(\Phi)$ the inputs differ in one bit, for the other S-box the inputs differ in two bits. For every three neighbouring S-boxes of Ψ there are only two possibilities for the S-box in $\Delta S(\Phi)$, whose inputs differ in only one bit. From a closer look at Table 4 it follows that $\Delta S(\Phi)$ must be neighbours and there are only two possible values for $\Delta S(\Psi)$ and $\Delta S(\Phi)$. From the *pairs xor distribution table* we find that the best combination for $\Psi \to \Phi$ has probability $\frac{12 \times 10 \times 4}{64^3}$. But then the probability for the 4-round iterative characteristic $P(4R) \leq \frac{1}{4^4} \times \frac{12 \times 10 \times 4}{64^3} < (\frac{1}{234})^2$.

Case B: The three S-boxes in $\Delta S(\Phi)$ are neighbours. From the proof of Theorem 3 we have $\Delta S(\Gamma) = \Delta S(\Psi)$. Then by Lemma 2 the outputs of each of the three neighbouring S-boxes in $\Delta S(\Phi)$ can differ in at most one bit, therefore the inputs must differ in at least two bits each by property 2. Then it follows from Table 5 that for each of the S-boxes in $\Delta S(\Gamma)$ the outputs must differ in two bits. For every triple of three neighbouring in $\Delta S(\Phi)$ there is only one possible way for the inputs to differ and only one possibility for $\Delta S(\Gamma)$. The best combination of $\Delta S(\Gamma)$ and $\Delta S(\Phi)$ gives a probability for the 4-round iterative characteristic $P(4R) \leq \frac{12 \times 12 \times 16 \times (8 \times 4)^2}{64^7} < (\frac{1}{234})^2$.

Case C: From Theorem 3 we have $\Delta S(\Gamma) = \Delta S(\Psi)$. The only possibility we have for a 4-round iterative characteristic of this kind is when $\Delta S(\Gamma) = \{S2, S3, S4\}$

and $\Delta S(\Phi) = \{S7\}$. The best combinations yields a probability for the 4-round iterative characteristic $P(4R) \leq \frac{1}{4^4} \times \frac{14 \times 8 \times 8}{64^3} < \left(\frac{1}{234}\right)^2$. $\qquad\square$

References

1. Eli Biham, Adi Shamir. *Differential Cryptanalysis of DES-like Cryptosystems.* Journal of Cryptology, Vol. 4 No. 1 1991.
2. Eli Biham, Adi Shamir. *Differential Cryptanalysis of Snefru, Khafre, REDOC-II, LOKI and Lucifer.* Extended abstract appears in Advances in Cryptology, proceedings of CRYPTO 91.
3. Eli Biham, Adi Shamir. *Differential Cryptanalysis of the full 16-round DES.* Technical Report # 708, Technion - Israel Institute of Technology.
4. Eli Biham, Adi Shamir. *Differential Cryptanalysis of Feal and N-Hash.* Extended abstract appears in Advances in Cryptology, proceedings of Euro-Crypt 91.
5. Y. Desmedt, J-J. Quisquater, M. Davio. *Dependence of output on input in DES: Small avalanche characteristics.* Advances in Cryptology: Proceedings of CRYPTO 84. Springer Verlag, Lecture Notes 196.
6. Kwangjo Kim. *Construction of DES-like S-boxes Based on Boolean Functions Satisfying the SAC.* To appear in the proceedings from ASIACRYPT'91, Lecture Notes, Springer Verlag.
7. Lars Ramkilde Knudsen. *Cryptanalysis of LOKI.* To appear in the proceedings from ASIACRYPT'91, Lecture Notes, Springer Verlag.
8. Xueija Lai, James L. Massey, Sean Murphy. *Markov Ciphers and Differential Cryptanalysis.* Advances in Cryptology - EUROCRYPT'91. Springer Verlag, Lecture Notes 547.
9. Lawrence Brown, Josef Pieprzyk, Jennifer Seberry. *LOKI - A Cryptographic Primitive for Authentication and Secrecy Applications.* Advances in Cryptology - AUSCRYPT '90. Springer Verlag, Lecture Notes 453, pp. 229-236, 1990.

DES is not a Group

Keith W. Campbell　and　Michael J. Wiener

Bell-Northern Research, P.O. Box 3511 Station C, Ottawa, Ontario, Canada, K1Y 4H7

Abstract. We prove that the set of DES permutations (encryption and decryption for each DES key) is not closed under functional composition. This implies that, in general, multiple DES-encryption is not equivalent to single DES-encryption, and that DES is not susceptible to a particular known-plaintext attack which requires, on average, 2^{28} steps. We also show that the size of the subgroup generated by the set of DES permutations is greater than 10^{2499}, which is too large for potential attacks on DES which would exploit a small subgroup.

1. Introduction

The Data Encryption Standard (DES) [3] defines a set of permutations on messages from the set $M = \{0, 1\}^{64}$. The permutations consist of encryption and decryption with keys from the set $K = \{0, 1\}^{56}$. Let $E_k: M \to M$ denote the encryption permutation for key k, and let E_k^{-1} be the corresponding decryption permutation. If the set of DES permutations were closed under functional composition, then for any two permutations t and u, there would exist some other permutation v such that $u(t(m)) = v(m)$ for all messages $m \in M$.

The question of whether the set of DES permutations is closed under functional composition is an important one because closure would imply that there exists a known-plaintext attack on DES that requires, on average, 2^{28} steps [4]. Furthermore, multiple encryption would be susceptible to the same attack because multiple encryption would be equivalent to single encryption.

Kaliski, Rivest, and Sherman developed novel cycling tests which gave evidence that the set of DES permutations is not closed [4]. However, their work relied upon randomness assumptions about either DES itself or a pseudo-random function $\rho: M \to K$ which was used in cycling experiments. Because of the randomness assumptions, it is difficult to use the results of their cycling tests to make any claims about the probability that DES is not closed.

We have developed our own DES cycling experiments which provide evidence that DES is not closed; this evidence does not rely upon randomness assumptions. Our cycling experiments are similar to those of Quisquater and Delescaille for finding DES collisions [7, 8]. Other recent related work is the switching closure tests of Morita, Ohta, and Miyaguchi [6].

Don Coppersmith has developed an approach to finding a lower bound on the size of the subgroup generated by the DES permutations [1]. He has shown this lower bound to be greater than the number of DES permutations, providing conclusive proof that DES is not closed.

Section 2 contains the new probabilistic argument against closure which relies upon the ability to find a set of four keys which quadruple-encrypt a particular plaintext message to a particular ciphertext message. Finding such four-key mappings can be done with an approach similar to finding DES collisions. In Section 3, we review previous work in collision finding and build up to the new method of finding four-key mappings. Section 4 contains further details on our experiments. In Section 5, we describe Don Coppersmith's approach to obtaining a lower bound on the size of the subgroup generated by the DES permutations, thereby proving that DES is not closed. We also discuss our results based on his approach.

2. Strong Evidence Against Closure

We begin with the hypothesis that the set of DES permutations is closed and search for a contradiction. Let S_p be the set of messages that can result from encrypting or decrypting a particular message p with any DES key. Because there are 2^{56} keys, S_p contains at most 2^{57} messages. From the hypothesis, S_p is also the set of all possible messages which can result when multiple permutations are applied to p. If a message $c \in M$ is selected at random, the probability that $c \in S_p$ is at most $2^{57}/2^{64} = 2^{-7}$. We selected 50 messages at random (by coin tossing), and for each random message c, we searched for a set of permutations which map p to c using $p=0$ in each case. In all 50 cases we found a set of four DES keys i, j, k, and l such that $E_l(E_k(E_j(E_i(p)))) = c$ (see Appendix). Therefore, $c \in S_p$ and the probability of this event occurring 50 times is at most $(2^{-7})^{50} = 2^{-350}$. Because this is an extremely unlikely occurrence, we must conclude that the original hypothesis is incorrect and the set of DES permutations is (almost certainly) not closed under functional composition.

The argument above does not rely upon any assumptions about the randomness of DES or any other function; the fact that four keys exist which map p to c for each randomly selected message c is sufficient to draw the conclusion. However, the method used to find the four keys in each case does rely upon randomness assumptions.

3. Collision Finding

The method used to find four keys which map one message to another is similar to the approach taken by Quisquater and Delescaille in finding DES collisions[1] [7]. In both cases a function $f:M \to M$ and an initial message x_0 are chosen which define the sequence $x_{i+1} = f(x_i)$ for $i = 0, 1, \dots$. Because M is finite, this sequence must eventually fall into a cycle. Unless x_0 is in the cycle, the sequence consists of a leader flowing into a cycle. The algorithms described by Sedgewick, Szymanski, and Yao [9] can be used to find the leader

---
[1] We have a DES collision when $E_i(m) = E_j(m)$ for some $m \in M$, and $i, j \in K$, $i \neq j$.

length λ and the cycle length μ. If $\lambda \neq 0$, this leads directly to finding a collision in f (i.e., $a, b \in M$ such that $f(a) = f(b)$, $a \neq b$, see Figure 1).

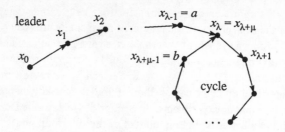

Figure 1. Leader and Cycle in a Sequence

DES Collisions

To find DES collisions, Quisquater and Delescaille used the function $f(x) = E_{g(x)}(m)$, where $g: M \to K$ takes a message and produces a key for DES encryption, and m is a fixed message. In this case, a collision in f is not necessarily a DES collision; if $f(a) = f(b)$, $a \neq b$, but $g(a) = g(b)$, then we have found a pseudo-collision where the keys are the same. Because there are fewer keys than messages, there can be at most $|K|$ distinct outputs from f. Assuming that DES is random and a suitable function g is selected, the probability of a collision in f leading to a DES collision is about $|K|/|M| = 2^{-8}$, and the expected time required to find a collision in f is on the order of $\sqrt{|K|} = 2^{28}$. Thus, the overall work factor in repeating this procedure until a DES collision is found is about $2^{28}/2^{-8} = 2^{36}$. This can be reduced somewhat using the method of distinguished points [7].

Two-Key Mapping

The method of finding DES collisions above was extended by Quisquater and Delescaille to find pairs of keys which double-encrypt a particular plaintext p to produce a particular ciphertext c [8]. In this case, collisions were found between two functions $f_1(x) = E_{g(x)}(p)$ and $f_0(x) = E_{g(x)}^{-1}(c)$. Given messages a, b such that $f_1(a) = f_0(b)$, $g(a)$ and $g(b)$ are a pair of keys with the desired property (i.e., $E_{g(b)}(E_{g(a)}(p)) = c$). To find a collision between f_1 and f_0, define the function f as follows:

$$f(x) = \begin{cases} f_1(x) & \text{if a particular bit of } x \text{ is set} \\ f_0(x) & \text{otherwise} \end{cases} \tag{1}$$

The particular bit that is used to choose between f_1 and f_0 is called the *decision bit*.

If DES is random, then we can expect collisions found in f to be collisions between f_1 and f_0 about half of the time. This increases the expected work factor from 2^{36} in the single-DES collision case to 2^{37} in this case.

Four-Key Mapping

The double-encryption collision finding above can be applied directly to the problem discussed in Section 2 of finding a set of permutations which map p to c. However, we improved upon this approach by searching for four keys rather than two. We chose different functions f_1 and f_0:

$$f_1(x) = E_{h(x)}(E_{g(x)}(p)) \quad \text{and} \quad f_0(x) = E_{h(x)}^{-1}(E_{g(x)}^{-1}(c)) \tag{2}$$

where functions g and h produce keys from messages, and the ordered pair $(g(x), h(x))$ is distinct for all $x \in M$. This approach doubles the number of encryptions which must be performed at each step of collision finding, but it eliminates the possibility of pseudo-collisions. The expected number of steps required to find a collision in f in this case is on the order of $\sqrt{|M|} = 2^{32}$. To compare this running time to the two-key mapping above, we should take into account that fact that this approach requires two DES operations at each step instead of one. Also, only about half of the collisions in f are collisions between f_1 and f_0. Thus, assuming that DES is random, the work factor in finding four keys with the required property is about 2^{34}, which is eight times faster than finding a two-key mapping. The speed-up may be less than a factor of eight if the method of distinguished points is used for finding two-key mappings.

4. Further Details on the Cycling Experiments

In the cycling experiments, four-key mappings were sought as described in section 3 using the functions f, f_1, and f_0 in equations (1) and (2). The functions g and h in equation (2) were selected for ease of implementation. In the DES document [3], keys are represented in 64 bits with every eighth bit (bits 8, 16, ..., 64) a parity bit,[1] leaving 56 independent bits. The function g produces a key from a message by converting every eighth bit into a parity bit. Function h produces a key from a message by shifting the message left one bit, and then converting every eighth bit into a parity bit. Note that the ordered pair $(g(x), h(x))$ is distinct for all $x \in M$ so that there is no possibility of pseudo-collisions.

As a test, a four-key mapping was sought for $p = c = 0$. This value of c is not one of the 50 randomly-selected values which contribute to the argument in section 2. Using bit number 30 as the decision bit and an initial message x_0=0123456789ABCDEF (hexadecimal) yielded a collision between f_1 and f_0 with the following results:

λ = 1143005696 (decimal)
μ = 2756683143 (decimal)
keys: 8908BF49D3DFA738, 10107C91A7BF4C73,
        4CEF086D6ED662AD, A7F7853737EAB057 (hexadecimal)

The results for the 50 random values of c are given in the Appendix. There were no additional values of c which were tried. This is important because failure for some values of c would greatly diminish the confidence in the conclusions drawn in section 2.

---

[2] In the DES document [3], bits of a message are numbered from 1 to 64 starting from the leftmost bit.

These experiments were conducted over a four-month period using the background cycles on a set of workstations. The average number of workstations in use over the four-month period was about ten, and in the end, more than 10^{12} DES operations were performed.

5. Conclusive Proof that DES is not Closed

In an as yet unpublished paper, Don Coppersmith described his latest work on finding a lower bound on the size of the subgroup, G, generated by the DES permutations [1]. He takes advantage of special properties of E_0 and E_1 (DES encryption with the all 0's and all 1's keys).

In earlier work [2], Coppersmith explained that the permutation E_1E_0 contains short cycles (of size about 2^{32}). This makes it practical to find the length of the cycle produced by repeatedly applying E_1E_0 to some starting message. Each of these cycle lengths must divide the order of E_1E_0. Therefore, the least common multiple of the cycle lengths for various starting messages is a lower bound on the order of E_1E_0. Also, the order of E_1E_0 divides the size of G. This makes is possible to get a lower bound on the size of G.

Coppersmith found the cycle lengths for 33 messages which proved that the size of G is at least 10^{277}. We have found the cycle lengths for 295 additional messages (see Table 2 in the Appendix). Combining our results with Coppersmith's yields a lower bound on the size of the subgroup generated by the DES permutations of 1.94×10^{2499}. This is greater than the number of DES permutations, which proves that DES is not closed. Also, meet-in-the-middle attacks on DES which would exploit a small subgroup [4] are not feasible.

It is interesting to note that in the course of investigating the cycle structure of weak and semi-weak DES keys in 1986 [5], Moore and Simmons published 5 cycle lengths from which one could have concluded that G has at least 2^{146} elements and that DES is not closed.

6. Conclusion

We have given probabilistic evidence as well as conclusive proof that DES is not a group. Furthermore, the subgroup generated by the DES permutations is more than large enough to prevent any meet-in-the-middle attacks which would exploit a small subgroup.

Acknowledgement

We would like to thank Alan Whitton for providing a large portion of our computing resources.

References

1. D. Coppersmith, "In Defense of DES", personal communication, July 1992 (This work was also described briefly in a posting to sci.crypt on Usenet News, 1992 May 18).

2. D. Coppersmith, "The Real Reason for Rivest's Phenomenon", *Advances in Cryptology - Crypto '85 Proceedings*, Springer-Verlag, New York, pp. 535-536.

3. *Data Encryption Standard*, Federal Information Processing Standards Publication 46, National Bureau of Standards, U.S. Department of Commerce, Washington, DC (1977 Jan. 15).

4. B.S. Kaliski, R.L. Rivest, and A.T. Sherman, "Is the Data Encryption Standard a Group? (Results of Cycling Experiments on DES)", *Journal of Cryptology*, vol. 1 (1988), no. 1, pp. 3-36.

5. J.H. Moore and G.J. Simmons, "Cycle Structure of the DES with Weak and Semi-weak Keys", *Advances in Cryptology - Crypto '86 Proceedings*, Springer-Verlag, New York, pp. 9-32.

6. H. Morita, K. Ohta, and S. Miyaguchi, "A Switching Closure Test to Analyze Cryptosystems", *Advances in Cryptology - Crypto '91 Proceedings*, Springer-Verlag, New York, pp. 183-193.

7. J.-J. Quisquater and J.-P. Delescaille, "How easy is collision search? Application to DES", *Advances in Cryptology - Eurocrypt 89 Proceedings*, Springer-Verlag, New York, pp. 429-434.

8. J.-J. Quisquater and J.-P. Delescaille, "How easy is collision search. New results and applications to DES", *Advances in Cryptology - Crypto '89 Proceedings*, Springer-Verlag, New York, pp. 408-413.

9. R. Sedgewick, T.G. Szymanski, and A.C. Yao, "The complexity of finding cycles in periodic functions", *Siam Journal on Computing*, vol. 11 (1982), no. 2, pp. 376-390.

Appendix: Results of Cycling

For each of 50 randomly selected messages c, Table 1 shows four DES keys i, j, k, and l such that $E_l(E_k(E_j(E_i(0)))) = c$. In each case, the initial message $x_0 = $0123456789ABCDEF was used. The DES keys in the table include eight parity bits as defined in the DES document [3]. The table also shows information from the collision search including the decision bit, the leader length λ, and the cycle length μ. All quantities are shown in hexadecimal except the decision bit, λ, and μ which are shown in decimal.

Table 2 lists the cycle lengths obtained by applying the $E_1 E_0$ permutation to various messages.

Table 1: Four-Key Mapping Results

ciphertext c	bit	λ	μ	key i	key j	key k	key l
D239854662E333D6	30	3089881971	1373508256	1C988A57AE5B1A6D	383315AE5EB637D9	754F0E80DC32C289	3B2607C16E196145
8611217BD3236C8A	30	1914494444	693463224	107658E66100061EF	20EFB0CDC219C1DC	7340CBB02FD9769E	B920E5D916ECBACE
0ADD04B7CDF1E742	30	6780759472	218463638	467C1C113CBDB085	8CF83B2370986208	51F16B0B324C0E08	A8FB5049BA78604
F30123D00EABAC2B	30	943553743	7310553453	07F1DE543D9D98F7	0EE3BFA87A3B31EC	C4A167E5BA26C449	E351B3F2DC1362A4
3A31BB9ABC6020F	30	4742565084	344544569	910E9DB0AE23103B	201C386246C72076	CE76646EC1255438	E63B32B6E0132A1C
7B5C224B59F14F2	30	2627610479	3114335933	131F49C8C1B9C11A	25AD91928013B034	8F04A26BA1C869E	46025134520E43CE
E4997DCBCAEBDB8F	30	5859776140	304193764	1562DA295E800E52	2AC7B652BC021FA7	89CE994B979F107	91FEC1162C94020E
FAA52FDBF51EAE92	30	20629979	4238943918	B68FB3B7C9E2FFD7F	6D1F76F83D5DFBFE	23FD832C5B29041C	76BF51DF9E926216
FA59DADF9170F1A7	30	2195424046	5907332685	98B58AC832DCAEA7	320B169267B95E4F	EC7FA2BF1E26C72F	0B6B8601C17CB0FD
79D117609B40BB21	30	433231912	2154862305	73ABDA4698D06E76	E657B68C31A1DCEC	15D60E0183FB62FB	02F1806DBADC5DA8
81B5D8A98B54D867	30	2552785502	1283047449	CE43FE26157033BC	9E85FE4C2AE36419	04E001D975BABA51	CBE5310EA826C2CE
751B8E4CCCB4F92C	30	2255145649	4136209653	2CBF6D135B8E367FD	5B7FDA25B5C4CDFB	97CB621C524C869E	75B5250B3DD01CD0
C6E4D4B4C74306D0	30	5247598183	2207575116	13107994329D808BA	2523F22A51AD1075	E9664A167AA13BA2	D389164638C2DC15
C775538756644E18	30	1923363275	3322757394	E92AEF709B3715EF	D057DFE0346E29DF	A4132F8F7085BA29	0EBF01DA401634DF
0BDE8BEED757013BB	30	6621189834	912287793	70948A077C5DFB52	E32A150DF8BAF4A7	1C7C02B5802F6BBF	07DC86C4DA077580
332D8679402816A4	30	248767409	6266969674	B0168C3704C82CBA	612F1A6E08975B75	0E990EB99860DE901	3794B5FE5DE9757A
D0B38FD1E7E1D031	30	3811140202	2843424010	461689D3856B6EFE	8C2F13A70BD6DFFD	837AC84676E0FF29E	86ADC113681C839E
668289C44AF40EF2	30	3619413642	1147320998	CE731CEA1AC01FEC	9DE53BD637923ED9	6296BFEBAD3EAF4	E5F7C48AA8E5C25B
AABEEEC736FF5B1B	30	3069331961	2005622212	07AB084A761A16FE	0E571394EC372FFD	BA04CE85BA4373B0	1FADD001A80752D9
2079AF9D6F2C2004	30	3906585734	1184537771	9ECB0BE376314F1D	3E9716C7EF619D79	FD2A2D0A084F4991	EAFD45A7AB8334F1
F8CF00BC81569693	30	193983657	2169973674	D97F6E51F40E6D7A	8FABF7AE46B9B3F1	0D588025D338043E	163DB6924C5E4319
644E441870C966C	27	846318690	328914116	C7D57A5723DC58F8	7637F201B3DC7A5B	C8EF8A1552CB85B5	29B3A2310IE61A10
9049A1BB57D5E5ED	27	1630497906	1441467245	BA9B7980D9EFBCAD	20A4752AF7544391	3E58A20510DA7B0	7089F7D332A83D91
C9B5D6BD2C8BAACA	27	3233207975	423171687	9152BA947A2AA149	7C6197C726DA3B0B	2C796D269BBC8632	7623A4FDB0E0E0C4
CE3A8663179E4B840	27	10304236841	1262357982	BFB0CBE313EC9D85	F7A2D07F6B8346C2	5267456201CE3720	C473AEC1D6C7E389
CD1D7C7655036088	27	2384563551	5311633846	DACE2F01C420ECF4	B69D5E028940D9EA	E310ECA464527920	98BF452A0216C4E5
CD24C9EC7971EB73	27	5122120604	261361938	929119DA7570C22A	252332B6EAE08554	EC454AF861CC18A	9D10319B8E580E9D5
205E69FBA5452EC6	27	10545326230	317843105	8C040B38A7CE46BA	190815704F9E8C76	57D0CE6461200D5D3	9DB62F3852A1F2CD
B7991B774FBDE001	27	2401212140	1891297161	643DF2ABBCBF64AB	CB79E6547A1FC854	8AE55E83AE89CC410	EF739402CD807F57
BB14D70990EFEC85	27	1975301710	5537484933	465773EF78CF4AE	8CAEAE7CEC1AEA5D	321C8A54072F8ACB	E654C80RBFB107C62
0A0BE3344B6A26CF1	24	6114881693	3979311181	FEE938C2B0BAA26B	FD5270B56275446D6	3B206233C802D3AB	2AB083F28C83E325
7F306674557B57C7	24	3329518115	5406967207	7C0B2CA88A3B5BCD	F8165D5415765B59B	016D5D73A743E59B	543B3B98551F726
7D4EC4D9D0D88AFB	24	6829918624	455335254	76A1BAAD80152F6D	FEF40755801295DDA	DFE529049801FEAD	92C72A3D1F3426D9
CD588CE10F137436	24	3325700738	674560323	8A985C2C1049BEC13	1631A7800834D925	CDAB9116F423F8C7	32DC44B0BF70377F
57B8BB0DFADB4B17	24	6955122751	337345258	1C3797FE83F1A861	3B6EFFE04E351C2	546107E61964C449	492F23ECFE5240D5
13F59D60E96745C1	24	7238487367	2567725571	9E76D6F2985D3D25	3DEFADE531B97A4A	E986677008A1EF4C	
9CCC5AA115D52D97	24	3122231806	936075576	E58985732AA88C32	CB130BE657521964	268C547A3D684CB0	
9C1C3C217847F73	24	2505406823	3622543567	9713F7D66BDCD54A	2F25ECAED59DAB97	67B94A627FE36EBE	
9223537B7C7391A4	24	2577190833	1693700608	73E5E5E6C43D75F4	E5C8C8D8979E9E9	925E45D9FEA783A8	
F584C208B15488FF	24	1398059397	1014634505	3E57AEA7F17C200E	7FAE5D4CE0F8431C	343EBCF8AEA213E3	
D88F5045AC15612D	24	3276162424	1167926168	B55D923DA4EAD068	68B9257949D6A1D0	86BA2CD646082C07	1A1FDFFD6D0008F1
79E11F13178IC081	24	3245567395	2355021803	F13E7CD37AB99D13	E07FF8AF4703825	C83EC43DA285C1FE	DA4516EAA2851602
6CD19C16143DB1B1	24	686670224	741307196	6DC6B294FA923157		6B19FD3152EA5DF2	649EE39E5143E07F
2760D762F03022C	24	5357109959	1933193087	C7C52DBF5D3D508	438CA7BC6BA4AB10	20A1ED06BA08CB	348C7F98A8752FF8
CB354D2FF5FF4048	24	2971721429	2992595032	D9CEED6219BCFD83	2EC9BE53D97D976	S1D5B5E943E5A207	9151CEE934158564
DE639517D229E809	21	4623799728	500111498	7610B0F452A479AD	8E2062EB974AF258	01F173656764975	017ABC73332CB0B
F24793142CC1A3B8	21	3617826299	60533389	FE2FFD672A4545	ADFD5EFDCE578A89	C1736BB57961BA9E	613834DABCB0DC4F
D16E1IE55A7D90EE	21	119490113	1525366355	A1C752DBF5D3D508	D32E3542662AE	340EEA3DFDF104BC	1A86F41F7F79025E
21110855E8A87EB3	21	4397358172	2763618490	EC949EDF292F89BA	B39D9BC4321AFB07	51D5B5E943E5A207	A86BDACE20F25102
42D93C3251BE47CD	18	1945651024			D92A3EBF515D1075	E07C79041331AE8F	F1BF3BD28998D646

Table 2: Cycles in E_1E_0

Fixed Point	Cycle Length
FCF4DFCDCD6258EF	28737542
2637A924F58B74BD	52726102
5FE79E047C375C9E	87605490
F86F776A3F29D215	120183041
453781698641582	123741142
EB147AC7621EA6DB	141528875
A22C41175610DD0A	157126532
964C03BF6D9484CE	180757910
9A8F18520C494C0F	181353093
FE5532899F4D01FC	204877793
D3C92F24ECD607A5	229430263
7B4E903C419FEE77	241491405
D1BF57C1681B0239	241970136
B1C537BD77E825CD	274132024
74AF3228EAA0ADE2	277651190
2F03BED91D7CB16E	286320467
63FD39D83034OD5F	311120314
8CB3ECBFADD205B3	337827436
F16941B1534C8CB	346375060
BAE2F389EDCDD00C	366197309
E8B79DDBA4FDBA13	370898345
23F50E3C63854946	382784102
F4FEB022D6662CC6	385833869
A382D83B6FB435C3	404923308*
9630287343696D5	417479850
63353539521C31D6	448409291*
811F3718D8F04175	467147934
6DE9D894880190A0	508130786
24D57E95080A4FB0	527729106
92B6FDDCEFE9CF2D	541798255
F2DFEBB3E98D9D5	543178224
DFFFD1E6420lDA17	549298502
B5E233FC14574CDA	559949983
022BED7A22E59128	572474003
4A7B7657BEE8666	607033653
FA2C0E43E4665530	628125220
C533E49F19CA472E	654423452
E72B8A2EAE9B13B	678517304
1883A14E567687FF	681583312
66A0CC36E6F3F100	709905971
B4E9F57BD0F8B679	726834017
D99C26BC6DA73936	766356532
172FEF9F90B6228D	767546084
9E57118I1C12E4DAB	794419263
8FA5A8261FC20EFA	805683389
FDB78F429EEADFA4	823007021
12C328347DF3EAE8	862573395
5FE93A859DAD6C29	870494059
FD30744EFF9EA757	883855821
ACE0A89879991A8F5	903017135
DAEF18D6317C75F0	935440566*
AA757AEB74AC08BC	954473685
F0591F59BD1C79D1	962933872
04968AFAB317659	1019170568
191CC6BFB3252119	1035340219
C4DEF2633D6B2BAD	1046106143
1F5A614311 5FA46B	1056029096
AC3C22BAF7113361	1078117118
FD4ADA2B652DAF14	1099417692
34BFC05A291EFCB8	1099384916
1DEB703B3971041A	1102596768
593D785FEECB2E11	1124554449
EC93D670OEF9D1E9	1139668928*
0BEF1110DC771C55	1160096502
9A28778C1832A029	1259919806
DF45B97314256B6F	1270969573
CB979FAD005CA52A	1288329310*
30AD6A3EB26D77B0	1295682916
106DB845E41BB505	1316780514
D95558388 74F07A8	1329512762
C75D3DEA4B3F8F92	1333813692
A798A0EC64F530D6	1362776543
A507CDD9A0E37CDD	1377253295
733C24355AF016C4	1408952249
582AE8818F89CED8	1411745523*
73D1BEF31FF43DE7	1440389551
272004C5A8DC8C1B	1452837055*
4E80AED88C4D7447	1456332586
54A3323DBC545563	1457951391
A3D1FAD47B65B2CE	1481121159
6107 23A4A638B148	1555624211
267CE6D2F57D4C4C	1572366534
116B35ABAC82B83D	1596684580
E5A7FB895DBB4283	1624144990
74ED56B5B009873	1646234340
77404 2FF322B933B	1658777926*
AE6E861A366EDCEE	1667794970
723CADDD7864D442	1720748098
74FCEF92A67710EB	1729629273
D5F67928825 9D405	1765832040*
293B88A916116A73	1772480044
719D8FD9CC2A871B	1802710702
C4B4504254122C8F	1840982002
A8C7D2F521679BC	1859355033
FAF0AE36AASF1EA5	1860438650
5928C2BFD514AED2	1869960235
8C034F89096 8F42A	1878340485
1B8EEB8441CDD382	1892427527
9A2569A0AEDB49D2	1916660837
D2A5D7A73197B4B	1950547180
AFA1ABBF4BB955DF	1951540803
16A1D35AC590E575	1960590858*
62BB3BD4C5E03810	1963575439
517B78 2B6B245EE8	1974439655
AF15F768E46CFF88	1975291199
494B23CCE0E156FB	1976289957*
147157851C2DA94E	1985676665
5EF2C597356007C5	2006244556
F0A1E4C1FA9CDB4C	2014317312
6B284B5BC26557D3	2014541822
4D1BBD29B150C61C	2035226896
A4C744R6AD127B55	2069824992
299ADCF37CE8CD3F	2071794071*
01DE680F4FAB48E2	2073876626*
E0F67CC670C2C14C	2096398889
BD31B8A755F45A9B	2135153368
CD95869B7FDDDB46	1135924274*
1AAC771E9380091A	2194367878
481DCC93A14C20EE	2204440708
F059B9059CEA918	2221853644
97BCA3E48ED809F	2279115448
9EA0A640426420C	2340054706*
0EE144EFFE5F7712	2351534544
D7A28C63755ElEEA	2369454965*
DB0F8C73A69DAFEE	2369547694
71D4FBDEDBF5A305	2371894158
C33A49F52102C5F	2441900413*
9DB78A73F7C9573	2446217335*
D098AF9B7C4BD43	2515072933
EB9J39F90FF47140	2515145939
E77FAFDCEAB4452C	2582506813
3EEA319352E87106	2600698023
DE1067C794525386	2606685976
0C5FCCE93C32AD03	2630972069
23423A96946D85BF	2708430383
B987BE4572C1E068	2717253722
23D81F45DCBC4201	2755233816*
A587CC140147FCEC	2761360957
927814FAFBF171E4	2821852324
9E2C4EC0ABB5B87E	2868112615
4B8A1375 4D14AC50	2942362723
3416B05300D49FA2	2986263853
BDEB9ECEEF8A7096	3052921261
036A9B4EAF272964	3094474831*
2D53A9077A3EF47E	3128640512
39D5A341EBF9EFD3	3166309170
3D82E8633EA0A272	3183868656
55004812 54EEBF97	3212100817*
5EDB8D27A34E44332	3246342391
7C287759OC8D2D5E	3273593348
F4078CF1D48F7F71	3311314857
E55FE3AD0FAE4FDD	3318474966
E57BE3AD6D715A1F	3335024550
94AAF6070F545BF9	3364883533
CFC33DB97378F208	3395916196
73B3A596976B3F3	3405347946
D706C0B5E854B9FF	3423707159
E3048AE05625DFFB	3483123062*
D315B52726BDA812	3488857882
FFBA18FD815BC327	3505126062
B5F4161C355A93C3	3513382457
392E5ODD868C7128	3545607921
13FC54C30DD99DCA	3553268870
FD1BEE163CC2766C	3618749492
400B31DA21AD6C96	3644910743
9B1I7F9FEC04FD6C	3682602304
AEDE1C85D97D4916	3756009149
1E44C92BAA43BAF7	3761758591*
0FB4F926485E31EE	3789936982
0432B0FF9CEFC9C5	3848300992
9BAFA768245461SF	3848492727*
3E029F79AE75FF8C	3936611694
4656FA9A8F05420C	4024232999
BE4FDD565F0FB789	4068954054*
F0071E5685175134	4113784876
3E9C67C6F84FE7EF	4148613660
759A08AD69EF31C4	4183043094
E959E8F2DC89E64E	4208755470*
5680CA94322106D	4246425419
A5C7B1B6667B772C	4249195877

Fixed Point	Cycle Length
028B303CE92C333B	4283087272
AE022AE9C4A52225	4298203540
63444CD11C18B4C4	4382270115
BACEA511BA41C759	4390335938
5B3F0EAE8D862D84	4459487784
1D7B02C8ACC7E53C	4508263560
CFF9F3C131CC7550	4580633338
EAC3E909F58A558D	4613073219*
E99B31CAA032D7C8	4624025139
E41EC962C6C47B65	4723147830
57EA50D8ADICEE18	4739063890
C8817AE4B6991EB	4784804293
683F326ECD48C5BB	4872065936
42401C77315C7B88	4894852081
4C041CA63D404722	4911410310
86O30IDFDA5F6CE9	4916166999
0A91867B20A0A878	4933454607
F8EDBCF899251808	4981750033
0D09225A6F23920C	4993175863
C8F29864F23C76CF	5061956573
65FF5031CC043066	5063489704
C8531CF9E8266298	5096034192
5E52B798C04A56A5	5147568304
DB430351EFF5A45	5153751028
B327F78B62127D5B	5253643840
5B040B741A69945A	5252632235
BDD52954BEB3CDB7	5338270753
7466CB0E05E47549	5375493367
192B8FB62A9A8B65	5400551559
46A3BA578D1DFF39	5435256032
F7F4B2A75E8129D3	5512472327
8493BA42C1AF97BB	5629649963
733DDF9357C79C33	5636606472
BF4CB1A6C45F2B1	5722528000
FCD084D25BEC96BBD	5805144356
7F8362AAAI649DC0	5831919016
A32B99FB3FC717587	5859853287
A7DE27C43B5C5C39	5958858003
20175B45BD4CA98D	5968358003
27C5CEA42FB889B07	5992136736
44CB24FB8F2EA3E	5992335770
2E6571EBF9FA00CB	6005957167
19642I0C0522DF27	6023557864
97E166C059F29C9B	6058340939
5358E006EAF28086	6075474474*

Fixed Point	Cycle Length
3FC814FE565204F7	6076137232
B82E80BEE4033A771	6174407692
2EC25679D6D58F5	6355464088*
C05680AD3C07F1B2	6403156820
A1643E70F40AC485	6411947449
7CBFC9F1EF594543	6423946064
EC7BE141D8F8E02A	6461094891
B172E3861497IBAB	6530104692
0A52F9F5B508535A	6541262041*
0D9337IB8D67C6B59	6571553375
F4A2EAE3410DIBDB	6641226295
44949ECE4983DBB	6667170278*
41CA463EA250A332	6787002094
3BCBB7F6683AA6D5	6795225973
605B16CFA01147D6	6951857282
4714ICA4E94E7215	6976824673
7598BE442B9F6882	6987647438
6897889FCC5D56D4	7073844641
A2314D0E2EBAF30D	7085878364
66CD3375E72003DC	7179626947
7B0CBCI763305D3	7255627009
68106C0A43FE6522	7430231952
A62DDB82EDB57529	7432217460
FEB3CBBEB0CD609	7441592579
3F1F8B74A69E06D0	7467140836
7B7269937CC60C95	7609598918
5AA9BC166BCBC1E0	7625629397
1BDAE54548288836	7661134106
92F24247E8C197C1	7778204234
19E963989 2EF9C12	7870418672
C34C33138F92846D	7978153130
4EDC7FDFA4EA9977	8000193283
5F02CB6AD214792	8063326246
4CFA478543ACE2B4	8170427064
F23078B46BAA7B88	8294313318
F68931B21E2E0A6	8295656075
F6F3D76136C84022	8421270154
A0FC512859DBC21C	8480871302
15E457C279CC7499	8515184617
39CDFF9507CFDBBD	8517167189
D67AD5F40B78CA8C	8547517623
4A56F2927725A424	8561303690
18D1B12D04887B83	8852280158
B434D0C7CEA94EB9	9041567214
78C1D96C74990310	9316341100

Fixed Point	Cycle Length
E067F0D748149AF2	9476168292
8F7E4EB0297787F8	9678698128
A3524F8F541E37A2	9705739403
0262CCE830A394BB	9711267022
9496D43FBCC91E87	9747304899
3FF08C6CED38A444	9769896281
ACBD4DA777D38BE3	9796615090
466AD2130B8AD2EC	9823918953
70A41ABD40C0EFA0	9836467612
7E9E33C139B2015E	9917373190
008BDCF4AEB733AFD	10006449361
33BC699F47851335	10064441381
062C933D766B0FA4	10076514201
CC2B00691E5EBE0E	10180552100
7F7322331744408C	10193525631
ED201340D9A2B4E8	10470707891
E6640686A6295AE	10479263238
99EC5A4CF7C3A0E9	10668733089
006DFE97E83F0FC6	10731024975
2786035A519568F8	10918119836
172B99FDDID74D0D	10990688763
0CEC1424726196B	11140433392
D3942LC50130C8CA	11162679154
94EBCA90F6428304	11240761345
532B1B06EA74A0B3	11260342500
9A31D7A5723C56B4	11294586603
92200FBA0FBDF66E	11407565190
842E9956E4B81920	11494443331
E686AFB65F2092F3	11893145004
C383AFB82A0EF481	12160327293
D5473BE134496315	12192580878
A0773F73D932371D	12742315020
0F7BCD33015A75D1	13004312584
413AE03B7A9AAA6F	13136649204
3F75D07BC68203AB	13548566368
09C0CC2DC8C31CEE	13564048I02
E7BBD0F753EB7080	13650787679
BC3F0AA77B7EC4B7	14283535135
9EBEF47C30CD9FA	14336899988
927A18722DBC03CA	14604244081
9197B53934E55BEF	15006473066
8E93415C528F9B68	15041961023
1E901C02D65BD8B5	15287551934
F73F93564 9B46D88	15298372664
2F6B824E5673266A	15874950095

Fixed Point	Cycle Length
E2F75B968FBECDF5	16062224185
086 5AED560BE868	16056667731
571A296A3C8BC81	16077856896
736E43159A4294F5	16201395230
DC951D638F8AEEB2	17174407494
7B443B3A7D272FA5	20737469521*
5232 9A83D7D8B6D1	21076207728
119EA346AFAAE345	21665705336
8BF8AF4A80AAF623	23510577127
6A0095E0DFB3309D	24142549973
D0401BB66BF30BB4	26928043663
A534E5B476385A1	27732705289
D3E42C9F9156E120	32908364861

* Starred entries were computed independently by Coppersmith. Taken in isolation they yield a lower bound of 1.16×10^{277}. The least common multiple of all the lengths listed is 1.94×10^{2499}.

A High-speed DES Implementation for Network Applications

Hans Eberle

Digital Equipment Corporation, Systems Research Center,
130 Lytton Ave, Palo Alto CA 94301, USA

Abstract. A high-speed data encryption chip implementing the Data Encryption Standard (DES) has been developed. The DES modes of operation supported are Electronic Code Book and Cipher Block Chaining. The chip is based on a gallium arsenide (GaAs) gate array containing 50K transistors. At a clock frequency of 250 MHz, data can be encrypted or decrypted at a rate of 1 GBit/second, making this the fastest single-chip implementation reported to date. High performance and high density have been achieved by using custom-designed circuits to implement the core of the DES algorithm. These circuits employ precharged logic, a methodology novel to the design of GaAs devices. A pipelined flow-through architecture and an efficient key exchange mechanism make this chip suitable for low-latency network controllers.

1 Introduction

Networking and secure distributed systems are major research areas at the Digital Equipment Corporation's Systems Research Center. A prototype network called Autonet with 100 MBit/s links has been in service there since early 1990 [14]. We are currently working on a follow-on network with link data rates of 1 GBit/s.

The work described here was motivated by the need for data encryption hardware for this new high-speed network. Secure transmission over a network requires encryption hardware that operates at link speed. Encryption will become an integral part of future high-speed networks.

We have chosen the Data Encryption Standard (DES) since it is widely used in commercial applications and allows for efficient hardware implementations. Several single-chip implementations of the DES algorithm exist or have been announced. Commercial products include the AmZ8068/Am9518 [1] with an encryption rate of 14 MBit/s and the recently announced VM007 with a throughput of 192 MBit/s [18].

An encryption rate of 1 GBit/s can be achieved by using a fast VLSI technology. Possible candidates are GaAs direct-coupled field-effect transistor logic (DCFL) and silicon emitter-coupled logic (ECL). As a semiconductor material GaAs is attractive because of the high electron mobility which makes GaAs circuits twice as fast as silicon circuits. In addition, electrons reach maximum velocity in GaAs at a lower voltage than in silicon, allowing for lower internal operating voltages, which decreases power consumption. These properties position

GaAs favorably with respect to silicon in particular for high speed applications. The disadvantage of GaAs technology is its immaturity compared with silicon technology. GaAs has been recognized as a possible alternative to silicon for over twenty years, but only recently have the difficulties with manufacturing been overcome. GaAs is becoming a viable contender for VLSI designs [8, 10] and motivated us to explore the feasibility of GaAs for our design.

In this paper, we will describe a new implementation of the DES algorithm with a GaAs gate array. We will show how high performance can be obtained even with the limited flexibility of a semi-custom design. Our approach was to use custom-designed circuits to implement the core of the DES algorithm and an unconventional chip layout that optimizes the data paths. Further, we will describe how encryption can be incorporated into network controllers without compromising network throughput or latency. We will show that low latency can be achieved with a fully pipelined DES chip architecture and hardware support for a key exchange mechanism that allows for selecting the key on the fly.

Section 2 of this paper outlines the DES algorithm. Section 3 describes the GaAs gate array that we used for implementing the DES algorithm. Section 4 provides a detailed description of our DES implementation. Section 5 shows how the chip can be used for network applications and the features that make it suitable for building low-latency network controllers. This section also includes a short analysis of the economics of breaking DES enciphered data. Finally, section 6 contains some concluding remarks.

2 DES Algorithm

The DES algorithm was issued by the National Bureau of Standards (NBS) in 1977. A detailed description of the algorithm can be found in [11, 13]. The DES algorithm enciphers 64-bit data blocks using a 56-bit secret key (not including parity bits which are part of the 64-bit key block). The algorithm employs three different types of operations: permutations, rotations, and substitutions. The exact choices for these transformations, i.e. the permutation and substitution tables are not important to this paper. They are described in [11]. As shown in Fig. 1, a block to be enciphered is first subjected to an initial permutation (IP), then to 16 iterations, or rounds, of a complex key-dependent computation, and finally to the inverse initial permutation (IP^{-1}). The key schedule transforms the 56-bit key into sixteen 48-bit partial keys by using each of the key bits several times.

Figure 2(a) shows an expanded version of the 16 DES iterations for encryption. The inputs to the 16 rounds are the output of IP and sixteen 48-bit keys $K_{1..16}$ that are derived from the supplied 56-bit key. First, the 64-bit output data block of IP is divided into two halves L_0 and R_0 each consisting of 32 bits. The outputs L_n and R_n of an iteration are defined by:

$$L_n = R_{n-1}$$
$$R_n = L_{n-1} \; XOR \; f(R_{n-1}, K_n)$$

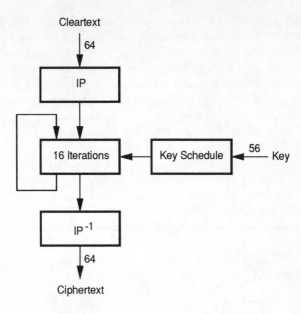

Fig. 1. Overview of the Data Encryption Standard

where n is in the range from 1 to 16. At the completion of the 16 iterations the two 32-bit words L_{16} and R_{16} are put together into a 64-bit block and used as the input to IP^{-1}.

Figure 2(b) represents the key scheduling algorithm for encryption. The 56-bit key first undergoes permuted choice 1 (PC1). The resulting 56 bits are divided into two 28-bit entities C_0 and D_0. The outputs of an iteration C_n and D_n are obtained by rotating C_{n-1} and D_{n-1} by one or two positions to the left, where n is in the range from 1 to 16. The number of left shifts at each iteration is a fixed part of the algorithm. After 16 rounds the two halves of the 56-bit key will have been shifted by 28 positions, i.e. C_{16} equals C_0 and D_{16} equals D_0. The key value K_n is obtained from C_n and D_n by choosing 48 bits of the available 56 bits according to permuted choice 2 (PC2).

Decryption and encryption use the same data path, and differ only in the order in which the key bits are presented to function f. That is, for decryption K_{16} is used in the first iteration, K_{15} in the second, and so on, with K_1 used in the 16th iteration. The order is reversed simply by changing the direction of the rotate operation performed on $C_{0..15}$ and $D_{0..15}$, that is, $C_{0..15}$ and $D_{0..15}$ are rotated to the left during encryption and rotated to the right during decryption.

Figure 3 describes the calculation of function f. First, the 32 bits of the right half R are permuted and expanded to 48 bits by the E bit-selection table (E). The expansion is achieved by repeating certain bits. The 48-bit result is then XORed with a 48-bit key value K obtained from the key schedule. Next, the 48-bit output of the XOR operation is split into blocks of 6 bits and delivered to eight substitution boxes $S_{1..8}$. Each S box implements a different nonlinear

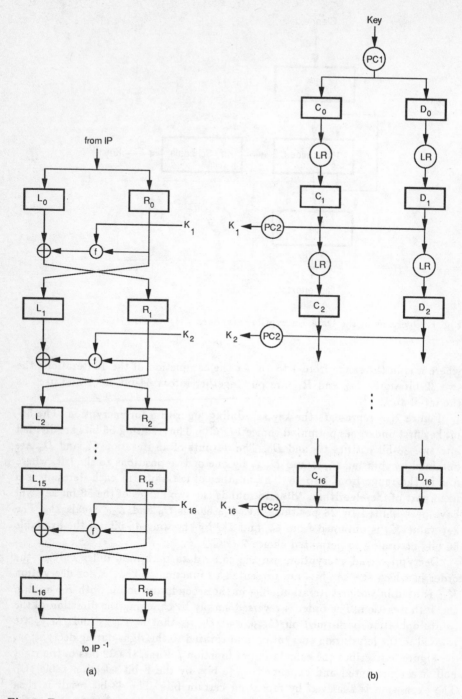

Fig. 2. Expanded Version of the 16 Iterations (a) and the Key Schedule (b) for Encryption

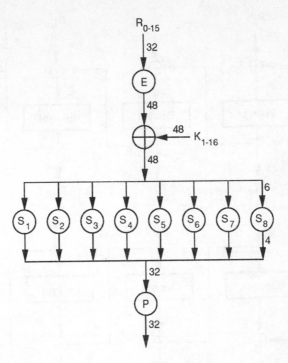

Fig. 3. Expanded Version of Function f

function yielding a 4-bit output block. Finally, the 32 bits produced by the S boxes undergo one more permutation function (P).

For enciphering data streams that are longer than 64 bits the obvious solution is to cut the stream into 64-bit blocks and encipher each of them independently. This method is known as Electronic Code Book (ECB) mode [12]. Since for a given key and a given plaintext block the resulting ciphertext block will always be the same, frequency analysis could be used to retrieve the original data. There exist alternatives to the ECB mode that use the concept of diffusion so that each ciphertext block depends on all previous plaintext blocks. These modes are called Cipher Block Chaining (CBC) mode, Cipher Feedback (CFB) mode, and Output Feedback (OFB) mode [12].

Our implementation complies with the NBS DES and supports ECB mode and CBC mode. We did not implement CFB and OFB modes because they are less useful in network applications. Figure 4 illustrates CBC mode. The plaintext p is split into 64-bit blocks $p = p_1 p_2 ... p_n$. The ciphertext block c_i is computed as:

$$c_i = DES_k(p_i \ XOR \ c_{i-1}) \ .$$

The resulting ciphertext is $c = c_1 c_2 ... c_n$. Knowing key k and c_0, which is also known as the initialization vector, the ciphertext can be deciphered by computing the plaintext block p_i as:

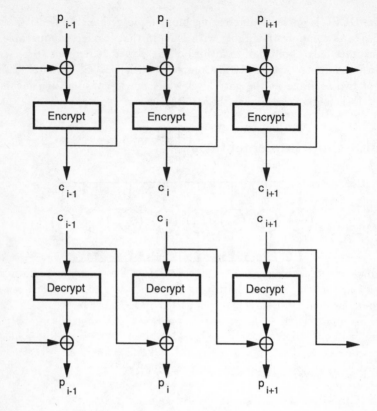

Fig. 4. Cipher Block Chaining

$$p_i = DES_k^{-1}(c_i) \; XOR \; c_{i-1} \; .$$

3 GaAs Gate Array

The DES chip is based on a FURY VSC15K gate array from Vitesse Semi-conductor [16]. It uses a 0.8 μm GaAs enhancement/depletion mode metal-semiconductor field-effect transistor (E/D-mode MESFET) process [17]. The array contains 50K transistors on a 8.1 mm by 7.1 mm die and can implement up to 15K unbuffered DCFL 2-input NOR gates. Of more interest to real applications, the array has the capacity for 4,000 buffered 2-input NOR gates or 1,500 D-flipflops.

Compared with silicon technologies, GaAs DCFL offers higher density than silicon ECL, which is the highest-performance bipolar silicon technology, but cannot yet compete with silicon CMOS, the densest silicon technology. Presently, the densest cell-based GaAs gate arrays offer up to 200K raw gates, while CMOS arrays can integrate up to 800K raw gates. It is worth noting that the density of GaAs DCFL is currently increasing more rapidly than the density of silicon CMOS. GaAs competes favorably with ECL in that it offers comparable speed, but consumes only about half to a third of the power. It remains to be seen how well GaAs competes with CMOS. Compared with CMOS, GaAs is faster by a factor of two to three at the gate level while power consumption favors GaAs only at clock frequencies higher than 100 MHz.

4 DES Chip Implementation

This section describes how we implemented the DES algorithm.

4.1 Organization

There are two ways to improve an algorithm's performance. One can choose a dense but slow technology such as silicon CMOS and increase performance by parallelizing the algorithm or flattening the logic. Alternatively, one can choose a fast but low-density technology such as silicon ECL or GaAs DCFL. The DES algorithm imposes limits on the former approach. As was shown in Fig. 4, the CBC mode of operation combines the result obtained by encrypting a block with the next input block. Since the result has to be available before the next block can be processed, it is impossible to parallelize the algorithm and operate on more than one block at a time. It is, however, possible to unroll the 16 rounds of Fig. 1 and implement all 16 iterations in sequence. Flattening the design in this manner will save the time needed to latch the intermediate results in a register on every iteration. Even though the density of CMOS chips is sufficient for doing this, the speed requirements of a 1 GBit/s CMOS implementation might still be challenging.

Since we wanted to use GaAs technology, we had to choose a different approach. The limited density of GaAs gate arrays forced us to implement only one of the 16 rounds and reuse it for all 16 iterations. Even without unrolling the 16 rounds, fitting the implementation into the available space and meeting the speed requirements was a major challenge. In order to achieve a data rate of 1 GBit/s, each block has to be processed in 64 ns, which corresponds to 4 ns per iteration or a clock rate of 250 MHz.

The register-level block diagrams for encryption and decryption are shown in Figures 5 and 6. The DES chip realizes a rigid 3-stage pipeline, that is, a block is first written into the input register I, is then moved into register LR, where it undergoes the 16 iterations of the cipher function f, and finally is written into the output register O.

The key schedule is formed by the master key register MK, which holds the encryption or decryption key, and a shift register CD, which supplies a different key value for each of the 16 iterations. Registers MK and CD can be written but not read by external circuitry. This is important since the security of a secret key system depends on the security of the keys. If the keys are compromised,

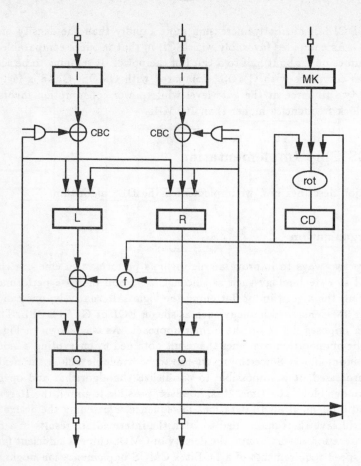

Fig. 5. Encryption

the whole system is. Once a key has been obtained, messages can be decoded or forged messages can be injected into the system.

The diagrams do not show the various permutations that must be applied to the data paths since these are accomplished solely with wiring.

Our implementation of the DES algorithm supports CBC mode. During encryption, a plaintext data block must be XORed with the previously encrypted block before it enters register LR of the encryption stage. During decryption, the decrypted block must be XORed with the previously encrypted block before it enters the output register O. In addition to the XOR gates, pipeline registers I' and I" are required during decryption in order to hold the encrypted version of a block. In ECB mode, blocks are not chained, that is, the CBC XOR gates are disabled.

A data path from the output register O to register CD allows for loading a key with a block from the data stream. The use of this feature will be explained in Sect. 5.1.

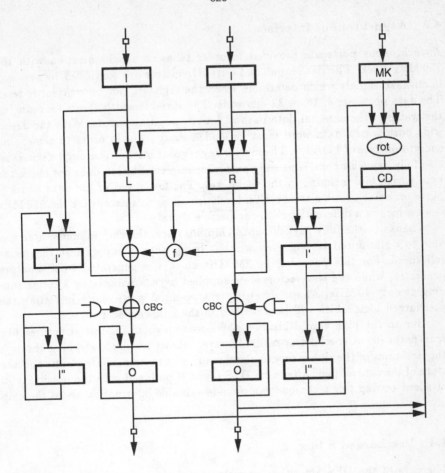

Fig. 6. Decryption

4.2 Implementation Characteristics

The implementation of the DES chip contains 480 flipflops, 2580 gates, and 8 PLAs. There are up to ten logic levels that have to be passed during the 4 ns clock period. The chip uses 84% of the transistors available in the VSC15K gate array. The high utilization is the result of a fully manual placement. Timing constraints further forced us to lay out signal wires partially by hand.

The chip's interface is completely asynchronous. The data ports are 8, 16, or 32 bits wide. A separate 7-bit wide port is available for loading the master key. Of the 211 available pins, 144 are used for signals and 45 are used for power and ground. With the exception of the 250 MHz clock, which is ECL compatible, all input and output signals are TTL compatible.

The chip requires power supply voltages of -2 V for the GaAs logic and 5 V for the TTL-compatible output drivers. The maximum power consumption is 8 W.

4.3   Asynchronous Interface

Asynchronous ports are provided in order to avoid synchronization with the 250 MHz clock. The data input and output registers are controlled by two-way handshake signals which determine when the registers can be written or read. The data ports are 8, 16, or 32 bits wide. The variable width allows for reducing the width of the external data path at lower operating speeds. With the 32-bit wide port, a new data word must be loaded every 32 ns in order to achieve an encryption rate of 1 GBit/s. The master key register is loaded through a separate, also fully asynchronous 7-bit wide port. Our implementation does not check the byte parity bits included in the 64-bit key. The low speed of the data and key ports makes it possible to use TTL-levels for all signals except for the 250 MHz clock which is a differential ECL-compatible signal.

Thanks to the fully asynchronous chip interface, the chip manufacturer was able to do at-speed testing even without being able to supply test vectors at full speed. For this purpose, the 250 MHz clock was generated by a separate generator, while the test vectors were supplied asynchronously by a tester running at only 40 MHz. At-speed testing was essential particularly in testing the precharged logic which will be described in the following section.

Due to the high chip utilization there was no room for test structures like scan-paths [9]. A special test mode, however, allows for single-stepping through the iterations of the cipher function and reading out intermediate results and the state of the control logic after each DES round. Combined with the possibility of at-speed testing this technique can provide valuable information about the chip internals.

4.4   Precharged S box

The core of the DES algorithm consists of eight substitution boxes (S boxes) which are part of the cipher function f in Fig. 2(a). Each S box computes a different boolean function with 6 inputs and 4 outputs. The most challenging and interesting part of the DES chip is to design and implement S boxes that are both fast and space-efficient.

The obvious implementation structure for the S boxes is a programmable logic array (PLA). In order to meet space and timing constraints, a precharged design using custom macros was chosen.

Precharging is a well-known design technique for silicon nMOS [5]. It offers the density of unbuffered gates and the speed of buffered gates. For FURY gate arrays, the difference in cell count between buffered logic versus unbuffered logic typically is a factor of four. The goal of precharged logic is to overcome the slow rise time of unbuffered gates that must drive large capacitive loads. The rise time of an unbuffered gate can be as much as ten times the fall time when driving a significant amount of metal because of the weak pullup transistors used in DCFL.

Figure 7(a) shows the basic building block of precharged NOR-NOR logic. The first-level gates have an extra input for the precharge signal, while the

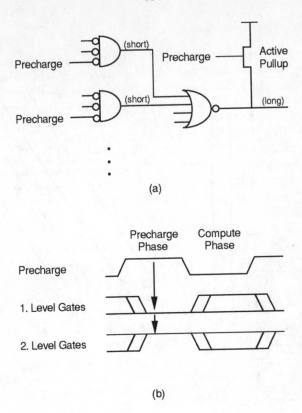

(a)

(b)

Fig. 7. Precharged NOR-NOR Logic (a) and Timing (b)

second-level gates have an active pullup connected to the output. As shown in Fig. 7(b), precharged logic operates in two phases: a precharge phase and a compute phase. During precharge, when the precharge signal is high, the outputs of the first-level gates are forced to a low level, while the active pullups will force the outputs of the second-level gates to a high level. During the compute phase, when the precharge signal is low, the outputs of the first-level gates stay low or go high while the outputs of the second-level gates stay high or go low. The first-level gates are placed adjacent to the second-level gates to make the rising edges of the first-level gates fast. The second-level gates are equipped with an active pullup to drive large capacitive loads. In a typical application several basic blocks are chained together. Notice that the slow low-to-high transitions for the second-level gates will occur in parallel during the precharge phase. During the compute phase, the long wires of the logic chain propagate only falling edges, which are fast. The penalty of this design technique is the time required for precharging. The precharge phase has to be long enough to charge the worst-case capacitance driven by any second-level gate. Therefore, the more levels of logic, the bigger the gain in performance.

The S box implementation shown in Fig. 8 contains two levels of precharged

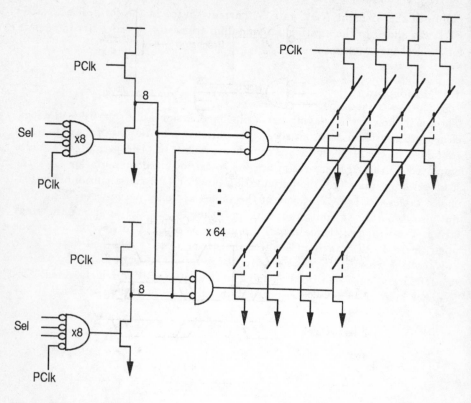

Fig. 8. Precharged S box

NOR-NOR logic: a 4-input NOR gate driving an inverter followed by a 2-input NOR gate driving from zero to four pulldown transistors. The row decoder uses two 3:8 decoders in order to save space. By using precharged logic, the S boxes occupy less than 10% of the die area. If standard macros were chosen, the S box implementation would require 5.5 times as many cells. An implementation with available macros would not have fit into the chosen gate array.

Contrary to the results obtained by analog simulations of the S box, the first implementation exhibited a discharge problem, which caused the chips to fail at high temperature. The discharge problem affected the last stage of the PLA structure in Fig. 8, which corresponds to a 32-bit wide NOR gate. The models of the pulldown transistors provided by the chip manufacturer basically ignored leakage currents. This caused the output of the PLA to drop from a high to a low level before the compute phase was over. Since leakage is proportional to temperature, the discharge problem was even worse at higher temperatures. The problem can be eliminated by lowering the voltage of the low level of the gates driving the 32-input NOR gate and thereby turning off the 32-input NOR gate harder. This requires a major change of the driving circuitry. Due to space constraints, we decided to improve the drop rate by simply changing the precharge pullup of the 32-input NOR gate. A current source in the form of a D-mode FET was added to the existing active pullup transistor in order to compensate for the leakage current.

4.5 Floorplan

The usual choices when laying out a pipelined design are to partition the logic either into register slices or bit slices. The various permutations of the data paths contained in the DES algorithm complicate this task. The permutation tables employed by the DES algorithm are the so-called initial permutation (IP), the E bit-selection table (E), the permutation function (P), and a pair of permuted-choice tables (PC1, PC2). Some of the tables not only permute the input bits but also duplicate or omit input bits and, thereby, expand or shrink the input string. The wiring of the data paths, however, is not as badly scrambled as one might fear. IP, IP^{-1}, and PC1 affect the wiring of the input and output pads only, not the wiring of the critical path, the iteration feedback loop. Fig. 9 shows one DES iteration. The wires belonging to the critical path are highlighted. This feedback loop contains two permuted data paths: permutations E and P.

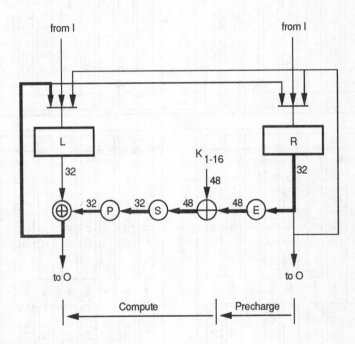

Fig. 9. One DES Iteration

While previous implementations have chosen a register-sliced layout [7, 15], we preferred a mixed strategy. As shown in Fig. 10, we first divide the design into blocks corresponding to the eight S boxes. We further subdivide each block into four bit slices each containing one bit of the left and the right half of registers I, I', I'', LR, and O. The register bits are laid out so that the wires connecting the outputs of the S boxes and the inputs of LR are as short as possible. Referring to Fig. 9, the only scrambled data path is permutation E which connects the outputs of R with the inputs of the XOR gate. These wires potentially have to go all the way across the chip. In our implementation, the longest of these wires is 6 mm long. The time to drive these wires is significant. However, driving these long wires happens at the beginning of a clock cycle and, therefore, coincides with the precharge phase. Thus, there is no data path with long wires that would contribute to the cycle time of the critical path.

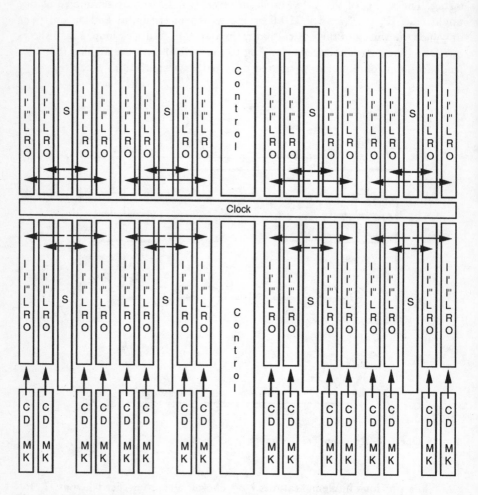

Fig. 10. DES Chip Floorplan

The key bits of register CD are laid out so that the wires connecting CD and the XOR gates are kept as short as possible. This scrambles the wiring of the key schedule (which implements two 28-bit wide registers that can be rotated by one or two bits either to the right or to the left). The timing of these wires is, however, not critical since the only logic this path contains is a multiplexer that implements the rotate function.

The control signals are generated in the middle columns of the chip. Drivers are duplicated; that is, there are separate drivers for each side of the chip in order to reduce the load and wire length and with it the propagation delay.

5 Applications

We now discuss applications of the DES chip, which is intended primarily for use in network controllers.

5.1 Low-latency Network Controller

Our implementation of the DES algorithm is tailored for high-speed network applications. This requires not only encryption hardware operating at link speed but also support for low-latency controllers. Operating at link data rates of 1 GBit/s requires a completely pipelined controller structure. Low latency can be achieved by buffering data in the controller as little as possible and by avoiding protocol processing in the controller. In this respect, the main features of the DES chip are a pipelined flow-through design and an efficient key exchange mechanism.

As described in the previous section, the chip is implemented as a rigid 3-stage pipeline with separate input and output ports. Each 64-bit data block is entered into the pipeline together with a command word. While the data block flows through the pipeline, the accompanying command instructs the pipeline stages which operations to apply to the data block. On a block-by-block basis it is possible to enable or disable encryption, to choose ECB or CBC mode, and to select the master key in MK or the key in CD. None of these commands causes the pipeline to stall. It is further possible to instruct the pipeline to load a block from the output register O into register CD. Typical usage of this feature is as follows: a data block is decrypted with the master key, is loaded into CD, and is then used for encrypting or decrypting subsequent data blocks. This operation requires a one-cycle delay slot; that is, the new key in CD cannot be applied to the data block immediately following.

The format of packets transmitted over the Autonet network efficiently uses the described architecture allowing for very low-latency controllers. The data flow of a packet transmission is as follows. With the help of a public key algorithm, a sender S and receiver R first exchange a key K that will subsequently be used for encrypting packets. Sender and receiver encrypt this key under their master keys and exchange the resulting values. Both store copies of $[K]_{MKS}$ and $[K]_{MKR}$ in their memories. MKS is the master key of S and MKR the master

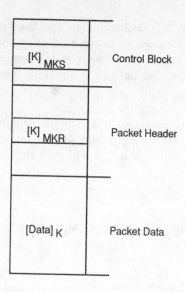

Fig. 11. Packet Format

key of R. Note that a plaintext version of K is not stored in either memory. The transmission of the actual data can now begin. The data flow through the sender's and receiver's DES chips is as follows.

Figure 11 shows the data that flows through the DES chip in the sender. First, a control block containing the key needed for encrypting the data part of the packet will be read from host memory and be presented to the sender's DES chip. The DES chip will decrypt [K]$_{MKS}$ and load the resulting key value K into key register CD. The control block will not be sent to the network since it contains only information required by the sender. Next, the header of the packet containing [K]$_{MKR}$ will pass through the DES chip without being manipulated, followed by the data, for which encryption and CBC mode are enabled. Both header and encrypted data will be sent over the network to the receiver.

When the header of the packet flows through the receiver's DES chip, [K]$_{MKR}$ will be picked out of the header, decrypted, and loaded into register CD. When the data part begins, decryption and CBC mode will be enabled. Note that in order to obtain key K, the receiver did not have to access memory or halt the DES pipeline.

5.2 Breaking DES

In 1979, Hellman published a paper with the title 'DES will be totally insecure within ten years' [6]. The controversy comes from the rather short length of the DES key, which could make an exhaustive search of the key space feasible [3, 4].

In 1977, Diffie and Hellman proposed a machine consisting of 1 million processors that would each be able to try 1 million keys per second. At an estimated cost of \$20M this machine would exhaust the key space in 20 hours [4]. In 1984, Hoornaert, Goubert, and Desmedt proposed a machine consisting of 25,000 devices that would each be able to try 1.13 million keys per second. At an estimated cost of \$1M this machine would exhaust the key space in about 4 weeks [7].

This section compares the length of time taken by our implementation to break DES with the time taken by two other popular implementations [1, 18]. We assume a known-plaintext cryptanalytic attack as described in [4]. The search starts out with one or several corresponding plaintext-ciphertext blocks, all encrypted under the same key. The attack is based on brute force in that key after key of the key space, which contains $2^{56} = 7.2 \times 10^{17}$ elements, is tried. Once the key is broken, messages can be forged or cryptograms for which the plaintext is not known can be read.

The data given in Table 1 illustrates the economics of breaking DES. As expected, the cost per GBit/s of decryption bandwidth and the time required for doing an exhaustive search drop with more recent implementations. The given duration for doing an exhaustive search assumes that one is willing to spend \$1M on DES chips alone. The necessary support circuitry might easily double that figure. The given cost per chip assumes quantities of thousands.

Part	Year	Technology	Data Rate	Cost/Chip	Cost/GBit/s	Exh. Search
Am9518	84	Silicon nMOS	14 MBit/s	\$19	\$1357	72 days
VM007	92	Silicon CMOS	192 MBit/s	\$170	\$885	47 days
GaAs DES	92	GaAs DCFL	1 GBit/s	\$300	\$300	16 days

Table 1. Cost of Breaking DES

For our implementation, it takes 16 days to try 2^{56} keys or an average of 8 days to find the key. With the separate key port our chip would be well suited for breaking DES in that the key could be easily changed every decryption cycle without stalling the pipeline. Moreover, the use of field-programmable gate arrays in our network controllers would easily allow for turning a network of controllers into a distributed machine for breaking DES. We believe that the full decryption bandwidth of 1 GBit/s per chip could be achieved without having to modify existing hardware. Therefore, a network of 10,000 machines each containing two DES chips to encrypt data full duplex at 1 GBit/s would exhaust the key space in 2 days and 16 hours.

Biham and Shamir recently showed that DES can be broken in less than the 2^{56} DES operations required for an exhaustive search [2]. The cryptanalytical attack consists of a data collection phase during which a pool of 2^{47} chosen plaintext blocks are encrypted and a data analysis phase which consists of 2^{37} DES-like operations. The proposed attack will not be further considered here since it cannot make use of existing DES implementations and since the practicability of the data collection phase is questionable.

6 Status and Conclusions

We began designing the DES chip in early 1989 and received the first prototypes at the beginning of 1991. The parts were logically functional, but exhibited electrical problems and failed at high temperature. A minor design change fixed this problem. In the fall of 1991, we received 25 fully functional parts that we plan to use in future high-speed network controllers.

With an encryption rate of 1 GBit/s, the design presented in this paper is the fastest DES implementation reported to date. Both ECB and CBC modes of operation are supported at full speed. This data rate is based on a worst case timing analysis and a clock frequency of 250 MHz. The fastest chips we tested run at 350 MHz or 1.4 GBit/s.

We have shown that a high-speed implementation of the DES algorithm is possible even with the limited flexibility of a semi-custom design. An efficient implementation of the S boxes offering both high performance and high density has been achieved with a novel approach to designing PLA structures in GaAs. An unconventional floorplan has been presented that eliminates long wires caused by permuted data bits in the critical path.

The architecture of the DES chip makes it possible to build very low-latency network controllers. A pipelined design together with separate fully asynchronous input and output ports allows for easy integration into controllers with a flow-through architecture. ECL levels are required only for the 250 MHz clock; TTL levels are used for all the data and control pins, thus providing a cost-effective interface even at data rates of 1 GBit/s. The provision of a data path for loading the key from the data stream allows for selecting the encryption or decryption key on the fly. These features make it possible to use encryption hardware for network applications with very little overhead.

References

1. Advanced Micro Devices: AmZ8068/Am9518 Data Ciphering Processor. Datasheet, July 1984
2. Biham, E., Shamir, A.: Differential Cryptanalysis of the Full 16-round DES. CRYPTO'92, Santa Barbara, August 16-20, 1992
3. Brassard, G.: Modern Cryptology. Lecture Notes in Computer Science, no. 325, Springer-Verlag, 1988
4. Diffie, W., Hellman, M.: Exhaustive cryptananlysis of the NBS Data Encryption Standard. Computer, vol. 10, no. 6, June 1977, pp. 74-84
5. Glasser, L., Dobberpuhl, D.: The Design and Analysis of VLSI Circuits. Addison-Wesley, 1988
6. Hellman, M.: DES will be totally insecure within ten years. IEEE Spectrum, vol. 16, July 1979, pp. 32-39
7. Hoornaert, F., Goubert, J. , Desmedt, Y.: Efficient hardware implementation of the DES. Advances in Cryptology: Proceedings of Crypto 84, Springer-Verlag, 1985, pp. 147-173

8. Lee, G., Donckels, B., Grey, A., Deyhimy, I.: A High Density GaAs Gate Array Architecture. CICC 1991: IEEE Custom Integrated Circuits Conference, San Diego, May 13-16, 1991, pp. 14.7.1-14.7.4

9. McCluskey, E.: Logic Design Principles. Prentice-Hall, 1986

10. Milutinovic, V., Fura, D.: Gallium Arsenide Computer Design. Computer Society Press of the IEEE, 1988

11. National Bureau of Standards: Data Encryption Standard. Federal Information Processing Standards Publication FIPS PUB 46-1, January 1988 (supersedes FIPS PUB 46, January 1977)

12. National Bureau of Standards: DES Modes of Operation. Federal Information Processing Standards Publication FIPS PUB 81, December 1980

13. National Bureau of Standards: Guidelines for Implementing and Using the NBS Data Encryption Standard. Federal Information Processing Standards Publication FIPS PUB 74, April 1981

14. Schroeder, M., Birrell, A., Burrows, M., Murray, H., Needham, R., Rodeheffer, T., Satterthwaite, E., Thacker, C.: Autonet: a High-speed, Self-configuring Local Area Network Using Point-to-point Links. Research Report 59, DEC Systems Research Center, Palo Alto, CA, 1990

15. Verbauwhede, I., Hoornaert, F., Vandewalle, J., De Man, H.: Security and Performance Optimization of a New DES Data Encryption Chip. IEEE Journal of Solid-State Circuits, Vol. 23, No. 3, June 1988, pp. 647-656

16. Vitesse Semiconductor Corporation: FURY Series Gate Array Design Manual. Version 3.0, June 1990

17. Vitesse Semiconductor Corporation: GaAs DCFL ASIC Design. Product Data Book: Application Note 7. 1991, pp. 7.30-7.35

18. VLSI Technology: VM007 Data Encryption Processor. Datasheet, October 1991 (Advance Information)

THRESHOLD SCHEMES WITH DISENROLLMENT

Bob Blakley[1], G.R. Blakley[2], A.H. Chan[3] and J.L. Massey[4]

[1] Entry Systems Division, IBM Corporation, Austin, TX 78758
[2] Department of Mathematics, Texas A&M University, College Station, TX
77843-3368
[3] College of Computer Science, Northeastern University, Boston, MA 02115.
Agnes Chan's work was supported by MITRE Sponsored Research Program.
[4] Swiss Federal Institute of Technology, Zurich 8092, Switzerland

Abstract. When a shadow of a threshold scheme is publicized, new shadows have to be reconstructed and redistributed in order to maintain the same level of security. In this paper we consider threshold schemes with disenrollment capabilities where the new shadows can be created by broadcasts through a public channel. We establish a lower bound on the size of each shadow in a scheme that allows L disenrollments. We exhibit three systems that achieve the lower bound on shadow size.

1 Introduction

In safeguarding a secret, there are many situations where two or more guardians provide more security than only one. Common examples can be found in safe deposit boxes and in the control of nuclear weapons. In these cases, two keys are needed to activate the control mechanism; the ability to exercise shared control is lost if either key is lost or either key's owner is incapacitated. To guard against such a loss, copies of keys or instructions may be made and distributed to different parties. However, increasing the number of distributed copies increases the risk of some copy being compromised, thus reducing the security of the system. By distributing "shadows" of a shared secret (which can be used as a key),threshold schemes allow shared control without risking compromise of the secret.

Let S be a secret which needs to be protected. The secret S is concealed among n different shadows in such a way that:

1. For some threshold $t, t \leq n$, called the "threshold size", any t shadows determine the secret S.
2. No $t - 1$ or fewer shadows uniquely determine the secret.

The secret S is secure against the collusion of any $t - 1$ or fewer owners of shadows, and the scheme is protected against the loss of any $n - t$ shadows.

Blakley[1] published a, (t, n) threshold scheme using hyperplanes. Shamir[7] proposed a threshold scheme using polynomials over a finite field. Various other schemes (using vector spaces, combinatorial designs, finite geometries and Reed-Solomon codes) exist [3, 4, 6, 9]. Schemes with the property that the disclosure of $t - 1$ or fewer shadows does not reveal any information about the secret are called *perfect* threshold schemes.

The disclosure of a shadow decreases the security against collusion of a threshold scheme since every $t - 1$ remaining shadows, together with the disclosed shadow, determine the secret. Thus, the threshold is reduced from t to $t - 1$. In order to maintain the same threshold t, the key must be changed and the shadows modified. One way to do this is to design a new (t, n) scheme where shadows are then distributed through secure channels. The security of the new system is not compromised if the new shadows are independent of the disclosed shadow. However, setting up the secure channels for distributing shadows can be expensive.

This paper considers schemes which distribute modifications to existing shadows through *insecure channels*. Such a scheme is said to have a *disenrollment capability*. Section 2 gives an information theoretic definition of threshold schemes with such a disenrollment capability and establishes a lower bound on the size of each shadow. Section 3 gives three examples of implementations that achieve the lower bound. The Brickell-Stinson Scheme[2] depends on the existence of a random number generator. The Nonrigid Hyperplane Scheme extends the original Blakley[1] Scheme to allow disenrollments. Finally, the Martin Scheme[5] makes use of threshold schemes with higher thresholds and reduces the cost of each public broadcast.

2 Information Theory and Lower Bound

A (t, n) threshold scheme distributes partially redundant shadows $S_1, ..., S_n$ among n users so that any t or more shadows uniquely determine the secret K. The random variable K representing the secret takes values in the space $\mathbb{K}$. The random variables $S_1, ..., S_n$ representing the shadows take values in a space $\mathbb{S}$. Using the entropy or "uncertainty" function $H(X)$ introduced by Shannon[8], we have the following definitions.

Definition 1. A (t, n) threshold scheme is a collection of random variables $(K, S_1, ..., S_n)$ such that for any $1 \leq i_1 < i_2 < ... < i_j \leq n$,

$$H(K|S_{i_1}, ..., S_{i_j}) = 0 \qquad \forall j \geq t, \tag{1}$$
$$H(K|S_{i_1}, ..., S_{i_j}) > 0 \qquad \forall j < t. \tag{2}$$

Condition (1) says that every set of t or more shadows determines the secret uniquely, whereas condition (2) indicates that the secret cannot be uniquely determined by fewer than t shadows. A (t, n) threshold scheme is said to be *perfect* if

$$H(K|S_{i_1}, ..., S_{i_j}) = H(K) \qquad \forall j < t. \tag{3}$$

Condition (3) says that knowledge of fewer than t shadows does not reduce one's uncertainty about the secret.

Let us consider the case where one shadow, say S_1, is disclosed or invalidated. In order to maintain the threshold level at t, a new secret key has to be chosen and new shadows have to be constructed. If information on the new shadows can be distributed through a public channel without compromising the secrecy of the new key, then such a (t, n) threshold scheme is said to have a 1-fold disenrollment capability. If $L+1$ secrets can be chosen so that, while disenrolling L shadows successively, the broadcast public messages do not compromise the secrecy of the new key, then such a (t, n) threshold scheme is said to have an L-fold disenrollment capability. An information-theoretic model of such a scheme is given below.

Let $K_0, K_1, ..., K_L$ denote the $L+1$ secrets. Let $S_1, ..., S_n$ represent the shadows, any t of which determine the original secret key K_0. Without loss of generality we may assume that S_i corresponds to the shadow that is invalidated at the i-th disenrollment, $i = 1, ..., L$. Let $P_1, ..., P_L$ denote the public messages that are broadcast successively at each disenrollment step. Note that each P_i may include informations obtained from the revealed shadows, $S_1, ..., S_i$.

Definition 2. A (t, n) threshold scheme with L-fold disenrollment capability is a collection of random variables $(K_0, K_1, ..., K_L, S_1, ..., S_n, P_1, ..., P_L)$ such that for each $i, i = 0, ..., L$,

$$H(K_i | \Delta_i(k), P_1, ..., P_i) = 0 \qquad \forall k \geq t, \qquad (4)$$

$$H(K_i | \Delta_i(k), P_1, ..., P_i, S_1, ..., S_i) > 0 \qquad \forall k < t, \qquad (5)$$

where $\Delta_i(k) = \{S_{i_1}, ..., S_{i_k}\} \subseteq \{S_{i+1}, S_{i+2}, ..., S_n\}$.

Definition 3. A (t, n) threshold scheme with L-fold disenrollment capability is said to be perfect if

$$H(K_i | \Delta_i(k), P_1, ..., P_i, S_1, ..., S_i) = H(K_i) \qquad \forall k < t. \qquad (6)$$

Let us assume that $H(K_i) = m$ bits. For a perfect (t, n) threshold scheme with L-fold disenrollment capability, conditions (4) and (6) can then be expressed in terms of mutual information as

$$I(K_i; \Delta_i(k), P_1, ..., P_i) = m \qquad \text{if} \quad k \geq t \qquad (7)$$

$$I(K_i; \Delta_i(k), P_1, ..., P_i, S_1, ..., S_i) = 0 \qquad \text{if} \quad k < t \qquad (8)$$

respectively, where we remind the reader that by definition,

$$I(X; Y) = H(X) - H(X|Y) = H(Y) - H(Y|X).$$

In order to minimize the cost of distributing shadows through secure channels, we wish to minimize the number of bits required to encode each shadow. It is conceivable that a (t, n) threshold scheme with higher disenrollment capability requires higher overhead for encoding the shadows. The following theorem shows that this is indeed the case by establishing a lower bound on the number of bits required to encode a shadow that grows linearly with the number L of disenrollments.

Theorem 4. *Let* $(K_0, K_1, ..., K_L, S_1, ..., S_n, P_1, ..., P_L)$ *be a perfect* (t, n) *threshold scheme with L-fold disenrollment capability. If $H(K_i) = m$, for $i = 0, ..., L$, then*

$$H(S_j) \geq (L+1)m \qquad \forall j = 1, ..., n.$$

To prove the theorem, we first establish that the knowledge of previous secret keys and the public messages, together with any $t - 1$ shadows, provides no information about the new secret.

Lemma 5. *For $L \geq i \geq 0$,*

$$I(K_i; K_0, K_1, ..., K_{i-1}, \Delta_i(k), P_1, ..., P_i, S_1, ..., S_i) = 0 \qquad \text{if } k \leq t - 1. \quad (9)$$

Proof. Recall from information theory that conditional mutual information is defined as $I(X; Y|Z) = H(X|Z) - H(X|Y, Z) = H(Y|Z) - H(Y|X, Z)$ and satisfies the identity $I(X, Y; Z) = I(X; Z) + I(Y; Z|X)$. Thus,

$$I(K_i; K_0, K_1, ..., K_{i-1}, \Delta_i(k), P_1, ..., P_i, S_1, ..., S_i)$$
$$= I(K_i; \Delta_i(k), P_1, ..., P_i, S_1, ..., S_i)$$
$$+ I(K_i; K_0, ..., K_{i-1}|\Delta_i(k), P_1, ..., P_i, S_1, ..., S_i).$$

If we can show that $I(K_i; K_0, ..., K_{i-1}|\Delta_i(k), P_1, ..., P_i, S_1, ..., S_i) = 0$ when $k \leq t - 1$, then (9) follows directly from (8). But

$$I(K_i; K_0, ..., K_{i-1}|\Delta_i(k), P_1, ..., P_i, S_1, ..., S_i) \leq H(K_i|\Delta_i(k), P_1, ..., P_i, S_1, ..., S_i)$$

and $H(K_i|\Delta_i(k), P_1, ..., P_i, S_1, ..., S_i) = 0$ by (4), so the desired result follows. □

We next observe the following identiy.

Lemma 6. *For $j \geq i + 1$,*

$$I(K_i; S_j|\Delta_i(t - 1), P_1, ..., P_i, K_0, .., K_{i-1}) = m.$$

Proof.

$$I(K_i; S_j|\Delta_i(t - 1), P_1, ..., P_i, K_0, ..., K_{i-1})$$
$$= I(K_i; S_j, \Delta_i(t - 1), P_1, ..., P_i, K_0, .., K_{i-1})$$
$$- I(K_i; \Delta_i(t - 1), P_1, ..., P_i, K_0, .., K_{i-1})$$
$$= I(K_i; \Delta_i(t), P_1, ..., P_i, K_0, ..., K_{i-1})$$
$$= m.$$

The second equality is obtained because $j \geq i + 1$ and thus joining S_j with $\Delta_i(t - 1)$ gives a set $\Delta_i(t)$ for use in (7), and by noticing that the second term in the previous equation is 0 from Lemma 5 because mutual information is nonnegative and $I(X; Y) \leq I(X; Y, Z)$. The last equality is obtained directly from Lemma 5. □

Proof of theorem. We first observe that for $j = 1, ..., n$, we may choose $S_j = S_{L+1}$. Thus $H(S_j) = H(S_{L+1})$ and we need to show only that $H(S_{L+1}) \geq (L+1)m$. Now,

$$
\begin{aligned}
&H(S_{L+1}) \\
&\geq H(S_{L+1}|\Delta_L(t-1)) \\
&\geq H(S_{L+1}|\Delta_L(t-1)) - H(S_{L+1}|P_1, ..., P_L, K_0, .., K_L, \Delta_L(t-1)) \\
&= I(P_1, ..., P_L, K_0, ..., K_L; S_{L+1}|\Delta_L(t-1)).
\end{aligned}
$$

If we can show that the last quantity is at least $(L+1)m$, then the theorem is proved. But

$$
\begin{aligned}
&I(P_1, ..., P_L, K_0, ..., K_L; S_{L+1}|\Delta_L(t-1)) \\
&= \sum_{i=1}^{L} I(P_i; S_{L+1}|\Delta_L(t-1), P_1, ..., P_{i-1}) \\
&\quad + \sum_{i=0}^{L} I(K_i; S_{L+1}|\Delta_L(t-1), P_1, ..., P_i, K_0, ..., K_{i-1}) \\
&\geq \sum_{i=0}^{L} I(K_i; S_{L+1}|\Delta_L(t-1), P_1, ..., P_i, K_0, ..., K_{i-1}) \\
&= (L+1)m
\end{aligned}
$$

where the last equality is obtained directly from Lemma 6. $\square$

We have shown that if a (t, n) threshold scheme can disenroll L participants, then each secret shadow must contain at least $(L+1)H(K_0)$ bits. In the next section we exhibit three examples of such threshold schemes where each shadow contains exactly $(L+1)H(K_0)$ bits.

3 Threshold Schemes with Disenrollment Capability

In this section we will exhibit three examples of perfect (n, t) threshold schemes that allow disenrollments and achieve the lower bound on shadow size established in the previous section

3.1 Brickell-Stinson Scheme[2]

Let $(K, S_1, ..., S_n)$ be a perfect (n, t) threshold scheme, where K represents the secret chosen from $\mathbb{K}$ and S_i represents a shadow chosen from $\$$. We further assume that $H(K) = m$. An (n, t) threshold scheme with L-fold disenrollment capability $(K_0, ..., K_L, \tilde{S}_1, ..., \tilde{S}_n, P_1, ..., P_L)$ can be constructed from $(K, S_1, ..., S_n)$ as follows:

- Each K_i represents a secret chosen uniformly from $\mathbb{K}$.

- Each $\tilde{S}_i$ represents a shadow $\tilde{S}_i = (S_i, R_{i,1}, ..., R_{i,L})$ where each $R_{i,j}$ is a random binary string of length m.
- When $\tilde{S}_i$ is invalidated, a new key K_i is chosen and associated with it are the new shadows $\{S_{i+1}^i, ..., S_n^i\}$ that are formed as specified by the original (n, t) threshold scheme. The public message P_i that is broadcast through the public channel is the union of messages of the type

$$\{R_{i+1,i} + S_{i+1}^i, R_{i+2,i} + S_{i+2}^i, ..., R_{n,i} + S_n^i\}.$$

Note that each $R_{i,j}$ is a random string and can be considered as a one-time pad that protects the shadow S_j^i; thus, $H(S_j^i) = H(S_j^i | P_i)$ and $H(K_i) = H(K_i | \Delta_i(k), P_1, ..., P_i)$ for $k < t$. Furthermore, it is easy to check that each shadow contains $(L + 1)m$ bits which is the lower bound given in Section 2. So, we have the following theorem.

Theorem 7. *The Brickell-Stinson scheme is a perfect (n, t) threshold scheme with L-fold disenrollment capability that achieves the lower bound, $H(S_j) = (L + 1)m$.*

3.2 Nonrigid Hyperplane Scheme

For simplicity we first consider the case where $L = t - 1$; the cases where $L \neq t - 1$ can be similarly designed and will be discussed later. Let $\mathbb{H}$ be the collection of all hyperplanes in a t-dimensional vector space E over $GF(q)$. The n hyperplanes represented by the rows of an n by $t + 1$ augmented matrix

$$A = \begin{bmatrix} a_{1,1} & a_{1,2} & \cdots & a_{1,t-1} & 1 & b_1 \\ a_{2,1} & a_{2,2} & \cdots & a_{2,t-1} & 1 & b_2 \\ \vdots & \vdots & \ddots & \vdots & \vdots & \vdots \\ a_{n,1} & a_{n,2} & \cdots & a_{n,t-1} & 1 & b_n \end{bmatrix} \tag{10}$$

must be in general orientation, that is, the unaugmented n by t matrix

$$U(A) = \begin{bmatrix} a_{1,1} & a_{1,2} & \cdots & a_{1,t-1} & 1 \\ a_{2,1} & a_{2,2} & \cdots & a_{2,t-1} & 1 \\ \vdots & \vdots & \ddots & \vdots & \vdots \\ a_{n,1} & a_{n,2} & \cdots & a_{n,t-1} & 1 \end{bmatrix} \tag{11}$$

must have the property that every one of its t by t submatrices is nonsingular. The intersection of the hyperplanes corresponding to any t or more rows of this matrix is a point v, whose first coordinate is the secret K_0. The intersection of hyperplanes corresponding to any collection of fewer than t rows must intersect in an affine subspace consisting of points which do not all share a common first coordinate. Equivalently, the vector $(10...0)$ must never appear as a row in the row reduced echelon form of any j by t submatrix of $U(A)$ given in (11) if $j < t$.

Let K_i correspond to the first coordinate of an arbitrarily chosen point v_i in the vector space E. Corresponds to every point v' in E, there is a translation of hyperplanes such that the new point of intersection is the point v'. Each shadow S_j is given by the j-th row of the matrix A in (10). Clearly, every shadow consists of $t \log_2 q$ bits, which is the lower bound given in Section 2. On revealing S_j, the public information P_j is the collection of translations of the unrevealed hyperplanes, that is, $\{c_{j,j+1}, c_{j,j+2}, ..., c_{j,n}\}$ such that the i-th newly translated hyperplane can be easily computed by converting the last entry in A to $b_i + c_{j,i}$.

Theorem 8. *The nonrigid hyperplane scheme is a perfect (n, t) threshold scheme with t-fold disenrollment capability that achieves the lower bound, $H(S_j) = t \log_2 q$.*

Proof. To show that the hyperplane scheme is a perfect (n, t) threshold scheme, we need to show that every key in $\mathbb{K}$ remains equally probable after each disenrollment. Let ℓ be a 1-dimensional subspace in E determined by $t-1$ hyperplanes in $\Delta_i(t-1)$, and let $\{v_0, \ldots, v_{i-1}\}$ be the chosen points in E that correspond to the known secrets $K_0, \ldots, K_{i-1}$ as defined above. For each each $j > i$, the translations of these chosen points given by $V = \{v_0, v_1 - (0, \ldots, c_{1,j}), \ldots, v_{i-1} - (0, \ldots, c_{i-1,j})\}$ must be contained in the hyperplane corresponding to participant j. Since $i \le t - 1$, for every point $p \in \ell$ and every $j > i$, there exists a hyperplane $H_j \in \mathbb{H}$ that contains the point p and the corresponding translated points in V. In other words, every $p \in \ell$ can be the chosen point v_i and every key can be the new secret. Thus, the entropy of every key remains the same and (6) is established. $\square$

In the case where the number of disenrollment L is less than $t-1$, we publish $t-1-L$ columns of the matrix $U(A)$ in (11) and still maintain the same perfect threshold scheme properties. If L is greater than $t-1$, then we use the additional columns to store informations about changing the orientation of each of the hyperplane after each disenrollment. Consider $L = t + x, x \ge 0$ and the matrix in (10) representing the shadows is then given by

$$
\begin{bmatrix}
a_{1,1} & a_{1,2} & \cdots & a_{1,t+x} & 1 & b_1 \\
a_{2,1} & a_{2,2} & \cdots & a_{2,t+x} & 1 & b_2 \\
\vdots & \vdots & \ddots & \vdots & \vdots & \vdots \\
a_{n,1} & a_{n,2} & \cdots & a_{n,t+x} & 1 & b_n
\end{bmatrix}
\tag{12}
$$

After i disenrollments, each new hyperplane is then given by $(a_{j,i_0}, \ldots, a_{j,i_{t-2}}, 1, b_j + c_{i,j})$ where $i_m = 1 + (i + m \mod t + x)$ and $c_{i,j}$ isthe corresponding broadcast translation. Such a scheme can be shown to be perfect by using similar arguments as above.

3.3 Martin Scheme[5]

Every $(n, t + i)$ threshold scheme, $i \ge 0$, can be used as an (n, t) threshold scheme by publishing i additional shadows from the shadow space $\$$. Thus, any t or more shadows together with the i published shadows can uniquely determine the secret. Based on the above notion, an (n, t) threshold scheme with L-fold disenrollment capability $(K_0, \ldots, K_L, \tilde{S}_1, \ldots, \tilde{S}_n, P_1, \ldots, P_L)$ can be constructed from $L + 1$ randomly chosen perfect $(n, t + L)$ threshold schemes $(K_i, S_1^i, \ldots, S_n^i), i = 0, \ldots, L$ as follows:

- Each K_i represents a secret chosen from the key space, $\mathbb{K}$.
- Each $\tilde{S}_i$ represents a shadow of the form $(S_i^0, S_i^1, \ldots, S_i^L)$ where each S_i^j is a shadow from the j-th $(n, t + L)$ threshold scheme, $(K_j, S_1^j, \ldots, S_n^j)$.
- When $\tilde{S}_i$ is invalidated, the new key K_i is used and associated with it, L additional "new" shadows have to be published. Among these L additional shadows are the revealed shadows, $S_1^i, S_2^i, \ldots, S_i^i$.

Since all the $L+1$ keys, $K_0, K_1, \ldots, K_L$, are independent of one another, the disclosures of K_j and $S_\ell^j, \ell \geq 1$, give no information on K_i, as long as $i \neq j$. However, the disclosed shadows, $S_1^i, \ldots, S_i^i$, together with $L + t - i$ other shadows can uniquely determine the key K_i. Thus, only $L - i$ additional shadows from $\mathbb{S}$ are needed to be broadcast through the public channel, and we have the following theorem,

Theorem 9. *The Martin scheme is a perfect (n, t) threshold scheme with L-fold disenrollment capability that achieves the lower bound, $H(\tilde{S}_i) = (L + 1)H(K_i)$.*

We can further modify the Martin Scheme to reduce the size of the public broadcast after each disenrollment. Specifically, we randomly choose an $(n, t + i)$ threshold scheme (instead of an (n, t_L) threshold scheme), for $0 \leq i \leq L$. After the i-th disenrollment, we use the i revealed shadows $S_1^i, \ldots, S_i^i$ as the additional shadows required to be published, thus reducing the size of the broadcast message.

4 Conclusion

We have established a lower bound on the initial overhead required for (n, t) threshold schemes that allow disenrollments and have given three examples of such implementations. We further modify the Martin Scheme to reduce the cost of broadcasting the public informations. An interesting open question remained to be solved is "What is the lower bound on the entropy of the public broadcast". We conjecture that the lower bound is given by

Conjecture. For $0 \leq i \leq L$,

$$H(P_i) \geq iH(K).$$

References

1. G.R. Blakley, *Safeguarding Cryptographic Keys.* Proceedings AFIPS 1979 Nat. Computer Conf. **48** (1979) 313–317
2. E.G. Brickell and D.R. Stinson, *oral communication.*

3. M. De Soete and K. Vedder, *Some New Classes of Geometric Threshold Schemes*, Proceedings EUROCRYPT 88 (1988).

4. E.D. Karnin, J.W. Greene and M.E. Hellman, *On Secret Sharing Systems*, IEEE Trans. on Information Theory, **IT-29** (1983), 35–41

5. K.M. Martin, *Untrustworthy Participants in Perfect Secret Sharing Schemes*, preprint

6. R.J. McEliece and D.V. Sarwarte, *On Sharing Secrets and Reed-Solomon Codes*, Communications of ACM **24** (Sept 1981), 583–584

7. A. Shamir, *How to Share a Secret*, Communications ACM **22** (Nov 1979), 612–613

8. C.E. Shannon, *Communication Theory of Secrecy Systems*, Bell System Technical Journal (1948), 656–715

9. D.R. Stinson and S.A. Vanstone, *A Combinatorial Approach to Threshold Schemes*, SIAM J. Disc. Math. **1–2** (1988), 230–236

Non-existence of homomorphic general sharing schemes for some key spaces
(Extended Abstract)

Yair Frankel[1]*, Yvo Desmedt[1]* and Mike Burmester[2]**

[1] EE & CS Department, University of Wisconsin–
Milwaukee, WI 53201, U.S.A.
[2] Dept. of Mathematics, Royal Holloway–University of London
Egham, Surrey TW20 OEX, U.K.

Abstract. Homomorphic threshold schemes were introduced by Benaloh and have found several applications. In this paper we prove that there do not exist perfect finite homomorphic general monotone sharing schemes for which the key space is a finite non-Abelian group (except for very particular access structures). This result is valid for the most general case, *e.g.*, if each participant receives shares from different sets and when these sets are not necessarily groups.
We extend the definition of homomorphic threshold scheme to allow that the homomorphic property is valid for two-operations. When the set of keys is a finite Boolean Algebra or a Galois field then there does not exist a perfect finite two-operation-homomorphic general sharing scheme.

1 Introduction

General sharing schemes [13] allow a distributor to distribute *shares* of a (secret) *key* to a set of participants (shareholders) A, such that when an authorized subset of participants, B, join their shares they can recompute the key. The set of authorized subsets (of A) is often called the *access structure* and denoted as Γ_A. A sharing scheme is *perfect* when a subset of B which is *not* in the access structure has no (additional) information about the (secret) key. Sharing schemes are a generalization of threshold schemes [5, 17] in which only subsets of A with cardinality greater than or equal to a threshold t are in the access structure.

Benaloh [3] introduced the concept of *homomorphic* sharing scheme. In such a scheme when a participant i has a share s_i of the key k and a share s'_i of the key k', then $s''_i = s_i \cdot s'_i$ is a share of $k * k'$. When all participants in B compute their s''_i from their shares s_i and s'_i and join these shares s''_i they can compute $k * k'$. The first application of homomorphic sharing schemes was used to set up secret ballot election schemes [3]. Recently, the interest in this

---

* This work has been supported by NSF Grant NCR-9106327 and INT-9123464. Part of this work was performed while visiting Royal Holloway–University of London.
** Part of this work was done while visiting the University of Wisconsin–Milwaukee.

topic was revived when it was used to set-up non-interactive threshold signature and threshold authentication schemes [8] (see also [7, 12, 16, 15]). With these schemes, the shares used to authenticate M (i.e., to calculate $h_k(M)$) should not provide a subset of shareholders not in the access structure with the ability to authenticate M' (i.e., to calculate $h_k(M')$). The non-interactive aspect was obtained by having each participant i, in B send $h'_{s_i}(M)$ to a combiner where s_i is the participant's share of the secret key k. From these $h'_{s_i}(M)$ the combiner could calculate $h_k(M)$ without finding k.

In [8], $h_k(\cdot)$ corresponded with RSA or an unconditionally secure authentication scheme. A natural open problem is whether this can be extended to any Boolean function. This would imply that shareholders could compute in a non-interactive way a deterministic function $h_k(I)$, where I is given and k is the secret, by just using their shares of k. In the context of [18, 11, 6] we could call this application a *non-interactive mental game* (observe that there is no random input to $h_k(I)$). Similar research has been done with secure circuit computation in an unconditionally secure model [2, 6] (which is also called a non-cryptographic model). Many schemes using the unconditionally secure model use a homomorphic sharing scheme with key space a finite Boolean Algebra (Galois field) for the XOR (addition) operation in the circuit (function) but the AND (multiplication) operation uses an interactive protocol. If a two-operation-homomorphic sharing scheme could be produced, then the communication complexity [10] of these schemes would be greatly improved. In this paper we prove that there is no perfect finite two-operation-homomorphic non-trivial monotone sharing scheme when the set of secrets is a finite Boolean Algebra or a Galois field, implying that the technique presented in [8] has a limited scope.

We also prove that there is no perfect finite sharing scheme when the set of keys is a finite non-Abelian group (except for very particular access structures). We remind the reader that the set, S_n, of permutations of n elements form a group (which is non-Abelian when $n \geq 3$) and that permutations are a key element in developing conventional cryptosystems (e.g., consult [14]).

2 Definitions and Notations

We now define a general sharing scheme [13].

Definition 1. Let K and A be sets respectively called the set of *keys* and the set of *participants*. For simplicity of notations we assume that $A = \{1, \ldots, |A|\}$. For each $i \in A$ we have the set S_i. Let C, called the set of *codewords* be a subset of $S_1 \times \ldots \times S_{|A|}$ and $f : C \to K$ be a function. Let $B = \{i_1, \ldots i_{|B|}\}$ where $B \subset A$. For each B we define $g_B : C \to S_{i_1} \times \ldots \times S_{i_{|B|}} : c = (s_1, \ldots, s_{|A|}) \to (s_{i_1}, \ldots, s_{i_{|B|}})$. We define the relation R_B between $g_B(C)$ and K as $R_B = \{(a, k) \mid \exists c \in C : a = g_B(c) \text{ and } k = f(c)\}$. When $\Gamma_A \subset \mathcal{P}(A)$ (the power set of A), then (Γ_A, K, C, f) is a general *monotone sharing scheme* if: $\forall B \in \Gamma_A : R_B$ is a function from $g_B(C)$ onto K. If so, we denote R_B by f_B. Observe that, from the definition of f_B, if $B \in \Gamma_A$ then $B' \in \Gamma_A$ for any B' with $A \supset B' \supset B$.

key	1's share	2's share		3's share	
0	0	0	0	0	0
0	0	0	1	1	0
1	0	1	0	1	1
1	0	1	1	0	1
1	1	0	0	1	0
1	1	0	1	0	0
0	1	1	0	0	1
0	1	1	1	1	1

Table 1. An example of C in a homomorphic sharing scheme.

Informally when a distributor gives shares of k he chooses a codeword c such that $f(c) = k$. Any subset B of A in the access structure can compute k uniquely using f_B on their shares which are elements of S_i. Due to [4] we allow the set of potential shares to be different. Observe that we allow that C is a proper subset of $S_1 \times \ldots \times S_{|A|}$, implying that given any $s_{i_1}, \ldots, s_{i_{|B|}}$ there does not necessarily exist a codeword c for which $g_B(c) = (s_{i_1}, \ldots, s_{i_{|B|}})$. However without affecting the generality we can assume that if $|B| = 1$ the above is satisfied. As an illustration we present in Table 1 a two out of three threshold scheme for which the set of keys is $K = Z_2(+)$. Observe that in this example $g_{\{2,3\}}(C)$ is a proper subset of $S_2 \times S_3$.

Definition 2. Let $B = \{i_1, \ldots, i_{|B|}\}$. Given a random variable **k**, on K, shares s_i are given (by the distributor) with a certain (known) probability distribution. If for all random variables **k** on K, all $B \notin \Gamma_A$, and all $(s_{i_1}, \ldots, s_{i_{|B|}}) \in g_B(C)$, $\text{prob}(\mathbf{k} = k \mid \mathbf{s}_{i_1} = s_{i_1}, \ldots, \mathbf{s}_{i_{|B|}} = s_{i_{|B|}}) = \text{prob}(\mathbf{k} = k)$, then the sharing scheme is *perfect*. If $\forall B \notin \Gamma_A : \forall (s_{i_1}, \ldots, s_{i_{|B|}}) \in g_B(C) : \forall k \in K : \exists c \in C : g_B(c) = (s_{i_1}, \ldots, s_{i_{|B|}})$ and $f(c) = k$ then we call the sharing scheme *weakly perfect*.

Definition 3. Let K have a binary operation "$*$" (so $K(*)$ is closed) and each S_i have a binary operation "$\cdot$". Let (Γ_A, K, C, f) be a sharing scheme. If f is a homomorphism then we have a homomorphic sharing scheme [3]. If K and the S_i have two binary operations and f is a homomorphism for both operations then we have a two-operation-homomorphic sharing scheme.

Note that Definition 3 implies that in a monotone homomorphic sharing scheme $C(\cdot)$ is closed.

Definition 4. A monotone sharing scheme (Γ_A, K, C, f) is finite if C is finite.

3 K is a finite non-Abelian group

We now prove that there is no weakly perfect finite homomorphic sharing scheme (except for very particular access structures) when the set of keys K is a finite non-Abelian group. *We observe that Benaloh's definition does not require that the set of shares S_i (which we allow to differ for each participant) have the same algebraic structure as K.*

To prove this we will first prove properties for an optimal sharing scheme.

Definition 5. When (Γ_A, K, C, f) is a sharing scheme, it is optimal if there does not exist a (Γ_A, K, C', f') sharing scheme such that $|C'| < |C|$.

Remark. It is possible that there are many optimal homomorphic threshold schemes.

In this section we will assume that the sharing scheme is finite. We will now prove that if each of the S_i is a not a quasigroup[3] then one can create a 'more' optimal sharing scheme.

Lemma 6. *In an optimal (weakly) perfect finite monotone homomorphic sharing scheme with key space K a finite group, each S_i and C are quasigroups.*

Proof. Suppose that $C(\cdot)$ is not a quasigroup. Then there exists a $c_1 \in C$ such that $c_1 \cdot C$ is a proper subset of C. Let $f(c_1) = k_1$. Since K is a finite group there exists an integer n such that $f(c_1)^n = k_1^{-1}$. Let $c_1' = \overbrace{c_1(c_1(\cdots(c_1 \cdot c_1)))}^{n \text{ times}}$ and $C' = c_1(c_1' \cdot C)$. Observe that $C' \subset c_1 \cdot C$ and that $f(c_1') = k_1^{-1}$.

We now create a 'more' optimal homomorphic sharing scheme using $C'(\circ)$. We define $x_1 \circ x_2 = c_1 \cdot (c_1' \cdot (x_1 \cdot x_2))$ for $x_1, x_2 \in C'$ and we use the restriction of f to C'. Clearly "$\circ$" is closed on C'. Because R_B is a function from $g_B(C)$ to K, for any $B \in \Gamma_A$, the restriction of R_B to $g_B(C')$ is a function. We now prove that the restriction of f_B is onto. Choose any $k' \in K$. Then there is a $c \in C$ such that $f_B(c) = k'$. Now $f_B(g_B(c_1 \cdot (c_1' \cdot c))) = f_B(g_B(c))$. Since $f(x_1 \circ x_2) = f(x_1 \cdot x_2)$, the new scheme is a homomorphic sharing scheme. Let $\bar{s} = (s_{i_1}, \ldots, s_{i_{|B|}}) \in g_B(C')$ and $B \notin \Gamma$. To prove that the new scheme remains perfect (if it was already), observe that

$$\text{prob}(\mathbf{k} = k, \bar{\mathbf{s}} = \bar{s}) = \sum_{\substack{s' \in g_B(C) \\ \bar{s} = g_B(c_1) \cdot (g_B(c_1') \cdot \bar{s}')}} \text{prob}(\mathbf{k} = k, \bar{\mathbf{s}}' = \bar{s}')$$

$$= \left(\text{prob}(\mathbf{k} = k) \cdot \sum_{\substack{s' \in g_B(C) \\ \bar{s} = g_B(c_1) \cdot (g_B(c_1') \cdot \bar{s}')}} \text{prob}(\bar{\mathbf{s}}' = \bar{s}') \right) = \text{prob}(\mathbf{k} = k) \cdot \text{prob}(\bar{\mathbf{s}} = \bar{s})$$

because the original scheme is perfect. Thus $C''(\circ)$ is a 'more' optimal scheme than $C(\cdot)$, contradicting the fact that C is optimal. $\qquad\square$

Lemma 7. *Consider a finite homomorphic monotone sharing scheme (Γ_A, K, C, f) with key space K a finite group. If the codeword space $C(\cdot)$ is a quasigroup then there exists a binary operation $\circ$ such that $C(\circ)$ has an identity element e and the function f remains a homomorphism. Thus each $S_i(\circ)$ has an identity.*

---

[3] In a finite quasigroup $\mathcal{S}(\cdot)$, for each $s \in \mathcal{S} : s \cdot \mathcal{S} = \mathcal{S} = \mathcal{S} \cdot s$

Proof. This follows from [1]. We briefly overview the proof. Let L_y be the mapping $L_y : b \to y \cdot b$ and R_z be the mapping $R_z : a \to a \cdot z$. Define $a \circ b = R_z^{-1}(a) \cdot L_y^{-1}(b)$ where $a, b \in C$. Then $C(\circ)$ is a quasigroup because $C(\cdot)$ is. Furthermore $e = y \cdot z$ is an identity element for $C(\circ)$. Indeed, $y \cdot z = R_z(y) = L_y(z)$, so that, $e \circ x = R_z^{-1}(y \cdot z) \cdot L_y^{-1}(x) = R_z^{-1}(R_z(y)) \cdot L_y^{-1}(x) = y \cdot L_y^{-1}(x) = L_y(L_y^{-1}(x)) = x$. Similarly, $x \circ e = R_z^{-1}(x) \cdot L_y^{-1}(y \cdot z) = R_z^{-1}(x) \cdot L_y^{-1}(L_y(z)) = R_z^{-1}(x) \cdot z = R_z(R_z^{-1}(x)) = x$.

We now prove that we can choose y and z in such a way that f remains a homomorphism. Choose $y, z \in C$ such that $f(z)^{-1} * f(y)^{-1} = 1 \in K$. Clearly for $a, b \in C$, $f(a \circ b) = f(R_z^{-1}(a) \cdot L_y^{-1}(b)) = f(R_z^{-1}(a)) * f(L_y^{-1}(b)) = f(a) * f(z)^{-1} * f(y)^{-1} * f(b) = f(a) * f(b)$, as can easily be verified. So, this scheme is homomorphic and if it is perfect it remains so since the set of codewords is the same. $\qquad\square$

From these lemmas we get the following result.

Theorem 8. *Consider a (weakly) perfect finite homomorphic monotone sharing scheme with key space K, a finite group, for which there exists a $B \in \Gamma_A$ such that $B = B' \cup B''$, with $B', B'' \notin \Gamma_A$ and $B' \not\subset B''$, $B'' \not\subset B'$. Then K is Abelian.*

Proof. From Lemma 6 and Lemma 7 it follows that there exists a homomorphic (weakly) perfect sharing scheme (Γ_A, K, C', f') for which C' is a quasigroup with identity. So we can assume that $(1, \ldots, 1) \in C$. Let $B = \{i_1, \ldots, i_{|B'|}, i_{|B'|+1}, \ldots, i_{|B|}\}$, with $B' = \{i_1, \ldots, i_{|B'|}\}$ and $B'' = \{i_{|B|-|B''|+1}, \ldots, i_{|B|}\}$. Since the scheme is weakly perfect for each $k' \in K$ then there exists a codeword $c' \in C$ such that $f(c') = k'$, $g_{B'}(c') = (1, \ldots, 1)$, and for each $k'' \in K$ there exists a codeword $c'' \in C$ such that $f(c'') = k''$ and $g_{B''}(c'') = (1, \ldots, 1)$. Observe that $g_B(c' \cdot c'') = g_B(c'' \cdot c')$, so because $B \in \Gamma_A$ we have $f(c' \cdot c'') = f(c'' \cdot c')$. Then, since f is a homomorphism, $f(c') * f(c'') = f(c'') * f(c')$. Thus K is Abelian (the conditions on B are crucial to allow that $f(c')$ is any k' and $f(c'')$ is any k''). $\qquad\square$

An example of an access structure that does not satisfy the above conditions is $A = \{1, 2, 3, 4\}$, $\Gamma_A = \mathcal{P}(A) \backslash \overline{\Gamma}_A$, where $\overline{\Gamma}_A = \{\emptyset, \{2\}, \{4\}, \{2, 4\}\}$. In general the above condition is not satisfied if the union of any two elements in $\overline{\Gamma}_A$ belongs to $\overline{\Gamma}_A$.

Corollary 9. *If (Γ_A, K, C, f) is a weakly perfect t-out-of-$|A|$ finite homomorphic threshold scheme and K is a finite non-Abelian group then $t = 1$.*

$\qquad\square$

Proof. Obvious.

4 Two operation sharing schemes

Let $K = K' \cup \{0\}$, where K' is a finite group and $0 \in K \backslash K'$ with $0 \cdot x = x \cdot 0 = 0$ for all $x \in K$. We call K a *finite group with zero*. Let us define the set $\tilde{\Gamma}_A = \{B \in \Gamma_A \mid \forall B' \underset{\neq}{\subset} B : B' \notin \Gamma_A\}$. A monotone sharing scheme for which $\{i\} \in \tilde{\Gamma}_A$ for all $i \in A$ is called *trivial* (for all practical purposes one can say that in such schemes each shareholder knows the key).

Theorem 10. *A non-trivial finite homomorphic monotone sharing scheme with key space K, a finite group with zero, cannot be weakly perfect.*

Proof. Let (Γ_A, K, C, f) be a non-trivial finite two-operation-homomorphic sharing scheme. Without loss of generality, we assume for all $i \in A$ that $g_{\{i\}}(C) = S_i$.

Let $B \in \tilde{\Gamma}_A$ and $|B| > 1$. We first prove that if $f(c) = 0$, then there exists an $i \in B$ such that $g_{\{i\}}(c) \cdot S_i \neq S_i$. Indeed suppose that for all $i \in B$ that $g_{\{i\}}(c) \cdot S_i = S_i$. This implies that $g_B(c) \cdot g_B(C) = g_B(C)$. Since f_B is a homomorphism $f_B(g_B(c) \cdot g_B(C)) = 0 * f_B(g_B(C)) = f_B(g_B(C)) = K$. So we have a contradiction.

For an optimal finite homomorphic non-trivial sharing scheme we now prove that if $s_i \cdot S_i \neq S_i$ then $f(c) = 0$ where $g_{\{i\}}(c) = s_i$. If $f(c) \neq 0$, then by Lemma 6 we can get a 'more' optimal scheme (replacing C by $c(c' \cdot C)$). This implies that the scheme is not weakly perfect. $\qquad\square$

Corollary 11. *In any finite two-operation-homomorphic sharing scheme with key space K, a Galois field, for every $c \in C$ and $B \in \Gamma_A$, there is a shareholder in B whose share uniquely determines the key. (Formally, $\forall B \in \Gamma_A : \forall c \in C : \exists i \in B : \exists k \in K : \mathrm{prob}(\mathbf{k} = k \mid \mathbf{s}_i = s_i) = 1.)$*

Proof. Observe that the multiplicative structure of a field is a group with zero. When the sharing scheme is trivial, the result is obvious. We now discuss non-trivial sharing schemes. It is now sufficient to consider $B \in \tilde{\Gamma}_A$ and $|B| > 1$. Denote the addition in C as $\oplus$, while we use $+$ in K. From the proof of the previous theorem (and because each $i \in B$ can simulate for his share the optimization technique[4] of Lemma 6) we see that for any $c \in C$ with key $k = 0$ there is a shareholder i in each $B \in \tilde{\Gamma}_A$ whose share s_i uniquely determines the key. That is $\mathrm{prob}(\mathbf{k} = 0 \mid \mathbf{s}_i = s_i) = 1$. We now use this fact.

Since K is a finite group over addition, for any $c \in C$ there exists a $c' \in C$ such that $f(c') = -f(c)$. Then $f(c \oplus c') = 0$. By Theorem 10, there exists an $i \in B$ such that $\mathrm{prob}(\mathbf{k} = 0 \mid \mathbf{s}_i = s_i) = 1$, where $s_i = g_{\{i\}}(c \oplus c')$. For any c'' with $g_{\{i\}}(c'') = g_{\{i\}}(c)$ we must have $f(c'') = f(c)$, since $g_{\{i\}}(c'' \oplus c') = s_i$ and $\mathrm{prob}(\mathbf{k} = 0 \mid \mathbf{s}_i = s_i) = 1$. Thus the value of $f(c)$ depends only on i's share. $\qquad\square$

Observe that Corollary 11 is not restricted to weakly perfect finite homomorphic sharing schemes. Moreover a stronger version of this corollary will be given in the final paper.

Corollary 12. *Let K be a finite Boolean algebra. In a non-trivial finite monotone sharing scheme with key space K which is two-operation-homomorphic (i.e., for the "AND"("$\cdot$") and "OR"("$+$") operations) cannot be weakly perfect.*

Proof. Let $B \in \tilde{\Gamma}_A$ and $|B| > 1$. Using a similar proof as in Lemma 6 one can prove that in an optimal sharing scheme $f(c) = 1$ implies that for all $i \in B$: $g_{\{i\}}(c) \cdot g_{\{i\}}(C) = g_{\{i\}}(C)$.

---

[4] We note that for this corollary we are only concerned that the algebraic properties, not the security properties, hold during the optimization.

Now, if $f(c) \neq 1$, then there exists an $i \in B$ such that $g_{\{i\}}(c) \cdot g_{\{i\}}(C) \neq g_{\{i\}}(C)$ by contradiction. Indeed, if $g_{\{i\}}(c) \cdot g_{\{i\}}(C) = g_{\{i\}}(C)$ for all $i \in B$ then $f_B(g_B(c) \cdot g_B(C)) = f(c) \cdot K = f_B(g_B(C)) = K$, but if $f(c) \cdot K = K$ then one can prove that $f(c) = 1$, contradiction. So, it is not weakly perfect. (To prove that $k \cdot K = K$ implies $k = 1$, observe that if $k \neq 1$, there exists a $k_1 \in K$ such that $k_1 + \bar{k} \neq k_1$ and moreover $k \cdot (k_1 + \bar{k}) = k \cdot k_1$.) $\qquad\square$

Corollary 13. *If the key space K of a weakly perfect t-out-of-$|A|$ finite two-operation-homomorphic threshold scheme is either a Boolean Algebra or a Galois field, then $t = 1$.*

5 Open problems

This paper demonstrates that finite homomorphic sharing schemes have limitations. However such schemes have already proved to be useful in several cryptographic applications [3, 7, 12, 16, 8, 15]. One wonders what more can be done with homomorphic sharing schemes.

Several examples can be constructed using algebraic structures which can be made homomorphic. For instance, it is obvious that a homomorphic sharing scheme exists for key space K and binary operation defined by $a * b = a$ for all $a, b \in K$. Thus a group structure is not required to make a homomorphic sharing scheme. A two-operation-homomorphic sharing scheme can be defined on a key space $K(+, *)$ where $K(+)$ is an Abelian group and the binary operation $*$ is defined as $a * b = 0$ for all $a, b \in K$. Thus there are rings in which two-operation-homomorphic sharing schemes can be made. Though these examples are trivial and do not seem to have cryptographic usefulness, they do introduce the following open problems:

1. What other (useful) algebraic structures have perfect homomorphic sharing schemes?
2. What other algebraic structures have two-operation-homomorphic sharing schemes?
3. What circuit evaluation can be made homomorphic (observe that it is possible to evaluate RSA in a homomorphic threshold way [8])?
4. How close to perfect can homomorphic sharing schemes be made for non-Abelian groups?
5. Are the share sets $S_i(\cdot)$ groups in an optimal homomorphic sharing scheme with key space a finite Abelian group, (so are the $S_i(\cdot)$ associative)?

6 Conclusion

Homomorphic sharing schemes have been useful in making secret ballot election [3], threshold signature, and threshold authentication [12, 16, 8, 15] schemes. This paper discusses the limitations of the concept of homomorphic sharing schemes by demonstrating that they cannot be used with non-interactive evaluation of Boolean circuits. Although, perfect homomorphic threshold exist for any Abelian group [9], this result does not extend to other groups. Several open problems have also been introduced.

Acknowledgment

We thank Bob Blakley for suggesting the open problem of whether perfect homomorphic sharing schemes exist over non-Abelian groups. The authors also wish to thank Moti Yung for discussions and references to papers related to two-operation-homomorphic sharing schemes.

References

1. Albert, A. A.: Non-associative algebras: Fundamental concepts and isotopy. Annals of Mathematics **43** (1942) 685–707
2. Ben-Or, M., Goldwasser, S., Wigderson, A.: Completeness theorems for non-cryptographic fault-tolerant distributed computation. In Proceedings of the twentieth annual ACM Symp. Theory of Computing, STOC (May 2–4, 1988) pp. 1–10
3. Benaloh, J. C.: Secret sharing homomorphisms: Keeping shares of a secret secret. In Advances in Cryptology, Proc. of Crypto '86 (Lecture Notes in Computer Science 263) (1987) A. Odlyzko, Ed. Springer-Verlag pp. 251–260
4. Benaloh, J. C., Leichter, J.: Generalized secret sharing and monotone functions. In Advances in Cryptology, Proc. of Crypto'88 (Lecture Notes in Computer Science 403) (1990) S. Goldwasser, Ed. Springer-Verlag pp. 27–35
5. Blakley, G. R.: Safeguarding cryptographic keys. In Proc. Nat. Computer Conf. AFIPS Conf. Proc. (1979) pp. 313–317
6. Chaum, D., Crépeau, C., Damgård, I.: Multiparty unconditionally secure protocols. In Proceedings of the twentieth annual ACM Symp. Theory of Computing, STOC (May 2–4, 1988) pp. 11–19
7. Desmedt, Y., Frankel, Y.: Threshold cryptosystems. In Advances in Cryptology – Crypto '89, Proceedings (Lecture Notes in Computer Science 435) (1990) G. Brassard, Ed. Springer-Verlag pp. 307–315
8. Desmedt, Y., Frankel, Y.: Shared generation of authenticators and signatures. In Advances in Cryptology – Crypto '91, Proceedings (Lecture Notes in Computer Science 576) (1992) J. Feigenbaum, Ed. Springer-Verlag pp. 457–469
9. Desmedt, Y., Frankel, Y.: Perfect zero-knowledge sharing schemes over any finite Abelian group. Presented at Sequences '91, June 17–22, 1991, Positano, Italy, to appear in: the Proceedings, Springer-Verlag 1991
10. Franklin, M., Yung, M.: Communication complexity of secure computation. In Proceedings of the twenty fourth annual ACM Symp. Theory of Computing, STOC (1992) pp. 699–710
11. Goldreich, O., Micali, S., Wigderson, A.: How to play any mental game. In Proceedings of the Nineteenth annual ACM Symp. Theory of Computing, STOC (May 25–27, 1987) pp. 218–229
12. Hwang, T.: Cryptosystems for group oriented cryptography. In Advances in Cryptology, Proc. of Eurocrypt '90 (Lecture Notes in Computer Science 473) (1991) I. Damgård, Ed. Springer-Verlag pp. 352–360

13. Ito, M., Saito, A., Nishizeki, T.: Secret sharing schemes realizing general access structures. In Proc. IEEE Global Telecommunications Conf., Globecom'87 (1987) IEEE Communications Soc. Press pp. 99–102
14. Kaliski, Jr., B. S., Rivest, R. L., Sherman, A. T.: Is the Data Encryption Standard a group? (results of cycling experiments on DES. Journal of Cryptology 1 (1988) 3–36
15. Laih, C.-S., Harn, L.: Generalized threshold cryptosystems. Presented at Asiacrypt'91, November 11–14, 1991, Fujiyoshida, Yamanashi, Japan, to appear in: Advances in Cryptology. Proc. of Asiacrypt'91 (Lecture Notes in Computer Science), Springer-Verlag
16. Pedersen, T. P.: A threshold cryptosystem without a trusted party. In Advances in Cryptology, Proc. of Eurocrypt '91 (Lecture Notes in Computer Science 547) (April 1991) D. W. Davies, Ed. Springer-Verlag pp. 522–526
17. Shamir, A.: How to share a secret. Commun. ACM **22** (1979) 612–613
18. Yao, A. C.: How to generate and exchange secrets. In The Computer Society of IEEE, 27th Annual Symp. on Foundations of Computer Science (FOCS) (1986) IEEE Computer Society Press pp. 162–167

An *l*-Span Generalized Secret Sharing Scheme

Lein Harn and Hung-Yu Lin

Computer Science Telecommunications Program
University of Missouri - Kansas City
Kansas City, MO 64110

Abstract. For some secret sharing applications, the secret reconstructed is not revealed to the participants, and therefore, the secret/shadows can be repeatedly used without having to be changed. But for other applications, in which the secret reconstructed is revealed to participants, a new secret must be chosen and its corresponding shadows must be regenerated and then secretly distributed to participants again, in order to enforce the same secret sharing policy. This is inefficient because of the overhead in the generation and distribution of shadows. In this paper, an *l*-span secret sharing scheme for the general sharing policy is proposed to solve the secret/shadows regeneration problem by extending the life span of the shadows from *1* to *l*, i. e., the shadows can be repeatedly used for *l* times to generate *l* different secrets.

I. Introduction

A secret sharing scheme is a method of hiding a secret among multiple shadows such that the secret can be retrieved by some subsets of these shadows but not by the others according to a given secret sharing policy. For example, Shamir's well-known (m,n)-threshold scheme [1] realizes the secret sharing policy in which any m, or more than m shadows, can reconstruct the secret. This sharing policy is far too simple for many applications because, implicitly, it assumes that every participant has equal privilege to the secret or every participant is equally trusted. Complicated sharing policies, in which participants have different privileges, can also be realized by other generalized secret sharing schemes [2, 3, 4]. One common feature among almost all secret sharing schemes is that once the reconstructed secret is exposed, a new secret must be chosen and its corresponding shadows must be regenerated and then secretly distributed to participants again, in order to enforce the same secret sharing policy. From life span aspects of the shadows, these traditional schemes are called 1-span secret sharing schemes.

Depending on applications, the secret can be reconstructed in a tamper-free device without revealing it to the participants. For such applications, the secret/shadows can be repeatedly used. But, for other applications, in which the secret reconstructed is revealed to participants, a new secret must be chosen and its corresponding shadows are then generated in order to enforce the same secret sharing policy. Such regeneration process is inefficient because of the overhead in the generation and distribution of shadows.

One previous work which tries to solve the shadow regeneration problem can be found in [5], but it deals with only traditional threshold schemes and the threshold value is decreased in proportion to the number of different secrets which have been revealed. In this paper, an l-span secret sharing scheme for the general sharing policy will be proposed to solve the secret/shadows regeneration problem by extending the life span of the shadows from 1 to l., i. e., the shadows can be repeatedly used for l times to generate l different secrets. Section II gives some definitions and Section III briefly reviews the scheme on which the proposed l-span generalized secret sharing scheme is based. The l-span generalized secret sharing scheme and an example are included in Section IV.

II. Definitions

Suppose a secret key k is to be shared according to a given secret sharing policy by a group of m participants $U = \{u_1, u_2, \ldots, u_m\}$. Each participant may be designated with a different privilege. A generalized secret sharing scheme is a method of breaking k into m pieces $k_1, k_2, \ldots, k_m$, with k_i secretly distributed to u_i such that

(1) if $\mathcal{A} \subseteq \mathcal{U}$ is a qualified subset of participants, called *positive access instance*, according to the secret sharing policy, then k can be reconstructed from shadows $\{k_i \mid u_i \in \mathcal{A}\}$.

(2) if $\mathcal{A} \subseteq \mathcal{U}$ is not a qualified subset of participants, called *negative access instance*, according to the secret sharing policy, then k cannot be reconstructed from $\{k_i \mid u_i \in \mathcal{A}\}$.

The set F of all positive access instances is called the positive access structure of the secret sharing policy and the set N of all negative access instances is called the negative access structure of the secret sharing policy. Suppose the positive access structure of a given sharing policy is F. The corresponding negative access structure is $N = 2^{\mathcal{U}} - F$.

III. Lin and Harn's Generalized Secret Sharing Scheme

The dealer first secretly selects two large primes, p and q, and publishes their product n=p∗q. Then it assigns a distinct prime p_j to each negative access instance N_j of $\mathcal{M}(N)$ and computes the tag t_i associated with participant u_i as

$$t_i = \prod_{u_i \in N_j} p_j,$$

where $\mathcal{M}(N)$ is the maximum set of the negative access structure.

The shadows assigned to the participants are computed as

$$k_i = k^{t_i} \mod n, \text{ for } i = 1, 2, ..., m, \text{ where } k \text{ is the secret.}$$

Each shadow k_i is then secretly distributed to participant u_i.

The dealer also publishes one pair of check values, t_c and k_c where

$$t_c = \prod_{N_j \in \mathcal{M}(N)} p_j.$$

and $k_c = k^{t_c} \pmod{n}$ for users' verification of the correctness of their received shadows.

The secret k can be reconstructed by any positive access instance according to the THEOREM 1 in [4]:

THEOREM 1. Given $k_1, k_2, e_1,$ and e_2 such that $k_1 = k^{e_1} \mod n$ and $k_2 = k^{e_2} \mod n, k^r \mod n$ can be easily computed if $\gcd(e_1, e_2) = r$.

IV. The l-Span Generalized Secret Sharing Scheme

In this l-span secret sharing scheme, the generation of tags associated with participants is the same as mentioned above. However, since there are multiple secrets corresponding to the same set of shadows, the choice of the secrets and the generation of shadows need to be modified.

First, the secret, k, is replaced by a sequence of secrets, s_j's, where

$$s_j = k^{t_c^{l-j}} \mod n, \text{ for } j = 1, 2, ..., l.$$

Note that each secret should be used only once to enforce the secret sharing policy and participants should reconstruct the secrets, $s_1, s_2, ..., s_l$, accordingly in order to obtain the maximum life span of k.

Then the shadows assigned to participants are computed as

$$k_i = k^{t_i^l} \mod n, \text{ for } i = 1, 2, ..., m.$$

Now suppose a positive access instance $\mathcal{A}$ wants to reconstruct secret s_j. Each participant $u_i \in \mathcal{A}$ computes

$$k_{i,j} = (k_i)^{(t_c/t_i)^{l}-j} \qquad \bmod n$$
$$= (k^{t_i^{l}})^{(t_c/t_i)^{l}-j} \qquad \bmod n$$
$$= (k)^{t_c^{l-j}t_i^{j}} \qquad \bmod n$$
$$= (k^{t_c^{l-j}})^{t_i^{j}} \qquad \bmod n$$
$$= (s_j)^{t_i^{j}} \qquad \bmod n,$$

and then submits it, instead of his shadow, k_i.

THEOREM 2. Any positive access instance $\mathcal{A}$ can reconstruct s_j, for $j = 1$, $2,..., l$.

<Proof> The greatest common divisor of t_i^{j}'s, for $u_i \in \mathcal{A}$, is 1, so s_j can be derived from $k_{i,j}$'s by Theorem 1.

Lemma 1. s_j's and $k_{i,j}$'s, $i = 1$ to m, can be derived from s_r, if $j < r \le l$.

<proof> Since $r-j > 0$, $s_j = k^{t_c^{l-j}} \bmod n$, $s_r = k^{t_c^{l-r}} \bmod n$, and modular exponentiation is an one-way function, we can derive s_j from s_r as

$$s_j = (s_r)^{t_c^{r-j}} \bmod n.$$

Similarly, we can derive $k_{i,j}$'s from s_r.

Lemma 2. s_j's and $k_{i,j}$'s, $i = 1$ to m, cannot be derived with knowledge of s_r, if $r < j \le l$

<proof> From RSA assumption in [4], i.e., the modular exponentiation is an one-way function.

THEOREM 3. No negative access instance can derives s_j unless some s_r, with $j < r$, have been revealed.

<proof> This theorem can be proved from Lemma 1, Lemma 2, and THEOREM 5 in [4].

Here we give an example to illustrate our idea.

EXAMPLE. Suppose there are four members in the system, Alice, Bob, Cathy, and David. The secret sharing policy is that either Alice and Bob working together, or Bob and Cathy working together, or Alice, Cathy, and David working together can reconstruct the secret. The positive access structure of this sharing policy can be represented as

$$F = (AB) \cup (BC) \cup (ACD).$$

The negative access structure is therefore the complement of the positive access structure and can be represented as

$$N = (AB'C'D') \cup (AB'C'D) \cup (AB'CD') \cup (A'BC'D') \cup (A'BC'D)$$
$$\cup (A'B'C'D') \cup (A'B'C'D) \cup (A'B'CD) \cup (A'B'CD').$$

By LEMMA 2-5 in reference [4], we can derive

$$\mathfrak{M}(N) = \mathfrak{M}(\mathfrak{M}(B'C') \cup \mathfrak{M}(B'D') \cup \mathfrak{M}(A'B') \cup \mathfrak{M}(A'C'))$$
$$= \mathfrak{M}(B'C') \cup \mathfrak{M}(B'D') \cup \mathfrak{M}(A'B') \cup \mathfrak{M}(A'C')$$
$$= \{\{Alice, David\} \{Alice, Cathy\} \{Cathy, David\} \{Bob, David\}\}.$$

This maximum set of the negative access structure tells that the secret key cannot be reconstructed either by Alice and David alone, or by Alice and Cathy alone, or by Cathy and David alone, or by Bob and David alone.

Now, the trusted key center selects two secret large primes, p and q, and publishes their product $n = p*q$. Then it selects $p_1, p_2, p_3,$ and p_4 as the public primes. These prime numbers can be chosen as small as possible. A secret key, k, is chosen from $[1, n-1]$. According to this l-span generalized secret sharing scheme, the secret keys to be shared are chosen as

$$s_j = k^{t_c^{l-j}} \bmod n, \text{ for } j = 1, 2, ..., l, \text{ where } t_c = p_1 p_2 p_3 p_4,$$

and the tags and the corresponding shadows associated with users are computed as

$$t_{Alice} = p_1 p_2, \qquad k_{Alice} = k^{p_1^l p_2^l} \bmod n,$$

$$t_{Bob} = p_4, \qquad k_{Bob} = k^{p_4^l} \bmod n,$$

$$t_{Cathy} = p_2 p_3, \qquad k_{Cathy} = k^{p_2^l p_3^l} \bmod n, \text{ and}$$

$$t_{David} = p_1 p_3 p_4, \qquad k_{David} = k^{p_1^l p_2^l p_4^l} \bmod n.$$

Now suppose Alice and Bob, which combination is a positive access instance, want to reconstruct s_j. Alice will present her shadow as

$$k_{Alice,j} = (k_{Alice})^{p_3^{l-j} p_4^{l-j}} = k^{p_1^l p_2^l p_3^{l-j} p_4^{l-j}} \bmod n$$

and Bob will present his shadow as

$$k_{Bob\ j} = (k_{Bob})^{p_1^{l-j} p_2^{l-j} p_3^{l-j}} = k^{p_1^{l-j} p_2^{l-j} p_3^{l-j} p_4^l} \bmod n.$$

By Euclid algorithm, since

$$\gcd(p_1^l p_2^l p_3^{l-j} p_4^{l-j}, p_1^{l-j} p_2^{l-j} p_3^{l-j} p_4^l) = p_1^{l-j} p_2^{l-j} p_3^{l-j} p_4^{l-j},$$

an integer pair (a, b) can be found such that

$$a * (p_1^l p_2^l p_3^{l-j} p_4^{l-j}) + b * (p_1^{l-j} p_2^{l-j} p_3^{l-j} p_4^l) = p_1^{l-j} p_2^{l-j} p_3^{l-j} p_4^{l-j}.$$

Therefore, the secret s_j can be reconstructed by computing

$$(k_{Alice,j})^a * (k_{Bob\ j})^b \bmod n$$

$$= (k)^{p_1^{l-j} p_2^{l-j} p_3^{l-j} p_4^{l-j}} \bmod n$$

$$= s_j$$

V. Conclusion

An l-span generalized secret sharing scheme is proposed in this paper. It allows secrets to be shared in a more efficient way in which same set of shadows can be used to reconstruct l different secrets. For applications in which the reconstructed secret must be revealed and the same secret sharing policy must still be enforced, it alleviates the overhead in the generation and distribution of shadows.

References

[1] A. Shamir, "How to Share a Secret", Communication ACM 22, 11, Nov. 1979, 612-613.

[2] M. Ito, A. Saito, and T. Nishizeki, "Secret Sharing Scheme Realizing General Access Structure", Proc. Glob. Com(1987).

[3] J. Benaloh, and J. Leichter, "Generalized Secret Sharing and Monotone Functions", Proc. Crypto '88, Springer-Verlag, 27-35.

[4] H. Y. Lin and L. Harn, "A Generalized Secret Sharing Scheme with Cheater Detection", Proc. Asiacrypt '91, Nov. 1991, Japan.

[5] C. S. Laih, L. Harn, J. Y. Lee and T. Hwang, "Dynamic Threshold Scheme Based on the Definition of Cross-Product in an N-Dimensional Linear Space", Proc. Crypto '89, Springer-Verlag, 286-297.

Provable Security Against Differential Cryptanalysis

Kaisa Nyberg [*1] and Lars Ramkilde Knudsen[2]

[1] Finnish Defence Forces, University of Helsinki, (on leave)[***]
[2] Aarhus University, DK-8000 Aarhus C.

1 Introduction

The purpose of this paper is to show that there exist DES-like iterated ciphers, which are provably resistant against differential attacks. The main result on the security of a DES-like cipher with independent round keys is Theorem 1, which gives an upper bound to the probability of r-round differentials, as defined in [3] and this upper bound depends only on the round function of the iterated cipher. Moreover, it is shown that there exist functions such that the probabilities of differentials are less than or equal to 2^{2-n}, where n is the length of the plaintext block. We also show a prototype of an iterated block cipher, which is compatible with DES and has proven security against differential attacks.

2 Differential Cryptanalysis of DES-like iterated ciphers

A DES-like cipher is a block cipher based on iterating a function, called **F**, several times. Each iteration is called a round. The input to each round is divided into two halves. The right half is fed into **F** together with a round key derived from a keyschedule algorithm. The output of **F** is added (modulo 2) to the left half of the input and the two halves are swapped except for the last round. The plaintext is the input to the first round and the ciphertext is the output of the last round.

Notation: Let the block size of the cipher be $2n$ and the size of the round key be m, $m \geq n$. Let $\mathbf{f} : GF(2)^m \rightarrow GF(2)^n$ and $E : GF(2)^n \rightarrow GF(2)^m$, an affine expansion mapping. Let L_i, R_i be the left and right halves of the input to the i'th round. Then $L_{i+1} = R_i$ and $R_{i+1} = \mathbf{f}(E(R_i) \oplus K_i) \oplus L_i$ and $\mathbf{F}(R_i, K_i) = \mathbf{f}(E(R_i) \oplus K_i)$.

In [1] Biham and Shamir introduced differential cryptanalysis of DES-like ciphers. In their attacks they make use of characteristics, which describe the behaviour of input and output differences for some number of consecutive rounds. The probability of a one-round characteristic is the conditional probability that given a certain difference in the inputs to the round we get a certain difference

* The work of the author on this project is supported by MATINE Board, Finland.
*** Current address: Prinz Eugen-Straße 18/6, A-1040 Vienna.

in the outputs of that round. Assume that in every round the inputs $E(R) \oplus K$ to **f** are independent and random. This assumption is satisfied if the round keys are uniformly random and independent. Then the probability of an r-round characteristic is obtained by multiplying the probabilities of the r one-round characteristics.

Lai and Massey [3] observed that for the success of differential cryptanalysis it is not necessary to fix the values of input and output differences for the intermediate rounds in a characteristic. They introduced the notion of *differentials*. The probability of an r-round differential is the conditional probability that given an input difference at the first round, the output difference at the r'th round will be some fixed value. Note that the probability of an r-round differential with input difference A and output difference B is the sum of the probabilities of all r-round characteristics with input difference A and output difference B. For $r \leq 2$ the probabilities for a differential and for the corresponding characteristic are equal, but in general the probabilities for differentials will be higher.

In order to make a successful attack on a DES-like iterated cipher by differential cryptanalysis the existence of good characteristics is sufficient. On the other hand to prove security against differential attacks for DES-like iterated ciphers we must ensure that there is no differential with a probability high enough to enable successful attacks.

The difference of two inputs $E(R) \oplus K$ and $E(R^*) \oplus K$ to **f** is $E(R) \oplus E(R^*)$. Since we assume E to be affine, the difference of two inputs depends only on the difference $R \oplus R^*$. Hence for DES-like ciphers the round probabilities of characteristics only depend on the intrinsic properties of **f**. Given $\mathbf{f} : GF(2)^m \rightarrow GF(2)^n$ denote

$$p_{max} = 2^{-m} max_\beta max_{\alpha \neq 0} \#\{X \in GF(2)^m | \mathbf{f}(X \oplus \alpha) \oplus \mathbf{f}(X) = \beta\}$$

That is, p_{max} is the highest probability for a non trivial one-round characteristic or differential.

Theorem 1 *It is assumed that in a DES-like cipher with* $\mathbf{f} : GF(2)^m \rightarrow GF(2)^n$ *the inputs to* **f** *at each round are independent and uniformly random. Then the probability of an r-round differential, $r \geq 4$, is less than or equal to $2p_{max}^2$.*

Proof: We shall first give the proof for $r = 4$. Let α_L and α_R be the left and right halves of the input difference at the first round and β_L and β_R be corresponding halves of the output difference at the last round. Either $\beta_L \neq 0$ or $\beta_R \neq 0$ or both. We shall give the proof in the case $\beta_L = 0$, $\beta_R \neq 0$, the other two cases are similar. We denote by $\Delta R(i)$ the right input differences to the i'th round, $i = 2, 3, 4$. Let $\beta_L = 0$ and $\beta_R \neq 0$. Then $\Delta R(4) = \beta_L = 0$ and $\Delta R(3) = \beta_R$. We separate between two cases: $\alpha_R \neq \beta_R$ and $\alpha_R = \beta_R$.

1. $\alpha_R \neq \beta_R$. Then $\Delta R(2) \neq 0$. For any given $\Delta R(2) \neq 0$ there is exactly one way of getting β_L, β_R from the input differences α_R and $\Delta R(2)$ at the second round, and the probability is less than or equal to p_{max}^2. Hence the probability of the four round differential is less than or equal to p_{max}^2.

2. $\alpha_R = \beta_R$. If $\Delta R(2) = 0$ it follows that the output difference from **F** at the

third round $\Delta \mathbf{F}(R(3)) = 0$, which happens with probability less than or equal to p_{max}, because $\Delta R(3) = \beta_R \neq 0$. Since $\alpha_R \neq 0$ we have

$$Prob(\Delta R(2) = 0 \,|\, \alpha_L, \alpha_R) \leq p_{max}$$

If $\Delta R(2) \neq 0$ the probability that $\Delta \mathbf{F}(R(3)) = \Delta R(2)$ is less than or equal to p_{max}. We also need to have $\Delta \mathbf{F}(R(2)) = 0$, which is true with probability less than or equal to p_{max}. So we obtain

$$Prob(\beta_L, \beta_R \,|\, \alpha_L, \alpha_R)$$
$$= \sum_{\Delta R(2)} Prob(\Delta R(2) \,|\, \alpha_L, \alpha_R) \, Prob(\beta_L, \beta_R \,|\, \alpha_L, \alpha_R, \Delta R(2))$$
$$= Prob(\Delta R(2) = 0 \,|\, \alpha_L, \alpha_R) \, Prob(\beta_L, \beta_R \,|\, \alpha_L, \alpha_R, \Delta R(2) = 0)$$
$$+ \sum_{\Delta R(2) \neq 0} Prob(\Delta R(2) \,|\, \alpha_L, \alpha_R) \, Prob(\beta_L, \beta_R \,|\, \alpha_L, \alpha_R, \Delta R(2))$$
$$\leq p_{max}^2 + \sum_{\Delta R(2) \neq 0} Prob(\Delta R(2) \,|\, \alpha_L, \alpha_R) \cdot p_{max}^2$$
$$\leq 2p_{max}^2$$

Let now $r > 4$. Then

$$Prob(\beta_L, \beta_R \,|\, \alpha_L, \alpha_R)$$
$$= \sum_{\Delta L(r-3), \Delta R(r-3)} [Prob(\Delta L(r-3), \Delta R(r-3) \,|\, \alpha_L, \alpha_R) \cdot$$
$$Prob(\beta_L, \beta_R \,|\, \alpha_L, \alpha_R, \Delta L(r-3), \Delta R(r-3))]$$

Since we assumed that the inputs to $\mathbf{f}$ are independent and uniformly random it follows from the proof for $r = 4$ that

$$Prob(\beta_L, \beta_R \,|\, \alpha_L, \alpha_R, \Delta L(r-3), \Delta R(r-3)) =$$
$$Prob(\beta_L, \beta_R \,|\, \Delta L(r-3), \Delta R(r-3)) \leq 2p_{max}^2$$

Thus $Prob(\beta_L, \beta_R \,|\, \alpha_L, \alpha_R) \leq 2p_{max}^2$. $\qquad \square$

If $\mathbf{f}$ is a permutation, then in every characteristic between two zero rounds there has to be at least two nonzero rounds and the following result can be proved.

Theorem 2 *It is assumed that the function $\mathbf{f}$ in a DES-like cipher is a permutation and that the inputs to $\mathbf{f}$ at each round are independent and uniformly random. Then the probability of an r-round differential for $r \geq 3$ is less than or equal to p_{max}^2.*

Proof: We give the proof for $r = 3$. The general case can then be proved like in the preceding theorem. Again we separate between three cases and use the same notation as before.

1. $\beta_L = 0$, $\beta_R \neq 0$. In this case the third round of each characteristic is a zero-round. At the second round the input difference $\Delta R(2) = \beta_R \neq 0$ results in

an output difference $\alpha \neq 0$ with probability less than or equal to p_{max}. At the first round we get the output difference $\alpha_L \oplus \beta_R \neq 0$ with probability less than or equal to p_{max} from the input difference $\Delta R(1) = \alpha_R \neq 0$. Hence $Prob(\beta_L, \beta_R \,|\, \alpha_L, \alpha_R) \leq p_{max}^2$.

2. $\beta_L \neq 0$, $\beta_R = 0$. Now the output difference at the third round equals $\Delta R(2)$ and it is different from zero. Given $\Delta R(2) \neq 0$ the probability of the third round is less than or equal to p_{max} and the same holds for the second round. Consequently

$$Prob(\beta_L, \beta_R \,|\, \alpha_L, \alpha_R)$$
$$= \sum_{\Delta R(2) \neq 0} Prob(\Delta R(2) \,|\, \alpha_L, \alpha_R) \, Prob(\beta_L, \beta_R \,|\, \alpha_L, \alpha_R, \Delta R(2))$$
$$\leq \sum_{\Delta R(2)} Prob(\Delta R(2) \,|\, \alpha_L, \alpha_R) \cdot p_{max}^2 \leq p_{max}^2$$

3. $\beta_L \neq 0$, $\beta_R \neq 0$. Assume first that $\Delta R(2) = 0$. Then for every characteristic the probability of the third round is less than or equal to p_{max}, the probability of the second round is one and the probability of the first round is less than or equal to p_{max}. Secondly, given $\Delta R(2) \neq 0$, the probability of the third round is less than or equal to p_{max} and the same is true for the second round. Hence $Prob(\beta_L, \beta_R \,|\, \alpha_L, \alpha_R) \leq p_{max}^2$ also in this case. $\qquad\square$

3 Almost perfect nonlinear permutations

For a mapping $\mathbf{f} : GF(2)^m \to GF(2)^n$ the lower bound for p_{max} is 2^{-n}. Mappings attaining this lower bound were investigated in [7], where they are called perfect nonlinear generalizing the definition of perfect nonlinearity given for Boolean functions in [6]. It was shown in [7] that perfect nonlinear mappings from $GF(2)^m \to GF(2)^n$ only exist for m even and $m \geq 2n$. Hence they can be adapted for use in DES-like ciphers only with expansion mappings that double the block length.

If the round function of a DES-like cipher does not involve any expansion, i.e. in the case when $\mathbf{f} : GF(2)^m \to GF(2)^n$ is a permutation, the trivial lower bound for p_{max} is 2^{1-n}, since then the difference

$$\mathbf{f}(\mathbf{x} + \mathbf{w}) + \mathbf{f}(\mathbf{x})$$

obtains half of the values in $GF(2)^n$ twice and never the other half of the values. We shall call the permutations with $p_{max} = 2^{1-n}$ *almost perfect nonlinear*. The purpose of this section is to show that such permutations exist.

Assume that $m = nd$, where m, n, d are all odd integers. In [8] permutations $\mathbf{f}$ in $GF(2^m) = GF(2^d)^n$ were constructed to satisfy the following property:

(P) Every nonzero linear combination of the components of $\mathbf{f}$ is a balanced quadratic form $\mathbf{x}^t \mathbf{C} \mathbf{x}$ in n indeterminates over $GF(2^d)$ with $rank(\mathbf{C} + \mathbf{C}^t) = n - 1$.

Indeed the following theorem holds.

Theorem 3 *Let* $\mathbf{f} : GF(2^d)^n \rightarrow GF(2^d)^n$ *be a permutation satisfying (P). Then* $p_{max} = 2^{d(1-n)}$.

For the sake of simplicity we shall give the proof in the case where $d = 1$ and $m = n$.

Lemma 1 *A quadratic form* $f(\mathbf{x}) = \mathbf{x}^t \mathbf{A} \mathbf{x}$ *in* n *indeterminates over* $GF(2)$ *is balanced if and only if* $f(\mathbf{w}) \neq 0$ *for the linear structure* $\mathbf{w}$ *of* f.

Recall that a linear structure $\mathbf{w}$ of $f : \mathbf{F}^n \rightarrow \mathbf{F}$ is a nonzero vector in $\mathbf{F}^n$ such that $f(\mathbf{x} + \mathbf{w}) + f(\mathbf{x})$ is constant. It was also shown in [8] that a quadratic form $f(\mathbf{x}) = \mathbf{x}^t \mathbf{A} \mathbf{x}$ in n indeterminates over $GF(2)$ with $rank(\mathbf{A} + \mathbf{A}^t) = n - 1$ has exactly one linear structure.
Proof of Lemma 1: Let

$$\varphi(x_1, \ldots\ldots, x_n) = x_1 x_2 + \ldots + x_{n-2} x_{n-1} + \delta x_n = \mathbf{x}^t \mathbf{C} \mathbf{x}$$

$\delta = 0$ or 1, be the quadratic forms to which all quadratic forms $f(\mathbf{x}) = \mathbf{x}^t \mathbf{A} \mathbf{x}$ with $rank(\mathbf{A} + \mathbf{A}^t) = n - 1$ are equivalent (see [4]). It means that there is a linear transformation $\mathbf{T}$ of coordinates such that $f(\mathbf{x}) = \varphi(\mathbf{T}\mathbf{x})$. Then $\mathbf{w}$ is a linear structure of f if and only if $\mathbf{T}\mathbf{w} = (0, 0, \ldots, 0, 1)$. Then f is balanced if and only if φ is balanced which is true if and only if $\delta = 1$. But $\delta = 1$ if and only if

$$f(\mathbf{w}) = \varphi(\mathbf{T}\mathbf{w}) = \varphi(0, \ldots, 0, 1) = 1.$$

$\square$

Lemma 2 *Let* $\mathbf{w} \in GF(2)^n$ *be not the linear structure of* $f : GF(2)^n \rightarrow GF(2)$, $f(\mathbf{x}) = \mathbf{x}^t \mathbf{A} \mathbf{x}$ *with* $rank(\mathbf{A} + \mathbf{A}^t) = n - 1$. *Then*

$$\mathbf{x} \mapsto f(\mathbf{x} + \mathbf{w}) + f(\mathbf{x})$$

is balanced.

Proof: It sufficies to show that

$$\varphi(\mathbf{x} + \mathbf{w}) + \varphi(\mathbf{x})$$

is balanced for every $\mathbf{w} \neq (0, \ldots, 0, 1)$. But this is true since

$\varphi(\mathbf{x} + \mathbf{w}) + \varphi(\mathbf{x}) =$
$(x_1 + w_1)(x_2 + w_2) + \ldots + (x_{n-2} + w_{n-2})(x_{n-1} + w_{n-1}) + x_n + w_n +$
$x_1 x_2 + \ldots + x_{n-2} x_{n-1} + x_n.$

is a non-constant affine or linear function for every $\mathbf{w} \neq (0, \ldots, 0, 1)$. $\square$

Lemma 3 *Let* $\mathbf{f} : GF(2)^n \rightarrow GF(2)^n$ *be a permutation with property (P). Then every nonzero vector* $\mathbf{w} \in GF(2)^n$ *is a linear structure of a nonzero linear combination of the components of* $\mathbf{f}$.

Proof: It sufficies to show that two different linear combinations of the components of $\mathbf{f}$ have different linear structures. Let $\mathbf{u}_1$ and $\mathbf{u}_2$ be nonzero vectors in $GF(2)^n$ and let $\mathbf{w}_1$ and $\mathbf{w}_2$ be the linear structures of $\mathbf{u}_1 \cdot \mathbf{f}$ and $\mathbf{u}_2 \cdot \mathbf{f}$, respectively. If $\mathbf{w}_1 = \mathbf{w}_2 = \mathbf{w}$ it follows that $\mathbf{w}$ is also the linear structure of $(\mathbf{u}_1 + \mathbf{u}_2) \cdot \mathbf{f}$. Since $\mathbf{u}_1 \cdot \mathbf{f}$ and $\mathbf{u}_2 \cdot \mathbf{f}$ are balanced it follows from Lemma 1 that

$$\mathbf{u}_1 \cdot \mathbf{f}(\mathbf{w}) = \mathbf{u}_2 \cdot \mathbf{f}(\mathbf{w}) = 1$$

and consequently

$$(\mathbf{u}_1 + \mathbf{u}_2) \cdot \mathbf{f}(\mathbf{w}) = 0.$$

If $\mathbf{u}_1 \neq \mathbf{u}_2$, then $(\mathbf{u}_1 + \mathbf{u}_2) \cdot \mathbf{f}$ is balanced. Thus by Lemma 1, $\mathbf{u}_1 = \mathbf{u}_2$. □
Now Theorem 3 for $d = 1$ is a consequence of the following

Theorem 4 Let $\mathbf{f} = (f_1, f_2, ... f_n)\colon GF(2)^n \to GF(2)^n$ be a permutation that satisfies (P). Then for every fixed nonzero difference $\mathbf{w} \in GF(2)^n$ of the inputs to $\mathbf{f}$, the differences of the outputs lie in an affine hyperplane of $GF(2)^n$ and are uniformly distributed there.

Proof: Let $\mathbf{w}$ be a nonzero input difference for $\mathbf{f}$. Then by Lemma 3 there is $\mathbf{v} \in GF(2)^n$, $\mathbf{v} \neq 0$, such that $\mathbf{w}$ is the linear structure of $\mathbf{v} \cdot \mathbf{f}$ and by Lemma 1

$$\mathbf{v} \cdot \mathbf{f}(\mathbf{x} + \mathbf{w}) + \mathbf{v} \cdot \mathbf{f}(\mathbf{x}) = 1$$

for all $\mathbf{x} \in GF(2)^n$.
Let $\mathbf{u}_1, ..., \mathbf{u}_{n-1}$ be linearly independent vectors in $GF(2)^n$ such that

$$\mathbf{v} \notin span\{\mathbf{u}_1, ..., \mathbf{u}_{n-1}\}$$

Then by Lemma 2 for every $\mathbf{u} \in span\{\mathbf{u}_1, ..., \mathbf{u}_{n-1}\}$ the function

$$\mathbf{x} \mapsto \mathbf{u} \cdot \mathbf{f}(\mathbf{x} + \mathbf{w}) + \mathbf{u} \cdot \mathbf{f}(\mathbf{x})$$

is balanced, which means (see [4]) that for every $(b_1, ..., b_{n-1}) \in GF(2)^{n-1}$ the system of equations

$$\mathbf{u}_i \cdot \mathbf{f}(\mathbf{x} + \mathbf{w}) + \mathbf{u}_i \cdot \mathbf{f}(\mathbf{x}) = b_i, \ i = 1, ..., n-1,$$

has 2 solutions $\mathbf{x} \in GF(2)^n$. Hence the system of n equations :

$$(2) \quad \begin{aligned} \mathbf{u}_i \cdot \mathbf{f}(\mathbf{x} + \mathbf{w}) + \mathbf{u}_i \cdot \mathbf{f}(\mathbf{x}) &= b_i, i = 1, ..., n-1, \\ \mathbf{v} \cdot \mathbf{f}(\mathbf{x} + \mathbf{w}) + \mathbf{v} \cdot \mathbf{f}(\mathbf{x}) &= b \end{aligned}$$

has 2 solutions if $b = 1$ and no solutions if $b = 0$. Every system of n equations

$$f_i(\mathbf{x} + \mathbf{w}) + f_i(\mathbf{x}) = a_i, \ i = 1, 2, ..., n.$$

is a linear transformation of (2), from which the claim follows.   □
By a similar argumentation one can prove the following generalization of Theorem 3.

Theorem 5 *Let* **f** *be a permutation in* $GF(2^d)^n$, *d and n odd, with property* (*P*) *and let* $f_1, ..., f_n$ *be the components of* **f** *with respect to some arbitrary fixed basis over* $GF(2^d)$. *Let* $l \leq n$ *and set* $\mathbf{h} = (f_1, f_2, ..., f_l)$. *Then* $p_{max} = 2^{d(1-l)}$ *for* **h**.

From the results in Section 2 we now obtain

Theorem 6 *Assume that in a DES-like cipher the function* **f** *is a mapping from* $GF(2)^m$ *to* $GF(2)^n$, $m \geq n$, *obtained from a permutation in* $GF(2)^n$ *with* (*P*) *by discarding* $m - n$ *output bits. Then* $p_{max} = 2^{1-n}$ *for* **f**. *Moreover, if* $m > n$, *then the probability of every r-round differential,* $r \geq 4$, *is less than or equal to* 2^{3-2n}, *assuming that the inputs to* **f** *are uniformly random and independent at each round. If* $m = n$, *the probability of every r-round differential,* $r \geq 3$, *is less than or equal to* 2^{2-2n}.

4 Examples of permutations with property (P)

Pieprzyk [9] observed that the permutations $\mathbf{f}(\mathbf{x}) = \mathbf{x}^{2^k+1}$ in $GF(2^n)$ with $gcd(k, n) = 1$, $1 \leq k < n$ and n odd are at a large distance from the linear mappings. We shall show that these permutations have property (P).
Let $\alpha_1, ..., \alpha_n$ be a basis in $GF(2^n)$ over $GF(2)$ and $\beta_1, ..., \beta_n$ be its dual basis. Let $\mathbf{x} = \sum_{i=1}^{n} x_i \alpha_i$, $x_i \in GF(2)$. Then the i'th component $f_i(\mathbf{x})$ of $\mathbf{f}(\mathbf{x})$ with respect to the basis $\alpha_1, ..., \alpha_n$ is

$$f_i(\mathbf{x}) = Tr(\beta_i \mathbf{x}^{2^k+1})$$
$$= Tr(\beta_i (\sum_{j=1}^{n} x_j \alpha_j)(\sum_{l=1}^{n} x_l \alpha_l)^{2^k})$$
$$= \sum_{j=1}^{n} \sum_{l=1}^{n} Tr(\beta_i \alpha_j \alpha_l^{2^k}) x_j x_l$$
$$= \sum_{j=1}^{n} \sum_{l=1}^{n} Tr(\gamma_i \alpha_j (\gamma_i \alpha_l)^{2^k}) x_j x_l$$

where $\gamma_i \in GF(2^n)$ is such that $\gamma_i^{2^k+1} = \beta_i$, $i = 1, 2, ..., n$.
Now it is straightforward to check that $Tr(\gamma_i \alpha_j (\gamma_i \alpha_l)^{2^k})$ is the jl'th entry in the matrix $\mathbf{A}_i = \mathbf{B}_i^t \mathbf{R}^k \mathbf{B}_i$ where

$$\mathbf{B}_i = \begin{pmatrix} \gamma_i \alpha_1 & \gamma_i \alpha_2 & .... & \gamma_i \alpha_n \\ (\gamma_i \alpha_1)^2 & (\gamma_i \alpha_2)^2 & .... & (\gamma_i \alpha_n)^2 \\ \cdot & \cdot & & \cdot \\ \cdot & \cdot & & \cdot \\ \cdot & \cdot & & \cdot \\ (\gamma_i \alpha_1)^{2^{n-1}} & (\gamma_i \alpha_2)^{2^{n-1}} & .... & (\gamma_i \alpha_n)^{2^{n-1}} \end{pmatrix}$$

is a $n \times n$ regular matrix over $GF(2^n)$ and

$$R = \begin{pmatrix} 0 & 1 & 0 & .... & 0 \\ 0 & 0 & 1 & .... & 0 \\ .. & .. & .. & .. & . \\ .. & .. & .. & .. & . \\ 0 & 0 & 0 & .... & 1 \\ 1 & 0 & 0 & .... & 0 \end{pmatrix}$$

is the cyclic shift for which $rank(\mathbf{R}^k + (\mathbf{R}^k)^t) = n - 1$ if $gcd(k, n) = 1$. Hence

$$f_i(\mathbf{x}) = \mathbf{x}^t \mathbf{A}_i \mathbf{x}$$

and

$$rank(\mathbf{A}_i + \mathbf{A}_i^t) = rank(\mathbf{B}_i^t(\mathbf{R}^k + (\mathbf{R}^k)^t)\mathbf{B}_i) = rank(\mathbf{R}^k + (\mathbf{R}^k)^t) = n - 1$$

over $GF(2^n)$. Thus $rank(\mathbf{A}_i + \mathbf{A}_i^t) = n - 1$ also over $GF(2)$, since the rank does not decrease when going to a subfield and it cannot be n. By the linearity of the trace function the same holds for every nonzero linear combination of the components f_i of $\mathbf{f}$. Moreover, since $\mathbf{f}$ is a permutation, they are all balanced, which completes the proof of property (P) for $\mathbf{f}$.

Matsumoto and Imai proposed in [5] a public key cryptosystem C^*, which is based on power polynomials $\mathbf{x}^{2^k+1}$. If the round function of an iterated DES-like cipher of block size 64 makes use of the mapping $\mathbf{x}^{2^k+1}$ as proposed below in Section 5, the description of the round function for efficient implementation would be less than the minimum size of the public key for C^* cryptosystem.

5 A prototype of a DES-like cipher for encryption

Let $\mathbf{g}(\mathbf{x}) = \mathbf{x}^3$ in $GF(2^{37})$. There are several efficient ways of implementing this power polynomial and each of them suggest a choice of a basis in $GF(2^{37})$. Let us fix a basis and discard five output coordinates. Then we have a function $\mathbf{f} : GF(2)^{37} \rightarrow GF(2)^{32}$. The 64-bit plaintext block is divided into two 32-bit halves L and R. The plaintext expansion is an affine mapping $E : GF(2)^{32} \rightarrow GF(2)^{37}$. Each round take a 32 bit input and a 37 bit key. The round function is $L \| R \mapsto R \| L \oplus \mathbf{f}(E(R) \oplus K)$.

In [2] Biham and Shamir introduced an improved differential attack on 16-round DES. This means, that in general for an r-round DES-like cipher the existence of an $(r-2)$-round differential with a too high probability may enable a successful differential attack. From Theorem 6 we have that every four and five round differential of this block cipher has probability less than or equal to 2^{-61}. Therefore we suggest at least six rounds for the block cipher. All round keys should be independent, therefore we need at least 222 key bits. This is equivalent to four DES keys, where all parity bits plus two other bits are discarded.

References

1. E. Biham, A. Shamir. *Differential Cryptanalysis of DES-like Cryptosystems.* Journal of Cryptology, Vol. 4 No. 1 1991.
2. E. Biham, A. Shamir. *Differential Cryptanalysis of the full 16-round DES.* Technical Report # 708, Technion - Israel Institute of Technology.
3. X. Lai, J. L. Massey, S. Murphy. *Markov Ciphers and Differential Cryptanalysis.* Advances in Cryptology - Eurocrypt '91. Lecture Notes in Computer Science 547, Springer Verlag.
4. R. Lidl, H. Niederreiter. *Finite Fields.* Encyclopedia of Mathematics and its applications, Vol. 20. Addison-Wesley, Reading, Massachusetts, 1983.
5. T. Matsumoto, H. Imai. *Public quadratic polynomial-tuples for efficient signature-verification and message-encryption.* Advances in Cryptology - Eurocrypt '88. Lecture Notes in Computer Science, Springer Verlag, 1989.
6. W. Meier, O. Staffelbach. *Nonlinearity criteria for cryptographic functions.* Proceedings of Eurocrypt '89, Springer Verlag 1990, 549-562.
7. K. Nyberg. *Perfect nonlinear S-boxes.* Advances in Cryptology - Proceedings of Eurocrypt '91. Lecture Notes in Computer Science 547, Springer Verlag.
8. K. Nyberg. *On the construction of highly nonlinear permutations.* Advances in Cryptology - Proceedings of Eurocrypt '92 (to appear).
9. J. Pieprzyk. *On bent permutations.* Technical Report CS91/11; The University of New South Wales.

Content-Addressable Search Engines and DES-like Systems

Peter C. Wayner

Computer Science Department
Cornell University
Ithaca, NY 14853

Abstract. A very simple parallel architecture using a modified version of content-addressable memory (CAM) can be used to cheaply and efficiently encipher and decipher data with DES-like systems. This paper will describe how to implement DES on these modified content-addressable memories at speeds approaching some of the better specialized hardware. This implementation is often much more attractive for system designers because the CAM can be reprogrammed to encrypt the data with other DES-like systems such as Khufu or perform system tasks like data compression or graphics.

The CAM memory architecture is also easily extendable to build a large scale engine for exhaustively searching the entire keyspace. This paper estimates that it will be possible to build a machine to test 2^{55} keys of DES in one day for \$30 million. This design is much less hypothetical than some of the others in the literature because it is based upon hardware that will be available off-the-shelf in the late end of 1992. The architecture of this key search machine is much more attractive to an attacker because it is easily reprogrammable to handle modified DES-like algorithms such as the UNIX password system or Khufu.

The original DES system was designed to be easily implemented in hardware [NBS77] and the current silicon manifestations of the cipher use modern processor design techniques to encipher and decipher information at about 1 to 30 megabits per second. Implementations of DES in software for standard CPUs, however, are markedly slower than specialized chips because many of the operations involved in DES are bit-level manipulations. As a result, many of the DES-like systems such as Merkle's Khufu [Mer90] were designed as replacements that could be easily implemented on conventional hardware.

There is one class of general architecture, however, that implements bit-level operations. The machines like the CM-1, CM-2 and CM-200 from Thinking Machines Corporation and the Maspar machine all have thousands of one-bit processors. The designers intended that a large number of processors would compensate for the deficiencies of the individual nodes.

Another example of this small architecture is now emerging from the labs of memory designers who are trying to build sophisticated content addressable memory (CAM). The individual processors of these machines are even weaker than the ones of the CM-1, but they can be packed very densely on a chip. The tiny processors have only a fraction of the memory of a CM-1 (42 bits versus thousands) and only a one dimensional interconnection network (vs. 12), but this is sufficient to implement DES. Most importantly, these restrictions allow a packing density (1024 processors per chip) that is significantly higher at a cheap price. (\$30-\$100 per chip)

Implementing the cipher on generalized parallel architectures like the CAM have one main advantage– cost. Many computer designers often find that the speed of a specialized DES chip is often not worth the price. Generalized, content-addressable machines, however, have many other applications and this makes them a good compromise for the system designer. The design presented here can be easily reprogrammed in software to encrypt with DES or any DES-like variant like Khufu. The hardware can also be used do data compression, data searches or even many different graphics operations.

This paper will describe how to implement the DES algorithm on this architecture and produce results that are on par with the middle range of the specialized hardware. The main contribution is not extremely fast encryption speeds. It is very fast speed coupled with software-level flexibility. Many other papers have offered flexible hardware designs [VHVM88, FMP85] that can be easily reworked to handle variants of DES, but none offer the flexibility of this system. Verbauwhede et al. [VHVM88] requires new silicon to be fabricated in all cases and the designs of Falfield et al. [FMP85] run internal microcode that can be easily reprogrammed to implement other slight variants of DES such as cipher-block chaining. However, new algorithms like Khufu, however, would require a new micro-code instruction set. The flexibility of this CAM based design is quite attractive to both the system designer and the brute-force attacker because it allows the hardware to be used for different purposes and different algorithms.

1 Content-Addressable Memory Machines

Standard memory maps an address to a value. Unfortunately, there are many applications when an algorithm needs to know which memory location holds a particular value. The only recourse is to search all the memory to find the value in question. Content-addressable memory is a hardware solution to this problem that will invert the search and provide the address holding a value in a single operation. This technique has been well-researched over the years and the book by Kohonen [Koh87] notes many approaches and summarizes some of the more salient aspects of this research. Several companies including AMD are making basic content-addressable memory modules.

Recently teams at Syracuse University (some publications include [Old86, OWN87, OSB87]), MIT and Cornell ([Bri90, WS89, Zip90]) have developed more sophicated and powerful implementations in silicon. These implementations allow the programmer to chain the result of several searches together in a simple fashion so that larger data structures and more complicated searches can be performed in hardware. Some of this hardware was originally intended to speed up logic programming, but many people have found surprising and interesting applications for the simple hardware. Oldfield and his team at Syracuse, for instance, are currently working on compressing data.

A company, Coherent Research Incorporated of Syracuse, New York, is building sophisticated content-addressable memory chips called the Coherent Processor for widespread use. This paper will use their chip as an example because it is commercially available, but there is no reason why the algorithms cannot be modified slightly for use on similar chips.

At the basic level, the Coherent Processor is a large, single dimensional array of very simple parallel processors. Each processor has 42 bits of memory ($W_i[0] \ldots W_i[41]$, the i denotes the processor number) and three one-bit registers (R_1, R_2 and R_3). It also has a processing unit that can execute instructions on the registers, transfer data between the registers and the memory, communicate with the two neighboring processors or match a value on the internal bus. The instructions are simple operations that read the three register bits of memory and store the result in one of the three. The match instructions can be used to simultaneously compare one 42-bit value against the entire array of processors. If there is a match, then the appropriate value is placed in a register.

The following table shows the basic Coherent Processor instructions and the number of clock cycles used to complete them.

1. *MATCH:* Simultaneously compare the 42 general bits at each processor with the values on a bus and store the result of this match in R_1. This is used to look up items quickly.

The match routine can include wild-card matches for individual bits so it is possible to match for strings of bits like "0000******11*****" (a "*" matches both a "0" and a "1"). If you want to move the value of bit $W_i[2]$ into R_3, then you would "match" a pattern with 1 in bit $W_i[2]$ and wild-card matches specified for the rest and store the result in R_3. If the value of bit $W_i[2]$ was 1 in a particular word, then the match would be successful and a 1 would be stored in R_3. If a zero was in bit $W_i[2]$, then the match would be unsuccessful and a zero would be stored. The values of the other columns would not be affected. Cost: 4 cycles.

2. *CALC:* Calculate a three-bit function of the three registers and store the result in a third register. Cost: 2 cycles.
3. *READ:* Take the result of a selected word and place it on the bus. This operation usually follows a MATCH operation. Cost: 3 cycles.
4. *WRITE:* Move the result from the bus into the selected word(s). Cost: 2 cycles.
5. *SHIFT:* The first registers of each word are interconnected. They can shift the bit in their register to adjacent words in one step. Cost: 2 cycles.
6. *WRITECOLUMN:* Moves a bit from a register into one of the 42 bits of memory. Cost: 2 cycles.

These commands can be strung together to manipulate data in simple and straight-forward methods.

2 Implementing Plain DES

There are three main operations involved in encrypting a block of 64 bits with the basic mode of the Data Encryption Standard known as the Electronic Code Book (ECB). They are 1) permuting the bits, 2) passing a 32-bit block through an s-box and 3) permuting the key structure. Each of these steps is easy to program on the Coherent Processor , in a large part because the architecture is so limited. Several features of the instruction set, however, make implementing the algorithm very easy.

Let the plaintext blocks of data be denoted, $B_1, \ldots, B_n$ and the individual bits of block B_i be $\{B_i[0] \ldots B_i[63]\}$. The key is K and the individual bits are $K[0] \ldots K[55]$.

There are sixteen rounds of encryption and the key scheduling algorithm chooses a 48-bit subset of key bits to be used on each round. Let $K^{(l)}[0] \ldots K^{(l)}[47]$ be the 48 bits used in round l. Each block of 64 bits is broken into two 32-bit halves (called B_L and B_R) and in each round the value of one of the halves is mixed with a subset of the key bits, passed through the s-box and then mixed with the right 32-bit half. More precisely, in each round:

$$B_L \leftarrow B_L \oplus f(E(B_R) \oplus K^{(l)}).$$

("$\oplus$"=XOR) Then B_L and B_R are exchanged. f is the s-box function that takes 48 bits and returns 32 and the $E()$ function is an "expansion" function that maps 32 bits into 48 bits so it can be combined with the 48 bits of key. Some bits of the input to E are used more than others.

The data to be encrypted is broken into 64-bit blocks and each block is stored in 32-bit halves in two adjacent 42 bit words in the array, W_i and W_{i+1}.

2.1 Permuting the Bits

At the beginning and the end of the encryption process, the 64 bits in the block are passed through a bit-wise permutation. This step is often considered the slowest part of many software implementations for general purpose machines and many people believe that it was included to slow down software implementations and force general CPUs to move bits one by one. The Coherent Processor must also move each bit one at a time, but at least this is the best that it can do. In practice, the large number of parallel processors makes up for the weakness.

Let the permutation be written as a set of cycles: $W_i[p_0] \rightarrow W_j[p_1] \rightarrow \ldots \rightarrow W_i[p_i] \rightarrow W_i[p_0]$. There are 64 bits to be exchanged, but they do not move in one cycle. The process can be accomplished by stringing together a chain of bit moving commands. When the bits to be exchanged are on different words, then the CAM must also execute a bit-passing command to swap the bit to the adjacent word. The work can be summarized in pseudo-code:

Move $W_i[p_0]$ into a bit .
for k:=1 to 63 do
    Move $W_i[p_k]$ into a bit.
    Move $W_i[p_{k-1}]$ into its destination.
    If $W_i[p_k]$ is on the wrong word,
        then pass it to the correct one.
Move $W_i[p_{63}]$ into $W_i[p_0]$.

There are only 32 bits that need to be shifted between words. It is possible to do this quickly. The next section which computes the values of the s-boxes is much more time intensive. The cost: 129 MATCH and WRITECOLUMN instructions, 32 SHIFT instructions. About 580 cycles.

2.2 Computing the S-boxes

The s-box are responsible for providing the non-linear mixing of the bits that is necessary to provide adequate security. At the highest level, the s-box is a function that maps 32 bits to 32 other bits. The s-boxes used in DES are, though, much simpler and they can be described as eight functions that take 6 out of the 32 bits and return four. Some bits are used more than others. These eight s-boxes can be further simplified into 32 functions that map six bits to one bit and this is the best level of abstraction to use when programming the Coherent Processor .

Meyer and Matyas [MM82] describe the design of the s-boxes in terms of *minterms*, which are roughly the same as clauses of boolean variables. An equation describing output of one bit of an s-box might look something like this:

$$B_i[1] \cdot \neg B_i[2] \cdot B_i[3] \cdot B_i[4] + B_i[1] \cdot \neg B_i[5] \cdot \neg B_i[6] + B_i[2] \cdot B_i[5]. \tag{1}$$

("." =boolean and, "+"=boolean or, "¬"=boolean not.) There are three minterms in the example and it is generally believed that the number of minterms in a minimal expression is one measure the complexity of the s-box. The recent papers by Biham and Shamir [BS91] and others , show that there are additional criterion that are more important. Meyer and Matyas note that there are 52 and 53 minterms in the description of each of the 8 s-boxes.

These minterm descriptions of the s-boxes can be directly converted into operations for the Coherent Processor . Each clause of variables to be ANDed together can be computed with a MATCH equation with appropriate set of ones for the variables in the clause, zeros for the negated variables in the clause and wildcards for the unrepresented variables. The expression from equation 1 can be encoded:

$$MATCH\ \text{"1011} * * * \ldots * **''} \to R_1$$
$$CALC R_1 \to R_2$$
$$MATCH\ \text{"1} * * * 00 * * \ldots * **''} \to R_1$$
$$CALC R_1 \cdot R_2 \to R_2$$
$$MATCH\ \text{"} * 1 * * 1 * * \ldots * **''} \to R_1$$
$$CALC R_1 \cdot R_2 \to R_1$$

$$(2)$$

This takes 6 cycles per minterm. At 53 minterms per s-box and 8 s-boxes per encryption round, this takes 2544 cycles per encryption round to calculate the values of the bits. It takes one SHIFT, one MATCH, one CALC and one COLUMNWRITE to XOR each of the 32 bits into the adjacent word. That is an additional 384 cycles for 2928 per encryption round. There are 16 rounds in DES, the permutations take 580 cycles and the overall encryption process takes 47,528 cycles.

2.3 Handling the Key

When the result of one of the 32 functions is computed it must be XOR-ed with the key and then passed to the adjacent word to be XOR-ed with the appropriate bit. The same key encrypts all the blocks at the same time and it can be included by XORing the key vector, $K^{(l)}$, into the match words. For instance, assume that "11001100 10101110 01001100 11100101" is the 48 bits of key being used in a round and the minterms from equation 1 define the s-box equations. Then the operations in example 2 become:

$$MATCH\ \text{"0111} * * * \ldots * **''} \to R_1$$
$$CALC R_1 \to R_2$$
$$MATCH\ \text{"0} * * * 11 * * \ldots * **''} \to R_1$$
$$CALC R_1 \cdot R_2 \to R_2$$
$$MATCH\ \text{"} * 0 * * 0 * * \ldots * **''} \to R_1$$
$$CALC R_1 \cdot R_2 \to R_1$$

$$(3)$$

The same key is used to encrypt or decrypt each block of data in the simple version of DES. There are 56 key bits, but only 48 of them are used during each of the 16 different rounds. The bits being used are maintained by the program running on the general machine that is driving the Coherent Processor . It selects the subset of 48 bits that are used in each encryption and modifies the s-box functions accordingly.

This method presupposes that the sixteen 48-bit subsets of the keys are precomputed and "compiled" into the code. This process is non-trivial and certain to cost some time. When the amount of data encrypted or decrypted per key change is large, then this "compilation" time is minimal. If the key is changed frequently,then there may be some impact on the encryption times. It is not likely to impact the overall throughput, however, if the CPU driving the CAM array is fast enough to interleave operations in between the various CAM instructions. This is not unreasonable because many of the CAM instructions take 2 to 4 cycles to complete. A modern pipelined RISC architecture should be able to complete the key scheduling instruction inbetween. A better understanding of the effects of this will need to wait until the software is completely implemented on a working system.

2.4 The Total Cost

The current version of the Coherent Processor will run at speeds up to 50 MHtz. If an encryption takes about 47,428 cycles, then each pair of words in the processor array can encrypt about 1,000 64-bit blocks per second. Writing a word into the array and reading it out takes 5 cycles in total. One chip of the current model has 1024 words or processors, so it can read in, encrypt and write out blocks of 32K in 52,548 cycles. This is equivalent to 31.2 megabits per second– something that is in line with the middle range of current DES chips. The Cryptech CRY12C102 data sheet reports that it runs at 22.5 megabits per second and the Pijnenburg PCC100 attains 20 megabits per second. Moreover, the Coherent Processor is designed to be easily expanded by linking together multiple copies of the chip and n chips will n times faster for small numbers of n. When there are hundreads or thousands of chips, the cost of writing and reading the information from the Coherent Processor becomes the limiting factor. Coherent Research reports that the new chip will cost about $100 per copy in small quantities and substantially less in large ones.

3 Exhaustive Attack on DES

When DES was introduced in 1977, some computer scientists protested that 56 bits were not sufficient because it would be possible to do an exhaustive search of the key space in a short amount of time using a massively parallel computer. In their book, Meyer and Matyas [MM82] discount that possiblity and predict that it would just not be physically possible to build the machine until the 1990's because there were too many physical limitations. Heat and power usage are two major barriers. Diffie and Hellman describe the design in detail and respond to these criticism in [DH77].

How easy would it be to build one today? Standard off-the-shelf encryption chips are plentiful and relatively cheap, but they require a second processor feeding them the keys and the test cases. Anyone who wants to build such a machine must undertake a project of building such a large array of distributed computers. This would require a large amount of custom design work. A truly dedicated attacker could even fabricate custom DES testing chips which have a built in circuit for incrementing the key by one bit and testing the result against another register. Only governments could afford a budget this large. Moreover, the slightest change in the algorithm would render this machine worthless.

Garon and Outerbridge calculated the approximate costs of designing such a machine and found that it would cost about $129,000 for a machine that would break DES within 1 year if the machine was built in 1990. [GO91]. They also say that a machine that could exhaustively search all the bits in one day for $46 million in 1990. This price would drop to $18 million in 1995. They assume that it is possible to build a node that encrypts 2 million key tests for $25 in 1990 in order to complete such a machine. They do not describe the details of how to design the board or manufacture it is sufficient quanties.

The Content Addressable Memory array chips, however, are designed to be built into large parallel arrays of chips. It is already possible to buy a board for a PC which has 64 chips of a previous model of the Coherent Processor . Large arrays should not be hard to create. Moreover, the algorithm is implemented in software, so the machine can also be used to attack many other subtle and not-so-subtle variations of DES.

What is the best way to do an exhaustive search with the current architecture of the Coherent Processor? The version described for simple encryption and decryption is able to work very quickly because it can encode the key in the stream of instructions fed to the Coherent Processor. This approach must be abandoned because an exhaustive search of the key space requires that each processing node must use a different key.

One alternative is to store the key bits in the 10 extra tag bits stored at each node. Two nodes are used to hold the two 32-bit half-blocks of each case, so there are up to 20 extra key bits which can be stored at each node. Let there be 2^n processors in the machine. That means there are 2^{n-1} potential keys that can be tested with each round because two nodes are used for each encryption. Assume that $n \leq 21$ and the problem does not overflow the physical space of the real machine. (Later versions of the architecture could have more free bits available.) At each pair of nodes, store a unique set of $n-1$ key bits. These bits will be used by this pair of nodes alone. The other $56 - (n-1)$ bits are shared by all the instances and they are encoded in the instruction stream as before.

At the beginning of each round of encryption, the local key bits must be XOR-ed into the appropriate half-block of bits before that half-block is passed through the s-boxes. These four or five instructions will XOR in the key bit K_i in to position B_j:

$$MATCH\,K_i \rightarrow R_1$$
$$SHIFT$$
$$MATCH\,B_j \rightarrow R_2$$
$$CALC\,R_1\,XOR\,R_2 \rightarrow R_2$$
$$WRITECOLUMN\,R_2 \rightarrow B_j$$

$$(4)$$

The SHIFT instruction is only necessary if the key bit is on the opposite node from the destination bit. This process is repeated at the end of the s-box calculation to remove the bit from the data. Only 48 of the 56 key bits are used at each round, but it is possible that up to $n-1$ of these bits will come from the bits stored locally. The operations in equation 4 take 16 cycles. They must be repeated $2n-2$ times for each round. The result takes $512n - 512$ extra cycles for each encryption. If a machine was built with a full complement of 2^{21} processors, then it would take 57,126 cycles to test 2^{20} potential keys. This step must be repeated 2^{36} times and the machine is capable of doing about 875 of these tests per second or about 76 million per day. Exhausting the entire space would require 904 days. If the well-known trick of exploiting symmetry in the keys is used to reduce the key space to 2^{55} keys, then one machine will test all in 452 days.

How much would such a machine cost? There are 2^{10} processors on a chip that will cost between \$30 and \$100. 2^{11} chips are necessary and this would cost between about \$60,000 and \$200,000. Control hardware would add additional \$10,000 to \$20,000. 45 machines would cost about \$3 million dollars and exhaustively search the space in 10 days. \$30 million would buy a machine that would search the space in 1 day with 450 machines. I'm assuming that volume discounts would apply at this scale and \$30 is a price that should apply at the end of 1992 when the chips become widely available.

Although this design is still hypothetical, it is much more real than some of the other designs available because the chip fabrication and design is already complete. The process of building a machine out of chips is not much different from connecting a large bank of memory up to a single processor. This paper does not pretend to addresss any of the important questions about heat and power dissipation. These could also affect the design and it is possible that my estimate of \$10,000 to \$20,000 for the support hardware is too low.

The standard assumptions about time and transistor density should apply to this model as well. It is entirely conceivable that we will see larger improvements in density and price of these machines in the near future because they are younger designs.

The UNIX password system uses a version of DES that was presumably modified to make it impossible to gang together a number of off-the-shelf DES chips and use the system to break UNIX passwords. This large machine, however, is not constrained by this modification or any other modification that re-arranges the pattern of s-boxes, permutations and mixing. The salting process used in the UNIX password operation is easy to express with extra bit swapping operations. The only problem with attacking systems like Merkle's Khufu is expressing the s-boxes as minterms. Incidentally, logic minimization is also easily handled by the Coherent Processor .

The availability of these systems puts even more pressure on the Unix password system. In 1989, Feldmeier and Karn [FK89] estimated that the UNIX password system was insecure for short alphanumeric passwords because a DEC 3100 could process about 1000 passwords per seconds. Given that each password needs 25 passes of DES, then it is possible to estimate that a Coherent Processor based processor will be able to test about 20,000 passwords per second per chip. If a basic Coherent Processor processor comes with between 8 to 64 chips, then it is easily possible to imagine computers with the ability to test between 160,000 and 1,280,000 tests per second. How fast could such a standard machine test all passwords made up of 6 alphanumeric characters ("A" – "Z", "a" – "z", "1" – "9")? Between about 3.75 days (8 chips) and about half a day (64 chips). A large scale machine with 2^{11} chips should be able to tackle passwords with 7 alphanumeric passwords in about one day.

4 DES with Modified Chaining

The last several sections described how to encrypt a large block of data in parallel using a simple DES with no feedback. A more robust version of DES feeds the result of encrypting each block into the key selection of the next block. Let $E_i = f(K, B_i)$ represents the ciphertext blocks. A feedback cipher sets $E_i = f(K, B_i \oplus E_{i-1})$ where "$\oplus$" represents boolean XOR. E_0 is set to a pre-arranged constant. This process is called Cipher Block Chaining (CBC).

The modification adds a great deal of strength to the plain DES because it reduces the redundancies that can developed if there is an 64 bit block that occurs often in the plaintext. The feedback mode ensures that a different value will permute each block and obscure the redundancy. It should be obvious that this system cannot be used when all the blocks are computed in parallel. Here is a modified version of chaining that can be implemented in parallel.

One solution is to exchange and XOR bits with neighbors at the end of certain rounds of encryption. In round 1, the left half of each block is used to compute the value XORed into the right half. After this, the left blocks are exchanged with the neighboring blocks and XOR'ed into the right halves of the neighboring block. This can be done with pseudo-code like this. W_i is the left half and W_{i+1} is the right half.

```
for k:=0 to 31 do
    MATCH W_i[k] → R_1
    CALC COPY R_1 → R_2
    SHIFT
    SHIFT
    CALC XOR R_1 R_2 → R_1
    WRITECOLUMN R_1 → W_{i+1}[k]
```

This command shifts one bit to the next pair of words over and XOR's it with the value of a neighboring block. It takes 16 cycles per bit to achieve this. This can be repeated as often as desired at the cost of slowing down the entire encryption. Doing this at the end of each round of encryption costs 8,192 cycles and this slows the encryption rate to 27.0 megabits per second. In this case, a change in block B_i will propigate through blocks B_i to B_{i+16} and effect their encrypted values. Arbitrarily complex shifting can be included as long as care is taken to ensure that the results can be reversed. If this step is done often in the process, it can effectively turns the encryption into one large block at a small decrease in speed.

5 Conclusion

This paper has described a simple architecture intended for information storage and retrieval that can also encrypt and decrypt messages faster than all but the best specialized chips. More importantly, the results are achieved in software so the process can be extended to other DES-like systems without refabricating the chips. The only problem is expressing the s-boxes so they can be implemented with minterms. This should make the chip much more desirable for many implementations of DES that require more flexibility than extreme speed.

Chips like the Coherent Processor also make it very easy to create a large-scale processor for exhaustive cryptanalysis of the key space because the chips were designed to be grouped together in a large array. The hypothetical machine described here is much different from the other machines described in the literature because it is both reprogrammable and substantially closer to being realized. Only a minimal amount of logic is necessary to turn the chips into machines that are able to handle DES and variants of DES like the UNIX password system or Khufu.

The flexible software structure also provides an easy method to test for broken chips. It is possible to load each line with a test vector, encrypt them in parallel and then test for failures with a MATCH instruction. Many of the earlier designs for large machines needed to build in a specific test function to maintain the system.

There are several changes to the Coherent Processor that would improve its ability to encrypt DES. Currently, the key is "compiled" into the program for the CAM and this may be a non-trivial event. If future versions of the architecture have more that 42 bits per word, then it could be practical to store the key locally and add the key in bit by bit as it is done in the brute force attack. Also, the current version of the Coherent Processor will only compute 3 bit functions. 4 or 5 bit functions may be quite practical and they would certainly speed the results of the process.

Working hardware is due in early 1993 and this will provide an opportunity to develop

6 Acknowledgements

The author would like to thank Chuck Stormon at Coherent Research for taking the time to teach me how to program the Coherent Processor and making many valuable comments about the structure of this paper. I would also like to thank Luke O'Connor for his comments on the structure of the paper and for providing the minterm representations of the s-box included here in the appendix. Richard Outerbridge also provided invaluable help and suggestions.

References

[Bri90]   Sharon Marie Britton. *8k-trit Database Accelerator with Error Detection.* PhD thesis, Massachusetts Institute of Technology, February 1990.

[BS91]   Eli Biham and Adi Shamir. Differential cryptanalysis of Snefru, Khafre, REDOC-II, LOKI, and lucifer. In *Crypto 91*, Santa Barbara, California, 1991.

[DH77]   Whitfield Diffie and Martin Hellman. Exhaustive cryptanalysis of the nbs data encryption standard. *Computer*, 10(6):74–84, 1977.

[FK89]   David C. Feldmeier and Philip R. Karn. Unix password security- ten years later. In
           G. Brassard, editor, *Advances in Cryptology: Proceedings of Crypto '89*, pages 44–63,
           New York City, Berlin, 1989. Springer-Verlag.

[FMP85]   Robert C. Fairfield, Alex Matusevich, and Joseph Plany. An lsi digital encryption pro-
           cessor. *IEEE Communication*, pages 23–27, July 1985.

[GO91]   Gilles Garon and Richard Outerbridge. Des watch: And examination of the sufficiency of
           the data encryption standard for financial institution's information security in the 1990's.
           Cryptologia, 15(3):177–193, July 1991.

[Koh87]   Teuvo Kohonen. *Content-Addressable Memories*. Springer-Verlag, Berlin, New York
           City, 1987.

[Mer90]   Ralph Merkle. Fast software encryption function. In A.J. Menezes and S.A. Van Stone,
           editors, *Crypto 90*, Berlin, New York City, 1990. Springer Verlag.

[MM82]   Carl H. Meyer and Stephen M. Matyas. *Cryptography: New Dimension in Computer
           Security*. John Wiley and Sons, New York, 1982.

[NBS77]   NBS. Data encryption standard (des). Technical report, National Bureau of Standards
           (US), Federal Information Processing Standards, Publication 46, National Technical In-
           formation Services, Springfield, Virginia, April 1977.

[Old86]   J.V. Oldfield. Logic programs and an experimental architecture for their execution.
           Procedings of the I.E.E.E. Part E, 133:163–167, 1986.

[OSB87]   J.V. Oldfield, Charles D. Stormon, and M.R. Brule. The application of vlsi content-
           addressable memories to the acceleration of logic programming systems. In *CompEuro
           87, VLSI and Computers*, pages 27–30, Hamburg, Germany, May 1987.

[OWN87]   J.V. Oldfield, R.D. Williams, and N.E.Wiseman. Content-addressable memories for stor-
           ing and processing recursively-divided images and trees. *Electronics Letters*, 23(6):262–
           263, 1987.

[ST79]   Robert Morris Sr. and Ken Thompson. Password security: A case history. *Communica-
           tions of the ACM*, 22:594–597, November 1979.

[VHVM88]   Ingrid Verbauwhede, Frank Hoornaert, Joos Vandewalle, and Hugo J. De Man. Security
           and performance optimization of a new des encryption chip. *IEEE Journal of Solid-State
           Circuits*, pages 647–656, June 1988.

[WS89]   John Wade and Charles Sodini. A ternary content-addressable search engine. *IEEE
           Journal of Solid-State Circuits*, 24(4):1003–1013, August 1989.

[Zip90]   Richard Zippel. Programming the data structure accelerator. In *Proceedings of
           Jerusalem Conference on Information, Technology*, Jerusalem, Israel, October 1990.

7 Appendix

Some minimal representations of the s-boxes provided by Luke O'Connor. S_1^2 represents the
function for the first bit of the second s-box. The "$+$" means means logical or. The logical
ands in each clause (implicant) are left out to save space. A variable with a bar over it "$\bar{x_6}$")
represents NOT x_6.

$S_1^1(x_1, x_2, x_3, x_4, x_5, x_6) = \bar{x_1}\bar{x_2}\bar{x_3}\bar{x_5}\bar{x_6} + \bar{x_1}x_3\bar{x_4}x_5\bar{x_6} + \bar{x_1}x_2x_4\bar{x_5}\bar{x_6} + \bar{x_1}x_2\bar{x_3}x_5\bar{x_6} + \bar{x_1}x_2x_3\bar{x_5}x_6 +$
$\bar{x_1}x_2\bar{x_3}\bar{x_5}x_6 + x_1\bar{x_2}x_4x_5\bar{x_6} + x_1\bar{x_2}x_3\bar{x_4}\bar{x_5}\bar{x_6} + \bar{x_2}x_3x_4x_5\bar{x_6} + x_1x_2\bar{x_3}\bar{x_5}\bar{x_6} + x_1\bar{x_3}x_4\bar{x_5}\bar{x_6} +$
$x_2\bar{x_4}x_5\bar{x_6} + \bar{x_2}x_3\bar{x_4}x_5x_6 + x_1\bar{x_2}x_3\bar{x_5}x_6 + x_1\bar{x_2}x_4x_5x_6 + x_1\bar{x_3}x_4x_5\bar{x_6} + x_2x_3\bar{x_4}\bar{x_5}x_6 + x_2x_4x_5x_6$

$S_2^1(x_1, x_2, x_3, x_4, x_5, x_6) = \bar{x_1}\bar{x_2}x_4\bar{x_5}\bar{x_6} + \bar{x_1}x_2x_3\bar{x_4}\bar{x_5}\bar{x_6} + \bar{x_1}x_2x_4x_5\bar{x_6} + \bar{x_1}x_2\bar{x_3}x_5\bar{x_6} +$
$\bar{x_1}\bar{x_2}x_4x_5x_6 + \bar{x_1}x_3x_4\bar{x_5} + \bar{x_1}x_2\bar{x_4}x_5x_6 + \bar{x_2}x_3\bar{x_5}\bar{x_6} + x_1\bar{x_2}x_4\bar{x_5} + \bar{x_2}x_3x_4x_5\bar{x_6} + x_1x_2\bar{x_3}x_5\bar{x_6} +$
$x_1x_2x_3x_4\bar{x_5} + \bar{x_2}x_3\bar{x_4}x_5x_6 + \bar{x_2}x_3\bar{x_4}\bar{x_5}x_6 + x_1\bar{x_3}x_4\bar{x_5} + x_1x_2x_4x_5x_6 + x_1x_3x_4x_5x_6$

$S_3^1(x_1, x_2, x_3, x_4, x_5, x_6) = \bar{x_1}\bar{x_2}x_4x_5\bar{x_6} + \bar{x_1}x_2\bar{x_3}x_5\bar{x_6} + \bar{x_1}x_2x_3x_4x_5\bar{x_6} + \bar{x_1}\bar{x_2}x_3x_4\bar{x_5}\bar{x_6} +$
$\bar{x_1}x_2x_3\bar{x_4} + \bar{x_1}\bar{x_2}x_4x_5x_6 + x_1x_2\bar{x_3}x_4 + x_1\bar{x_2}x_4\bar{x_5}x_6 + \bar{x_2}x_3x_4\bar{x_5}x_6 + x_1\bar{x_2}x_3x_5\bar{x_6} + x_1x_2\bar{x_3}x_4x_5 +$
$x_1x_2\bar{x_4}\bar{x_5}\bar{x_6} + x_1x_3\bar{x_4}x_5\bar{x_6} + x_1\bar{x_2}\bar{x_3}x_4x_5x_6 + x_1\bar{x_2}x_4x_5x_6 + x_1x_2x_2\bar{x_4}\bar{x_5}x_6 + x_2\bar{x_3}x_5x_6 + x_1x_2x_3\bar{x_5}x_6 +$
$x_2x_3x_4\bar{x_5}x_6$

$S_4^1(x_1, x_2, x_3, x_4, x_5, x_6) = \bar{x_1}\bar{x_2}x_3x_4\bar{x_6} + \bar{x_1}x_3\bar{x_4}x_5\bar{x_6} + \bar{x_1}x_2x_3x_5\bar{x_6} + \bar{x_1}\bar{x_2}\bar{x_3}x_4x_5\bar{x_6} +$
$\bar{x_1}\bar{x_2}x_4\bar{x_5} + \bar{x_1}x_2\bar{x_3}x_4x_5x_6 + x_1x_2x_3\bar{x_4} + \bar{x_1}x_3x_4\bar{x_5}x_6 + x_1\bar{x_2}\bar{x_3}x_4x_5\bar{x_6} + x_1x_3\bar{x_4}\bar{x_5}\bar{x_6} + x_1\bar{x_2}x_3x_4x_5 +$

$$x_1x_2\bar{x}_3x_4\bar{x}_6 + x_2\bar{x}_4\bar{x}_5\bar{x}_6 + x_1x_2\bar{x}_5\bar{x}_6 + x_1\bar{x}_2x_3x_5x_6 + \bar{x}_2x_3x_4x_6 + x_1\bar{x}_3\bar{x}_4\bar{x}_5x_6 + x_1x_2\bar{x}_3\bar{x}_4x_6 +$$
$$x_1x_2\bar{x}_3\bar{x}_5 + x_1x_3x_4x_5x_6$$

$$S_1^2(x_1,x_2,x_3,x_4,x_5,x_6) = \bar{x}_1x_2\bar{x}_3\bar{x}_5\bar{x}_6 + \bar{x}_1\bar{x}_2x_3\bar{x}_4\bar{x}_5\bar{x}_6 + \bar{x}_1\bar{x}_3x_4\bar{x}_5\bar{x}_6 + \bar{x}_1\bar{x}_3x_4x_5\bar{x}_6 +$$
$$\bar{x}_1\bar{x}_2x_3\bar{x}_5\bar{x}_6 + \bar{x}_1\bar{x}_2x_3x_4\bar{x}_6 + \bar{x}_2x_3\bar{x}_4\bar{x}_5\bar{x}_6 + x_1\bar{x}_2x_3\bar{x}_5\bar{x}_6 + x_1\bar{x}_3\bar{x}_4x_5\bar{x}_6 +$$
$$x_1\bar{x}_2\bar{x}_3\bar{x}_5\bar{x}_6 + x_2x_3\bar{x}_4\bar{x}_5x_6 + x_2x_3x_4x_5\bar{x}_6 + \bar{x}_2\bar{x}_3x_4x_5\bar{x}_6 + x_1\bar{x}_2\bar{x}_3\bar{x}_5x_6 + x_1\bar{x}_2x_4x_5x_6 +$$
$$x_2\bar{x}_3\bar{x}_4\bar{x}_5x_6 + x_2x_3x_4x_5x_6 + x_2x_3x_4\bar{x}_5x_6 + x_1x_2x_4x_5x_6$$

$$S_2^2(x_1,x_2,x_3,x_4,x_5,x_6) = \bar{x}_1x_2\bar{x}_4\bar{x}_5\bar{x}_6 + \bar{x}_1\bar{x}_2x_3x_5\bar{x}_6 + \bar{x}_1x_2\bar{x}_3\bar{x}_5x_6 + \bar{x}_1\bar{x}_2x_3x_5x_6 + \bar{x}_1\bar{x}_2x_3x_4x_6 +$$
$$\bar{x}_1\bar{x}_2x_4x_5 + \bar{x}_1x_2\bar{x}_4\bar{x}_5x_6 + \bar{x}_1x_3x_4\bar{x}_5 + \bar{x}_1x_3x_4x_5x_6 + x_1\bar{x}_2x_4x_5\bar{x}_6 + x_1\bar{x}_2x_4\bar{x}_5\bar{x}_6 + x_1x_2\bar{x}_3\bar{x}_5\bar{x}_6 +$$
$$x_1x_2x_4x_5\bar{x}_6 + x_1\bar{x}_2\bar{x}_3x_4\bar{x}_5\bar{x}_6 + x_1x_2x_4\bar{x}_5x_6 + x_1x_2\bar{x}_3x_5x_6 + x_1x_3x_4x_5x_6 + x_1x_3x_4\bar{x}_5x_6$$

$$S_3^2(x_1,x_2,x_3,x_4,x_5,x_6) = \bar{x}_1x_2x_3\bar{x}_4 + \bar{x}_1x_2x_3\bar{x}_5\bar{x}_6 + \bar{x}_1\bar{x}_2\bar{x}_3x_4x_5\bar{x}_6 + \bar{x}_1x_2\bar{x}_3x_4x_5\bar{x}_6 +$$
$$\bar{x}_1\bar{x}_2x_4\bar{x}_5 + \bar{x}_1\bar{x}_3x_4x_5x_6 + \bar{x}_1x_3x_4\bar{x}_5x_6 + x_1\bar{x}_2\bar{x}_3x_5\bar{x}_6 + x_1\bar{x}_2\bar{x}_3x_4\bar{x}_6 + x_2x_3x_4x_5\bar{x}_6 + \bar{x}_2x_3\bar{x}_4\bar{x}_5 +$$
$$x_1\bar{x}_3x_4x_5\bar{x}_6 + x_1x_2x_3x_5\bar{x}_6 + x_1x_2x_3x_4\bar{x}_6 + x_2x_3x_4x_5x_6 + \bar{x}_2x_3x_5x_6 + x_1x_2\bar{x}_3x_4x_6 + x_1\bar{x}_3x_4\bar{x}_5x_6 +$$
$$x_2x_3x_4\bar{x}_5x_6$$

$$S_4^2(x_1,x_2,x_3,x_4,x_5,x_6) = \bar{x}_1\bar{x}_2x_4x_5\bar{x}_6 + \bar{x}_1x_3\bar{x}_4x_6 + x_1\bar{x}_2\bar{x}_3x_5\bar{x}_6 + \bar{x}_1x_3x_4\bar{x}_5x_6 + \bar{x}_1\bar{x}_2x_3x_5x_6 +$$
$$\bar{x}_1x_2x_4\bar{x}_5x_6 + x_1\bar{x}_2x_4\bar{x}_6 + x_1x_2x_3x_5 + \bar{x}_2x_3x_4x_5x_6 + \bar{x}_2\bar{x}_4\bar{x}_5x_6 + x_1x_3\bar{x}_4x_5x_6 +$$
$$x_1\bar{x}_3x_4\bar{x}_5x_6 + x_2x_3\bar{x}_4x_5x_6$$

$$S_1^3(x_1,x_2,x_3,x_4,x_5,x_6) = \bar{x}_1x_2x_4\bar{x}_5\bar{x}_6 + x_1x_2\bar{x}_3x_4x_5 + \bar{x}_1x_3x_4x_5\bar{x}_6 + x_1x_2x_3\bar{x}_4x_5 + x_1x_2x_3x_4x_5\bar{x}_6 +$$
$$\bar{x}_1x_2\bar{x}_3\bar{x}_4x_5 + \bar{x}_1\bar{x}_2x_4x_5x_6 + \bar{x}_1x_3x_4x_5x_6 + \bar{x}_1x_2x_4x_5x_6 + \bar{x}_1x_2x_3x_5x_6 + \bar{x}_2x_3x_4x_5x_6 + x_1\bar{x}_2x_4x_5\bar{x}_6 +$$
$$x_1x_3x_4\bar{x}_5\bar{x}_6 + x_1x_3x_4x_5\bar{x}_6 + x_1x_3\bar{x}_4x_5\bar{x}_6 + x_1x_2x_3x_4\bar{x}_5\bar{x}_6 + x_1\bar{x}_2x_4x_5x_6 + x_1x_2x_4\bar{x}_5x_6 +$$
$$x_2\bar{x}_3x_4x_5x_6 + x_1x_3x_4\bar{x}_5x_6 + x_2x_3x_4\bar{x}_5x_6 + x_1x_2x_3x_4x_5x_6$$

$$S_2^3(x_1,x_2,x_3,x_4,x_5,x_6) = \bar{x}_1x_2x_3\bar{x}_5x_6 + \bar{x}_1x_2x_4x_5\bar{x}_6 + x_1x_2x_4x_5\bar{x}_6 + \bar{x}_1x_2\bar{x}_3x_4x_6 + \bar{x}_1x_2\bar{x}_4x_5x_6 +$$
$$x_1x_2x_3\bar{x}_4 + \bar{x}_1x_2x_3\bar{x}_5x_6 + \bar{x}_1x_3x_4\bar{x}_5x_6 + x_1\bar{x}_2x_3\bar{x}_5x_6 + x_1\bar{x}_2x_4x_5x_6 + x_1x_2x_3\bar{x}_5x_6 + x_1x_2x_4x_5\bar{x}_6 +$$
$$x_1x_3x_4\bar{x}_5x_6 + x_1x_2x_3x_4\bar{x}_5x_6 + x_1x_2\bar{x}_3x_5x_6 + x_1x_2x_4x_5x_6 + x_1x_3x_4x_5x_6$$

$$S_3^3(x_1,x_2,x_3,x_4,x_5,x_6) = \bar{x}_1\bar{x}_2x_4x_5\bar{x}_6 + \bar{x}_1x_3x_4x_5\bar{x}_6 + \bar{x}_1x_3x_4\bar{x}_5\bar{x}_6 + x_1x_3x_4x_5x_6 + \bar{x}_2x_3x_4x_5\bar{x}_6 +$$
$$\bar{x}_1x_2x_3\bar{x}_4x_5x_6 + x_1\bar{x}_2x_4x_5\bar{x}_6 + \bar{x}_2x_3x_4x_4\bar{x}_6 + x_1x_2\bar{x}_3x_4\bar{x}_5 + x_2x_3x_4x_5x_6 + x_1x_3\bar{x}_4x_5\bar{x}_6 +$$
$$x_2x_3\bar{x}_4x_5x_6 + x_2x_3x_4x_5x_6 + x_1x_3x_4x_5\bar{x}_6 + x_1x_2\bar{x}_3x_4\bar{x}_5x_6$$

$$S_4^3(x_1,x_2,x_3,x_4,x_5,x_6) = \bar{x}_1x_2x_4x_5\bar{x}_6 + \bar{x}_1x_2x_3x_5\bar{x}_6 + x_1x_2\bar{x}_3x_5\bar{x}_6 + \bar{x}_1x_2x_3x_5x_6 + \bar{x}_1x_2\bar{x}_4x_5x_6 +$$
$$x_1\bar{x}_2x_4x_5x_6 + \bar{x}_1x_2x_3x_5x_6 + x_1x_3x_4\bar{x}_5\bar{x}_6 + x_1x_2x_3x_4x_5\bar{x}_6 + \bar{x}_2x_3x_4x_5\bar{x}_6 + \bar{x}_2x_3x_4\bar{x}_5\bar{x}_6 +$$
$$x_2x_3x_4\bar{x}_6 + x_1x_2x_3x_4x_5x_6 + x_1\bar{x}_2x_3x_5x_6 + x_1x_2\bar{x}_3x_5x_6 + x_1x_2x_3x_4x_6$$

$$S_1^4(x_1,x_2,x_3,x_4,x_5,x_6) = \bar{x}_1x_2\bar{x}_3x_4x_5 + x_1\bar{x}_2x_4x_5\bar{x}_6 + \bar{x}_1x_3x_4x_5\bar{x}_6 + x_1x_2x_3\bar{x}_4x_6 + \bar{x}_1x_3x_4x_5\bar{x}_6 +$$
$$\bar{x}_1x_2x_3\bar{x}_5x_6 + \bar{x}_1x_3x_4x_5x_6 + \bar{x}_1x_2x_3x_4x_6 + \bar{x}_2x_3x_4\bar{x}_5\bar{x}_6 + x_1x_2\bar{x}_4x_5\bar{x}_6 + x_1x_2x_3x_5x_6 + x_1x_3\bar{x}_4x_5\bar{x}_6 +$$
$$x_1x_2\bar{x}_3x_4x_5 + x_1x_2x_3x_4\bar{x}_5\bar{x}_6 + \bar{x}_2x_3\bar{x}_4x_4x_5x_6 + x_1\bar{x}_2x_3\bar{x}_5x_6 + x_1x_3x_4x_5x_6 + x_1x_2\bar{x}_4x_5x_6 +$$
$$x_2x_4x_5x_6$$

$$S_2^4(x_1,x_2,x_3,x_4,x_5,x_6) = \bar{x}_1x_2\bar{x}_3x_5x_6 + \bar{x}_1x_3x_4x_5\bar{x}_6 + \bar{x}_1x_2x_3\bar{x}_4x_6 + \bar{x}_1x_3x_4x_5x_6 + \bar{x}_1x_3x_4x_5x_6 +$$
$$\bar{x}_1x_2x_3x_4x_5 + x_2x_3x_4x_5\bar{x}_6 + x_1x_3x_4x_5x_6 + x_1x_2x_4x_5\bar{x}_6 + x_1x_3x_4x_5\bar{x}_6 + x_1x_2x_3x_4x_6 + x_1x_2x_3x_5x_6 +$$
$$x_2x_3x_4x_5x_6 + x_1x_2x_3x_4x_5 + x_2x_3x_4x_5x_6 + x_1x_2x_4x_5x_6 + x_1x_2x_3x_5x_6$$

$$S_3^4(x_1,x_2,x_3,x_4,x_5,x_6) = \bar{x}_1\bar{x}_2x_3\bar{x}_5x_6 + \bar{x}_1\bar{x}_2x_4x_5x_6 + \bar{x}_1x_2\bar{x}_4x_5x_6 + x_1\bar{x}_3x_4x_5\bar{x}_6 + x_1x_2x_3x_4x_6 +$$
$$\bar{x}_1x_3x_4x_5\bar{x}_6 + x_1\bar{x}_2x_3x_5 + \bar{x}_1x_3x_4\bar{x}_5x_6 + x_1x_2\bar{x}_3x_4\bar{x}_5x_6 + \bar{x}_2x_3x_4x_5x_6 + x_2x_3x_4x_5x_6 + x_2x_3x_4x_5x_6 + x_1x_2x_4x_5x_6 +$$
$$x_1x_3x_4x_5\bar{x}_6 + x_1\bar{x}_2x_4x_5x_6 + x_1x_3x_4x_5x_6$$

$$S_4^4(x_1,x_2,x_3,x_4,x_5,x_6) = \bar{x}_1x_3x_4x_5\bar{x}_6 + \bar{x}_1x_2x_3\bar{x}_5x_6 + x_1x_2x_4x_5x_6 + \bar{x}_1x_2x_3\bar{x}_5x_6 + \bar{x}_1x_2x_4x_5x_6 +$$
$$x_1x_2\bar{x}_3x_4x_5x_6 + \bar{x}_1x_2x_3x_4\bar{x}_5 + \bar{x}_1x_3x_4x_5x_6 + x_2x_3x_4\bar{x}_5x_6 + x_1x_2x_3\bar{x}_5x_6 + x_1x_2\bar{x}_3x_4\bar{x}_6 +$$
$$x_1x_3x_4\bar{x}_5x_6 + x_2\bar{x}_4\bar{x}_5x_6 + x_1x_2x_4x_5x_6 + x_2x_3x_4x_5x_6 + x_1x_2x_3x_4x_5 + x_1x_3x_4x_5x_6 + x_1x_2x_3x_4x_6 +$$
$$x_1x_3x_4x_5x_6$$

$$S_1^5(x_1,x_2,x_3,x_4,x_5,x_6) = \bar{x}_1x_2x_4x_5\bar{x}_6 + x_1x_2x_4x_5\bar{x}_6 + x_1x_2\bar{x}_4x_5\bar{x}_6 + \bar{x}_1x_3x_4x_5x_6 + x_1x_2x_4\bar{x}_5x_6 +$$
$$\bar{x}_1x_2x_3x_4x_5x_6 + \bar{x}_1x_3x_4x_5 + x_1x_2x_4x_5x_6 + x_1x_2x_3x_4x_6 + x_1x_2x_3x_4x_6 + x_1x_2x_3x_5x_6 + x_2x_3x_4x_5x_6 +$$
$$\bar{x}_2x_3x_4x_6 + x_1x_2x_3x_5x_6 + x_1x_2x_3x_5 + x_1x_3x_4x_5x_6 + x_2x_3x_4x_5x_6 + x_1x_2x_3x_4x_5x_6$$

$$S_2^5(x_1,x_2,x_3,x_4,x_5,x_6) = \bar{x}_1x_2x_3x_4\bar{x}_5\bar{x}_6 + \bar{x}_1x_3x_4x_5\bar{x}_6 + \bar{x}_1x_2x_3x_4x_5\bar{x}_6 + \bar{x}_1x_2x_3x_4x_5 +$$
$$\bar{x}_1x_2x_3\bar{x}_5x_6 + x_1\bar{x}_2x_4x_5x_6 + \bar{x}_1x_2x_3\bar{x}_5x_6 + x_1x_2x_3\bar{x}_5x_6 + \bar{x}_1x_2x_3x_4x_5x_6 + x_1x_2x_3x_4x_5 +$$
$$x_1\bar{x}_2x_3x_4x_5\bar{x}_6 + x_1x_3x_4x_5x_6 + x_1x_2x_3x_5x_6 + x_2x_3x_4x_5x_6 + x_1x_2x_4x_5x_6 +$$
$$x_1x_2x_3x_4x_6 + x_2x_3x_4x_5x_6 + x_2x_3x_4x_5x_6 + x_1x_2x_4x_5x_6 + x_2x_3x_4x_5x_6 + x_1x_2x_4x_5x_6 + x_1x_2x_3x_4x_5x_6$$

$$S_3^5(x_1,x_2,x_3,x_4,x_5,x_6) = \bar{x}_1x_2x_4x_5x_6 + \bar{x}_1x_2x_3x_6 + x_1x_2x_3x_4 + \bar{x}_1x_3x_4x_5x_6 + \bar{x}_1x_2x_4x_5x_6 +$$

$$\bar{x}_1\bar{x}_3x_4\bar{x}_5x_6+x_1\bar{x}_2\bar{x}_3x_5\bar{x}_6+x_1x_2\bar{x}_4\bar{x}_5+x_1\bar{x}_3\bar{x}_4x_5\bar{x}_6+x_1x_2x_3\bar{x}_5\bar{x}_6+\bar{x}_2\bar{x}_3\bar{x}_4\bar{x}_5x_6+x_1\bar{x}_2\bar{x}_3x_4x_5+$$
$$\bar{x}_2x_3\bar{x}_4x_5x_6+x_1\bar{x}_2x_3x_4\bar{x}_5+x_1x_2\bar{x}_3\bar{x}_4x_6+x_2x_3\bar{x}_4\bar{x}_5x_6+x_2x_3x_4x_5x_6$$

$$S_4^5(x_1,x_2,x_3,x_4,x_5,x_6)=\bar{x}_1x_2\bar{x}_3x_4\bar{x}_6+\bar{x}_1x_3\bar{x}_4\bar{x}_5x_6+\bar{x}_1x_2x_4\bar{x}_5x_6+\bar{x}_1\bar{x}_2\bar{x}_4x_5x_6+\bar{x}_1\bar{x}_2x_3x_4x_6+$$
$$\bar{x}_1x_2\bar{x}_3x_6+x_1\bar{x}_2x_4\bar{x}_5x_6+x_1x_3\bar{x}_4x_5x_6+x_1\bar{x}_2x_4\bar{x}_5\bar{x}_6+\bar{x}_2x_3x_4\bar{x}_5x_6+x_1x_2\bar{x}_3\bar{x}_4\bar{x}_6+x_2x_3x_5\bar{x}_6+$$
$$\bar{x}_3x_4x_5\bar{x}_6+x_1x_3x_4x_5\bar{x}_6+x_1\bar{x}_2\bar{x}_4\bar{x}_5x_6+x_1x_2\bar{x}_3x_5+x_1x_2x_3x_4x_6+x_1x_4x_5\bar{x}_6$$

$$S_1^6(x_1,x_2,x_3,x_4,x_5,x_6)=\bar{x}_1\bar{x}_2x_4x_5\bar{x}_6+\bar{x}_1x_2\bar{x}_3\bar{x}_4x_5\bar{x}_6+\bar{x}_1x_3\bar{x}_4\bar{x}_5\bar{x}_6+\bar{x}_1x_3x_4x_5\bar{x}_6+$$
$$\bar{x}_1\bar{x}_2\bar{x}_3\bar{x}_4x_6+x_1\bar{x}_2\bar{x}_3x_4x_6+\bar{x}_1x_3\bar{x}_4x_5\bar{x}_6+x_1\bar{x}_2x_4x_5\bar{x}_6+\bar{x}_2x_3x_5\bar{x}_6+x_1x_2\bar{x}_3x_4x_5\bar{x}_6+x_1x_3x_4x_5\bar{x}_6+$$
$$x_1x_3x_4\bar{x}_5+x_1\bar{x}_2x_4x_5x_6+x_1\bar{x}_2x_3\bar{x}_5x_6+\bar{x}_2x_3x_4\bar{x}_5x_6+x_1x_2\bar{x}_3\bar{x}_4x_6+x_2x_3x_4x_5x_6$$

$$S_2^6(x_1,x_2,x_3,x_4,x_5,x_6)=\bar{x}_1\bar{x}_2\bar{x}_3\bar{x}_4\bar{x}_5\bar{x}_6+\bar{x}_1x_2\bar{x}_4x_5\bar{x}_6+\bar{x}_1x_3x_4x_5\bar{x}_6+\bar{x}_1x_2x_3\bar{x}_5\bar{x}_6+$$
$$\bar{x}_1\bar{x}_2x_4x_5x_6+x_1\bar{x}_2x_3\bar{x}_4x_6+x_1\bar{x}_2x_3x_5x_6+x_1x_2\bar{x}_3\bar{x}_5x_6+\bar{x}_1x_3x_4\bar{x}_5\bar{x}_6+x_1x_2\bar{x}_3x_5x_6+\bar{x}_2x_3\bar{x}_4\bar{x}_5\bar{x}_6+$$
$$x_1x_2\bar{x}_3\bar{x}_5\bar{x}_6+x_1\bar{x}_3x_4\bar{x}_5\bar{x}_6+x_1x_2x_3\bar{x}_5\bar{x}_6+x_1\bar{x}_2x_3\bar{x}_4x_6+x_1\bar{x}_3x_4x_5x_6+\bar{x}_2x_3\bar{x}_4x_5x_6+$$
$$x_1x_2\bar{x}_3x_4\bar{x}_5+x_1x_2\bar{x}_3x_5x_6+x_2\bar{x}_3x_4x_5x_6+x_1x_2x_3\bar{x}_4x_5x_6+x_1x_2x_4x_5x_6$$

$$S_3^6(x_1,x_2,x_3,x_4,x_5,x_6)=\bar{x}_1\bar{x}_2\bar{x}_3x_4\bar{x}_6+x_1\bar{x}_2x_4x_5\bar{x}_6+x_1x_3x_4x_5\bar{x}_6+x_1x_2x_3\bar{x}_4x_6+\bar{x}_1x_3x_4x_5\bar{x}_6+$$
$$x_1x_3x_4x_5x_6+\bar{x}_1x_2x_4\bar{x}_5x_6+x_1\bar{x}_2x_3x_4\bar{x}_5+\bar{x}_1x_2x_3x_4\bar{x}_5x_6+x_1x_2\bar{x}_3x_4x_5+\bar{x}_2x_3x_4\bar{x}_5\bar{x}_6+$$
$$x_1x_2\bar{x}_3x_4\bar{x}_5\bar{x}_6+x_1x_3x_4x_5\bar{x}_6+x_1x_2\bar{x}_3x_4\bar{x}_5+\bar{x}_1x_2x_3x_4\bar{x}_5x_6+x_1x_2x_3x_4\bar{x}_6+x_2x_3x_4x_5\bar{x}_6+$$
$$\bar{x}_2\bar{x}_3x_4x_5x_6+x_1\bar{x}_2x_4x_5x_6+x_1x_2x_3x_4x_6+x_2\bar{x}_3x_4x_5x_6+x_1\bar{x}_3x_4x_5x_6+x_2x_3x_4x_5x_6+x_1x_2\bar{x}_3x_4x_5x_6+x_1x_2\bar{x}_3x_4x_5x_6$$

$$S_4^6(x_1,x_2,x_3,x_4,x_5,x_6)=\bar{x}_1x_2x_3\bar{x}_4\bar{x}_5+\bar{x}_1x_2x_4\bar{x}_5+x_1x_2\bar{x}_3x_4x_5+\bar{x}_1\bar{x}_2x_3x_4\bar{x}_6+\bar{x}_1\bar{x}_3x_4x_5+$$
$$\bar{x}_1x_2\bar{x}_4x_5+x_1\bar{x}_2\bar{x}_3x_5\bar{x}_6+\bar{x}_2x_3x_4\bar{x}_5\bar{x}_6+x_1\bar{x}_2x_4x_5\bar{x}_6+x_1x_2\bar{x}_4x_5\bar{x}_6+\bar{x}_2x_3\bar{x}_4x_5\bar{x}_6+x_2x_3x_4\bar{x}_5\bar{x}_6+$$
$$x_1\bar{x}_2\bar{x}_4x_5x_6+\bar{x}_2x_3\bar{x}_5x_6+x_1x_2\bar{x}_3x_5x_6+x_1x_2x_4x_5x_6$$

$$S_1^7(x_1,x_2,x_3,x_4,x_5,x_6)=\bar{x}_1\bar{x}_3\bar{x}_4x_5\bar{x}_6+\bar{x}_1\bar{x}_2x_3\bar{x}_5x_6+\bar{x}_1x_2\bar{x}_3x_4\bar{x}_5x_6+\bar{x}_1x_2\bar{x}_4x_5\bar{x}_6+$$
$$\bar{x}_1x_3\bar{x}_4\bar{x}_5x_6\bar{x}_6+\bar{x}_1x_2x_3x_5x_6+\bar{x}_1x_3\bar{x}_4x_5x_6+\bar{x}_1x_2x_3x_4\bar{x}_5x_6+x_1\bar{x}_2\bar{x}_3x_4+\bar{x}_2x_3\bar{x}_4\bar{x}_5x_6+\bar{x}_2x_4x_5\bar{x}_6+$$
$$x_1x_2\bar{x}_3\bar{x}_4x_6+x_1x_2x_3x_4\bar{x}_5x_6+x_1\bar{x}_2x_3x_5x_6+\bar{x}_2x_3x_4x_5x_6+x_1x_3x_4x_5+x_1\bar{x}_2x_4x_5x_6+x_2\bar{x}_3x_4x_5x_6+$$
$$x_1x_2x_4\bar{x}_5x_6+x_1x_2x_4x_5x_6$$

$$S_2^7(x_1,x_2,x_3,x_4,x_5,x_6)=\bar{x}_1x_2x_3\bar{x}_5x_6+\bar{x}_1x_3x_4x_5+\bar{x}_1\bar{x}_2x_4\bar{x}_5+x_1x_2\bar{x}_3\bar{x}_5x_6+\bar{x}_1x_2x_3x_5x_6+$$
$$x_1\bar{x}_2x_3x_5\bar{x}_6+x_1x_2\bar{x}_3\bar{x}_5\bar{x}_6+\bar{x}_2x_4x_5\bar{x}_6+\bar{x}_2\bar{x}_3x_4x_5x_6+x_1x_2\bar{x}_3x_4\bar{x}_5x_6+x_1x_2\bar{x}_4x_5\bar{x}_6+x_1\bar{x}_2x_3\bar{x}_5x_6+$$
$$x_1\bar{x}_2x_3x_5x_6+x_1x_2\bar{x}_3x_5x_6+x_1x_2x_3\bar{x}_4\bar{x}_5x_6+x_2x_4x_5x_6$$

$$S_3^7(x_1,x_2,x_3,x_4,x_5,x_6)=\bar{x}_1x_2\bar{x}_3x_5x_6+\bar{x}_1x_2x_3\bar{x}_4\bar{x}_5x_6+\bar{x}_1x_3x_4x_5\bar{x}_6+\bar{x}_1x_2x_4x_5\bar{x}_6+$$
$$\bar{x}_1x_2x_3\bar{x}_4x_5\bar{x}_6+\bar{x}_1x_2x_3x_4+x_1x_2\bar{x}_4x_6+\bar{x}_1x_3x_4x_5x_6+x_1\bar{x}_2x_3\bar{x}_5x_6+x_1\bar{x}_2x_4\bar{x}_5x_6+\bar{x}_2\bar{x}_3x_4\bar{x}_5\bar{x}_6+$$
$$x_1x_2\bar{x}_3\bar{x}_4x_6+x_1\bar{x}_3x_4x_5\bar{x}_6+x_1x_3x_4x_5x_6+x_1\bar{x}_2\bar{x}_3\bar{x}_4x_6+\bar{x}_2x_3x_4x_5x_6+x_1x_2\bar{x}_3x_4x_5x_6+$$
$$x_2x_3x_3\bar{x}_4x_6+x_1x_3x_4x_5x_6$$

$$S_4^7(x_1,x_2,x_3,x_4,x_5,x_6)=\bar{x}_1\bar{x}_2\bar{x}_3x_4x_5\bar{x}_6+\bar{x}_1x_2x_3\bar{x}_5x_6+\bar{x}_1x_2x_4x_5\bar{x}_6+\bar{x}_1x_3\bar{x}_4x_5\bar{x}_6+$$
$$\bar{x}_1x_3x_4x_5x_6\bar{x}_6+x_1\bar{x}_2x_3\bar{x}_5x_6+x_1\bar{x}_2x_3x_4x_6+x_1\bar{x}_2x_4x_5\bar{x}_6+\bar{x}_1x_3x_4x_5x_6+x_1x_3x_4x_5x_6+x_1x_2\bar{x}_3x_5\bar{x}_6+$$
$$x_1x_2\bar{x}_3x_4x_6+x_1x_2x_4x_5x_6+x_1x_3x_4x_5\bar{x}_6+x_1x_3x_4x_5\bar{x}_6+x_1x_3x_4x_5x_6+\bar{x}_2x_3x_4x_5x_6+x_1x_2x_3x_4\bar{x}_5x_6+$$
$$x_1x_2\bar{x}_3x_4x_5x_6+x_1x_2\bar{x}_3x_4x_6+x_2x_3x_4x_5x_6+x_1x_2x_3x_5x_6+x_1x_2x_3x_4\bar{x}_5$$

$$S_1^8(x_1,x_2,x_3,x_4,x_5,x_6)=\bar{x}_1x_2x_4\bar{x}_5\bar{x}_6+\bar{x}_1x_3x_4x_5\bar{x}_6+\bar{x}_1x_2\bar{x}_3x_5x_6+\bar{x}_1x_2x_3x_5x_6+\bar{x}_1x_2x_3\bar{x}_4x_5x_6+$$
$$\bar{x}_1x_3x_4x_5x_6+\bar{x}_1x_2x_3x_4x_5x_6+\bar{x}_1x_2x_3\bar{x}_4\bar{x}_5+x_1\bar{x}_2x_4x_5\bar{x}_6+\bar{x}_2x_3x_4x_5\bar{x}_6+\bar{x}_2x_3x_4\bar{x}_5\bar{x}_6+$$
$$x_1x_2\bar{x}_3x_4\bar{x}_6+x_1\bar{x}_3x_4x_5\bar{x}_6+x_1x_2\bar{x}_3x_4x_6+x_2\bar{x}_3x_4x_5\bar{x}_6+x_1\bar{x}_2x_3x_5x_6+x_1\bar{x}_2x_4x_5\bar{x}_6+x_2x_3x_4x_5\bar{x}_6+$$
$$x_1x_2\bar{x}_3x_4x_6+x_1\bar{x}_3x_4x_5x_6+x_1x_3x_4x_5x_6$$

$$S_2^8(x_1,x_2,x_3,x_4,x_5,x_6)=\bar{x}_1x_3x_4x_5\bar{x}_6+\bar{x}_1x_3x_4\bar{x}_5\bar{x}_6+\bar{x}_1x_2x_4x_5\bar{x}_6+\bar{x}_1x_3x_4x_5x_6+\bar{x}_1x_2x_4\bar{x}_5x_6+$$
$$\bar{x}_1x_3x_4\bar{x}_5x_6+\bar{x}_2x_3\bar{x}_4\bar{x}_5x_6+x_1\bar{x}_2x_4x_5\bar{x}_6+\bar{x}_2x_3x_4x_5x_6+x_1x_2\bar{x}_3x_5\bar{x}_6+x_2x_3\bar{x}_5\bar{x}_6+\bar{x}_2\bar{x}_3x_4\bar{x}_5\bar{x}_6+$$
$$x_1\bar{x}_2x_4x_5x_6+x_1\bar{x}_2x_3\bar{x}_4x_5x_6+\bar{x}_2x_3x_4x_5x_6+x_2\bar{x}_3x_4x_6+x_2\bar{x}_4x_5x_6+x_1x_2x_3x_4\bar{x}_5$$

$$S_3^8(x_1,x_2,x_3,x_4,x_5,x_6)=\bar{x}_1x_2\bar{x}_3x_5\bar{x}_6+\bar{x}_1\bar{x}_2x_4\bar{x}_5+\bar{x}_1\bar{x}_2x_3\bar{x}_5+\bar{x}_1x_2\bar{x}_3x_4+\bar{x}_1x_2x_4x_5+$$
$$\bar{x}_1x_3x_4x_5x_6+x_1\bar{x}_2x_3\bar{x}_4x_6+x_1x_2x_3x_4\bar{x}_6+x_2x_3x_4x_5\bar{x}_6+x_1x_2x_3\bar{x}_4x_6+x_1x_2\bar{x}_4x_5x_6+x_1\bar{x}_2\bar{x}_3x_4x_6+$$
$$\bar{x}_2x_3x_4x_5x_6+x_1\bar{x}_3x_4\bar{x}_5x_6+x_1x_2x_3\bar{x}_5x_6+x_2x_3x_4x_5x_6$$

$$S_4^8(x_1,x_2,x_3,x_4,x_5,x_6)=\bar{x}_1\bar{x}_2x_3x_5\bar{x}_6+\bar{x}_1x_2x_3x_4\bar{x}_6+\bar{x}_1x_2\bar{x}_3x_4x_5+x_1x_2\bar{x}_3x_4\bar{x}_5\bar{x}_6+$$
$$\bar{x}_1x_3x_4x_5\bar{x}_6+\bar{x}_1\bar{x}_2\bar{x}_3x_5\bar{x}_6+\bar{x}_1\bar{x}_2x_4x_5x_6+x_1\bar{x}_2x_3x_5x_6+\bar{x}_1x_3x_4x_5x_6+\bar{x}_2x_3\bar{x}_4x_5\bar{x}_6+x_1\bar{x}_2x_3x_5+$$
$$x_1x_3\bar{x}_4x_5\bar{x}_6+x_1\bar{x}_3x_4x_5\bar{x}_6+x_2x_3\bar{x}_4\bar{x}_5\bar{x}_6+x_1x_2x_3\bar{x}_4+x_1x_2x_3\bar{x}_5x_6+x_1x_3\bar{x}_4x_5x_6+x_1x_3x_4x_5x_6$$

FFT-Hash-II is not yet Collision-free

Serge Vaudenay

LIENS * , 45 rue d'Ulm, 75230 Paris cedex 05, France

Abstract. In this paper, we show that the FFT-Hash function proposed by Schnorr [2] is not collision free. Finding a collision requires about 2^{24} computation of the basic function of FFT. This can be done in few hours on a SUN4-workstation. In fact, it is at most as strong as a one-way hash function which returns a 48 bits length value. Thus, we can invert the proposed FFT hash-function with 2^{48} basic computations. Some simple improvements of the FFT hash function are also proposed to try to get rid of the weaknesses of FFT.

History

The first version of FFT-Hashing was proposed by Schnorr during the rump session of Crypto'91 [1]. This function has been shown not to be collision free at Eurocrypt'92 [3]. An improvement of the function has been proposed the same day [2] without the weaknesses discovered. However, FFT-Hashing has still some other weaknesses as it is proved in this paper.

1   FFT-Hash-II, Notations

The FFT-hash function is built on a basic function $< . >$ which takes one 128-bits long hash block H and one 128-bits long message block M, and return a 128-bits long hash block $< H, M >$. The hash value of n message blocks $M_1, \ldots, M_n$ is $< \ldots << H_0, M_1 >, M_2 >, \ldots, M_n >$ where H_0 is a constant given in hexadecimal by :

$$H_0 = 0123\ 4567\ 89ab\ cdef\ fedc\ ba98\ 7654\ 3210$$

The basic function is defined by two one-to-one functions Rec and FT2 on the set $(\mathrm{GF}_p)^{16}$ where $p = 2^{16}+1$. The concatenation HM defines 16 16-bits numbers which represents 16 numbers in GF_p between 0 and $p-2$. $(\mathrm{Rec} \circ \mathrm{FT2} \circ \mathrm{Rec})(HM)$ defines 16 numbers of GF_p. The last 8 numbers taken modulo 2^{16} are the result $< H, M >$.

---

* The *Laboratoire d'Informatique de l'Ecole Normale Supérieure* is a research group affiliated with the CNRS

We define the following notations :

$$A(M) = H_0 M$$
$$B(M) = \text{Rec}(A(M))$$
$$C(M) = \text{FT2}(B(M))$$
$$D(M) = \text{Rec}(C(M))$$

So, $< H_0, M >$ is the last 8 numbers of $D(M)$ taken modulo 2^{16}. We define X_i the i-th number of X (from 0 to 15), and $X[i, j]$ the list of the i-th to the j-th number of X.

If $x_i \in \text{GF}_p$, $i = 0, \ldots, 15$, we define $y_{-3} = x_{13}$, $y_{-2} = x_{14}$, and $y_{-1} = x_{15}$. Then, following Schnorr :

$$y_i = x_i + y_{i-1}^* y_{i-2}^* + y_{i-3} + 2^i \tag{1}$$

where $y^* = 1$ if $y = 0$ and $y^* = y$ otherwise. Then, we let :

$$\text{Rec}(x_0, \ldots, x_{15}) = y_0, \ldots, y_{15}$$

If $x_i \in \text{GF}_p$, $i = 0, \ldots, 7$, we define :

$$y_j = \sum_{i=0}^{7} \omega^{ij} x_i$$

where $\omega = 2^4$. Then, we define $FT(x_0, \ldots, x_7) = y_0, \ldots, y_7$.

If $x_i \in \text{GF}_p$, $i = 0, \ldots, 15$, we define $y_0, y_2, \ldots, y_{14} = FT(x_0, x_2, \ldots, x_{14})$ and $y_1, y_3, \ldots, y_{15} = FT(x_1, x_3, \ldots, x_{15})$. Then, we define $\text{FT2}(x_0, \ldots, x_{15}) = y_0, \ldots, y_{15}$.

2 Basic Remarks

If we want to find a collision to the hash function, we may look for a pair (x, x') of two 128-bits strings such that $< H_0, x >=< H_0, x' >$. In fact, we will look for x and x' such that $D(x)[8, 15] = D(x')[8, 15]$.

First, we notice that we have necessarily $C(x)[11, 15] = C(x')[11, 15]$. In one direction, we show that $C(x)_i = C(x')_i$ for $i = 11, \ldots, 15$. This is due to the equation :

$$C_i = D_i - D_{i-1}^* D_{i-2}^* - D_{i-3} - 2^i$$

Conversely, if we have both $C(x)[11, 15] = C(x')[11, 15]$ and $D(x)[8, 10] = D(x')[8, 10]$, then we have $D(x)[8, 15] = D(x')[8, 15]$.

Moreover, we notice on the equation 1 that $B(x)[0, 7]$ is a function of $x[5, 7]$ only. Let us denote :

$$B(x)[0, 7] = g(x[5, 7])$$

Finally, we notice that $FT2$ is a linear function.

3 Breaking *FFT*

3.1 Outlines

If we get a set of 3.2^{24} strings x such that $C(x)[11, 15]$ is a particular string R chosen arbitrarily[2], we will have a collision on $D(x)[8, 10]$ with probability 99% thanks to the birthday paradox. We will describe an algorithm which gives some x with the definitively chosen R for any $x[5, 7] = abc$.

Given $abc = x[5, 7]$, we can compute $B(x)[0, 7] = g(abc)$. If we denote $y = B(x)[8, 15]$, the following equation is a linear equation in y ;

$$FT2(g(abc)y)[11, 15] = R \tag{2}$$

We can define a function ϕ_R and three vectors U_e, U_o, U'_e such that :

$$(2) \iff \exists \lambda, \lambda', \mu \quad y = \phi_R(abc) + \lambda U_e + \lambda' U'_e + \mu U_o$$

(see section 3.2).

Finally, the system :

$$\begin{cases} x[5, 7] = abc \\ C(x)[11, 15] = R \end{cases}$$

is equivalent to the system :

$$\begin{cases} x[5, 7] = abc \\ y = \phi_R(abc) + \lambda U_e + \lambda' U'_e + \mu U_o \\ H_0 x = \text{Rec}^{-1}(g(abc)y) \end{cases}$$

Which is equivalent to :

$$\begin{cases} y = \phi_R(abc) + \lambda U_e + \lambda' U'_e + \mu U_o \\ y_{13} = a + y_{12}^* y_{11}^* + y_{10} + 2^{13} \\ y_{14} = b + y_{13}^* y_{12}^* + y_{11} + 2^{14} \\ y_{15} = c + y_{14}^* y_{13}^* + y_{12} + 2^{15} \\ x[5, 7] = abc \\ x[0, 4] = \text{Rec}^{-1}(g(abc)y)[8, 12] \end{cases} \tag{3}$$

Is we substitute y by the expression of the first equation in the other equations, we obtain a system of three equations of three unknown λ, λ', μ. This system can be shown linear in λ and λ' by a good choice of U_e, U_o and U'_e. Then, this system can have some solutions only if the determinant, which is a degree 2 polynomial in μ is 0. This can gives some μ. Then, the number of (λ, λ') is almost always unique. For more details, see section 3.3.

Finally, this gives 0 or 2 solutions x, with an average number of 1 for a given abc. If we try $1 \le a < p$, $1 \le b \le 768$ and $c = 2$, we have 3.2^{24} abc.

---

[2] For the collisions found in this paper, R is the image of my phone number by *FT2*.

3.2 Solving (2)

The function $X \longmapsto FT2(X)[11, 15]$ is linear, and has a kernel of dimension 3. If we define :

$$U = (0, 0, 0, 0, 4081, 256, 1, 61681)$$
$$U' = (0, 0, 0, 0, 65521, 4352, 1, 0)$$

we notice that :

$$FT(U) = (482, 56863, 8160, 57887, 7682, 0, 0, 0)$$
$$FT(U') = (4337, 61202, 65503, 544, 61170, 3855, 0, 0)$$

Let us introduce the following notation :

$$(x_0, \ldots, x_7) \times (y_0, \ldots, y_7) = (x_0, y_0, \ldots, x_7, y_7)$$

We have $FT2(X \times Y) = FT(X) \times FT(Y)$. Thus, we can can define :

$$U_e = U \times 0$$
$$U_o = 0 \times U$$
$$U'_e = U' \times 0$$

So, we have :

$$U_e = (0, 0, 0, 0, 0, 0, 0, 0, 4081, 0, 256, 0, 1, 0, 61681, 0)$$
$$U_o = (0, 0, 0, 0, 0, 0, 0, 0, 0, 4081, 0, 256, 0, 1, 0, 61681)$$
$$U'_e = (0, 0, 0, 0, 0, 0, 0, 0, 65521, 0, 4352, 0, 1, 0, 0, 0)$$

These vectors are a base of the kernel of $X \longmapsto FT2(X)[11, 15]$.

If M denotes the matrix of FT, we can write it using four 4×4 blocks :

$$M = \begin{pmatrix} M_{11} & M_{12} \\ M_{21} & M_{22} \end{pmatrix}$$

If x and y are two vectors of 4 elements, we have :

$$FT(xy)[4, 7] = 0 \iff y = -M_{22}^{-1} M_{21} x$$

Let us define :

$$N = -M_{22}^{-1} M_{21} = \begin{pmatrix} 65281 & 4335 & 289 & 61170 \\ 3823 & 8992 & 53012 & 65248 \\ 8447 & 61748 & 56545 & 4335 \\ 4369 & 57090 & 3823 & 256 \end{pmatrix}$$

Now, if x and y are two vectors of 8 elements, we have :

$$FT2(xy)[8, 15] = 0 \iff y = Nx^0 \times Nx^1$$

Where $x = x^0 \times x^1$. Let us define :

$$\phi_R(abc) = 0(Nx^0 \times Nx^1 + y^0)$$

where $g(abc) = x^0 \times x^1$ and $R = FT2(0y^0)[11, 15]$ for an arbitrary y^0 (one's phone number for instance). Then, $\phi_R(abc)$ is a vector which begins by $g(abc)$, and such that $FT2(\phi_R(abc))$ ends by a constant vector R.

So, we have :

$$(2) \iff \exists \lambda, \lambda', \mu \quad y = \phi_R(abc) + \lambda U_e + \lambda' U'_e + \mu U_o$$

3.3 Solving (3)

If we hope that no y_i $(i = 11, 12, 13, 14)$ is equal to 0 (we may ultimately test this condition, and forget the solutions y which do not pass this test, but this will be very rare), the system :

$$
\begin{cases}
y = \phi_R(abc) + \lambda U_e + \lambda' U'_e + \mu U_o \\
y_{13} = a + y^*_{12} y^*_{11} + y_{10} + 2^{13} \\
y_{14} = b + y^*_{13} y^*_{12} + y_{11} + 2^{14} \\
y_{15} = c + y^*_{14} y^*_{13} + y_{12} + 2^{15} \\
x[5, 7] = abc \\
x[0, 4] = \operatorname{Rec}^{-1}(g(abc)y)[8, 12]
\end{cases}
$$

imply :

$$z_{13} + \mu = a + (z_{12} + \lambda + \lambda')(z_{11} + 256\mu) + z_{10} + 256\lambda + 4352\lambda' + 2^{13}$$

$$z_{14} + 61681\lambda = b + (z_{13} + \mu)(z_{12} + \lambda + \lambda') + (z_{11} + 256\mu) + 2^{14}$$

$$z_{15} + 61681\mu = c + (z_{14} + 61681\lambda)(z_{13} + \mu) + (z_{12} + \lambda + \lambda') + 2^{15}$$

where $z = \phi_R(abc)$. If we define :

$$a' = a + z_{12}z_{11} + z_{10} + 2^{13} - z_{13}$$

$$b' = b + z_{13}z_{12} + z_{11} + 2^{14} - z_{14}$$

$$c' = c + z_{14}z_{13} + z_{12} + 2^{15} - z_{15}$$

we have :

$$
\begin{pmatrix}
z_{11} + 256\mu + 256 & z_{11} + 256\mu + 4352 & a' - (1 - 256z_{12})\mu \\
z_{13} + \mu - 61681 & z_{13} + \mu & b' + (256 + z_{12})\mu \\
61681(z_{13} + \mu) + 1 & 1 & c' - (61681 - z_{14})\mu
\end{pmatrix}
\begin{pmatrix}
\lambda \\
\lambda' \\
1
\end{pmatrix} = 0
$$

This is a linear system of unknown λ and λ'. If this system has an equation, which determinant has to be 0.

3.4 Discussion

This condition may be sufficient in most of the cases. The determinant should be a degree 3 polynomial. However, the coefficient of μ^3 is the determinant of the following matrix :

$$
\begin{pmatrix}
256 & 256 & (1 - 256z_{12}) \\
1 & 1 & -(256 + z_{12}) \\
61681 & 0 & (61681 - z_{14})
\end{pmatrix}
$$

which is 0 since the first line is 256 time the second.

The coefficient of μ^2 is 0 with probability $1/p$, this is rare. In this case, we have one solution if the equation has a degree one, and zero or p solutions in the other cases.

μ has to satisfy a degree 2 equation. If the discriminant is different from 0, it has a square root with probability 50%. So, we have two different μ or no solution with probability 50%, and a single solution with probability $1/p$.

For each μ, we are likely to have a uniq solution (λ, λ'). However, it is possible to have 0 or p solutions, but it is rare. So, for each solution (λ, λ', μ), we can 'compute y in the system (3), then x. Finally, we have zero or two solutions x in almost all cases.

3.5 Reduction of the Function FFT

To sum up, we have a function f_R such that for a given abc :

$$f_R(abc) = \{D(x)[8, 10]; x[5, 7] = abc \land C(x)[11, 15] = R\}$$

$f_R(abc)$ is a list of 0 or 2 $D(x)[8, 10]$ for each x such that $x[5, 7] = abc$ and $C(x)[11, 15] = R$. The average of number of x is 1, so f_R is almost a function.

The function f_R is a kind of reduction of FFT since a collision for f_R gives a collision for FFT. We can use the birthday paradox with f_R to get some collision. The expected complexity is $O(2^{24})$.

We can invert FFT with f_R to. If we are looking for x such that $D(x)[8, 15] = z$, we can compute $R = \text{Rec}^{-1}(z)[11, 15]$ and look for abc such that $f_R(abc) = z[0, 2]$. The complexity is 2^{48}. Then, we get the x required.

4 Finding Collisions with the Birthday Paradox

If we suppose that f_R is like a real random function, the probability that a set $\{f_R(x_i)\}$ for k different x_i have k elements is next to :

$$e^{-\frac{k^2}{2n}}$$

where n is the cardinality of the image of f_R, when k is next to $\sqrt{n}$. So, with $n = 2^{48}$ and $k = 3.2^{24}$, the probability is 1%.

Two collisions have been found in 24 hours by a SUN4 workstation with $k = 3.2^{24}$ different x. With the choice :

$$R = 5726\ 17fc\ b115\ c5c0\ a631$$

We got :

$$FFT(17b3\ 2755\ 4e52\ b915\ 2218\ 1948\ 00a8\ 0002) =$$
$$FFT(9c70\ 504e\ 834c\ b15c\ f404\ 94e2\ 02a7\ 0002) =$$
$$0851\ 393d\ 37c9\ 66e3\ d809\ d806\ 5e8c\ 05b8$$

and :

$$FFT(8ccc\ 23a4\ 086d\ fbb9\ 85f4\ 70b2\ 029e\ 0002) =$$
$$FFT(9d53\ 45ae\ 3286\ ada7\ 8c77\ 9877\ 02b4\ 0002) =$$
$$10e5\ 49f5\ 9df0\ d91b\ 0450\ afcc\ fba4\ 2063$$

Conclusion

The main weakness of FFT-Hash-II are described in section 2. First, the beginning of the computation depends on too few information of the input : $B(x)[0, 7]$ is a function of $x[5, 7]$. Second, the output allows to compute too much information of the computations in FFT : $D(x)[8, 15]$ allows to compute $C(x)[11, 15]$. The connection between $B(x)$ and $C(x)$ is linear, this makes our attack possible.

To get rid of the first weakness, we might mix H_0 and x in $A(x)$ before applying Rec. Similarly, the result of $< H_0, x >$ should be the set of $D(x)_{2i+1}$ instead of the right side.

Acknowledgment

I am happy to thank JEAN-MARC COUVEIGNES, ANTOINE JOUX, ADI SHAMIR and JACQUES STERN from the *Groupe de Recherche en Complexité et Cryptographie* for any advices. I owe a lot of time to JACQUES BEIGBEDER, RONAN KERYELL and all the *Service des Prestations Informatiques* for hardware and software advices. Finally, I should thank *France Telecom* to have given to me a phone number which hid so many collisions.

References

1. C. P. Schnorr. FFT-Hashing : An Efficient Cryptographic Hash Function. Presented at the rump session ot the CRYPTO'91 Conference (unpublished)
2. C. P. Schnorr. FFT-Hash II, Efficient Cryptographic Hashing. Presented at the EUROCRYPT'92 Conference (unpublished)
3. T. Baritaud, H. Gilbert, M. Girault. FFT Hashing is not Collision-free. Presented at the EUROCRYPT'92 Conference (unpublished)

Springer-Verlag
and the Environment

We at Springer-Verlag firmly believe that an international science publisher has a special obligation to the environment, and our corporate policies consistently reflect this conviction.

We also expect our business partners – paper mills, printers, packaging manufacturers, etc. – to commit themselves to using environmentally friendly materials and production processes.

The paper in this book is made from low- or no-chlorine pulp and is acid free, in conformance with international standards for paper permanency.

Lecture Notes in Computer Science

For information about Vols. 1–670
please contact your bookseller or Springer-Verlag

Vol. 708: C. Laugier (Ed.), Geometric Reasoning for Perception and Action. Proceedings, 1991. VIII, 281 pages. 1993.

Vol. 709: F. Dehne, J.-R. Sack, N. Santoro, S. Whitesides (Eds.), Algorithms and Data Structures. Proceedings, 1993. XII, 634 pages. 1993.

Vol. 710: Z. Ésik (Ed.), Fundamentals of Computation Theory. Proceedings, 1993. IX, 471 pages. 1993.

Vol. 711: A. M. Borzyszkowski, S. Sokołowski (Eds.), Mathematical Foundations of Computer Science 1993. Proceedings, 1993. XIII, 782 pages. 1993.

Vol. 712: P. V. Rangan (Ed.), Network and Operating System Support for Digital Audio and Video. Proceedings, 1992. X, 416 pages. 1993.

Vol. 713: G. Gottlob, A. Leitsch, D. Mundici (Eds.), Computational Logic and Proof Theory. Proceedings, 1993. XI, 348 pages. 1993.

Vol. 714: M. Bruynooghe, J. Penjam (Eds.), Programming Language Implementation and Logic Programming. Proceedings, 1993. XI, 421 pages. 1993.

Vol. 715: E. Best (Ed.), CONCUR'93. Proceedings, 1993. IX, 541 pages. 1993.

Vol. 716: A. U. Frank, I. Campari (Eds.), Spatial Information Theory. Proceedings, 1993. XI, 478 pages. 1993.

Vol. 717: I. Sommerville, M. Paul (Eds.), Software Engineering – ESEC '93. Proceedings, 1993. XII, 516 pages. 1993.

Vol. 718: J. Seberry, Y. Zheng (Eds.), Advances in Cryptology – AUSCRYPT '92. Proceedings, 1992. XIII, 543 pages. 1993.

Vol. 719: D. Chetverikov, W.G. Kropatsch (Eds.), Computer Analysis of Images and Patterns. Proceedings, 1993. XVI, 857 pages. 1993.

Vol. 720: V.Mařík, J. Lažanský, R.R. Wagner (Eds.), Database and Expert Systems Applications. Proceedings, 1993. XV, 768 pages. 1993.

Vol. 721: J. Fitch (Ed.), Design and Implementation of Symbolic Computation Systems. Proceedings, 1992. VIII, 215 pages. 1993.

Vol. 722: A. Miola (Ed.), Design and Implementation of Symbolic Computation Systems. Proceedings, 1993. XII, 384 pages. 1993.

Vol. 723: N. Aussenac, G. Boy, B. Gaines, M. Linster, J.-G. Ganascia, Y. Kodratoff (Eds.), Knowledge Acquisition for Knowledge-Based Systems. Proceedings, 1993. XIII, 446 pages. 1993. (Subseries LNAI).

Vol. 724: P. Cousot, M. Falaschi, G. Filè, A. Rauzy (Eds.), Static Analysis. Proceedings, 1993. IX, 283 pages. 1993.

Vol. 725: A. Schiper (Ed.), Distributed Algorithms. Proceedings, 1993. VIII, 325 pages. 1993.

Vol. 726: T. Lengauer (Ed.), Algorithms – ESA '93. Proceedings, 1993. IX, 419 pages. 1993

Vol. 727: M. Filgueiras, L. Damas (Eds.), Progress in Artificial Intelligence. Proceedings, 1993. X, 362 pages. 1993. (Subseries LNAI).

Vol. 728: P. Torasso (Ed.), Advances in Artificial Intelligence. Proceedings, 1993. XI, 336 pages. 1993. (Subseries LNAI).

Vol. 729: L. Donatiello, R. Nelson (Eds.), Performance Evaluation of Computer and Communication Systems. Proceedings, 1993. VIII, 675 pages. 1993.

Vol. 730: D. B. Lomet (Ed.), Foundations of Data Organization and Algorithms. Proceedings, 1993. XII, 412 pages. 1993.

Vol. 731: A. Schill (Ed.), DCE – The OSF Distributed Computing Environment. Proceedings, 1993. VIII, 285 pages. 1993.

Vol. 732: A. Bode, M. Dal Cin (Eds.), Parallel Computer Architectures. IX, 311 pages. 1993.

Vol. 733: Th. Grechenig, M. Tscheligi (Eds.), Human Computer Interaction. Proceedings, 1993. XIV, 450 pages. 1993.

Vol. 734: J. Volkert (Ed.), Parallel Computation. Proceedings, 1993. VIII, 248 pages. 1993.

Vol. 735: D. Bjørner, M. Broy, I. V. Pottosin (Eds.), Formal Methods in Programming and Their Applications. Proceedings, 1993. IX, 434 pages. 1993.

Vol. 736: R. L. Grossman, A. Nerode, A. P. Ravn, H. Rischel (Eds.), Hybrid Systems. VIII, 474 pages. 1993.

Vol. 737: J. Calmet, J. A. Campbell (Eds.), Artificial Intelligence and Symbolic Mathematical Computing. Proceedings, 1992. VIII, 305 pages. 1993.

Vol. 738: M. Weber, M. Simons, Ch. Lafontaine, The Generic Development Language Deva. XI, 246 pages. 1993.

Vol. 739: H. Imai, R. L. Rivest, T. Matsumoto (Eds.), Advances in Cryptology – ASIACRYPT '91. X, 499 pages. 1993.

Vol. 740: E. F. Brickell (Ed.), Advances in Cryptology – CRYPTO '92. Proceedings, 1992. X, 593 pages. 1993.

Vol. 741: B. Preneel, R. Govaerts, J. Vandewalle (Eds.), Computer Security and Industrial Cryptography. Proceedings, 1991. VIII, 275 pages. 1993.

Vol. 742: S. Nishio, A. Yonezawa (Eds.), Object Technologies for Advanced Software. Proceedings, 1993. X, 543 pages. 1993.

Vol. 743: S. Doshita, K. Furukawa, K. P. Jantke, T. Nishida (Eds.), Algorithmic Learning Theory. Proceedings, 1992. X, 260 pages. 1993. (Subseries LNAI)

Vol. 744: K. P. Jantke, T. Yokomori, S. Kobayashi, E. Tomita (Eds.), Algorithmic Learning Theory. Proceedings, 1993. XI, 423 pages. 1993. (Subseries LNAI)

Vol. 745: V. Roberto (Ed.), Intelligent Perceptual Systems. VIII, 378 pages. 1993. (Subseries LNAI)

Vol. 746: A. S. Tanguiane, Artificial Perception and Music Recognition. XV, 210 pages. 1993. (Subseries LNAI)

Vol. 747: M. Clarke, R. Kruse, S. Moral (Eds.), Symbolic and Quantitative Approaches to Reasoning and Uncertainty. Proceedings, 1993. X, 390 pages. 1993.

Vol. 748: R. H. Halstead Jr., T. Ito (Eds.), Parallel Symbolic Computing: Languages, Systems, and Applications. Proceedings, 1992. X, 419 pages. 1993.

Vol. 751: B. Jähne, Spatio-Temporal Image Processing. XII, 208 pages. 1993.

Vol. 753: L. J. Bass, J. Gornostaev, C. Unger (Eds.), Human-Computer Interaction. Proceedings, 1993. X, 388 pages. 1993.